Microorganisms for Sustainability

Volume 33

Series Editor

Naveen Kumar Arora, Environmental Microbiology, School for Environmental Science, Babasaheb Bhimrao Ambedkar University, Lucknow, Uttar Pradesh, India

Microorganisms perform diverse roles on our planet most of which are important to make earth a habitable and sustainable ecosystem. Many properties of microorganisms are being utilized as low input biotechnology to solve various problems related to the environment, food security, nutrition, biodegradation, bioremediation, sustainable agriculture, bioenergy and biofuel, bio-based industries including microbial enzymes/ extremozymes, probiotics etc. The book series covers all the wider aspects and unravels the role of microbes towards achieving a sustainable world. It focuses on various microbial technologies related to sustenance of ecosystems and achieving targets of Sustainable Development Goals. Series brings together content on microbe based technologies for replacing harmful chemicals in agriculture, green alternatives to fossil fuels, use of microorganisms for reclamation of wastelands/ stress affected regions, bioremediation of contaminated habitats, biodegradation purposes. Volumes in the series also focus on the use of microbes for various industrial purposes including enzymes, extremophilic microbes and enzymes, effluent treatment, food products.

The book series is a peer reviewed compendium focused on bringing up contemporary themes related to microbial technology from all parts of the world, at one place for its readers, thereby ascertaining the crucial role of microbes in sustaining the ecosystems.

More information about this series at http://www.springer.com/series/14379

Naveen Kumar Arora • Brahim Bouizgarne
Editors

Microbial BioTechnology for Sustainable Agriculture Volume 1

Springer

Editors
Naveen Kumar Arora
Department of Environmental Science
Babasaheb Bhimrao Ambedkar University
Lucknow, Uttar Pradesh, India

Brahim Bouizgarne
Department of Biology
Ibn Zohr University
Agadir, Morocco

ISSN 2512-1901 ISSN 2512-1898 (electronic)
Microorganisms for Sustainability
ISBN 978-981-16-4842-7 ISBN 978-981-16-4843-4 (eBook)
https://doi.org/10.1007/978-981-16-4843-4

This Springer imprint is published by the registered company Springer Nature Singapore Pte Ltd.
The registered company address is: 152 Beach Road, #21-01/04 Gateway East, Singapore 189721, Singapore

Preface

In the era of environmental pollution and land degradation, achieving the targets of sustainable agriculture has become a big challenge. Sustainable agriculture should lead to economic viability, be ecologically sound, enrich the farming communities, families, and farmers' lives, and above all ensure food security without polluting the agro-ecosystems. To meet the demands of sustainable food production, the study and understanding of the rhizosphere and the associated plant growth-promoting microorganisms (PGPMs) would be one of the most important aspects for future. The plant-associated microbiome affects plant growth and modifies the responses to pathogens, environmental stresses, pollutants, and nutrient cycling in the soil. To achieve the goals of sustainability through microbial technology, intricate mechanistic details of action against abiotic stresses, pollutants, and pathogen attack would be highly helpful in developing novel and efficient biostimulants, biofertilizers, biodegraders, biopesticides, and other bioinoculants.

Plant growth-promoting organisms have been gaining attention to achieve the targets of climate-smart agricultural practices. Amidst the extreme climatic variations and presence of xenobiotic compounds and pollutants, plants evolve through modifying the microbial assemblages and adjusting the signaling and molecular responses for better plant-microbe associations. These beneficial microbes in turn are tremendous "stress therapists," offering limitless facilities to plants, improving their health, and elevating the fertility and productivity of soils. The establishment of plant-microbe associations involves a nexus of signaling pathways and mechanisms through biomolecules from both the partners. The microbial assemblages have the potential to protect the plant from various stresses and enhance their biomass, growth rate, and productivity by direct and indirect mechanisms. Therefore, studies about these interactions are the basis of microbial technology without which the application facets would be an incomplete story. The rhizosphere is an excellent example of community-level co-occurrence of microbes; however, niche sharing or niche exclusion between the groups or species directly fine-tunes the growth, nutrition, health, and stress resistivity in plants. For example, actinomycetes are often found to be associated with arbuscular mycorrhizal fungi (AMF); AMF being present both inside and outside of plant roots influence other microorganisms in the rhizosphere, niche sharing between non-mycorrhizal fungi and bacteria. Successful recruitment or establishment of microbial populations in the rhizosphere results in the accomplishment of various functions to improve the health and fitness of plants in diverse or even adverse conditions. The direct mechanisms of plant growth promotion include production of phytohormones, nutrient cycling, and water availability, while indirect approaches are protection against pathogens, production of antibiotics, niche exclusion of unwanted microbes, and inducing systemic resistance in plants; additionally the third type of "conditional mechanisms" include production of secondary metabolites or initiation of biochemical and molecular responses according to abiotic stresses faced by plants and microbes. The latter mechanism of action induces the production of exopolysaccharides, osmolytes, antioxidants, biosurfactants, quorum sensing molecules (specific for stress mitigation), or the role of PGPMs in immobilization and degradation of toxic metals or pollutants.

The book *Microbial BioTechnology for Sustainable Agriculture* (Volume 1) covers the mechanistic aspects related to plant-microbe interactions. The book consists of 18 chapters from experts all across the globe covering various microbial communities and their diverse roles ranging from rhizosphere interactions and cross-talks, nutrient cycling to mitigation of biotic/abiotic stresses (such as salinity and drought), and biodegradation of xenobiotic and recalcitrant pollutants. The book includes chapters that discuss heterogeneous groups of microorganisms found in the rhizosphere or living as endophytes and explains how their presence affects the sustainability of agricultural systems. The volume will inform about the latest developments and mechanistic insights of action of plant microbiome in promoting plant growth and soil remediation. The role of soil microbes which can be conserved as novel PGPMs, such as actinomycetes and yeasts, is also explained in the volume.

The editors are thankful to all the authors for channeling their knowledge and contributing their research in this volume. The editors are also thankful to the team at Springer Nature, specifically Ms. Aakanksha Tyagi, Senior Commissioning Editor (Books), Life Sciences, India, and Dr. Naren Aggarwal, Editorial Director-Books, Medicine and Life Sciences, Asia, for their support.

NKA also expresses gratitude towards Prof. Sanjay Singh, Vice-Chancellor, BBA University, Lucknow, UP, India, for the support. NKA would also like to thank his research scholars, namely Tahmish Fatima, Isha Mishra, Priyanka Verma, Priya Mishra, Ankita Bhattarcharya, Chanda, and Dr. Jitendra Mishra, for helping in editing of the manuscripts.

We hope that the reader will find the book useful and helpful in the generation of new ideas in the area of sustainable agriculture. We also hope that the volume will be a step towards achieving the targets of sustainable development goals (#SDG2 and #SDG15).

Lucknow, Uttar Pradesh, India — Naveen Kumar Arora
Agadir, Morocco — Brahim Bouizgarne

Contents

About the Editors

Naveen Kumar Arora PhD in Microbiology, Fellow of International Society of Environmental Botanists (FISEB), Professor and Dean of the School of Earth and Environmental Sciences, Babasaheb Bhimrao Ambedkar University (A Central University), Lucknow, Uttar Pradesh, India, is a renowned researcher in the field of Environmental Microbiology and Biotechnology. His specific area of research is plant-microbe interactions, sustainable agriculture, and environment. He has more than 100 research articles published in premium international journals and several articles published in magazines and dailies. He is editor of 35 books, published by Springer Nature. He is the Series Editor of the book series "Microorganisms for Sustainability" published by Springer Nature. He is the Editor in Chief of the journal *Environmental Sustainability* also published by Springer Nature. He is a member of several national and international societies, on the editorial board of 04 journals, and a reviewer of several international journals. He has 01 patent to his name. He has successfully completed 06 R&D projects sanctioned by funding agencies such as DST, DBT, CSIR, UPCST, and UGC and presently working on 03 projects funded by the Ministry of Science and Technology, New Delhi, India. He has delivered invited and keynote talks in conferences and seminars around the globe; some of these include international conferences in the USA, Portugal, China, Turkey, Morocco, Uzbekistan, Vietnam. Eleven scholars have been awarded PhD under his guidance. He has been awarded by national and international bodies and societies for the research and academic work. Although an academician and researcher by profession he has a huge obsession for the wildlife and its conservation and has authored a book, *Splendid Wilds*. He is President of the Society for Conservation of Wildlife and Secretary of the Society for Environmental Sustainability (website: www.ses-india.org).

Brahim Bouizgarne obtained his PhD in 2005 at Cadi Ayyad University, Marrakesh, Morocco, in Microbial and Plant Biotechnology. He worked at Ibn Zohr University (Agadir, Morocco) as assistant professor in 2007. Currently, Brahim BOUIZGARNE is a full professor (Professor of Higher Education PES) in Microbial and Plant Biotechnology, and he is head of the research team Microbial Biotechnologies (Laboratory of Plant Biotechnology), at the Faculty of Sciences, Ibn Zohr University. His specific area of research is soil and agriculture microbiology, actinobacteria taxonomy, rhizosphere microbiology, plant-microbe interactions, phytopathology, plant growth-promoting rhizobacteria (PGPR), and biofertilizers and biopesticide development.

Brahim Bouizgarne supervises 6 PhD candidates and has published 13 papers and 4 chapters with Springer. He is a reviewer of several international journals. He contributed to six research projects and supervises two research projects.

He is also a member of the Moroccan Association of Biotechnology and Protection of Natural Resources "Microbiona." He was the chair of the fourth edition of Microbiod 4 "Microbial Biotechnology for Sustainable Development," Agadir, 2019, and served as a member of scientific committees of other conferences.

For more information: https://www.researchgate.net/Brahim_Bouizgarne

Nutrient Acquisition and Soil Fertility: Contribution of Rhizosphere Microbiomes in Sustainable Agriculture

1

Satyavir S. Sindhu, Anju Sehrawat, Manisha Phour, and Rakesh Kumar

Abstract

Recent application of modern agricultural biotechnologies and use of high-yielding crop varieties provided sufficient crop yields and food for human population. But, the major constraint to enhance crop yield and food production is the availability of various nutrients in the soil, which include nitrogen, phosphorus, potassium, sulphur, zinc and other micronutrients. Currently, chemical fertilizers and plant growth regulators are widely utilized for obtaining high crop yields. However, the injudicious application of fertilizers along with other agrochemicals has resulted in environmental pollution along with deleterious effects on beneficial microflora and fauna. Recently, soil-inhabiting beneficial microorganisms are screened and exploited for use as biofertilizers to enhance soil fertility and crop yield with reduced application of chemical fertilizers. Thus, the use of microbial inoculants has emerged as novel agrobiotechnology for attaining sustainable agricultural production systems. These beneficial microorganisms contribute immensely towards management of plant nutrients in the soil by way of nitrogen fixation, solubilization of phosphorus, potassium and zinc along with other nutrients. Other beneficial characteristics of microbial inoculants include phytohormones production, inhibition of phytopathogens' growth, bioremediation of pollutants and heavy metals and amelioration of abiotic as well as biotic stresses. However, the efficacies of microbial inoculants in improving crop yield under field conditions remain variable under varied farming situations in different agro-ecosystems. Recently, genetic engineering techniques are being employed to improve the beneficial traits in plants and microorganisms to improve nutrient availability, soil fertility and crop yield. In

S. S. Sindhu (✉) · A. Sehrawat · M. Phour · R. Kumar
Department of Microbiology, CCS Haryana Agricultural University, Hisar, Haryana, India

N. K. Arora, B. Bouizgarne (eds.), *Microbial BioTechnology for Sustainable Agriculture Volume 1*, Microorganisms for Sustainability 33,
https://doi.org/10.1007/978-981-16-4843-4_1

addition, identification of effective microbial inoculants and their persistence in soil and quality of these inoculants is a never-ending process for harnessing desirable impacts on crop productivity. Considering the importance of beneficial microorganisms in biogeochemical cycling of nutrients, various mechanisms involved in improving nutrients availability are reviewed for increasing food production.

Keywords

Microbial inoculants · Nutrient acquisition · Rhizospheric microorganisms · Phytohormones · Induced systemic resistance

1.1 Introduction

The growth of crop plants is dependent on the supply of appropriate amount of nutrients, adequate amount of moisture and favourable environmental conditions. Under the current intensive crop production systems in agriculture, the nutrient supply capacity of soil is being exhausted continuously due to the uptake of available nutrients by plants, soil erosion and land degradation. Therefore, farmers apply excessive amounts of agrochemicals in the soil for improving crop growth (Newbould 1989). This injudicious use of chemical fertilizers adversely affected the soil health and resulted in public health hazards (Tilman et al. 2002; Meena et al. 2017). For ensuring the availability of essential nutrients to plants in a sustainable manner and to reduce the use of potentially deleterious agrochemicals, integrated nutrient management system is needed (Maurya et al. 2014; Sindhu et al. 2019a). Therefore, applications of microbial inoculants as biofertilizers have emerged as potential tools for improving agriculture productivity and for restoration of sustainability in agricultural systems (Gomiero et al. 2011; Fasusi et al. 2021; Zhang et al. 2021). Rhizosphere-inhabiting microorganisms possessing the ability of nutrient solubilization and release of plant growth-promoting (PGP) substances may be used for inoculation of crop plants to improve nutrient supply system for higher crop production (Fig. 1.1).

Some soil or rhizosphere-inhabiting nitrogen (N_2)-fixing bacteria develop symbiotic and non-symbiotic associations with plants resulting in improved plant growth and yield of various leguminous, cereal as well as oilseed crops (Afzal and Bano 2008; Yadav et al. 2010; Chaudhary and Sindhu 2016). Other plant growth-promoting rhizobacteria (PGPR) were found to solubilize insoluble phosphate into soluble available forms in the soil by secreting organic acids leading to stimulation of plant growth (Khan et al. 2007; Wani et al. 2007; Sindhu et al. 2014a). Similarly, several potassium-solubilizing microorganisms (KSMs) were characterized, which caused release of soluble potassium from insoluble rocks and silicate materials (Parmar and Sindhu 2013, 2019). Besides this, degradation of plant residues and leaf litre by lignocellulosic fungi, i.e. *Trichoderma harzianum*, *Polyporus ostriformis* and *Phanerochaete chrysosporium*, and bacteria including species of

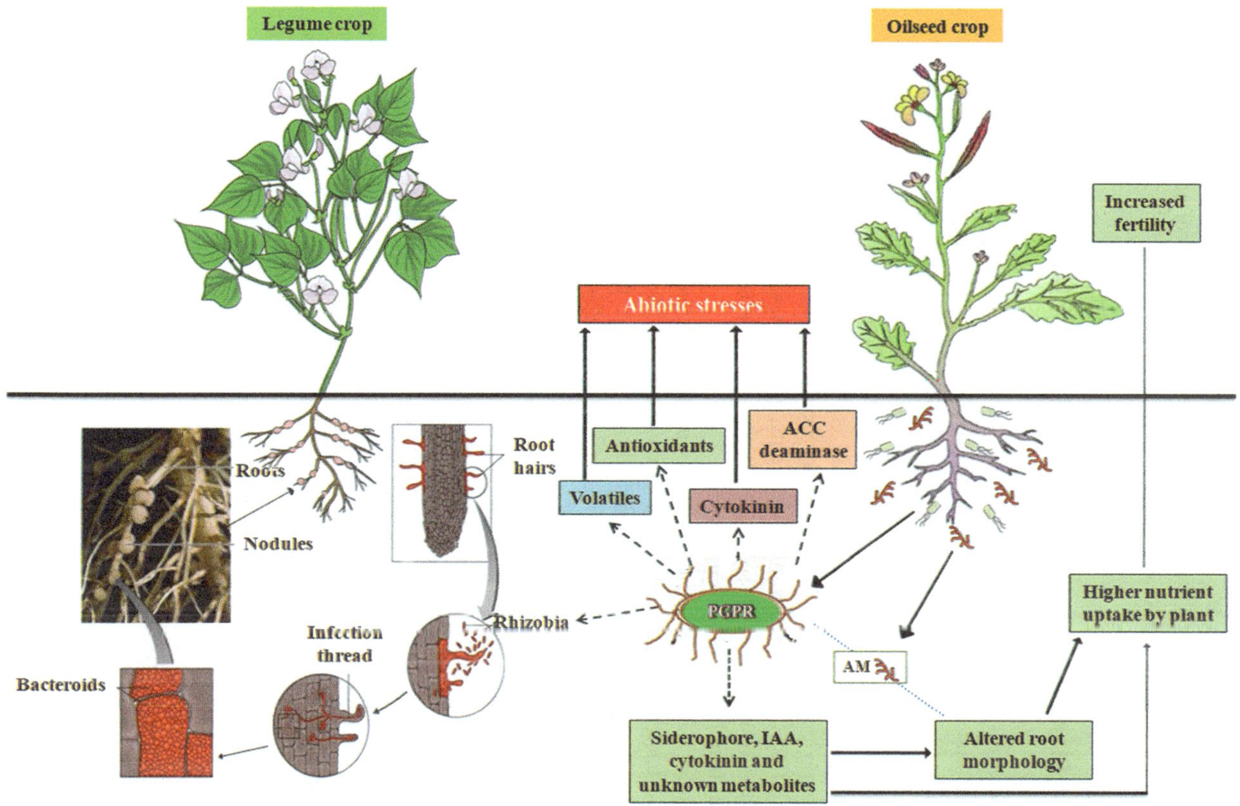

Fig. 1.1 Beneficial attributes of plant growth-promoting rhizobacteria causing improvement in plant biomass and crop productivity

Pseudomonas, *Cellulomonas*, *Sporocytophaga* and *Streptomyces* was demonstrated to enhance nutrient availability for plants (Goyal and Sindhu 2011; Verma et al. 2014). Thus, PGPR and beneficial fungi are being applied as biofertilizers for enhancing nutrient use efficiency along with improvement in productivity of agricultural crops under greenhouse and field conditions (Gravel et al. 2007; Naiman et al. 2009; Sindhu and Sharma 2020).

1.2 Plants and Their Associated Microbial Community

The compositions of microbial communities and physiology of the plant along with physical and chemical characteristics of soil are some important factors, which influence growth of plants (Ryan et al. 2009; Lareen et al. 2016). Selective enrichment of microbiomes in the rhizosphere for beneficial traits including mineralization/solubilization of nutrients, pathogens' growth inhibition and degradation of pollutants and pesticides may result in stimulation of plant biomass resulting in better crop productivity (Kennedy and Smith 1995; Chaparro et al. 2012; Sindhu et al. 2019b). In agricultural ecosystems, plant-associated microbes acquire root exudates from plants (Sasse et al. 2018), and in return, these microbes provide plant utilizable nutrients to their hosts.

Besides beneficial microbes, various plant pathogens including smallest viroids, viruses, bacteria, fungi and nematodes were found to cause many important plant diseases leading to significant reductions in crop yield. Major bacterial plant pathogens include *Pseudomonas syringae*, *Pseudomonas corrugata*, *Ralstonia solanacearum*, *Corynebacterium michiganensis*, *Agrobacterium tumefaciens*, *Xanthomonas oryzae*, *Xanthomonas campestris*, *Erwinia amylovora*, etc. (Mansfield et al. 2012). Soil-borne fungal pathogens include *Gaeumannomyces graminis* (take-all disease of wheat), *Rhizoctonia solani* (root and stem rots), *Neovossia indica* (Karnal bunt disease in wheat), *Pythium* spp. (damping off), *Armillaria mellea* (honey fungus), *Fusarium oxysporum* (wilt in chickpea) and *F. oxysporum f. sp. cubense* (wilt in banana) (Doehlemann et al. 2017). The most destructive fungal pathogens in Brassicaceae group have been characterized as *Alternaria brassicicola* that causes dark spot disease (Pochon et al. 2012). Rice blast disease caused by fungal pathogen *Magnaporthe grisea* has been found to affect rice cultivation drastically (Talbot and Foster 2001).

1.2.1 Beneficial Microbes

Beneficial soil microorganisms improve nutrient availability; suppress various plant diseases; provide tolerance to adverse environmental conditions such as salt stress, drought stress and weed infestation; and also help in plant defence response through secretion of various secondary metabolites (Sindhu and Sharma 2020) (Fig. 1.1). Some of the soil microbes also contribute towards decomposition of organic waste and detoxification of toxic substances such as pesticides (Nihorimbere et al. 2011;

Aislabie Jand and Deslippe 2013; Ma et al. 2016). Bacterial genera including *Azospirillum*, *Rhizobium*, *Azotobacter*, *Arthrobacter*, *Bacillus*, *Pseudomonas*, etc. show the capability to fix atmospheric nitrogen, thereby enhancing the fertility of soil (Mishra and Arora 2012; Sangwan et al. 2012; Chaudhary et al. 2013; Arora 2015). Some cyanobacterial genera, viz. *Nostoc*, *Chlorogloeopsis*, *Calothrix*, *Plectonema*, *Anabaena*, *Cylindrospermum* and *Anabaenopsis*, are known to produce indole acetic acid (IAA) phytohormone in free-living and symbiotic associations in addition to nitrogen fixation (Karthikeyan et al. 2007; Manjunath et al. 2011; Rossi and Philippis 2015). Many of the beneficial microbes solubilize bound form of nutrients (P, K, Zn, etc.) for utilization of the plants. Growth inhibition of phytopathogens resulting in suppression of diseases is caused by antagonistic microbes through production of antibiotics, siderophores, wall-degrading hydrolytic enzymes such as chitinase and by induction of systemic resistance (ISR) (Vejan et al. 2016; Sehrawat and Sindhu 2019). Moreover, some PGPR strains also play an important role in amelioration of various environmental and biotic stresses under stressed environments (Khare and Arora 2011; Kang et al. 2014; Sindhu and Sharma 2019; Phour and Sindhu 2020).

1.2.1.1 Legume-Rhizobium Symbiosis

Legumes comprise of over 20,000 species, and some legumes including soybean, common bean and chickpea are important constituents of diets for human beings. Legumes establish symbiotic association with rhizobia and fix atmospheric inert N_2 to ammonia in the nodules (Sindhu and Dadarwal 1995; Sindhu et al. 2021). Contributions of annual nitrogen input in agricultural systems by biological nitrogen fixation (BNF) have been estimated to be about 21.45 Tg, in which major contribution has been reported by soybean rhizobia accounting to 16.4 Tg (Herridge et al. 2008). Nitrogen fixation in diazotrophic bacteria is performed with nitrogenase enzyme, which consists of two metalloprotein components, i.e. dinitrogenase reductase (Fe protein) and dinitrogenase (MoFe protein). Using mutational and complementation analysis and transcriptional profiling, various genes involved in nitrogen fixation (*nif* and *fix* genes) have been identified in free-living *Klebsiella pneumoniae*, *Azotobacter chroococcum*, *Azospirillum brasilense* and symbiotic rhizobia (Fischer 1994). Nitrogenase enzyme is highly sensitive to high concentration of oxygen (O_2), which causes irreversible damage to nitrogenase activity in diazotrophic bacteria (Wang et al. 1984; Khadem et al. 2010). However, this O_2 paradox has been resolved under symbiotic conditions in root nodules by synthesis of leghemoglobin/phytoglobins (Berger et al. 2018; Pucciariello et al. 2019).

Significant increases in the root and shoot weight, number of pods and yield of various legume crops were found by inoculation of effective strains of rhizobia under pot house and field conditions (Thies et al. 1991; Sindhu and Dadarwal 1992; Franche et al. 2009). Furthermore, coinoculations of rhizobia with PGPR strains were found to increase nodulation, nitrogen fixation, plant biomass and yield of various legumes (Sindhu et al. 1999; Goel et al. 2002; Chaudhary and Sindhu 2017; Zhang et al. 2021). Genes involved in symbiosis can be horizontally transferred from rhizobia to other soil-inhabiting bacteria, allowing development of novel symbionts with the capability to develop effective association with incompatible legumes growing in diverse soils (Andrews et al. 2018).

1.2.1.2 Plant and PGPR Interactions

PGPRs have been characterized from diverse habitats for their subsequent use for increasing the plant growth through various direct or indirect mechanisms (Vessey 2003; Sindhu et al. 2010; Tabassum et al. 2017). For example, insoluble bound form of phosphate in the soil/rhizosphere may be solubilized by phosphorus-solubilizing bacteria through secretion of various organic acids, i.e. gluconic acid, citric acid, oxalic acid and malic acid (Khan et al. 2010; Parmar and Sindhu 2019). Production of antibiotics, hydrolytic enzymes, hydrogen cyanide, siderophores and induction of systemic resistance (ISR) in plants contributed towards killing of pathogens by various antagonistic bacteria (Sindhu et al. 1999; Sharma et al. 2018a; Sehrawat and Sindhu 2019). Besides this, synthesis and secretion of plant hormones by PGPRs were found to exert strong influences on crop productivity even under adverse soil conditions (Malik and Sindhu 2011; Etesami and Maheshwari 2018; Khan et al. 2020). Some PGPRs produce biofilm that help in survival, propagation and functioning of microbial community in rhizosphere niches under stressed conditions (Barthe et al. 2015). Biofilm-producing photoautotropic cyanobacterial strain *Anabaena torulosa* was found to increase nitrogen fixation of rhizobia under in vitro conditions and its coinoculation of wheat with phosphorus-solubilizing bacteria (PSB) caused enhancement of phosphorus uptake under pot house conditions (Swarnalakshmi et al. 2013).

1.2.1.3 Plant AM Fungi Association

Mycorrhizal symbiosis has been reported in about 250,000 species of plants worldwide (Smith and Read 2008). Plant-mycorrhizal fungus interactions are classified as ectomycorrhiza, endomycorrhiza and ectendomycorrhiza, according to the colonization of fungal mycelium in the root structure (Bonfante and Perotto 2000; Brundrett 2004; Smith and Read 2008). Arbuscular mycorrhiza (AM) represents the most studied plant-mycorrhizal association among the endomycorrhiza with a great economic impact worldwide (Smith et al. 2009; Balestrini et al. 2015). AM fungi establish symbiosis with more than 80% of land plants, including most agricultural crop species, and help in increasing host plant nutrition under low-nutrient-availability soil conditions (Smith and Smith 2011; Corradi and Bonfante 2012). AM fungi have an obligate life cycle, which receives photosynthates from host plant, and in turn supply various nutrients including phosphorus, potassium and water for the host plants (Smith and Smith 2011; Ferrol et al. 2019). After penetration of AM fungi into root cortex cells, arbuscules are produced by dichotomous branching of hyphal tips, which are the key structures indicative of symbiotic effect (Luginbuehl and Oldroyd 2017). In most of the inoculation experiments, AM fungi were found to increase plant growth and crop productivity even under heat and drought stress conditions (Wiebe et al. 2015; Wang et al. 2017).

1.3 Mechanisms of Growth Promotion by PGPR

Plant growth-promoting microorganisms are usually isolated from the rhizosphere soil using either dilution plate method or soil enrichment technique. Subsequently, purified cultures are screened for different growth-promoting characteristics for their subsequent use as biocontrol agents or biofertilizers (Glick 1995; Reed and Glick 2013; Santoyo et al. 2016; Dahiya et al. 2021) (Fig. 1.2). Currently, several microbial strains have been commercialized to increase plant growth and crop yield under pot house or field conditions (Sehrawat and Sindhu 2019; Sindhu and Sharma 2020).

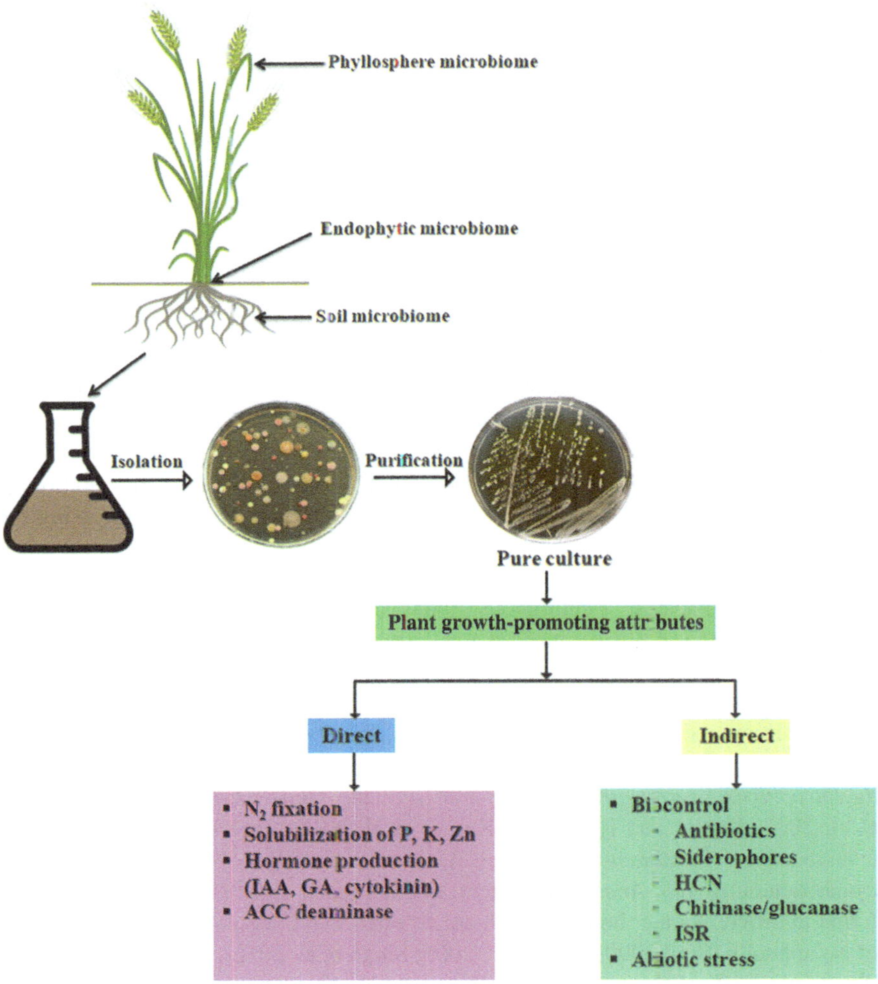

Fig. 1.2 Isolation and purification of beneficial microorganisms from rhizosphere soil and their subsequent screening for different plant growth promoting attributes

1.3.1 Role of Plant Microbiome in Nutrient Acquisition

Nitrogen, phosphorous, potassium, iron and zinc are the major nutrients required for optimal growth of crop plants, and many agricultural soils are deficient in either one or more of these nutrients. For increasing crop yields, farmers apply sufficient/excessive amounts of chemical fertilizers to obviate this nutrient problem (Newbould 1989). However, the injudicious use of chemical fertilizers results in depletion of nonrenewable resources and causes environmental and public health hazards. Therefore, alternative means of providing nitrogen, phosphorus, potassium and zinc in the soil include biogeochemical cycling of nutrients by rhizosphere-inhabiting microorganisms (Marschner 2011; Sindhu et al. 2019a, b; Sindhu and Sharma 2020).

1.3.1.1 Nitrogen

Nitrogen (N) is often the most important nutrient needed for plant growth and crop productivity. Nitrogen in the soil is usually present in the unavailable form either as inert or organic form; and it is absorbed by the plants normally either in ammonium or nitrate form (Smil 1999; Mikkelsen and Hartz 2008). Nitrogen plays role in the formation of amino acids, vitamins (biotin, thiamine, niacin, riboflavin) and nucleic acids (Kalra and Sharma 2015; Sharma et al. 2016). Besides this, nitrogen is essential for plant cell division and photosynthesis, and its involvement in utilization of carbohydrates has been reported to affect energy reactions in the plant (Schaetzl and Anderson 2005). Symptoms of nitrogen deficiency include stunted growth and yellowing of leaves. Crops growing in nitrogen-deficient soils have retarded plant growth, which subsequently affect yield and quality of agri-produce (Mikkelsen and Hartz 2008; Hayat et al. 2010).

Some prokaryotes including bacteria and archaea possess the ability to reduce atmospheric inert nitrogen into plant utilizable ammoniacal form using the nitrogenase enzyme (Sindhu and Dadarwal 1995; Sindhu 2004). Various nitrogen-fixing organisms are generally categorized either as free-living, associative or symbiotic N_2-fixing bacteria. Leguminous plants develop symbiotic association with members of the family Rhizobiaceae and form nitrogen-fixing nodules on the roots or stem of plants (Sindhu and Dadarwal 1995; Ahemad and Khan 2012; Sindhu et al. 2021). Other categories of nitrogen-fixing bacteria include free-living, associative and endophytic microorganisms such as species of *Azotobacter*, *Azospirillum*, *Azoarcus*, *Acetobacter nitrogenifigens*, *Bacillus*, *Clostridium pasteurianum*, *Gluconacetobacter diazotrophicus*, *Herbaspirillum*, *Pseudomonas* and some methanogens (Saravanan et al. 2007; Kumar and Rao 2012; Kumar et al. 2017). Free-living photosynthetic cyanobacteria (*Anabaena*, *Nostoc*, *Aulosira*, *Calothrix*, etc.) or symbiotic cyanobacteria (*Azolla-Anabaena* system) or blue green algae (BGA) are commonly found in different water bodies, which are reported to fix about 4–6 billion kilograms of N_2 annually (Smil 1999; Song et al. 2005; Singh et al. 2016). Estimations revealed that nearly 13% of the total fertilization on arable land is carried out by biological nitrogen fixation (Anglade et al. 2015; Sindhu et al. 2019a).

1.3.1.2 Phosphorus

Another major growth-limiting nutrient in the agriculture soils is phosphorous (P). The availability of P is limited in the soil due to its poor solubility and quick fixation in soil (Illmer and Schinner 1995). The P content in average soil is about 0.05% (w/w). Depending upon the neutral or acidic nature of the soil, the insoluble forms of phosphorus in soil include tricalcium phosphate $(Ca_3PO_4)_2$, aluminium phosphate (Al_3PO_4) or iron phosphate (Fe_3PO_4). Phosphorus availability and uptake from the soil affect many biological processes including photosynthesis, respiration, energy storage and transfer, cell division and enlargement (Shukla et al. 2014). Phosphorus has been demonstrated to play a significant role in early root formation and growth of roots, leading to improvement in the quality of fruits, vegetables and grains (Kadiri et al. 2013).

P-solubilizing microorganisms inhabit different soil ecosystems and are involved in conversion of bound insoluble phosphate into soluble form (Khan et al. 2013; Sindhu et al. 2014a). Various species of *Arthrobacter*, *Aerobacter*, *Azotobacter*, *Bacillus*, *Enterobacter*, *Erwinia*, *Flavobacterium*, *Micrococcus*, *Pseudomonas*, *Rhodococcus* and *Pantoea* have been identified as efficient phosphate-solubilizing bacteria (Sindhu et al. 2009; Richardson and Simpson 2011). The prominent fungal species involved in solubilization of bound phosphorus include *Aspergillus*, *Arthrobotrys oligospora*, *Penicillium*, *Fusarium*, *Trichoderma*, *Eupenicillium*, *Talaromyces*, yeast, etc. (Pradhan and Shukla 2005; Rawat et al. 2015; Saxena et al. 2016). Production of organic acids, chelation of mineral ions or decrease of pH has been established as the dominant mechanism for phosphorus solubilization (Goldstein 1994; Oliveira et al. 2009; Sindhu et al. 2014a). Previous studies have reported that inoculation of many phosphate-solubilizing bacteria, i.e. species of *Bacillus*, *Enterobacter*, *Burkholderia* and *Pseudomonas*, can enhance the phosphorus uptake and content of plants leading to enhancement of plant biomass and aid in increasing the crop yield (Oteino et al. 2013; Anand et al. 2016; Parastesh et al. 2019). Coinoculation of PSB and PGPR strains, i.e. *Bacillus polymyxa*, *Rhizobium* and *Pseudomonas fluorescens*, in wheat showed significant enhancement in nutrient uptake over the single inoculation of PGPR (*Rhizobium* or *Pseudomonas fluorescens*), PSB (*Bacillus polymyxa*) or uninoculated plants (Singh and Singh 2012; Jaybhay et al. 2017). However, inconsistent responses were observed due to inoculation of phosphorus-solubilizing bacterial strains under field conditions, owing to the different agronomic, climatic and management conditions in the field.

1.3.1.3 Potassium

The third major essential macronutrient potassium (K) plays a prominent role in protein synthesis, carbohydrate metabolism and translocation of starch in plants (Read et al. 2006; Bahadur et al. 2014). Potassium activates various enzymes and improves quality of seeds and fruits. In case of K deficiency, the susceptibility of plants increases towards wilting, necrosis, chlorosis and brown spotting, and more damage is observed due to exposure to the frost and heat (Jagessar and Allen 2012). Majority of potassium in soil (98%) exists in the unavailable mineral forms (Sparks

1987), and feldspars, muscovite, biotite, phlogopite and micas constitute the biggest pool of bound potassium in soils (Barre et al. 2007; Nishanth and Biswas 2008). The availability of K to crop plants in soils has decreased recently due to crop uptake, runoff, leaching and soil erosion (Sheng and Huang 2002). Therefore, under recent intensive agriculture system, the demand of potassium for agricultural crops is growing steadily at 3–3.5% per annum (Manning 2010).

The unavailable mineral forms of potassium are solubilized by K-solubilizing microorganisms (KSM) using several mechanisms, which include acid hydrolysis, secretion of soluble/insoluble molecules and secretion of polymers for chelation of mineral elements (Sindhu et al. 2014b; Parmar and Sindhu 2019). Several soil microorganisms such as *Acidithiobacillus ferrooxidans*, *Bacillus mucilaginosus*, *Bbacillus circulans*, *Bbacillus edaphicus*, *Bbacillus megaterium*, *Burkholderia* spp., *Paenibacillus* sp. and *Pseudomonas* spp. have been characterized to release potassium from K-bearing minerals in soil (Sheng et al. 2002; Basak and Biswas 2012; Teotia et al. 2016; Parmar and Sindhu 2019). Furthermore, inoculation with different strains of KSMs have been found to exert profound stimulatory effects on the development and growth of plants under different agro-ecosystems (Zeng et al. 2012; Parmar and Sindhu 2013; Sindhu et al. 2016; Yasin et al. 2016). For example, inoculation of potassium-solubilizing *Mesorhizobium* sp., *Paenibacillus* sp. and *Arthrobacter* sp., obtained from rape rhizospheric soil, into crops growing in K-limited soils was demonstrated to increase available K content resulting into improved growth, vigour, biomass yield and K uptake in ryegrass (Sindhu et al. 2014b; Xiao et al. 2017). Currently, several bacteria with efficient K solubilization efficiency are being extensively used as biological K fertilizers for agricultural sustainability (Zhang and Kong 2014; Nath et al. 2017; Sarkar et al. 2017).

1.3.1.4 Zinc

Another vital micronutrient required for optimum plant growth and plant metabolism is zinc (Zn) (Hughes and Poole 1991). Zn is adsorbed by plants as divalent cation (Zn^{2+}). It is estimated that about 30% of the cultivated soils are deficient in Zn worldwide (Suzuki et al. 2006). Several Zn-solubilizing bacteria have been isolated from agricultural soil, which include *Pseudomonas aeruginosa*, *Burkholderia*, *Bacillus subtilis*, *Acinetobacter*, *Enterobacter aerogenes*, *Mycobacterium brisbanense*, *Stenotrophomonas maltophilia*, *Thiobacillus thiooxidans* and *Xanthomonas retroflexus* (Jha et al. 2013; Kumari et al. 2016; Sindhu et al. 2019b). Similarly, many fungi belonging to *Aspergillus* and *Penicillium* sp. (Saravanan et al. 2003) and ericoid mycorrhizae, *Oidiodendron maius* Cd8, were reported to solubilize insoluble Zn compounds like zinc oxide, zinc carbonate and zinc sulphide (Martino et al. 2003). Zinc-solubilizing microorganisms were reported to solubilize Zn through production of various organic acids, protons and chelated ligands (Cunningham and Kuiack 1992; Sindhu et al. 2019b). Application of Zn-solubilizing bacterium *Bacillus* sp. as biofertilizer was demonstrated to improve Zn uptake by the field crops in soil ecosystems (Mahdi et al. 2010).

1.3.1.5 Other Nutrients

Plants also need other micronutrients in small quantities including calcium (Ca), magnesium (Mg), sulphur (S), iron (Fe), chlorine (Cl), manganese (Mn), zinc (Zn), copper (Cu), boron (B) and molybdenum (Mo) (Hochmuth and Hanlon 2010). These micronutrients participate in various enzymatic reactions and hormonal processes, strengthen the cell wall structure and contribute towards protection of the plants against heat stress and various diseases (Maathuis 2009). The deficiency of any of the macro- or micronutrients in soil adversely affects the growth and development of the entire plant (White and Brown 2010). Therefore, some of the rhizosphere-inhabiting bacteria have been identified, which improve the availability of these micronutrients for proper growth of plants. For example, inoculation of *P. fluorescens* strain C7 (having the ability to synthesize Fe-pyoverdine complex) increased iron content of the plant tissues and improved growth of *Arabidopsis thaliana* plants (Vansuyt et al. 2007). Similarly, inoculation of siderophore-producing *Pseudomonas* strain GRP3 on *Vigna radiata* increased the content of iron, chlorophyll a and chlorophyll b, along with decline in chlorotic symptoms in comparison to uninoculated plants (Sharma et al. 2003). Different bacteria including *Pseudomonas*, *Klebsiella*, *Salmonella*, *Enterobacter*, *Serratia* and *Comamonas* were found to mobilize sulphate-esters (Hummerjohann et al. 2000). Inoculation of arbuscular mycorrhizal (AM) fungi to maize and clover plants resulted in enhanced uptake of sulphur (Gray and Gerdemann 1973). Similarly, sulphur uptake was enhanced in tomato by inoculation of AM fungi (Cavagnaro et al. 2006). Allen and Shachar-Hill (2009) reported that inoculation of *Glomus intraradices* on transformed carrot roots enhanced the uptake of reduced forms of sulphur in vitro.

1.3.2 Phytohormone Production by Plant Microbiome and Nutrient Uptake

Different phytohormones like indole acetic acid (IAA), gibberellins, cytokinins and ethylene are produced by rhizospheric microorganisms (Kloepper et al. 2007; Malik and Sindhu 2008; Saleem et al. 2018). These PGPR-released hormones have been demonstrated to increase root surface area and number of root tips in many plants (Han et al. 2005; Khan et al. 2020) and significantly influence the crop growth and food production (Etesami and Maheshwari 2018; Khan et al. 2020). Besides this, phytohormone production by several microbes has been reported to alter plant physiology differentially, resulting in diverse outcomes that vary from pathogenesis to stimulation of plant growth (Spaepen 2015; Adolfsson et al. 2017).

Various strains of bacteria and fungi including *Pseudomonas*, *Agrobacterium*, *Rhizobium*, *Bradyrhizobium*, *Azospirillum*, *Botrytis*, *Aspergillus* and *Rhizopus* were found to produce auxins (Costacurta and Vanderleyden 1995; Jangu and Sindhu 2011). Inoculation of IAA-producing *Azospirillum* sp. resulted in improvement and growth of wheat (*Triticum aestivum* L.) plants (Spaepen et al. 2008; Baudoin et al. 2010). Besides this, production of IAA by some rhizobia was reported to function in symbiotic signalling that enhanced the nodule number and nitrogen fixation leading to enhancement of legume plant growth (Alikhani et al. 2010; Puente et al. 2018; Buhian and Bensmihen 2018). Moreover, the over-production of IAA in *Rhizobium leguminosarum* bv. *viciae* strain was caused by introduction of IAA biosynthetic pathway, and the constructed strains induced more efficient nodulation on *Vicia hirsuta* leading to significant effects on symbiotic parameters (Camerini et al. 2008), whereas differential response on nodulation and plant biomass enhancement was observed by coinoculation of IAA-producing *Pseudomonas* sp. with *Rhizobium* in green gram and black gram under chillum jar conditions (Jangu and Sindhu 2011). Similarly, a higher number of nodules were formed on *Phaseolus vulgaris* cultivar DOR364 by coinoculation of *A. brasilense* strain Sp245 with *Rhizobium etli* strain CNPAF512 (Remans et al. 2008). On the other hand, a less number of nodules were produced by coinoculation of an *Azospirillum* mutant strain having the capacity to produce only 10% of the IAA of the wild-type strain. These results indicated that IAA production by several rhizosphere bacteria contribute significantly in establishment and functioning of effective symbiosis.

Some PGPR isolates also possessed the ability to solubilize phosphate and produced ammonia along with production of IAA, and their inoculation showed stimulation of chickpea plant growth (Yadav et al. 2010). Similarly, accumulation of specific zeatin riboside (ZR)-like and isopentenyl adenosine (iPA)-like compounds was observed in both root and shoot of tobacco on colonization with arbuscular mycorrhizal fungi *G. intraradices* (Shaul-Keinan et al. 2002). Similarly, the levels of IAA and iPAs were significantly increased in litchi (*Litchi chinensis* Sonn.) seedlings by AM fungal inoculation, which induced root morphological alterations (Yao et al. 2005). Ectomycorrhizal fungi were also demonstrated to synthesize IAA-like compounds (Strzelczyk et al. 1997). On the other hand, it is also interesting that plant hormones can regulate the development of AM fungal symbiosis. For example, auxins were found to play a key role in promoting colonization and arbuscule formation (Ludwig-Muller and Guther 2007; Etemadi et al. 2014; Liao et al. 2018). Reduced arbuscule abundance was observed due to inoculation of abscisic acid (ABA)-deficient mutant (Sitiens) in tomato in comparison to increased AM colonization and arbuscule formation with wild-type fungi, suggesting that ABA plays a significant role in plant-fungal symbiosis (Herrera-Medina et al. 2007).

Gibberellins synthesized by *Gibberella fujikuroi* were firstly discovered to cause disease development on rice plants (Kurosawa 2003). Subsequently, various bacterial genera including *Azospirillum* sp., *Rhizobium* sp., *Acetobacter diazotrophicus*, *Herbaspirillum seropedicae*, *Bacillus* sp. and *Fusarium moniliforme* were reported to produce gibberellin-like substances (Bottini et al. 2004; Meleigy and Khalaf 2009). Other hormones produced by rhizosphere bacteria, i.e. jasmonic acid, salicylic acid and ethylene, were found to alter the hormonal signalling in plants, when exposed to different abiotic and biotic stresses, resulting into induction of systemic resistance (Zamioudis and Pieterse 2012). In the last decade, nitrogen oxide (NO) release in the rhizosphere has been demonstrated to act as the second messenger in IAA signalling pathway that affects different plant developmental processes (Molina-Favero et al. 2008). *A. brasilense* was demonstrated to synthesize NO by different aerobic pathways, and it promoted root branching in tomato (Molina-Favero et al. 2008). Similarly, NO production in *Sinorhizobium meliloti* was found to cause induction of root nodule organogenesis in *Medicago truncatula* roots (Pii et al. 2007).

1.3.3 Role of Plant Microbiome in Tolerance to Abiotic and Biotic Stresses

Soil or rhizosphere-inhabiting microbes possess the capabilities to ameliorate abiotic and biotic stresses in plants (Gopalakrishnan et al. 2015; Sehrawat and Sindhu 2019; Nabi et al. 2019; Sindhu and Sharma 2019; Phour and Sindhu 2020). Several microorganisms have been characterized including species of *Pseudomonas*, *Bacillus*, *Burkholderia*, *Azotobacter*, *Enterobacter*, *Methylobacterium*, *Serratia* and *Trichoderma*, which contributed significantly towards mitigation of different kinds of abiotic as well as biotic stresses (Atieno et al. 2012; Sorty et al. 2016; Meena et al. 2017) (Fig. 1.3). For example, inoculation of *Burkholderia phytofirmans* strain PsJN to wheat plants ameliorated the drought stress and improved the chlorophyll content, increased photosynthetic rate, water-use efficiency and grain yield in comparison to uninoculated control plants under field conditions (Naveed et al. 2014a). Similarly, enhancement of shoot and root biomass, leaf area and photosynthetic efficiency of maize plants was observed under drought stress conditions on coinoculation of *B. phytofirmans* and *Enterobacter* sp. strain FD17 in comparison to uninoculated plants (Naveed et al. 2014b). In similar studies, improved uptake of essential plant nutrients along with enhanced accumulation of antioxidants and osmolytes was obtained on inoculation of *T. harzianum* on Indian mustard (*Brassica juncea*) under saline conditions (Ahmad et al. 2015). Enhanced salinity tolerance was observed in rice (*Oryza sativa* L.) due to inoculation of *Pseudomonas* sp., and exopolysaccharide production and better root-colonizing capability of the inoculant strain were correlated with alleviation of salt stress (Sen and Chandrasekhar 2014). In another experiment, exopolysaccharide-producing thermotolerant *Pseudomonas putida* strain NBR10987 was demonstrated to play a role in amelioration of drought stress (Srivastava et al. 2008). Similarly, increased tolerance to low non-freezing temperatures was reported due to higher accumulation of stress-related proteins and metabolites by rhizosphere microorganisms (Theocharis et al. 2012).

Another group of microbes including microbial endophytes also imparted tolerance towards abiotic stresses. For example, enhanced salinity tolerance in *Miscanthus sinensis* was observed on inoculation of nitrogen-fixing consortium consisting of anaerobic *Clostridium* and *Enterobacter* sp. (Ye et al. 2005). Inoculated plants showed more plant biomass even at 100 mM NaCl treatment than the uninoculated plants. In another study, common endophyte of sugarcane *Gluconacetobacter diazotrophicus* was found to tolerate high sucrose concentrations (Velázquez-Hernández et al. 2011). Expression of enzyme levansucrase, which hydrolysed sucrose to glucose and fructose, was involved for transport of sucrose

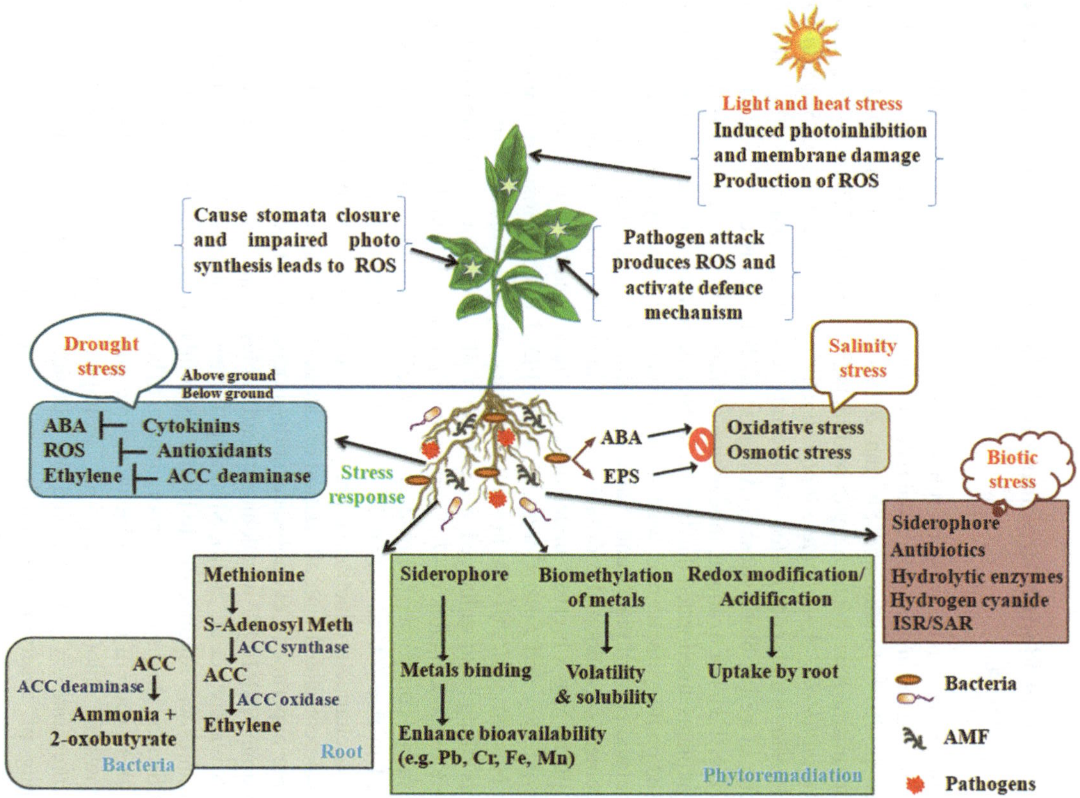

Fig. 1.3 Amelioration of different stresses (abiotic and biotic) and bioremediation of pollutants by beneficial microbes

into the bacterial cell and subsequent metabolism. Mutation of the gene encoding levansucrase reduced the production of enzyme resulting into decreased tolerance to desiccation and sodium chloride, suggesting the involvement of levansucrase enzyme as an osmoprotectant.

Besides abiotic stresses, plant growth and crop yield are also strongly influenced by biotic stresses caused by infestation of pathogenic microorganisms, insects, nematodes and weeds. Pathogenic microorganisms usually weaken or destroy the plant tissue and adversely affect the growth and yield of crop plants. Diseases caused by plant pathogens are estimated to cause about 20–40% yield losses annually worldwide in various cereal and legume crops (Oerke 2006). Recently, several microorganisms including bacteria, actinomycetes, fungi, viruses and nematodes have been isolated from the rhizosphere of crop plants, which showed antagonistic activity and suppressed the various diseases of agricultural crops (Mendes et al. 2011; Sehrawat and Sindhu 2019). Species of *Agrobacterium*, *Bacillus*, *Burkholderia*, *Enterobacter*, *Pseudomonas* and *Serratia* showed antagonism towards various plant pathogens and suppressed different diseases in several crops (Sehrawat and Sindhu 2019). Similarly, many bacterial strains, viz. *Bacillus thuringiensis*, *Bacillus cereus*, *Brevibacillus laterosporus*, *Pseudomonas entomophila*, *Serratia entomophila* and *S. maltophilia*, showed high mortality against specific insects compared to conventional insecticides (Vega and Kaya 2012; Sindhu et al. 2017). Production of secondary metabolites, i.e. antibiotics, toxins, siderophores, lipopeptides, hydrogen cyanide, volatile organic compounds and hydrolytic enzymes, contributed towards inhibition of phytopathogens and insects (Sindhu and Dadarwal 2001a; Mishra and Arora 2018; Sharma et al. 2018b). Interestingly, many of the biocontrol agents also possessed plant growth-promoting attributes, and the inoculation of these rhizobacterial strains caused stimulation of plant biomass and crop yield (Sindhu et al. 1999, 2002; Sahu and Sindhu 2011).

1.3.4 Induced Systemic Resistance to Pathogens Via Plant-Microbe Associations

Activation of the dormant defence mechanisms in plants leads to development of resistance in response to either infection of pathogen or attack by the insect (Pieterse et al. 2012, 2014; Walters et al. 2013). Some PGPR strains such as *Bacillus*, *Pseudomonas* and *Azospirillum* were reported to induce systemic resistance (ISR). Various metabolites/ISR-inducing agents produced by rhizobacteria included lipopolysaccharides; cyclic lipopeptides; siderophores; 2,4-diacetylphloroglucinol antibiotic; and 2,3-butanediol and acetoin (Lugtenberg and Kamilova 2009). Additionally, production of ethylene and jasmonic acid was found to induce resistance in plants against various pathogens (Lugtenberg and Kamilova 2009; Glick 2012).

Evidences showed that beneficial soil-inhabiting microorganisms hijack plant hormone signalling pathways to suppress the host defences (Pieterse et al. 2012; Sharifi and Ryu 2018; Tyagi et al. 2018). This phenomenon is illustrated in symbiotic association of legumes with rhizobia, where the defence reactions of the plants were quickly suppressed resulting into development of successful symbiotic association (Guinel 2015). Besides this, siderophore pseudobactin was found effective in inducing ISR (Meziane et al. 2005). In similar studies, *B. phytofirmans* PsJN was demonstrated to induce resistance to grey mould in grapevines (Theocharis et al. 2012). Upon invasion by pathogens, beneficial ISR-eliciting microbes have been found to sensitize the whole plant for faster activation of defence responses, and this phenomenon is termed as priming (Choudhary et al. 2007; Berendsen et al. 2012; Martinez-Medina et al. 2016). Moreover, exposure of plant roots to a particular population density of bacteria, i.e. about 10^5–10^7 colony-forming units per gram of root, was found to induce systemic resistance (Jankiewicz and Koltonowicz 2012; Bakker et al. 2013; Pieterse et al. 2014). Thus, higher microbial density obtained in the rhizosphere (varying from 10^8 to 10^9 bacteria per gram soil) due to the release of plant root exudates may be contributing towards specific stimulation or repression of particular microbial community (Berendsen et al. 2012; Pii et al. 2016; Stringlis et al. 2018).

1.3.5 Microbe-Induced Bioremediation and Increased Nutrient Uptake

Excessive concentration of toxic metals and pollutants in the rhizosphere causes stunted growth of plants due to excessive accumulation of contaminants. Various physico-chemical and biological techniques are used to reduce the heavy metal contaminants from soil (Glick 2010; Ullah et al. 2015). But, biological remediation using various microorganisms is considered as the most suitable, low-cost and eco-friendly technology for removal of toxic metals (Doble and Kumar 2005; Hassan et al. 2017). Under stressed conditions, complex toxic substances are converted to simple and nontoxic compounds resulting in higher survival rate of microbes (Thassitou and Arvanitoyannis 2001; Mustapha and Halimoon 2015) (Fig. 1.3). Metal sequestration or metal chelation was increased by lowering pH of the soil due to production of organic acids by certain soil microbes (Mishra et al. 2017). Besides this, complex formation capacity of metals with siderophores, other metabolites and transporter proteins from bacteria lowers the availability of metal contaminants in the soil (Rajkumar et al. 2010; Ahemad 2012; Ullah et al. 2015). In addition, some bacteria, viz. *Pseudomonas*, *Microbacterium*, *Verrucomicrobia* and *Actinobacteria*, along with some fungi including *Lewia* sp., and mycorrhizal fungi were reported to contribute towards rhizoremediation (Kawasaki et al. 2012; Yang et al. 2016). These microbial strains altered the mobility and bioavailability of metals, thereby resulting in increased uptake of metals by plants. Moreover, novel gene modulation technique employed to overexpress target genes or using gene insertion may result in construction of the desired transgenic microbes, which can serve as an alternative with higher-throughput capacity of metal sequestration and translocation (Singh et al. 2011). Therefore, inoculation with beneficial microbes for soil bioremediation has been found beneficial resulting into increased plant growth under climate-changing scenarios (Nautiyal et al. 2013; Tiwari et al. 2016).

Mycorrhizal fungi also have the potential to accumulate heavy metals and may enhance the uptake of diverse heavy metal ions significantly (Kumar et al. 2016a; Shukla et al. 2016). The extensive mycelial mats of vesicular arbuscular mycorrhiza (VAM) help the plants to acquire nutrients, particularly phosphorus, and their association is well recognized for growth promotion of various crop plants (Smith and Read 1997; Saxena et al. 2016). The heavy metal dislocation from the available nutrient pool by VAM fungi at the interface of soil and plant is responsible for removal of heavy metals. Thus, VAM fungi play a prominent role in alleviation of heavy metal stress (Leyval et al. 1997; Kumar et al. 2016a).

Other filamentous fungi including species of *Trichoderma*, *Penicillium* and *Aspergillus* have also been demonstrated to tolerate metal stress (Ezzouhri et al. 2009; Oladipo et al. 2017). The negatively charged functional groups of fungus cell wall possess metal-binding properties and play an important role in amelioration of metal stress (Ong et al. 2017). Thus, inoculation of *Aspergillus niger* (Coreño-Alonso et al. 2014) and *Trichoderma* sp. in chickpea plants was demonstrated to cause effective reduction in the metal concentration in the soil (Tripathi et al. 2013, 2017). Besides this, local soil microflora growing in the contaminated soil may also facilitate increased accumulation of heavy metals by the plant (Dikshit et al. 2013). Therefore, application of rhizospheric microflora and filamentous fungi has been reported to magnify the detoxification process by both elimination and immobilization mechanisms.

1.4 Inoculation Effect of Beneficial Microbes for Improving Plant Growth

Various beneficial microbes including *Azotobacter*, *Azospirillum* and rhizobia are currently being used or tested as inoculants on different crops under greenhouse or field conditions (Dadarwal et al. 1997; Herridge et al. 2008; Tiwari and Singh 2017; Kumar et al. 2016b). Thilakarathna and Raizada (2017) performed screening of rhizobial inoculants on soybean, and diverse *Bradyrhizobium* and *Sinorhizobium* isolates including some indigenous strains were found highly effective inoculants under field conditions. However, the symbiotic effectiveness of these rhizobial inoculants varied depending on legume genotypes, legume and rhizobia interactions, formulation of inoculant and its application methods (Sindhu and Dadarwal 1999, 2001a; Denton et al. 2017). Application of rhizobia with phosphorus and manure was found to increase the yield of common bean and soybean by twofold and threefold, respectively (Rurangwa et al. 2018). Moreover, the yield of subsequent crop maize also increased by about 6-fold and 2.5-fold following common bean and soybean, respectively.

Many PGPRs were reported to improve plant growth and crop yield by increasing nutrient uptake, producing plant hormones and suppressing soil-borne diseases (Sindhu and Sharma 2020). Inoculation of six phosphorus-solubilizing strains including three strains of *Bacillus* and one each of *Serratia*, *Arthrobacter* and *Pantoea* to rapeseed plants enhanced the final crop yield by 21–44% even without P application under field conditions (Valetti et al. 2018). Some of the PGPR strains possessed multiple beneficial activities, and their inoculation may cause a synergistic effect leading to enhanced crop production. For example, *Bacillus* strain M-3, *Burkholderia* strain OSU-7 and *Pseudomonas* strain BA-8 showed the capability of phosphorus solubilization along with production of IAA and cytokinin (Aslanta et al. 2007). The improved nutrient status and production of plant hormones caused by inoculation of these PGPR strains increased the fruit yield by 73.7, 88.2 and 137.5%, respectively, in young apple trees. Similarly, inoculation of *A. chroococcum* and *Bradyrhizobium* sp., having the ability to solubilize insoluble phosphate along with secretion of IAA and production of siderophores, was found to stimulate growth of *V. radiata* (Ahmad et al. 2016). In similar studies, inoculation of *P. fluorescens* strain N21.4, having the capability to produce siderophores and chitinases, to blackberry roots resulted in enhanced plant growth and number of fruits along with increase in promotion of total phenolics, flavonols and epicatechins/catechins metabolites (Garcia-Seco et al. 2015). Thus, applications of diverse microbial inoculants may also enhance the product quality and promote the synthesis of functional secondary metabolites along with improvement in the productivity of agricultural crops.

Inoculation with filamentous fungi and mycorrhizal fungi has also been found to increase the growth and yield of different agriculture and horticultural crops. More pronounced growth-promoting effects were observed with inoculation of AM fungal inocula. However, the production of AM fungal inocula is produced only in symbiosis with plants, and thus massive culture production is carried out in association with transformed hairy roots (Gianinazzi and Vosátka 2004; Berruti et al. 2016). In an interesting experiment, inoculation with a mixed AM fungal inocula caused a significant improvement in the fruit quality in terms of sweetness index to titratable acidity ratio, but no stimulatory effect on plant growth of tomato (in terms of fruit fresh weight and fruit number) was observed under field conditions (Bona et al. 2017). In some cases, the native AM fungal community has been found equally effective to the exogenously introduced AM fungi in promoting crop growth under field conditions (Panja and Chaudhuri 2004). Native AM fungal inoculation was found to increase tomato and carrot yields by 26% and 300%, respectively, in a field experiment infected with nematodes (Affokpon et al. 2011).

For obtaining better crop productivity, sometimes coinoculation of more than one group of microbes with different beneficial activities is recommended for practical purposes. Thus, coinoculation of rhizobia and PGPRs or AM fungi with rhizobia or different PGPRs may produce synergistic effects leading to improved plant growth and crop yield of legume crops under pot house or field conditions (Sindhu et al. 2002; Sivaramaiah et al. 2007). For example, combined inoculation of *Bradyrhizobium* spp. and *A. brasilense* significantly increased the soybean yield in comparison to single inoculation of *Bradyrhizobium* spp. in a field experiment (Hungria et al. 2015). Similarly, the application of rhizobia with cyanobacterium *Anabaena laxa* and *Trichoderma* sp. showed promotive effects on nodulation, nitrogen fixation and crop yield of pea, chickpea and lentil (Babu et al. 2015) In another study, 20% increase in wood yield of non-nodulating legume *Schizolobium parahyba* var. *amazonicum* was observed by coinoculation of AM fungi and PGPRs (Cely et al. 2016). Thus, dual inoculation strategies have shown more stimulation of plant growth and increase in crop yield than single inoculation. However, appropriate synergistic combinations are the prerequisite for selection of effective microbial strains for a particular crop species (Goel et al. 2002; Sivaramaiah et al. 2007; Chaudhary and Sindhu 2016). Besides these, the inoculation of different PGPR strains have been reported to improve the availability of nitrogen, phosphorus and potassium, and their growth-promoting effects on various crops are presented in Table 1.1.

In some cases, several constraints have been found to limit crop growth under field conditions in agricultural systems, and the inoculation of multifunctional PGPR strains does not improve root and shoot growth, yield and quality of agri-produce. Another factor is regarding screening of PGPR strains for beneficial traits. Growth conditions of microbes are generally different under laboratory and greenhouse conditions, which affect the survival and functioning of inoculated microorganisms under field conditions (Sindhu et al. 1995; Sindhu and Dadarwal 2000).

1.5 Engineering of Rhizosphere to Enhance Soil Fertility and Promotion of Plant Growth

Due to exponential increase in the world population, food production has to be enhanced from the available productive land by using genetic engineering technologies (Sharma et al. 2018a). Thus, rhizosphere microbiome engineering has emerged as a novel approach, which aims to alter plant root exudation patterns in the rhizosphere for attracting beneficial microorganisms enabling the plants to achieve enhanced biomass and crop yield within limited resources (Bais et al. 2004; Ryan et al. 2009; Sasse et al. 2018). Moreover, the transfer and expression of specific genes cloned from beneficial microbes into transgenic plants may cause production/excretion of specific nutritional compound leading to stimulation of beneficial microorganisms (Reddy et al. 2002; Marasco et al. 2012; Savka et al. 2013; Mueller and Sachs 2015). Recently, manipulation of host plants, engineering of metabolic pathways and specific expression of microbial genes are the important strategies for rhizosphere modification, contributing towards enhanced crop productivity (Berendsen et al. 2012; Quiza et al. 2015; Reinhold-Hurek et al. 2015).

1.5.1 Genetic Engineering of Plants for Enhancing Nutrient Uptake

Genetic engineering techniques have been successfully applied for enhancing particular gene expression in plants and/or increasing activities of various membrane

Table 1.1 Inoculation effect of PGPR as biofertilizers and their PGP activities

Types of biofertilizer	Bacterial strain	Plant growth-promoting activity	Effect on plant productivity	References
Nitrogen-fixing bacteria	*Azotobacter* sp. strain Avi2 (MCC 3432)	Nitrogen fixation, production of IAA, siderophore	Increased vegetative and reproductive growth of rice	Banik and Dangar (2019)
	Herbaspirillum seropedicae (strain ZAE94)	Nitrogen fixation, production of siderophore, IAA	Enhanced mineral uptake and increased yield of maize	Ávila et al. (2020)
	Azotobacter chroococcum, *Azotobacter vinelandii*	Nitrogen fixation, P solubilization, production of NH_3, HCN, IAA	Increased root length, shoot length, leaf number and chlorophyll content of maize plantlets	Jain et al. (2021)
	Rhizobium meliloti	Nitrogen fixation, production of siderophore and chitinase	Enhanced nitrogen efficiency, vegetative growth and yield of peanut	Mondal et al. (2020)
Phosphorus-solubilizing bacteria	*Alcaligenes faecalis* sub sp. *faecalis* str. S8	P solubilization, IAA, chitinolytic, proteolytic and pectinolytic enzymes, hydrogen cyanide	Increased plant height, root length and fresh root weight	Abdallah et al. (2016)
	Bacillus subtilis LK14	P solubilization, production of IAA	Enhanced host plant's nutrient uptake and amelioration of biotic and abiotic stresses	Khan et al. (2016)
	R. leguminosarum	P solubilization	Under drought stress, increased vegetative growth and yield performance of soybean	Igiehon et al. (2019)
Potassium-solubilizing bacteria	*Acidithiobacillus ferrooxidans*	Potassium solubilization	Increased oil and yield of pumpkin	Ansari et al. (2017)
Coinoculation of beneficial bacteria	*Azotobacter* (N_2 fixer), phosphate-solubilising bacteria (PSB) and potash-mobilizing bacteria (KMB)	Nitrogen fixation, P solubilization, K mobilization, IAA	Increased yield components and yield of wheat, improved soil nutrients balance, increased rhizosphere microbial activity	Game et al. (2020)
	Azospirillum lipoferum, *Azotobacter chroococcum*	Nitrogen fixation, IAA production	Increased vegetative growth, chlorophyll content, K^+/Na^+ ratio, increased various ROS enzyme activities of salt-affected maize plants	Latef et al. (2020)

transporter proteins, which are involved in absorbing soluble nutrients from the rhizosphere in several crop species. For example, nitrogen is adsorbed from the soil in the form of either NH_4^+ or NO_3^- through root-specific transporters (Masclaux-Daubresse et al. 2010). Similarly, phosphate transporter (PT) proteins of crop plants are involved in absorption of soluble form of Pi (orthophosphate) from the soil (López-Arredondo et al. 2014). Potassium is another important nutrient for the growth of plants, and K^+ absorption from soils is carried out by using K^+ channels in plant roots (Ashley et al. 2006; Wang and Wu 2013). However, the complicated plant-microbe interactions, availability of nutrients in the soil and environmental conditions deserve more attention for the uptake of different nutrients.

Attempts have been made to develop self-fertilizing crops for nitrogen, but limited success is achieved mainly due to the inability of nitrogenase expression in the absence of an oxygen protection system in eukaryotes (Bohlool et al. 1992; Dixon et al. 1997). However, nodule-like structures or pseudonodules were induced in wheat and rice crops with the help of lytic enzymes or hormones treatment, which showed nitrogenase activity and $^{15}N_2$ incorporation (Cocking et al. 1994). But, the expression of nitrogenase activity was less than 1% of the value observed in legumes. Similarly, induction of 'para nodules' was observed on wheat by inoculation with rhizobia or *Azospirillum*, following treatment with the 2,4-dichlorophenoxyacetic acid herbicide or with indole acetic acid and naphthaleneacetic acid auxins (Tchan and Kennedy 1989). The low level of nitrogenase activity was found only after inoculation with *Azospirillum*, and there was little evidence for bacterial infection. Thus, further understanding of the signal-exchange mechanisms between plants and microorganisms is needed for achieving the expression of active nitrogenase in cereal crops (Dixon et al. 1997; Shantharam and Mattoo 1997).

Another approach involves plant breeding strategies to alter the genetic make-up of plants for creating preferred rhizosphere communities (O'Connell et al. 1996; Sindhu et al. 2018). Plant genotypes could be screened among germplasms, which may restrict nodulation by ineffective indigenous rhizobia. Working on similar hypothesis, experiments were targeted for breeding of red clover with improved nitrogen fixation capability, and genotypes were constructed that maintained its superiority in nodulation against various *R. leguminosarum* bv. *trifolii* strains (Nutman 1984). During symbiotic interaction of pea and *R. leguminosarum*, it was observed that specific compatibility between nodulation gene *nod*X of *R. leguminosarum* bv. *viciae* strain TOM and *sym*2 of *Pisum sativum* cv. Afghanistan prevented nodulation by indigenous rhizobia, and only the inoculant strains was permitted to make nodules. The genetic manipulations of pea cultivar (Trapper) by transfer of *sym*2 gene and *nod*X transfer in rhizobial strains were found to improve symbiotic parameters under field studies (Fobert et al. 1991). Using a similar approach, symbiotic nitrogen fixation was improved in soybean-*Bradyrhizobium japonicum* symbiosis involving genetic modification of both micro- and macro-symbiont (Cregan et al. 1989; Sadowsky et al. 1991). Thus, development of legume cultivars with broad-spectrum effectiveness for nodulation and nitrogen fixation may obviate the requirement for repeated inoculation of legumes (Brockwell and Bottomley 1995).

Another interesting approach involves construction of transgenic plants having enhanced H^+ efflux and release of organic acids from roots. For example, enhanced exudation of citric acid was obtained by transferring a bacterial citrate synthase gene in tobacco roots, which improved the phosphate availability to the plant roots (Lopez-Bucio et al. 2000). Interestingly, enhanced leaf and fruit biomass were obtained by plantation of transgenic citrate-overproducing tobacco plants under P-limiting conditions. Similarly, alteration in the release of plant hormones, i.e. ethylene, may control root formation by regulating auxin transport within the root tip (Swarup et al. 2007). In another study, the secretion of rosmarinic acid (a caffeic acid ester) was stimulated by infection of *Pythium* fungus on *Ocimum basilicum* (Bais et al. 2002). Rosmarinic acid showed antimicrobial activity against pathogenic fungi in hairy root cultures to prevent further infection. In line with this, the infection of pathogenic bacterium *P. syringae* pv. tomato Pst DC 3000 on *Arabidopsis* was found to enhance colonization of beneficial *B. subtilis* strain FB17 (Rudrappa et al. 2008) that stimulated secretion of L-malic acid by roots. In another study, aphid attack at the root of *Capsicum annum* was reported to increase population of beneficial *B. subtilis*, which resulted into lowering of the pathogen *Ralstonia solanacearum* populations (Lee et al. 2012). Thus, manipulation of the plant-microbe signalling in response to attack of pathogens or insects on plants may result in release of specific metabolites in root exudates that may selectively stimulate population of plant growth-promoting microbes.

1.5.2 Manipulation of Beneficial Microbes

Bioengineering of microbial communities also presents a unique opportunity for improving crop productivity (Berg et al. 2014). While hundreds of bacterial strains possessing many beneficial traits have been identified (Köberl et al. 2015; Sehrawat and Sindhu 2019), engineering of a microbial community represents a significant challenge. Thus, successes have been achieved in transferring of nitrogen-fixation *nif*H gene from *Paenibacillus* to a plant growth-promoting *Bacillus* spp. (Kim and Timmusk 2013). Similarly, microbes have been constructed which possessed the capability to produce high concentrations of plant hormones (Arkhipova et al. 2005). Genes encoding antibiotics, siderophores or HCN may be transferred to other *Bacillus* species resulting into enhanced biocontrol activity against plant pathogens (Köberl et al. 2015).

Recently, considerable efforts have been made to improve the efficiency of biological nitrogen fixation in diazotrophic organisms to reduce our overall dependence on nitrogenous fertilizers (Dixon et al. 1997; Sindhu and Dadarwal 2001b; Schmidt et al. 2017). Considering the host-specific restriction of nodulation among *Rhizobium*-legume associations (Brewin 1991), attempts were made to extend the host range of rhizobia to other related bacteria. For example, symbiotic (*sym*) megaplasmid (pRme41b) of *R. meliloti* was transferred into *A. tumefaciens*

(Kondorosi et al. 1982), and ineffective nodule-like deformations were induced by these transconjugates on alfalfa roots. Using a similar approach, protoplast fusion approach was used to construct strains between *Rhizobium* sp. *Vigna* and *Rhizobium* sp. *Cicer* (Sindhu and Dadarwal 1993). The recombinants developed effective nodules on green gram, but ineffective pseudonodules were formed on chickpea under sterilized chillum jar conditions. Addition of multiple copies of *nif*A gene from *K. pneumoniae* was found to confer increased nodulation competitiveness to the recombinant strains of *R. meliloti* (Sanjuan and Olivares 1991). Similarly, mobilization of host-specific *nod*FEGHPQ genes of *R. meliloti* into *R. leguminosarum* bv. *trifolii* or bv. *viciae* caused nodule formation onto non-host alfalfa plants by these transconjugants (Putnoky and Kondorosi 1986). However, nodulation on normal hosts, i.e. white clover and vetch, was strongly inhibited by these transconjugants (Debelle et al. 1988; Faucher et al. 1989). In similar studies, transfer of nodulation gene *nod*Z was made to *R. leguminosarum* bv. *viciae*, and the transconjugates produced fucosylated Nod signals leading to formation of nodules even on *Macroptilium* (Lopez-Lara et al. 1996). These results suggested that substitutions or modifications of the nodulation genes of rhizobia may result in different decorations on nodulation (Nod) factors resulting in broadening of the host range of rhizobia.

The manipulation or transfer of nitrogen fixation (*nif*) genes of the nitrogenase enzyme complex in diazotrophic bacteria may further improve the N_2 fixation efficiency. Thus, increasing the NifA production (transcriptional activator of other *nif* genes) in diazotrophic bacteria could enhance the expression of the whole N_2-fixing system (Szeto et al. 1990). Using this approach, *R. meliloti* strains were constructed with enhanced *nif*A gene products, and their inoculation caused 7–15% increase in alfalfa plant biomass in comparison to wild-type parent strain under greenhouse conditions (Williams et al. 1990). On similar observations, enhanced rates of nitrogen fixation were observed in some diazotrophs, i.e. *Azotobacter vinelandii*, *Rhizobium phaseoli* and *Azorhizobium sesbania*, due to possession of more than one copy of the *nif*H gene (encoding dinitrogenase reductase) (Quinto et al. 1985; Jacobson et al. 1986). The multiple copies of *nif*H gene in these rhizobia could be the basis for enhanced nitrogen fixation due to increased turnover rate of electrons to MoFe protein. Dicarboxylic acid utilization provides energy for nitrogen fixation, and therefore recombinant strains of *R. meliloti* and *B. japonicum* showed 15% increase in the rate of N_2 fixation in comparison to wild-type strains (Ronson et al. 1990). Mutations in *hup* genes (confers hydrogenase activity to recycle evolved hydrogen during nitrogen fixation) were performed in Hup$^+$ *B. japonicum* strains (Merberg and Maier 1983) or *Rhizobium* sp. strains (Sindhu and Dadarwal 1992) for increasing the expression of hydrogenase activity. As expected, increased dry matter yields in green gram and black gram were obtained by the inoculation of the *Rhizobium* mutants having enhanced hydrogenase activity. However, sometimes the genetic manipulated strains may not perform effectively under field conditions. For example, increased rates of N_2 fixation were demonstrated by recombinant strains of *R. meliloti* and *B. japonicum* having increased expression of *nif*A and *dct*A genes under lab conditions. But, increase in N_2 fixation or yield improvement was not obtained by inoculation of the same constructs under the field conditions (Ronson et al. 1990).

Using recombinant DNA technology, attempts were made to improve phosphorus solubilization efficacy of phosphate-solubilizing bacteria by genetic transfer of genes (*mps* genes) involved in mineral phosphate solubilization (Rodriguez et al. 2006). For example, transfer of a PQQ synthase gene from *Erwinia herbicola* was found to improve the mineral phosphate solubilization in PGPR strains (Rodrıguez et al. 2001). The recombinant plasmid containing PQQ synthase gene from *E. coli* was transferred to PGPR strains of *Burkholderia cepacia* and *P. aeruginosa*. Production of larger clearing halo zones of tricalcium phosphate solubilization by several ex-conjugants indicated the heterologous expression of this gene in the recombinant PGPR strains. However, the expression of the *mps* genes in different hosts depended on genetic background of the recipient strain, the copy number of the indigenous plasmids and their metabolic interactions (Sindhu et al. 2009). In another study, phosphate solubilization and crop yield of *Medicago* plants were enhanced by inoculation of *S. meliloti* strain, which was genetically manipulated to overexpress PhoT protein (phosphate uptake ABC transporter permease protein) and phoC protein (phosphate uptake ABC transporter ATP-binding protein) (Carmen and Roberto 2010).

1.6 Future Prospects and Conclusion

Soil fertility and nutrient management are critical components to ensure food security for ever-increasing world population and for environmental sustainability (Alley and Vanlauwe 2009; Zahedi 2016; Kumar et al. 2017). However, the intensified agricultural patterns using high input of agrochemicals, i.e. fertilizers and pesticides, lead to environmental pollution and public health hazards along with huge economic loss. In sustainable agriculture approach, beneficial soil microorganisms have been characterized, and their inoculation caused an increase in soil fertility, suppression of the plant diseases and alleviation of biotic and abiotic stresses leading to growth enhancement and yield increase of various crops (Nyamangara et al. 2011; Etesami and Maheshwari 2018; Sindhu and Sharma 2020). It is established that rhizobial strains isolated from soil receiving high input of nitrogenous fertilizers are not symbiotically effective as compared with those strains isolated from soil receiving the low N level (Peoples et al. 1995; Reid et al. 2011). Similarly, high application of agrochemicals (e.g. phosphate, fungicide) adversely affected AM fungal colonization and functioning (Zhang et al. 2016, 2018). Recently, sequencing of a large number of microbial genomes and identification of specific genes provided a powerful tool to enhance the synthesis and release of PGP metabolites by the beneficial microbes (Bakker et al. 2012). The gene transfer and gene modification using recombinant DNA technology are routinely used for increasing the efficiency of microbial inoculants in relation to nitrogen fixation; solubilization of mineral potassium, phosphate and zinc; improving production of hormones; or enhancing the synthesis/release of different organic acids in the rhizosphere. However, different biotic and abiotic stress factors affect the survival of microbes along with detrimental effects on the functioning of genes involved in plant growth promotion. Therefore, understanding of the plant-microbe signalling, modification of rhizosphere population, microbial diversity, root colonization ability, mechanisms of action, formulation of inoculants and their application under field conditions are the major challenges for eco-friendly sustainable agricultural systems (Sehrawat et al. 2020). Furthermore, the use of climate-smart farming techniques and engineering of crop plants for better nutrient efficiency and disease control may contribute towards solving critical problems for managing sustainable environmental system for food security.

References

Abdallah RAB, Trabelsi BM, Nefzi A, Khiareddine HJ, Remadi MD (2016) Isolation of endophytic bacteria from *Withania somnifera* and assessment of their ability to suppress Fusarium wilt disease in tomato and to promote plant growth. J Plant Pathol Microbiol 7:352

Adolfsson L, Nziengui H, Abreu IN, Šimura J, Beebo A, Herdean A et al (2017) Enhanced secondary-and hormone metabolism in leaves of arbuscular mycorrhizal *Medicago truncatula*. Plant Physiol 175(1):392–411

Affokpon A, Coyne DL, Lawouin L, Tossou C, Agbede RD, Coosemans J (2011) Effectiveness of native West African arbuscular mycorrhizal fungi in protecting vegetable crops against root-knot nematodes. Biol Fertil Soils 47(2):207–217

Afzal A, Bano A (2008) *Rhizobium* and phosphate solubilizing bacteria improve the yield and phosphorus uptake in wheat (*Triticum aestivum*). Int J Agric Biol 10:85–88

Ahemad M (2012) Implication of bacterial resistance against heavy metals in bioremediation: a review. J Inst Integr Omics Appl Biotechnol 3:39–46

Ahemad M, Khan MS (2012) Ecological assessment of biotoxicity of pesticides towards plant growth promoting activities of pea (*Pisum sativum*) specific *Rhizobium* sp. strain MRP1. Emir J Food Agric 24:334–343

Ahmad P, Hashem A, Abd-Allah EF, Alqarawi AA, John R, Egamberdieva D, Gucel S (2015) Role of *Trichoderma harzianum* in mitigating NaCl stress in Indian mustard (*Brassica juncea* L) through antioxidative defense system. Front Plant Sci 6:868

Ahmad E, Zaidi A, Khan MS (2016) Effects of plant growth promoting rhizobacteria on the performance of green gram under field conditions. Jordan J Biol Sci 9(2):79–88

Aislabie Jand A, Deslippe JR (2013) Soil microbes and their contribution to soil services. In: Dymond JR (ed) Ecosystem services in New Zealand—conditions and trends. Manaaki Whenua Press, Lincoln, New Zealand, pp 143–161

Alikhani HA, Etesami H, Mohammadi L (2010) Effect of superior indole-3-acetic acid producing rhizobia and the combination with Ag and L-tryptophan on wheat growth indices under *in vitro* conditions. J Food Agric Environ 8(3–4):949–954

Allen JW, Shachar-Hill Y (2009) Sulfur transfer through an arbuscular mycorrhiza. Plant Physiol 149:549–560

Alley MM, Vanlauwe B (2009) The role of fertilizers in integrated plant nutrient management. International Fertilizer Industry Association, Paris

Anand K, Kumari B, Mallick MA (2016) Phosphate solubilizing microbes: an effective and alternative approach as biofertilizers. Int J Pharm Sci 8(2):37–40

Andrews M, De Meyer S, James EK, Stępkowski T, Hodge S, Simon MF et al (2018) Horizontal transfer of symbiosis genes within and between rhizobial genera: occurrence and importance. Gene 9(7):321

Anglade J, Billen G, Garnier J (2015) Relationships for estimating N_2 fixation in legumes: incidence for N balance of legume-based cropping systems in Europe. Ecosphere 6(3):1–24

Ansari MH, Hashemabadi D, Kaviani B (2017) Effect of cattle manure and sulphur on yield and oil composition of pumpkin (*Cucurbita pepo* var. *Styriaca*) inoculated with *Thiobacillus thiooxidans* in calcareous soil. Commun Soil Sci Plant Anal 48:2103–2118

Arkhipova TN, Veselov SU, Melentiev AI, Martynenko EV, Kudoyarova GR (2005) Ability of bacterium *Bacillus subtilis* to produce cytokinins and to influence the growth and endogenous hormone content of lettuce plants. Plant Soil 272(1):201–209

Arora NK (2015) Plant microbes symbiosis: applied facets. Springer, Dordrecht, p 381

Ashley MK, Grant M, Grabov A (2006) Plant responses to potassium deficiencies: a role for potassium transport proteins. J Exp Bot 57:425–436

Aslanta R, Cakmakc R, Sahin F (2007) Effect of plant growth promoting rhizobacteria on young apple tree growth and fruit yield under orchard conditions. Sci Hortic 111(4):371–377

Atieno M, Hermann L, Okalebo R, Lesueur D (2012) Efficiency of different formulations of *Bradyrhizobium japonicum* and effect of co-inoculation of *Bacillus subtilis* with two different strains of *Bradyrhizobium japonicum*. World J Microbiol Biotechnol 28:2541–2550

Ávila JS, Ferreira JS, Santos JS, Rocha PAD, Baldani VL (2020) Green manure, seed inoculation with *Herbaspirillum seropedicae* and nitrogen fertilization on maize yield. Rev Bras Eng Agrícola Ambient 24:590–595

Babu S, Prasanna R, Bidyarani N, Nain L, Shivay YS (2015) Synergistic action of PGP agents and *Rhizobium* spp. for improved plant growth, nutrient mobilization and yields in different leguminous crops. Biocatal Agric Biotechnol 4(4):456–464

Bahadur I, Meena VS, Kumar S (2014) Importance and application of potassic biofertilizer in Indian agriculture. Int Res J Biol Sci 3:80–85

Bais HP, Walker TS, Schweizer H, Vivanco J (2002) Root specific elicitation and antimicrobial activity of rosmarinic acid in hairy root cultures of *Ocimum basilicum*. Plant Physiol Biochem 40:983–995

Bais HP, Park SW, Weir TL, Callaway RM, Vivanco JM (2004) How plants communicate using the underground information superhighway. Trends Plant Sci 9(1):26–32

Bakker M, Manter D, Sheflin A, Weir T, Vivanco J (2012) Harnessing the rhizosphere microbiome through plant breeding and agricultural management. Plant Soil 360:1–13

Bakker PAHM, Doornbos RF, Zamioudis C, Berendsen RL, Pieterse CMJ (2013) Induced systemic resistance and the rhizosphere microbiome. Plant Pathol J 29:136–143

Balestrini R, Lumini E, Borriello R, Bianciotto V (2015) Plant–soil biota interactions. In: Paul EA (ed) Soil microbiology, ecology and biochemistry. Elsevier, London, pp 311–338

Banik A, Dangar TK (2019) Application of rice (*Oryza sativa* L.) root endophytic diazotrophic *Azotobacter* sp. strain AviZ (MCC 3432) can increase rice yield under green house and field condition. Microbiol Res 219:56–65

Barre P, Velde B, Catel N, Abbadie L (2007) Soil-plant potassium transfer: impact of plant activity on clay minerals as seen from X-ray diffraction. Plant Soil 292:137–146

Barthe A, Ten-Hage L, Lamy A, Rols JL, Leflaive J (2015) Resilience of aggregated microbial communities subjected to drought: small-scale studies. Microb Ecol 70(1):9–20

Basak B, Biswas D (2012) Modification of waste mica for alternative source of potassium: evaluation of potassium release in soil from waste mica treated with potassium solubilizing bacteria (KSB). Lambert Academic, Saarbrücken. ISBN: 978-3659298424

Baudoin E, Lerner A, Mirza MS, El Zemrany H, Prigent-Combaret C, Jurkevich E et al (2010) Effects of *Azospirillum brasilense* with genetically modified auxin biosynthesis gene *ipd*C upon the diversity of the indigenous microbiota of the wheat rhizosphere. Res Microbiol 161(3): 219–226

Berendsen RL, Pieterse CM, Bakker PA (2012) The rhizosphere microbiome and plant health. Trends Plant Sci 17:478–486

Berg G, Grube M, Schloter M, Smalla K (2014) Unraveling the plant microbiome: looking back and future perspectives. Front Microbiol 5:148

Berger A, Brouquisse R, Pathak PK, Hichri I, Singh I, Bhatia S et al (2018) Pathways of nitric oxide metabolism and operation of phytoglobins in legume nodules: missing links and future directions. Plant Cell Environ 41:2057–2068

Berruti A, Lumini E, Balestrini R, Bianciotto V (2016) Arbuscular mycorrhizal fungi as natural biofertilizers: let's benefit from past successes. Front Microbiol 6:1559

Bohlool BB, Ladha JK, Garrity DP, George T (1992) Biological nitrogen fixation for sustainable agriculture: a perspective. Plant Soil 141(1–2):1–11

Bona E, Cantamessa S, Massa N, Manassero P, Marsano F, Copetta A et al (2017) Arbuscular mycorrhizal fungi and plant growth-promoting pseudomonads improve yield, quality and nutritional value of tomato: a field study. Mycorrhiza 27(1):1–11

Bonfante P, Perotto S (2000) Outside and inside the roots: cell to-cell interactions among arbuscular mycorrhizal fungi, bacteria and host plants. In: Podila GK, Douds DD (eds) Current advances in mycorrhizae research. APS Press, St. Paul, pp 141–155

Bottini R, Cassán F, Piccoli P (2004) Gibberellin production by bacteria and its involvement in plant growth promotion and yield increase. Appl Microbiol Biotechnol 65(5):497–503

Brewin NJ (1991) Development of the legume root nodule. Annu Rev Cell Biol 7:191–226

Brockwell J, Bottomley PJ (1995) Recent advances in inoculant technology and prospects for the future. Soil Biol Biochem 27:683–697

Brundrett M (2004) Diversity and classification of mycorrhizal associations. Biol Rev 79:473–495

Buhian WP, Bensmihen S (2018) Nod factor regulation of phytohormone signaling and homeostasis during rhizobia-legume symbiosis. Front Plant Sci 9:1247

Camerini S, Senatore B, Lonardo E, Imperlini E, Bianco C, Moschetti G et al (2008) Introduction of a novel pathway for IAA biosynthesis to rhizobia alters vetch root nodule development. Arch Microbiol 190(1):67–77

Carmen B, Roberto D (2010) Improvement of phosphate solubilization and *Medicago* plant yield by an indole-3-acetic acid overproducing strain of *Sinorhizobium meliloti*. Appl Environ Microbiol 76:4626–4632

Cavagnaro TR, Jackson LE, Six J, Ferris H, Goyal S, Asami D et al (2006) Arbuscular mycorrhizas, microbial communities, nutrient availability, and soil aggregates in organic tomato production. Plant Soil 282:209–225

Cely MV, Siviero MA, Emiliano J, Spago FR, Freitas VF, Barazetti AR et al (2016) Inoculation of *Schizolobium parahyba* with mycorrhizal fungi and plant growth-promoting rhizobacteria increases wood yield under field conditions. Front Plant Sci 7:1708

Chaparro JM, Sheflin AM, Manter DK, Vivanco JM (2012) Manipulating the soil microbiome to increase soil health and plant fertility. Biol Fertil Soils 48:489–499

Chaudhary SR, Sindhu SS (2016) Growth stimulation of clusterbean (*Cyamopsis tetragonoloba*) by coinoculation with rhizosphere bacteria and *Rhizobium*. Legume Res 39(6):1003–1012

Chaudhary D, Sindhu SS (2017) Amelioration of salt stress in chick pea (*Cicer arietinum* L.) by inoculation of ACC deaminase containing rhizospheric bacteria with *Mesorhizobium* strains. Legume Res 40(1):80–86

Chaudhary D, Narula N, Sindhu SS, Behl RK (2013) Plant growth stimulation of wheat (*Triticum aestivum* L.) by inoculation of salinity tolerant *Azotobacter* strains. Physiol Mol Biol Plants 19(4):515–519

Choudhary DK, Prakash A, Johri BN (2007) Induced systemic resistance (ISR) in plants: mechanism of action. Indian J Microbiol 47:289–297

Cocking EC, Webster G, Batchelor CA, Davey MR (1994) Nodulation of non-legume crops: a new look. Agro-Industry Hi-Tech, pp 21–24

Coreño-Alonso A, Solé A, Diestra E, Esteve I, Gutiérrez-Corona JF, López GR et al (2014) Mechanisms of interaction of chromium with *Aspergillus niger* var. *tubingensis* strain Ed8. Bioresour Technol 158:188–192

Corradi N, Bonfante P (2012) The arbuscular mycorrhizal symbiosis: origin and evolution of a beneficial plant infection. PLoS Pathog 8(4):e1002600

Costacurta A, Vanderleyden J (1995) Synthesis of phytohormones by plant-associated bacteria. Crit Rev Microbiol 21(1):1–18

Cregan PB, Keyser HH, Sadowsky MJ (1989) Host plant effect on nodulation and competitiveness of the *Bradyrhizobium japonicum* serotype strains constituting serocluster 123. Appl Environ Microbiol 55:2532–2536

Cunningham JE, Kuiack C (1992) Production of citric and oxalic acid and solubilization of calcium phosphate by *Penicillium billai*. Appl Environ Microbiol 58:1451–1458

Dadarwal LR, Yadav LS, Sindhu SS (1997) Biofertilizer production technology: prospects. In: Dadarwal KR (ed) Biotechnological approaches in soil microorganisms for sustainable crop production. Scientific Publishers, Jodhpur, pp 149–191

Dahiya A, Kumar R, Sindhu SS (2021) Microbial endophytes mediated phosphorus solubilization: sustainable approach to improve soil fertility and plant growth. In: Maheshwari DK, Dheeman S (eds) Endophytes: mineral nutrient management, vol 3. Springer, Cham, pp 35–75

Debelle F, Maillet F, Vasse J, Rosenberg C, de Billy F, Truchet G et al (1988) Interference between *Rhizobium meliloti* and *Rhizobium trifolii* nodulation genes: Genetic basis of *R. meliloti* dominance. J Bacteriol 170:5718–5727

Denton MD, Phillips LA, Peoples MB, Pearce DJ, Swan AD, Mele PM et al (2017) Legume inoculant application methods: effects on nodulation patterns, nitrogen fixation, crop growth and yield in narrow-leaf lupin and faba bean. Plant Soil 419(1–2):25–39

Dikshit A, Shukla SK, Mishra RK (2013) Plant growth promotion by plant growth promoting rhizobacteria (PGPR) with nanomaterials in current agricultural scenario. Lap Lambert Academic Publishing. ISBN 978- 3-659-36774-8

Dixon R, Cheng Q, Shen GF, Day A, Day MD (1997) *nif* genes and expression in chloroplasts: prospects and problems. Plant Soil 194:193–203

Doble M, Kumar A (2005) Biotreatment of industrial effluents. Butterworth-Heinemann, Oxford, pp 19–38

Doehlemann G, Ökmen B, Zhu W, Sharon A (2017) Plant pathogenic fungi. In: Heitman J, Howlett BJ, Crous PW, Stukenbrock EH, James TY, Gow NAR (eds) The Fungal Kingdom, pp 701–726

Etemadi M, Gutjahr C, Couzigou JM, Zouine M, Lauressergues D, Timmers A et al (2014) Auxin perception is required for arbuscule development in arbuscular mycorrhizal symbiosis. Plant Physiol 166(1):281–292

Etesami H, Maheshwari DK (2018) Use of plant growth promoting rhizobacteria (PGPRs) with multiple plant growth promoting traits in stress agriculture: action mechanisms and future prospects. Ecotoxicol Environ Saf 156:225–246

Ezzouhri L, Castro E, Moya M, Espinola F, Lairini K (2009) Heavy metal tolerance of filamentous fungi isolated from polluted sites in Tangier, Morocco. Afr J Microbiol Res 3:35–48

Fasusi OA, Cruz C, Babalola OO (2021) Agricultural sustainability: microbial biofertilizers in rhizosphere management. Agriculture 11:163

Faucher C, Camut S, Denarie J, Truchet G (1989) The *nod*H and *nod*Q host range genes of *Rhizobium meliloti* behave as virulence genes in *R. leguminosarum* bv. *viciae* and determine changes in the production of plant-specific extracellular signals. Mol Plant Microbe Interact 2: 291–300

Ferrol N, Azcón-Aguilar C, Pérez-Tienda J (2019) Arbuscular mycorrhizas as key players in sustainable plant phosphorus acquisition: an overview on the mechanisms involved. Plant Sci 280:441–447

Fischer HM (1994) Genetic regulation of nitrogen fixation in rhizobia. Microbiol Rev 58:352–386

Fobert PR, Roy N, Nash JH, Iyer VN (1991) Procedure for obtaining efficient root nodulation of a pea cultivar by a desired *Rhizobium* strain and preempting nodulation by other strains. Appl Environ Microbiol 57:1590–1594

Franche C, Lindström K, Elmerich C (2009) Nitrogen-fixing bacteria associated with leguminous and non-leguminous plants. Plant Soil 321(1-2):35–59

Game BC, Ilhe BM, Pawar VS, Khandagale PP (2020) Effect of *Azotobacter*, phosphate solubilizing bacteria and potash mobilizing bacteria inoculants on productivity of wheat (*Triticum aestivum* L.). Int J Curr Microbiol App Sci 9(3):2800–2807

Garcia-Seco D, Zhang Y, Gutierrez-Mañero FJ, Martin C, Ramos-Solano B (2015) Application of *Pseudomonas fluorescens* to blackberry under field conditions improves fruit quality by modifying flavonoid metabolism. PLoS One 10(11):e0142639

Gianinazzi S, Vosátka M (2004) Inoculum of arbuscular mycorrhizal fungi for production systems: science meets business. Can J Bot 82(8):1264–1271

Glick BR (1995) The enhancement of plant growth by free-living bacteria. Can J Microbiol 41(2): 109–117

Glick BR (2010) Using soil bacteria to facilitate phytoremediation. Biotechnol Adv 28:367–374

Glick BR (2012) Plant growth-promoting bacteria: mechanisms and applications. Scientifica 2012: 963401

Goel AK, Sindhu SS, Dadarwal KR (2002) Stimulation of nodulation and plant growth of chickpea (*Cicer arietinum*) by *Pseudomonas* spp. antagonistic to fungal pathogens. Biol Fertil Soils 36: 391–396

Goldstein AH (1994) Involvement of the quinoprotein glucose dehydrogenase in the solubilization of exogenous phosphates by Gram-negative bacteria. In: Torriani-Gorini A, Yagil E, Silver S (eds) Phosphate in microorganisms: cellular and molecular biology. ASM Press, Washington, DC, pp 197–203

Gomiero T, Pimental D, Paoletti MG (2011) Is there a need for more sustainable agriculture? CRC Crit Rev Plant Sci 30:6–23

Gopalakrishnan S, Sathya A, Vijayabharathi R, Varshney RK, Gowda CL, Krishnamurthy L (2015) Plant growth promoting rhizobia: challenges and opportunities. 3 Biotech 5(4):355–377

Goyal S, Sindhu SS (2011) Composting of rice straw using different inocula and analysis of compost quality. Microbiol J 1:126–138

Gravel V, Antoun H, Tweddell R (2007) Growth stimulation and fruit yield of greenhouse tomato plants by inoculation with *Pseudomonas putida* or *Trichoderma atroviride*: possible role of indole acetic acid (IAA). Soil Biol Biochem 39:1968–1977

Gray L, Gerdemann J (1973) Uptake of sulphur-35 by vesicular-arbuscular mycorrhizae. Plant Soil 39:687–689

Guinel FC (2015) Ethylene, a hormone at the center-stage of nodulation. Front Plant Sci 6:1121

Han J, Sun L, Dong X, Cai Z, Sun X, Yang H et al (2005) Characterization of a novel plant growth-promoting bacteria strain *Delftia tsuruhatensis* HR4 both as a diazotroph and a potential biocontrol agent against various plant pathogens. Syst Appl Microbiol 28(1):66–76

Hassan TU, Bano A, Naz I (2017) Alleviation of heavy metals toxicity by the application of plant growth promoting rhizobacteria and effects on wheat grown in saline sodic field. Int J Phytoremed 19:522–529

Hayat R, Ali S, Amara U, Khalid R, Ahmed I (2010) Soil beneficial bacteria and their role in plant growth promotion. Ann Microbiol 60:579–598

Herrera-Medina MJ, Steinkellner S, Vierheilig H, Ocampo Bote JA, Garciá Garrido JM (2007) Abscisic acid determines arbuscule development and functionality in the tomato arbuscular mycorrhiza. New Phytol 175:554–564

Herridge DF, Peoples MB, Boddey RM (2008) Global inputs of biological nitrogen fixation in agricultural systems. Plant Soil 311(1–2):1–18

Hochmuth GJ, Hanlon EA (2010) Commercial vegetable fertilization principles. Fla Coop Ext Serv Cir SL 319. http://edis.ifas.ufl.edu/pdffiles/cv/cv00900.pdf

Hughes MN, Poole RK (1991) Metal speciation and microbial growth—the hard (and soft) facts. J Gen Microbiol 137(4):725–734

Hummerjohann J, Laudenbach S, Rétey J, Leisinger T, Kertesz MA (2000) The sulfur-regulated arylsulfatase gene cluster of *Pseudomonas aeruginosa*, a new member of the cys regulon. J Bacteriol 182(7):2055–2058

Hungria M, Nogueira MA, Araujo RS (2015) Soybean seed coinoculation with *Bradyrhizobium* spp. and *Azospirillum brasilense*: a new biotechnological tool to improve yield and sustainability. Am J Plant Sci 6(06):8–11

Igiehon NO, Babalola OO, Aremu BR (2019) Genomic insights into plant growth promoting rhizobia capable of enhancing soybean germination under drought stress. BMC Microbiol 19: 159

Illmer PA, Schinner F (1995) Solubilization of inorganic calcium phosphates solubilization mechanisms. Soil Biol Biochem 27:257–263

Jacobson MR, Premakumar R, Bishop PE (1986) Transcriptional regulation of nitrogen fixation by molybdenum in *Azotobacter vinelandii*. J Bacteriol 167:480–486

Jagessar RC, Allen R (2012) Phytochemical screening and atomic absorption spectroscopic studies of solvent type extract from leaves of *Terminalia catappa*, (almond). Nat Appl Sci 3(3):17–26

Jain D, Sharma J, Kaur G, Bhojiya AA, Chauhan S, Sharma V et al (2021) Phenetic and molecular diversity of nitrogen fixating plant growth promoting *Azotobacter* isolated from semiarid regions of India. Bio Med Res Int 2021:6686283

Jangu OP, Sindhu SS (2011) Differential response of inoculation with indole acetic acid producing *Pseudomonas* sp. in green gram (*Vigna radiata* L.) and black gram (*Vigna mungo* L.). Microbiol J 1(5):159–173

Jankiewicz U, Koltonowicz M (2012) The involvement of *Pseudomonas* bacteria in induced systemic resistance in plants. Appl Biochem Microbiol 48:244–249

Jaybhay SA, Taware SP, Varghese P (2017) Microbial inoculation of *Rhizobium* and phosphate solubilizing bacteria along with inorganic fertilizers for sustainable yield of soybean [*Glycine max* (L.) Merrill]. J Plant Nutr 40(15):2209–2216

Jha A, Saxena J, Sharma V (2013) An investigation on phosphate solubilization potential of agricultural soil bacteria as affected by different phosphorus sources, temperature, salt and pH. Commun Soil Sci Plant Anal 44(16):2443–2458

Kadiri DD, Gorle N, Peetala KVR, Peela S (2013) Isolation, screening and identification of phosphate solubilising bacteria from different regions of Visakhapatnam and Araku Valley. Int J Adv Biotechnol Res 4(4):518–526

Kalra VP, Sharma PK (2015) Agro-climatic indices of fodder maize *Zea mays* L in relation to nutritional status. Int J Farm Sci 5(2):7–14

Kang SM, Khan AL, Waqas M, You YH, Kim JH, Kim JG et al (2014) Plant growth-promoting rhizobacteria reduce adverse effects of salinity and osmotic stress by regulating phytohormones and antioxidants in *Cucumis sativus*. J Plant Interact 9(1):673–682

Karthikeyan N, Prasanna RL, Nain L, Kaushik BD (2007) Evaluating the potential of plant growth promoting cyanobacteria as inoculants for wheat. Eur J Soil Biol 43:23–30

Kawasaki A, Watson ER, Kertesz MA (2012) Indirect effects of polycyclic aromatic hydrocarbon contamination on microbial communities in legume and grass rhizospheres. Plant Soil 358:169–182

Kennedy AC, Smith KL (1995) Soil microbial diversity and the sustainability of agricultural soils. Plant Soil 170(1):75–86

Khadem AF, Pol A, Jetten MSM, Op den Camp HJM (2010) Nitrogen fixation by verruco microbial methanotroph *Methylacidiphilum fumariolicum* Sol V. Microbiology 156:1052–1059

Khan MS, Zaidi A, Wani PA (2007) Role of phosphate solubilizing microorganisms in sustainable agriculture: a review. Agron Sustain Dev 27:29–43

Khan MS, Zaidi A, Ahemad M, Oves M, Wani PA (2010) Plant growth promotion by phosphate solubilizing fungi: current perspective. Arch Agron Soil Sci 56(1):73–98

Khan MS, Ahmad E, Zaidi A, Oves M (2013) Functional aspect of phosphate-solubilizing bacteria: importance in crop production. In: Maheshwari DK, Saraf M, Aeron A (eds) Bacteria agrobiol: crop productivity. Springer, Berlin, pp 237–265

Khan AL, Halo BA, Elyassi A, Ali S, Al-Hosni K, Hussain J (2016) Indole acetic acid and ACC deaminase from endophytic bacteria improves the growth of *Solanum lycopersicum*. Electron J Biotechnol 21:58–64

Khan N, Bano A, Ali S, Babar Md A (2020) Crosstalk among phytohormones from planta and PGPR under biotic and abiotic stresses. Plant Growth Regul 90:189–203

Khare E, Arora NK (2011) Physiologically stressed cells of fluorescent *Pseudomonas* EKi as better option for bioformulation development for management of charcoal rot caused by *Macrophomina phaseolina* in field conditions. Curr Microbiol 62(6):1789–1793

Kim SB, Timmusk S (2013) A simplified method for gene knockout and direct screening of recombinant clones for application in *Paenibacillus polymyxa*. PLoS One 8(6):e68092

Kloepper JW, Gutierrez-Estrada A, McInroy JA (2007) Photoperiod regulates elicitation of growth promotion but not induced resistance by plant growth-promoting rhizobacteria. Can J Microbiol 53(2):159–167

Köberl M, White RA, Erschen S, El-Arabi TF, Jansson JK, Berg G (2015) Draft genome sequence of *Streptomyces* sp. strain Wb2n-11, a desert isolate with broad-spectrum antagonism against soilborne phytopathogens. Genome Announc 3(4):e00860–e00815

Kondorosi A, Kondorosi E, Pankhurst CE, Broughton WJ, Banfalvi Z (1982) Mobilization of *Rhizobium meliloti* megaplasmid carrying nodulation and nitrogen fixation genes in other rhizobia and *Agrobacterium*. Mol Gen Genet 188:433–439

Kumar SRS, Rao KVB (2012) Biological nitrogen fixation: a review. Int J Adv Life Sci 1:1–6

Kumar R, Mishra RK, Qidwai A, Shukla SK, Mishra V, Pandey A et al (2016a) Detoxification and tolerance of heavy metals in plants. In: Ahmad P (ed) Plant metal interaction: emerging remediation techniques. Elsevier, Amsterdam, pp 335–360

Kumar A, Patel JS, Bahadur I, Meena VS (2016b) The molecular mechanisms of KSMs for enhancement of crop production under organic farming. In: Meena VS, Maurya BR, Verma JP, Meena RS (eds) Potassium solubilizing microorganisms for sustainable agriculture. Springer, New Delhi, pp 61–75

Kumar A, Meena VS, Maurya BR, Raghuwanshi R, Bisht JK, Pattanayak A (2017) Towards the biological nitrogen fixation and nitrogen management in legume under sustainable agriculture. Appl Soil Ecol 117

Kumari KS, Devi SNP, Vasandha S (2016) Zinc solubilizing bacterial isolates from the agricultural fields of Coimbatore, Tamil Nadu, India. Curr Sci 110(2):196–205

Kurosawa E (2003) Experimental studies on the nature of the substance secreted by the "bakanae" fungus. Lancet Infect Dis 3(3):125–126

Lareen A, Burton F, Schafer P (2016) Plant root-microbe communication in shaping root microbiomes. Plant Mol Biol 90:575–587

Latef HA, Alhmad MFA, Zakir A (2020) Inoculation with *Azospirillum lipoferum* or *Azotobacter chroococcum* reinforces maize growth by improving physiological activities under saline conditions. J Plant Growth Regul 39:1293–1308

Lee B, Lee S, Ryu CM (2012) Foliar aphid feeding recruits rhizosphere bacteria and primes plant immunity against pathogenic and non-pathogenic bacteria in pepper. Ann Bot 110(2):281–290

Leyval C, Turnau K, Haselwandter K (1997) Effect of heavy metal pollution on mycorrhizal colonization and function—physiological, ecological and applied aspects. Mycorrhiza 7:139–153

Liao D, Wang S, Cui M, Liu J, Chen A, Xu G (2018) Phytohormones regulate the development of arbuscular mycorrhizal symbiosis. Int J Mol Sci 19(10):3146

López-Arredondo DL, Leyva-González MA, González-Morales SI López-Bucio J, Herrera-Estrella L (2014) Phosphate nutrition: improving low-phosphate tolerance in crops. Annu Rev Plant Biol 65:95–123

Lopez-Bucio J, de la Vega OM, Guevara-Garcia A, Herreera-Estrella I (2000) Enhanced phosphate uptake in transgenic tobacco plants that over-produce citrate. Nat Biotechnol 18:450–453

Lopez-Lara IM, Blok-Tip L, Quinto C, Garcia ML, Bloemberg GV, Lamers GEM et al (1996) *nodZ* of *Bradyrhizobium* extends the nodulation host range of *Rhizobium* by adding a fucosyl residue to nodulation factors. Mol Microbiol 21:397–408

Ludwig-Muller J, Guther M (2007) Auxins as signals in arbuscular mycorrhiza formation. Plant Signal Behav 2:194–196

Luginbuehl LH, Oldroyd GE (2017) Understanding the arbuscule at the heart of endomycorrhizal symbioses in plants. Curr Biol 27(17):952–963

Lugtenberg B, Kamilova F (2009) Plant-growth-promoting rhizobacteria. Annu Rev Microbiol 63: 541–556

Ma Y, Oliveira RS, Freitas H, Zhang C (2016) Biochemical and molecular mechanisms of plant-microbe-metal interactions: relevance for phytoremediation. Front Plant Sci 7:918

Maathuis FJM (2009) Physiological functions of mineral macronutrients. Curr Opin Plant Biol 12: 250–258

Mahdi SS, Dar SA, Ahmad S, Hassan GI (2010) Zinc availability—a major issue in agriculture. Res J Agric Sci 3(3):78–79

Malik DK, Sindhu SS (2008) Transposon-derived mutants of *Pseudomonas* strains altered in indole acetic acid production: effect on nodulation and plant growth in green gram (*Vigna radiata* L.). Physiol Mol Biol Plants 14(4):315–320

Malik DK, Sindhu SS (2011) Production of indole acetic acid by *Pseudomonas* sp.: effect of coinoculation with *Mesorhizobium* sp. *cicer* on nodulation and plant growth of chickpea (*Cicer arietinum*). Physiol Mol Biol Plants 17(1):25–32

Manjunath M, Prasanna R, Shama P, Nain L, Singh R (2011) Developing PGPR consortia using novel genera *Providencia* and *Alcaligenes* along with cyanobacteria for wheat. Arch Agron Soil Sci 57:873–887

Manning DAC (2010) Mineral sources of potassium for plant nutrition. Rev Agron Sustain Dev 30: 281–294

Mansfield J, Genin S, Magori S, Citovsky V, Sriariyanum M, Ronald P et al (2012) Top 10 plant pathogenic bacteria in molecular plant pathology. Mol Plant Pathol 13(6):614–629

Marasco R, Rolli E, Ettoumi B, Vigani G, Mapelli F, Borin S et al (2012) A drought resistance-promoting microbiome is selected by root system under desert farming. PLoS One 7:e48479

Marschner H (2011) Mineral nutrition of higher plants. Academic Press, London

Martinez-Medina A, Flors V, Heil M, Mauch-Mani B, Pieterse CMJ, Pozo MJ et al (2016) Recognizing plant defense priming. Trends Plant Sci 21:818–822

Martino E, Perotto S, Parsons R, Gadd GM (2003) Solubilization of insoluble inorganic zinc compounds by ericoid mycorrhizal fungi derived from heavy metal polluted sites. Soil Biol Biochem 35:133–141

Masclaux-Daubresse C, Daniel-Vedele F, Dechorgnat J, Chardon F, Gaufichon L, Suzuki A (2010) Nitrogen uptake, assimilation and remobilization in plants: challenges for sustainable and productive agriculture. Ann Bot 105(7):1141–1157

Maurya BR, Meena VS, Meena OP (2014) Influence of inceptisol and alfisol's potassium solubilizing bacteria (KSB) isolates on release of K from waste mica. Vegetos 27:181–187

Meena KK, Sorty AM, Bitla UM, Choudhary K, Gupta P, Pareek A et al (2017) Abiotic stress responses and microbe-mediated mitigation in plants: the omics strategies. Front Plant Sci 8:172

Meleigy SA, Khalaf MA (2009) Biosynthesis of gibberellic acid from milk permeate in repeated batch operation by a mutant *Fusarium moniliforme*, cells immobilized on loofa sponge. Bioresour Technol 100(1):374–379

Mendes R, Kruizt M, de Bruizn I, Dekkers E, van der Voort M, Schneider JH (2011) Deciphering the rhizosphere microbiome for disease suppressive bacteria. Science 332:1097–1100

Merberg D, Maier RJ (1983) Mutants of *Rhizobium japonicum* with increased hydrogenase activity. Science 220:1064–1065

Meziane H, Van der Sluis I, Van Loon LC, Höfte M, Bakker PA (2005) Determinants of *Pseudomonas putida* WCS358 involved in inducing systemic resistance in plants. Mol Plant Pathol 6:177–185

Mikkelsen R, Hartz TK (2008) Nitrogen sources for organic crop production. Better Crops 92:16–19

Mishra S, Arora NK (2012) Evaluation of rhizospheric *Pseudomonas* and *Bacillus* as biocontrol tool for *Xanthomonas campestris* pv *campestris*. World J Microbiol Biotechnol 28(2):693–702

Mishra J, Arora NK (2018) Secondary metabolites of fluorescent pseudomonads in biocontrol of phytopathogens in sustainable agriculture. Appl Soil Ecol 125:35–45

Mishra J, Singh R, Arora NK (2017) Alleviation of heavy metal stress in plants and remediation of soil by rhizosphere microorganisms. Front Microbiol 8:1706

Molina-Favero C, Creus CM, Simontacchi M, Puntarulo S, Lamattina L (2008) Aerobic nitric oxide production by *Azospirillum brasilense* Sp245 and its influence on root architecture in tomato. Mol Plant Microbe Interact 21(7):1001–1009

Mondal M, Skalicky M, Garai S, Hossain A, Sarkar S, Banerjee H et al (2020) Supplementing nitrogen in combination with *Rhizobium* inoculation and soil mulch in peanut (*Arachis hypogaea* L.) production system Part II. Effect on phenology, growth, yield attributes, pod quality, profitability and nitrogen use efficiency. Agronomy 10:1513

Mueller UG, Sachs JL (2015) Engineering microbiomes to improve plant and animal health. Trends Microbiol 23:606–617

Mustapha MU, Halimoon N (2015) Screening and isolation of heavy metal tolerant bacteria in industrial effluent. Proc Environ Sci 30:33–37

Nabi RBS, Tayade R, Hussain A, Kulkarni KP, Imran QM, Mun B-G et al (2019) Nitric oxide regulates plant responses to drought, salinity and heavy metal stress. Environ Exp Bot 161:120–133

Naiman AD, Latronico A, Garca de Salamone IE (2009) Inoculation of wheat with *Azospirillum brasilense* and *Pseudomonas fluorescens*: impact on the production and culturable rhizosphere microflora. Eur J Soil Biol 45:44–51

Nath D, Maurya BR, Meena VS (2017) Documentation of five potassium-and phosphorus-solubilizing bacteria for their K and P-solubilization ability from various minerals. Biocatal Agric Biotechnol 10:174–181

Nautiyal CS, Srivastava S, Chauhan PS, Seem K, Mishra A, Sopory SK (2013) Plant growth-promoting bacteria *Bacillus amyloliquefaciens* NBRISN13 modulates gene expression profile of leaf and rhizosphere community in rice during salt stress. Plant Physiol Biochem 66:1–9

Naveed M, Hussain MB, Zahir ZA, Mitter B, Sessitsch A (2014a) Drought stress amelioration in wheat through inoculation with *Burkholderia phytofirmans* strain PsJN. Plant Growth Regul 73: 121–131

Naveed M, Mitter B, Reichenauer TG, Wieczorek K, Sessitsch A (2014b) Increased drought stress resilience of maize through endophytic colonization by *Burkholderia phytofirmans* PsJN and *Enterobacter* sp. FD17. Environ Exp Bot 97:30–39

Newbould P (1989) Use of nitrogen fertilizer in agriculture: where do we go sustainably and ecologically? Plant Soil 115:297–311

Nihorimbere V, Ongena M, Smargiassi M, Thonart P (2011) Beneficial effect of the rhizosphere microbial community for plant growth and health. Biotechnol Agron Soc Environ 2:327–337

Nishanth D, Biswas DR (2008) Kinetics of phosphorus and potassium release from rock phosphate and waste mica enriched compost and their effect on yield and nutrient uptake by wheat (*Triticum aestivum*). Bioresour Technol 99(9):3342–3353

Nutman PS (1984) Improving nitrogen fixation in legumes by plant breeding: The relevance of host selection experiments in red clover (*Trifolium pretense* L.) and subterraneum clover (*T. subterraneum* L.). Plant Soil 82:285–301

Nyamangara J, Makarimayi E, Masvaya EN, Zingore S, Delve RJ (2011) Effect of soil fertility management strategies and resource-endowment on spatial soil fertility gradients, plant nutrient uptake and maize growth at two smallholder areas, North-Western Zimbabwe. S Afr J Plant Soil 28(1):1–10

O'Connell KP, Goodman RM, Handelsman J (1996) Engineering the rhizosphere: a expressing a bias. Trends Biotechnol 14:83–86

Oerke EC (2006) Crop losses to pests. J Agric Sci 44(1):31–43

Oladipo OG, Awotoye CO, Olayinka A, Bezuidenhout CC, Maboeta MS (2017) Heavy metal tolerance traits of filamentous fungi isolated from gold and gemstone mining sites. Braz J Microbiol 49:29–37

Oliveira CA, Alves VMC, Marriel IE, Gomes EA, Scotti MR, Carneiro NP et al (2009) Phosphate solubilizing microorganisms isolated from rhizosphere of maize cultivated in an oxisol of the Brazilian Cerrado Biome. Soil Biol Biochem 41:1782–1787

Ong GH, Ho XH, Shamkeeva S, Fernando MS, Shimen A, Wong LS (2017) Biosorption study of potential fungi for copper remediation from Peninsular Malaysia. Remediat J 27:59–63

Oteino N, Culhane J, Germaine KJ, Ryan D, Brazil D, Dowling DN (2013) Screening of large collections of plant associated bacteria for effective plant growth promotion and colonization. In: Proceeding of positive plant microbial interactions: their role in maintaining sustainable agricultural and natural ecosystems. Association of Applied Biologists (AAB) Conference, North Linconshire, pp 13–18

Panja BN, Chaudhuri S (2004) Exploitation of soil arbuscular mycorrhizal potential for AM-dependent mandarin orange plants by pre-cropping with mycotrophic crops. Appl Soil Ecol 26(3):249–255

Parastesh F, Alikhani HA, Etesami H (2019) Vermicompost enriched with phosphate–solubilizing bacteria provides plant with enough phosphorus in a sequential cropping under calcareous soil conditions. J Clean Prod 221:27–37

Parmar P, Sindhu SS (2013) Potassium solubilization by rhizosphere bacteria: influence of nutritional and environmental conditions. J Microbiol Res 3:25–31

Parmar P, Sindhu SS (2019) The novel and efficient method for isolating potassium solubilizing bacteria from rhizosphere soil. Geomicrobiol J 36(2):130–136

Peoples MB, Herridge DF, Ladha JK (1995) Biological nitrogen fixation: an efficient source of nitrogen for sustainable agricultural production? In: Ladha JK, Peoples MB (eds) Management of biological nitrogen fixation for the development of more productive and sustainable agricultural systems. Springer, Dordrecht, pp 3–28

Phour M, Sindhu SS (2020) Exploiting plant growth promoting bacteria for salt tolerance and crop productivity in mustard. Appl Soil Ecol 149:103518

Pieterse CM, Van der Does D, Zamioudis C, Leon-Reyes A, Van Wees SC (2012) Hormonal modulation of plant immunity. Annu Rev Cell Dev Biol 28:489–521

Pieterse CM, Zamioudis C, Berendsen RL, Weller DM, Van Wees SC, Bakker PA (2014) Induced systemic resistance by beneficial microbes. Annu Rev Phytopathol 52:347–375

Pii Y, Crimi M, Cremonese G, Spena A, Pandolfini T (2007) Auxin and nitric oxide control indeterminate nodule formation. BMC Plant Biol 7:21

Pii Y, Borruso L, Brusetti L, Crecchio C, Cesco S, Mimmo T (2016) The interaction between iron nutrition, plant species and soil type shapes the rhizosphere microbiome. Plant Physiol Biochem 99:39–48

Pochon S, Terrasson E, Guillemette T, Iacomi-Vasilescu B, Georgeault S, Juchaux M et al (2012) The *Arabidopsis thaliana-Alternaria brassicicola* pathosystem: a model interaction for investigating seed transmission of necrotrophic fungi. Plant Methods 8(1):16

Pradhan N, Shukla LB (2005) Solubilization of inorganic phosphates by fungi isolated from agriculture soil. Afr J Biotechnol 5:850–854

Pucciariello C, Boscari A, Tagliani A, Brouquisse R, Perata P (2019) Exploring legume-rhizobia symbiotic models for waterlogging tolerance. Front Plant Sci 10:578

Puente ML, Gualpa JL, Lopez GA, Molina RM, Carletti SM, Cassan FD (2018) The benefits of foliar inoculation with *Azospirillum brasilense* in soybean are explained by an auxin signaling model. Symbiosis 76(1):41–49

Putnoky P, Kondorosi A (1986) Two gene clusters of *Rhizobium meliloti* code for early essential nodulation functions and a third influences nodulation efficiency. J Bacteriol 167:881–887

Quinto C, de la Vega H, Flores M, Leemans J, Cevallos MA, Pardo MA et al (1985) Nitrogenase reductase: a functional multigene family in *Rhizobium phaseoli*. Proc Natl Acad Sci 82:1170–1174

Quiza L, St-Arnaud M, Yergeau E (2015) Harnessing phytomicrobiome signaling for rhizosphere microbiome engineering. Front Plant Sci 6:507

Rajkumar M, Ae N, Prasad MNV, Freitas H (2010) Potential of siderophore-producing bacteria for improving heavy metal phytoextraction. Trends Biotechnol 28(3):142–149

Rawat J, Sanwal P, Saxena J (2015) Effect of *Aspergillus niger* K7 inoculation and biochar amendment on the growth and yield of *Brassica nigra* plants. In: Proceedings of the 1st National conference on recent advances in sciences and technology (NCRAST-2015) towards the mechanisms of nutrient solubilization and fixation in soil system, vol 1, pp 73–81

Read JJ, Reddy KR, Johnie N, Jenkins JN (2006) Yield and fiber quality of upland cotton as influenced by nitrogen and potassium nutrition. Eur J Agron 24(3):282–290

Reddy PM, James EK, Ladha JK (2002) Nitrogen fixation in rice. In: Leigh GJ (ed) Nitrogen fixation at the Millenium. Elsevier, Amsterdam, pp 421–445

Reed MLE, Glick BR (2013) Applications of plant growth-promoting bacteria for plant and soil systems. In: Gupta VK, Schmoll M, Maki M, Tuohy M, Mazutti MA (eds) Applications of microbial engineering. Taylor Francis, Enfield, CT, pp 181–229

Reid DE, Ferguson BJ, Gresshoff PM (2011) Inoculation-and nitrate-induced CLE peptides of soybean control NARK-dependent nodule formation. Mol Plant Microbe Interact 24(5): 606–618

Reinhold-Hurek B, Bünger W, Burbano CS, Sabale M, Hurek T (2015) Roots shaping their microbiome: global hotspots for microbial activity. Annu Rev Phytopathol 53:403–424

Remans R, Beebe S, Blair M, Manrique G, Tovar E, Rao I et al (2008) Physiological and genetic analysis of root responsiveness to auxin-producing plant growth-promoting bacteria in common bean (*Phaseolus vulgaris* L.). Plant Soil 302(1–2):149–161

Richardson AE, Simpson RJ (2011) Soil microorganisms mediating phosphorus availability update on microbial phosphorus. Plant Physiol 156(3):989–996

Rodrıguez H, Gonzalez T, Selman G (2001) Expression of a mineral phosphate solubilizing gene from *Erwinia herbicola* in two rhizobacterial strains. J Biotechnol 84(2):155–161

Rodriguez H, Fraga R, Gonzalez T, Bashan Y (2006) Genetics of phosphate solubilization and its potential applications for improving plant growth promoting bacteria. Plant Soil 287:15–21

Ronson CW, Bosworth A, Genova M, Gudbrandsen S, Hankinson T, Kwaitowski R et al (1990) Field release of genetically engineered *Rhizobium meliloti* and *Bradyrhizobium japonicum* strains. In: Gresshoff PM, Roth LE, Stacey G, Newton WE (eds) Nitrogen fixation: achievements and objectives. Chapman and Hall, New York, pp 397–403

Rossi F, Philippis RD (2015) Role of cyanobacterial exopolysaccharides in phototrophic biofilms and in complex microbial mats. Life 5(2):1218–1238

Rudrappa T, Czymmek KJ, Pare PW, Bais HP (2008) Root-secreted malic acid recruits beneficial soil bacteria. Plant Physiol 148:1547–1556

Rurangwa E, Vanlauwe B, Giller KE (2018) Benefits of inoculation, P fertilizer and manure on yields of common bean and soybean also increase yield of subsequent maize. Agr Ecosyst Environ 261:219–229

Ryan PR, Dessaux Y, Thomashow LS, Weller DM (2009) Rhizosphere engineering and management for sustainable agriculture. Plant Soil 321:363–383

Sadowsky MJ, Cregan PB, Gottfert M, Sharma A, Gerhold D, Rodriquez-Quinones F et al (1991) The *Bradyrhizobium japonicum nol*A gene and its involvement in the genotype-specific nodulation of soybeans. Proc Natl Acad Sci U S A 88:637–641

Sahu GK, Sindhu SS (2011) Disease control and plant growth promotion of green gram by siderophore producing *Pseudomonas* sp. Res J Microbiol 6:735–749

Saleem AR, Brunetti C, Khalid A, Della Rocca G, Raio A, Emiliani G et al (2018) Drought response of *Mucuna pruriens* (L.) inoculated with ACC deaminase and IAA producing rhizobacteria. PLoS One 13(2):e0191218

Sangwan VP, Sindhu SS, Dahiya OS, Kharub RPS (2012) Improvement of wheat (*Triticum aestivum* L.) yield under field conditions by inoculation of microbial strains. Microbiol J 2: 86–95

Sanjuan J, Olivares J (1991) Multicopy plasmids carrying the *Klebsiella pneumoniae nif*A gene enhances *Rhizobium meliloti* nodulation competitiveness on alfalfa. Mol Plant Microbe Interact 4:365–369

Santoyo G, Moreno-Hagelsieb G, del Carmen O-MM, Glick BR (2016) Plant growth-promoting bacterial endophytes. Microbiol Res 183:92–99

Saravanan VS, Subramoniam SR, Raj SA (2003) Assessing *in vitro* solubilization of different zinc solubilizing bacterial (ZBS) isolates. Braz J Microbiol 34:121–125

Saravanan VS, Madhaiyan M, Osborne J, Thangaraju M, Sa TM (2007) Ecological occurrence of *Gluconacetobacter diazotrophicus* and nitrogen-fixing *Acetobacteraceae* members: their possible role in plant growth promotion. Microb Ecol 55:130–140

Sarkar A, Saha M, Meena VS (2017) Plant beneficial rhizospheric microbes (PBRMs): prospects for increasing productivity and sustaining the resilience of soil fertility. In: Meena V, Mishra P, Bisht J, Pattanayak A (eds) Agriculturally important microbes for sustainable agriculture. Springer, Singapore, pp 3–29

Sasse J, Martinoia E, Northen T (2018) Feed your friends: do plant exudates shape the root microbiome? Trends Plant Sci 23(1):25–41

Savka MA, Dessaux Y, McSpadden Gardener BB, Mondy S, Kohler PRA, de Bruijn FJ et al (2013) The "biased rhizosphere" concept and advances in the omics era to study bacterial competitiveness and persistence in the phytosphere. In: de Bruijn FJ (ed) Molecular microbial ecology of the rhizosphere, vol 1 and 2. Wiley, Hoboken, pp 1145–1161

Saxena J, Rawat J, Sanwal P (2016) Enhancement of growth and yield of *Glycine max* plants with inoculation of phosphate solubilizing fungus *Aspergillus niger* K7 and biochar amendment in soil. Commun Soil Sci Plant Anal 47:2334–2347

Schaetzl RJ, Anderson S (2005) Soils: genesis and geomorphology. Cambridge University Press, Cambridge, pp 741–790

Sehrawat A, Phour M, Sindhu SS, Glick BR (2020) Interkingdom signaling in plant-rhizomicrobe interactions for sustainable agriculture. Microbiol Res 241:126589. https://doi.org/10.1016/j.micres.2020.126589

Schmidt JE, Weese DJ, Lau JA (2017) Long-term agricultural management does not alter the evolution of a soybean–*Rhizobium* mutualism. Ecol Appl 27(8):2487–2496

Sehrawat A, Sindhu SS (2019) Potential of biocontrol agents in plant disease control for improving food safety. Defence Life Sci J 4(4):220–225

Sen S, Chandrasekhar CN (2014) Effect of PGPR on growth promotion of rice (*Oryza sativa* L.) under salt stress. Asian J Plant Sci Res 4:62–67

Shantharam S, Mattoo AK (1997) Enhancing biological nitrogen fixation: An appraisal of current and alternative technologies for N input into plants. Plant Soil 194:205–216

Sharifi R, Ryu CM (2018) Sniffing bacterial volatile compounds for healthier plants. Curr Opin Plant Biol 44:88–97

Sharma A, Johri BN, Sharma AK, Glick BR (2003) Plant growth-promoting bacterium *Pseudomonas* sp. strain GRP3 influences iron acquisition in mung bean (*Vigna radiata* L. Wilzeck). Soil Biol Biochem 35(7):887–894

Sharma A, Shankhdhar D, Shankhdhar SC (2016) Potassium-solubilizing microorganisms: mechanism and their role in potassium solubilization and uptake. In: Meena VS, Maurya BR, Verma JP, Meena RS (eds) Potassium solubilizing microorganisms for sustainable agriculture. Springer, New Delhi, pp 203–219

Sharma R, Sindhu S, Sindhu SS (2018a) Bioinoculation of mustard (*Brassica juncea* L.) with beneficial rhizobacteria: a sustainable alternative to improve crop growth. Int J Curr Microbiol App Sci 7(5):1375–1386

Sharma R, Sindhu S, Sindhu SS (2018b) Suppression of Alternaria blight disease and plant growth promotion of mustard (*Brassica juncea* L.) by antagonistic rhizosphere bacteria. Appl Soil Ecol 129:145–150

Shaul-Keinan O, Gadkar V, Ginzberg I, Grünzweig JM, Chet I, Elad Y et al (2002) Hormone concentrations in tobacco roots change during arbuscular mycorrhizal colonization with *Glomus intraradices*. New Phytol 154(2):501–507

Sheng XF, Huang WY (2002) Mechanism of potassium release from feldspar affected by the strain NBT of silicate bacterium. Acta Pedol Sin 39:863–871

Sheng XF, He LY, Huang WY (2002) The conditions of releasing potassium by a silicate dissolving bacterial strain NBT. Agric Sci China 1:662–666

Shukla R, Sharma YK, Shukla AK (2014) Molecular mechanism of nutrient uptake in plants. Int J Curr Res Acad Res 2(12):142–154

Shukla SK, Mishra RK, Pandey M, Mishra V, Pathak A, Pandey A et al (2016) Land reformation using plant growth promoting rhizobacteria with context to heavy metal contamination. In: Ahmad P (ed) Plant metal interaction: emerging remediation techniques. Elsevier, Amsterdam, pp 499–530

Sindhu SS (2004) Biological nitrogen fixation. In: Narwal SS, Dahiya SS, Singh JP (eds) Research methods in plant sciences. Allelopathy, soil analysis. Scientific Publishers, Jodhpur, pp 326–342

Sindhu SS, Dadarwal KR (1992) Symbiotic effectivity of cowpea miscellany *Rhizobium* mutants having increased hydrogenase activity. Indian J Microbiol 32:411–416

Sindhu SS, Dadarwal KR (1993) Broadening of host range infectivity in cowpea miscellany *Rhizobium* by protoplast fusion. Indian J Exp Biol 31:521–528

Sindhu SS, Dadarwal KR (1995) Molecular biology of nodule development and nitrogen fixation in *Rhizobium*-legume symbiosis. In Srivastava HS, Singh RP (eds) Nitrogen nutrition in higher plants. Associated Publishing Company, New Delhi, pp 57–129

Sindhu SS, Dadarwal KR (1999) Improvement of biological nitrogen fixation by genetic manipulations to increase crop productivity. In: Sihag RC, Patel RK, Singh M (eds) Recent advances in biosciences for sustainable food security. Acad Agric Res Edu Manag, pp 75–86

Sindhu SS, Dadarwal KR (2000) Competition for nodulation among rhizobia in legume-*Rhizobium* symbiosis. Indian J Microbiol 40(4):211–246

Sindhu SS, Dadarwal KR (2001a) Chitinolytic and cellulolytic *Pseudomonas* sp. antagonistic to fungal pathogens enhances nodulation by *Mesorhizobium* sp. *Cicer* in chickpea. Microbiol Res 156(4):353–358

Sindhu SS, Dadarwal KR (2001b) Genetic manipulation of rhizobia to improve nodulation and nitrogen fixation in legumes. In Yadav AK, Motsara MR, Ray Choudhary S (eds) Recent Advances in Biofertilizer Technology. Society for Promotion and Utilization of Resources and Technology, New Delhi, pp 1–97

Sindhu SS, Sharma R (2019) Amelioration of biotic stress by application of rhizobacteria for agriculture sustainability. In: Sayyed RZ, Tabassum B (eds) Plant growth promoting rhizobacteria for sustainable stress management. Microorganisms for sustainability. Springer, Singapore, pp 111–168

Sindhu SS, Sharma R (2020) Plant growth promoting rhizobacteria (PGPR): a sustainable approach for managing soil fertility and crop productivity. In: Malik DK, Rathi M, Kumar R, Bhatia D (eds) Microbes for humankind and applications. Daya Publishing House, New Delhi, pp 97–130

Sindhu SS, Mor S, Dadarwal KR, Tauro P (1995) Rhizobial lipo-oligosaccharide nodulation factors: a novel class of signal molecules involved in microbe-plant interactions specificity. In: Proceedings of the International Satellite symposium—complex carbohydrates and advances in biosciences. Soc Biosci Pub, pp 521–562

Sindhu SS, Goel AK, Dadarwal KR (1999) Antagonistic effect of *Pseudomonas* sp. on pathogenic fungi and enhancement of growth of green gram (*Vigna radiata*). Biol Fertil Soils 29:62–68

Sindhu SS, Parmar N, Suneja S, Goel AK, Dadarwal KR (2002) Plant growth promoting effect of *Pseudomonas* sp. on coinoculation with *Mesorhizobium* sp. *Cicer* strain under sterile and wilt sick soil conditions. Appl Soil Ecol 19:57–64

Sindhu SS, Verma MK, Mor S (2009) Molecular genetics of phosphate solubilization in rhizosphere bacteria and its role in plant growth promotion. Nova Science, Hauppauge, NY, pp 199–228

Sindhu SS, Jangu OP, Sivaramaiah N (2010) Genetic engineering of diazotrophic bacteria to improve nitrogen fixation for sustainable agriculture. In: Sayyed RZ, Patil AS (eds) Biotechnology: emerging trends. Scientific Publishers, Jodhpur, pp 73–112

Sindhu SS, Parmar P, Phour M (2014a) Nutrient cycling: Potassium solubilization by microorganisms and improvement of crop growth. In: Parmar N, Singh A (eds) Geomicrobiology and biogeochemistry. Soil biology, vol 39. Springer, Berlin, Heidelberg, pp 175–198

Sindhu SS, Phour M, Choudhary SR, Chaudhary D (2014b) Phosphorus cycling: prospects of using rhizosphere microorganisms for improving phosphorus nutrition of plants. In: Parmar N, Singh A (eds) Geomicrobiology and biogeochemistry. Soil biology, vol 39. Springer, Berlin, Heidelberg, pp 199–237

Sindhu SS, Parmar P, Phour M, Sehrawat A (2016) Potassium-solubilizing microorganisms (KSMs) and its effect on plant growth improvement. In: Meena VS, Maurya BR, Verma JP, Meena RS (eds) Potassium solubilizing microorganisms for sustainable agriculture. Springer, Singapore, pp 171–185

Sindhu SS, Sehrawat A, Sharma R, Khandelwal A (2017) Biological control of insect pests for sustainable agriculture. In: Adhya T, Mishra B, Annapurna K, Verma D, Kumar U (eds) Advances in soil microbiology: recent trends and future prospects. Microorganisms for sustainability, vol 4. Springer, Singapore, pp 189–218

Sindhu SS, Khandelwal A, Phour M, Sehrawat A (2018) Bioherbicidal potential of rhizosphere microorganisms for ecofriendly weed management. In: Meena V (ed) Role of rhizospheric microbes in soil. Springer, Singapore, pp 331–376

Sindhu SS, Sharma R, Sindhu S, Sehrawat A (2019a) Soil fertility improvement by symbiotic rhizobia for sustainable agriculture. In: Panpatte D, Jhala Y (eds) Soil fertility management for sustainable development. Springer, Singapore, pp 101–166

Sindhu SS, Sharma R, Sindhu S, Phour M (2019b) Plant nutrient management through inoculation of zinc-solubilizing bacteria for sustainable agriculture. In: Giri B, Prasad R, Wu QS, Varma A (eds) Biofertilizers for sustainable agriculture and environment. Springer, Cham, pp 173–201

Sindhu D, Sindhu S, Yadav SK (2021) Data mining and phylogenetic analysis of *nif*A gene sequences using nodule-forming *Azorhizobium* strain. Int J Emerg Trends Technol Comput Sci 10:1–8

Singh S, Singh JP (2012) Effect of organic and inorganic nutrient sources on some soil properties and wheat yield. J Indian Soc Soil Sci 60(3):237–240

Singh JS, Abhilash PC, Singh HB, Singh RP, Singh DP (2011) Genetically engineered bacteria: an emerging tool for environmental remediation and future research perspectives. Gene 480:1–9

Singh JS, Kumar A, Rai AN, Singh DP (2016) Cyanobacteria: a precious bio-resource in agriculture, ecosystem, and environmental sustainability. Front Microbiol 7:529

Sivaramaiah N, Malik DK, Sindhu SS (2007) Improvement in symbiotic efficiency of chickpea (*Cicer arietinum*) by coinoculation of *Bacillus* strains with *Mesorhizobium* sp. *Cicer*. Indian J Microbiol 47(1):51–56

Smil V (1999) Nitrogen in crop production: an account of global flows. Global Biogeochem Cycles 13:647–662

Smith SE, Read DJ (1997) Mycorrhizal symbiosis, 2nd edn. Academic Press, London, p 605

Smith SE, Read DJ (2008) Mycorrhizal symbiosis, 3rd edn. Academic Press, San Diego, CA

Smith SE, Smith FA (2011) Roles of arbuscular mycorrhizas in plant nutrition and growth: new paradigms from cellular to ecosytems scales. Annu Rev Plant Biol 62:227–250

Smith FA, Grace EJ, Smith SE (2009) More than a carbon economy: nutrient trade and ecological sustainability in facultative arbuscular mycorrhizal symbioses. New Phytol 182:347–358

Song T, Martensson L, Eriksson T, Zheng W, Rasmussen U (2005) Biodiversity and seasonal variation of the cyanobacterial assemblage in a rice paddy field in Fujian, China. The Fed Eur Mat Soc Microbiol Ecol 123:54131–54140

Sorty AM, Meena KK, Choudhary K, Bitla UM, Minhas PS, Krishnani KK (2016) Effect of plant growth promoting bacteria associated with halophytic weed (*Psoralea corylifolia* L.) on

germination and seedling growth of wheat under saline conditions. Appl Biochem Biotechnol 180:872–882
Spaepen S (2015) Plant hormones produced by microbes. In: Lugtenberg B (ed) Principles of plant-microbe interactions. Springer, Cham, pp 247–256
Spaepen S, Dobbelaere S, Croonenborghs A, Vanderleyden J (2008) Effects of *Azospirillum brasilense* indole-3-acetic acid production on inoculated wheat plants. Plant Soil 312:15–23
Sparks DL (1987) Potassium dynamics in soils. Adv Soil Sci 6:1–63
Srivastava S, Yadav A, Seem K, Mishra S, Chaudhary V, Nautiyal CS (2008) Effect of high temperature on *Pseudomonas putida* NBRI0987 biofilm formation and expression of stress sigma factor RpoS. Curr Microbiol 56(5):453–457
Stringlis IA, Yua K, Feussnerb K, de Jonge R, Van Bentum S, Van Verk MC et al (2018) MYB72-dependent coumarin exudation shapes root microbiome assembly to promote plant health. Proc Natl Acad Sci 115:5213–5222
Strzelczyk E, Dahm H, Pachlewski R, Rozycki H (1997) Production of indole compounds by the ectomycorrhizal fungus *Catharellus cibarius*. Fr. Pedobiology 41:402–411
Suzuki M, Takahashi M, Tsukamoto T, Watanabe S, Matsuhashi S, Yazaki J et al (2006) Biosynthesis and secretion of mugineic acid family phytosiderophores in zinc-deficient barley. Plant J 48:85–97
Swarnalakshmi K, Prasanna R, Kumar A, Pattnaik S, Chakravarty K, Shivay YS et al (2013) Evaluating the influence of novel cyanobacterial biofilmed biofertilizers on soil fertility and plant nutrition in wheat. Eur J Soil Biol 55:107–116
Swarup R, Perry P, Hagenbeek D, van der Straeten D, Beemster GT, Sandberg G et al (2007) Ethylene upregulates auxin biosynthesis in *Arabidopsis* seedlings to enhance inhibition of root cell elongation. Plant Cell 19(7):2186–2196
Szeto W, Kwiatkowski R, Cannon FC, Ronson CW (1990) The enhancement of symbiotic nitrogen fixation in *Bradyrhizobium japonicum*. In: Abstracts of fifth international symposium on the molecular genetics of plant-microbe interactions. Interlaken, Switzerland, p 152
Tabassum B, Khan A, Tariq M, Ramzan M, Khan MSI, Shahid N et al (2017) Bottlenecks in commercialisation and future prospects of PGPR. Appl Soil Ecol 12[illegible]:102–117
Talbot NJ, Foster AJ (2001) Genetics and genomics of the rice blast fungus *Magnaporthe grisea*: developing an experimental model for understanding fungal diseases of cereals. Adv Bot Res 34:263–287
Tchan YT, Kennedy IR (1989) Possible nitrogen-fixing root nodules induced in non-legumes. Agric Sci 2:57–59
Teotia P, Kumar V, Kumar M, Shrivastava N, Varma A (2016) Rhizosphere microbes: potassium solubilization and crop productivity-present and future aspects. In: Meena VS, Maurya BR, Verma JP, Meena RS (eds) Potassium solubilizing microorganisms for sustainable agriculture. Springer, New Delhi, pp 315–325
Thassitou P, Arvanitoyannis I (2001) Bioremediation: a novel approach to food waste management. Trends Food Sci Technol 12:185–196
Theocharis A, Bordiec S, Fernandez O, Paquis S, Dhondt-Cordelier S, Baillieul F et al (2012) *Burkholderia phytofirmans* PsJN primes *Vitis vinifera* L. and confers a better tolerance to low nonfreezing temperatures. Mol Plant-Microbe Interact 25(2):241–249
Thies JE, Singleton PW, Bohlool BB (1991) Influence of size of indigenous rhizobial populations on establishment and symbiotic performance of introduced rhizobia on field-grown legumes. Appl Environ Microbiol 57(1):19–28
Thilakarathna MS, Raizada MN (2017) A meta-analysis of the effectiveness of diverse rhizobia inoculants on soybean traits under field conditions. Soil Biol Biochem 105:177–196
Tilman D, Cassman KG, Matson PA, Naylor R, Polasky S (2002) Agricultural sustainability and intensive productive practices. Nature 148:671–677
Tiwari P, Singh JS (2017) A plant growth promoting rhizospheric *Pseudomonas aeruginosa* strain inhibits seed germination in *Triticum aestivum* (L) and *Zea mays* (L). Microbiol Res 8:1–7

Tiwari S, Lata C, Chauhan PS, Nautiyal CS (2016) *Pseudomonas putida* attunes morpho-physiological, biochemical and molecular responses in *Cicer arietinum* L. during drought stress and recovery. Plant Physiol Biochem 99:108–117

Tripathi P, Singh PC, Mishra A, Chaudhry V, Mishra S, Tripathi RD et al (2013) *Trichoderma* inoculation ameliorates arsenic induced phytotoxic changes in gene expression and stem anatomy of chickpea (*Cicer arietinum*). Ecotoxicol Environ Saf 89:8–14

Tripathi P, Singh PC, Mishra A, Srivastava S, Chauhan R, Awasthi S et al (2017) Arsenic tolerant *Trichoderma* sp. reduces arsenic induced stress in chickpea (*Cicer arietinum*). Environ Pollut 223:137–145

Tyagi S, Mulla SI, Lee KJ, Chae JC, Shukla P (2018) VOCs-mediated hormonal signaling and crosstalk with plant growth promoting microbes. Crit Rev Biotechnol 38:1277–1296

Ullah A, Heng S, Munis MFH, Fahad S, Yang X (2015) Phytoremediation of heavy metals assisted by plant growth promoting (PGP) bacteria: a review. Environ Exp Bot 117:28–40

Valetti L, Iriarte L, Fabra A (2018) Growth promotion of rapeseed (*Brassica napus*) associated with the inoculation of phosphate solubilizing bacteria. Appl Soil Ecol 132:1–10

Vansuyt G, Robin A, Briat JF, Curie C, Lemanceau P (2007) Iron acquisition from Fe-pyoverdine by *Arabidopsis thaliana*. Mol Plant Microbe Interact 20(4):441–447

Vega FE, Kaya HK (2012) Insect pathology. Academic, New York

Vejan P, Abdullah R, Khadiran T, Ismail S, Nasrulhaq Boyce A (2016) Role of plant growth promoting rhizobacteria in agricultural sustainability—a review. Molecules 21(5):573

Velázquez-Hernández ML, Baizabal-Aguirre VM, Cruz-Vázquez F, Trejo-Contreras MJ, Fuentes-Ramírez LE, Bravo-Patiño A et al (2011) *Gluconacetobacter diazotrophicus* levansucrase is involved in tolerance to NaCl, sucrose and desiccation, and in biofilm formation. Arch Microbiol 193(2):137–149

Verma N, Wati L, Sindhu SS, Goyal S (2014) Organic waste utilization for production of compost and biofuel. In: Nagpal R, Kumar A, Singh R (eds) Microbes in the service of mankind: tiny bugs with huge impact. JBC Press, New Delhi, pp 95–124

Vessey JK (2003) Plant growth promoting rhizobacteria as biofertilizers. Plant Soil 255(2):571–586

Walters DR, Ratsep J, Havis ND (2013) Controlling crop diseases using induced resistance: challenges for the future. J Exp Bot 64(5):1263–1280

Wang Y, Wu WH (2013) Potassium transport and signaling in higher plants. Annu Rev Plant Biol 64:451–476

Wang Z-C, Burns A, Watt GD (1984) Complex formation and O_2 sensitivity of *Azotobacter vinelandii* nitrogenase and its component proteins. Biochemistry 24:214–221

Wang N, Feng Z, Zhou Y, Zhu H, Yao Q (2017) External hyphae of *Rhizophagus irregularis* DAOM 197198 are less sensitive to low pH than roots in arbuscular mycorrhizae: evidence from axenic culture system. Environ Microbiol Rep 9(5):649–657

Wani PA, Khan MS, Zaidi A (2007) Synergistic effect of the inoculation with nitrogen-fixing and phosphate- solubilizing rhizobacteria on performance of field grown chickpea. J Plant Nutr Soil Sci 70(2):283–287

White PJ, Brown PH (2010) Plant nutrition for sustainable development and global health. Ann Bot 105(7):1073–1080

Wiebe K, Lotze-Campen H, Sands R, Tabeau A, van der Mensbrugghe D, Biewald A et al (2015) Climate change impacts on agriculture in 2050 under a range of plausible socioeconomic and emissions scenarios. Environ Res Lett 10(8):085010

Williams MK, Beynon JL, Ronson CW, Cannon FC (1990) Abstracts of fifth international symposium on the molecular genetics of plant-microbe interactions. Interlaken, Switzerland, p 152

Xiao Y, Wang X, Chen W, Huang Q (2017) Isolation and identification of three potassium-solubilizing bacteria from rape rhizospheric soil and their effects on ryegrass. Geomicrobiol J 34(10):873–880

Yadav J, Verma JP, Tiwari KN (2010) Effect of plant growth promoting rhizobacteria on seed germination and plant growth chickpea (*Cicer arietinum* L.) under in vitro conditions. Bio Forum Int J 2(2):15–18
Yang Y, Liang Y, Han X, Chiu T-Y, Ghosh A, Chen H, Tang M (2016) The roles of arbuscular mycorrhizal fungi (AMF) in phytoremediation and tree-herb interactions in Pb contaminated soil. Sci Rep 6:20469
Yao Q, Zhu HH, Chen JZ (2005) Growth responses and endogenous IAA and iPAs changes of litchi (*Litchi chinensis* Sonn.) seedlings induced by arbuscular mycorrhizal fungal inoculation. Sci Hortic 105:145–151
Yasin M, Munir I, Faisal M (2016) Can *Bacillus* spp. enhance K^+ uptake in crop species. In: Meena VS, Maurya BR, Verma JP, Meena RS (eds) Potassium solubilizing microorganisms for sustainable agriculture. Springer, New Delhi, pp 163–170
Ye B, Saito A, Minamisawa K (2005) Effect of inoculation with anaerobic nitrogen-fixing consortium on salt tolerance of *Miscanthus sinensis*. Soil Sci Plant Nutr 51(2):243–249
Zahedi H (2016) Growth-promoting effect of potassium-solubilizing microorganisms on some crop species. In: Meena VS, Maurya BR, Verma JP, Meena RS (eds) Potassium solubilizing microorganisms for sustainable agriculture. Springer, New Delhi, pp 31–42
Zamioudis C, Pieterse CM (2012) Modulation of host immunity by beneficial microbes. Mol Plant Microbe Interact 25(2):139–150
Zeng X, Liu X, Tang J, Hu S, Jiang P, Li W et al (2012) Characterization and potassium-solubilizing ability of *Bacillus circulans* Z 1–3. Adv Sci Lett 10:173–176
Zhang C, Kong F (2014) Isolation and identification of potassium-solubilizing bacteria from tobacco rhizospheric soil and their effect on tobacco plants. Appl Soil Ecol 82:18–25
Zhang L, Xu M, Liu Y, Zhang F, Hodge A, Feng G (2016) Carbon and phosphorus exchange may enable cooperation between an arbuscular mycorrhizal fungus and a phosphate-solubilizing bacterium. New Phytol 210(3):1022–1032
Zhang L, Feng G, Declerck S (2018) Signal beyond nutrient, fructose, exuded by an arbuscular mycorrhizal fungus triggers phytate mineralization by a phosphate solubilizing bacterium. ISME J 12(10):2339–2351
Zhang J, Cook J, Nearing JT, Zhang J, Raudonis R, Glick BR et al (2021) Harnessing the plant microbiome to promote the growth of agricultural crops. Microbiol Res 2021:126690

Role of Plant Growth-Promoting Rhizobacteria in Combating Abiotic and Biotic Stresses in Plants

2

Muhammad Zahid Mumtaz, Maqshoof Ahmad, Khadija Mehmood, Ahsan Sattar Sheikh, Arif Malik, Azhar Hussain, Sajid Mehmood Nadeem, and Zahir Ahmad Zahir

Abstract

Global climate change accelerates the coincidence of a variety of abiotic stresses, viz., salinity, drought, flooding, high and low temperature, and biotic stresses, viz. phytopathogens which degrade agricultural productivity. In such circumstances, plant growth-promoting rhizobacteria (PGPR) are eco-friendly and sustainable candidates to combat these stresses. Several PGPR with the ability to support plant growth under various stressed conditions have been commercialized. The current chapter is mainly restricted to beneficial effects of PGPR on plant growth and development under environmental and biotic stresses. It begins with the description of various abiotic and biotic stress factors affecting plant growth and their tolerance achieved by both physiological and molecular mechanisms of adaptation. The use of PGPR helps ameliorate these stresses in rhizosphere by using several mechanisms and has beneficial effects on plant growth after efficiently colonizing the root surface. Plant growth stimulation through PGPR is the net result of multiple mechanisms of action that may be activated simultaneously. Such bacteria are more likely to be used for stress

M. Z. Mumtaz · A. S. Sheikh · A. Malik
Institute of Molecular Biology and Biotechnology, The University of Lahore, Lahore, Pakistan

M. Ahmad (✉) · K. Mehmood · A. Hussain
Department of Soil Science, The Islamia University of Bahawalpur, Bahawalpur, Pakistan

S. M. Nadeem
University of Agriculture Faisalabad, Sub-Campus Burewala-Vehari, Burewala, Vehari, Punjab, Pakistan

Z. A. Zahir
Institute of Soil and Environmental Sciences, University of Agriculture Faisalabad, Faisalabad, Punjab, Pakistan

N. K. Arora, B. Bouizgarne (eds.), *Microbial BioTechnology for Sustainable Agriculture Volume 1*, Microorganisms for Sustainability 33,
https://doi.org/10.1007/978-981-16-4843-4_2

tolerance to fulfill the need for food production under extreme environmental conditions. The bacterial inoculants also enhance nutrient uptake and crop growth. They are also involved in biocontrol so they may be good supplements to chemical fertilizers and agrochemicals. This chapter discusses the potential and key mechanisms used by PGPR under stress conditions for sustainable agricultural productivity followed by their prospects.

Keywords

ACC-deaminase activity · Beneficial bacteria · Environmental sustainability · Integrated stress · Phytohormones · Plant tolerance

2.1 Introduction

Global warming leads to abiotic and biotic stresses that are affecting crop productivity all over the globe. Abiotic stresses include salinity, drought, flooding, heavy metals, heat, and cold stress, while biotic stresses are caused by the attack of insects, pests, and microbial infection. Such abiotic and biotic stresses are collectively known as integrated stresses (Flynn 2003; Brown et al. 2017). Abiotic stresses are also known as environmental stresses that drastically reduce plant growth by disrupting metabolic and biochemical attributes of a plant (Suzuki et al. 2014). Abiotic stresses induce plant responses through modification in cellular metabolism, reduction in plant growth, and modulation in biochemical attributes (Singh 2015). Many studies have illustrated the effect of selective abiotic stresses such as salinity (Mumtaz et al. 2018; Sapre et al. 2021), heavy metals (Thakare et al. 2021), drought (Mumtaz et al. 2020; Ansari et al. 2021), flooding (Bal and Adhya 2021), heat (Sarkar et al. 2021), and cold (Vega-Celedón et al. 2021) on the growth of different plants.

Microorganisms can be handy in mitigation of abiotic and biotic stresses. Beneficial bacteria which constitute up to 2–5% of the total bacterial population in soil are beneficial for plant growth even under adverse environments (Nazli et al. 2020). Millions of microorganisms colonizing the rhizosphere form a complex biological community with a plant root system and stimulate plant growth through their metabolic activities (Schmidt et al. 2014; Zvinavashe et al. 2021). Plant growth promoting bacteria (PGPR) improve plant nutrient availability and help in sustained crop yields (Hayat et al. 2010) through efficient nutrient solubilization and mobilization, providing biocontrol against phytopathogens, improving soil structure and quality by sequestrating toxic salt ions, and even degrading various xenobiotic compounds (Braud et al. 2009; Koskey et al. 2021). The application of such inoculants is a promising strategy for the amelioration of various stresses and improving plant health (Ahmad et al. 2013a). Bioaugmentation of PGPR modulates environmental stresses through direct and indirect mechanisms that induce systemic tolerance (Yang et al. 2009). Production of phytohormones and 1-amino-cyclopropane-1-carboxylic acid (ACC) deaminase activity by various PGPR has a crucial role in plant growth under environmental as well as biotic stresses. It activates the radicular system and promotes root elongation through the development of apical meristem and lateral roots, thus improving plant access to soil nutrients (Vessey 2003; Liu et al. 2021). In the current chapter, we have discussed the status of various abiotic and biotic stresses and their effects on plant growth. We have also discussed the potential of PGPR for remediation of these stresses. Abiotic stress induced by salt, drought, flooding, heat, and cold factors and biotic stress due to fungal, bacterial, and nematode pathogen have been reviewed. Various direct and indirect mechanisms used by PGPR for amelioration of these stresses have also been discussed.

2.2 Plant Responses to Stress

Plant stress is a condition in which plants are developed in non-ideal circumstances for their growth. It could be due to various external biological and non-biological factors that lead to a direct effect on plant growth. Biological stresses are also known as biotic stress, observed on plants due to plant interactions with pathogenic bacteria, fungi, virus, nematodes, weeds, and/or insects. Non-biological stress could be due to physical factors such as deficiency or toxicity of nutrients, high or low temperature, limited water availability or excess of water conditions, etc. (McKersie and Lesheim 1994). Non-biological factors result in non-favorable conditions to plants and enforce them under a state of stresses known as abiotic or environmental stresses. Mostly, abiotic stresses have a huge impact on plants by causing up to a 50% reduction in growth. Both biotic and abiotic stresses could affect plant growth for short or long periods depending on the exposure level (Mosa et al. 2017). Plants are exposed to abiotic and biotic stress factors, collectively called stressors that deviate plant's optimal growth conditions. Stressors damage plant growth up to certain tolerance limits and, after that, may cause death. Stressors are interconnected to each other and affect plants by causing osmotic stress, ion imbalance, disrupting plant metabolism, production of ethylene, reactive oxygen species (ROS), and cell homeostasis. Many groups of genes have also shown changes in their expression patterns due to various stressors (Mittler 2006; Ramegowda et al. 2020; Kusale et al. 2021; Sofy et al. 2021).

Plants use a stress-sensing mechanism that depends on plant species and type of stress. Plant response to stress can be categorized into alarming, resistance, exhaustion, and regeneration (Mosa et al. 2017). In initial stress exposure, distinguishable changes from optimal plant growth in terms of functional deviation, declining vitality, and increase in catabolism take place. If stressors continue, these affect plant growth and productivity, and plant responds through stress adaptation, repair, and hardening processes. To minimize damage caused by various stresses, plant activates a complex signaling cascade of defense and specific ion channels; generates ROS and phytohormones, and expresses various stress-related genes that improve plant stress tolerance (Ramegowda et al. 2020; Tanveer and Ahmed 2020; Salvi et al. 2021). Every plant species has a certain limit for resistance against stresses that cannot be extended through external modification. Further increase in stress dosage beyond the plant resistance will cause loss of vitality, severe cell damage, and plant death (Mosa et al. 2017). Before complete senescence, plants can regenerate their physiological standards that allow plants to survive under various stresses. However, if stressors cause permanent damage, then it will not recover. A plant can survive under stress conditions but with limited growth and reproduction ability. These responses usually involve physiological-, biochemical-, and molecular-level actions (Kranner et al. 2010; Brunetti et al. 2019; Ansari et al. 2021). Plant response to stress can be modified/improved through exogenous application of various chemicals, phytohormones, and/or by modification of growing conditions. Rhizosphere microbiome engineering has also been studied as an eco-friendly, economical, and sustainable approach by the soil microbiologists.

2.3 Strategies for Stress Amelioration

Several strategies have been proposed by plant scientists to ameliorate stress. These strategies are comprehensively reviewed and available in literature (Glick et al. 2007; Saharan and Nehra 2011; Nadeem et al. 2014). Certain chemicals are used to reduce the impact of stresses on plant growth. These chemicals can reduce the negative impact of ethylene on plant growth by inhibiting its excess production under stress. Important chemicals reported in the literature are silver ions (Ag^{+}), cobalt ions (Co^{2+}), and aminoethoxyvinylglycine (AVG) which improve plant growth by suppressing ethylene production (Coupland and Jackson 1991; Mckeon et al. 1995; Kim and Mulkey 1997). Savvides et al. (2016) reported that plant priming with chemical agents, hydrogen peroxide, melatonin, polyamines, sodium hydrosulfide, and sodium nitroprusside, can improve multiple-stress tolerance. Priming with γ-amino butyric acid and β-amino butyric acid also induces promising tolerance in plants against abiotic stresses (Vijayakumari et al. 2016). Soaking seed with these priming agents shows greater stress tolerance by improving metabolic events and shows early seed germination and seedling emergence as compared to non-primed seeds (Ashraf et al. 2018). These chemicals also improve antioxidative enzymes, proline, malondialdehyde, and improve stress-responsive proteins (Thakur et al. 2019; Buttar et al. 2020).

The use of chemicals in agriculture is, however, expensive and deteriorates the plant, soil, and human health (Dodd et al. 2004; Pahalvi et al. 2021). These chemicals can increase crop yield but reduce farm profitability. Scientists have also raised certain environmental issues due to the persistence and toxicity of these chemicals for the natural habitat (Ahmadi et al. 2009). The use of osmoprotectants and potassium to reduce the negative effects has also been reported. However, the use of chemicals cannot reduce the negative effects of ion toxicity caused by salinity or root desiccation under osmotic stress. Although plants employ some specific mechanisms to cope with stresses, the presence of beneficial microbes can be helpful to combat the negative effects of stresses on plant growth. Rhizosphere engineering by using PGPR can therefore be exploited as a useful strategy to ameliorate the negative effects on plant growth. It could be the most important approach to remediate multiple stresses for sustainable agriculture. Recently, such beneficial bacteria have been acknowledged by scientists for their capability to improve crop productivity as well as raise stress tolerance in various crops (Goyal et al. 2019; Njeru and Koskey 2021).

2.4 Role of Plant Growth-Promoting Rhizobacteria to Combat Stresses in Plants

Plants are associated with soil microorganisms in a symbiotic relationship; thus there is a strong plant-microbe interaction in the rhizosphere. Such symbiotic free-living soil microorganisms inhabiting the rhizosphere of many plant species having diverse beneficial effects on host plants are known as PGPR. Their interactions with plants are commercially explored and have biotechnological applications for sustainable agriculture to improve plant health in an eco-friendly manner (Gonzalez et al. 2015; Gouda et al. 2018). Biological activities of soil microbiome in the ecosystem make it vigorous for sustainable agriculture production. The PGPR dwell in plant roots and improve plant growth through various mechanisms including nitrogen (N) fixation (Kuan et al. 2016; Aasfar et al. 2021), solubilization of phosphorus (P) (Selvakumar et al. 2009; Kusale et al. 2021); potassium (K) (Ahmad et al. 2016) and zinc (Zn) solubilization (Mumtaz et al. 2017, 2019); production of siderophores (Gouda et al. 2018); ACC-deaminase (Misra et al. 2017; Ansari et al. 2021; Vega-Celedón et al. 2021); and phytohormones, e.g. indole-3-acetic acid (IAA) (Kim et al. 2017; Khan et al. 2020a, b; Kusale et al. 2021), cytokinins (CK) (Liu et al. 2013; Ilyas et al. 2020), gibberellins (GA) (Gutiérrez-Mañero et al. 2001; Khan et al. 2020a, b), abscisic acid (ABA) (Belimov et al. 2014; Kang et al. 2019), hydrogen cyanide (HCN) (Kremer and Souissi 2001), and antibiotics and/or lytic enzymes (Gouda et al. 2018; Migunova and Sasanelli 2021). They also possess more specific plant growth-promoting (PGP) traits that can ameliorate/reduce environmental stresses such as heavy metals, salinity, and drought stress and biological control of phytopathogens (Egamberdieva and Lugtenberg 2014). Such PGPR are classified into two main types: (1) the extracellular PGPR that can colonize the rhizoplane and (2) intracellular PGPR that colonize inside the root cortex or nodule cells (Bhattacharyya and Jha 2012).

The PGPR possess several PGP characteristics that are involved in enhancing plant physiology under normal as well as stress conditions through various mechanisms (Ahmad et al. 2011; Zakry et al. 2012; Asghari et al. 2020). They produce volatile organic compounds (VOCs) and enzymes for biocontrol of phytopathogens and neutralize abiotic and biotic stresses. However, different PGPR have different modes of action which differ under certain circumstances according to their host plant, abiotic and biotic factors, plant defense and developmental stages, and presence of competition for microflora (Garcia et al. 2015). They can support plant growth and development through enhancing nutrient availability by nitrogen fixation, solubilization of minerals, mineralization of nutrients, and production of phytohormones and siderophores (Bhardwaj et al. 2014). Such mechanisms improve plant growth directly and may contribute according to the ability of microbial strains. In the presence of PGPR, improved individual ion fluxes at the root surface enrich nutrient uptake (Gouda et al. 2018). The PGPR may indirectly improve plant growth and development by producing suppressive compounds that prevent the phytopathogen attack on plants (Singh and Jha 2015; Migunova and Sasanelli 2021). This process actively helps plants grow under biotic stresses (Akhgar et al. 2014). In this mechanism, PGPR contribute through the production of chitinases, cellulases, proteases, and other hydrolytic enzymes and production of antibiotics, siderophores, VOCs, exopolysaccharides (EPS), and induced systemic resistance (ISR) in response to a plant pathogen (Gupta et al. 2015). The occurrence of abiotic and biotic stresses drastically affects crop productivity over the globe. So, there is a need to explore unique PGP characteristics of soil microbiome under extreme environmental conditions, their relations with crop plants, genetic diversity, and successful application methods for increased agriculture production (Grover et al. 2011).

2.4.1 Abiotic Stress

External conditions of either physical or biological nature that unfavorably disturb plant growth are known as stresses. Physical stresses, viz., salinity, drought, waterlogging, heat, cold and heavy metal contamination, etc., are termed as abiotic stresses (Brown et al. 2017), while biological agents like microbial and insect and pest attacks are known as biotic stresses (Flynn 2003). According to an estimate, between 2003 and 2013, 78 types of natural disasters caused 140 billion dollars worth of losses in developing countries (FAO 2015). Abiotic stresses impose their hazardous effects on plant growth by altering the metabolic, biochemical, and physiological attributes of plant growth (Suzuki et al. 2014; Sapre et al. 2021). Abiotic stresses initiate plant responses in terms of cellular metabolism, changes in growth and yield attributes, modulation in metabolites and biochemical attributes, and alteration in gene expression (Singh 2015; Sofy et al. 2021).

Abiotic stresses impair photosynthetic machinery in plants such as photosystem I and II, stomatal conductance, carbon (C) fixation, electron transport, and ATP-generating system (Nouri et al. 2015; Chatterjee et al. 2020; Sharma et al. 2020). These may modify the structure of chloroplasts and thylakoid complexes, cause changes in metabolites concentration, and generate ROS (Gan et al. 2019). For example, salinity stress causes loss of photosynthetic pigments and stomatal closure due to reduced nutrient uptake (Stepien and Kłbus 2006). It also reduces the carbon dioxide (CO_2) availability and carbon (C) fixation that generate ROS, thus causing cellular damage through oxidation of lipids, proteins, and nucleic acids (Lopez-Raez 2016). Abiotic stresses adversely affect plant growth by decreasing water availability that causes an imbalance between enzymatic and non-enzymatic reactions taking place in chloroplasts, mitochondria, and electron transport reactions (Koyro et al. 2012; Adak et al. 2020). Various crop plants differ in their ability to withstand most of the abiotic stresses. The plant tolerates abiotic stresses through genetic and physiological adaptive mechanisms and interacts symbiotically with endophytic and rhizospheric bacteria. Many studies have demonstrated genetic adaptation to abiotic stresses including molecular mechanism, identification of overexpressed genes, and development of stress-tolerant transgenic plants (Zang et al. 2017; Datir et al. 2019; Woo et al. 2020). Several studies have also reported plant-microbe interactions in abiotic stress tolerance. Pre-treatment of plant seed or seedlings with PGPR can improve plant survival during abiotic stresses which are being discussed in subsequent sections for each abiotic stress.

2.4.1.1 Salt Stress

Agriculture is moving into regions with lower rainfall and saline soil that occupy 6% of the world's land area due to the increased need for food production (Munns 2005). High salinity is one of the major abiotic factors that adversely affect crop growth, development, and productivity (Mumtaz et al. 2018). It converts fertile fields to marginal lands and causes desertification. Soil salinization is the buildup of dissolved salts in agricultural soils over a prolonged time. It may arise due to weathering of rocks and minerals, precipitation, wind-borne salts, the influx of seawater, and cultivation operations, viz. unnecessary irrigation, land clearing, and insufficient drainage. The dryland salinity originated due to salt deposition in the topsoil through water evaporation (Rengasamy 2002, 2010; Pitman and Läuchli 2002; Shrivastava and Kumar 2015). Salinity effects are more harmful under the raised groundwater table, arid, and coastal areas, and only salt-tolerant plants called halophytes can grow (Rengasamy 2002; Flowers and Colmer 2008). The salinity-sensitive plants called glycophytes eventually die as their physiological and biochemical attributes are susceptible to salt-induced stress (Hajiboland et al. 2018).

Salinity stress hinders crop production by causing various physiological and metabolic changes. It suppresses plant growth through osmotic stress and ion toxicity (Rozema and Flowers 2008; Rahnama et al. 2010). Osmotic stress due to higher salt buildup in soil and plants causes a reduction in water absorption ability, accelerated water loss from leaves, and various physiological changes, viz. nutrient imbalance, membrane interruption, inability to reduce ROS, and alteration in antioxidant enzymes (Munns and Tester 2008; Rahnama et al. 2010; Sagar et al. 2020). Salinity causes accumulation of sodium (Na^+) and chloride (Cl^-) ions in plant tissues that cause severe ion imbalance and physiological disorders. Accumulation of higher Na^+ contents retard the uptake of potassium ions (K^+) that is necessary for plant growth and development and may cause plant death (James et al. 2011). Due to salt stress, oxidative stress elevates the production of ROS which leads to oxidative damages in proteins, lipids, and DNA and interrupts vital cellular functions of plants (Xie et al. 2014; Kusale et al. 2021).

Plants tolerate high salt stress by developing various physiological adaptations *viz.* ion homeostasis, transport, and compartmentalization, synthesis of osmoprotectants, compatible solutes, polyamines, nitric oxide, and activation of antioxidant enzymes and hormone modulation (Gupta and Huang 2014; Dubey and Verma 2019; Sofy et al. 2021). There is a need to apply an attenuated approach to remediate the effects of salt stress on crops. Due to reduced agricultural production under salinity stress, the production of salinity-resistant crops is a desirable scientific achievement. For successful saline soil reclamation, effective soil management practices are needed (Munns and Tester 2008). The sustainable management practices to facilitate future agriculture production provide more support regarding crop production. Various plant growth-promoting microorganisms (PGPM) play a beneficial role to promote agricultural productivity in an ecofriendly manner (Rodriguez and Redman 2008). Such PGPM associated with the stressed plant augment to adapt their microenvironments (Paul and Lade 2014). Various crops under salinity-stressed conditions respond to PGPM that improves plant growth. Such biotechnology can help to develop salinity-tolerant technologies for growing crops. Augmented PGPR colonizes plant's rhizospheric soil and alleviates stress in crops. This remediation strategy could be successful due to a wide range of salinity-tolerant microbes, their associated interactions with plants, potential PGP metabolites, and prospective application methods.

The PGPR have been investigated under salt-induced stress for their role in improving the efficiency of photosynthesis, ion homeostasis, and plant-water relations regulated by a signaling network that mitigates the stress (Wang et al. 2016; Mahadik and Kumudini 2020; Taj and Challabathula 2021). They optimize water uptake and stomatal opening by improving the transpiration rate and hydraulic conductivity. Marulanda et al. (2010) reported that inoculation with *Bacillus megaterium* improved root hydraulic conductivity in maize under 2.59 dS m^{-1} salt stress that correlated with increased ZmPIP expression. *Arabidopsis thaliana* treated with *Bacillus subtilis* encourages proline production with the expression of proBA genes that conferred salt tolerance to withstand the initial osmotic shocks (Chen et al. 2007). In another study, *Bacillus amyloliquefaciens* strain SN13 inoculated in rice augmented salt stress of 200 mM NaCl through upregulation of stress-related genes (Nautiyal et al. 2013). Upadhyay and Singh (2015) reported that inoculation with *Bacillus aquimaris* promoted total sugar contents in wheat that related to increased NPK and biomass accumulation and reduction of Na^+ in leaves. del Amor and Cuadra-Crespo (2012) reported that pepper plants inoculated with *Azospirillum brasilense* and *Pantoea dispersa* improved salinity resistance in pepper and recorded higher plant dry matter correlated with enhanced stomatal conductance and photosynthetic rate. The PGPR induces accumulation of osmoprotectants and compatible solutes, viz., glycine, proline, and trehalose having more potency than their linked host plants under salt stress (Arif and Ghoul 2018; Taj and Challabathula 2021). These compatible solutes are absorbed by plant roots that maintain osmotic balance. The bean crop overexpressed trehalose-6-phosphate gene when co-inoculated by *Paenibacillus polymyxa* and *Rhizobium tropici* that promoted nodulation and N accumulation. The upregulation of stress tolerance genes through *R. tropici* and *P. polymyxa* induced salinity tolerance in beans (Figueiredo et al. 2008).

The PGPR trap cations in their EPS matrix and limit salt uptake by plants that improve their root structure and regulate ion affinity transporter's expression. They relieve nutrient imbalance by reducing Na^+ and Cl^- influx and increased exchange of macronutrients and micronutrients through mineralizing the nutrients, lowering rhizospheric pH, and producing siderophores (Nadeem et al. 2016; Etesami and Alikhani 2019; Ji et al. 2020). They maintain ion homeostasis by boosting activity of K^+ transporters, improving K^+:Na^+ ratios, decreasing Na^+ and Cl^- buildup, and increasing Na^+ exclusion. Maize inoculation with IAA-producing *Azotobacter* strains C5 and C9 showed improved Na^+ exclusion and K^+ uptake (Rojas-Tapias et al. 2012). Under 200 mM NaCl stress, *Puccinellia tenuiflora* inoculated with *B. subtilis* strain GB03 exhibited reduced Na^+ buildup (Niu et al. 2016). PGPR produce exogenous phytohormones and enzymes that may help ameliorate salt stress. During salinity stress, plants also synthesize phytohormones in the response of plant-microbe relations (Dodd et al. 2010; Kusale et al. 2021).

Several PGPR can produce phytohormones, viz., IAA, CK, GA, and ABA, proliferate roots hairs, and enhance uptake of water and nutrients under salt stress. Salinity stress reduces the endogenous IAA and ABA levels in plants which can be maintained at their normal level through PGPR inoculation. For example, *B. amyloliquefaciens* strain SQR9 stimulated ABA production of inoculated plants under salinity stress (Chen et al. 2016). Similarly, Kim et al. (2014) inoculated *Enterobacter* sp. strain EJ01 in *Arabidopsis thaliana* under salt stress and reported increased expression of salt stress-responsive genes. They also reported that strain EJ01-inoculated plant showed increase in ROS scavenging activities through production of ascorbate peroxidase (APX) under salinity stress. PGPR produce CK under salt stress; however their role in salt stress tolerance is still under question (Arkhipova et al. 2005). Egamberdieva (2009) inoculated wheat with CK-producing *Pseudomonas* strains TSAU22, TSAU6, and TSAU20 that enhanced wheat growth under salt stress. PGPR modulate ABA biosynthesis under salinity stress and support plant growth. In another study, *Dietzia natronolimnaea* strain STR1 induced salt-tolerance in wheat by modulating ABA signaling cascade and upregulation of salt stress-induced gene. These strains contributed to antioxidant enzymes and proline content gene expression that enhanced salt tolerance (Bharti et al. 2016). Cucumber inoculation with strains of *Burkholderia*, *Promicromonospora*, and *Acinetobacter* showed a significant increase in biomass, water potential, and decreased electrolyte leakage at 120 mM NaCl stress. These strains showed downregulation of ABA and increased levels of salicylic acid and GA contents in inoculated plant (Kang et al. 2014). Similarly, improvement in maize growth has also been reported on compacted saline-sodic soil through seed inoculation with *Pseudomonas syringae* and *Pseudomonas fluorescens* (Zafar-ul-Hye et al. 2018).

Under salinity stress, ethylene levels increase, which causes inhibition in auxin transcription and constrain plant growth. PGPR produce ACC-deaminase that inhibit the biosynthesis of ethylene in plants. Soil bacteria produce ACC-deaminase enzyme that converts ACC to ammonia and a-ketobutyrate and enhance salt tolerance and promote plant growth (Glick et al. 2007). Tomato seedlings inoculated with *Pseudomonas putida* strain UW4 grown under 90 mM NaCl stress showed increased shoot growth due to the expression of upregulated Toc GTPase that import stress response proteins (Yan et al. 2014). Okra inoculation with *Enterobacter* sp. strain UPMR18 ameliorated salt stress due to escalating antioxidant activities (Habib et al. 2016). Nadeem et al. (2009) reported salt tolerance in maize due to co-inoculation of ACC-deaminase-producing *Pseudomonas* and *Enterobacter* spp. The ACC-deaminase- and IAA-producing PGPR can ameliorate the negative effects of salt stress effectively. The accumulation of IAA stimulates the ACC synthase genes, resulting in a higher ACC amount that leads to ethylene production. The ACC-deaminase-producing PGPR break down the excess ACC and lower the toxic level of ethylene, and simultaneously IAA production promotes plant growth (Glick 2012; Ahmad et al. 2013b). In another study, the combined use of ACC-deaminase-containing PGPR strains *Agrobacterium fabrum* and *B. amyloliquefaciens* along with biochar improved wheat productivity under drought-stressed conditions (Danish and Zafar-ul-Hye 2019; Zafar-ul-Hye et al. 2019).

The IAA- and ACC-deaminase-producing *Arthrobacter* sp. and *Bacillus* sp. inoculation promoted proline contents under osmotic stress and revealed downregulation of stress-inducible genes in sweet pepper (Sziderics et al. 2007). In another study, Panwar et al. (2016) inoculated IAA- and ACC-deaminase-producing *P. dispersa* strain PSB3 in chickpea under 150 mM NaCl stress and reported significant improvement in plant biomass and grain yield attributes due to reduced Na^+ uptake, electrolyte leakage, and enhanced K^+ uptake. The extracellular secretions from PGPR could also manipulate signaling pathways and alleviate salt stress (Bhattacharyya et al. 2015; Zhou et al. 2016; Kumawat et al. 2020). PGPR produce EPS responsible for bacterial attachment to soil particles. EPS improves soil structure and water-holding capacity by binding soil particles (Upadhyay et al. 2011). Qurashi and Sabri (2012) inoculated chickpea with *Halomonas variabilis* strain HT1 and *Planococcus rifietoensis* strain RT4 and reported salt tolerance by improving plant growth and soil aggregation. *P. polymyxa* ANM59 and *Paenibacillus* sp. ANM76 have also been reported as potential bioinoculants to improve abiotic stress tolerance of crop plants (Ahmad et al. 2019a). The mechanisms of action adopted by PGPR to mitigate salt stress are summarised in Fig. 2.1 and Table 2.1.

2.4.1.2 Drought Stress

Drought is a condition of an uncommon drop in soil moisture content due to a prolonged period of low rainfall. It is the utmost devastating abiotic stress that restricts crop growth and productivity worldwide. The arable lands affected by drought have doubled during the period from 1970 to 2000 and is expected to increase further by 2050, posing serious threat to crop productivity (Vinocur and Altman 2005). In the last four decades, drought caused reduction in cereal yield by up to 10% and is estimated to affect crop production in over 50% of cultivated land by 2050 (Lesk et al. 2016). Drought may be of moderate, short, extremely severe, and prolonged duration (Bottner et al. 1995). It causes a reduction in plant growth by altering physiological traits of a plant (Rahdari and Hoseini 2012; Mumtaz et al. 2020). Growth reduction due to drought stress has been described in cereal crops such as barley (Hellal et al. 2018), maize (Avramova et al. 2015), rice (Mumtaz et al. 2020), wheat (Kosová et al. 2016), etc. Among growth attributes, drought commonly causes a reduction in water content and fresh weight of plants (Jaleel et al. 2009; Meenakshi et al. 2019). Due to limited water availability, drought causes a reduction in nutrient availability and transportation (Selvakumar et al. 2012). Free radicals induced under drought stress cause oxidative stress that affects antioxidant defense and ROS production. The ROS generation damages lipid peroxidation and degradation of membrane, protein, and lipids in plants (Hendry 2005; Chiappero et al. 2019). Drought stress causes a reduction in chlorophyll content,

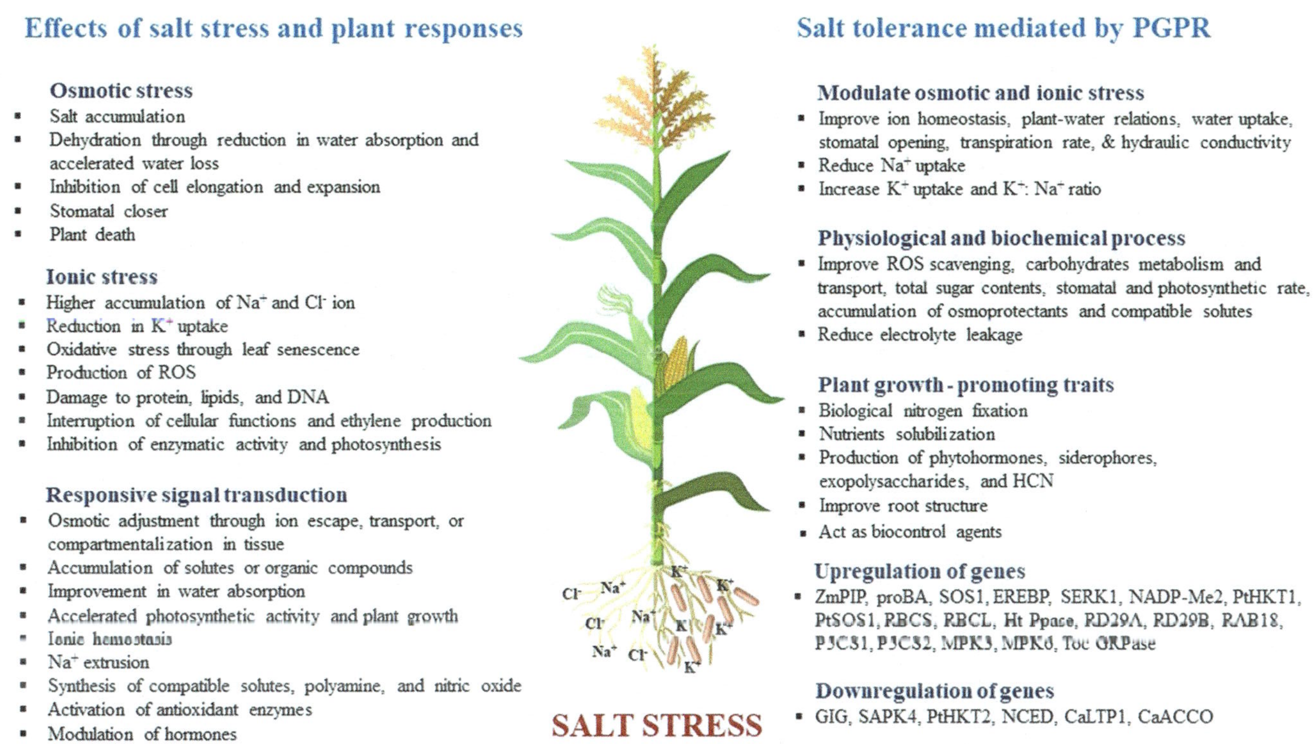

Fig. 2.1 Proposed salt-tolerance mechanisms mediated by PGPR. Rhizobacteria colonize the root surface and produce extracellular polysaccharides to ameliorate salinity stress. They manipulate phytohormones level and improve nutritional balance and photosynthetic activity through various physiological and biochemical processes

Table 2.1 The use of PGPR to mitigate abiotic stresses in plants

PGPR	Plants	Findings	References
Salt stress			
Klebsiella variicola (SURYA6)	*Triticum aestivum*	This strain showed a potential to ameliorate salinity stress by producing various metabolites including phytase, organic acid, P solubilization, siderophore, IAA, ACC-deaminase, SOD, CAT, and GPX	Kusale et al. (2021)
Pseudomonas azotoformans (CHB 1107)	*Solanum lycopersicum*	This strain possesses ACC-deaminase gene and showed a potential to ameliorate saline stress by increasing shoot and root dry weights and K^+/Na^+ ratio and reducing Na^+ uptake in tomato plants	Liu et al. (2021)
Acinetobacter bereziniae (IG 2), *Enterobacter ludwigii* (IG 10), *Alcaligenes faecalis* (IG 27)	*Pisum sativum*	Application of these strains ameliorated salt stress by modulated chlorophyll and proline contents, total soluble sugars, electrolyte leakage, antioxidant enzymatic activities, and reduced electrolyte leakage and H_2O_2 content	Sapre et al. (2021)
Bacillus subtilis, *Pseudomonas fluorescens*	*P. sativum*	These endophytic bacteria promoted salt stress tolerance in pea through increasing SOD, CAT, POD, glutathione reductase, proline contents, IAA, GA, MSI, photosynthetic pigments, K^+ uptake and expressing antioxidant enzyme genes and reducing Na^+ accumulation	Sofy et al. (2021)
Staphylococcus sciuri (*ET101*)	*Lens esculentum*, *Oryza sativa*	Strain ET101 increased transpiration rate, stomatal conductance, intracellular CO_2, RWC, proline, and glycine betaine in rice and tomato to ameliorate the salinity stress	Taj and Challabathula (2021)
Glutamicibacter sp. (YD01)	*O. sativa*	Strain YD01 was tolerant to 10% salt stress and possessed ACC-deaminase	Ji et al. (2020)

(continued)

Table 2.1 (continued)

PGPR	Plants	Findings	References
		activity and IAA production. Upon inoculation in rice, it enhanced salt tolerance by regulating ethylene production, ACC-oxidase, and K^+ uptake and reducing Na^+ accumulation and electrolyte leakage	
Rhizobium sp. (LSMR-32), *Enterococcus mundtii* (LSMRS-3)	*Vigna radiata*	These strains were positive for IAA, EPS, siderophores production, P-solubilization, biofilm formation, and salt tolerance. Their co-inoculation reported an increase in nodulation, leghemoglobin, phosphatase, dehydrogenase, proline, SOD, CAT, POD, K+ uptake, and K^+/Na^+ ratio and reduced Na^+ accumulation	Kumawat et al. (2020)
Pseudomonas sp. (SPF-33, SPF-37)	*Eleusine coracana*	The IAA, ACC-deaminase, siderophore-producing, and biofilm-forming strains promoted antioxidant enzymatic activity, proline, phenolics, flavonoids, protein contents, seed germination, vigor index, and plant height and reduced lipid peroxidation and H_2O_2 production in finger millet under salt stress	Mahadik and Kumudini (2020)
B. subtilis (NBRI 28B, NBRI 33 N), *Bacillus safensis* (NBRI 12 M)	*Zea mays*	Salt-tolerant strains produced metabolites such as ACC-oxidase and ACC-synthase that ameliorated salinity stress in maize through promoting defense enzymes, chlorophyll, proline, and soluble sugar contents	Misra and Chauhan (2020)
Pseudomonas argentinensis (HMM57),	*Brassica juncea*	These strains showed their ability to produce IAA, aminolevulinic acid	Phour and Sindhu (2020)

(continued)

Table 2.1 (continued)

PGPR	Plants	Findings	References
Pseudomonas azotoformans (JMM15)		synthase, and ACC utilization activity and reported increase in seed germination and growth of Indian mustard seedlings up to 12 dS m^{-1} salt stress	
Enterobacter sp. (PR14)	*O. sativa*, *Panicum miliaceum*	Strain PR14 having multiple PGP traits including IAA, ACC-deaminase, P solubilization, SOD, CAT, and GPX promoted alkaline and saline stress tolerance in rice and millets through promoting seed germination, seedling vigor, and shoot length	Sagar et al. (2020)
Bacillus siamensis (PM13), *Bacillus* sp. (PM15), *Bacillus methylotrophicus* (PM19)	*T. aestivum*	Salt-tolerant *Bacillus* strains showed IAA, ACC-deaminase, and EPS production. Their inoculation in wheat seedling enhanced seed germination, plant growth, and chlorophyll contents under high salt stress conditions	Amna et al. (2019)
Jeotgalicoccus huakuii (NBRI 13E)	*Z. mays*	Maize plant treated with abiotic stress-tolerant strain NBRI 13E showed salt stress tolerance through modulating defense enzymes, chlorophyll, proline, and soluble sugar contents and improving yield parameters	Misra et al. (2019)
Bacillus megaterium (A12)	*S. lycopersicum*	Inoculated plants showed higher shoot length, dry biomass, and yield under salt stress. This strain showed a reduction in ethylene production and promoted the production of APX, CAT, and SOD and WUE in plants to ameliorate salt stress	Aslam et al. (2018)
Zhihengliuella flava (F-9), *S. jettensis* (F-11), *B.*	*Z. mays*	These strains were positive for IAA, ACC-deaminase, P solubilization, and	Aslam and Ali (2018)

(continued)

Table 2.1 (continued)

PGPR	Plants	Findings	References
megaterium (F-58), *S. arlettae* (F-71)		biofilm formation. Inoculation of these strains in maize seedling mitigated salinity stress and promoted shoot and root length	
Glutamicibacter halophytocola (KLBMP 5180)	*Limonium sinense*	Halotolerant PGP strain KLBMP 5180 significantly promoted chlorophyll, proline, flavonoids, K^+, Ca^{2+} contents, and antioxidative enzymes and reduced Na^+ and MDA concentration	Qin et al. (2018)
Klebsiella sp. (IG 3)	*Avena sativa*	Plant inoculated with strain IG 3 reported higher shoot and root length, dry weight, relative water content, proline content, electrolyte leakage, MDA, and other antioxidant enzymatic activities	Sapre et al. (2018)
Klebsiella sp. (SBP-8)	*T. aestivum*	Strain SBP-8 promoted proline content, total soluble sugars and protein contents, SOD, CAT, and POX and reduces MDA in wheat under salt stress	Singh and Jha (2017)
Drought stress			
P. azotoformans FAP5	*T. aestivum*	This strain was capable to produce EPS, IAA, and P solubilization and possessed ACC-deaminase and biofilm-forming genes under water stress conditions. The inoculated wheat plants showed increased plant growth, photosynthetic pigment, physiological attributes, and antioxidative enzymatic activities under drought stress	Ansari et al (2021)
Azotobacter chroococcum (Ac), *Azotobacter brasilense* (Ab)	*Mentha pulegium*	These strains reduced the adverse effects of water stress on physiological and biochemical attributes and secondary metabolites in plants	Asghari et al. (2020)

(continued)

Table 2.1 (continued)

PGPR	Plants	Findings	References
Bacillus licheniformis, *P. fluorescens*, *A. brasilense*	*Chenopodium quinoa*	Seed-based inoculation of PGPR mitigated drought stress by improving crop growth, net assimilation rate, WUE, chlorophyll and phenolic contents, grain yield, protein, and P and K contents in grains	Aslam et al. (2020)
B. subtilis, *Azospirillum brasilense*	*T. aestivum*	These strains showed the production of EPS, osmolytes, and stress-induced phytohormones, viz., IAA, GA, CK, and ABA. They promoted seed germination, seedling vigor, shoot, and root length, leaf area, chlorophyll, carotenoid, proline, amino acid, sugar, and protein contents, SOD, CAT, and POD activities under drought stress	Ilyas et al. (2020)
B. licheniformis, *Bacillus haynesii*, *Bacillus paralicheniformis*	*O. sativa*	PGPR inoculation alleviated drought stress in rice by increasing biomass accumulation and grain yield and reducing SOD, CAT, guaiacol peroxidase, and enzymatic activities	Joshi et al. (2020)
A. calcoaceticus (EU-LRNA-72)	*Setaria italica*	This P-solubilizing strain effectively mitigated the negative effects of drought on foxtail millet by increasing glycine, betaine, proline, and sugar accumulation and decreasing lipid peroxidation	Kour et al. (2020)
B. subtilis (GOT9)	*A. thaliana*	It enhanced drought and salt tolerance in *Arabidopsis* through increasing transcripts of various drought stress- and salt stress-inducible genes	Woo et al. (2020)
Pseudomonas moraviensis	*T. aestivum*	This strain along with biogas slurry improved stomatal and sub-stomatal conductance, transpiration, photosynthetic rates, RWC, CAT, APX, plant	Yaseen et al. (2020)

(continued)

Table 2.1 (continued)

PGPR	Plants	Findings	References
		height, and grain yield under drought stress	
P. putida (KT2440, *P. fluorescens* (Pf1)	*Z. mays*	They promoted root length, protein, chlorophyll content, and plant biomass through downregulating drought stress-responsive gene, viz., cold-related dehydrin 410 gene, lipoxygenase genes, and OPR7 genes under drought stress	Ahmad et al. (2019b)
P. fluorescens (WCS417), *B. amyloliquefaciens* (GB03)	*Mentha piperita*	Their inoculation mitigated drought stress by increasing phenolic content, enzymatic activities, and reduced membrane lipid peroxidation under drought conditions	Chiappero et al. (2019)
B. amyloliquefaciens	*T. aestivum*	ACC-deaminase-producing strain along with biochar improved chlorophyll contents photosynthetic and transpiration rate, grain yield, and macronutrient concentrations in grains under drought stress	Danish and Zafar-ul-Hye (2019)
Bacillus sp. (12D6), *Enterobacter* sp. (16i)	*T. aestivum*, *Z. mays*	These phytohormones, IAA, and salicylic acid-profiling PGPR strains significantly promoted root branching, root length, diameter, and surface area in wheat and maize	Jochum et al. (2019)
B. subtilis, *Bacillus thuringiensis*, *B. megaterium*	*Cicer arietinum*	A consortium of these strains and plant growth regulator significantly enhanced the metabolic activities including chlorophyll, protein, sugar, proline, lipid peroxidation, accumulation of riboflavin, L-asparagine, aspartate, glycerol, nicotinamide, 3-hydroxy-3-methyglutarate, nicotinamide, 4-hydroxy-	Khan et al. (2019)

(continued)

Table 2.1 (continued)

PGPR	Plants	Findings	References
		methylglycine, and arginine and decreased the activities of CAT, APOX, POD, and SOD under drought stress	
Paenibacillus beijingensis (BJ-18), *Bacillus* sp. (L-56)	*T. aestivum*, *Cucumis sativus*	These strains along with super absorbent polymer promoted seed germination, plant growth, urease, sucrose, dehydrogenase activities, and downregulated expression of genes involved in ROS scavenging, ethylene biosynthesis, stress response, salicylic acid, and transcription activation in plants under drought stress	Li et al. (2019)
Ochrobactrum pseudogrignonense (RJ12), *Pseudomonas* sp. (RJ15), *B. subtilis* (RJ46)	*Vigna mungo*, *P. sativum*	These strains were capable of N fixation, P solubilization, production of IAA, HCN, siderophores, and ACC-deaminase activity, and their consortium offered drought stress tolerance through promoting seed germination, root and shoot length, dry weight, ROS scavenging enzymes, cellular osmolytes, RWC, and chlorophyll contents	Meenakshi et al. (2019)
P. moraviensis	*T. aestivum*	This strain inoculation along with biogas slurry promoted grain and biological yield and N, P, and K contents in straw and grains of wheat under water drought stress	Yaseen et al. (2019)
Genus of *Bacillus*, *Enterobacter*, *Moraxella*, *Pseudomonas*	*T. aestivum*	IAA-producing PGPR significantly improved shoot and spike length, seed weight, number of tillers, and spikelets under drought stress	Raheem et al. (2018)
Azospirillum sp. (AZ39)	*Z. mays*	This strain was capable to produce siderophore, N	García et al. (2017)

(continued)

Table 2.1 (continued)

PGPR	Plants	Findings	References
		fixation, ACC-deaminase activity, IAA, tolerant to osmotic, and salt stress and responded against water deficit condition	
P. putida (GAP-P45)	*Arabidopsis thaliana*	Inoculated plants showed delayed and prolonged upregulation of proline biosynthesis genes, viz., OAT, P5CS1, and P5CR, as well as proline catabolism genes, viz., PDH1 and P5CDH, were positively correlated with better growth and physiology of plants under drought stress	Ghosh et al. (2017)
Flooding stress			
Microbacterium sp. (AR-ACC2)	*O. sativa*	The ACC-deaminase utilizing strain significantly increased seed germination, seedling vigor index, growth attributes, and reduced ethylene production under submerged condition	Bal and Adhya (2021)
Pseudomonas veronii (KJ)	*Sesamum indicum*	This acdS gene-coded strain mitigated waterlogging stress by increasing chlorophyll content, shoot length, root length, and biomass of plant	Ali et al. (2018)
P. veronii (KJ)	*A. thaliana*	Waterlogging stress and salt stress were ameliorated in mutant *A. thaliana* by inoculation of acdS coding strain KJ	Jung et al. (2018)
K. variicola (AY13)	*Glycine max*	IAA-producing strain AY13 initiated adventitious root and promoted plant growth and chlorophyll contents under flooding stress	Kim et al. (2017)
P. putida (WT)	*Rumex palustris*	ACC-deaminase-producing strain ameliorated the waterlogging stress by altering plant hormonal balance	Ravanbakhsh et al. (2017)

(continued)

Table 2.1 (continued)

PGPR	Plants	Findings	References
Rhizobium sp. (IRBG74)	*O. sativa*, *Sesbania cannabina*	Strain IRBG74 possessed β-glucuronidase and green fluorescent protein altered the synthesis of the rhamnose-containing O-antigen in nodulation and symbiotic N fixation in *S. cannabina* as well as involved in rice plant growth promotion under flooded stress	Mitra et al. (2016)
P. fluorescens (REN1)	*O. sativa*	ACC-deaminase- and IAA-producing strains REN1 promoted root elongation and endophytic root colonization in rice under flooded stress conditions	Etesami et al. (2014)
Serratia odorifera (CC7), *Aerococcus viridans* (CK3)	*T. aestivum*	These ameliorated the negative effects of waterlogging stress through increasing root and shoot length and dry root and shoot weight of wheat	Bangash et al. (2013)
Pseudomonas sp. UW4	*Cucumis sativus*	This strain ameliorated the waterlogged stress by regulating responsive proteins and metabolism pathways	Li et al. (2013)
Achromobacter xylosoxidans (d2), *S. ureilytica* (Bac5), *Herbaspirillum seropedicae* (Oci9), *Ochrobactrum rhizosphaerae* (Oci13)	*Ocimum sanctum*	ACC-deaminase-producing PGPR showed a potential to protect plant from flooding-induced negative stresses including increasing production of ethylene, lipid peroxidation, reduction in chlorophyll content, and nutrient uptake	Barnawal et al. (2012)
A. chroococcum, *Azospirillum* spp., *Pseudomonas* spp., *B. subtilis*	*Brassica napus*	Seed inoculation and foliar application of biofertilizers alleviated waterlogging stress on canola by improving SOD, CAT, and POX activities and reducing lipid peroxidation and ethylene production	Habibzadeh et al. (2012)

(continued)

Table 2.1 (continued)

PGPR	Plants	Findings	References
Mesorhizobium cicero (LMS-1)	*Cicer arietinum*	This mutant strain with an exogenous ACC-deaminase gene showed increased nodulation and plant growth of chickpea	Nascimento et al. (2012)
Heat stress			
B. safensis	*T. aestivum*	Seed primed with this strain showed heat tolerance by improved production of defensive cell organelles, mitigation of ROS by regulating redox enzymes and antioxidants, expression of HSPs and modulation of putrescine biosynthetic genes	Sarkar et al. (2021)
B. cereus (TCR17), *Providencia rettgeri* (TCR21), *Myroides odoratimimus* (TCR22)	*Sorghum bicolor*	The chromium-reducing thermotolerant strains were capable to produce siderophores, IAA, and P-solubilization and induced heat tolerance in sorghum by reducing oxidative stress, improving antioxidant status, and reducing proline and malondialdehyde contents	Bruno et al. (2020)
Brevibacterium linens (RS16)	*Eucalyptus grandis*	Foliar application of ACC-deaminase producing strain RS16 reduced the immediate reduction in photosystem II, net assimilation rate, and stomatal conductance caused reduction in isoprene emissions, and induced expression of DnaK genes under short time heat stress	Chatterjee et al. (2020)
B. cereus (SA1)	*G. max*	Strain SA1 was able to produce IAA, GA, and organic acids and showed amelioration of heat stress in soybean by improving biomass, chlorophyll, salicylic acid, ascorbic acid, SOD, POD, glutathione, and reducing ABA	Khan et al. (2020a)

(continued)

Table 2.1 (continued)

PGPR	Plants	Findings	References
B. cereus (SA1)	*S. lycopersicum*	Combined application of strain SA1 and exogenous humic acid improved plant biomass and chlorophyll contents, salicylic acids, APX, SOD, glutathione, and nutrient uptake, and reduced ABA under heat stress	Khan et al. (2020b)
B. cereus	*S. lycopersicum*	ACC-deaminase and EPS producing strains significantly reduced heat stress effects by promoting shoot and root length, fresh and dry weight, and leaf surface area	Mukhtar et al. (2020)
B. tequilensis (SSB07)	*S. max*	The strain counteracted heat stress on soybean by increasing shoot length, biomass, photosynthetic pigment, jasmonic acid, salicylic acid, and down-regulation of stress-responsive ABA	Kang et al. (2019)
Paraburkholderia phytofirmans (PsJN)	*L. esculentum*	This strain alleviated the harmful effect of high temperature by improving plant growth and physiological attributes of tomatoes	Issa et al. (2018)
B. safensis, *Ochrobactrum pseudogrignonense*	*T. aestivum*	Seed priming with these strains improved cell viability, chlorophyll contents, antioxidative response, redox enzymes, osmolytes, and reduced photosynthetic damage, ROS production, and membrane damage due to heat stress	Sarkar et al. (2018)
B. amyloliquefaciens (UCMB5113), *A. brasilense* (NO40)	*T. aestivum*	Seed inoculation with these strains improved heat tolerance in wheat by reducing ROS generation, alteration in the metabolome, and activating heat shock transcription factors.	Abd El-Daim et al. (2014)

(continued)

Table 2.1 (continued)

PGPR	Plants	Findings	References
Cold stress			
Pseudomonas sp. (TmR5a), *Curtobacterium* sp. (BmP22c)	*S. lycopersicum*	The consortia application of psychrotolerant strains possessing ACC-deaminase gene, and nitrogenase reductase gene, biocontrol activities, IAA, P-solubilization promoted tomato plant growth under cold stress conditions	Vega-Celedón et al. (2021)
Genus of *Pseudomonas*, *Serratia*, *Staphylococcus*	*Deschampsia antarctica*, *Colobanthus quitensis*	Cold tolerant ACC-degrading bacteria were isolated from the rhizosphere, phyllosphere, and endosphere of Antarctic vascular plants and could ameliorate cold stress in plants	Araya et al. (2020)
Pseudomonas spp.	*S. lycopersicum*	Psychrotolerant PGP strain reported increasing tomato germination and plantlets under low-temperature stress	Tapia-Vázquez et al. (2020)
P. fragi, *P. chlororaphis*, *P. fluorescens*, *P. proteolytica*, *B. frigoritolerans*	*Phaseolus vulgaris*	Psychrotolerant bacterial strains capable to produce ACC-deaminase activity, production of extracellular proteins, stimulation of SOD, CAT, POD, and glutathione reductase improved the cold resistance in bean seedlings through decreasing freezing injury, ice nucleation, lipid peroxidation, and reduction in ROS level	Tiryaki et al. (2019)
Bacillus spp. (CJCL2, RJGP41)	*T. aestivum*	The genetic screening of psychrophilic strain revealed the presence of genes responsible for cold stress tolerance, membrane transport, signal transduction, and osmotic regulation under cold stress. Wheat plants inoculated with this strain responded to cold stress by regulating ABA, lipid peroxidation, and proline accumulation	Zubair et al. (2019)

(continued)

Table 2.1 (continued)

PGPR	Plants	Findings	References
Genus of *Bacillus*, *Paenarthrobacter*, *Pseudomonas*, *Paenibacillus*, *Cupriavidus*	*Lepidium meyenii*	Psychrotrophic PGPR strains were positive for IAA production, P-solubilization, tolerant to varying concentrations of Cd and Pb, and promoted seed germination exposed to low temperature and heavy metal toxicity	Ortiz-Ojeda et al. (2017)
Arthrobacter agilis (L77)	–	Hydrolytic enzymes producing PGP strain L77 genome under cold stress revealed diverse genes involved in metabolism and cold shock variations and diverse polysaccharides biosynthesis	Singh et al. (2016)
Genus of *Arthrobacter*, *Flavimonas*, *Flavobacterium*, *Massilia*, *Pedobacter*, *Pseudomonas*	*S. lycopersicum*	Psychrotolerant bacteria caused an increase in seed germination, plant growth, decrease in membrane damage, induced antioxidant enzymes, and proline synthesis in tomato under chilling stress	Subramanian et al. (2016)
Genus of *Streptomyces*, *Arthrobacter*, *Paenibacillus*	*Thylacospermum ceaspitosum*	Bacterial community and structure of cold desert revealed that actinobacteria dominated cultivable parts of a community. A more diverse bacterial community was associated with finer soil derived from easily weathering and cushion plants in a xeric environment	Řeháková et al. (2015)
Pseudomonas spp.	*T. aestivum*	Inoculation of wheat with *Pseudomonas* strains ameliorated cold stress by increasing chlorophyll, anthocyanin, proline, amino acids, phenolics, RWC, and starch contents and reducing electrolyte leakage and Na^+/K^+ ratio	Mishra et al. (2011)

photosynthetic and transpirational rates, and stomatal conductance in various crops (Ohashi et al. 2006; Mumtaz et al. 2020).

Drought as multi-dimensional stress affects various subcellular compartments, cell organs, and whole plant levels (Horn and Chapman 2012). Drought destructively upsets the quantity and quality of plants. Drought stress alleviation is important to fulfill food demand. To address this global challenge, there is a dire need to develop improved crop germplasm and technologies to increase water use efficiency (WUE) (Passioura 2007). Recently, the application of beneficial microorganisms has got attention for the mitigation of drought by improving WUE (Marulanda et al. 2009; Ngumbi and Kloepper 2016). Under drought stress, the plant-associated bacterial community in the rhizosphere alleviates drought stress and improves plant health by improving the resistance in plants against stressors (Cherif et al. 2015; Aslam et al. 2020; Kour et al. 2020). The PGPR in the water-limited rhizosphere with repeated dry periods is more stress-tolerant and increases plant growth as compared to PGPR population of the well-irrigated rhizosphere (Mayak et al. 2004). Drought tolerance mediated by PGPR could be through the production of phytohormones, viz., IAA, ABA, and GA, production of ACC-deaminase to reduce ethylene production, and induced drought tolerance through the production of metabolites, viz., siderophore and EPS (Timmusk and Nevo 2011; Meenakshi et al. 2019; Ansari et al. 2021).

Plants produce phytohormones such as IAA, ABA, GA, and CK which play important role in drought stress alleviation (Egamberdieva 2013; Fahad et al. 2015; Kumar et al. 2019a). PGPR are reported to produce these phytohormones under drought stress that can improve plant growth (Khan et al. 2020c). The IAA-producing PGPR alleviate drought stress by improving lateral root development and water uptake (Mantelin and Touraine 2004). For example, Dimkpa et al. (2009) reported that *Azospirillum* (IAA producing) promoted drought stress tolerance in plants. Nitric oxide (NO, signaling molecule) produced by *A. brasilense* inoculation in tomato showed induction of IAA pathway and development of adventitious roots. Its inoculation in maize grown under drought stress also improved water content in comparison to uninoculated control plants (Molina-Favero et al. 2008). Wheat treated with *A. brasilense* strain Sp245 improved drought stress tolerance through increased grain yield; Mg, K, and Ca concentration; water content; and lowering volumetric cell wall elasticity (Creus et al. 2004). The IAA-producing *Bacillus thuringiensis* inoculation in *Lavandula dentata* showed drought stress amelioration and improved plant metabolic activities, nutrition, and physiology (Armada et al. 2014). GA- and ABA-producing PGPR also help plants withstand drought stress. Soybean inoculation with *P. putida* strain H-2-3 (GA producing) showed an increase in plant growth under drought stress (Sang-Mo et al. 2014). Similarly, inoculation of maize with *Azospirillum lipoferum* ameliorated drought stress in plants (García et al. 2017). The ABA is a stress hormone, and its biosynthesis is induced during cellular dehydration under water deficit conditions. The inoculation of CK-producing *B. subtilis* in *Platycladus orientalis* improved drought stress alleviation by increasing ABA levels (Egamberdieva et al. 2017).

Drought stress enhances the production of ethylene that causes retarded plant growth (Hardoim et al. 2008). The ACC-deaminase-producing bacteria sequester and degrade plant ACC and improve plant growth, supply N and energy, and reduce the deleterious effect of ethylene in plants (Glick 2005; Nadeem et al. 2017; Danish and Zafar-ul-Hye 2019). Mayak et al. (2004) inoculated ACC-deaminase-producing *Achromobacter piechaudii* in tomato and pepper and reported increase in plant growth under drought stress by reducing ethylene production. The ACC-deaminase-containing PGPR enhance grain yield and N accumulation in grains and repair depressed nodulation in plants (Dodd et al. 2004; Danish et al. 2020). Inoculation with ACC-deaminase-producing *P. fluorescens* in *Pisum sativum* under drought stress showed longer roots due to increased water uptake (Zahir et al. 2008). *Variovorax paradoxus* strain 5C-2 improved pea productivity through increase in WUE due to ABA production in drought stress (Belimov et al. 2009). PGPR promoted wheat growth due to better water and nutrient acquisition over uninoculated control (Shakir et al. 2012; Ansari et al. 2021). Sharma et al. (2013) reported that co-inoculation of *Bacillus*, *Pseudomonas*, and *Mesorhizobium* strains (ACC-deaminase producing) in *Cicer arietinum* under drought conditions promoted seed germination and seedling biomass over uninoculated control (Hui and Kim 2013).

2.4.1.3 Flooding Stress

Waterlogging and submergence are abiotic stresses, collectively called flooding stresses. Flooding stress is a global issue that influences plant composition and productivity (Phukan et al. 2016). Under flooding conditions, soil becomes completely saturated with water and develops an anaerobic environment. Plant roots become hypoxic which impairs the growth and development of the plants (Nishiuchi et al. 2012; Glick 2014). According to an estimate, flooding affected nearly 13% of the country's land and nearly 10% of agricultural lands globally (Cramer et al. 2011). Flooding stress causes yield loss up to 15–80% in various crops (Patel et al. 2014).

Under flooding conditions, poorly adapted plant species are severely affected by excess water that results in 10^4-fold slower oxygen diffusion as compared to air (Armstrong and Drew 2002). Oxygen-deficient conditions lead to higher production of plant hormone "ethylene" that increases ROS production, destruction of macromolecules, and inhibition of photochemical and ultimately causes plant death (Grichko and Glick 2001a; Glick 2005; Ahmed et al. 2006). Under flooding conditions, plants exhibit symptoms of chlorosis, necrosis, and low productivity (Paul et al. 2016). Various wetland plant species are highly effective due to a combination of various physiological adaptations to withstand flooding stress conditions (Ravanbakhsh et al. 2017). Such plants acclimatize flooding stress through a physical method including escape from submerged conditions, prevention of oxygen deficiency via internal aeration, anoxia tolerance, and avoiding oxidative damage (Armstrong et al. 1994; Gibbs and Greenway 2003; Voesenek et al. 2003). Under flooding stress, plant roots produce a larger amount of ethylene that modifies phytohormone cascade of ABA, GA, and IAA to encourage the emergence of adventitious rooting in plants (Vidoz et al. 2010; Dawood et al. 2016; Ali et al. 2020). Root development also requires ethylene that persuades the development of ROS in epidermal cells leading to cell death by dropping antioxidant activity (Steffens and Sauter 2009; Yamauchi et al. 2011).

Under flooding stress, ACC-deaminase-producing bacteria play a key role in reducing excess ethylene production by catabolizing ACC to ammonia and α-ketobutyric acid (Ali and Kim 2018; Sasidharan et al. 2018; Bal and Adhya 2021). Limited literature is available to describe PGPR's role in flooding stress mainly through their ability to produce ACC-deaminase. Transgenic tomato developed through ACC-deaminase expressing transcriptional control of *rolD* promoter from *Agrobacterium rhizogenes* and 35*S* constitutive expression was evaluated under flooding stress (Grichko and Glick 2001b). Development of transgenic plants with the insertion of *rolD* promoter showed ACC-deaminase expression and increased flooding tolerance by improving seedling biomass, epinasty, leaf chlorophyll, and protein content and reducing the ethylene production (Grichko and Glick 2001b). Under flooding stress, ACC synthase is produced in a higher amount in the plant tissues that induces the production of ACC-deaminase by bacteria. Symbiotic bacteria can reduce ethylene stress up to 60–90% by producing ACC-deaminase enzyme (Saleem et al. 2007; Jung et al. 2018; Ali et al. 2018). Grichko and Glick (2001a) inoculated tomato with *Enterobacter cloacae* (strains UW4 and CAL2) and *P. putida* (strains pRKACC) and reported that the expression of ACC-deaminase genes in these strains promoted substantial flooding stress tolerance by improving plant growth and physiological attributes (Grichko and Glick 2001a).

Farwell et al. (2007) inoculated ACC-deaminase-producing *P. putida* strain UW4 in ACC-deaminase-producing transgenic canola and non-transformed canola under flooding and nickel (Ni) stress. They reported that transgenic canola inoculated with strain UW4 showed better shoot biomass as compared to non-transformed canola under flooding conditions (Farwell et al. 2007). Barnawal et al. (2012) investigated the effect of *Achromobacter xylosoxidans* strain Fd2, *Herbaspirillum seropedicae* strain Oci9, *Ochrobactrum rhizosphaerae* strain Oci13, and *Serratia ureilytica* strain Bac5 on *Ocimum sanctum*. They reported that bacterial treatment alleviated waterlogging stress and reduced ethylene production, lipid peroxidation, proline contents, and improved chlorophyll content and nutrient uptake. Strain Fd2 exhibited maximum waterlogging tolerance with 46.5% higher growth and herb yield and a 53% reduction in ACC concentration over uninoculated waterlogged plants (Barnawal et al. 2012). Habibzadeh et al. (2012) treated canola plants with biofertilizers through foliar and seed application and reported their effect on physiological and morphological attributes of canola under flooding conditions. The biofertilizers were AAP (strains of *Azotobacter*, *Azospirillum*, and *Pseudomonas*) and APB (strains of *Azospirillum*, *Pseudomonas*, *Bacillus*). Both foliar and seed treatment with these biofertilizers significantly alleviated flooding stress by improving superoxide dismutase (SOD), catalase (CAT), and peroxidases (POD) activities and by reducing ethylene and lipid peroxidation (Habibzadeh et al. 2012).

Etesami et al. (2014) reported that rice inoculation with endophytic ACC-deaminase-producing *P. fluorescens* strain REN1 promoted root growth under flooding stress. The *Streptomyces* sp. strain GMKU inoculation in mung bean alleviated flooding stress by improving shoot and root growth and chlorophyll content and reducing ethylene formation (Jaemsaeng et al. 2018). Similarly, Ali et al. (2018) reported that *Sesamum indicum* inoculation with *Pseudomonas veronii* strain KJ showed flooding stress mitigation by improving plant growth and physiological attributes. Production of IAA by PGPR induces adventitious root growth that also helps plant maintain their growth and ameliorate flooding stress. Rice seedling inoculation with IAA-producing strain REN1 also showed better root development under flooding conditions (Etesami et al. 2014). Similarly, Kim et al. (2017) inoculated *Klebsiella variicola* strain AY-13 in soybean under waterlogged conditions and reported its potential to induce adventitious root growth, improved plant growth, and flooding tolerance (in soybean). Strain AY-13 was able to produce IAA up to 84.27 ± 3.55 μg mL^{-1} as analyzed through gas chromatography-mass spectrometry selected ion monitoring (GC-MS/SIM) as reported by Kim et al. (2017).

2.4.1.4 Heat Stress

Temperature is a primary factor affecting plant growth and development (Hatfield and Prueger 2015). Each plant species requires a certain temperature regime characterized as a minimum, maximum, and optimal temperature range for its growth and productivity (Hatfield et al. 2011). Global warming is becoming a major issue worldwide due to elevated carbon dioxide that is causing economic losses. It is predicted that greenhouse gas emission is causing a steady increase in average temperature worldwide. The increase in average temperature linked with climate changes usually aggravates the heat stress in agriculture (Carbonell-Bojollo et al. 2019; Zhang et al. 2019). Heat stress negatively affects plant growth and development by causing changes in plant morphology and physiology (Prasad et al. 2008). Repeated extremely high-temperature events have caused phenological disturbances in different plant species (Li et al. 2017). In the early crop growth stages, heat stress inhibits seed germination and seedling's emergence (Wahid et al. 2008). It can damage photosynthetic pigments, carbon metabolism, and organic solute transport (Hasanuzzaman et al. 2013; Balal et al. 2016). El-Daim et al. (2014) reported that temperature at 45 °C caused mortality of wheat seedlings up to 80–95%. The impact of heat stress on the reproductive phase of crops generally leads to variation in mineral nutrition, deactivation of the immune system, and the successive generation of ROS (Zhang et al. 2016; Lanza and dos Reis 2021). Heat stress elevation disrupts the rate of photosynthesis, stomatal conductance, and transpiration (Fahad et al. 2016). The activity of photosystem II (PSII) was inhibited due to heat stress that ultimately reduced photosynthetic rate which is the major cause of the inhibition of chlorophyll production (Chen et al. 2007; Killi et al. 2020).

Plants tolerate heat stress by altering osmotic pressure, antioxidants, plant hormone level, and regulation of HSPs. Plant regulates the production of certain plant hormones, such as ABA, GA, JA, and IAA, to improve heat resistance (Abdelrahman et al. 2017; Khan et al. 2020c). Among these phytohormones, IAA and GA improve plant development processes, while JA and ABA can alleviate the heat stress (Ahammed et al. 2016). Hydrogen sulfide production also improves resistance to heat stress in plants (Chen et al. 2016). The HSPs protect the natural protein and improve protein stability, which in turn improve the thermal tolerance in the plant (Wei et al. 2021). The accumulation of HSP in terms of ZmHSP17.2 in corn and OsHSP26 in rice showed a strong correlation in heat stress tolerance (Lee et al. 2007; Liu et al. 2013; Sarkar et al. 2021). Maize leaves under heat stress expressed sHSP17.2, sHSP17.4, and sHSP26, while these proteins were also regulated by ABA (Hu et al. 2010). HSFA1 is a major regulator of signal perception, transduction, and control of the expression of stress-sensitive genes in tomato plants. The expression of HSP26, HSP17.4, and HSP17.2 also played an important role in heat tolerance in corn (Asada 2006). A multifunctional cytoplasmic protein known as eukaryotic elongation factor 1α (eEF1A) is responsible for the heat resistance in tomatoes (Momčilović et al. 2016). Plants also increase the activity of antioxidant enzymes, including alpha-tocopherol; glutathione; ascorbic acid; CAT, SOD, APX, and POD; and glutathione reductase, to repair the oxidative damage caused by heat stress (Asada 2006; Khan et al. 2020a, b).

Plant inoculation with PGPR improves heat tolerance by enhancing the photosynthetic rate, water use efficiency, antioxidant enzymatic activity, and uptake of water and nutrients (Zhu et al. 2012; Maya and Matsubara 2013; Kang et al. 2019; Sarkar et al. 2021). The literature on the mechanism of heat stress tolerance due to bacterial inoculation is very limited; however, studies are available to support that bacterial augmentation can ameliorate heat stress by producing ACC-deaminase, EPS, and phytohormones which are involved in the expression of HSP. Application of ACC-deaminase-producing PGPR under heat stress could ease unfavorable conditions. Production of ACC-deaminase by PGPR reduces the inhibitory effect of stress hormone "ethylene" generated under heat stress through degrading ACC into ammonia and α-ketobutyrate (Barnawal et al. 2012). The detailed mechanism of conversion of ACC into ammonia and α-ketobutyrate by the action of ACC-deaminase produced by PGPR has already been reported in the above abiotic stress sections of this chapter. The PGPR uses a similar mechanism of action in all abiotic stress to reduce the lethal concentration of ethylene. Recently, Mukhtar et al. (2020) inoculated PGP *Bacillus cereus* having the ability to produce ACC-deaminase up to 0.76 to C0.9 μM/mg protein/h in tomato under heat stress. They reported a significant improvement in growth and physiological and biochemical attributes of tomato plants due to inoculation with EPS-producing, ACC-deaminase-containing *B. cereus*. Wu and Yang (2019) reported that ethylene-mediated signaling under heat stress was involved in the reduction of oxidative damage, maintenance of chlorophyll content, and enhancement of thermo-tolerance in rice seedlings. They also reported that HSP and ethylene-signaling-related genes were involved in complex network regulation that conferred thermo-tolerance to rice seedlings.

Srivastava et al. (2008) reported that high-temperature (40 °C up to 5 days) tolerant *P. putida* strain NBRI0987 isolated from chickpea rhizosphere was affected with drought stress. Strain NBRI0987 was able to tolerate heat stress due to its ability of biofilm formation and overexpression of sigma factor genes. Ali et al. (2009)

inoculated sorghum with thermo-tolerant *Pseudomonas* sp. strain AKM-P6 under elevated temperatures (47–50 °C for day and 30–33 °C for a night). They reported that inoculation with strain AKM-P6 ameliorated the elevated temperature stress through well-established growth on the root surface and promoted root biomass, shoot biomass, amino acids, sugar, proline, chlorophyll, and induced synthesis of protein (Ali et al. 2009). El-Daim et al. (2014) reported the heat tolerance in wheat through inoculation with *B. amyloliquefaciens* strain UCMB5113 and *A. brasilense* strain NO40. They reported that inoculation with these strains improved heat tolerance through activating ascorbate-glutathione redox cycle, reducing generation of ROS and cell damage (El-Daim et al. 2014).

Ali et al. (2011) investigated heat tolerance in wheat through inoculation with thermo-tolerant PGPR *P. putida* strain AKMP7. They reported that strain AKMP7 promoted growth of wheat in terms of root length, shoot length, biomass, tiller, and spikelet number and promoted grain formation through reducing membrane injury and improving the activity of SOD, APX, and CAT and various metabolites like chlorophyll, proline, protein, and carbohydrates (Ali et al. 2011). The heat shock protein could be a key factor in maintaining the functional conformation of the protein under heat (Wang et al. 2004). Among HSP, HSP17.8 is highly upregulated in wheat seedlings under heat stress; however, the response to bacterial treatment may vary between different cultivars and the bacterial strains. Interestingly, the expression of HsfA3 was downregulated in *Bacillus*-treated wheat seedlings compared to untreated heat-loaded seedlings (Suzuki et al. 2008). In *Arabidopsis*, HsfA3 improved the production of several HSPs (Ikeda and Ohme-Takagi 2009) and regulated two independent pathways, suggesting that they play an integrated role in stress management (Liu and Charng 2013). El-Daim et al. (2014) reported that inoculation of wheat with these strains *B. amyloliquefaciens* strain UCMB5113 and *A. brasilense* strain NO40 improved heat tolerance through the transcript level of a stress-related gene. They reported that the preactivation of heat shock transcription factors had an important role in heat stress tolerance (El-Daim et al. 2014).

2.4.1.5 Cold Stress

Low temperature is one of the most important abiotic stresses also known as cold or chilling stress. It limits crop productivity by damaging various physiological and biochemical cell functioning with visible symptoms (Ruelland and Zachowski 2010). Cold stress slows down the plant metabolism by damaging cellular macromolecules and membrane function, clotting of the cell membrane, and alteration in lipid composition and induces synthesis of compatible solutes, proteins, and carbohydrate metabolism (Welti et al. 2002; Baek and Skinner 2003). Chilling stress forms extracellular ice crystals leading to frozen dehydration and concentration of cellular juice and causes several other mechanical effects (Mishra et al. 2012).

Adaptation to cold stress tolerance encompasses a range of biochemical, molecular, and metabolic processes (Zhu et al. 2007; Ritonga and Chen 2020). Plants in the first attempt try to resist changes in cellular metabolism due to lower temperatures. Further, plant regulatory mechanisms are activated and restore levels and normal flow of metabolites (Fernie et al. 2005). They maintain homeostasis to gain tolerance to freezing and implying gene expression (Cook et al. 2004). The cold stress adaptation process involves changes in gene expression, membrane lipids, protein content, soluble proteins, sugar, proline, total phenols, deoxygenase, low leakage of ions from the cell membrane; changes in the accumulation of anthocyanins; and production of metabolites, including osmotic protective agents, and allows osmosis (Kasuga et al. 2004; Barka et al. 2006; Tiryaki et al. 2019). Plant cells sense cold stress by cold-responsive membrane hardening induced by genes that contribute to cold acclimation of alfalfa and *Brasicca napus* (Sangwan et al. 2001).

A major portion of the earth is covered with a low-temperature environment that acts as a reservoir of cold-tolerant (psychrophilic) bacteria having a low metabolic activity (Kumar et al. 2019b). Various psychrophilic bacterial strains including *Acinetobacter rhizosphaerae*, *Arthrobacter nicotianae*, *Bacillus* sp., *Bordetella* sp., *Brevundimonas terrae*, *Exiguobacterium acetylicum*, *Mesorhizobium* sp., *Pseudomonas cedrina*, *P. fluorescens*, *P. putida*, *Providencia* sp., *Pseudomonas corrugata*, *Pseudomonas fragi*, *Rhizobium leguminosarum*, *Serratia marcescens*, *Sinorhizobium meliloti*, *Stenotrophomonas*, etc. having the ability to increase P and iron (Fe) availability under cold stress by producing siderophore and solubilizing P were reported previously (Prevost et al. 1999, 2003; Pandey et al. 2002, 2006; Katiyar and Goel 2004; Negi et al. 2005; Trivedi and Sa 2008; Gulati et al. 2009; Selvakumar et al. 2008, 2009; Verma et al. 2015; Yadav et al. 2015; Zubair et al. 2019; Araya et al. 2020; Tapia-Vázquez et al. 2020; Vega-Celedón et al. 2021). Recently, Awasthi et al. (2019), Gautam et al. (2019b), Kumar et al. (2019b), and Tiryaki et al. (2019) also reported cold-tolerant bacteria including *Arthrobacter humicola*, *Brevibacillus invocatus*, *Brevibacterium frigoritolerans*, *Pseudomonas chlororaphis*, *P. fluorescens*, *Pseudomonas fragi*, *Pseudomonas proteolytica*, *Pseudomonas helmanticensis*, *Pseudomonas koreensis*, *Pseudomonas mandelii*, and *Viridibacillus arenosi* isolated from various cold regions.

The psychrophilic bacteria alleviate cold injury in plants using various mechanisms. Such bacteria could be efficient enough to synthesize antifreeze proteins and have a strong affinity for the promotion of root growth (Sun et al. 1995). Cold-tolerant *Bradyrhizobium japonicum* was reported for its ability to promote nodule formation and nitrogen fixation in soybean under cold stress (Mishra et al. 2009). Barka et al. (2006) reported improvement in cold tolerance in grapevine seedlings due to inoculation with *Burkholderia phytofirmans* that reduce electrolyte leakage. Inoculation with endophytic *B. phytofirmans* strain PsJN improved cold stress tolerance through the process of antioxidant scavenging, modulation in carbohydrate metabolism, and trehalose metabolism in grapevine (Fernandez et al. 2012; Theocharis et al. 2012). Mishra et al. (2011) and Barka et al. (2006) reported that psychrophilic bacterial inoculation also improved sugar metabolism pathway, photosynthetic activity, total phenolic contents, and activation of genes related to C-repeat binding factor. Inoculation with strains of *Azospirillum*, *Bacillus*, and *Raoultella* spp. improved cold tolerance in wheat and barley by reducing freezing injury and sustaining antioxidant enzyme activity (Turan et al. 2013). Tomato inoculated with *Pseudomonas frederiksbergensis* strain OS251 and *Pseudomonas vancouverensis* strain OB155 showed improved cold tolerance due to enhanced cold-acclimatized gene expression, antioxidant activity, proline synthesis, and reduction in membrane damage (Subramanian et al. 2015, 2016). Production of phytohormones by *Serratia nematodiphila* improved cold tolerance in pepper seedling (Kang et al. 2015). Inoculation with strains of *Bacillus* and *Pseudomonas* spp. in *Phaseolus vulgaris* has been reported to improve cold tolerance in plants by regulating freezing injury, ice-nucleating activity, lipid peroxidation, and activities of SOD, CAT, and POD (Tiryaki et al. 2019). So, it is concluded that plant interaction with cold-tolerant bacteria under cold stress environment can sustainably engineer cold stress (Kushwaha et al. 2020).

2.4.2 Biotic Stress

The PGPR help in improving the plant growth by suppressing the growth of phytopathogens, and weeds thus acting as a biocontrol and bioherbicide agents (Adetunji et al. 2019; Khan et al. 2020c). They use the indirect mechanisms of action to promote plant growth through alleviating biotic stresses due to growth of deleterious organisms (Arya et al. 2018). PGPR as biocontrol agents are environment-friendly supplements to treat plant diseases and to reduce the use of chemicals in agriculture (Gerhardson 2002). They control pathogens through pathogen exclusion based on bacterial capacity to colonize faster and effectively and by reducing the nutrient availability to harmful pathogens (Table 2.2). PGPR, for example, limit Fe availability to pathogenic microbes by production of siderophores that chelate Fe, making it unavailable for use by pathogens (Singh et al. 2011). They also produce a heterogeneous group of organic compounds called antibiotics which are lethal to various microorganisms even at low concentrations (Haas and Defago 2005). They reduce the pathogen attack by provoking a defense response in plants known as induced systemic resistance (ISR) involving synthesis of various defense metabolites that modify physical and biochemical properties of host without causing disease (Bent 2006; Bukhat et al. 2020).

PGPR produce antibiotics that are bioactive metabolites having the ability to inhibit growth of disease-causing agents. Under stress, PGPR secrete antibiotics having varying degrees of specificity and mode of action (Pathak et al. 2017). *Pseudomonas* sp., for example, can produce antibiotics including 2,4-diacetylphloroglucinol, amphisin, phenazine, phenzine, pyoluteorin, pyrrolnitrin, tensin, and tropolone (Compant et al. 2005; Showkat 2012). Many *Pseudomonas* strains produce pyrrolnitrin antibiotic which is a broad-spectrum antifungal compound. Similarly, in another study, Perneel et al. (2008) reported that *Pseudomonas aeruginosa* and *P. chlororaphis* can produce phenazine, which is a heterocyclic, broad-spectrum antibiotic used in biocontrol against phytopathogens. Two broad-spectrum antibiotics called kanosamine and zwittermicin are produced by *B. cereus* that act against damping-off of the alfalfa seedlings (Lozano et al. 2016). Iturins with β-amino fatty acid produced by strains of *Bacillus* sp. exhibit strong antifungal activity and rely predominantly on the ability to augment membrane permeability due to strong interaction with the phospholipids and sterols

Table 2.2 Use of PGPR to mitigate biotic stresses in plants

Biocontrol agent	Plant	Phytopathogen	Findings	References
P. peoriae (RP20, RP51, RP62, RP51), *Paenibacillus* sp. (RP31)	*R. pseudoacacia*, *D. triangulare*, *O. semicastrata*, *C. arietinum*, *A. crassicarpa*, *A. implexa*	*F. graminearum*, *M. oryzae*, *R. solani*, *S. sclerotiorum*, *B. cinerea*	These nodule-inhabiting strains showed an antagonistic effect against phytopathogenic fungi by producing multiple hydrolytic enzymes, siderophores, and lipopeptide fusaricidins	Ali et al. (2021)
Bacillus sp. (BS-Z15)	Cotton	*V. dahliae*	BS-Z15 possessed various genes coding for antagonism or synthesizing mycosubtilin, chitinases, glycoside hydrolases, antibiosis, and biofilm production	Chen et al. (2021)
P. polymyxa (N179)	Pear, Turnip cabbage	*E. amylovora*, *X. campestris*	Strain N179 suppressed phytopathogen in pear fruit and turnip cabbage	Fallahzadeh-Mamaghani et al. (2021)
S. fulvissimus (A12), *S. venezuelae* (A30), *S. anulatus* (A34), *P. donghuensis* (P17), *Pseudomonas* sp. (P3)	Onion	*P. penetrans*, *F. oxysporum* (C1-1), *Ilyonectria* spp. (C2-1), *I. macrodidyma* (C1-1),	Results revealed that application of PGPR caused a reduction in the growth of phytopathogen through delaying production of nematicidal lytic enzymes, viz., chitinases and proteases and form biofilms	Marin-Bruzos et al. (2021)
Bacillus sp., *T. harzianum*	Rice	*M. oryzae* (MG01)	Metagenome sequencing explored the antagonistic diversity, distribution, and abundance of microflora to phytopathogen strain MG01	Prasannakumar et al. (2021)

(continued)

Table 2.2 (continued)

Biocontrol agent	Plant	Phytopathogen	Findings	References
B. velezensis EB14	*Populus* sp.	*S. musiva*	Strain EB14 possessed several gene clusters coding for the biosynthesis of antimicrobial compounds	Sachin et al. (2021)
B. halotolerans (JZ7)	Chinese jujube	*F. oxysporum*	This strain showed antagonistic activity by producing acetoin, 2,3-butanediol, and fenretinide	Wang et al. (2021)
B. amyloliquefaciens (PMB04, PMB05)	Strawberry	*C. gloeosporioides* (SC01)	These strains intensified plant immunity through the generation of ROS and callose deposition against conidia and reduced the occurrence of anthracnose on strawberries	Wu et al. (2021)
B. velezensis (8-4)	Potato scab	*P. foveat*, *R. solani*, *F. avenaceum*, *C. coccodes*	This strain showed antagonistic activity against potato pathogen	Cui et al. (2020)
C. muytjensii (JZ38)		*P. infestans*	Strain JZ38 revealed its ability to grow under a range of abiotic stresses due to the presence of genes coding for phytohormone	Eida et al. (2020)
B. subtilis (RH5)	Rice	*R. solani*	This strain produced defense-related enzymes, siderophores, hydrolytic enzymes, viz., chitinase, protease, cellulase, xylanase activity, and antimicrobial peptide biosynthetic genes (bacylisin, surfactin, and fengycin) in rice plants against *R. solani*	Jamali et al. (2020)
S. plymuthica (A294), *E. amnigenus* (A167), *R. aquatilis* (H145), *S. rubidaea* (H440), *S. rubidaea* (H469)	Potato tubers	*Pectobacterium* spp., *Dickeya* spp.	Consortium powder formulation of these strains reduced the severity and incidence of soft rot disease	Maciag et al. (2020)

Bacillus, *Pseudomonas*, *Sphingomonas* spp.	*D. officinale*	*S. rolfsii*	These strains produced antibiotic-encoding genes, viz., surfactin, phenazine, and iturin A	Shen et al. (2020)
B. xiamenensis (PM14)	*S. officinarum*	*C. falcatum*, *F. moniliforme*, *F. oxysporum*, *P. splendens*, *R. solani*, *M. phaseolina*	The results revealed the production of antioxidative enzymes and proline by strain PM14 involved in biocontrol activity	Xia et al. (2020)
B. velezensis	Maize	*T. funiculosus*, *P. oxalicum*, *F. verticillioides*	It showed biological control by producing lipopeptide antibiotic genes (bioA, bmyB, ituC, fenD, srfAA, srfAB, yngG, and yndJ)	Yang et al. (2020)
B. velezensis (D747, QST713)	Olive trees	*X. fastidiosa* (ST53)	Strains D747 and QST713 were registered as a biocontrol agent to control olive quick decline syndrome	Zicca et al. (2020)
K. oxytoca (D1/3), *E. indicum* (D1/8), *B. cereus* (D1/17, D1/18),	Tomato	*P. aphanidermatum*	These strains were capable of controlling the damping-off of tomato	Al-Hussini et al. (2019)
L. plantarum (CC100, PM411, TC92)	Kiwifruit, Prunes, Strawberry	*P. syringae*, *X. arboricola*, *X. fragariae*	These strains showed biocontrol against phytopathogen due to chitosan, acibenzolar-S-methyl, kasugamycin, and production of lactic acid	Daranas et al. (2019)
B. amyloliquefaciens (KU_2S1, $R_2S_{(1)}$, $RG1_{(3)}$, $AG1_{(7)}$)	Tomato	*C. michiganensis*	These strains showed antagonistic activity against the bacterial canker of tomato	Gautam et al. (2019a)

(continued)

Table 2.2 (continued)

Biocontrol agent	Plant	Phytopathogen	Findings	References
P. parafulva (JBCS1880)	Soybean	*X. axonopodis*, *B. glumae*	Strain JBCS1880 showed a potential for antagonism by producing novel lipopeptide	Kakembo and Lee (2019)
B. amyloliquefaciens (SW-34)	Ginseng plants	*B. cinerea*	This strain produced broad-spectrum antibiotics	Sun et al. (2019)
Streptomyces sp. (N2)	Rice	*R. solani*	This bacterium produces novel antifungalmycin N2	Wu et al. (2019)

(Alvarez et al. 2012). Microbial HCN is involved in biocontrol as well as sequestration of metal ions and consequently increases the availability of nutrients which is beneficial both for the rhizobacteria and plant host (Rijavec and Lapanje 2016; Meena et al. 2020).

Microbial production of siderophores has a central role in the development of a plant (Beneduzi et al. 2012). Siderophores are one of the strongest carriers of Fe^{3+} in combination with enterobactin (Saharan and Nehra 2011). The production of siderophores is affected by the type of iron ion; pH; enough availability of N, C, and P; and other trace elements (Compant et al. 2005). Pseudobactin or pyoverdine have been reported to be produced by *P. fluorescens*, which have a high affinity for sequestration of Fe^{2+}. The ferripyoverdins (a complex of pyoverdine with Fe^{3+}) are reported to be combined with the receptors of the outer membrane and transported to the cytoplasm and ultimately get reduced to Fe^{2+} (Haas and Defago 2005). The plant growth is directly increased by production of the siderophore as they make Fe available for plants. It has been reported that PGPR also consumes Fe for effective rhizospheric colonization. PGPR produce lytic enzymes to degrade cells of the pathogens (Compant et al. 2005). The synergistic effects of PGPR metabolites such as antibiotics and lytic enzymes are responsible for the degradation of cell walls of microbial pathogen and weeds (Dunne et al. 1998).

In response to phytopathogens, PGPR help plants stimulate a defense mechanism having a hypersensitive reaction called ISR (Compant et al. 2005). Numerous *Bacillus* sp. trigger ISR in biotic stress that improves plant growth through ultrastructural changes and cytochemical alteration in plants (Kloepper et al. 2004). The bacteriocins produced by PGPR are different from narrow-spectrum antibiotics that can kill closely related bacterial strains (Beneduzi et al. 2012). The bio-surfactants are considered to be effective for the biocontrol of phytium species (Perneel et al. 2008). For example, *P. fluorescens* is capable of protecting plants from disease-causing pathogenic fungi. It secretes a varying degree of exoproducts through antibiosis, phytohormone activities, lytic activities, and antimicrobial compounds to be considered as a primary mechanism to suppress disease in root area (Péchy-Tarr et al. 2008). Such PGPR play a major role in biocontrol by producing a wide range of secondary metabolites including the production of volatile compounds, antibiotics, siderophores, HCN, and other disease-suppressing compounds (Showkat 2012).

Numerous microorganisms have been reported as potential bioherbicides. PGPR are the most suitable biocontrol agents as compared to other microorganisms due to their rapid growth and genetic variation via mutagenesis (Johnson et al. 1996; Li et al. 2003). Various strains of *P. fluorescens* have shown their benefits as well as inhibitory effects on plants (Banowetz et al. 2008). Their suppressive effects were reported due to production of extracellular metabolites that suppress plant germination (Banowetz et al. 2008). Growth inhibition of various grassy weeds including downy brome was observed through inoculation with *P. fluorescens* strain D7 (Kennedy et al. 2001). These strains showed plant growth inhibition due to combined production of extracellular peptides and lipopolysaccharides (Gurusiddaiah et al. 1994). However, their mechanism for weed growth suppression has not yet been described. *P. fluorescens* strain WH6 showed growth inhibition of 21 monocot and 8 dicot plant species; however it was unable to affect the germination of hybrid corn (Banowetz et al. 2008). Its ability to suppress weed germination was reported due to production of germination arrest factor (GAF) composed of 4-formylaminooxy-L-vinylglycine and oxyvinylglycines due to synthesis of their regulatory genes (Banowetz et al. 2008; McPhail et al. 2010; Halgren et al. 2013; Okrent et al. 2014). The strain BRG100 showed suppression of *Setaria viridis* due to the production of extracellular metabolites (pseudophomin A and B) having phytotoxic effects (Caldwell et al. 2012).

Pseudomonas strains produced phytotoxins including phenazine-1-carboxylic acid, 2-aminophenoxazone, and 2-aminophenol that have allelopathic effects against plant growth (Adetunji et al. 2019). The strains of *Xanthomonas campestris* were also reported as a potential weed control agent. Its strain JT-P482 showed specific control against grasses *Poa annua* and *Poa attenuate* (Imaizumi et al. 1997). The strain LVA-987 also showed weed control against horseweed (Boyette and Hoagland 2015). No information on phytotoxic compounds produced by these strains is available in support of its application as a bioherbicide. Such strains spread in their host plant through colonizing plant xylem. They interact with host plant through their receptor proteins that can recognize pathogen-associated molecular patterns that activates plant defense responses in terms of production of ROS and programmed cell death (Guy et al. 2013; Dugé de Bernonville et al. 2014). Such allelopathic bacteria may also produce enzymes and broad-spectrum antibiotics that have inhibitory effects on plant growth (Abbas et al. 2017). Production of elevated IAA levels by PGPR also possesses allelopathic effects (Sarwar and Kremer 1995). The HCN production by rhizobacteria is a major suppressor of weed growth (Kremer and Souissi 2001).

Recent developments in bioherbicides support the potential use of HCN-producing bacterial strains along with multiple PGP traits for improving crop growth by suppression of weeds. In a recent study, Dar et al. (2020) reported *Pseudomonas* strains with potential *Pseudomonas* for both potential to control *Phalaris minor* and *Avena fatua* in wheat. They reported four *Pseudomonas* strains having HCN-producing ability along with multiple PGP traits such as P solubilization, ACC-deaminase activity, siderophores production, and oxidase activity during in vitro studies. In the bioassay, they reported these strains as strong phytotoxic for lettuce seedling having a mortality rate of up to 73.3%. Consortia of these strains were more effective than sole inoculation in the biocontrol of *Pectoralis minor* and *Avena fatua* seedlings. In addition to biocontrol of weeds, these strains were also effective in improving wheat growth. They argued that the use of these strains as bioherbicide may improve human and environmental health. So, inoculation with allelopathic bacteria could support the existing herbicides that not only control weeds but also alleviate the allelopathic effects of weeds. The mechanisms of biotic stress amelioration by PGPR have been summarized in Fig. 2.2.

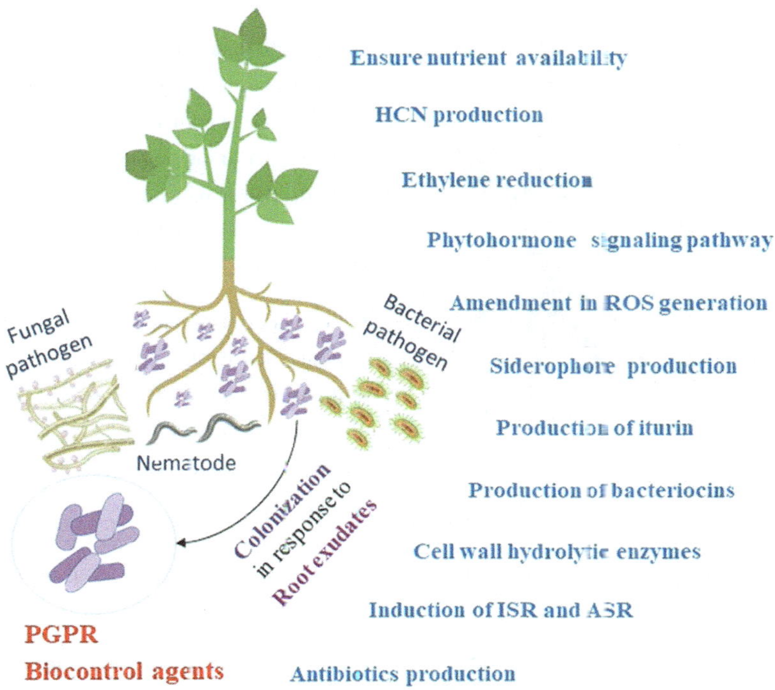

Fig. 2.2 Mechanisms of biocontrol by plant growth-promoting rhizobacteria

2.4.3 Integrated Stress Amelioration

Abiotic and biotic stresses due to climate change are a major cause of productivity loss in major crops (Anderson et al. 2020). The effects of abiotic and biotic stresses on plant growth and their amelioration by PGPR have already been discussed in the above sections. Under field conditions, crops face multiple stresses which drastically decline crop productivity (Pandey et al. 2017; Hussain et al. 2018). Managing abiotic and biotic stresses with competent and cost-effective approaches is a major challenge. Such approaches including the development of stress-tolerant varieties and application of chemical fertilizers and pesticides are cost-intensive. Recently, the application of PGPR to ameliorate various environmental and biotic stresses has been revealed as a cost-effective and efficient approach in integrated stress management (Grover et al. 2011). We have proposed a possible mechanism of action in the amelioration of integrated stresses (Fig. 2.3).

As described in the above sections, the PGPR having the ability to produce ACC-deaminase, phytohormones, antioxidant enzymes, siderophores, EPS, VOCs, and solubilization of nutrients can ameliorate all the abiotic stresses. Similarly, in

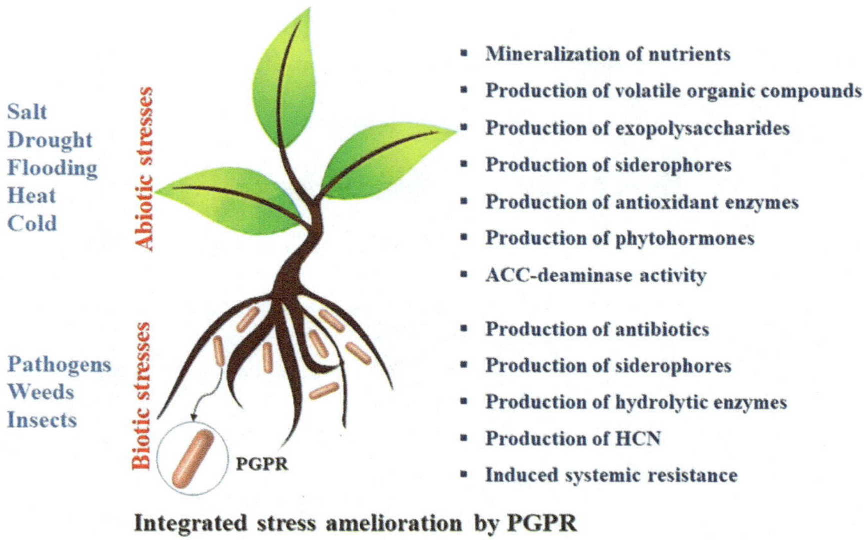

Fig. 2.3 Proposed mechanisms of action for integrated stress amelioration by plant growth-promoting rhizobacteria

biotic stress control, the HCN-producing PGPR along with production of antibiotics, lytic enzymes, siderophores, and ISR are effective (Fig. 2.3).

The PGPR with multiple PGP traits can be effective in integrated stress amelioration (ISA). As a combination of stresses enhance the ethylene production, ACC-deaminase-producing PGPR with multiple PGP traits can be more effective in ISA. For example, recent studies reported the effectiveness of ACC-deaminase-producing PGPR in ISA (Barnawal et al. 2017; Srinivasan et al. 2017; Saikia et al. 2018; Orozco-Mosqueda et al. 2020; Li et al. 2020). Some mechanisms of stress mitigation by PGPR including the production of siderophores, nutrient mineralization, production of VOCs, and EPS are not well understood in all abiotic stresses; however, these metabolic activities have been reported to be effective in various environmental stresses (Talaat and Shawky 2017; Khan and Bano 2019). PGPR release secondary metabolites including antibiotics, siderophores, hydrolytic enzymes, and HCN that mitigate the biotic stresses in terms of infection by pathogens and allelopathic effects of weed as well as help improve crop productivity (Jha 2018). As reported by Dar et al. (2020), HCN-producing rhizobacteria with multiple PGP traits can be strong candidates for integrated stress amelioration.

2.5 Conclusion and Prospects

Under natural field conditions, plants face several abiotic and biotic stresses which negatively affect plant growth. These stresses result in the enhanced biosynthesis of ethylene and damage the plant physiological and biochemical attributes. The application of PGPR is well reported for mitigation of these stresses. These beneficial rhizobacteria directly mitigate integrated stresses by metabolic activities in terms of production of phytohormones, ACC-deaminase, siderophores, and EPS and ensure nutrient availability through improving their solubility. PGPR have an indirect role as well in stress alleviation by producing antibiotics, HCN, and hydrolytic enzymes and induce ISR that help plant tolerate biotic stresses.

Improvement in PGPR population and diversity through rhizosphere engineering or use of microbial inoculants and/or their metabolites should be practiced for better agroecosystem and soil health. There is a need to study interactive effects of various stresses on a plant to know about impacts of combined stresses on crop production. Future research is needed to evaluate the role of PGPR under multiple stress conditions.

Future research should also be focused on understanding the mechanisms involved in PGPR-induced stress tolerance in crop plants. There is also a need to investigate why certain PGPR with specific PGP traits could not induce stress tolerance in crop plants under soil (field) and environmental conditions. Strategies should be developed to study and improve plant-microbe interactions using bioinformatics, modeling tools, and other molecular techniques to enhance crop stress tolerance and for sustainable agroecosystems.

References

Aasfar A, Bargaz A, Yaakoubi K, Hilali A, Bennis I, Zeroual Y, Meftah Kadmiri I (2021) Nitrogen fixing azotobacter species as potential soil biological enhancers for crop nutrition and yield stability. Front Microbiol 12:354. https://doi.org/10.3389/fmicb.2021.628379

Abbas T, Zahir ZA, Naveed M (2017) Bioherbicidal activity of allelopathic bacteria against weeds associated with wheat and their effects on growth of wheat under axenic conditions. Bio Control 62(5):719–730

Abd El-Daim IA, Bejai S, Meijer J (2014) Improved heat stress tolerance of wheat seedlings by bacterial seed treatment. Plant Soil 379(1):337–350

Abdelrahman M, El-Sayed M, Jogaiah S, Burritt DJ, Tran LSP (2017) The "STAY-GREEN" trait and phytohormone signaling networks in plants under heat stress. Plant Cell Rep 36(7): 1009–1025

Adak MK, Sarkar B, De AK, Saha I, Ghosh A (2020) Alleviation of abiotic stress by nonconventional plant growth regulators in plant physiology. In: Rakshit A, Singh HB, Singh AK, Singh US, Fraceto L (eds) New frontiers in stress management for durable agriculture. Springer, Singapore, pp 197–211

Adetunji CO, Oloke JK, Bello OM, Pradeep M, Jolly RS (2019) Isolation, structural elucidation and bioherbicidal activity of an eco-friendly bioactive 2-(hydroxymethyl) phenol, from *Pseudomonas aeruginosa* (C1501) and its ecotoxicological evaluation on soil. Environ Technol Innov 13: 304–317

Ahammed GJ, Li X, Zhou J, Zhou YH, Yu JQ (2016) Role of hormones in plant adaptation to heat stress. In: Ahammed GJ, Yu JQ (eds) Plant hormones under challenging environmental factors. Springer, Dordrecht, pp 1–21

Ahmad M, Zahir ZA, Asghar HN, Asghar M (2011) Inducing salt tolerance in mung bean through co-inoculation with *Rhizobium* and PGPR containing ACC-deaminase. Can J Microbiol 57(7): 578–589

Ahmad M, Zahir ZA, Khalid M, Nazli F, Arshad M (2013a) Efficacy of *Rhizobium* and *Pseudomonas* strains to improve physiology, ionic balance and quality of mung bean under salt-affected conditions on farmer's fields. Plant Physiol Biochem 63:170–176

Ahmad M, Zahir ZA, Nazli F, Akram F, Arshad M, Khalid M (2013b) Effectiveness of halo-tolerant, auxin producing *Pseudomonas* and *Rhizobium* strains to improve osmotic stress tolerance in mung bean (*Vigna radiata* L.). Braz J Microbiol 44(4):1341–1348

Ahmad M, Nadeem S, Naveed M (2016) Potassium-solubilizing bacteria and their application in agriculture. In: Meena VS, Maurya BR, Verma JP, Meena RS (eds) Potassium solubilizing microorganisms for sustainable agriculture. Springer, New Delhi, pp 293–313

Ahmad M, Naseer I, Hussain A, Mumtaz MZ, Mustafa A, Hilger TH, Zahir ZA, Xu M (2019a) Appraising endophyte-plant symbiosis for improved growth, nodulation, nitrogen fixation and abiotic stress tolerance: an experimental investigation with chickpea (*Cicer arietinum* L.). Agronomy 9:621. https://doi.org/10.3390/agronomy9100621

Ahmad I, Zaib S, Alves PC, Luthe DS, Bano A, Shakeel SN (2019b) Molecular and physiological analysis of drought stress responses in *Zea mays* treated with plant growth promoting rhizobacteria. Biol Plant 63:536–547

Ahmadi N, Mibus H, Serek M (2009) Characterization of ethylene-induced organ abscission in F1 breeding lines of miniature roses (*Rosa hybrida* L.). Postharvest Biol Technol 52:260–266

Ahmed S, Nawata E, Sakuratani T (2006) Changes of endogenous ABA and ACC, and their correlations to photosynthesis and water relations in mungbean (*Vigna radiata* L.) Wilczak cv. KPS1 during waterlogging. Environ Exp Bot 57(3):278–284

Akhgar R, Arzanlou M, Bakker PA, Hamidpour M (2014) Characterization of 1-aminocyclopropane-1-carboxylate (ACC) deaminase-containing *Pseudomonas* sp. in the rhizosphere of salt-stressed canola. Pedosphere 24:161–468

Al-Hussini HS, Al-Rawahi AY, Al-Marhoon AA, Al-Abri SA, Al-Mahmooli IH, Al-Sadi AM, Velazhahan R (2019) Biological control of damping-off of tomato caused by *Pythium aphanidermatum* by using native antagonistic rhizobacteria isolated from Omani soil. J Plant Pathol 101(2):315–322

Ali S, Kim WC (2018) Plant growth promotion under water: decrease of waterlogging-induced ACC and ethylene levels by ACC deaminase-producing bacteria. Front Microbiol 9:1096

Ali SZ, Sandhya V, Grover M, Kishore N, Rao LV, Venkateswarlu B (2009) *Pseudomonas* sp. strain AKM-P6 enhances tolerance of sorghum seedlings to elevated temperatures. Biol Fertil Soils 46:45–55

Ali SZ, Sandhya V, Grover M, Linga VR, Bandi V (2011) Effect of inoculation with a thermotolerant plant growth promoting *Pseudomonas putida* strain AKMP7 on growth of wheat (*Triticum* spp.) under heat stress. J Plant Interact 6(4):239–246

Ali S, Khan MA, Kim WC (2018) *Pseudomonas veronii* KJ mitigates flood stress-associated damage in *Sesamum indicum* L. Appl Biol Chem (5):575–585

Ali Q, Shahid S, Nazar N, Hussain AI, Ali S, Chatha SA, Perveen R, Naseem J, Haider MZ, Hussain B, Hussain SM (2020) Use of phytohormones in conferring tolerance to environmental stress. In: Hasanuzzaman M (ed) Plant ecophysiology and adaptation under climate change: mechanisms and perspectives II. Springer, Singapore, pp 245–355

Ali M, Lou Y, Hafeez R, Li X, Hossain A, Xie T, Lin L, Li B, Yin Y, Yan J, An Q (2021) Functional analysis and genome mining reveal high potential of biocontrol and plant growth promotion in nodule-inhabiting bacteria within *Paenibacillus polymyxa* complex. Front Microbiol 11:3627

Alvarez F, Castro M, Principe A, Borioli G, Fischer S, Mori G, Jofre E (2012) The plant-associated *Bacillus amyloliquefaciens* strains MEP218 and ARP23 capable of producing the cyclic lipopeptides iturin or surfactin and fengycin are effective in biocontrol of sclerotinia stem rot disease. J Appl Microbiol 112(1):159–174

Amna DBU, Sarfraz S, Xia Y, Kamran MA, Javed MT, Sultan T, Munis MF, Chaudhary HJ (2019) Mechanistic elucidation of germination potential and growth of wheat inoculated with exopolysaccharide and ACC-deaminase producing *Bacillus* strains under induced salinity stress. Ecotoxicol Environ Saf 183:109466

Anderson R, Bayer PE, Edwards D (2020) Climate change and the need for agricultural adaptation. Curr Opin Plant Biol 56:197–2020. https://doi.org/10.1016/j.pbi.2019.12.006

Ansari FA, Jabeen M, Ahmad I (2021) *Pseudomonas azotoformans* FAP5, a novel biofilm-forming PGPR strain, alleviates drought stress in wheat plant. Int J Environ Sci Technol 3:1–6

Araya MA, Valenzuela T, Inostroza NG, Maruyama F, Jorquera MA, Acuña JJ (2020) Isolation and characterization of cold-tolerant hyper-acc-degrading bacteria from the rhizosphere, endosphere, and phyllosphere of Antarctic vascular plants. Microorganisms 8(11):1788

Arif F, Ghoul M (2018) Halotolerance of indigenous fluorescent Pseudomonads in the presence of natural osmoprotectants. Annu Res Rev Biol 24(4):1–11

Arkhipova T, Veselov S, Melentiev A, Martynenko EV, Kudoyarova GR (2005) Ability of bacterium *Bacillus subtilis* to produce cytokinins and to influence the growth and endogenous hormone content of lettuce plants Plant Soil 272(1–2):201–209

Armada E, Roldan A, Azcon R (2014) Differential activity of autochthonousbacteria in controlling drought stress in native Lavandula and Salvia plants species under drought conditions in natural arid soil. Microb Ecol 67:410–420

Armstrong W, Drew MC (2002) Root growth and metabolism under oxygen deficiency. In: Waisel Y, Eshel A, Kafkafi U (eds) Plant roots: the hidden half, 3rd edn. Marcel Dekker, New York, pp 729–761

Armstrong W, Brändle R, Jackson MB (1994) Mechanisms of flood tolerance in plants. Acta Bot Neerl 43(4):307–358

Arya B, Komala BR, Sumalatha NT, Surendra GM, Gurumurthy PE (2018) PGPR induced systemic tolerance in plant. Int J Curr Microbiol App Sci 7:453–462

Asada K (2006) Production and scavenging of reactive oxygen species in chloroplasts and their functions. Plant Physiol 141(2):391–396

Asghari B, Khademian R, Sedaghati B (2020) Plant growth promoting rhizobacteria (PGPR) confer drought resistance and stimulate biosynthesis of secondary metabolites in pennyroyal (*Mentha pulegium* L.) under water shortage condition. Sci Hort 263:109132

Ashraf MA, Akbar A, Askari SH, Iqbal M, Rasheed R, Hussain I (2018) Recent advances in abiotic stress tolerance of plants through chemical priming: an overview. In: Rakshit A, Singh HB (eds) Advances in seed priming. Springer, Singapore, pp 51–79

Aslam F, Ali B (2018) Halotolerant bacterial diversity associated with *Suaeda fruticosa* (L.) forssk. Improved growth of maize under salinity stress. Agronomy 8:131

Aslam H, Ahmad SR, Anjum T, Akram W (2018) Native halotolerant plant growth promoting bacterial strains can ameliorate salinity stress on tomato plants under field conditions. Int J Agric Biol 20:315–322

Aslam MU, Raza MA, Saleem MF, Waqas M, Iqbal R, Ahmad S, Haider I (2020) Improving strategic growth stage-based drought tolerance in quinoa by rhizobacterial inoculation. Commun Soil Sci Plant Anal 51(7):853–868

Avramova V, AbdElgawad H, Zhang Z, Fotschki B, Casadevall R, Vergauwen L, Knapen D, Taleisnik E, Guisez Y, Asard H, Beemster GT (2015) Drought induces distinct growth response, protection, and recovery mechanisms in the maize leaf growth zone. Plant Physiol 169(2): 1382–1396

Awasthi S, Sharma A, Saxena P, Yadav J, Pandiyan K, Kumar M, Singh A, Chakdar H, Bhowmik A, Kashyap PL, Srivastava AK (2019) Molecular detection and in silico characterization of cold shock protein coding gene (cspA) from cold adaptive *Pseudomonas koreensis*. J Plant Biochem Biotechnol 28:405–413

Baek KH, Skinner DZ (2003) Alteration of antioxidant enzyme gene expression during cold acclimation of near-isogenic wheat lines. Plant Sci 165(6):1221–1227

Bal HB, Adhya TK (2021) Alleviation of submergence stress in rice seedlings by plant growth-promoting rhizobacteria with ACC-deaminase activity. Front Sust Food Syst 5:36

Balal RM, Shahid MA, Javaid MM, Iqbal Z, Anjum MA, Garcia-Sanchez F, Mattson NS (2016) The role of selenium in amelioration of heat-induced oxidative damage in cucumber under high temperature stress. Acta Physiol Plant 38(6):158. https://doi.org/10.1007/s11738-016-2174-y

Bangash N, Khalidi A, Mahmood T, Siddique TM (2013) Screening rhizobacteria containing ACC deaminase for growth promotion of wheat under water stress. Pak J Bot 45(SI):91–96

Banowetz GM, Azevedo MD, Armstrong DJ, Halgren AB, Mills DI (2008) Germination-Arrest Factor (GAF): biological properties of a novel, naturally-occurring herbicide produced by selected isolates of rhizosphere bacteria. Biol Control 46:380390

Barka EA, Nowak J, Clément C (2006) Enhancement of chilling resistance of inoculated grapevine plantlets with a plant growth-promoting rhizobacterium, *Burkholderia phytofirmans* strain PsJN. Appl Environ Microbiol 72(11):7246–7252

Barnawal D, Bharti N, Maji D, Chanotiya CS, Kalra A (2012) 1-Aminocyclopropane-1-carboxylic acid (ACC) deaminase-containing rhizobacteria protect *Ocimum sanctum* plants during waterlogging stress via reduced ethylene generation. Plant Physiol Biochem 58:227–235

Barnawal D, Bharti N, Pandey SS, Pandey A, Chanotiya CS, Kalra A (2017) Plant growth-promoting rhizobacteria enhance wheat salt and drought stress tolerance by altering endogenous phytohormone levels and TaCTR1/TaDREB2 expression. Physiol Plant 161(4):502–514

Belimov AA, Dodd IC, Hontzeas N, Theobald JC, Safronova VI, Davies WJ (2009) Rhizosphere bacteria containing 1-aminocyclopropane-1-carboxylatedeaminase increase yield of plants grown in drying soil via both local andsystemic hormone signaling. New Phytol 181:413–423

Belimov AA, Dodd IC, Safronova VI, Dumova VA, Shaposhnikov AI, Ladatko AG, Davies WJ (2014) Abscisic acid metabolizing rhizobacteria decrease ABA concentrations in planta and alter plant growth. Plant Physiol Biochem 74:84–91

Beneduzi A, Ambrosini A, Passaglia LMP (2012) Plant growth-promoting rhizobacteria (PGPR): Their potential as antagonists and biocontrol agents. Genet Mol Biol 35(4):1044–1051

Bent E (2006) Induced systemic resistance mediated by plant growth-promoting rhizobacteria (PGPR) and fungi (PGPF). In: Tusuz S, Bent E (eds) Multigenic and induced systemic resistance in plants. Springer, Boston, MA, pp 225–258

Bhardwaj D, Ansari MW, Sahoo RK, Tuteja N (2014) Biofertilizers function as key player in sustainable agriculture by improving soil fertility, plant tolerance and crop productivity. Microb Cell Fact 3(66):1–10

Bharti N, Pandey SS, Barnawal D, Patel VK, Kalra A (2016) Plant growth promoting rhizobacteria *Dietzia natronolimnaea* modulates the expression of stress responsive genes providing protection of wheat from salinity stress. Sci Rep 6:34768. https://doi.org/10.1038/srep34768

Bhattacharyya PN, Jha DK (2012) Plant growth-promoting rhizobacteria (PGPR): emergence in agriculture. World J Microbiol Biotechnol 28:1327–1350

Bhattacharyya D, Garladinne M, Lee YH (2015) Volatile indole produced by rhizobacterium *Proteus vulgaris* JBLS202 stimulates growth of *Arabidopsis thaliana* through auxin, cytokinin, and brassinosteroid pathways. J Plant Growth Regul 34:158–168

Bottner P, Couteaux MM, Vallejo VR (1995) Soil organic matter in Mediterranean-type ecosystems and global climatic changes: a case study-thesoils of the Mediterranean basin. In: Jose M, Oechel WC (eds) Global change and mediterranean-type ecosystems. Springer-Verlag, New York, pp 306–325

Boyette CD, Hoagland RE (2015) Bioherbicidal potential of *Xanthomonas campestris* for controlling *Conyza canadensis*. Biocontrol Sci Technol 25:229–237

Braud A, Jézéquel K, Bazot S, Lebeau T (2009) Enhanced phytoextraction of an agricultural Cr and Pb-contaminated soil by bioaugmentation with siderophore-producing bacteria. Chemosphere 74:280–286

Brown AJP, Cowen LE, di Pietro A, Quinn J (2017) Stress adaptation. Fungal Kingdom 1:463–485

Brunetti C, Tattini M, Guidi L, Velikova V, Ferrini F, Fini A (2019) An integrated overview of physiological and biochemical responses of *Celtis australis* to drought stress. Urban For Urban Green 46:126480. https://doi.org/10.1016/j.ufug.2019.126480

Bruno LB, Karthik C, Ma Y, Kadirvelu K, Freitas H, Rajkumar M (2020) Amelioration of chromium and heat stresses in Sorghum bicolor by Cr^{6+} reducing-thermotolerant plant growth promoting bacteria. Chemosphere 244:125521

Bukhat S, Imran A, Javaid S, Shahid M, Majeed A, Naqqash T (2020) Communication of plants with microbial world: exploring the regulatory networks for PGPR mediated defense signaling. Microbiol Res 1:126486

Buttar ZA, Wu SN, Arnao MB, Wang C, Ullah I, Wang C (2020) Melatonin suppressed the heat stress-induced damage in wheat seedlings by modulating the antioxidant machinery. Plants 9(7): 809

Caldwell CJ, Hynes RK, Boyetchko SM, Korber DR (2012) Colonization and bioherbicidal activity on green foxtail by *Pseudomonas fluorescens* BRG100 in a pesta formulation. Can J Microbiol 58:1–9

Carbonell-Bojollo R, Veroz-Gonzalez O, Ordoñez-Fernandez R, Moreno-Garcia M, Basch G, Kassam A, Repullo-Ruiberriz de Torres MA, Gonzalez-Sanchez EJ (2019) The effect of conservation agriculture and environmental factors on CO_2 emissions in a rainfed crop rotation. Sustainability 11(14):3955. https://doi.org/10.3390/su11143955

Chatterjee P, Kanagendran A, Samaddar S, Pazouki L, Sa TM, Niinemets Ü (2020) Influence of *Brevibacterium linens* RS16 on foliage photosynthetic and volatile emission characteristics upon heat stress in *Eucalyptus grandis*. Sci Total Environ 700:134453

Chen M, Wei H, Cao J, Liu R, Wang Y, Zheng C (2007) Expression of *Bacillus subtilis* proBA genes and reduction of feedback inhibition of proline synthesis increases proline production and confers osmotolerance in transgenic Arabidopsis. J Biochem Mol Biol 40:396–403

Chen L, Liu Y, Wu G, Veronican NK, Shen Q, Zhang N, Zhang R (2016) Induced maize salt tolerance by rhizosphere inoculation of *Bacillus amyloliquefaciens* SQR9. Physiol Plant 158: 34–44

Chen ZY, Abuduaini X, Mamat N, Yang QL, Wu MJ, Lin XR, Wang R, Lin RR, Zeng WJ, Ning HC, Zhao HP (2021) Genome sequencing and functional annotation of *Bacillus* sp. strain BS-Z15 isolated from cotton rhizosphere soil having antagonistic activity against *Verticillium dahliae*. Arch Microbiol 5:1–1

Cherif H, Marasco R, Rolli E, Ferjani R, Fusi M, Soussi A, Mapelli F, Blilou I, Borin S, Boudabous A, Cherif A (2015) Oasisdesert farming selects environment-specific date palm root endophyticcommunities and cultivable bacteria that promote resistance to drought. Environ Microbiol Rep 7:668–678

Chiappero J, del Rosario CL, Alderete LG, Palermo TB, Banchio E (2019) Plant growth promoting rhizobacteria improve the antioxidant status in *Mentha piperita* grown under drought stress leading to an enhancement of plant growth and total phenolic content. Ind Crop Prod 139: 111553

Compant S, Duffy B, Nowak J, Clément C, Barka EA (2005) Use of plant growth-promoting bacteria for biocontrol of plant diseases: principles, mechanisms of action, and future prospects. Appl Environ Microbiol 71(9):4951–4959

Cook D, Fowler S, Fiehn O, Thomashow MF (2004) A prominent role for the CBF cold response pathway in configuring the low-temperature metabolome of Arabidopsis. Proc Natl Acad Sci 101(42):15243–15248

Coupland D, Jackson MB (1991) Effects of mecoprop (an auxin analogue) on ethylene evolution and epinasty in two biotypes of stellaria media. Ann Bot 68:167–172

Cramer GR, Urano K, Delrot S, Pezzotti M, Shinozaki K (2011) Effects of abiotic stress on plants: a systems biology perspective. BMC Plant Biol 11(1):163. https://doi.org/10.1186/1471-2229-11-163

Creus CM, Sueldo RJ, Barassi CA (2004) Water relations and yield in *Azospirillum*-inoculated wheat exposed to drought in the field. Can J Bot 82:273–281

Cui L, Yang C, Wei L, Li T, Chen X (2020) Isolation and identification of an endophytic bacteria *Bacillus velezensis* 8-4 exhibiting biocontrol activity against potato scab. Biol Control 141: 104156

Danish S, Zafar-ul-Hye M (2019) Co-application of ACC-deaminase producing PGPR and timber-waste biochar improves pigments formation, growth and yield of wheat under drought stress. Sci Rep 9(1):1–3

Danish S, Zafar-ul-Hye M, Mohsin F, Hussain M (2020) ACC-deaminase producing plant growth promoting rhizobacteria and biochar mitigate adverse effects of drought stress on maize growth. PLoS One 15(4):e0230615

Dar A, Zahir ZA, Asghar HN, Ahmad R (2020) Preliminary screening of rhizobacteria for biocontrol of little seed canary grass (*Phalaris minor* Retz.) and wild oat (*Avena fatua* L.) in wheat. Can J Microbiol 66:368–376

Daranas N, Roselló G, Cabrefiga J, Donati I, Francés J, Badosa E, Spinelli F, Montesinos E, Bonaterra A (2019) Biological control of bacterial plant diseases with *Lactobacillus plantarum* strains selected for their broad-spectrum activity. Ann Appl Biol 174(1):92–105

Datir SS, Kochle M, Jindal S (2019) Molecular approaches and salt tolerance mechanisms in leguminous plants. In: Akhtar MS (ed) Salt stress, microbes, and plant interactions: mechanisms and molecular approaches. Springer, Singapore, pp 49–67

Dawood T, Yang X, Visser EJ, Te Beek TA, Kensche PR, Cristescu SM, Lee S, Floková K, Nguyen D, Mariani C, Rieu I (2016) A coopted hormonal cascade activates dormant adventitious root primordia upon flooding in *Solanum dulcamara*. Plant Physiol 170:2351–2364

del Amor FM, Cuadra-Crespo P (2012) Plant growth-promoting bacteria as a tool to improve salinity tolerance in sweet pepper. Funct Plant Biol 39:82–90

Dimkpa C, Weinand T, Asch F (2009) Plant-rhizobacteria interactions alleviateabiotic stress conditions. Plant Cell Environ 32:1682–1694

Dodd IC, Belimov AA, Sobeih WY, Safronova VI, Grierson D, Davies WJ (2004) Will modifying plant ethylene status improve plant productivity in water-limited environments. In: Handbook and Abstracts for the 4th International Science Congress, Brisbane, Australia, p 134

Dodd IC, Zinovkina NY, Safronova VI, Belimov AA (2010) Rhizobacterial mediation of plant hormone status. Ann Appl Biol 157:361–379

Dubey A, Verma AK (2019) Salinity stress-dependent coordination of metabolic networks in relation to salt tolerance in plants. In: Giri B, Verma A (eds) Microorganisms in saline environments: strategies and functions. Springer, Cham, pp 401–422

Dugé de Bernonville T, Noël LD, SanCristobal M, Danoun S, Becker A, Soreau P, Arlat M, Lauber E (2014) Transcriptional reprogramming and phenotypical changes associated with growth of *Xanthomonas campestris* pv. campestris in cabbage xylem sap. FEMS Microbiol Ecol 89:527–541

Dunne C, Moënne-Loccoz Y, McCarthy J, Higgins P, Powell J, Dowling DN, O'Gara F (1998) Combining proteolytic and phloroglucinol-producing bacteria for improved biocontrol of Pythium-mediated damping-off of sugar beet. Plant Pathol 47(3):299–307

Egamberdieva D (2009) Alleviation of salt stress by plant growth regulators and IAA producing bacteria in wheat. Acta Physiol Plant 31:61–864

Egamberdieva D (2013) The role of phytohormone producing bacteria inalleviating salt stress in crop plants. In: Miransari M (ed) Biotechnological techniques of stress tolerance in plants. Stadium Press LLC, Houston, TX, pp 21–39

Egamberdieva D, Lugtenberg B (2014) Use of plant growth-promoting rhizobacteria to alleviate salinity stress in plants. PGPR to alleviate salinity stress on plant growth. In: Miransari M (ed) Use of microbes for the alleviation of soil stresses. Springer, New York, pp 73–96

Egamberdieva D, Wirth SJ, Alqarawi AA, AbdAllah EF, Hashem A (2017) Phytohormones and beneficial microbes: essential components for plants to balance stress and fitness. Front Microbiol 8:2104. https://doi.org/10.3389/fmicb.2017.02104

Eida AA, Bougouffa S, L'Haridon F, Alam I, Weisskopf L, Bajic VB, Saad MM, Hirt H (2020) Genome insights of the plant-growth promoting bacterium *Cronobacter muytjensii* JZ38 with volatile-mediated antagonistic activity against *Phytophthora infestans*. Front Microbiol 11:369

El-Daim IAA, Bejai S, Meijer J (2014) Improved heat stress tolerance of wheat seedlings by bacterial seed treatment. Plant Soil 379(1–2):337–350

Etesami H, Alikhani HA (2019) Halotolerant plant growth-promoting fungi and bacteria as an alternative strategy for improving nutrient availability to salinity-stressed crop plants. In: Kumar M, Estesami H, Kumar V (eds) Saline soil-based agriculture by halotolerant microorganisms. Springer, Singapore, pp 103–146

Etesami H, Hosseini HM, Alikhani HA (2014) Bacterial biosynthesis of 1-aminocyclopropane-1-caboxylate (ACC) deaminase, a useful trait to elongation and endophytic colonization of the roots of rice under constant flooded conditions. Physiol Mol Biol Plants 20(4):425–434

Fahad S, Hussain S, Bano A, Saud S, Hassan S, Shan D, Khan FA, Khan F, Chen Y, Wu C, Tabassum MA (2015) Potential role of phytohormones and plant growth-promoting rhizobacteria in abiotic stresses: consequences for changing environment. Environ Sci Pollut Res 22:4907–4921

Fahad S, Hussain S, Saud S, Hassan S, Tanveer M, Ihsan MZ, Shah AN, Ullah A, Khan F, Ullah S, Alharby H (2016) A combined application of biochar and phosphorus alleviates heat-induced adversities on physiological, agronomical and quality attributes of rice. Plant Physiol Biochem 103:191–198

Fallahzadeh-Mamaghani V, Golchin S, Shirzad A, Mohammadi H, Mohamadivand F (2021) Characterization of *Paenibacillus polymixa* N179 as a robust and multifunctional biocontrol agent. Biol Control 154:104505

FAO Food and Agriculture Organization of the United Nations (2015) The impact of natural hazards and disasters on agriculture and food security and nutrition: a call for action to build resilient livelihoods. FAO, Rome

Farwell AJ, Vesely S, Nero V, Rodriguez H, McCormack K, Shah S, Dixon DG, Glick BR (2007) Tolerance of transgenic canola plants (*Brassica napus*) amended with plant growth-promoting bacteria to flooding stress at a metal-contaminated field site. Environ Pollut 147(3):540–545

Fernandez O, Vandesteene L, Feil R, Baillieul F, Lunn JE, Clément C (2012) Trehalose metabolism is activated upon chilling in grapevine and might participate in *Burkholderia phytofirmans* induced chilling tolerance. Planta 236(2):355–369

Fernie AR, Geigenberger P, Stitt M (2005) Flux an important, but neglected, component of functional genomics. Curr Opin Plant Biol 8(2):174–182

Figueiredo MV, Burity HA, Martínez CR, Chanway CP (2008) Alleviation of drought stress in the common bean (*Phaseolus vulgaris* L.) by coinoculation with *Paenibacillus polymyxa* and *Rhizobium* tropici. Appl Soil Ecol 40:182–188

Flowers TJ, Colmer TD (2008) Salinity tolerance in halophytes. New Phytol 1:945–963

Flynn P (2003) Biotic vs. abiotic—distinguishing disease problems from environmental stresses. Hortic Home Pest News 489:22

Gan P, Liu F, Li R, Wang S, Luo J (2019) Chloroplasts- beyond energy capture and carbon fixation: tuning of photosynthesis in response to chilling stress. Int J Mol Sci 20(20):5046. https://doi.org/10.3390/ijms20205046

Garcia FP, Menendez E, Rivas R (2015) Role of bacterial bio fertilizers in agriculture and forestry. AIMS Bioeng 2:183–205

García JE, Maroniche G, Creus C, Suárez-Rodríguez R, Ramirez-Trujillo JA, Groppa MD (2017) In vitro PGPR properties and osmotic tolerance of different *Azospirillum* native strains and their effects on growth of maize under drought stress. Microbiol Res 202:21–29

Gautam S, Chauhan A, Sharma R, Sehgal R, Shirkot CK (2019a) Potential of *Bacillus amyloliquefaciens* for biocontrol of bacterial canker of tomato incited by *Clavibacter michiganensis* ssp. michiganensis. Microb Pathog 130:196–203

Gautam N, Sharma P, Rana JC, Singh M (2019b) Plant growth promoting traits of a novel psychrotrophic bacterium *Virdibacillus arenosi* PH15 isolated from rhizosphere of *Podophyllum hexandrum*. R Acad J Med Plants 7(1):013–019

Gerhardson B (2002) Biological substitutes for pesticides. Trends Biotechnol 20:338–343

Ghosh D, Sen S, Mohapatra S (2017) Modulation of proline metabolic gene expression in Arabidopsis thaliana under water-stressed conditions by a drought-mitigating *Pseudomonas putida* strain. Ann Microbiol 67(10):655–668

Gibbs J, Greenway H (2003) Mechanisms of anoxia tolerance in plants. I. Growth, survival and anaerobic catabolism. Funct Plant Biol 30:1–47

Glick BR (2005) Modulation of plant ethylene levels by the bacterial enzyme ACC deaminase. FEMS Microbiol Lett 251(1):1–7

Glick BR (2012) Plant growth-promoting bacteria: mechanisms and applications. Scientifica 2012: 963401. https://doi.org/10.6064/2012/963401

Glick BR (2014) Bacteria with ACC deaminase can promote plant growth and help to feed the world. Microbiol Res 169(1):30–39

Glick BR, Cheng Z, Czarny J, Duan J (2007) Promotion of plant growth by ACC deaminase-producing soil bacteria. Eur J Plant Pathol 119:329–339

Gonzalez AJ, Larraburu EE, Llorente BE (2015) *Azospirillum brasilense* increased salt tolerance of Jojoba during in vitro rooting. Ind Crop Prod 76:41–48

Gouda S, Kerry RG, Das G, Paramithiotis S, Shin HS, Patra JK (2018) Revitalization of plant growth promoting rhizobacteria for sustainable development in agriculture. Microbiol Res 206: 131–140

Goyal D, Prakash O, Pandey J (2019) Rhizospheric microbial diversity: an important component for abiotic stress management in crop plants toward sustainable agriculture. In: Singh JS, Singh DP (eds) New and future developments in microbial biotechnology and bioengineering. Elsevier, Amsterdam, pp 115–134

Grichko VP, Glick BR (2001a) Amelioration of flooding stress by ACC deaminase-containing plant growth-promoting bacteria. Plant Physiol Biochem 39(1):11–17

Grichko VP, Glick BR (2001b) Flooding tolerance of transgenic tomato plants expressing the bacterial enzyme ACC deaminase controlled by the 35S, rolD or PRB-1b promoter. Plant Physiol Biochem 39(1):19–25

Grover M, Ali SZ, Sandhya V, Rasul A, Venkateswarlu B (2011) Role of microorganisms in adaptation of agriculture crops to abiotic stresses. World J Microbiol Biotechnol 27(5): 1231–1240

Gulati A, Vyas P, Rahi P, Kasana RC (2009) Plant growth-promoting and rhizosphere-competent *Acinetobacter rhizosphaerae* strain BIHB 723 from the cold deserts of the Himalayas. Curr Microbiol 58:371–377

Gupta B, Huang B (2014) Mechanism of salinity tolerance in plants: physiological, biochemical, and molecular characterization. Int J Genomics 2014:701596. https://doi.org/10.1155/2014/701596

Gupta G, Parihar SS, Ahirwar NK, Snehi SK, Singh V (2015) Plant growth promoting Rhizobacteria (PGPR): current and future prospects for development of sustainable agriculture. J Microbiol Biochem 7:96–102

Gurusiddaiah S, Gealy DR, Kennedy AC, Ogg AG (1994) Isolation and characterization of metabolites from *Pseudomonas fluorescens* D7 for control of downy brome (*Bromus tectorum*). Weed Sci 42:492–501

Gutiérrez-Mañero FJ, Ramos-Solano B, Probanza AN, Mehouachi JR, Tadeo F, Talon M (2001) The plant-growth-promoting rhizobacteria *Bacillus pumilus* and *Bacillus licheniformis* produce high amounts of physiologically active gibberellins. Physiol Plant 111(2):206–211

Guy E, Lautier M, Chabannes M, Roux B, Lauber E, Arlat M, Noël LD (2013) xopAC-triggered immunity against *Xanthomonas* depends on *Arabidopsis* receptor-like cytoplasmic kinase genes PBL2 and RIPK. PLoS One 8:e73469. https://doi.org/10.1371/journal.pone.0073469

Haas D, Defago G (2005) Biological control of soil-borne pathogens by fluorescent Pseudomonads. Nat Rev Microbiol 3:307–319

Habib SH, Kausar H, Saud HM (2016) Plant growth-promoting rhizobacteria enhance salinity stress tolerance in okra through ROS-scavenging enzymes. Biomed Res Int 2016:6284547. https://doi.org/10.1155/2016/6284547

Habibzadeh F, Sorooshzadeh A, Pirdashti H, Modarres Sanavy SA (2012) A comparison between foliar application and seed inoculation of biofertilizers on canola (*Brassica napus* L.) grown under waterlogged conditions. Aust J Crop Sci 6(10):1435–1440

Hajiboland R, Bahrami-Rad S, Akhani H, Poschenrieder C (2018) Salt tolerance mechanisms in three Irano-Turanian Brassicaceae halophytes relatives of *Arabidopsis thaliana*. J Plant Res 131:1029–1046

Halgren A, Maselko M, Azevedo M, Mills D, Armstrong D, Banowetz G (2013) Genetics of germination-arrest factor (GAF) production by *Pseudomonas fluorescens* WH6: identification of a gene cluster essential for GAF biosynthesis. Microbiology 159:36–45

Hardoim PR, Overbeek LSV, Elsas JDV (2008) Properties of bacterialendophytes and their proposed role in plant growth. Trends Microbiol 16:463–471

Hasanuzzaman M, Nahar K, Gill SS, Fajita M (2013) Drought stress responses in plants, oxidative stress, and antioxidant defense. In: Tuteja N, Gill SS (eds) Climate change and plant abiotic stress tolerance. Wiley-VCH Verlag GmbH & Co. KGaA, Weinheim, pp 209–250

Hatfield JL, Prueger JH (2015) Temperature extremes: effect on plant growth and development. Weather Clim Extremes 10:4–10

Hatfield JL, Boote KJ, Kimball BA, Ziska LH, Izaurralde RC, Ort D, Thomson AM, Wolfe D (2011) Climate impacts on agriculture: implications for crop production. Agron J 103(2):351–370

Hayat R, Ali S, Amara U, Khalid R, Ahmad I (2010) Soil beneficial bacteria and their role in plant growth promotion: a review. Ann Microbiol 60:579–598

Hellal FA, El-Shabrawi HM, El-Hady MA, Khatab IA, El-Sayed SA, Abdelly C (2018) Influence of PEG induced drought stress on molecular and biochemical constituents and seedling growth of Egyptian barley cultivars. J Genet Eng Biotechnol 16(1):203–212

Hendry GA (2005) Oxygen free radical process and seed longevity. Seed Sci J 3:141–147

Horn PJ, Chapman KD (2012) Lipidomics in tissues, cells and subcellular compartments. Plant J 70(1):69–80

Hu X, Li Y, Li C, Yang H, Wang W, Lu M (2010) Characterization of small heat shock proteins associated with maize tolerance to combined drought and heat stress. J Plant Growth Regul 29(4):455–464

Hui LJ, Kim SD (2013) Induction of drought stress resistance by multifunctional PGPR *Bacillus licheniformis* K11 in Pepper. Plant Pathol J 29:201–208

Hussain HA, Hussain S, Khaliq A, Ashraf U, Anjum SA, Men S, Wang L (2018) Chilling and drought stresses in crop plants: implications, cross talk, and potential management opportunities. Front Plant Sci 9:393. https://doi.org/10.3389/fpls.2018.00393

Ikeda M, Ohme-Takagi M (2009) A novel group of transcriptional repressors in Arabidopsis. Plant Cell Physiol 50:970–975

Ilyas N, Mumtaz K, Akhtar N, Yasmin H, Sayyed RZ, Khan W, Enshasy HAE, Dailin DJ, Elsayed EA, Ali Z (2020) Exopolysaccharides producing bacteria for the amelioration of drought stress in wheat. Sustainability 12(21):8876

Imaizumi S, Nishino T, Miyabe K, Fujimori T, Yamada M (1997) Biological control of annual bluegrass (*Poa annua* L.) with a Japanese isolate of *Xanthomonas campestris* pv. *Poae* (JT-P482). Biol Control 8:7–14

Issa A, Esmaeel Q, Sanchez L, Courteaux B, Guise JF, Gibon Y, Ballias P, Clément C, Jacquard C, Vaillant-Gaveau N, Aït Barka E (2018) Impacts of *Paraburkholderia phytofirmans* strain PsJN on tomato (*Lycopersicon esculentum* L.) under high temperature. Front Plant Sci 9:1397

Jaemsaeng R, Jantasuriyarat C, Thamchaipenet A (2018) Positive role of 1-aminocyclopropane-1-carboxylate deaminase-producing endophytic *Streptomyces* sp. GMKU 336 on flooding resistance of mung bean. Agric Nat Resour 52(4):330–334

Jaleel CA, Manivannan P, Wahid A (2009) Drought stress in plants: a review on morphologicalcharacteristics and pigments composition. Int J Agric Biol 11:100–105

Jamali H, Sharma A, Srivastava AK (2020) Biocontrol potential of *Bacillus subtilis* RH5 against sheath blight of rice caused by *Rhizoctonia solani*. J Basic Microbiol 60(3):268–280

James RA, Blake C, Byrt CS, Munns S (2011) Major genes for Na^+ exclusion, Nax1 and Nax2 (wheat HKT1;4 and HKT1;5), decrease Na^+ accumulation in bread wheat leaves under saline and waterlogged conditions. J Exp Bot 62:2939–2947

Jha Y (2018) Induction of anatomical, enzymatic, and molecular events in maize by PGPR under biotic stress. In: Meena VS (ed) Role of rhizospheric microbes in soil. Springer, Singapore, pp 125–141

Ji J, Yuan D, Jin C, Wang G, Li X, Guan C (2020) Enhancement of growth and salt tolerance of rice seedlings (*Oryza sativa* L.) by regulating ethylene production with a novel halotolerant PGPR strain *Glutamicibacter* sp. YD01 containing ACC deaminase activity. Acta Physiol Plant 42:42. https://doi.org/10.1007/s11738-020-3034-3

Jochum MD, McWilliams KL, Borrego EJ, Kolomiets MV, Niu G, Pierson EA, Jo YK (2019) Bioprospecting plant growth-promoting rhizobacteria that mitigate drought stress in grasses. Front Microbiol 10:2106

Johnson DR, Wyse DL, Jones KJ (1996) Controlling weeds with phytopathogenic bacteria. Weed Technol 10:621–624

Joshi B, Chaudhary A, Singh H, Kumar PA (2020) Prospective evaluation of individual and consortia plant growth promoting rhizobacteria for drought stress amelioration in rice (*Oryza sativa* L.). Plant Soil 457(1):225–240

Jung H, Ali S, Kim JY, Kim WC (2018) Transgenic Arabidopsis expressing acdS gene of *Pseudomonas veronii*-KJ alleviate the adverse effects of salt and water-logging stress. Plant Breed Biotechnol 6(3):221–232

Kakembo D, Lee YH (2019) Analysis of traits for biocontrol performance of *Pseudomonas parafulva* JBCS1880 against bacterial pustule in soybean plants. Biol Control 134:72–81

Kang SM, Khan AL, Waqas M, You YH, Kim JH, Kim JG, Hamayun M, Lee IJ (2014) Plant growth-promoting rhizobacteria reduce adverse effects of salinity and osmotic stress by regulating phytohormones and antioxidants in *Cucumis sativus*. J Plant Interact 9:673–682

Kang SM, Khan AL, Waqas M, You YH (2015) Gibberellin-producing *Serratia nematodiphila* PEJ1011 ameliorates low temperature stress in *Capsicum annum* L. Eur J Soil Biol 68:85–93

Kang SM, Khan AL, Waqas M, Asaf S, Lee KE, Park YG, Kim AY, Khan MA, You YH, Lee IJ (2019) Integrated phytohormone production by the plant growth-promoting rhizobacterium *Bacillus tequilensis* SSB07 induced thermotolerance in soybean. J Plant Interact 14(1):416–423

Kasuga M, Miura S, Shinozaki K, Yamaguchi-Shinozaki K (2004) A combination of the Arabidopsis DREB1A gene and stress-inducible rd29A promoter improved drought-and low-temperature stress tolerance in tobacco by gene transfer. Plant Cell Physiol 45(3):346–350

Katiyar V, Goel R (2004) Siderophore mediated plant growth promotion at low temperature by mutant of fluorescent pseudomonad. Plant Growth Regul 42:239–244

Kennedy AC, Johnson BN, Stubbs TL (2001) Host range of a deleterious rhizobacterium for biological control of downy brome. Weed Sci 49:792–797

Khan N, Bano A (2019) Rhizobacteria and abiotic stress management. In: Syed RZ, Arora NK, Reddy MS (eds) Plant growth promoting rhizobacteria for sustainable stress management. Springer, Singapore, pp 65–80

Khan N, Bano A, Rahman MA, Guo J, Kang Z, Babar MA (2019) Comparative physiological and metabolic analysis reveals a complex mechanism involved in drought tolerance in chickpea (*Cicer arietinum* L.) induced by PGPR and PGRs. Sci Rep 9(1):1–9

Khan MA, Asaf S, Khan AL, Jan R, Kang SM, Kim KM, Lee IJ (2020a) Extending thermotolerance to tomato seedlings by inoculation with SA1 isolate of *Bacillus cereus* and comparison with exogenous humic acid application. PLoS One 15(4):e0232228

Khan MA, Asaf S, Khan AL, Jan R, Kang SM, Kim KM, Lee IJ (2020b) Thermotolerance effect of plant growth-promoting *Bacillus cereus* SA1 on soybean during heat stress. BMC Microbiol 20(1):1–4

Khan N, Bano A, Ali S, Babar MA (2020c) Crosstalk amongst phytohormones from planta and PGPR under biotic and abiotic stresses. Plant Growth Regul 90:189–203

Killi D, Raschi A, Bussotti F (2020) Lipid peroxidation and chlorophyll fluorescence of photosystem II performance during drought and heat stress is associated with the antioxidant capacities of C3 sunflower and C4 maize varieties. Int J Mol Sci 21(14):4846

Kim SY, Mulkey TJ (1997) Effect of ethylene antagonists on auxin-induced inhibition of intact primary root elongation in maize (*Zea mays* L.). J Plant Biol 40:256–260

Kim K, Jang YJ, Lee SM, Oh BT, Chae JC, Lee KJ (2014) Alleviation of salt stress by *Enterobacter* sp. EJ01 in tomato and Arabidopsis is accompanied by up-regulation of conserved salinity responsive factors in plants. Mol Cells 37:109–117

Kim AY, Shahzad R, Kang SM, Seo CW, Park YG, Park HJ, Lee IJ (2017) IAA-producing *Klebsiella variicola* AY13 reprograms soybean growth during flooding stress. J Crop Sci Biotechnol 20(4):235–242

Kloepper JW, Ryu CM, Zhang S (2004) Induced systemic resistance and promotion of plant growth by *Bacillus* spp. Phytopathology 94(11):1259–1266

Koskey G, Mburu SW, Awino R, Njeru EM, Maingi JM (2021) Potential use of beneficial microorganisms for soil amelioration, phytopathogen biocontrol, and sustainable crop production in smallholder agroecosystems. Front Sustain Food Syst 5:606308. https://doi.org/10.3389/fsufs

Kosová K, Urban MO, Vítámvás P, Prášil IT (2016) Drought stress response in common wheat, durum wheat, and barley: transcriptomics, proteomics, metabolomics, physiology, and breeding for an enhanced drought tolerance. In: Hossain MA, Wani SH, Bhattacharjee S, Burritt DJ, Tran LSP (eds) Drought stress tolerance in plants, vol 2. Springer, Cham, pp 277–314

Kour D, Rana KL, Yadav AN, Sheikh I, Kumar V, Dhaliwal HS, Saxena AK (2020) Amelioration of drought stress in Foxtail millet (*Setaria italica* L.) by P-solubilizing drought-tolerant microbes with multifarious plant growth promoting attributes. Environ Sustain 3(1):23–34

Koyro HW, Ahmad P, Geissler N (2012) Abiotic stress responses in plants: an overview. In: Ahmad P, Prasad NV (eds) Environmental adaptations and stress tolerance of plants in the era of climate change. Springer, New York, pp 1–28

Kranner I, Minibayeva FV, Beckett RP, Seal CE (2010) What is stress? concepts, definitions and applications in seed science. New Phytol 188:655–673

Kremer RJ, Souissi T (2001) Cyanide production by rhizobacteria and potential for suppression of weed seedling growth. Curr Microbiol 43:182–186

Kuan KB, Othman R, Rahim KA, Shamsuddin ZH (2016) Plant growth-promoting rhizobacteria inoculation to enhance vegetative growth, nitrogen fixation and nitrogen remobilisation of maize under greenhouse conditions. PLoS One 11(3):0152478. https://doi.org/10.1371/journal.pone.0152478

Kumar A, Patel JS, Meena VS, Srivastava R (2019a) Recent advances of PGPR based approaches for stress tolerance in plants for sustainable agriculture. Biocatal Agric Biotechnol 29:101271. https://doi.org/10.1016/j.bcab.2019.101271

Kumar S, Suyal DC, Yadav A, Shouche Y, Goel R (2019b) Microbial diversity and soil physiochemical characteristic of higher altitude. PLoS One 14(3):e0213844. https://doi.org/10.1371/journal.pone.0213844

Kumawat KC, Sharma P, Nagpal S, Gupta RK, Sirari A, Nair RM, Bindumadhava H, Singh S (2020) Dual microbial inoculation, a Game changer?–Bacterial biostimulants with multifunctional growth promoting traits to mitigate salinity stress in Spring Mungbean. Front Microbiol 11:600576. https://doi.org/10.3389/fmicb.2020.600576

Kusale SP, Attar YC, Sayyed RZ, Malek RA, Ilyas N, Suriani NL, Khan N, El Enshasy HA (2021) Production of plant beneficial and antioxidants metabolites by *Klebsiella variicola* under salinity stress. Molecules 26(7):1894

Kushwaha P, Kashyap PL, Kuppusamy P (2020) Microbes for cold stress resistance in plants: mechanism, opportunities, and challenges. In: Goel R, Soni R, Suyal DC (eds) Microbiological advancements for higher altitude agro-ecosystems and sustainability. Springer, Singapore, pp 269–292

Lanza MGDBR, dos Reis AR (2021) Roles of selenium in mineral plant nutrition: ROS scavenging responses against abiotic stresses. Plant Physiol Biochem 164:27–43. https://doi.org/10.1016/j.plaphy.2021.04.026

Lee DG, Ahsan N, Lee SH, Kang KY, Bahk JD, Lee IJ, Lee BH (2007) A proteomic approach in analyzing heat-responsive proteins in rice leaves. Proteomics 7(18):3369–3383

Lesk C, Rowhani P, Ramankutty N (2016) Influence of extreme weather disasters on global crop production. Nature 529:84–87

Li YQ, Sun ZL, Zhuang XF, Xu L, Chen SF, Li MZ (2003) Research progress on microbial herbicides. Crop Prot 22:247–252

Li J, McConkey BJ, Cheng Z, Guo S, Glick BR (2013) Identification of plant growth-promoting bacteria-responsive proteins in cucumber roots under hypoxic stress using a proteomic approach. J Proteomics 84:119–131

Li Z, Long R, Zhang T, Wang Z, Zhang F, Yang Q, Kang J, Sun Y (2017) Molecular cloning and functional analysis of the drought tolerance gene MsHSP70 from alfalfa (*Medicago sativa* L.). Int J Plant Res 130(2):387–396

Li Y, Shi H, Zhang H, Chen S (2019) Amelioration of drought effects in wheat and cucumber by the combined application of super absorbent polymer and potential biofertilizer. Peer J 7:e6073

Li X, Sun P, Zhang Y, Jin C, Guan C (2020) A novel PGPR strain *Kocuria rhizophila* Y1 enhances salt stress tolerance in maize by regulating phytohormone levels, nutrient acquisition, redox potential, ion homeostasis, photosynthetic capacity and stress-responsive genes expression. Environ Exp Bot 174:104023. https://doi.org/10.1016/j.envexpbot.2020.104023

Liu HC, Charng YY (2013) Common and distinct functions of Arabidopsis class A1 and A2 heat shock factors in diverse abiotic stress responses and development. Plant Physiol 163(1):276–290

Liu F, Xing S, Ma H, Du Z, Ma B (2013) Cytokinin-producing, plant growth-promoting rhizobacteria that confer resistance to drought stress in *Platycladus orientalis* container seedlings. Appl Microbiol Biotechnol 97(20):9155–9164

Liu CH, Siew W, Hung YT, Jiang YT, Huang CH (2021) 1-Aminocyclopropane-1-carboxylate (ACC) Deaminase gene in *Pseudomonas azotoformans* is associated with the amelioration of salinity stress in tomato. J Agric Food Chem 69(3):913–921

Lopez-Raez JA (2016) How drought and salinity affect arbuscular mycorrhizal symbiosis and strigolactone biosynthesis? Planta 243:1375–1385

Lozano GL, Holt J, Ravel J, Rasko DA, Thomas MG, Handelsman J (2016) Draft genome sequence of biocontrol agent *Bacillus cereus* UW85. Genome Announc 4(5):5–6

Maciag T, Krzyzanowska DM, Jafra S, Siwinska J, Czajkowski R (2020) The Great Five-an artificial bacterial consortium with antagonistic activity towards *Pectobacterium* spp. and *Dickeya* spp.: formulation, shelf life, and the ability to prevent soft rot of potato in storage. Appl Microbiol Biotechnol 104(10):4547–4561

Mahadik S, Kumudini BS (2020) Enhancement of salinity stress tolerance and plant growth in finger millet using fluorescent Pseudomonads. Rhizosphere 15:100226

Mantelin S, Touraine B (2004) Plant growth-promoting rhizobacteria and nitrateavailability: impacts on root development and nitrate uptake. J Exp Bot 55:27–34

Marin-Bruzos M, Grayston SJ, Forge T, Nelson LM (2021) Isolation and characterization of streptomycetes and pseudomonad strains with antagonistic activity against the plant parasitic nematode *Pratylenchus penetrans* and fungi associated with replant disease. Biol Control 158: 104599

Marulanda A, Barea JM, Azcon R (2009) Stimulation of plant growth and drought tolerance by native microorganisms (AM Fungi and Bacteria) from dry environments: mechanisms related to bacterial effectiveness. J Plant Growth Regul 28:115–124

Marulanda A, Azcón R, Chaumont F, Ruiz-Lozano JM, Aroca R (2010) Regulation of plasma membrane aquaporins by inoculation with a *Bacillus megaterium* strain in maize (*Zea mays* L.) plants under unstressed and salt-stressed conditions. Planta 232:533–543

Maya MA, Matsubara YI (2013) Influence of arbuscular mycorrhiza on the growth and antioxidative activity in cyclamen under heat stress. Mycorrhiza 23(5):381–390

Mayak S, Tirosh T, Glick BR (2004) Plant growth-promoting bacteria confer resistance in tomato plants to salt stress. Plant Physiol Biochem 42(6):565–572

McKeon TA, Fernandez-Maculet JC, Yang SF (1995) Biosynthesis and metabolism of ethylene. In: Davies PJ (ed) Plant hormones physiology, biochemistry and molecular biology. Kluwer Academic, Dordrecht, pp 118–139

McKersie BD, Lesheim Y (1994) Stress and stress coping in cultivated plants, 1st edn. Springer, Dordrecht, p 256

McPhail KL, Armstrong DJ, Azevedo MD, Banowetz GM, Mills DI (2010) 4-Formylaminooxyvinylglycine, an herbicidal germination-arrest factor from *Pseudomonas* rhizosphere bacteria. J Nat Prod 73:1853–1857

Meena M, Swapnil P, Divyanshu K, Kumar S, Tripathi YN, Zehra A, Marwal A, Upadhyay RS (2020) PGPR-mediated induction of systemic resistance and physiochemical alterations in plants against the pathogens: current perspectives. J Basic Microbiol 60(10):828–861

Meenakshi, Annapurna K, Govindasamy V, Ajit V, Choudhary DK (2019) Mitigation of drought stress in wheat crop by drought tolerant endophytic bacterial isolates. Vegetos 32(4):486–493

Migunova VD, Sasanelli N (2021) Bacteria as biocontrol tool against phytoparasitic nematodes. Plants 10(2):389

Mishra PK, Mishra S, Bisht SC, Selvakumar G, Kundu S, Bisht JK, Gupta HS (2009) Isolation, molecular characterization and -growth-promotion activities of a cold tolerant bacterium *Pseudomonas* sp. NARs9 (MTCC9002) from the Indian Himalaya. Biol Res 42(3):305–313

Mishra PK, Bisht SC, Ruwari P, Selvakumar G, Joshi GK, Bisht JK, Bhatt JC, Gupta HS (2011) Alleviation of cold stress in inoculated wheat (*Triticum aestivum* L.) seedlings with psychrotolerant *Pseudomonads* from NW Himalayas. Arch Microbiol 193(7):497–513

Mishra PK, Bisht SC, Bisht JK, Bhatt JC (2012) Cold-tolerant PGPRs as bioinoculants for stress management. In: Maheshwari DK (ed) Bacteria in agrobiology: stress management. Springer, Berlin, Heidelberg, pp 95–118

Misra S, Chauhan PS (2020) ACC deaminase-producing rhizosphere competent *Bacillus* spp. mitigate salt stress and promote *Zea mays* growth by modulating ethylene metabolism. 3 Biotech 10:119. https://doi.org/10.1007/s13205-020-2104-y

Misra S, Dixit VK, Khan MH, Mishra SK, Dviwedi G, Yadav S, Lehri A, Chauhan PS (2017) Exploitation of agro-climatic environment for selection of 1-aminocyclopropane-1-carboxylic acid (ACC) deaminase producing salt tolerant indigenous plant growth promoting rhizobacteria. Microbiol Res 205:25–34

Misra S, Dixit VK, Mishra SK, Chauhan PS (2019) Demonstrating the potential of abiotic stress-tolerant *Jeotgalicoccus huakuii* NBRI 13E for plant growth promotion and salt stress amelioration. Ann Microbiol 69(4):419–434

Mitra S, Mukherjee A, Wiley-Kalil A, Das S, Owen H, Reddy PM, Ané J-M, James EK, Gyaneshwar P (2016) A rhamnose-deficient lipopolysaccharide mutant of *Rhizobium* sp. IRBG74 is defective in root colonization and beneficial interactions with its flooding-tolerant hosts *Sesbania cannabina* and wetland rice. J Exp Bot 67(19):5869–5884

Mittler R (2006) Abiotic stress, the field environment and stress combination. Trends Plant Sci 11: 15–19

Molina-Favero C, Creus CM, Simontacchi M, Puntarulo S, Lamattina L (2008) Aerobic nitric oxide production by *Azospirillum brasilense* Sp245 and its influence on root architecture in tomato. Mol Plant Microbe Interact 2:1001–1009

Momčilović I, Pantelić D, Zdravković-Korać S, Oljača J, Rudić J, Fu J (2016) Heat-induced accumulation of protein synthesis elongation factor 1A implies an important role in heat tolerance in potato. Planta 244(3):671–679

Mosa KA, Ismail A, Helmy M (2017) Introduction to plant stresses. In: Mosa KA, Ismail A, Helmy M (eds) Plant stress tolerance. Springer, Cham, pp 1–19

Mukhtar T, Smith D, Sultan T, Seleiman MF, Alsadon AA, Ali S, Chaudhary HJ, Solieman TH, Ibrahim AA, Saad MA (2020) Mitigation of heat stress in *Solanum lycopersicum* L. by ACC-deaminase and exopolysaccharide producing *Bacillus cereus*: effects on biochemical profiling. Sustainability 12(6):2159. https://doi.org/10.3390/su12062159

Mumtaz MZ, Ahmad M, Jamil M, Hussain T (2017) Zinc solubilizing *Bacillus* spp. potential candidates for biofortification in maize. Microbiol Res 202:51–60

Mumtaz MZ, Saqib M, Abbas G, Akhtar J, Qamar ZU (2018) Genotypic variation in rice for grain yield and quality as affected by salt-affected field conditions. J Plant Nutr 41(2):233–242

Mumtaz MZ, Barry KM, Baker AL, Nichols DS, Ahmad M, Zahir ZA, Britz ML (2019) Production of lactic and acetic acids by *Bacillus* sp. ZM20 and *Bacillus cereus* following exposure to zinc oxide: a possible mechanism for Zn solubilization. Rhizosphere 12:100170. https://doi.org/10.1016/j.rhisph.2019.100170

Mumtaz MZ, Saqib M, Abbas G, Akhtar J, Qamar ZU (2020) Drought stress impairs grain yield and quality of rice genotypes by impaired photosynthetic attributes and K nutrition. Ric Sci 27(1): 5–9

Munns R (2005) Genes and salt tolerance: bringing them together. New Phytol 167:645–663

Munns R, Tester M (2008) Mechanisms of salinity tolerance. Annu Rev Plant Biol 59:651–681

Nadeem SM, Zahir ZA, Naveed M, Arshad M (2009) Rhizobacteria containing ACC-deaminase confer salt tolerance in maize grown on salt-affected fields. Can J Microbiol 55:1302–1309

Nadeem SM, Ahmad M, Zahir ZA, Javaid A, Ashraf M (2014) The role of mycorrhizae and plant growth promoting rhizobacteria (PGPR) in improving crop productivity under stressful environments. Biotechnol Adv 32:429–448

Nadeem SM, Ahmad M, Naveed M, Imran M, Zahir ZA, Crowley DE (2016) Relationship between *in vitro* characterization and comparative efficacy of plant growth-promoting rhizobacteria for improving cucumber salt tolerance. Arch Microbiol 198(4):379–387

Nadeem SM, Imran M, Naveed N, Khan MY, Ahmad M, Zahir ZA, Crowley DE (2017) Synergistic use of biochar, compost and plant growth-promoting rhizobacteria for enhancing cucumber growth under water deficit conditions. J Sci Food Agric 97:5139–5145

Nascimento F, Brígido C, Alho L, Glick BR, Oliveira S (2012) Enhanced chickpea growth-promotion ability of a *Mesorhizobium* strain expressing an exogenous ACC deaminase gene. Plant Soil 353:221–230

Nautiyal CS, Srivastava S, Chauhan PS, Seem K, Mishra A, Sopory S (2013) Plant growth-promoting bacteria *Bacillus amyloliquefaciens* NBRISN13 modulates gene expression profile of leaf and rhizosphere community in rice during salt stress. Plant Physiol Biochem 66:1–9

Nazli F, Najm-ul-Seher, Khan MY, Jamil M, Nadeem SM, Ahmad M (2020) Soil microbes and plant health. In: Haq IU, Ijaz S (eds) Plant disease management strategies for sustainable agriculture through traditional and modern approaches. Sustainability in plant and crop protection, vol 13. Springer, Cham, pp 111–135

Negi YK, Kumar J, Garg SK (2005) Cold-tolerant fluorescent *Pseudomonas* isolates from Garhwal Himalayas as potential plant growth promoting and biocontrol agents in pea. Curr Sci 89:2151–2156

Ngumbi E, Kloepper J (2016) Bacterial-mediated drought tolerance: current and future prospects. Appl Soil Ecol 105:109–125

Nishiuchi S, Yamauchi T, Takahashi H, Kotula L, Nakazono M (2012) Mechanisms for coping with submergence and waterlogging in rice. Rice 5:2. https://doi.org/10.1186/1939-8433-5-2
Niu SQ, Li HR, Paré PW, Aziz M, Wang SM, Shi H, Li J, Han QQ, Guo SQ, Li J, Guo Q (2016) Induced growth promotion and higher salt tolerance in the halophyte grass *Puccinellia tenuiflora* by beneficial rhizobacteria. Plant Soil 407:217–230
Njeru EM, Koskey G (2021) Using beneficial microorganisms to promote sustainable crop production and resilience of smallholder agroecosystems to changing climate. In: Mallappa VKH, Shirur M (eds) Climate change and resilient food systems. Springer, Singapore, pp 287–314
Nouri MZ, Moumeni A, Komatsu S (2015) Abiotic stresses: insight into gene regulation and protein expression in photosynthetic pathways of plants. Int J Mol Sci 16:20392–20416
Ohashi Y, Nakayama N, Saneoka H, Fujita K (2006) Effects of drought stress on photosynthetic gas exchange, chlorophyll fluorescence and stem diameter of soybean plants. Biol Plant 50(1): 138–141
Okrent RA, Halgren AB, Azevedo MD, Chang JH, Mills DI, Maselko M, Armstrong DJ, Banowetz GM, Trippe KM (2014) Negative regulation of germination-arrest factor production in *Pseudomonas fluorescens* WH6 by a putative extracytoplasmic function sigma factor. Microbiology 160:2432–2442
Orozco-Mosqueda MC, Glick BR, Santoyo G (2020) ACC deaminase in plant growth-promoting bacteria (PGPB): an efficient mechanism to counter salt stress in crops. Microbiol Res 235: 126439. https://doi.org/10.1016/j.micres.2020.126439
Ortiz-Ojeda P, Ogata-Gutiérrez K, Zúñiga-Dávila D (2017) Evaluation of plant growth promoting activity and heavy metal tolerance of psychrotrophic bacteria associated with maca (*Lepidium meyenii* Walp.) rhizosphere. AIMS Microbiol 3(2):279
Pahalvi HN, Rafiya L, Rashid S, Nisar B, Kamili AN (2021) Chemical fertilizers and their impact on soil health. In: Dar GH, Bhat RA, Mehmood MA, Hakeem KR (eds) Microbiota and biofertilizers, vol 2. Springer, Cham, pp 1–20
Pandey A, Palni LM, Mulkalwar P, Nadeem M (2002) Effect of temperature on solubilization of tricalcium phosphate by *Pseudomonas corrugata*. J Sci Ind Res 61:457–460
Pandey A, Trivedi P, Kumar B, Palni LM (2006) Characterization of a phosphate solubilizing and antagonistic strain of *Pseudomonas putida* (B0) isolated from a sub-alpine location in the Indian central Himalaya. Curr Microbiol 53(2):102–107
Pandey P, Irulappan V, Bagavathiannan MV, Senthil-Kumar M (2017) Impact of combined abiotic and biotic stresses on plant growth and avenues for crop improvement by exploiting physio-morphological traits. Front Plant Sci 8:537. https://doi.org/10.3389/fpls.2017.00537
Panwar M, Tewari R, Gulati A (2016) Indigenous salt-tolerant rhizobacterium *Pantoea dispersa* (PSB3) reduces sodium uptake and mitigates the effects of salt stress on growth and yield of chickpea. Acta Physiol Plant 38:278. https://doi.org/10.1007/s11738-016-2284-6
Passioura J (2007) The drought environment: physical, biological and agricultural perspectives. J Exp Bot 58:113–117
Patel PK, Singh AK, Tripathi N, Yadav D, Hemantaranjan A (2014) Flooding: abiotic constraint limiting vegetable productivity. Adv Plant Agric Res 1(3):96–103
Pathak R, Shrestha A, Lamichhane J, Gauchan D (2017) PGPR in biocontrol mechanisms and roles in disease suppression. Int J Agron Agric Res 11(1):69–80
Paul D, Lade H (2014) Plant-growth-promoting rhizobacteria to improve crop growth in saline soils: a review. Agron Sustain Dev 34:737–752
Paul MV, Iyer S, Amerhauser C, Lehmann M, van Dongen JT, Geigenberger P (2016) Oxygen sensing via the ethylene response transcription factor RAP2 12 affects plant metabolism and performance under both normoxia and hypoxia. Plant Physiol 172(1):141–153
Péchy-Tarr M, Bruck DJ, Maurhofer M, Fischer E, Vogne C, Henkels MD, Donahue KM, Grunder J, Loper JE, Keel C (2008) Molecular analysis of a novel gene cluster encoding an insect toxin in plant-associated strains of *Pseudomonas fluorescens*. Environ Microbiol 10(9): 2368–2386

Perneel M, D'Hondt L, De Maeyer K, Adiobo A, Rabaey K, Höfte M (2008) Phenazines and biosurfactants interact in the biological control of soil-borne diseases caused by *Pythium* spp. Environ Microbiol 10(3):778–788

Phour M, Sindhu SS (2020) Amelioration of salinity stress and growth stimulation of mustard (*Brassica juncea* L.) by salt-tolerant *Pseudomonas* species. Appl Soil Ecol 149:103518

Phukan UJ, Mishra S, Shukla RK (2016) Waterlogging and submergence stress: affects and acclimation. Crit Rev In Biotech 36(5):956–966

Pitman MG, Läuchli A (2002) Global impact of salinity and agricultural ecosystems. In: Pitman MG, Lauchi A (eds) Salinity: environment-plants-molecules. Springer, Dordrecht, pp 3–20

Prasad PVV, Staggenborg SA, Ristic Z (2008) Impacts of drought and/or heat stress on physiological, developmental, growth, and yield processes of crop plants. Response of crops to limited water: understanding and modeling water stress effects on plant growth processes. In: Ahuja LR, Reddy VR, Saseendran SA, Yu Q (eds) Advances in agricultural systems modeling, vol 1, pp 301–355

Prasannakumar MK, Netravathi LM, Mahesh HB, Parivallal PB, Puneeth ME, Sathish A, Pramesh D, Middha SK, Das AJ, Rohit BS (2021) Comparative metagenomic analysis of rice soil samples revealed the diverse microbial population and biocontrol organisms against plant pathogenic fungus *Magnaporthe oryzae*. 3 Biotech 11(5):1–1

Prevost D, Drouin P, Antoun H (1999) The potential use of cold-adapted rhizobia to improve symbiotic nitrogen fixation in legumes cultivated in temperate regions. In: Margesin R, Schinner F (eds) Biotechnological applications of cold-adapted organisms. Springer, Berlin, Heidelberg, pp 161–176

Prevost D, Drouin P, Laberge S, Bertrand A, Cloutier J, Lévesque G (2003) Cold-adapted rhizobia for nitrogen fixation in temperate regions. Can J Bot 81(12):1153–1161

Qin S, Feng WW, Zhang YJ, Wang TT, Xiong YW, Xing K (2018) Diversity of bacterial microbiota of coastal halophyte Limonium sinense and amelioration of salinity stress damage by symbiotic plant growth-promoting actinobacterium *Glutamicibacter halophytocola* KLBMP 5180. Appl Environ Microbiol 84(19):e01533

Qurashi AW, Sabri AN (2012) Bacterial exopolysaccharide and biofilm formation stimulate chickpea growth and soil aggregation under salt stress. Braz J Microbiol 43(3):1183–1191

Rahdari P, Hoseini SM (2012) Drought stress, a review. Int J Agron Plant Prod 3:443–446

Raheem A, Shaposhnikov A, Belimov AA, Dodd IC, Ali B (2018) Auxin production by rhizobacteria was associated with improved yield of wheat (*Triticum aestivum* L.) under drought stress. Arch Agron Soil Sci 64(4):574–587

Rahnama A, James RA, Poustini K, Munns R (2010) Stomatal conductance as a screen for osmotic stress tolerance in durum wheat growing in saline soil. Funct Plant Biol 37(3):255–263

Ramegowda V, Da Costa MV, Harihar S, Karaba NN, Sreeman SM (2020) Abiotic and biotic stress interactions in plants: a cross-tolerance perspective. In: Hossain MA, Liu F, Huang B (eds) Priming-mediated stress and cross-stress tolerance in crop plants. Academic Press, San Diego, CA 267-302

Ravanbakhsh M, Sasidharan R, Voesenek LA, Kowalchuk GA, Jousset A (2017) ACC deaminase-producing rhizosphere bacteria modulate plant responses to flooding. J Ecol 105(4):979–986

Řeháková K, Chroňáková A, Krištůfek V, Kuchtová B, Čapková K, Scharfen J, Čapek P, Doležal J (2015) Bacterial community of cushion plant *Thylacospermum ceaspitosum* on elevational gradient in the Himalayan cold desert. Front Microbiol 6:304

Rengasamy P (2002) Transient salinity and subsoil constraints to dryland farming in Australian sodic soils: an overview. Aust J Exp Agric 42:351–361

Rengasamy P (2010) Soil processes affecting crop production in salt-affected soils. Funct Plant Biol 37(7):613–620

Rijavec T, Lapanje A (2016) Hydrogen cyanide in the rhizosphere: not suppressing plant pathogens, but rather regulating availability of phosphate. Front Microbiol 7:1–14

Ritonga FN, Chen S (2020) Physiological and molecular mechanism involved in cold stress tolerance in plants. Plants 9(5):560

Rodriguez R, Redman R (2008) More than 400 million years of evolution and some plants still can't make it on their own: plant stress tolerance via fungal symbiosis. J Exp Bot 59:1109–1114

Rojas-Tapias D, Moreno-Galván A, Pardo-Díaz S, Obando M, Rivera D, Bonilla R (2012) Effect of inoculation with plant growth-promoting bacteria (PGPB) on amelioration of saline stress in maize (*Zea mays*). Appl Soil Ecol 61:264–272

Rozema J, Flowers T (2008) Ecology: crops for a salinized world. Science 322:1478–1480

Ruelland E, Zachowski A (2010) How plants sense temperature. Environ Exp Bot 69:225–232

Sachin N, Tsang A, Shaanker RU, Dayanandan S (2021) Genome Sequence Resource of *Bacillus velezensis* EB14, a native endophytic bacterial strain with biocontrol potential against the poplar stem canker causative pathogen, *Sphaerulina musiva*. Phytopathology 1:PHYTO09200433A

Sagar A, Sayyed RZ, Ramteke PW, Sharma S, Marraiki N, Elgorban AM, Syed A (2020) ACC deaminase and antioxidant enzymes producing halophilic *Enterobacter* sp. PR14 promotes the growth of rice and millets under salinity stress. Physiol Mol Biol Plants 26(9):1847–1854

Saharan BS, Nehra V (2011) Plant growth promoting rhizobacteria: a critical review. Life Sci Med Res 21(1):1–29

Saikia J, Sarma RK, Dhandia R, Yadav A, Bharali R, Gupta VK, Saikia R (2018) Alleviation of drought stress in pulse crops with ACC deaminase producing rhizobacteria isolated from acidic soil of Northeast India. Sci Rep 8(1):1–16

Saleem M, Arshad M, Hussain S, Bhatti AS (2007) Perspective of plant growth promoting rhizobacteria (PGPR) containing ACC deaminase in stress agriculture. J Ind Microbiol Biotechnol 34(10):635–648

Salvi P, Manna M, Kaur H, Thakur T, Gandass N, Bhatt D, Muthamilarasan M (2021) Phytohormone signaling and crosstalk in regulating drought stress response in plants. Plant Cell Rep 22: 1–25

Sang-Mo K, Radhakrishnan R, Khan AL, Min-Ji K, Jae-Man P, Bo-Ra K, Dong-Hyun S, In-Jung L (2014) Gibberellin secreting rhizobacterium, *Pseudomonas putida* H-2-3 modulates the hormonal and stress physiology of soybean to improve the plant growth under saline and drought conditions. Plant Physiol Biochem 84:115–124

Sangwan V, Foulds I, Singh J, Dhindsa RS (2001) Cold-activation of *Brassica napus* BN115 promoter is mediated by structural changes in membranes and cytoskeleton, and requires Ca^{2+} influx. Plant J 27(1):1–12

Sapre S, Gontia-Mishra I, Tiwari S (2018) *Klebsiella* sp. confers enhanced tolerance to salinity and plant growth promotion in oat seedlings (*Avena sativa*). Microbiol Res 206:25–32

Sapre S, Gontia-Mishra I, Tiwari S (2021) Plant growth-promoting rhizobacteria ameliorates salinity stress in pea (*Pisum sativum*). J Plant Growth Regul. https://doi.org/10.1007/s00344-021-10329-y

Sarkar J, Chakraborty B, Chakraborty U (2018) Plant growth promoting rhizobacteria protect wheat plants against temperature stress through antioxidant signalling and reducing chloroplast and membrane injury. J Plant Growth Regul 37(4):1396–1412

Sarkar J, Chakraborty U, Chakraborty B (2021) High-temperature resilience in Bacillus safensis primed wheat plants: A study of dynamic response associated with modulation of antioxidant machinery, differential expression of HSPs and osmolyte biosynthesis. Environ Exp Bot 182: 104315

Sarwar M, Kremer RJ (1995) Enhanced suppression of plant growth through the production of L-tryptophan derived compounds by deleterious rhizobacteria. Plant Soil 172:261–269

Sasidharan R, Hartman S, Liu Z, Martopawiro S, Sajeev N, van Veen H, Yeung E, Voesenek LA (2018) Signal dynamics and interactions during flooding stress. Plant Physiol 176(2): 1106–1117

Savvides A, Ali S, Tester M, Fotopoulos V (2016) Chemical priming of plants against multiple abiotic stresses: mission possible? Trends Plant Sci 21(4):329–340

Schmidt R, Köberl M, Mostafa A, Ramadan EM, Monschein M, Jensen KB, Bauer R, Berg G (2014) Effects of bacterial inoculants on the indigenous microbiome and secondary metabolites of chamomile plants. Front Microbiol 5:64. https://doi.org/10.3389/fmicb.2014.00064

Selvakumar G, Kundu S, Joshi P, Nazim S, Gupta AD, Mishra PK, Gupta HS (2008) Characterization of a cold tolerant plant growth-promoting bacterium *Pantoea dispersa* 1A isolated from a sub-alpine soil in the North Western Indian Himalayas. World J Microbiol Biotechnol 24(7): 955–960

Selvakumar G, Joshi P, Nazim S, Mishra P, Bisht J, Gupta H (2009) Phosphate solubilization and growth promotion by *Pseudomonas fragi* CS11RH1 (MTCC 8984), a psychrotolerant bacterium isolated from a high altitude Himalayan rhizosphere. Biologia 64(2):239–245

Selvakumar G, Panneerselvam P, Ganeshamurthy AN (2012) Bacterial mediated alleviation of abiotic stress in crops. In: Maheshwari DK (ed) Bacteria in agrobiology: stress management. Springer-Verlag, Berlin, Heidelberg, pp 205–224

Shakir MA, Asghari B, Arshad M (2012) Rhizosphere bacteria containing ACC deaminase conferred drought tolerance in wheat grown under semi-arid climate. Soil Environ 31:108–112

Sharma P, Khanna V, Kumar P (2013) Efficacy of aminocyclopropane-1-carboxylic acid (ACC)-deaminase producing rhizobacteria in ameliorating water stress in chickpea under axenic conditions. Afr J Microbiol Res 7:5749–5757

Sharma S, Joshi J, Kataria S, Verma SK, Chatterjee S, Jain M, Pathak K, Rastogi A, Brestic M (2020) Regulation of the Calvin cycle under abiotic stresses: an overview. In: Tripathi DK, Singh VP, Ramawat N (eds) Plant life under changing environment. Academic Press, San Diego, CA, pp 681–717

Shen YC, Korkor NL, Xiao R, Pu Q, Hu M, Zhang SS, Kong DD, Zeng G, Hu XF (2020) Antagonistic activity of combined bacteria strains against southern blight pathogen of *Dendrobium officinale*. Biol Control 151:104291

Showkat S (2012) Biological control of *Fusarium oxysporum* and *Aspergillus* sp. by *Pseudomonas fluorescens* isolated from wheat rhizosphere soil of Kashmir. IOSR Int J Pharm Biol Sci 1(4): 24–32

Shrivastava P, Kumar R (2015) Soil salinity: a serious environmental issue and plant growth promoting bacteria as one of the tools for its alleviation. Saudi J Biol Sci 22(2):123–131

Singh JS (2015) Biodiversity: current perspective. Clim Chang Environ Sustain 3(1):71–72

Singh RP, Jha PN (2015) Molecular identification and characterization of rhizospheric bacteria for plant growth promoting ability. Int J Curr Biotechnol 3:12–18

Singh RP, Jha PN (2017) Analysis of fatty acid composition of PGPR Klebsiella sp. SBP-8 and its role in ameliorating salt stress in wheat. Symbiosis 73(3):213–222

Singh JS, Pandey VC, Singh DP (2011) Efficient soil microorganisms: a new dimension for sustainable agriculture and environmental development. Agric Ecosyst Environ 140:339–353

Singh RN, Gaba S, Yadav AN, Gaur P, Gulati S, Kaushik R, Saxena AK (2016) First high quality draft genome sequence of a plant growth promoting and cold active enzyme producing psychrotrophic *Arthrobacter agilis* strain L77. Stand Genomic Sci 11(1):1–9

Sofy MR, Aboseidah AA, Heneidak SA, Ahmad HR (2021) ACC deaminase containing endophytic bacteria ameliorate salt stress in *Pisum sativum* through reduced oxidative damage and induction of antioxidative defense systems. Environ Sci Pollut Res. https://doi.org/10.1007/s11356-021-13585-3

Srinivasan R, Mageswari A, Subramanian P, Maurya VK, Sugnathi C, Amballa C, Sa T, Gothandam KM (2017) Exogenous expression of ACC deaminase gene in psychrotolerant bacteria alleviates chilling stress and promotes plant growth in millets under chilling conditions. Indian J Exp Biol 55(7):463–468

Srivastava S, Yadav A, Seem K, Mishra S, Chaudhary V, Nautiyal CS (2008) Effect of high temperature on *Pseudomonas putida* NBRI0987 biofilm formation and expression of stress sigma factor RpoS. Curr Microbiol 56(5):453–457

Steffens B, Sauter M (2009) Epidermal cell death in rice is confined to cells with a distinct molecular identity and is mediated by ethylene and H_2O_2 through an autoamplified signal pathway. Plant Cell 21:184–196

Stepien P, Kłbus G (2006) Water relations and photosynthesis in *Cucumis sativus* L. leaves under salt stress. Biol Plant 50(4):610. https://doi.org/10.1007/s10535-006-0096-z

Subramanian P, Mageswari A, Kim K, Lee Y, Sa T (2015) Psychrotolerant endophytic *Pseudomonas* sp. strains OB155 and OS261 induced chilling resistance in tomato plants (*Solanum Lycopersicum* Mill.) by activation of their antioxidant capacity. Mol Plant Microbe Interact 28:1073–1081

Subramanian P, Kim K, Krishnamoorthy R, Mageswari A, Selvakumar G, Sa T (2016) Cold stress tolerance in psychrotolerant soil bacteria and their conferred chilling resistance in tomato (*Solanum lycopersicum* Mill.) under low temperatures. PLoS One 11(8):e0161592. https://doi.org/10.1371/journal.pone.0161592

Sun X, Griffith M, Pasternak JJ, Glick BR (1995) Low temperature growth, freezing survival, and production of antifreeze protein by the plant growth promoting rhizobacterium *Pseudomonas putida* GR12-2. Can J Microbiol 41(9):776–784

Sun Z, Yang LM, Han M, Han ZM, Yang L, Cheng L, Yang X, Lv ZL (2019) Biological control ginseng grey mold and plant colonization by antagonistic bacteria isolated from rhizospheric soil of *Panax ginseng* Meyer. Biol Control 138:104048

Suzuki N, Bajad S, Shuman J, Shulaev V, Mittler R (2008) The transcriptional co-activator MBF1c is a key regulator of thermotolerance in Arabidopsis thaliana. J Biol Chem 283(14):9269–9275

Suzuki N, Rivero RM, Shulaev V, Blumwald E, Mittler R (2014) Abiotic and biotic stress combinations. New Phytol 203:32–43

Sziderics AH, Rasche F, Trognitz F, Sessitsch A, Wilhelm E (2007) Bacterial endophytes contribute to abiotic stress adaptation in pepper plants (*Capsicum annuum* L.). Can J Microbiol 53: 1195–1202

Taj Z, Challabathula D (2021) Protection of photosynthesis by halotolerant Staphylococcus Sciuri ET101 in tomato (*Lycoperiscon esculentum*) and rice (*Oryza sativa*) plants during salinity stress: Possible interplay between carboxylation and oxygenation in stress mitigation. Front Microbiol 11:547750. https://doi.org/10.3389/fmicb.2020.547750

Talaat NB, Shawky BT (2017) Microbe-mediated induced abiotic stress tolerance responses in plants. In: Singh DP, Singh HB, Prabha R (eds) Plant-microbe interactions in agro-ecological perspectives. Springer, Singapore, pp 101–133

Tanveer M, Ahmed HAI (2020) ROS signalling in modulating salinity stress tolerance in plants. In: Hasanuzzaman M, Tanveer M (eds) Salt and drought stress tolerance in plants, signaling networks and adaptive mechanisms. Springer, Cham, pp 299–314

Tapia-Vázquez I, Sánchez-Cruz R, Arroyo-Domínguez M, Lira-Ruan V, Sánchez-Reyes A, del Rayo S-CM, Padilla-Chacón D, Batista-García RA, Folch-Mallol JL (2020) Isolation and characterization of psychrophilic and psychrotolerant plant-growth promoting microorganisms from a high-altitude volcano crater in Mexico. Microbiol Res 232:126394

Thakare M, Sarma H, Datar S, Roy A, Pawar P, Gupta K, Pandit S, Prasad R (2021) Understanding the holistic approach to plant-microbe remediation technologies for removing heavy metals and radionuclides from soil. Curr Res Biotechnol 3:84–98. https://doi.org/10.1016/j.crbiot.2021.02.004

Thakur M, Sharma P, Anand A (2019) Seed priming-induced early vigor in crops: an alternate strategy for abiotic stress tolerance. In: Hasanuzzaman M, Fotopoulos V (eds) Priming and pretreatment of seeds and seedlings: implication in plant stress tolerance and enhancing productivity in crop plants. Springer, Singapore, pp 163–180

Theocharis A, Bordiec S, Fernandez O, Paquis S, Dhondt-Cordelier S, Baillieul F, Clément C, Barka EA (2012) *Burkholderia phytofirmans* PsJN primes *Vitis vinifera* L. and confers a better tolerance to low nonfreezing temperatures. Mol Plant Microbe Interact 25(2):241–249

Timmusk S, Nevo E (2011) Plant root associated biofilms. In: Maheshwari DK (ed) Bacteria in agrobiology: plant nutrient management, vol 3. Springer, Verlag, Berlin, pp 285–300

Tiryaki D, Ihsan A, Okkes A (2019) Psychrotolerant bacteria isolated from the leaf apoplast of coldadapted wild plants improve the cold resistance of bean (*Phaseolus vulgaris* L.) under low temperature. Cryobiology 86:111–119

Trivedi P, Sa T (2008) *Pseudomonas corrugate* (NRRL B-30409) mutants increased phosphate solubilisation, organic acid production, and plant growth at lower temperatures. Curr Microbiol 56:140–144

Turan M, Gulluce M, Cakmak R, Sahin F (2013) Effect of plant growth-promoting rhizobacteria strain on freezing injury and antioxidant enzyme activity of wheat and barley. J Plant Nutr 36: 731–748

Upadhyay SK, Singh DP (2015) Effect of salt-tolerant plant growthpromoting rhizobacteria on wheat plants and soil health in a saline environment. Plant Biol 17:288–293

Upadhyay SK, Singh JS, Singh DP (2011) Exopolysaccharide-producing plant growth-promoting rhizobacteria under salinity condition. Pedosphere 21:214–222

Vega-Celedón P, Bravo G, Velásquez A, Cid FP, Valenzuela M, Ramírez I, Vasconez IN, Álvarez I, Jorquera MA, Seeger M (2021) Microbial diversity of psychrotolerant bacteria isolated from wild flora of andes mountains and patagonia of chile towards the selection of plant growth-promoting bacterial consortia to alleviate cold stress in plants. Microorganisms 9(3):538

Verma P, Yadav AN, Khannam KS, Panjiar N, Kumar S, Saxena AK, Suman A (2015) Assessment of genetic diversity and plant growth promoting attributes of psychrotolerant bacteria allied with wheat (*Triticum aestivum*) from the northern hills zone of India. Ann Microbiol 65(4): 1885–1899

Vessey JK (2003) Plant growth promoting rhizobacteria as biofertilizers. Plant and Soil 255(2): 571–586

Vidoz ML, Loreti E, Mensuali A, Alpi A, Perata P (2010) Hormonal interplay during adventitious root formation in flooded tomato plants. Plant J 63:551–562

Vijayakumari K, Jisha KC, Puthur JT (2016) GABA/BABA priming: a means for enhancing abiotic stress tolerance potential of plants with less energy investments on defence cache. Acta Physiol Plant 38(9):230. https://doi.org/10.1007/s11738-016-2254-z

Vinocur B, Altman A (2005) Recent advances in engineering plant tolerance to abiotic stress: achievements and limitations. Curr Opin Biotechnol 16:123–132

Voesenek LA, Benschop JJ, Bou J, Cox MC, Groeneveld HW, Millenaar FF, Vreeburg RA, Peeters AJ (2003) Interactions between plant hormones regulate submergence-induced shoot elongation in the flooding-tolerant dicot *Rumex palustris*. Ann Bot 91:205–211

Wahid A, Sehar S, Perveen M, Gelani S, Basra SM, Farooq M (2008) Seed pretreatment with hydrogen peroxide improves heat tolerance in maize at germination and seedling growth stages. Seed Sci Technol 36(3):633–645

Wang W, Vinocur B, Shoseyov O, Altman A (2004) Role of plant heat-shock proteins and molecular chaperones in the abiotic stress response. Trends Plant Sci 9(5):244–252

Wang Q, Dodd IC, Belimov AA, Jiang F (2016) Rhizosphere bacteria containing 1-aminocyclopropane-1-carboxylate deaminase increase growth and photosynthesis of pea plants under salt stress by limiting Na^+ accumulation. Funct Plant Biol 43(2):161–172

Wang X, Xiao C, Ji C, Liu Z, Song X, Liu Y, Li C, Yan D, Li H, Qin Y, Liu X (2021) Isolation and characterization of endophytic bacteria for controlling root rot disease of Chinese jujube. J Appl Microbiol 130(3):926–936

Wei Y, Zhu B, Liu W, Cheng X, Lin D, He C, Shi H (2021) Heat shock protein 90 co-chaperone modules fine-tune the antagonistic interaction between salicylic acid and auxin biosynthesis in cassava. Cell Rep 34(5):108717

Welti R, Li W, Li M, Sang Y, Biesiada H, Zhou HE, Rajashekar CB, Williams TD, Wang (2002) Profiling membrane lipids in plant stress responses role of phospholipase Dα in freezing-induced lipid changes in Arabidopsis. J Biol Chem 277(35):31994–32002

Woo OG, Kim H, Kim JS, Keum HL, Lee KC, Sul WJ, Lee JH (2020) *Bacillus subtilis* strain GOT9 confers enhanced tolerance to drought and salt stresses in *Arabidopsis thaliana* and *Brassica campestris*. Plant Physiol Biochem 148:359–367

Wu YS, Yang CY (2019) Ethylene-mediated signaling confers thermotolerance and regulates transcript levels of heat shock factors in rice seedlings under heat stress. Bot Stud 60(1):1–12

Wu ZM, Yang Y, Li KT (2019) Antagonistic activity of a novel antifungalmycin N2 from *Streptomyces* sp. N2 and its biocontrol efficacy against *Rhizoctonia solani*. FEMS Microbiol Lett 366(3):fnz018

Wu YM, Chen X, Wang F, Hsiao CY, Yang CY, Lin ST, Wu LH, Chen YK, Liang YS, Lin YH (2021) *Bacillus amyloliquefaciens* strains control strawberry anthracnose through antagonistic activity and plant immune response intensification. Biol Control 157:104592

Xia Y, Farooq MA, Javed MT, Kamran MA, Mukhtar T, Ali J, Tabassum T, ur Rehman S, Munis MF, Sultan T, Chaudhary HJ (2020) Multi-stress tolerant PGPR *Bacillus xiamenensis* PM14 activating sugarcane (*Saccharum officinarum* L.) red rot disease resistance. Plant Physiol Biochem 151:640–649

Xie HT, Wan ZY, Li S, Zhang Y (2014) Spatiotemporal production of reactive oxygen species by NADPH oxidase is critical for tapetal programmed cell death and pollen development in Arabidopsis. Plant Cell 26:2007–2023

Yadav AN, Singh RN, Sachan SG, Kaushik R (2015) Prospecting cold deserts of north western Himalayas for microbial diversity and plant growth promoting attributes. J Biosci Bioeng 119(6):683–693

Yamauchi T, Rajhi I, Nakazono M (2011) Lysigenous aerenchyma formation in maize root is confined to cortical cells by regulation of genes related to generation and scavenging of reactive oxygen species. Plant Signal Behav 6:759–761

Yan J, Smith MD, Glick BR, Liang Y (2014) Effects of ACC deaminase containing rhizobacteria on plant growth and expression of Toc GTPases in tomato (*Solanum lycopersicum*) under salt stress. Botany 92:775–781

Yang J, Kloepper JW, Ryu CM (2009) Rhizosphere bacteria help plants tolerate abiotic stress. Trends Plant Sci 14:1–4

Yang F, Zhang R, Wu X, Xu T, Ahmad S, Zhang X, Zhao J, Liu Y (2020) An endophytic strain of the genus *Bacillus* isolated from the seeds of maize (*Zea mays* L.) has antagonistic activity against maize pathogenic strains. Microb Pathog 142:104074

Yaseen R, Zafar-ul-Hye M, Hussain M (2019) Integrated application of ACC-deaminase containing plant growth promoting rhizobacteria and biogas slurry improves the growth and productivity of wheat under drought stress. Int J Agric Biol 21:869–878

Yaseen R, Aziz O, Saleem MH, Riaz M, Zafar-ul-Hye M, Rehman M, Ali S, Rizwan M, Nasser Alyemeni M, El-Serehy HA, Al-Misned FA (2020) Ameliorating the drought stress for wheat growth through application of ACC-deaminase containing rhizobacteria along with biogas slurry. Sustainability 12(15):6022

Zafar-ul-Hye M, Nasir A, Aon M, Hussain S, Ahmad M, Naz I (2018) Seed inoculation with *Pseudomonas fluorescens* and *Pseudomonas syringae* enhanced maize growth in a compacted saline-sodic soil. Phyton 87:25–31

Zafar-ul-Hye M, Danish S, Abbas M, Ahmad M, Munir TM (2019) ACC Deaminase producing PGPR *Bacillus amyloliquefaciens* and *Agrobacterium fabrum* along with biochar improve wheat productivity under drought stress. Agronomy 9:343. https://doi.org/10.3390/agronomy9070343

Zahir ZA, Munir A, Asghar HN, Shahroona B, Arshad M (2008) Effectiveness of rhizobacteria containing ACC-deaminase for growth promotion of peas (*P. sativum*) under drought conditions. J Microbiol Biotechnol 18:958–963

Zakry FA, Shamsuddin ZH, Rahim KA, Zakaria ZZ, Rahim AA (2012) Inoculation of *Bacillus sphaericus* UPMB-10 to young oil palm and measurement of its uptake of fixed nitrogen using the N isotope dilution technique. Microbial Environ 27:257–262

Zang X, Geng X, Wang F, Liu Z, Zhang L, Zhao Y, Tian X, Ni Z, Yao Y, Xin M, Hu Z (2017) Overexpression of wheat ferritin gene TaFER-5B enhances tolerance to heat stress and other abiotic stresses associated with the ROS scavenging. BMC Plant Biol 17(1):14. https://doi.org/10.1186/s12870-016-0958-2

Zhang CX, Fu GF, Yang XQ, Yang YJ, Zhao X, Chen TT, Zhang XF, Jin QY, Tao LX (2016) Heat stress effects are stronger on spikelets than on flag leaves in rice due to differences in dissipation capacity. J Agron Crop Sci 202(5):394–408

Zhang X, Rong X, Cai M, Meng Q (2019) Collaborative optimization of emissions and abatement costs for air pollutants and greenhouse gases from the perspective of energy structure: an empirical analysis in Tianjin. Sustainability 11:3872. https://doi.org/10.3390/su11143872

Zhou C, Ma Z, Zhu L, Xiao X, Xie Y, Zhu J, Wang J (2016) Rhizobacterial strain *Bacillus megaterium* BOFC15 induces cellular polyamine changes that improve plant growth and drought resistance. Int J Mol Sci 17:976. https://doi.org/10.3390/ijms17060976

Zhu J, Dong CH, Zhu JK (2007) Interplay between cold-responsive gene regulation, metabolism and RNA processing during plant cold acclimation. Curr Opin Plant Biol 10(3):290–295

Zhu XC, Song FB, Liu SQ, Liu TD, Zhou X (2012) Arbuscular mycorrhizae improve photosynthesis and water status of *Zea mays* L. under drought stress. Plant Soil Environ 58(4):186–191

Zicca S, De Bellis P, Masiello M, Saponari M, Saldarelli P, Boscia D, Sisto A (2020) Antagonistic activity of olive endophytic bacteria and of *Bacillus* spp. strains against *Xylella fastidiosa*. Microbiol Res 236:126467

Zubair M, Hanif A, Farzand A, Sheikh TM, Khan AR, Suleman M, Ayaz M, Gao X (2019) Genetic screening and expression analysis of psychrophilic *Bacillus* spp. reveal their potential to alleviate cold stress and modulate phytohormones in wheat. Microorganisms 7(9):337

Zvinavashe AT, Mardad I, Mhada M, Kouisni L, Marelli B (2021) Engineering the plant microenvironment to facilitate plant-growth-promoting microbe association. J Agric Food Chem. https://doi.org/10.1021/acs.jafc.1c00138

Endophytic Seed Bacteria: A Relevant Pool of Microorganisms with the Ability to Promote Plant Growth

3

M. L'Hoir, B. Nasslahsen, H. Ferhout, Y. Prin, A. Smouni, V. Walker, and R. Duponnois

Abstract

According to the Food and Agricultural Organization (FAO), "global agricultural production needs to increase by 70% and double in developing countries by 2050 to meet the demand of 9 billion people". The major challenge all countries are

M. L'Hoir and B. Nasslahsen contributed equally to this work.

M. L'Hoir
IRD, Laboratoire des Symbioses Tropicales & Méditerranéennes, Campus International de Baillarguet TA A-82/J, Montpellier Cedex 5, France

Ovinalp Fertilisation, Quartier de la tour du Puy, Ribiers, France

B. Nasslahsen
IRD, Laboratoire des Symbioses Tropicales & Méditerranéennes, Campus International de Baillarguet TA A-82/J, Montpellier Cedex 5, France

Agro-nutrition, Carbonne, France

H. Ferhout
Agro-nutrition, Carbonne, France

Y. Prin
CIRAD, Laboratoire des Symbioses Tropicales & Méditerranéennes, Campus International de Baillarguet TA A-82/J, Montpellier Cedex 5, France

A. Smouni
Laboratoire de Physiologie et Biotechnologie Végétale, Faculté Des Sciences de Rabat, Université Mohammed V, Rabat, Morocco

V. Walker
Ovinalp Fertilisation, Quartier de la tour du Puy, Ribiers, France

R. Duponnois (✉)
IRD, Laboratoire des Symbioses Tropicales & Méditerranéennes, Campus International de Baillarguet TA A-82/J, Montpellier Cedex 5, France
e-mail: robin.duponnois@ird.fr

N. K. Arora, B. Bouizgarne (eds.), *Microbial BioTechnology for Sustainable Agriculture Volume 1*, Microorganisms for Sustainability 33,
https://doi.org/10.1007/978-981-16-4843-4_3

currently facing in the world is food security. It is connected to two inseparable components, namely, land and water. In a context of agro-ecological transition, many actors including those of research are mobilized to develop sustainable solutions. These include the use of biofertilizers of microbial origin. Microbial biofertilizers are a sustainable alternative to currently widely used inputs. In contrast to these, biofertilizers ensure plant productivity and conservation of water, soil and biodiversity resources. Moreover, their use does not impact human health. Plants can be described as complex micro-ecosystems to which the different habitats are exploited by a wide variety of bacteria. Endophytes are non-pathogenic organisms that perform part or all of their life cycle within plant tissues. Endophytic bacteria are likely to have beneficial effects similar to the known effects of plant growth-promoting rhizobacteria (PGPR). A number of studies evoke the interest of endophytic bacteria, but only a few bring up the topic of seed endophytic bacteria. Hence, seeds are essential tools for agricultural production, as well as potential supports for endophytic bacteria. This chapter aims to highlight the potentialities of the seed-borne endophytic bacteria for agricultural use due to their remarkable abilities (survival in extreme conditions, motility, competitive advantage, vertical transmission, contribution to germination, growth-promoting effects, biocontrol abilities). The examples mentioned for the different kind of cultures (market gardening, big cultures, forage crops, niche market cultures) demonstrate the relevance of agricultural purposes.

Keywords

Endophytic bacteria · Biofertilizers · Biopesticides · Agro-ecology · PGPR

3.1 Introduction

According to the data given by the Food and Agricultural Organization (FAO), an increase of the world agricultural production is a must to ensure nine billion people nutrition (FAO 2021). The major challenge all countries are currently facing in the world is food security. It is connected to two inseparable components, namely, land and water. This may seem alarmist, but what will happen in 2050, in a context of competition in food resources and water, population pressure and climate change? The increase in productivity per hectare and per worker in the context of the Green Revolution has certainly been effective, but this increase has reached limits. For about 60 years, the production methods have mutated according to the following criteria: (1) a varietal selection of high-yielding field crop plants, (2) mono-specific crops to optimize the productivity of agrosystems, (3) systematic use of chemical fertilizers and pesticides, and (4) massive use of water resources in irrigation systems. Nevertheless, it is associated with agrosystems whose limits are recognized (Suman et al. 2016). These cultural practices have been responsible for numerous degradations in the biological mechanisms governing the productivity and stability of agroecosystems and the services associated with them. Beyond the environmental impacts, there are technical problems (Aktar et al. 2009). In relation to intensive agricultural systems, phenomena of resistance and soil fertility decline have developed: soils are often deficient in macro- and micro-nutrients and are also reservoirs of phytopathogenic microorganisms such as those belonging to the genera *Fusarium*, *Pythium* and *Phytophthora* (Dixon and Tilston 2010; Weil and Brady 2016) calling into question the sustainability of these systems. In addition, the socio-economic impacts are blatant: the issue of the health of farmers and consumers is threatened (Douglas et al. 2018), and the farmers' malaise is a proven fact (Daghagh Yazd et al. 2019). It is imperative to change current agricultural systems in a sustainable way: as stated in the 2011 FAO Synthesis Report, agriculture must be "the backbone of economic development, provide environmental services and finally be at the heart of the fight against rural poverty". In order to meet this challenge, numerous initiatives have been undertaken to define farming itineraries combining respect for the environment and productivity compatible with the food needs of the world's population, in a context of global change. Among the "tools" likely to satisfy the rules of modern agriculture is the valorization of microbial resources associated with plants that are involved in the mineral nutrition of plant species and which facilitate their development in soils deficient in different macro-elements (i.e. nitrogen and phosphate). In context of agro-ecological transition (FAO 2015), many actors including those of research are mobilized to develop sustainable solutions. These include the use of biofertilizers of microbial origin (Glick 2014). These correspond to living microorganisms that have beneficial effects on the plant species with which they interact (Suman et al. 2016). Their effects may relate to improved plant nutrition and resistance to biotic and abiotic stresses (Johnson et al. 2005; Lugtenberg and Kamilov 2009). Microbial biofertilizers are a sustainable alternative to currently widely used inputs (Bhardwaj et al. 2014). In contrast to these, biofertilizers ensure plant productivity and conservation of water, soil and biodiversity resources. Moreover, their use does not impact human health. Despite numerous scientific studies showing the potential of these microorganisms, their application in the field remains rare. Indeed, biofertilizer formulations are not controlled and do not guarantee sufficient development conditions for microorganisms (Dobbelaere et al. 2001). The result is (1) a short life span of biofertilizers, (2) poor survival of the microorganisms, and (3) a high degree of contamination (Suman et al. 2015). This is due in particular to the limits of the forms of inoculation, namely, (1) the stresses of biotic origin (competition with soil-dependent microorganisms occupying the soil ecological niche) and (2) stress of abiotic origin (temperature variations, water conditions, pollution, soil nutrient status, soil nature, soil porosity) (Dobbelaere et al. 2001). In this context, current research focuses on alternative inoculation methods that would improve the viability of microorganisms (Suman et al. 2016). In this respect, the potential of endophytic bacteria is of growing interest (Johnston-Monje and Raizada 2011). Plants can be described as complex micro-ecosystems to which the different habitats are exploited by a wide variety of microorganisms (McInroy and Kloepper 1991). These habitats are not limited to the outer surfaces of plants on which most epiphytic microbes reside but also include internal tissues in which multiple microorganisms have the ability to penetrate and survive (Lodewyckx et al. 2002). Within these micro-ecosystems, microbial species including bacteria and fungi (Fisher et al. 1992) are able to interact and establish a balance. Endophytes are non-pathogenic organisms that perform part or all of their life cycle within plant tissues (Rosenblueth and Martínez-Romero 2006). Endophytic bacteria are likely to have beneficial effects similar to the known effects of plant growth-promoting rhizobacteria (PGPR) (Kloepper et al. 1991; Höflich et al. 1994). Bacterial endophytes are beneficial for their host plant: Cankar et al. (2005) present these sorts of microorganisms as potential biofertilizers to enable sustainable and renewable crop production. Indeed, endophyte have the ability to enhance plant growth and protect them from abiotic and biotic stresses. A number of studies evoke the interest of endophytic bacteria, but only a few bring up the topic of seed endophytic bacteria. Seeds are essential tools for agricultural production, as well as potential supports for endophytic bacteria (Xu et al. 2014). Since these plant organs are involved in different stages of the plant reproductive process, these habitats potentially represent reservoirs of microbial biodiversity that can promote the development of their host plant (Kristin and Miranda 2013; Truyens et al. 2015). Colonization from within the grain is advantageous for endophytic bacteria. In fact, endophytic seed bacteria are not subject to competition; they are linked to the large number of bacteria around the host plant (Rosenblueth and Martínez-Romero 2006). In addition, endophytic bacteria are protected against stresses and other rhizosphere bacteria (Hallmann et al. 1997). This opens up new avenues for the introduction of microorganisms into seeds, to make biofertilizers more effective (Khalaf and Raizada 2018). In close interaction with plants, endemic bacteria evolve in the optimal development environment. Thus, the development of inoculation of these microbes seems conceivable to overcome the technical constraints related to biofertilizers (Karthik et al. 2017) and a real commercial and scientific infatuation.

3.2 Microbial Resources: Plant Localization and Functions

3.2.1 Context

Nearly all eukaryotic organisms have the ability to develop interactions with minute life forms that are termed "microorganisms" (Taylor et al. 2005; Berg et al. 2014). A wide range of these microorganisms can conquer some of the external and internal parts of both plants and animals (Moënne-Loccoz et al. 2015). These tight interactions played an important role in the formation of plant and animal kingdom due to their ability to control and modulate trophic networks and biogeochemical cycles, regulate ecosystem productivity and determine the ecology and health of plant and animal partners (Hossain et al. 2017). Plants are constantly in contact with microbes through water via water cycle or through wind, soil and air (Rastogi et al. 2012); these small-size entities play an important role in their performance and survival (Berg et al. 2014) and have the ability to colonize a large range of plant tissues and compartments (Hossain et al. 2017). They can promote the plant growth by controlling a vast array of host plant hormones (Glick 2014; Bilal et al. 2018). Hence, some recent researches have proven that symbiotic microorganisms including bacteria and fungi can impact plant growth and plant tolerance to different abiotic and biotic stresses and manipulation of the plant microbiome has the potential to improve the yield of important crops (Rolli et al. 2015; Owen et al. 2015). These capacities are driven by their abilities to enhance nutrient uptake and phytohormone production as well as to reprogram plant gene expression, through differential activation of plant signalling pathways (Owen et al. 2015; Glick 2014). Different groups of bacteria and fungi interact with higher plants.

This review adopts this broad definition and includes only bacteria.

3.2.2 The Diversity of Plant Microbiota Origin

Plants are in a constant selection of their microbial host from different surrounding microbial reservoirs (Hardoim 2015). Plant bacterial communities are not randomly established; they are controlled by many rules and factors. Plant microbiota is influenced by the soil type, plant immune system, plant development stage, season and host genotype and species (Hassani et al. 2018), and the bacterial microbiota of plants is dominated by three major phyla (*Proteobacteria*, *Actinobacteria*, and *Bacteroidetes*) (Hassani et al. 2018). Plant microbiota has different origins: soil, phyllosphere that is the leaf surface, flowers that are also called the anthosphere, the spermosphere that refers to the external surface of germinated seeds or the carposphere and plant fruit (Hardoim 2015).

3.2.3 Plant Endophytic Bacteria

Several bacteria taxa can penetrate into plant tissues, but many factors control the colonization and the transmission of endophytes in plants. The quota of plant resources and the colonization ability of the bacteria are the most determinant (Compant et al. 2019). A considerable variation in the ecology of endosphere and phyllosphere bacteria is noticed. Most endophytes spread systemically via the xylem to distinct compartments of the plant like stem, leaves and fruits; otherwise they can also penetrate the plant tissues through aerial parts of the plant such as flowers and fruits (Wallace et al. 2018; Compant et al. 2019).

3.2.4 Functions of Endophytes

Plant-associated microbes include beneficial, neutral and pathogenic microorganisms. The plant growth-promoting bacteria (PGPB), which are the beneficial microbes, can improve and promote the plant growth by either producing some phytohormones like auxin, cytokinin and gibberellin (Herrera et al. 2016) or secreting enzymes like ACC deaminase to reduce the ethylene level, the hormone of stress in plants. The PGPB can also improve the nutrient uptake by phosphate solubilization, nitrogen fixation or iron sequestration (Rascovan et al. 2016; Olanrewaju et al. 2017). A few studies also showed that some plant bacteria can have the ability to enhance host resistance against pathogen infection (Herrera et al. 2016; Olanrewaju et al. 2017; De Vrieze et al. 2018).

3.2.4.1 PGPB and Phytohormones

Khan et al. (2011) describe the phytohormones as chemical compounds that are produced in one part and exert an effect in another part influencing the physiological and biochemical processes of the plants (Khan et al. 2011). Phytohormones are critical for plant growth and development; they can regulate the protective responses of plants against both biotic and abiotic stresses (Table 3.1) by means of coordinated synergistic or antagonistic signals controlling downstream stress responses (Schmelz et al. 2003; Khan et al. 2011).

Auxin

Auxins interfere in many important functions such as geotropism, phototropism, vascular tissue differentiation, apical dominance, root initiation, cell division and stem and root elongation (Grobelak et al. 2015). The main auxin in plants that controls many important physiological processes is the indole-3-acetic acid (IAA) (Lüthen et al. 1999; Karadeniz et al. 2006). Contrary to other auxins, such as indole-3-butyric acid (IBA) and phenyl acetic acid (PAA) also identified in plants (Normanly 1997), little is known about their physiological function. The IAA production by microbial isolates varies depending on different species and strains and is controlled by the availability of substrate(s). It plays an important and critical role in plant growth promotion because the bacterial IAA producers have the capacity to input IAA into the plant's auxin pool-generating modifications in the plant tissues (Finnie and Van Staden 1985; Karadeniz et al. 2006). IAA can be synthetized using multiple pathways that differ sometimes in the same organism (Karadeniz et al. 2006).

Cytokinin

Cytokinin is a phytohormone that regulates apical dominance, cell division, root elongation, seed germination, xylem and chloroplast differentiation, flower and fruit development, nutritional signalling, leaf senescence and plant-pathogen interactions (Sakakibara 2006; De Rybel et al. 2016). Cytokinin is produced in the plant roots and transported by translocation through the xylem (De Rybel et al. 2016). Some microorganisms have the capacity to synthetize cytokinins (Arshad and Frankenberger 1991) and supply the soil to raise its cytokinin content and promote plant growth (Arshad and Frankenberger 1991; Spaepen 2015). Some studies also showed that cytokinin can regulate stomatal opening and restrict damaging foliar water loss by controlling stomatal closure (Arkhipova et al. 2005).

Table 3.1 The ability of some phytohormone-producing microorganisms to decrease plant stress responses (Egamberdieva et al. 2017a)

Microorganisms	Phytohormone	Host plant, abiotic stress	References
Pseudomonas sp., *Bacillus* sp.	IAA	*Sulla carnosa* (Desf.), salt stress	Hidri et al. (2016)
Bacillus licheniformis	IAA	*Triticum aestivum* L., salt stress	Singh and Jha (2016)
Bacillus subtilis, *Arthrobacter* sp.	IAA	*Triticum aestivum* L., salt stress	Upadhyay et al. (2012)
Pseudomonas putida, *Bacillus megaterium*	IAA	*Trifolium repens*, drought stress	Marulanda et al. (2009)
Marinobacterium sp., *Pseudomonas* sp., *Rhizobium* sp., *Sinorhizobium* sp.	IAA	*Triticum aestivum* L., salt stress	Sorty et al. (2016)
Serratia plymuthica, *Stenotrophomonas rhizophila*, *Pseudomonas fluorescens*, *Pseudomonas extremorentialis*	IAA	*Cucumis sativus*, salt stress	Egamberdieva et al. (2011)
Acinetobacter faecalis, *Bacillus cereus*, *Enterobacter hormaechei*, *Pantoea agglomerans*	IAA	*Triticum aestivum* L., salt stress	Egamberdieva et al. (2008)
Curtobacterium flaccumfaciens, *Ensifer garamanticus*	IAA	*Hordeum vulgare*, salt stress	Cardinale et al. (2015)
Streptomyces coelicolor, *Streptomyces geysiriensis*	IAA	*Triticum aestivum* L., salt stress	Yandigeri et al. (2012)
Bacillus subtilis	IAA	*Acacia gerrardii* Benth., salt stress	Hashem et al. (2016)
Pseudomonas sp.	IAA	*Zea mays*, salt and heat stresses	Mishra et al. (2017)
Serratia sp.	IAA	*Cicer arietinum* L., nutrient stress	Zaheer et al. (2016)
Achromobacter xylosoxidans	IAA	*Brassica juncea*, Cu stress	Ma et al. (2008)
Pseudomonas putida	IAA	*Glycine max* (L.) Merr., salt stress	Egamberdieva et al. (2017b)
Leifsonia sp., *Bacillus* sp.	IAA	*Zea mays*, Cd stress	Ahmad et al. (2016)

(continued)

Table 3.1 (continued)

Microorganisms	Phytohormone	Host plant, abiotic stress	References
Burkholderia sp.	IAA	*Solanum lycopersicum* L., Cd stress	Dourado et al. (2013)
Bacillus subtilis	IAA	*Brassica juncea* L., Ni stress	Zaidi et al. (2006)
Bacillus megaterium	IAA	*Vinca rosea* L., Ni stress	Khan et al. (2017)
Achromobacter xylosoxidans, *Bacillus pumilus*	SA	*Helianthus annuus*, drought stress	Forchetti et al. (2010)
Serratia marcescens	SA	*Zea mays*, salt stress	Lavania and Nautiyal (2013)
Micrococcus luteus	CK	*Zea mays*, drought stress	Raza and Faisal (2013)
Arthrobacter sp., *Bacillus* sp., *Azospirillum* sp.	CK	*Glycine max* (L.) Merr., salt stress	Naz et al. (2009)
Bacillus subtilis	CK	*Platycladus orientalis*, drought stress	Liu et al. (2013)
Aspergillus fumigatus	GA	*Glycine max* (L.) Merr., salt stress	Khan et al. (2011)
Azospirillum lipoferum	GA	*Triticum aestivum* L., drought stress	Creus et al. (2004)
Phoma glomerata, *Penicillium* sp.	GA	*Cucumis sativus*, drought stress	Waqas et al. (2012)
Bacillus amyloliquefaciens	ABA	*Oryza sativa* L., salt stress	Shahzad et al. (2017)
Bacillus licheniformis, *Pseudomonas fluorescens*	ABA	*Vitis vinifera* L., water stress	Salomon et al. (2014)
Trichoderma asperellum	IAA, GA, ABA	*Cucumis sativus*, salt stress	Zhao and Zhang (2015)
Bacillus aryabhattai	IAA, GA, ABA	*Glycine max* (L.) Merr., heat stress	Park et al. (2017)

Gibberellin

Gibberellin is a phytohormone, which is able to stimulate growth and activate an important growth process that includes stem elongation, seed germination, flowering, fruit setting that improves photosynthesis rate and chlorophyll content

(Hedden and Thomas 2012; Halo et al. 2015; Olanrewaju et al. 2017). Genera of bacteria that produce GAs are *Achromobacter xylosoxidens*, *Gluconacetobacter diazotrophicus*, *Acinetobacter calcoaceticus*, *Rhizobia*, *Azotobacter* spp., *Bacillus* spp., *Herbaspirillum seropedicae* and *Azospirillum* spp. (Tsukanova et al. 2017). Gibberellins can induce shoot growth and development and also root growth, root hair abundance, inhibition of floral bud differentiation in woody angiosperms, regulation of vegetative and reproductive bud dormancy and delay of senescence in many organs of a range of plant species (Nelson and Steber 2018).

3.2.4.2 Enzyme's Secretion

ACC Deaminase

All plants, in different development stages, are affected by ethylene. The synthesis of ethylene is usually affected by the concentration of different phytohormones, temperature, nutrient availability, light and the presence or not of any biotic or abiotic stress (Gamalero and Glick 2015). The presence of a high concentration of ethylene in plants is a response to different stresses such as the presence of metals, high temperature, water deficit or excess and pathogen presence (Ali et al. 2014). Bacteria expressing 1-aminocyclopropane-1-carboxylic acid (ACC) deaminase have the capacity to lower the impact on plants of different stresses of biotic and abiotic kinds and enhance plant tolerance to the stresses. These PGPB reduce the stress by lowering the concentration of ethylene via hydrolysing ACC by the enzyme ACC deaminase (Gamalero and Glick 2015; Raghuwanshi and Prasad 2018). A lot of recent studies proved that the inoculation of plants with ACC deaminase producing PGPR can provide a higher resistance to different biotic stresses and abiotic stresses such as salinity, drought and flood against various pathogens (Mallick et al. 2018; Saikia et al. 2018; Gupta and Pandey 2019).

3.2.4.3 PGPB and Nutrient Uptake

Nitrogen Fixation

To ensure a good nutrition, plants require generally 16 nutrients (C, H, O, N, P, K, Ca, Mg, S, B, Cl, Cu, Fe, Mo, Mn and Z) that are taken up either by atmosphere or from soil in different concentrations (Marschner 2011). The most vital components such as protein or nucleic acids are composed of nitrogen (Scherer-Lorenzen et al. 2008). But the major problem is that more than 78% of the atmospheric nitrogen is not used by plants due to the non-assimilable form which is dinitrogen (N_2). To be used and available for plants, it must be converted into ammonia. Plants do not have such mechanism in their cells (Haynes et al. 1993). Some bacterial strains have the ability to fix nitrogen and convert it into ammonia. Two types of nitrogen-fixing bacteria are available; the symbiotic and free-living bacteria (Saharan and Nehra 2011). The symbiotic bacteria can be *Rhizobium*. They are strictly bound to the leguminous type of plants and attracted towards plant roots. They form nodules to fix up to 80% of needed nitrogen; the free-living bacteria like *Cyanobacteria* or

Azoarcus (Graham and Vance 2000) survive on plant residues and are able to fix 20% of needed nitrogen (Dash et al. 2015).

Phosphorous Solubilization

Among the essential nutrients taken up from soil, is phosphorus (Lavania and Nautiyal 2013). Due to its insolubility, the availability of phosphorous depends on the quality of soil (Yadav and Tarafdar 2001). Many researchers found that due to the presence of various beneficial bacterial species in soil, hydroxyapatite, di-calcium phosphate, rock phosphate and tricalcium phosphate are converted into soluble forms of phosphorous (Chen et al. 2006). These microbes use different mechanisms to solubilize phosphate, including acid production and enzyme release, allowing plant growth enhancement. Some of the important phosphate-solubilizing bacterial genera are *Rhizobium*, *Acinetobacter*, *Pseudomonas*, *Arthrobacter*, *Burkholderia*, *Enterobacter* and *Bacillus* (Mehmood et al. 2018).

Iron Sequestration

One of the foremost important nutrients for plant growth is iron, due to the role it plays in several plant development mechanisms. Ferric ions are the iron forms taken up by plants from soil through transport mechanisms. PGPB have the capability to produce siderophores to compete and attain ferric ions (Fe^{3+}) from the surroundings under iron scarcity. Siderophores are small iron-chelating compounds that are released by some beneficial microorganisms. They significantly enhance plant biomass and chlorophyll content (Gull and Hafeez 2012; Mehmood et al. 2018).

3.2.4.4 PGPB and Stress Tolerance

Agricultural crops are exposed to many abiotic and biotic stresses that reduce crop yield (Atkinson and Urwin 2012). Different pathogenic microorganisms such as fungi, bacteria, viruses and nematodes attack plants. They induce biotic stresses, while abiotic stresses are more induced by drought, extreme temperatures, air pollution, salinity and moisture (Mehmood et al. 2018).

To protect their development and under stress conditions, plants recourse to different physiological and metabolic changes (Han and Lee 2005). Plant inoculation with PGPB eases their stress; these microorganisms not only enhance their ability to uptake nutrients from soil, but they can also enhance their ability to tolerate heavy metals or extreme weather conditions (Khan 2005; Rojas-Tapias et al. 2012; Majeed et al. 2018). Many studies have shown the positive action of the PGPB under biotic and abiotic stresses. For example, soybean crop is sensitive to soil salinity, nutrient and water uptake, nodule formation and ultimately yield which is reduced under salt-stressed conditions (Han and Lee 2005). The ability to tolerate heavy metals makes Pseudomonads good candidates under field conditions for phytoremediation (Chacko et al. 2009). Drought stress tolerance in crops has been enhanced by PGPR (Kim et al. 2012; Vurukonda et al. 2016). PGPB showing ACC deaminase activity enhance growth and salt tolerance of red pepper seedlings (*Capsicum annuum* L.) by regulating stress ethylene synthesis (Siddikee et al. 2011).

3.2.5 Localization of Endophytic Bacteria in Different Plant Organs

3.2.5.1 Leaf's Endophytes and Their Potential Use as PGP

Bacteria have the ability to colonize different plant organs and tissues including plant leaves. These microorganisms can improve the plant growth and enhance its development. Many studies demonstrated the potential of using leaves' endophytic bacteria as biofertilizers due to their ability to promote the plant growth. Gagne-Bourgue et al. (2013) isolated and characterized indigenous endophytic bacteria associated with leaves of switchgrass (*Panicum virgatum* L.) cultivars; the study showed wide bacterial diversity with a high plant growth potential. Another study in 2010 by Bhore et al. described the isolation and identification of the bacterial endophytes from the leaves of Sambung Nyawa [*Gynura procumbens* (Lour.) Merr.] and their ability to enhance plant development due to production of cytokinin-like compounds. Leaves' endophytic bacterial strains can also act as biocontrol agents, conferring a protection against plant pathogens. This was demonstrated in a study in 2006 by Shiomi et al. where the suppression of plant diseases due to the action of endophytic microorganisms isolated from leaves has been proven. This experiment under controlled conditions involved endophytic bacteria isolated from leaves and branches of *Coffea arabica* L. and *Coffea robusta* L. The endophytic bacterial isolates tested proved to be effective in inhibiting spore germination and/or rust development, with values above 50%. The endophytic isolates were identified as *Bacillus lentimorbus* Dutky and *Bacillus cereus* (Shiomi et al. 2006).

3.2.5.2 Root Endophytic Bacteria Potentialities

The variation of endophytic bacterial communities depends on many criteria such as plant genotypes, field conditions, plant nutrient status, soil type and cultivation practices. Endophytic bacteria colonize different plant compartments and can be detected in different plant tissues like the root cortex, phloem and xylem. Root emergence sites can be considered as the perfect spot allowing the bacterial colonization. The emergence of the lateral roots forms an entrance to the epidermis, cortex and pericycle allowing the bacteria to penetrate the plant cells and be transported to different plant organs such as leaves and roots (Sprent and De Faria 1989; Agarwhal and Shende 1987), because root cracks allow the leakage of plant metabolites, they become sites that attract bacteria (Hallmann et al. 1997). Many studies suggested that root hairs play an important and decisive role in plant root colonization by beneficial bacteria (Rincón et al. 2005; Mattos et al. 2008), but lateral roots can also be an entrance for the bacteria (Huang 1986). To succeed the plant colonization, bacteria must have the motility and chemotaxis and be able to produce cell wall-degrading products and lipopolysaccharides. The importance of motility and polysaccharide production to facilitate the roots' bacterial penetration has been proven in a study including bacterial species like *Alcaligenes faecalis* and *Azospirillum brasilense* (Bashan and Holguin 1993; You et al. 1995) facilitating the penetration of those microbes. Chemotaxis is the key step of bacterial colonization of roots of several plant species (Zhulin et al. 1988; Turnbull et al. 2001). Chemotaxis and motility play many important roles in the symbiotic interactions of rhizobia with their hosts

(Currier and Strobel 1977). Motility and chemotaxis are two important characteristics of PGPR and play an important role in the interaction process (Gupta Sood 2003). Bacterial root colonization has so many beneficial effects on plant development and growth (Kloepper et al. 1989) using different mechanisms (Etesami and Maheshwari 2018). These bacteria colonize the internal plant tissues without any negative influence on their host. It has been proven in different researches that transplanting plant species in the absence of bacteria is notoriously difficult (Leifert et al. 1989; Hardoim et al. 2008), demonstrating the role of bacteria in plant growth under stressful conditions.

3.2.5.3 Root Endophytic Bacterial Diversity and Their Potential as PGP

The endophytic bacteria colonizing the plant's roots have a very wide range of diversity, and many studies showed this variousness. Li et al. (2010) studied the community structure and diversity of endophytic bacteria in reed (*Phragmites australis*) roots growing in the Beijing Cuihu Wetland, China, using the 16S rRNA library technique. They found that *Pleomorphomonas*, *Azospirillum*, *Aeromonas*, *Dechloromonas*, *Desulfovibrio* and *Sulfurospirillum* were the dominant bacteria genera. These bacteria have important roles improving the plant metabolism. Therefore, this study shows that within the various bacterial communities found in reed roots, endophytic strains could have a great potential for improving phytotherapy on reed wetlands (Li et al. 2010). In 1998, the diversity of bacteria associated with the roots of canola plants grown at three field locations in Saskatchewan, Canada, was investigated by Germida et al. (1998). They selected and identified randomly over 520 bacteria including 220 root endophytic bacteria. Four genera were identified: *Bacillus*, *Flavobacterium*, *Micrococcus* and *Rathayibacter* (Germida et al. 1998). Han et al. (2009) showed that the population diversity of culturable bacteria is abundant in the root domains of moso bamboo plants and that obvious differences exist among the rhizospheric, rhizoplane and endophytic bacterial communities. Abedinzadeh et al. (2019) characterize in their study a culturable rhizosphere and endophytic bacterial isolated from both rhizosphere soil and roots of maize plant with two types of irrigation: industrial and municipal wastewater. The results illustrated that both rhizosphere isolates and endophytic ones had various plant growth-promoting characteristics in terms of both the number and the production amount of these characteristics. A substantial number of the bacterial isolates (both endophytic isolates and rhizosphere isolates) were tolerant to heavy metals (multi-metal-resistant bacteria). Both endophytic isolates and rhizosphere ones showed remarkable resistance to salinity and tolerance to heavy metal presence. Another study in 2008, by Sheng and his team, showed two lead (Pb)-resistant endophytic bacteria from rape roots grown in heavy metal-contaminated soils. They found that during assays an increase in root elongation of inoculated rape seedlings was compared to the control plants. Strain G16 produced IAA, siderophores and ACC deaminase. Increases in biomass production and total Pb uptake in the bacteria-inoculated plants were obtained compared to the control. The two strains could colonize the root interior and rhizosphere soil of rape after root inoculation. In another study, Cherif et al. (2015) found that date palm roots included endophytic communities that have the capacity to promote plant growth under drought conditions. Endophytic bacterial flora isolated from root and stem tissues of black pepper (*Piper nigrum* L.) was identified as an effective antagonistic endophyte for biological control of *Phytophthora* foot rot in black pepper (Aravind et al. (2009). To deal with soil contamination, 11 cadmium-tolerant bacterial strains were isolated from the root zone of Indian mustard (*Brassica juncea* L. Czern.) seedlings grown in Cd-supplemented soils. The bacteria isolated increased the plant tolerance to other metals including Zn, Cu, Ni and Co. The isolated strains including *Variovorax paradoxus*, *Rhodococcus* sp. and *Flavobacterium* sp. were able to stimulate root elongation of *B. juncea* seedlings in the presence and absence of toxic Cd concentrations. Some of the strains have also the capacity to produce auxins or siderophores. The isolated bacteria can be used as inoculants to improve plant growth under the presence of toxic Cd concentrations and for the development of plant-inoculant systems useful for phytoremediation of polluted soils (Belimov et al. 2005). Most root microbiota is horizontally transferred, and microbiome is derived from the soil which contains a diverse range of microorganisms (Fierer 2017). However, bacteria can also be transmitted vertically via seeds. Seeds represent an important source of microorganisms that proliferate in the plant root system (Liu et al. 2012). In this regard, seed endophytic bacteria constitute a potential vector of PGPB (Bergna et al. 2018).

3.3 Potentialities of Seed Endophytic Bacteria

The growing interest for seed endophytic bacteria is due to their intrinsic characteristics. Indeed, seed endophytic bacteria are not typical: they have their own abilities and differ from endophytes from other parts of the plants (Compant et al. 2010; Truyens et al. 2015; Lopez et al. 2018).

3.3.1 Seed Endophytic Bacteria Can Withstand the Conditions of the Particular Ecological Niche: Seed

Seed endophytic bacteria are able to survive in peculiar conditions: seeds are not a stable and rich ecological niche. During seed maturation, there is an accumulation of starch and a drastic decline in water content. The consequence is that they have to deal with dehydration and high osmotic pressure (Ebeltagy et al. 2000; Nicholson et al. 2000; Mano et al. 2006; Compant et al. 2010; Lopez et al. 2018). In order to survive in such conditions, they may have the trait of endospore formation, which protects them from the changes that occur inside seeds (Compant et al. 2011). As seeds can be in dormancy for a long time, they have to survive during this period (Nelson 2004; Geisen et al. 2017) and to deal with starvation (Nicholson et al. 2000; Compant et al. 2010; Truyens et al. 2015). Endophytic seed bacteria have also a particular mode of nutrition because of the uncommon trophic resources of seeds. They possess a cortege of enzymes such as amylase to utilize starch as source of carbon (Mano et al. 2006) and phytase to use phytate as a source of phosphate (López-López et al. 2010).

3.3.1.1 Motility

As seeds harden, seed endophytic bacteria have the ability to migrate through the plant and colonize other tissues (Johnston-Monje and Raizada 2011; Truyens et al. 2015). In this regard, Ebeltagy et al. (2000) and Okunishi et al. (2005) found out that almost all the seed endophytic bacteria from rice were motile. They can thus migrate from the different plant tissues to the seeds and vice versa. Of these endophytes, some use xylem vessels. They are endowed with a flagellum facilitating their mobility and assisted by ascending water flows guided notably by the phenomenon of evapotranspiration (Compant et al. 2010). Others use intercellular spaces and are then equipped with enzymes capable of degrading cell walls (Dong et al. 1994).

3.3.1.2 Competitive Advantage

As endophytic seed bacteria are seed-borne, they assure their presence in plants and are the first communities present in the seedlings (Truyens et al. 2014), which represent a competitive advantage comparing to the other bacteria present in the environment. Some studies proved the transmission from seed to mature tissues (Ferreira et al. 2008; Ringelberg et al. 2012).

3.3.1.3 Vertical Transmission

Seed endophytic bacteria have also the potential to persist across generations, through the process of vertical transmission (Shahzad et al. 2018). There is still a lack of evidence to confirm it, but some studies have already proved it for different species: Hardoim et al. (2012) for rice, Gagne-Bourgue et al. (2013) for switchgrass seeds, Truyens et al. (2015) for *Arabidopsis thaliana* and Cope-Selby et al. (2017) for *Miscanthus*.

3.3.1.4 Contribution to Germination

As seed endophytic bacteria are seed-borne, they are therefore the first communities present at the beginning of the seedling establishment. Consequently, these bacteria may play an important role during both germination and seedling development. Some studies showed that seed endophytic bacteria facilitate seed germination in soil (Rodríguez et al. 2018), and others assume that these bacteria are beneficial for both germination and seedling establishment (Truyens et al. 2015; Khalaf and Raizada 2018). In this regard, the beneficial effects were shown in the cases of rice (Mano et al. 2006; Kaga et al. 2009), eucalyptus (Ferreira et al. 2008) and maize (Rijavec et al. 2007). Seed-borne bacteria are sometimes indispensable for the germination process. Their removal by surface sterilization and antibiotic treatments inhibits drastically the seedling growth in the case of rice seeds (Verma et al. 2017). One study also showed that seed endophytic bacteria might play a role to release seed dormancy. Goggin et al. (2015) showed that bacteria help release seed dormancy through production of cytokinins for ryegrass. Seed endophytic bacteria also share beneficial traits similar to other bacteria beneficial for plants. Indeed, they have growth-promoting effects through their nutritional functions (Weyens et al. 2009) and hormone production (Herrera et al. 2016). Moreover, they display defensive functions as well (Weyens et al. 2009). They can also struggle against abiotic stresses such as metal pollution (Mastretta et al. 2009).

3.3.1.5 Growth-Promoting Effects

Numerous studies showed the growth-promoting effects of seed endophytic bacteria on plants such as rice (Hardoim et al. 2012), cactus (Puente et al. 2009), tomato (Xu et al. 2014) and wheat (Herrera et al. 2016). They can display this property through the improvement of nutrition of their plant host or through the hormone production.

Nutritional Functions

Seed endophytic bacteria improve the nutrition of their host (Puente et al. 2009; Johnston-Monje and Raizada 2011; Truyens et al. 2015; Khalaf and Raizada 2016; Mitter et al. 2016; Khalaf and Raizada 2018). This improvement is linked with not only the involvement of nitrogen fixation and nutrient mobilization (Weyens et al. 2009; Herrera et al. 2016) but also the production of enzymes and secondary metabolites important for plant nutrition (Shahzad et al. 2018). The nutrient mobilization includes phosphate solubilization (Verma et al. 2001; López-López et al. 2010; Johnston-Monje and Raizada 2011; Gagne-Bourgue et al. 2013; Xu et al. 2014; Chimwamurombe et al. 2016) iron acquisition (Weyens et al. 2009; Xu et al. 2014; Truyens et al. 2014; Chimwamurombe et al. 2016) and nitrogen fixation (Ebeltagy et al. 2000; Verma et al. 2001; Gagne-Bourgue et al. 2013; Zawoznik et al. 2014; Xu et al. 2014; Chimwamurombe et al. 2016).

Hormone Production

Seed-borne endophytic bacteria have the potential to produce various phytohormones (Shahzad et al. 2018). For example, these bacteria can produce IAA (Ebeltagy et al. 2000; Xu et al. 2014; Truyens et al. 2014; Zawoznik et al. 2014; Chimwamurombe et al. 2016), cytokinins (Weyens et al. 2009). They can also suppress stress ethylene production by ACC deaminase activity (Weyens et al. 2009; Johnston-Monje and Raizada 2011; Xu et al. 2014; Chimwamurombe et al. 2016).

3.3.1.6 Defensive Functions

Seed endophytic bacteria have defensive functions through various means. They can compete with pathogens for space and nutrients and produce hydrolytic enzymes. They also have the ability to act as biocontrol agents via antifungal and antibiotic activities. Eventually, they can induce plant defence mechanisms (Weyens et al. 2009).

Competition for Space and Nutrients

Because seed endophytic bacteria evolve in the same ecological niches as some pathogens, they compete with them. Moreover, the production of siderophores leads to competitions with harmful microorganisms for the acquisition of iron (Loper and Henkels 1999).

Antibiotic and Antifungal Activities

Seed-borne bacteria have the ability to produce antimicrobial compounds (Johnston-Monje and Raizada 2011; Truyens et al. 2015; Mitter et al. 2016; Khalaf and Raizada 2016; Shahzad et al. 2018; Khalaf and Raizada 2018). These may be volatile antimicrobial compounds (Mukhopadhyay et al. 1996), and can also involve lipopeptides such as surfactin, iturin and mycobacillin (Gagne-Bourgue et al. 2013). Seed-borne endophytic bacteria own a cortege of hydrolytic enzymes useful for biocontrol such as chitinolytic enzymes (Mukhopadhyay et al. 1996). The effects against pathogenic fungi are shown in several studies involving rice (Mukhopadhyay et al. 1996; Cottyn et al. 2001; Bacilio-Jiménez et al. 2001; Ruiza et al. 2011), pumpkins (Fürnkranz et al. 2012), peanut (Sobolev et al. 2013), tomato (Sundaramoorthy and Balabaskar 2013), maize (Gond et al. 2015), wheat (Herrera et al. 2016) and cucurbits (Khalaf and Raizada 2018).

Induction of Plant Defence Mechanisms

Seed-borne bacteria have the ability to trigger induced systemic resistance (ISR) and therefore to reduce plant susceptibility to pathogen attack (Weyens et al. 2009). They can release quorum-sensing molecules such as acetoin (Johnston-Monje and Raizada 2011). Quorum-sensing molecules may participate by the induction of plant gene expression for defence (Mathesius et al. 2003). They can release substances and natural elicitors, such as hydrogen peroxide (Pitzschke 2016).

3.3.1.7 Reduction of Pollutant Phytotoxicity

Seed endophytes can reduce cadmium phytotoxicity: Mastretta et al. (2009) showed it for tobacco plants. They may also have properties of phytoremediation. Truyens et al. (2014) assume this phenomenon in the case of *Agrostis capillaris* with the phytoremediation of cadmium.

3.3.2 Some Examples That Demonstrate the Potential of Seed-Borne Endophytic Bacteria for Agricultural Use

The potential of seed endophytic bacteria is still unexploited for agricultural use. Some studies concerning different kinds of cultures have been led to and shown several benefits in order to improve crop yield. On one hand, the interest of seed endophytic bacteria was assessed for market gardening. Market gardening requires a large number of inputs such as pesticides and fertilizers. The studies mentioned below show how much the seed endophytic bacteria are interesting to replace non-sustainable agricultural inputs. They concern the following crops: tomatoes, beans, pumpkins, peanuts and cucurbits. On the other hand, other studies prove their use interest for big crops such as rice, wheat, barley, maize and soya, and reveal their key role for agricultural transition. Moreover, the effect of these bacteria has been evaluated on forage crops, more precisely on agrostis. Finally, peculiar crops, which constitute niche market, are mentioned and reveal seed endophytic bacteria applications for agricultural use. The concerned cultures are quinoa and cactus.

3.3.2.1 The Potential of Seed Endophytic Bacteria for Market Gardening

Tomato

Two studies related to tomato have been found (Xu et al. 2014; Lopez et al. 2018). They conduce not only to plant growth-promoting properties of entophytic bacteria originated from seeds but also to their biocontrol ability. Xu et al. (2014) have isolated 84 culturable endophytic bacteria from tomato seeds of four commercial varieties. Concerned isolated strains showed many plant growth-promoting traits such as (1) hormone production (37% of the isolates produced IAA), (2) phosphate solubilization (37% of the isolates), (3) siderophore production (24% of the isolates), nitrogen fixation (85% of the isolates) and (4) enzyme activity: 6% of the isolates showed ACC deaminase activity. The present study also assessed the effect of these strains on plant growth promotion and led to positive results: this experiment showed the ability of one strain to enhance the growth of tomato seedlings and as a consequence the potential of the concerned *B. subtilis* endophytic strain for industrial application. Lopez et al. (2018) have isolated bacteria and have studied plant growth promotion traits. The strains they found exhibited beneficial traits such as IAA production, siderophore production and phosphate solubilization. Moreover, they assessed not only antifungal activities from endophytic bacteria and in vivo antagonism but also the effect of cell-free supernatant of endophytic bacteria against fungal pathogens. The results led to the conclusion that seed endophytic bacteria are able to synthetize antifungal substances and have the ability to deter plant pathogens.

Beans

Two publications in relation to seed endophytic bacteria have been found on beans. Both of them show the potential of these bacteria to contribute to plant growth promotion and plant protection. Chimwamurombe et al. (2016) isolated 123 isolates from marama bean (*Tylosema esculentum*). They screened these strains for plant growth-promoting activities which revealed the ability of IAA production. The strains exhibit not only ACC deaminase activity, siderophore production, endoglucanase activity, protease activity, phosphate solubilization and nitrogen fixation but also production of N-acylhomoserine lactones, components which are involved in successful colonization (Compant et al. 2010) and play a role in plant defences (Schuhegger et al. 2006). These bacterial activities may contribute to plant growth promotion and protect plant against pathogens. The lytic activities such as protease are useful for plant protection. López-López et al. (2010) have studied the bacterial endophytic community of *Phaseolus vulgaris* cultivars. The study has permitted to conclude that these bacteria have the ability to solubilize phytate. This property enables to improve plant iron nutrition.

Pumpkins

The study of Fürnkranz et al. (2012) explains the ability of seed endophytic bacteria of Styrian oil pumpkin (*Cucurbita pepo* L. subsp. pepo var. styriaca Greb) for biocontrol. In vitro antagonism tests against *D. bryoniae* showed positive results. Moreover, Fürnkranz et al. (2012) also reported 43 isolates that have the ability to inhibit the growth of the bacterial pumpkin pathogens *Pectobacterium carotovorum*, *Pseudomonas viridiflava* and *Xanthomonas cucurbitae*. They concluded that there can be promising use of these strains for agricultural purposes in relation to pumpkin crop protection against pathogens.

Peanut

Sobolev et al. (2013) made a survey on seed endophytic bacteria on peanut culture. The aim of the study included the characterization of strains that present an activity against the pathogen *A. flavus*. The results showed that some bacteria showed an ability to deter the phytopathogen.

Cucurbits

Khalaf and Raizada (2018) focused on the antagonist ability of seed endophytic bacteria from diverse cultivated cucurbits. They isolated 169 bacterial endophytes from cucumbers, melons and pumpkins. They affected different tests including dual culture assays against *Rhizoctonia solani*, *Fusarium graminearum*, *Phytophthora capsici* and *Pythium aphanidermatum* and tests in plants for antagonism against a foliar pathogen, which is the causative agent of cucurbit powdery mildew. Then, tests in vitro for assessing secretion of volatile organic compounds were realized, followed by the evaluation of extracellular ribonuclease activity, which play a role in plant defences against pathogens. It was found that the majority of the strains presented strong antagonistic properties. Seventy percent of the isolates have the ability to be effective against the phytopathogens. In addition, 67% of the strains produced organic volatile compounds, and 62% showed extracellular ribonuclease activity. All the results led to the belief in the potential of seed endophytic bacteria to suppress diseases issues linked to cucurbit crops.

3.3.2.2 The Potential of Seed Endophytic Bacteria for the Productivity of the Main Food Crops

Rice

Seed endophytic bacteria from rice have been studied in several studies, including Mukhopadhyay et al. (1996), Ebeltagy et al. (2000), Verma et al. (2001), Cottyn et al. (2001), Feng et al. (2005) and Mano et al. (2006). All the studies reveal the potential of seed endophytic bacteria isolated from rice cultivars, including not only plant growth promotion but also the biocontrol aspect. Feng et al. (2005) focused on *Pantoea agglomerans* rice endophytic bacterium. This bacterium has beneficial properties for plant growth promotion as nitrogen fixation and production of phytohormones (IAA, abscisic acid, gibberellic acid and cytokinin). Ebeltagy et al. (2000) have isolated endophytic bacteria originated from different tissues and organs of rice plants, including seeds. They found similarity to Feng et al. (2005) some beneficial traits such as nitrogen fixation and IAA production. They also discovered other specific traits including pectinase and cellulose excretion, osmotic resistance and capsule formation. Verma et al. (2001) obtained similar results with positive results concerning cellulase and pectinase activity, IAA production and nitrogen fixation. Cottyn et al. (2001) isolated 428 bacteria. Their studies focused on the biocontrol properties. They found out that 4% of their isolates showed in vitro antifungal activity against *Rhizoctonia solani* and *Pyricularia grisea*. Mukhopadhyay et al. (1996) obtained results concerning biocontrol properties of isolates. They found that some bacteria they have isolated showed a high antifungal ability against the pathogens *R. solani*, *Pythium myriotylum*, *Gaeumannomyces graminis* and *Heterobasidion annosum*. More recently, this aspect was reinforced by the study of Ruiza et al. (2011). They found out that the 39 isolated seed-borne bacteria of rice exhibited probiotic activities and antifungal activity against *Curvularia* sp., *F. oxysporum* var. *radices-lycopersici* and *Pythium ultimum* in dual culture plates. These isolates also demonstrated plant growth-promoting traits such as nitrogen fixation, phosphate solubilization and production of IAA.

Maize

Studies relative to maize seed endophytic bacteria showed their potential for farming by the ability to express plant growth-promoting traits and show biocontrol activity. On one hand, the publication from Johnston-Monje and Raizada (2011) revealed that the 26 different genera of bacteria isolated from maize seeds exhibit plant growth-promoting traits (plant growth stimulation, nitrogen fixation, phosphate solubilization, iron sequestration, production of the phytohormones auxin and acetoin, possess ACC deaminase, pectinase and cellulase activities). They also discovered the ability of endophytes to show antagonism towards pathogens: a majority showed antibiosis against bacteria or yeast. Gond et al. (2015) worked particularly on the biocontrol traits of such endophytic bacteria: they evaluated bacteria for antifungal compound production by the evaluation of lipopeptide production. They also tested these isolates for induction of defence gene expression. It was found that the lipopeptide extracts produced by the isolates exhibited antifungal activity against *Fusarium moniliforme* and that one isolate showed the induction of pathogenesis-relative genes in relation to plant defence against fungal pathogens.

Wheat

Herrera et al. (2016) underlined the role of endophytic strains from wheat seeds by studying plant growth properties and antifungal activities against *F. graminearum*. More precisely, nitrogen fixation, IAA, phosphate solubilization and siderophore production were assessed, and all the traits presented at least one beneficial trait. The in vitro antagonism against *F. graminearum* and the biofilm production in relation to biocontrol activity were also tested and conduced to confirm the biocontrol activity of these strains.

Barley

Zawoznik et al. (2014) isolated eight isolates from barley seeds and screened them for nitrogen fixation and IAA production. They found out that the majority of the strains were able to fix nitrogen and that all of them were capable to produce the phytohormone aforementioned.

Soy Beans

Oehrle et al. (2000) focused on seed-borne endophyte from soya seeds. They did not focus on plant growth-promoting traits nor on biocontrol properties. They study the role of seed-borne microbes on root colonization by the symbiont *B. japonicum*. It was found that soybean microbiome can affect the symbiosis of nodulative bacteria.

3.3.2.3 The Potential of Seed Endophyte Bacteria for Forage Crops

Agrostis

Truyens et al. (2014) led studies on a forage crop, *Agrostis capillaris*. Seed endophytic bacteria from seeds of contaminated soils exhibited some traits such as ACC deaminase activity (78.4%), production of siderophores (42.4%), phytohormones biosynthesis (IAA, 82.6%; acetoin, 43.4%) and metal tolerance.

3.3.2.4 The Potential of Seed Endophytic Bacteria for Niche Market Cultures

Quinoa

Pitzschke (2016) made a study on seed-borne endophytes in quinoa. They tested the seed-borne endophytes for catalase and amylase activity, and the results were positive. Moreover, it was found that these microorganisms have high superoxide levels. They concluded that seed endophytic bacteria have a potential for agriculture use.

Cactus

Puente et al. (2009) discovered the remarkable abilities of endophytic bacteria in cactus seeds. They found out that these bacteria are necessary for the seed to develop. The endophytic bacteria have the property to promote the growth of giant cactus seedlings.

Tobacco

Mastretta et al. (2009) have isolated seed endophytic bacteria from *Nicotiana tabacum*. After having positive results for cadmium tolerance for many strains, they tested the isolates for plant growth promotion with and without cadmium stress. Generally, the strains permit an increase in biomass production. The potential of seed endophytic bacteria is notingly due to their vertical transmission trait which remains unclear but constitutes a potentially interesting trait for commercial use.

3.3.3 Vertical Transmission of Seed Entophytic Bacteria

Plants contain a number of microbes that constitute the plant microbiome. Among this microbiome, bacteria play an important role. These bacteria can be transmitted by different ways: on one hand, the transmission can be exerted in a horizontal way by the environment (via soil, atmosphere and insects), on the other hand, bacteria can be transmitted in a vertical manner (via pollen and seeds) (Frank et al. 2017). Until now, the knowledge of their route of colonization remains unclear (Hardoim et al. 2012; Klaedtke et al. 2016; Rodríguez et al. 2018). In their review, Shade et al. (2017) have reported three principal ways of transmission to constitute the seed microbiome; (1) through the xylem or nonvascular tissues of the mother plant, (2) the stigma of the mother plant and (3) contamination through contact with microorganisms.

3.3.3.1 Context

In their review, Truyens et al. (2015) underline the peculiar potential of seed endophytic bacteria because they may be transmitted across successive generations. Soil is generally considered as the main source of endophytic bacteria. Rodríguez et al. (2018) explained that the assumption of heritability of bacterial endophytes from seed to seed lacks evidence. To that extent, Vandenkoornhuyse et al. (2015) assumed that the weight of vertical transmission relatively to horizontal transmission is still unclear. Shade et al. (2017) pointed out the fact that this weight was relying on the plant species and fluctuant. Sanchez-Canizares et al. (2017) pointed out that even if the part of contribution of seed microbiome to constitute the future plant microbiome is small compared to environment contribution, seed microbiome does play a role for development of plant microbiome. Frank et al. (2017) have reported the beneficial effects of seed endophytic bacteria, and regarding this, they suggested that such mutualistic interactions may be transmitted vertically via seeds (Ewald 1989). Vertical transmission may favour mutualism because endophytic bacterial survival relies on the fitness of its host plant (Truyens et al. 2015). At the beginning of the twenty first century, the vertical transmission of endophytic symbionts colonizing host plants with the seeds was proved with the biological model of grass species. Bacteria were not considered earlier, but fungal species such as *Neotyphodium coenophialum* were involved (Clay and Schardl 2002; Selosse et al. 2004). Specific studies concerning bacteria were done later.

Transmission from Seed to Mature Tissues

There are several studies which prove the transmission of seed microbiome to mature tissues. Ferreira et al. showed in 2008 the vertical transmission of seed endophytic bacteria from seeds to seedlings. Their biological model was *Eucalyptus* sp. In 2012, Ringelberg et al. have confirmed this fact with another biological model, wheatgrass. They studied bacterial endophyte communities of two distinct wheatgrass varieties. In order to have a vision of bacterial communities, they used culturable and direct rDNA PCR amplification techniques. They compared the communities of leaves, seeds and roots. They affected the evaluation in two distinct media, and found out that cultures in both media showed the possible translocation of bacteria (*Actinobacteria*, *Firmicutes*, *Gammaproteobacteria*) from seeds to mature plant tissues.

Flower to Seed Transmission

One study went further and proved vertical inheritance of bacteria from flower to seed (Mitter et al. 2017). Mitter et al. (2017) succeeded to prove it by introducing the endophyte *Paraburkholderia phytofirmans* PsJN to the flowers of parent plants and by evaluating its inclusion to seed microbiome. The results permit to demonstrate that the introduction of endophytic bacteria to the flower brought modifications to seed endophytic bacteria communities in wheat.

Seed to Seed Transmission

The vertical transmission of seed microbiome has been proved for several species. Gagne-Bourgue et al. (2013) have succeeded to prove the vertical transmission of endophytic bacteria seed to seed in switchgrass seeds. In switchgrass plants some endophytic bacteria were reported which were also present in seeds collected a year earlier. Hardoim et al. (2012) investigated the transmission of seed endophytic bacteria from seed to seed by assessing the endophytic bacterial community encompassing two rice generations. To do that, they used two methods: cultivation-dependent and cultivation-independent methods. It was found that 45% of the bacterial community found in the first seed generation was also found in the second generation. Truyens et al. (2013) studied the changes in the population of seed bacteria of trans-generationally Cd-exposed *A. thaliana*. Their results suggested that there is a selective process of endophytes across generations that shape microbiome according to bacterial properties and environmental stress factors endured by the mother plant. Cope-Selby et al. (2017) wanted to study the extent of vertical transmission of endophytic bacteria in *Miscanthus* seed via the seed. They found out that vertical transmission is potentially a major source of the microbiome of *Miscanthus* seeds.

Conservation of Seed Microbiome Through Generations Improving Plant Fitness

This vertical transmission has interest for plant fitness. Some studies suggest that the advantageous microbiome is selected across generations to improve the survival of plant species depending on the biotope, suggesting a coevolution of seed microbiome in the long term. Several arguments suggest this selection: (1) conservation of seed microbiome through generation in the long term (Johnston-Monje and Raizada 2011), (2) low diversity of seed endophytic bacteria compared to bacteria from other parts of plants (Adam et al. 2018; Bergna et al. 2018) and (3) the shape of seed microbiome depending on the environmental conditions (Bergna et al. 2018; Adam et al. 2018).

Conservation of Seed Microbiome Through Generation in the Long Term
Johnston-Monje and Raizada (2011) made researches about the conservation of endophytic species from wild ancestor of maize, teosinte, to the domesticated plant genus *Zea*. They studied ten maize varieties for its endophytic communities by culturing, cloning and DNA fingerprinting. Despite of the variation of seed endophyte composition between the varieties, they noted a "core microbiota" of seed endophytic bacteria that is conserved from teosinte to modern cultivated varieties of maize.

Low Diversity of Seed Endophytic Bacteria Compared to Bacteria from Other Parts of Plants
Some studies (Bergna et al. 2018; Adam et al. 2018) show that seed microbiome has a lower diversity than the microbiome of the other parts of the plant. Bergna et al. (2018) found that soil, root endosphere and rhizosphere were the habitats with the highest bacterial diversity, whereas seeds harbour selective bacterial communities. Adam et al. (2018) discovered similar results: contrary to rhizospheric microbiomes, seed microbiomes showed a low diversity.

The shape of seed microbiome depending on the environmental conditions (Bergna et al. 2018; Adam et al. 2018). Bergna et al. (2018) provides a better understanding of its origin. These studies focussed on: (1) to know the main drivers of the microbiome composition, (2) what the impact of soil quality on the seed microbiome is and (3) how the transmission of PGPB in the seed across generations works. It was found that plant habitat and genotype have its own microbiome and its own PGPB (plant growth-promoting bacteria). Studies also pointed out a selection process that conserves beneficial traits across generations depending on the biotope: seeds grown under field conditions included endophytic bacteria with biocontrol traits, whereas seeds of the successive generation grown in a pathogen-free and with a scarcity of nutrient harboured a majority of bacteria that have the ability to improve the nutrition of plants. These results show that the environment has the power to shape the seed microbiome and to improve the fitness of plant depending on its needs. Adam et al. (2018) also found out that soil exert an influence on seed microbiome.

3.3.3.2 The Importance of the Plant Genotype

Recently, studies permit to assume the link between plant genotype and seed microbiome, suggesting vertical transmission and coevolution between plants and seed microbiome. Adam et al. (2018) studied the *Cucurbita pepo* seed microbiome and its shaping agents. They evaluated bacterial communities of seeds and rhizospheres of 14 Styrian oil pumpkin genotypes in comparison to bulk soil. Their results permit to assume that there is a strong influence of the *C. pepo* genotype on the composition of the seed microbiome, which reinforces the vertical transmission aspect. Berg and Raaijmakers (2018) insisted that factors, as plant genotype and management practices, have also a strong influence on seed microbiome. They particularly insisted on the phenomenon of plant domestication, which is an important driver of seed microbiota, proving that vertical transmission is effective. In this regard, Rybakova et al. (2017) studied indigenous microbiota associated with the seed of several cultivars of oilseed rape. They discovered that cultivars had different microbiomes and that it (the microbiome) acted differently according to the concerned cultivar to inoculation, reinforcing the assumption that plant genotype is linked to seed microbiota and confirming the potential existence of vertical transmission of seed microbiota.

3.3.3.3 Possible Ways of Vertical Transmission

In their review, Truyens et al. (2014) detail the ways bacterial endophytes can get in, which are multiple. Indeed, seed microbiota may be transmitted: (1) from the vegetative parts of the plant to the seed via vascular connections, (2) via gametes (3) via future reproductive meristems.

3.3.4 Horizontal Transmission of Seed Endophytic Bacteria

Even if there are proofs of the role of vertical transmission of the seed microbiome, horizontal transmission of seed endophytic bacteria still seems to play the major role for seed microbiome constitution. The key component of construction of seed endophytic bacteria communities seems to be soil. Sanchez-Canizares et al. (2017) affirmed that soil is the main source of microbiome. This assumption is especially reinforced by the study of Hardoim et al. (2012). The team as mentioned above made a study on dynamics of seed-borne rice endophytes. They conducted an experiment to know what the factors are that shape the endophytic bacterial community structure. They found out that soil type is the major effector of the bacterial endophytes. The soil is not the only environmental part which influences the seed microbiome, Berg and Raaijmakers (2018) insisted that the rest of the environment and management practices have also a strong influence on seed microbiome. In this regard, Klaedtke et al. (2016) studied the influence of host genotypes and terroir on the structure of the seed microbiota by using the metabarcoding method. They analysed the results from different seed endophytic communities of five bean cultivars originated from farms. It was found that there was a significant effect of the location (i.e. the farm site) on the structure of both the bacterial and fungal communities: the location is estimated to be responsible for 12.2% and 39.7% of variance in bacterial and fungal diversity respectively, across samples. On the contrary, host genotype was not a determinant factor of variance in microbial assemblage structure. They concluded that seed endophytic bacterial communities are determined by the terroir, which is a driver for selection of seed microbiota. The environment thus plays a significant role. In this regard, Rezki et al. (2018) described in their study this "ecological drift". They focused on the assembly of seed-associated microbial communities within and across successive radish generations. Their work consisted in the evaluation of the seed microbiome of radish across successive generations by assessing its structure and composition to have a better understanding of the involvement of seeds in the transmission of plant bacteria. The results showed that seed microbial communities have a low heritability across plant generations. It was found that "ecological drift" is a factor for the assembly of seed microbiome.

3.4 Seed Endophytic Bacteria as Promising Tools for Agricultural Purposes

The seed-borne endophytic bacteria present a potential for agricultural use due to their remarkable abilities summarized in the figure (Fig. 3.1). The examples mentioned before for the different kinds of crop systems (market gardening, big crops, forage crops, niche market cultures) demonstrate the relevance of endophytes for agricultural purposes. Numerous scientists assume that seed-borne endophytic bacteria are promising biofertilizers (Ruiza et al. 2011; Truyens et al. 2015; Khalaf and Raizada 2016; Khalaf and Raizada 2018). Berg et al. (2017) and Khalaf and Raizada (2018) insisted on their biocontrol potential. The characteristic of vertical transmission constitute a growing interest (Johnston-Monje and Raizada 2011). Bergna et al. (2018) showed that seed can be considered as primary vehicle of PGPB transmission. Consequently, this is a potential strategy to maintain beneficial bacteria, and the design of seed treatments seems promising. Bacilio-Jiménez et al. (2001), Nelson (2004) and Mitter et al. (2017) highlighted on the competitive advantage of seed-borne endophytic bacteria for inoculation. Bacilio-Jiménez et al. (2001) assumed that these kinds of bacteria are strongly competitive and have the potential to be as competitive as bacteria introduced as inoculants. This is because seed-borne endophytic bacteria have a privileged interaction with their plant host (Nelson 2004): they escape from competition because of the fact that they are from the beginning inside plant tissues. In this regard, Shade et al. (2017) qualify them as a starting point and an endpoint for community assembly. Seed endophytic bacteria seem thereby to be promising and efficient biofertilizers. Some scientists assume that seeds could play the role of support for inoculation by manipulating its microbiome (Berg and Raaijmakers 2018; Adam et al. 2018). In this regard, Berg and Raaijmakers (2018) considered that it would be interesting to join breeding and introduction of beneficial microbes with seeds as vectors of inoculation. In addition, bacteria in seeds might be less impacted by environmental constraints and mechanical perturbation (Mitter et al. 2017). Since seed microbiome improves the germination conditions, plants grown from seeds would present the advantages (López-López et al. 2010). This method could permit to increase targeted inoculant populations by pre-inoculation, with time, work and cost savings. For example, in Brazilian farming systems, the cultivation of eucalyptus has increased considerably. The development of eucalyptus is mainly conditioned by the endophytic fungi which act directly on the increase of eucalyptus growth. Cloned seeds inoculated with fungi improved yields (Vitorino et al. 2016). Adam et al. (2018) found out that although there is a small difference between the *C. pepo* L. subsp. pepo var. styriaca genotypes, cultivar-dependent differences were found in the seed microbiome. They suggest that it could be possible to select the cultivar with its associated beneficial microbiome. Mitter et al. (2017) went further and succeeded to manipulate

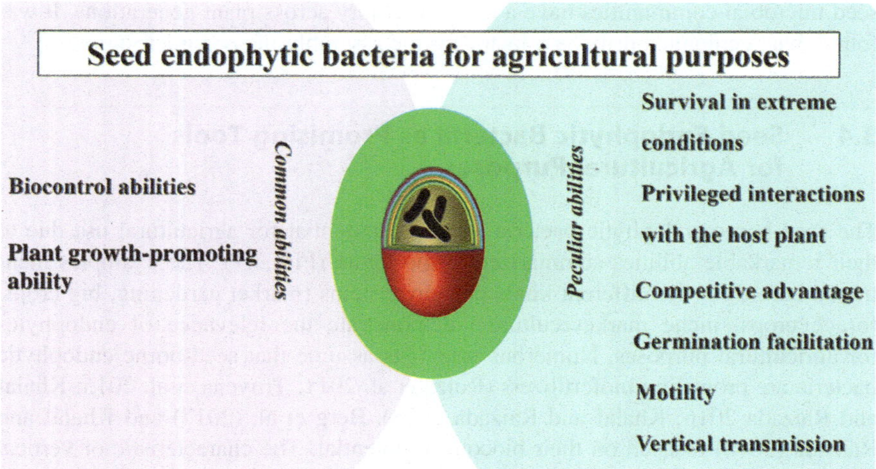

Fig. 3.1 Seed endophytic bacteria and their traits for agricultural purposes

the seed microbiome in a targeted way. They conceived a way to introduce targeted and directed traits into plants in one generation without plant breeding strategy and thereby saving time. They demonstrated that the modulation of an optimized seed microbiome was possible. They developed a new approach to modulate seed microbiome by introducing beneficial strains into seeds and thereby to optimize plant development. They introduced the endophyte *P. phytofirmans* PsJN (a powerful plant growth promoter, Mitter et al. 2013) to the flower of parent plant and demonstrated to include it in progeny seed microbiome. The aim of the experiment was to induce its vertical inheritance to the offspring generation. Experiment showed that this modulation of seed microbiome can be useful for agricultural ends. The endophyte mentioned above was introduced in both monocotyledon and dicotyledonous plant species (maize, wheat, soy, pepper), and it succeeded in all the cases. This modulation resulted in important modifications to seed microbiome composition. The modulation of the plant microbiome creates thereby new halobionts (Turner et al. 2013) and determines plant fitness (Jefferson 1994) by modulating plant's traits. The conception of novel seed-based microbiomes is thereby possible. Determining plant beneficial traits by modulating seed microbiome seems to be a feasible strategy to substitute abusive use of fertilizers and pesticides.

3.5 Conclusion

Seed endophytic bacteria are of growing interest because they have particular characteristics compared to bacteria from other plant compartments, known for their biofertilizing properties. Among these particular characteristics, it is possible to mention:

1. The resistance of these bacteria to difficult abiotic conditions.
2. The competitive advantage of these microbes.
3. Vertical transmission which would allow endophytic seed bacteria to be transmitted from generation to generation.

In connection with this ability of vertical transmission, some scientists consider that the seeds are the primary vehicles of beneficial bacteria. There is a selection of the procession of microbiome important for the host plant from generation to generation in order to increase the latter's fitness. Seeds are thus essential agronomic tools with an under-exploited potential. Indeed, they are not only a microbiological gold mine, but they are also a means of making bacterial inoculation viable and feasible for agronomic purposes. In this regard, it is not only possible to operate an alternative seed design for effective inoculation but also to manipulate the seed microbiome so as to introduce targeted agronomic traits.

References

Abedinzadeh M, Etesami H, Alikhani HA (2019) Characterization of rhizosphere and endophytic bacteria from roots of maize (*Zea mays* L.) plant irrigated with wastewater with biotechnological potential in agriculture. Biotechnol Rep 21:e0030

Adam E, Bernhart M, Müller WH, Berg G (2018) The *Cucurbita pepo* seed microbiome: genotype-specific composition and implication for breeding. Plant Soil 422(1):35–49

Agarwhal S, Shende ST (1987) Tetrazolium reducing microorganisms inside the root of *Brassica* species. Curr Sci 56:187–188

Ahmad I, Akhtar MJ, Asghar HN, Ghafoor U, Shahid M (2016) Differential effects of plant growth-promoting rhizobacteria on maize growth and cadmium uptake. J Plant Growth Regul 35(2): 303–315

Aktar W, Sengupta D, Chowdhury A (2009) Impact of pesticides use in agriculture: their benefits and hazards. Interdiscip Toxicol 2(1):1

Ali S, Charles TC, Glick BR (2014) Amelioration of high salinity stress damage by plant growth-promoting bacterial endophytes that contain ACC deaminase. Plant Physiol Biochem 80:160–167

Aravind R, Kumar A, Eapen SJ, Ramana KV (2009) Endophytic bacterial flora in root and stem tissues of black pepper (*Piper nigrum* L.) genotype: isolation, identification and evaluation against Phytophthora capsica. Lett Appl Microbiol 48(1):58–64

Arkhipova TN, Veselov SU, Melentiev AI, Martynenko EV, Kudoyarova GR (2005) Ability of bacterium *Bacillus subtilis* to produce cytokinins and to influence the growth and endogenous hormone content of lettuce plants. Plant Soil 272(1):201–209

Arshad M, Frankenberger WT (1991) Microbial production of plant hormones. Plant Soil 133:1–8

Atkinson NJ, Urwin PE (2012) The interaction of plant biotic and abiotic stresses: from genes to the field. J Exp Bot 63(10):3523–3543

Bacilio-Jiménez M, Aguilar-Flores S, DelValle MV, Pérez A, Zepeda A, Zenteno E (2001) Endophytic bacteria in rice seeds inhibit early colonization of roots by *Azospirillum brasilense*. Soil Biol Biochem 33(2):167–172

Bashan Y, Holguin G (1993) Anchoring of Azospirillum brasilense to hydrophobic polystyrene and wheat roots. Microbiology 139(2):379–385

Belimov AA, Hontzeas N, Safronova VI, Demchinskaya SV, Piluzza G, Bullitta S, Glick BR (2005) Cadmium-tolerant plant growth-promoting bacteria associated with the roots of Indian mustard (*Brassica juncea* L. Czern.). Soil Biol Biochem 37(2):241–250

Berg G, Raaijmakers JM (2018) Saving seed microbiomes. ISME J 12(5):1167–1170

Berg G, Grube M, Schloter M, Smalla K (2014) Unraveling the plant microbiome: looking back and future perspectives. Front Microbiol 5:148

Berg G, Köberl M, Rybakova D, Müller H, Grosch R, Smalla K (2017) Plant microbial diversity is suggested as the key to future biocontrol and health trends. FEMS Microbiol Ecol 93(5)

Bergna A, Cernava T, Rändler M, Grosch R, Zachow C, Berg G (2018) Tomato seeds preferably transmit plant beneficial endophytes. Phytobiomes J 2(4):183–193

Bhardwaj D, Ansari MW, Sahoo RK, Tuteja N (2014) Biofertilizer function as key player in sustainable agriculture by improving soil fertility, plant tolerance and crop productivity. Microb Cell Fact 13(1):1–10

Bhore SJ, Ravichantar N, Loh CY (2010) Screening of endophytic bacteria isolated from leaves of Sambung Nyawa [*Gynura procumbens* (Lour.) Merr.] for cytokinin-like compounds. Bioinformation 5(5):191

Bilal S, Shahzad R, Khan AL, Kang SM, Imran QM, Al-Harrasi A, Lee IJ (2018) Endophytic microbial consortia of phytohormones-producing fungus *Paecilomyces formosus* LHL10 and bacteria *Sphingomonas* sp. LK11 to (*Glycine max* L.) regulates physio-hormonal changes to attenuate aluminum and zinc stresses. Front Plant Sci 9:1273

Cankar K, Kraigher H, Ravnikar M, Rupnik M (2005) Bacteial endophytes from seeds of Norway spruce (*Picea abies* L. karst). FEMS Microbiol Lett 244(2):341–345

Cardinale M, Ratering S, Suarez C, Montoya AMZ, Geissler-Plaum R, Schnell S (2015) Paradox of plant growth promotion potential of rhizobacteria and their actual promotion effect on growth of barley (*Hordeum vulgare* L.) under salt stress. Microbiol Res 181:22–32

Chacko S, Ramteke PW, John SA (2009) Amidase from plant growth promoting rhizobacterium. Afr J Bacteriol Res 1(4):046–050

Chen YP, Rekha PD, Arun AB, Shen FT, Lai WA, Young CC (2006) Phosphate solubilizing bacteria from subtropical soil and their tricalcium phosphate solubilizing abilities. Appl Soil Ecol 34(1):33–41

Cherif H, Marasco R, Rolli E, Ferjani R, Fusi M, Soussi A, Cherif A (2015) Oasis desert farming selects environment-specific date palm root endophytic communities and cultivable bacteria that promote resistance to drought. Environ Microbiol Rep 7(4):668–678

Chimwamurombe PM, Grönemeyer JL, Reinhold-Hurek B (2016) Isolation and characterization of culturable seed-associated bacterial endophytes from gnotobiotically grown Marama bean seedlings. FEMS Microbiol Ecol 92(6):fiw083

Clay K, Schardl CL (2002) Evolutionary origins and ecological consequences of endophyte symbiosis with grasses. Am Nat 160:99–127

Compant S, Clément C, Sessitsch A (2010) Plant growth-promoting bacteria in the rhizo-and endosphere of plants: their role, colonization, mechanisms involved and prospects for utilization. Soil Biol Biochem 42(5):669–678

Compant S, Mitter B, Coli-Mull JG, Gangl H, Sessitsch A (2011) Endophytes of grapevine flowers, berries, and seeds: identification of cultivable bacteria, comparison with other plant parts, and visualization of niches of colonization. Microb Ecol 62:188–197

Compant S, Samad A, Faist H, Sessitsch A (2019) A review on the plant microbiome: ecology, functions, and emerging trends in microbial application. J Adv Res 19:29–37

Cope-Selby N, Cookson A, Squance M, Donnison I, Flavemm R, Farrar K (2017) Endophytic bacteria in Miscanthus seed: implications for germination, vertical inheritance of endophytes, plant evolution and breeding. GCB Bioenergy 9(1):57–77

Cottyn B, Regalado E, Lanoot B, De Cleene M, Mew TW, Swings J (2001) Bacterial populations associated with rice seed in the tropical environment. Phytopathology 91(3):282–292

Creus CM, Sueldo RJ, Barassi CA (2004) Water relations and yield in *Azospirillum* inoculated wheat exposed to drought in the field. Can J Bot 82(2):273–281

Currier WW, Strobel GA (1977) Chemotaxis of *rhizobium spp.* to a glycoprotein produced by birdsfoot trefoil roots. Science 96(288):434–436

Daghagh Yazd S, Wheeler SA, Zuo A (2019) Key risk factors affecting farmers' mental health: a systematic review. Int J Environ Res Public Health 16(23):4849

Dash NP, Kumar A, Singh PK (2015) Cyanobacteria, pesticides and rice interaction. Biodivers Conserv 24:995–1005
De Rybel B, Mähönen AP, Helariutta Y, Weijers D (2016) Plant vascular development: from early specification to differentiation. Nat Rev Mol Cell Biol 17(1):30
De Vrieze M, Germanier F, Vuille N, Weisskopf L (2018) Combining different potato-associated *Pseudomonas* strains for improved biocontrol of *Phytophthora infestans*. Front Microbiol 9: 2573
Dixon GR, Tilston EL (2010) Soil-borne pathogens and their interactions with the soil environment. In: Dixon G, Tilston E (eds) Soil microbiology and sustainable crop production. Springer, Berlin, pp 197–271
Dobbelaere S, Croonenborghs A, Thys A, Ptacek D (2001) Agronomically important crops to inoculation with *Azospirillum*. Aust J Plant Physiol 28(9):871–879
Dong Z, Canny MJ, McCully ME, Roboredo MR, Cabadilla CF, Ortega E, Rodes R (1994) A nitrogen-fixing endophyte of sugarcane stems (a new role for the apoplast). Plant Physiol 105(4):1139–1147
Douglas P, Robertson S, Gay R, Hansell AL, Gant TW (2018) A systematic review of the public health risks of bioaerosols from intensive farming. Int J Hyg Environ Health 221(2):134–173
Dourado MN, Martins PF, Quecine MC, Piotto FA, Souza LA, Franco MR, Azevedo RA et al (2013) Burkholderia sp. SCMS54 reduces cadmium toxicity and promotes growth in tomato. Ann Appl Biol 163(3):494–507
Ebeltagy A, Nishioka K, Suzuki H, Sato T, Sato Y, Morisaki H et al (2000) Isolation and characterization of endophytic bacteria from wild and traditionally cultivated rice varieties. Soil Sci Plant Nutr 46(3):617–629
Egamberdieva D, Kamilova VS, Gafurova L, Kucharova Z, Lugtenberg B (2008) High incidence of plant growth-stimulating bacteria associated with the rhizosphere of wheat grown on salinated soil in Uzbekistan. Environ Microbiol 10(1):1–9
Egamberdieva D, Kucharova Z, Davranov K, Berg G, Makarova N, Azarova T et al (2011) Bacteria able to control foot and root rot and to promote growth of cucumber in salinated soils. Biol Fertil Soils 47(2):197–205
Egamberdieva D, Wirth SJ, Alqarawi AA, Abd Allah EF, Hashem A (2017a) Phytohormones and beneficial microbes: essential components for plants to balance stress and fitness. Front Microbiol 8:2104
Egamberdieva D, Wirth S, Jabborova D, Räsänen LA, Liao H (2017b) Coordination between *Bradyrhizobium* and *Pseudomonas* alleviates salt stress in soybean through altering root system architecture. J Plant Interact 12(1):100–107
Etesami H, Maheshwari DK (2018) Use of plant growth promoting rhizobacteria (PGPRs) with multiple plant growth promoting traits in stress agriculture: action mechanisms and future prospects. Ecotoxicol Environ Safe 156:225–246
Ewald PW (1989) Transmission modes and evolution of the parasitism-mutualism continuum. Ann N Y Acad Sci 503(1):295–306
FAO (2015) Promotion of fruit and vegetables for health, report of the Pacific regional workshop. FAO, Rome
FAO (2021) Résumé de L'État de la sécurité alimentaire et de la nutrition dans le monde 2021. Transformer les systèmes alimentaires pour que la sécurité alimentaire, une meilleure nutrition et une alimentation saine et abordable soient une réalité pour tous. FAO, Rome
Feng Y, Shen D, Song W (2005) Rice endophyte *Pantoe agglomerans* YS19 promotes host plant growth and affects allocations of host photosynthates. J Appl Microbiol 100(5):938–945
Ferreira A, Quecine MC, Lacava PT, Oda S, Azevedo JL, Araujo WL (2008) Diversity of endophytic bacteria from Eucalyptus species seeds and colonization of seedlings by *Pantoea agglomerans*. FEMS Microbiol Lett 287(1):8–14
Fierer N (2017) Embracing the unknown: disentangling the complexities of the soil microbiome. Nat Rev Microbiol 15(10):579–590

Finnie JF, Van Staden J (1985) Effect of seed weed concentrate and applied hormones on in vitro cultured tomato roots. J Plant Physiol 120:215–222

Fisher PJ, Petrini O, Lappi SHM (1992) The distribution of some fungal and bacterial endophytes in maize (*Zea mays* L.). New Phytol 122(2):299–305

Forchetti G, Masciarelli O, Izaguirre MJ, Alemano S, Alvarez D, Abdala G (2010) Endophytic bacteria improve seedling growth of sunflower under water stress, produce salicylic acid, and inhibit growth of pathogenic fungi. Curr Microbiol 61(6):485–493

Frank AC, Guzman JPS, Shay JE (2017) Transmission of bacterial endophytes. Microorganisms 5(4):70

Fürnkranz M, Lukesch B, Müller H, Huss H, Grube M, Berg G (2012) Microbial diversity inside pumpkins: microhabitat-specific communities display a high antagonistic potential against phytopathogens. Microb Ecol 63(2):418–428

Gagne-Bourgue F, Aliferis KA, Seguin P, Rani M, Samson R, Jabaji S (2013) Isolation and characterization of indigenous endophytic bacteria associated with leaves of switchgrass (*Panicum virgatum* L.) cultivars. J Appl Microbiol 114(3):836–853

Gamalero E, Glick BR (2015) Bacterial modulation of plant ethylene levels. Plant Physiol 169(1): 13–22

Geisen S, Kostenko O, Cnossen MC, ten Hooven FC, Vreš B, van der Putten WH (2017) Seed and root endophytic fungi in a range expanding and a related plant species. Front Microbiol 8:1645

Germida JJ, Siciliano SD, Renato de Freitas J, Seib AM (1998) Diversity of root-associated bacteria associated with field-grown canola (*Brassica napus* L.) and wheat (*Triticum aestivum* L.). FEMS Microbiol Ecol 26(1):43–50

Glick BR (2014) Bacteria with ACC deaminase can promote plant growth and help to feed the world. Microbiol Res 169(1):30–39

Goggin DE, Emery RJN, Kurepin LV, Powles SB (2015) A potential role for endogenous microflora in dormancy release, cytokinin metabolism and the response to fluridone in *Lolium rigidum* seeds. Ann Bot 115(2):293–301

Gond SK, Bergen MS, Torres MS, White JF (2015) Endophytic Bacillus spp. produce antifungal lipopeptides and induce host defense gene expression in maize. Microbiol Res 172:79–87

Graham PH, Vance CP (2000) Nitrogen fixation in perspective: an overview of research and extension needs. Field Crop Res 65(2–3):93–106

Grobelak A, Napora A, Kacprzak M (2015) Using plant growth-promoting rhizobacteria (PGPR) to improve plant growth. Ecol Eng 84:22–28

Gull M, Hafeez FY (2012) Characterization of siderophore producing bacterial strain *Pseudomonas fluorescens* Mst 8.2 as plant growth promoting and biocontrol agent in wheat. Afr J Microbiol Res 6(33):6308–6318

Gupta S, Pandey S (2019) Unravelling the biochemistry and genetics of ACC deaminase-an enzyme alleviating the biotic and abiotic stress in plants. Plant Gene 18:100175

Gupta Sood S (2003) Chemotactic response of plant-growth-promoting bacteria towards roots of vesicular-arbuscular mycorrhizal tomato plants. FEMS Microbiol Ecol 45(3):219–227

Hallmann J, Quadt-Hallmann A, Mahaffee WF, Kloepper JW (1997) Bacterial endophytes in agricultural crops. Can J Microbiol 43(10):895–914

Halo BA, Khan AL, Waqas M, Al-Harrasi A, Hussain J, Ali L, Lee IJ (2015) Endophytic bacteria (*Sphingomonas* sp. LK11) and gibberellin can improve *Solanum lycopersicum* growth and oxidative stress under salinity. J Plant Int 10(1):117–125

Han HS, Lee KD (2005) Physiological responses of soybean-inoculation of Bradyrhizobium japonicum with PGPR in saline soil conditions. Res J Agric Biol Sci 1(3):216–221

Han J, Xia D, Li L, Sun L, Yang K, Zhang L (2009) Diversity of culturable bacteria isolated from root domains of moso bamboo (*Phyllostachys edulis*). Microb Ecol 58(2):363–373

Hardoim PR (2015) Heading to the origins-Rice microbiome as functional extension of the host. J Rice Res 3(2):1–3

Hardoim PR, van Overbee LS, van Elsas JD (2008) Properties of bacterial endophytes and their proposed role in plant growth. Trends Microbiol 16(10):463–471

Hardoim PR, Hardoim CCP, Overbeek LS, van Elsas JD (2012) Dynamics of seed-borne rice endophytes on early plant growth stages. PLoS One 7(2):e30438

Hashem A, Abd Allah EF, Alqarawi A, Al-Huqail AA, Wirth S, Egamberdieva D (2016) The interaction between arbuscular mycorrhizal fungi and endophytic bacteria enhances plant growth of *Acacia gerrardii* under salt stress. Front Plant Sci 7:1089

Hassani MA, Durán P, Hacquard S (2018) Microbial interactions within the plant holobiont. Microbiome 6(1):1–17

Haynes RJ, Martin RJ, Goh KM (1993) Nitrogen fixation, accumulation of soil nitrogen and nitrogen balance for some field-grown legume crops. Field Crop Res 35(2):85–92

Hedden P, Thomas SG (2012) Gibberellin biosynthesis and its regulation. Biochem J 444(1):11–25

Herrera SD, Grossi C, Zawoznik M, Groppa MD (2016) Wheat seeds harbour bacterial endophytes with potential as plant growth promoters and biocontrol agents of *Fusarium graminearum*. Microbiol Res 186:37–43

Hidri R, Barea JM, Mahmoud OM, Abdelly C, Azcón R (2016) Impact of microbial inoculation on biomass accumulation by *Sulla carnosa* provenances, and in regulating nutrition, physiological and antioxidant activities of this species under non-saline and saline conditions. J Plant Physiol 201:28–41

Höflich G, Wiehe W, Kühn G (1994) Plant growth stimulation by inoculation with symbiotic and associative rhizosphere mechanisms. Experientia 50(10):897–905

Hossain MM, Sultana F, Islam S (2017) Plant growth-promoting fungi (PGPF): phytostimulation and induced systemic resistance. In: Singh D, Singh H, Prabha R (eds) Plant-microbe interactions in agro-ecological perspectives. Springer, Singapore, pp 135–191

Huang JS (1986) Ultrastructure of bacterial penetration in plants. Annu Rev Phytopathol 24(1): 141–157

Jefferson R (1994) The Hologenome. Agriculture, environment and the developing world: a future of PCR. Cold Spring Harbor Laboratory Press, NewYork, p 411

Johnson DL, Anderson DR, McGrath SP (2005) Soil microbial response during the phytoremediation of a PAH contaminated soil. Soil Biol Biochem 37(12):2334–2336

Johnston-Monje D, Raizada MN (2011) Conservation and diversity of seed associated endophytes in Zea across boundaries of evolution, ethnography and ecology. PLoS One 6(6):e20396

Kaga H, Mano H, Tanaka F, Watanabe A, Kaneko S et al (2009) Rice seeds as sources of endophytic bacteria. Microb Environ 24:0904220080

Karadeniz A, Topcuoğlu ŞF, Inan S (2006) Auxin, gibberellin, cytokinin and abscisic acid production in some bacteria. World J Microbiol Biotechnol 22(10):1061–1064

Karthik M, Pushpakanth P, Krishnamoorthy R, Senthilkumar M (2017) Endophytic bacteria associated with banana cultivars and their inoculation effect on plant growth. J Hortic Sci Biotechnol 92(6):568–576

Khalaf EM, Raizada MN (2016) Taxonomic and functional diversity of cultured seed associated microbes of the cucurbit family. BMC Microbiol 16(1):1–16

Khalaf EM, Raizada MN (2018) Bacterial seed endophytes of domesticated cucurbits antagonize fungal and oomycete pathogens including powdery mildew. Front Microbiol 9:42

Khan AG (2005) Role of soil microbes in the rhizospheres of plants growing on trace metal contaminated soils in phytoremediation. J Trace Elem Med Biol 18(4):355–364

Khan AL, Hamayun M, Kim YH, Kang SM, Lee JH, Lee IN (2011) Gibberellins producing endophytic *Aspergillus fumigatus sp.* LH02 influenced endogenous phytohormonal levels isoflavonoids production and plant growth in salinity stress. Process Biochem 46(2):440–447

Khan WU, Ahmad SR, Yasin NA, Ali A, Ahmad A, Akram W (2017) Application of *Bacillus megaterium* MCR-8 improved phytoextraction and stress alleviation of nickel in *Vinca rosea*. Int J Phytoremediation 19(9):813–824

Kim YC, Glick BR, Bashan Y, Ryu CM (2012) Enhancement of plant drought tolerance by microbes. In: Aroca R (ed) Plant responses to drought stress. Springer, Berlin, Heidelberg, pp 383–413

Klaedtke S, Jacques MA, Raggi L, Préveaux A, Bonneau S, Negri V, Chable V, Barret M (2016) Terroir is a key driver of seed_associated microbial assemblages. Environ Microbiol 18(6): 1792–1804

Kloepper JW, Lifshitz R, Zablotowicz RM (1989) Free-living bacterial inocula for enhancing crop productivity. Trends Biotechnol 7(2):39–44

Kloepper JW, Tipping EM, Lifshitz R (1991) Plant growth promotion mediated by bacterial rhizosphere colonizers. In: Keister DL, Cregan PB (eds) The rhizosphere and plant growth. Beltsville Symposia in Agricultural Research, vol 14. Springer, Dordrecht, pp 315–326

Kristin A, Miranda H (2013) The root microbiota—a fingerprint in the soil? Plant Soil 370(1): 671–686

Lavania M, Nautiyal C (2013) Solubilization of tricalcium phosphate by temperature and salt tolerant *Serratia marcescens* NBRI1213 isolated from alkaline soils. Afr J Microbiol Res 7(34):4403–4413

Leifert C, Waites WM, Nicholas JR (1989) Bacterial contaminants of micropropagated plant cultures. J Appl Bacteriol 67(4):353–361

Li YH, Zhu JN, Zhai ZH, Zhang Q (2010) Endophytic bacterial diversity in roots of *Phragmites australis* in constructed Beijing Cuihu wetland (China). FEMS Microbiol Lett 309(1):84–93

Liu S, Zuo L, Xu Y, Zou W (2012) Song study on diversity of endophytic bacterial communities in seeds of hybrid maize and their parental lines. Arch Microbiol 194(12):1001–1101

Liu F, Xing S, Ma H, Du Z, Ma B (2013) Cytokinin-producing, plant growth-promoting rhizobacteria that confer resistance to drought stress in *Platycladus orientalis* container seedlings. Appl Microbiol Biotechnol 97(20):9155–9164

Lodewyckx C, Vangronsveld J, Porteous F, Moore ERB, Taghavi S, Mezgeay M, Van der Lelie D (2002) Endophytic bacteria and their potential applications. Crit Rev Plant Sci 21(6):583–606

Loper J, Henkels MD (1999) Utilization of heterologous siderophores enhances levels of iron available to *Pseudomonas putida* in the rhizosphere. Appl Environ Microbiol 65(12): 5357–5363

Lopez SMY, Pastorino GN, Franco MEE, Medina R, Lucenttini CG, Saparrat MCN, Balatti PA (2018) Microbial endophytes that live within the seeds of two tomato hybrids cultivated in Argentina. Agronomy 8(8):136

López-López A, Rogel MA, Ormeño-Orillo E, Martínez-Romero J, Martínez-Romero E (2010) *Phaseolus vulgaris* seed-borne endophytic community with novel bacterial species such as *rhizobium endophyticum* sp. nov. Syst Appl Microbiol 33(6):322–327

Lugtenberg B, Kamilov F (2009) Plant-growth-promoting rhizobacteria. Annu Rev Microbiol 63: 541–556

Lüthen H, Claussen M, Böttger M (1999) Growth: progress in auxin research. In: Esser K, Kadereit JW, Lüttge U, Runge M (eds) Progress in botany, vol 60. Springer, Berlin, Heidelberg, pp 315–340

Ma Y, Rajkumar M, Fritas H (2008) Inoculation of plant growth promoting bacterium *Achromobacter xylosoxidans* strain Ax10 for the improvement of copper phytoextraction by *Brassica juncea*. J Environ Manage 90(2):831–837

Majeed A, Muhammad Z, Ahmad H (2018) Plant growth promoting bacteria: role in soil improvement, abiotic and biotic stress management of crops. Plant Cell Rep 37(12):1599–1609

Mallick I, Bhattacharyya C, Mukherji S, Dey D, Sarkar SC, Mukhopadhyay UK, Ghosh A (2018) Effective rhizoinoculation and biofilm formation by arsenic immobilizing halophilic plant growth promoting bacteria (PGPB) isolated from mangrove rhizosphere: a step towards arsenic rhizoremediation. Sci Total Environ 610:1239–1250

Mano H, Tanaka F, Watanabe A, Kaga H, Okunishi S, Morisaki H (2006) Culturable surface and endophytic bacterial flora of the maturing seeds of rice plants (*Oryza sativa*) cultivated in a paddy field. Microbes Environ 21(2):86–100

Marschner H (2011) Marschner's mineral nutrition of higher plants. Academic Press, San Diego, CA

Marulanda A, Barea JM, Azcón R (2009) Stimulation of plant growth and drought tolerance by native microorganisms (AM fungi and bacteria) from dry environments: mechanisms related to bacterial effectiveness. J Plant Growth Regul 28(2):115–124

Mastretta C, Taghavi S, Van der Lelie D, Mengoni A, Galardi F, Gonnelli C et al (2009) Endophytic bacteria from seeds of *Nicotiana tabacum* can reduce cadmium phytotoxicity. Int J Phytoremediation 11(3):251–267

Mathesius U, Mulders S, Go M, TeplitskI M, Caetano-Anollés G, Rolfe BG, Bauer WD (2003) Extensive and specific responses of a eukaryote to bacterial quorum-sensing signals. Proc Natl Acad Sci U S A 100(3):1444–1449

Mattos KA, Pádua VL, Romeiro A, Hallack LF, Neves BC, Ulisses TM, Mendonça-Previato L (2008) Endophytic colonization of rice (*Oryza sativa* L.) by the diazotrophic bacterium *Burkholderia kururiensis* and its ability to enhance plant growth. An Acad Bras Cienc 80(3): 477–493

McInroy JA, Kloepper JW (1991) Novel bacterial taxa inhabiting internal tissues of sweet corn and cotton. In: Ryder MH, Stephens PM, Bowen GD (eds) Improving plant productivity with rhizosphere bacteria. CSIRO, Adelaide, p 190

Mehmood U, Inam-ul-Haq M, Saeed M, Altaf A, Azam F, Hayat S (2018) A brief review on plant growth promoting rhizobacteria (PGPR): a key role in plant growth promotion. Plant Prot 2(2): 77–82

Mishra SK, Khan MH, Misra S, Dixit KV, Khare P, Srivastava S et al (2017) Characterization of *Pseudomonas* spp. and *Ochrobactrum* sp. isolated from volcanic soil. Antonie Van Leeuwenhoek 110(2):253–270

Mitter B, Petric A, Chain SG, Trognitz F, Nowak J, Compant S, Sessitsch A et al (2013) Genome analysis, ecology, and plant growth promotion of the endophyte *Burkholderia phytofirmans* strain PsJN. Mol Microb Ecol Rhizosphere 1:865–874

Mitter B, Pfaffenbichler N, Sessitsch A (2016) Plant-microbe partnerships in 2020. J Microbial Biotechnol 95(5):635–640

Mitter B, Pfaffenbichler N, Flavell R, Compant S, Antonielli L, Petric A, Berninger T, Muhammad N, Sheibani-Tezeji R, Von Maltzahn G, Sessitsch A (2017) A new approach to modify plant microbiomes and traits by introducing beneficial bacteria at flowering into progeny seeds. Front Microbiol 8:11

Moënne-Loccoz Y, Mavingui P, Combes C, Normand P, Steinberg C (2015) Microorganisms and biotic interactions. In: Bertrand JC, Caumette P, Lebaron P, Matheron R, Normand P, Sime-Ngando T (eds) Environmental microbiology: fundamentals and applications. Springer, Dordrecht, pp 395–444

Mukhopadhyay NK, Garrison NK, Hinton DM, Bacon CW, Khush GS, Peck HD, Datta N (1996) Identification and characterization of bacterial endophytes of rice. Mycopathologia 134(3): 151–159

Naz I, Bano A, Ul-Hassan T (2009) Isolation of phytohormones producing plant growth promoting rhizobacteria from weeds growing in Khewra salt range, Pakistan and their implication in providing salt tolerance to soybean (*Glycine max* L.). Afr J Biotechnol 8(21):5762–5766

Nelson E (2004) Microbial dynamics and interactions in the spermosphere. Annu Rev Phytopathol 42:271–309

Nelson SK, Steber CM (2018) Gibberellin hormone signal perception: down-regulating DELLA repressors of plant growth and development. Annu Rev Online:153–187

Nicholson W, Munakata N, Horneck G, Melosh H, Setlow P (2000) Resistance of Bacillus endospores to extreme terrestrial and extraterrestrial environments. Microbiol Mol Boil Rev 64(3):548–572

Normanly J (1997) Auxin metabolism. Physiol Plant 100:431–442

Oehrle NW, Karr DB, Kremer RJ, Emerich DW (2000) Enhanced attachment of *Bradyrhizobium japonicum* to soybean through reduced root colonization of internally seedborne microorganisms. Can J Microbiol 46(6):600–606

Okunishi S, Sako K, Mano H, Imamura A, Morisaki H (2005) Bacterial flora of endophytes in the maturing seed of cultivated rice (*Oryza sativa* L.). Microbes Environ 20(3):168–177

Olanrewaju OS, Glick BR, Babalola OO (2017) Mechanisms of action of plant growth promoting bacteria. World J Microbiol Biotechnol 33(11):1–16

Owen D, Williams AP, Griffith GW, Withers PJ (2015) Use of commercial bio-inoculants to increase agricultural production through improved phosphrous acquisition. Appl Soil Ecol 86: 41–54

Park YG, Mun BG, Kang SM, Hussain A, Shahzad R, Seo CW et al (2017) *Bacillus aryabhattai* SRB02 tolerates oxidative and nitrosative stress and promotes the growth of soybean by modulating the production of phytohormones. PLoS One 12:e0173203

Pitzschke A (2016) Developmental peculiarities and seed-borne endophytes in quinoa: omnipresent, robust bacilli contribute to plant fitness. Front Microbiol 7:2–12

Puente ME, Li CY, Bashan Y (2009) Endophytic bacteria in cacti seeds can improve the development of cactus seedlings. Environ Exp Bot 66(3):402–408

Raghuwanshi R, Prasad JK (2018) Perspectives of rhizobacteria with ACC deaminase activity in plant growth under abiotic stress. In: Root biology, pp 303–321

Rascovan N, Carbonetto B, Perrig D, Díaz M, Canciani W, Abalo M, Vazquez MP (2016) Integrated analysis of root microbiomes of soybean and wheat from agricultural fields. Sci Rep 6(1):1–12

Rastogi G, Sbodio A, Tech JJ, Suslow TV, Coaker GL, Leveau JH (2012) Leaf microbiota in an agroecosystem: spatiotemporal variation in bacterial community composition on field-grown lettuce. ISME J 6(10):1812–1822

Raza A, Faisal M (2013) Growth promotion of maize by desiccation tolerant *Micrococcus luteus*-chp37 isolated from Cholistan desert, Pakistan. Aust J Crop Sci 7(11):1693–1698

Rezki S, Campion C, Simoneau P, Jacques MA, Shade A, Barret M (2018) Assembly of seed-associated microbial communities within and across successive plant generations. Plant Soil 422(1):67–79

Rijavec T, Lapanje A, Dermastia M, Rupnik M (2007) Isolation of bacterial endophytes from germinated maize kernels. Can J Microbiol 53(6):802–808

Rincón A, Ruiz-Díez B, García-Fraile S, García JAL, Fernández-Pascual M, Pueyo JJ, de Felipe MR (2005) Colonisation of *Pinus halepensis* roots by *Pseudomonas fluorescens* and interaction with the ectomycorrhizal fungus *Suillus granulatus*. FEMS Microbiol Ecol 51(3):303–311

Ringelberg D, Foley K, Reynolds CM (2012) Bacterial endophyte communities of two wheatgrass varieties following propagation in different growing media. Can J Microbiol 58(1):67–80

Rodríguez CE, Mitter B, Barret M, Sessitsch A, Compant S (2018) Commentary: seed bacterial inhabitants and their routes of colonization. Plant Soil 422(1):129–134

Rojas-Tapias DF, Bonilla RR, Dussán J (2012) Effect of inoculation with plant growth-promoting bacteria on growth and copper uptake by sunflowers. Water Air Soil Pollut 223(2):643–654

Rolli E, Marasco R, Vigani G, Ettoumi B, Mapelli F, Deangelis ML, Daffonchio D (2015) Improved plant resistance to drought is promoted by the root-associated microbiome as a water stress-dependent trait. Environ Microbiol 17(2):316–331

Rosenblueth M, Martínez-Romero E (2006) Bacterial endophytes and their interactions with hosts. Mol Plant Microbe Interact 19(8):827–837

Ruiza D, Agaras B, de Werrab P, Wall LG, Valverde C (2011) Characterization and screening of plant probiotic traits of bacteria isolated from rice seeds cultivated in Argentina. J Microbiol 49(5):902–912

Rybakova D, Mancinelli R, Wikstrom M, Birch-Jensen AS, Postma J, Ehlers RU, Goertz S, Berg G (2017) The structure of the *Brassica napus* seed microbiome is cultivar-dependent and affects the interactions of symbionts and pathogens. Microbiome 5(1):104

Saharan BS, Nehra V (2011) Plant growth promoting rhizobacteria: a critical review. Life Sci Med Res 21(1):30

Saikia J, Sarma RK, Dhandia R, Yadav A, Bharali R, Gupta VK, Saikia R (2018) Alleviation of drought stress in pulse crops with ACC deaminase producing rhizobacteria isolated from acidic soil of Northeast India. Sci Rep 8(1):1–16

Sakakibara H (2006) Cytokinins: activity, biosynthesis, and translocation. Annu Rev Plant Biol 57: 431–449

Salomon MV, Bottini R, de Souza Filho GA, Cohen AC, Moreno D, Gil M et al (2014) Bacteria isolated from roots and rhizosphere of *Vitis vinifera* retard water losses, induce abscisic acid accumulation and synthesis of defense-related terpenes in in vitro cultured grapevine. Physiol Plant 51(4):359–374

Sanchez-Canizares C, Jorrin C, Polle PS, Tkacz A (2017) Understanding the holonbiont : the interdependence of plants and their microbiome. Curr Opin Microbiol 38:188–196

Scherer-Lorenzen M, Venterink HO, Buschmann H (2008) Nitrogen enrichment and plant invasions: the importance of nitrogen-fixing plants and anthropogenic eutrophication. In: Nentwig W (ed) Biological invasions. Ecological studies (analysis and synthesis), vol 193. Springer, Berlin, Heidelberg, pp 163–180

Schmelz EA, Engelberth J, Alborn HT, O'donnell P, Sammons M, Toshima H, Tumlinson JH (2003) Simultaneous analysis of phytohormones, phytotoxins, and volatile organic compounds in plants. Proc Natl Acad Sci 100(18):10552–10557

Schuhegger R, Ihring A, Gantner S, Bahnweg G, Knappe C, Vogg G, Langebartels C (2006) Induction of systemic resistance in tomato by N-acyl-L-homoserine lactone-producing rhizosphere bacteria. Plant Cell Environ 29(5):909–918

Selosse MA, Baudoin E, Vandenkoornhuyse P (2004) Symbiotic microorganisms, a key for ecological success and protection of plants. C R Biol 327(7):639–648

Shade A, Jacques MA, Barret M (2017) Ecological patterns of seed microbiome diversity, transmission, and assembly. Curr Opin Microbiol 37:15–22

Shahzad R, Khan AL, Bilal S, Waqas M, Kang SM, Lee IJ (2017) Inoculation of abscisic acid-producing endophytic bacteria enhances salinity stress tolerance in *Oryza sativa*. Environ Exp Bot 13:68–77

Shahzad R, Khan AL, Bilal S, Asaf S, Lee IJ (2018) What is there in seeds? Vertically transmitted endophytic resources for sustainable improvement in plant growth. Front Plant Sci 9:24

Sheng XF, Xia JJ, Jiang CY, He LY, Qian M (2008) Characterization of heavy metal-resistant endophytic bacteria from rape (*Brassica napus*) roots and their potential in promoting the growth and lead accumulation of rape. Environ Pollut 156(3):1164–1170

Shiomi HF, Silva HSA, Melo ISD, Nunes FV, Bettiol W (2006) Bioprospecting endophytic bacteria for biological control of coffee leaf rust. Sci Agric 63(1):32–39

Siddikee MA, Glick BR, Chauhan PS, Jong Yim W, Sa T (2011) Enhancement of growth and salt tolerance of red pepper seedlings (*Capsicum annuum* L.) by regulating stress ethylene synthesis with halotolerant bacteria containing 1-aminocyclopropane-1-carboxylic acid deaminase activity. Plant Physiol Biochem 49(4):427–434

Singh RP, Jha PN (2016) A halotolerant bacterium *Bacillus licheniformis* HSW-16 augments induced systemic tolerance to salt stress in wheat plant (*Triticum aestivum*). Front Plant Sci 7: 1890

Sobolev VS, Orner VA, Arias RS (2013) Distribution of bacterial endophytes in peanut seeds obtained from axenic and control plant material under field conditions. Plant Soil 371(1): 367–376

Sorty AM, Meena KK, Choudhary K, Bitla UM, Minhas PS, Krishnani KK (2016) Effect of plant growth promoting bacteria associated with halophytic weed (*Psoralea corylifolia* L.) on germination and seedling growth of wheat under saline conditions. Appl Biochem Biotechnol 180(5): 872–882

Spaepen S (2015) Plant hormones produced by microbes. In: Lugtenberg B (ed) Principles of plant-microbe interactions. Springer, Cham, pp 247–256

Sprent JI, De Faria SM (1989) Mechanisms of infection of plants by nitrogen fixing organisms. In: Skinner FA, Boddey RM, Fendrik I (eds) Nitrogen fixation with non-legumes. Developments in plant and soil sciences, vol 35. Springer, Dordrecht, pp 3–11

Suman A, Yadav AN, Verma P (2016) Endophytic microbes in crops: diversity and beneficial impact for sustainable agriculture. In: Singh D, Singh H, Prabha R (eds) Microbial inoculants in sustainable agricultural productivity. Springer, New Delhi, pp 117–143

Sundaramoorthy S, Balabaskar P (2013) Evaluation of combined efficacy of *Pseudomonas fluorescens* and *Bacillus subtilis* in managing tomato wilt caused by *Fusarium oxysporum f.* sp. *lycopersici* (Fol). Plant Pathol J (Faisalabad) 12(4):154–161

Taylor MW, Schupp PJ, De Nys R, Kjelleberg S, Steinberg PD (2005) Biogeography of bacteria associated with the marine sponge *Cymbastela concentrica*. Environ Microbiol 7(3):419–433

Truyens S, Weyens N, Cuypers A, Vangronsveld J (2013) Changes in the population of seed bacteria of transgenerationally cd-exposed *Arabidopsis thaliana*. Plant Biol 15(6):971–981

Truyens S, Jambon I, Croes S, Janssen J, Weyens N, Mench M, Vangronsveld J et al (2014) The effect of long-term cd and Ni exposure on seed endophytes of *Agrostis capillaris* and their potential application in phytoremediation of metal-contaminated soils. Int J Phytoremediation 16(7–8):643–659

Truyens S, Weyens N, Cuypers A, Vangronsveld J (2015) Bacterial seed endophytes : genera, vertical transmission and interaction with plants. Environ Microbiol Rep 7(1):40–50

Tsukanova KA, Chebotar VK, Meyer JJM, Bibikova TN (2017) Effect of plant growth-promoting Rhizobacteria on plant hormone homeostasis. S Afr J Bot 113:91–102

Turnbull GA, Morgan JAW, Whipps JM, Saunders JR (2001) The role of motility in the in vitro attachment of *Pseudomonas putida* PaW8 to wheat roots. FEMS Microbiol Ecol 35(1):57–65

Turner T, James EK, Poole PS (2013) The plant microbiome. Genome Biol 14(6):1–10

Upadhyay SK, Singh JS, Saxena AK, Singh DP (2012) Impact of PGPR inoculation on growth and antioxidant status of wheat under saline conditions. Plant Biol 14(4):605–611

Vandenkoornhuyse P, Quaiser A, Duhamel M, Le Van A, Dufresne A (2015) The importance of the microbiome of the plant holobiont. New Phytol 206(4):1196–1206

Verma SC, Ladha JK, Tripathi AK (2001) Evaluation of plant growth promoting and colonization ability of endophytic diazotrophs from deep water rice. J Biotechnol 91(2–3):127–141

Verma SK, Kingsley K, Irizarry I, Bergen M, Kharwar RN, White JF Jr (2017) Seed-vectored endophytic bacteria modulate development of rice seedlings. J Appl Microbiol 122(6): 1680–1691

Vitorino LC, Bessa LA, Carvalho LG, Silva FG (2016) Growth promotion mediated by endophytic fungi in cloned seedlings of *Eucalyptus grandis* x *Eucalyptus urophylla* hybrids. Afr J Biotechnol 15(48):2729–2738

Vurukonda SSKP, Vardharajula S, Shrivastava M, SkZ A (2016) Enhancement of drought stress tolerance in crops by plant growth promoting rhizobacteria. Microbiol Res 184:13–24

Wallace J, Laforest-Lapointe I, Kembel SW (2018) Variation in the leaf and root microbiome of sugar maple (*Acer saccharum*) at an elevational range limit. Peer J 14:6e5293

Waqas M, Khan AL, Kamran M, Hamayun M, Kang SM, Kim YH, Lee IJ et al (2012) Endophytic fungi produce gibberellins and indoleacetic acid and promotes host-plant growth during stress. Molecules 17(9):10754–10773

Weil RR, Brady NC (2016) The nature and properties of soils. Pearson Press, New York

Weyens N, van der Lelie D, Taghavi S, Vangronsveld J (2009) Phytoremediation: plant–endophyte partnerships take the challenge. Curr Opin Biotechnol 20(2):248–254

Xu M, Sheng J, Chen L, Men Y, Gan L, Guo S, Shen L (2014) Bacterial community compositions of tomato (*Lycopersicum esculentum* mill.) seeds and plant growth promoting activity of ACC deaminase producing *Bacillus subtilis* (HYT-12-1) on tomato seedlings. World J Microbiol Biotechnol 30(3):835–845

Yadav RS, Tarafdar JC (2001) In fluence of organic and inorganic phosphorus supply on the maximum secretion of acid phosphatase by plants. Biol Fertil Soils 34:140–143

Yandigeri MS, Meena KK, Singh D, Malviya N, Singh DP, Solanki MK, Arora DK et al (2012) Drought-tolerant endophytic actinobacteria promote growth of wheat (*Triticum aestivum*) under water stress conditions. Plant Growth Regul 68(3):411–420
You CB, Lin M, Fang XJ, Song W (1995) Attachment of *Alcaligenes* to rice roots. Soil Biol Biochem 27(4–5):463–466
Zaheer A, Mirza BS, Mclean JE, Yasmin S, Shah TM, Malik KA, Mirza MS (2016) Association of plant growth-promoting *Serratia* spp. with the root nodules of chickpea. Res Microbiol 167(6): 510–520
Zaidi S, Usmani S, Singh BR, Musarrat J (2006) Significance of *Bacillus subtilis* strain SJ-101 as a bioinoculant for concurrent plant growth promotion and nickel accumulation in *Brassica juncea*. Chemosphere 64(6):991–997
Zawoznik MC, Vázquez S, Díaz Herrera SM, Groppa M (2014) Search for endophytic diazotrophs in barley seeds. Braz J Microbiol 45(2):621–625
Zhao L, Zhang YQ (2015) Effects of phosphate solubilization and phytohormone production of *Trichoderma asperellum* Q1 on promoting cucumber growth under salt stress. J Integr Agric 14(8):1588–1597
Zhulin IB, Tretyakova SE, Ignatov VV (1988) Chemotaxis of *Azospirillum brasilense* towards compounds typical of plant root exudates. Folia Microbiol 33(4):277–280

4 Microorganisms as Biocontrol Agents of Pests and Diseases

Muriel da Silva Folli-Pereira, Amanda Azevedo Bertolazi, Juliana Garlet, Lucas Zanchetta Passamani, Sávio Bastos de Souza, Fernanda Senna Ferreira Costa, and Alessandro Coutinho Ramos

Abstract

Plant diseases and insect pests are among the main factors that limit the world agroforestry production, and possible control measures of phytopathogens and insects include the use of chemical pesticides, although major concerns about their effects on human health and the environment limit their use. Biological control of pests and plant diseases through biocontrol agents, such as microorganisms, has been increasingly used, especially as a eco-friendly approach to the environment, and meeting demand of food products free from agrochemical residues. These microorganisms can be easily isolated by conventional methods and used mainly due to their ease of cultivation. Thus, this chapter presents the main microbial biocontrol agents of pests and plant diseases, and the mechanisms associated, as well as a discussion about the economic impacts of biological control.

Keywords

Biological control · In vitro screening · Pest-insects · Phytopathogens · Mechanisms · Economic impacts

M. da Silva Folli-Pereira (✉) · J. Garlet
Universidade do Estado de Mato Grosso—Unemat, Av. Perimetral Rogério Silva, Alta Floresta, MT, Brazil
e-mail: muriel.folli@unemat.br

A. A. Bertolazi · S. B. de Souza · F. S. F. Costa · A. C. Ramos
Biotechnology and Environmental Microbiology Lab, Universidade Vila Velha (UVV), Vila Velha, ES, Brazil

L. Z. Passamani
FAESA Centro Universitário, Vitória, ES, Brazil

N. K. Arora, B. Bouizgarne (eds.), *Microbial BioTechnology for Sustainable Agriculture Volume 1*, Microorganisms for Sustainability 33,
https://doi.org/10.1007/978-981-16-4843-4_4

4.1 Introduction

Estimates from the United Nations demonstrate an accelerated population growth for the coming years, in which the world population is likely to exceed eight billion inhabitants in 2024 and 9.5 billion by 2050 (ONU 2012). Certainly, this population growth will increase the demand for food, which traditionally, makes intensive use of pesticides for the control of pests and diseases, creating environmental and human health problems.

In the same proportion as the demand for food grows, so does society's concern about the impacts of pesticides use on the environment. Agricultural sector has been increasingly pressured to adopt more sustainable production systems that result in food production without the use of pesticides (Morandi and Bettiol 2009).

The use of genetically modified plants or even plant cultivars that are resistant to pests and/or diseases is a widely used method for controlling pests and diseases (Havey 2018), although it does not work well for all plant species because these germplasms are not widely used in commercial cultivars. For some plants, disease control through the use of pesticides has been inefficient, due to the resistance accumulated by the wide application of a limited number of registered molecules, which has favoured the use of other methods like the use of biological control agents (BCAs) (Rivera-Méndez et al. 2020). Despite certain control effects, the problems of plant resistance to pesticides, environmental pollution and human health are becoming more serious and have increasingly drawn the world's attention. To solve these problems, several studies have been carried out to research and explore biopesticides produced by involving endophytic microorganisms, to control plant diseases (Cui et al. 2019).

The application of fungicides to control root pathogens in the soil has a questionable efficacy, also resulting in the accumulation of toxic compounds and compromising the survival and diversity of soil microorganisms, which are useful for plant development and soil quality (Rojo et al. 2007; Arora 2018) and can cause pathogen resistance (Selim et al. 2014; Gholami et al. 2019). In this context, sustainable alternatives for disease control, such as BCAs, are necessary. The interaction of microorganisms used as BCAs with host plants results in important processes that lead to the suppression and/or elimination of phytopathogens, such as by the production of antibiotics, parasitism/predation, production of cell wall degrading enzymes or lytic enzymes, competition for nutrients or space and establishment of induced resistance in the plant (Chernin et al. 1995; Cohen-Kupiec and Chet 1998; Léon-Kloosterziel et al. 2005; Verma et al. 2019).

Lopes (2009) highlights that the mass production of microorganisms to control pests and diseases, enabling the supply of large amounts of these agents, is essential for the evolution of this control method. Many biological control programs with microorganisms use the flood strategy, in which the control agent must be available in large quantities. Even for inoculative strategies, which occur with a certain periodicity, the availability of a microbial product in the most favourable times for application is essential. In this context, the market for biological products for the control of pests and agricultural diseases grew by more than 70% in 2019 in Brazil, for example, moving around R$ 464 million in 2018. The Brazilian result is considered the most expressive in the sector's history and exceeds the percentage presented by the international market (Brazil 2019).

4.2 Microorganisms with Biotechnological Potential for Biological Control of Pest-Insects and Phytopathogens

Insect pests are one of the main challenges in cultivated ecosystems, affecting the production costs, and considering that a large percentage of agricultural and forest plantations are found in the tropical regions, the incidence of these insects increases considerably. The search for forms of insect pests' control and regulation that are less aggressive to the environment, with less use of chemical insecticides, is gaining, and the increase in organic production systems also stimulates the development of new control alternatives (Van Driesche and Bellows 1996; Fontes et al. 2020). One of the control options with the least environmental impact is the biological control, which can be defined as "the purposeful utilization of introduced or resident living organisms, to suppress the activities and populations of one or more plant pathogens" (Pal and Gardener 2006). Thus, the use of pathogens (microorganisms) in the biological control of insect pests has increased significantly in integrated pest management programs.

The microorganisms that cause disease in insects are called entomopathogens. First reports of insect diseases appeared almost 5000 years ago, when bees and silkworms were exploited by Egyptians and Chinese (Alves 1998). One of the major milestones in the control of insect pests with entomopathogens was the discovery of the bacterium *Bacillus thuringiensis* Berliner (1915) (*Eubacteriales*: *Bacillaceae*) in the early twentieth century. From this discovery, the microbial control of insects had great advances, reaching the Era of Genetic Engineering, giving rise to transgenic plants that involve strains of *B. thuringiensis* (Lopes 2009). The main microorganisms used in the control of insect pests are bacteria, fungi, viruses and nematodes, and all groups have products commercialized in many countries. Microbial insecticides have the entomopathogenic microorganisms in their formulation as an active biological compound, and their insecticidal action comes from the organism itself or a substance it produces. Its main advantage is greater selectivity and less or no toxicity when compared to conventional chemical insecticides.

Despite their potential use in sustainable disease management, the success of these microorganisms and their interaction with other methods worldwide, BCAs have not been widely used for most agricultural crops (Subedi et al. 2020). However, studies involving plant-microorganism-microorganism interactions have been increasing in recent years and have stood out as an effective strategy to increase agricultural productivity in a sustainable way using BCAs of phytopathogens (Table 4.1). Strategies for the biological control of plant diseases that use beneficial microorganisms, whether endophytic, rhizobacteria or residing in the phylloplane (e.g. *Pseudomonas*, *Trichoderma* and *Bacillus* species), are considered ecologically correct for the management of plant diseases (Ayukea et al. 2017). These microorganisms can act by direct antagonism, by inducing resistance, or even, in many cases, by both ways, concomitantly (Morandi and Bettiol 2009). They provide several advantages for plants such as growth promotion and suppression and/or reduction of disease (Smith et al. 1999).

4.2.1 Bacteria

The main species of known entomopathogenic bacteria belong to the families *Bacillaceae*, *Pseudomonadaceae*, *Enterobacteriaceae*, *Streptococcaceae* and *Micrococcaceae*. However, the genus *Bacillus* Cohn (1872), of the *Bacillaceae* family, mainly the species *B. thuringiensis* (Bt) and its subspecies and varieties, have been the most used BCA of insect pests in different cultivated systems. According to ISAAA (International Service for the Acquisition of Agri-Biotech Applications) (2020), Bt is easily grown by fermentation, and for the past 50 years, it has been used as an insecticide by farmers around the world. Organic agriculture has benefited from the Bt insecticide, since it is one of the few products allowed in this cultivation system. Thus, considering the importance of the genus *Bacillus* for the biological control of insects, Fig. 4.1 shows the main species, variety and subspecies of this genus used in the control of insect pests and which insect groups are controlled with these agents.

B. thuringiensis is a Gram positive, aerobically or optionally anaerobic bacterium, naturally found in the soil, and can be kept in latency for significant periods in adverse conditions. To be effective in controlling insects, this entomopathogenic bacteria needs to be in contact with the insect, and often needs to be ingested, thus producing endotoxins in the digestive system, leading the insect to death. In the sporulation phase, these bacteria synthesize proteins that accumulate on the periphery of spores in the form of crystals in one of the poles of the cell (Peferöen 1997). These crystals are composed of one or more *Cry* proteins, which are highly toxic and specific to a certain group of insects. When ingested by the insect, *B. thuringiensis* produces *Cry* proteins which are activated in the alkaline condition of the intestine and pierce the midgut, leaving the insect unable to feed, so it dies within a few days (ISAAA 2020). This bacterium has a great biological diversity and can synthesize different toxic proteins to different pests. Being highly specialized, it is used mainly to control defoliating caterpillars, especially the subspecies *B. thuringiensis kurstaki*, known as Btk (Herrero et al. 2001; Usta 2013). Products that use Bt as an active compound have been available in the market since the 1970s, successfully controlling several insects of the order Lepidoptera, Diptera and Coleoptera. According to Wang et al. (2020), most strains of *Bacillus* associated with plants are plant growth-promoting rhizobacteria (PGPR), which provide new possibilities for sustainable agriculture, because this interaction results in plants resistant to a wide range of fungal and bacterial phytopathogens through various mechanisms.

Several species of PGPR of the genera *Azotobacter*, *Azospirillum*, *Pseudomonas* and *Bacillus* have been studied for their ability to biocontrol plant diseases (Rai and Nabti 2017; Boukerma et al. 2017; Gouda et al. 2018). In addition, several strains of

Table 4.1 Examples of screenings performed for biological control agents of plant pathogens

Biocontrol agent	Target phytopathogen	References
Acaulospora terricola	*Fusarium oxysporum* f. sp. *herbemontis*	Brum et al. (2012)
Acremonium obclavatum	*Puccinia arachidis*	Sathiyabama and Balasubramanian (2017)
Alternaria sp.	*Colletotrichum acutatum*	Landum et al. (2016)
Ampelomyces quisqualis	*Erysiphales* ssp.	Angeli et al. (2013)
Choiromyces aboriginum	*Rhizoctonia solani, Sclerotium rolfsii* *Rhizoctonia cerealis* *Fusarium graminearum* *F. oxysporum* f. sp. *perniciosum* *F. oxysporum* f. sp. *vasinfectum*	Cao et al. (2009)
Bacillus atrophaeus	*Colletotrichum gloeosporioides*	Guardado-Valdivia et al. (2018)
Bacillus subtilis	*Fusarium* ssp. *Ralstonia solanacearum*	Yu et al. (2011) and Djaya et al. (2019)
Bacillus pumilus	*Sclerotinia sclerotiorum* *Paecilomyces lilacinus*	Cavalcanti et al. (2020) and Zou et al. (2007)
Bacillus amyloliquefaciens	*Clavibacter michiganensis*	Gautam et al. (2019)
Bacillus licheniformis	*Phytophthora capsici* *Meloidogyne* ssp.	Li et al. (2020) and Chen and Dickinson (2004)
Bacillus velezensis	*Phoma foveat* *R. solani* *Fusarium averaceum* *Colletotrichum coccodes* *Phytophthora nicotianae* *Botrytis cinerea*	Cui et al. (2019), Guo et al. (2020), Jiang et al. (2018) and Jiang et al (2019)
Bacillus thuringiensis	*S. sclerotiorum* *Plutella xylostella*	Wang et al (2020)
Epicoccum nigrum	*Fusarium verticillioides,* *Ceratocystis paradoxa* *Pectobacterium carotovora* subsp. *atrosepticum*	Favaro et al. (2012) and Bagy et al. (2019)
Pasteuria penetrans	*Meloidogyne* ssp.	Dickson et al. (1992)
Pseudomonas chlororaphis	*Erwinia carotovora* ssp. *Carotovora*	Han et al. (2006)
Pseudomonas fluorescens	*R. solani* *Macrophomina phaseolina*	Shanmugaiah et al. (2009)
Pseudomonas aeruginosa	*B. cinerea*	Audenaert et al. (2002)
Pseudomonas putida	*B. cinerea* *R. solanacearum* *Xanthomonas oryzae* pv. *oryzae* *X. oryzae* pv. *oryzicola* *Xanthomonas citri* subsp. *citri*	Ongena et al. (2005) and Sun et al (2017)

(continued)

Table 4.1 (continued)

Biocontrol agent	Target phytopathogen	References
Pseudomonas parafulva	*Xanthomonas axonopodis* pv. *glycines* *Burkholderia glumae*	Kakemboa and Leea (2019)
Pseudomonas brassicacearum *Pseudomonas protegens* *Pseudomonas vranovensis*	*Ralstonia pseudosolanacearum*	Subedi et al. (2020)
Rhizopycnis vagum	*R. solani* *Corynespora cassiicola* *C. acutatum* *Phytophthora infestans* *F. oxysporum* *S. rolfsii*	Anisha et al. (2018)
Streptomyces lydicus	Soil pathogens	Gardener and Fravel (2002)
Streptomyces griseoviridis	*Fusarium* ssp., *Alternaria brassicicola*, *Phomopsis* ssp., *Botrytis* ssp., *Phytophthora* ssp., *Phytium* ssp.	
Streptomyces jietaisiensis	*Meloidogyne incognita*	Ruanpanun and Nimnoi (2020)
Streptomyces thermocarboxydus	*Glomerella cingulata*	Marian et al. (2020)
Streptomyces rochei	*F. oxysporum* f. sp. *lycopersici*	Kanini et al. (2013)
Trichoderma harzianum	*Gaeumannomyces graminis* var. *tritici* *B. cinerea*, *R. solani* and *Pythium ultimum* *Pythium irregulare* *S. sclerotiorum* *Crinipellis perniciosa* *M. phaseolina* *Pyrenophora teres*	Almassi et al. (1991), Dickinson et al. (1989), Vinale et al. (2006), Vinale et al. (2009), Marco et al. (2003), Khalili et al. (2016) and Paulina et al. (2019)
Trichoderma koningii, *Trichoderma aureoviride*	*Bipolaris sorokiniana*, *F. oxysporum*, *Phytophthora cinnamomi*, *Pythium middletonii*, *R. solani*	Vinale et al. (2014)
Trichoderma viride	*S. rolfsii*, *R. solani*, *P. ultimum*, *Fusarium solani* *M. phaseolina*	Awad et al. (2018) and Shanmugaiah et al. (2009)
Trichoderma longibrachiatum	*P. ultimum* *P. teres*	Migheli et al. (1998) and Paulina et al. (2019)
Trichoderma hamatum	*S. sclerotiorum*	Jones et al. (2014)

(continued)

Table 4.1 (continued)

Biocontrol agent	Target phytopathogen	References
Trichoderma atroviride	Soil pathogens in pastures	Kandula et al. (2015)
Trichoderma asperellum	*Meloidogyne javanica* *Sclerotium cepivorum* *Phytophthora megakarya*	Sharon et al. (1993), Rivera-Méndez et al. (2020) and Mbarga et al. (2020)
Trichoderma citrinoviride	*Cylindrocarpon destructans* *R. solani* *B. cinerea* *Alternaria panax* *C. destructans* *Phytophthora cactorum* *Pythium* spp. *B. cinerea*	Park et al. (2018)
Trichoderma gamsii	*Epicoccum nigrum*, *Scytalidium lignicola* *Pleospora herbarum* *Fusarium flocciferum*	Chen et al. (2016)

Erwinia, *Stenotrophomonas*, *Serratia*, *Burkholderia*, *Streptomyces*, *Actinomyces*, *Debaryomyces*, *Acinetobacter*, *Enterobacter*, *Escherichia* and *Ralstonia* were evaluated as antagonists of several phytopathogens, demonstrating potential efficiency as BCAs (Saddler 2005; Meziane et al. 2006; Bencheqroun et al. 2007; Messiha et al. 2007; Taqarort et al. 2008; Saravanakumar et al. 2008; Xue et al. 2009; Liu et al. 2010; Ramadasappa et al. 2012; Ramesh and Phadke 2012; Wei et al. 2013; Dukare et al. 2018). Of all the genera mentioned above, the species of *Pseudomonas* and *Bacillus* have been the most studied in recent years, and several results indicate that they can effectively control several plant diseases (Gao et al. 2018; Jiang et al. 2018; Prakash and Arora 2021).

Currently, there are more than 400 Bt-based formulations that have been registered in the market, and most of them contain viable insecticidal proteins and spores. Products formulated with Bt can be found in different formulations such as wettable powder, concentrated suspension, emulsifiable concentrate and dispersible granules (Usta 2013). In Brazil, wettable powder and concentrated suspension are the main formulations sold, and it is possible to find 17 registered commercial products based on *Bacillus* in the country (AGROFIT 2020a, ABCBio 2020). The great revolution that Bt brought was the possibility to produce transgenic plants resistant to insect pests. At the end of 2018, it was estimated that 23.7 million hectares were planted with crops containing the Bt gene in more than 40 countries, with the main cultivated Bt crops being cotton, (*Gossypium hirsutum* L.), corn (*Zea mays* L.) and soy (*Glycine max* (L) Merrill), and all crops were developed to be resistant to the attack of defoliating caterpillars (ISAAA 2020).

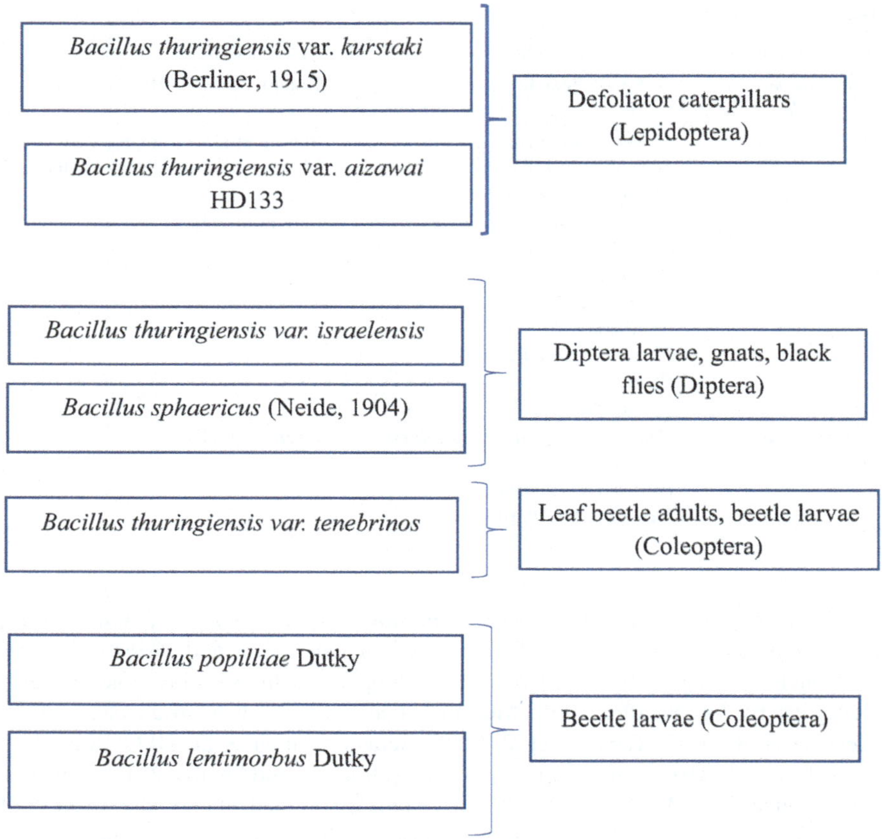

Fig. 4.1 Main species and varieties of the genus *Bacillus* used to control insect pests (Moraes et al. 1998; Silva 2000; ISAAA 2020; AGROFIT 2020a; ABCBio 2020)

4.2.2 Fungi

Fungi are eukaryotic microorganisms whose cell walls contain chitin and/or cellulose and glucans, and most of them grow in filamentous form, producing hyphae that, when grouped, are called mycelium. The order *Hypocreales* (*Ascomycota*: *Sordariomycetes*) has the largest number of entomopathogenic genera known as *Beauveria* Vuill. (*Cordycipitaceae*), *Metarhizium* Sorokin (*Clavicipitaceae*), *Isaria* Pers. (*Cordycipitaceae*), *Verticillium* Nees von Esenbeck (1816) (Clavicipitaceae), *Lecanicillium* W. Gams & Zare. (*Cordycipitaceae*) and *Nomuraea* Maubl. (*Clavicipitaceae*) (Usta 2013; Valadares-Inglis et al. 2020).

There are frequent changes in the taxonomic classification of entomopathogenic fungi, and, with the help of molecular techniques, the phylogenetic position of some groups of fungi has been determined with greater precision, even for identification of species. For instance, *Metarhizium anisopliae* (Metschn.) Sorokin, one of the main species used in insect control, was considered a complex of species that are being studied and identified. The first report of a fungi as a causative agent of disease in insects was in 1835 with *Beauveria bassiana* (Bals.-Criv.) Vuill., and between 1878 and 1888, efficiency of *M. anisopliae* as a microbial agent of pests' control was determined. Since then, more than 700 species of entomopathogenic fungi have been reported to cause disease in insects (Valadares-Inglis et al. 2020). But, despite that, the species *B. bassiana* and *M. anisopliae* are still the main fungi used in biological control programs, with an infinity of registered formulations and products. Figure 4.2 shows the main groups of insects controlled by these microorganisms. Another fungal species that also deserves to be highlighted is the microsporidium *Paranosema locustae* (Canning), which, until 2003, was known as *Nosema locustae* (Canning) and is a mandatory parasite in orthopterans and, among them, especially grasshoppers (Orthoptera). There are studies reporting its use in the control of grasshoppers in Argentina, Brazil, China, the United States and other countries (Feng et al. 2015; Valadares-Inglis et al. 2020). It should be noted that *Microsporidia* is a phylum of unicellular parasitic spore-forming fungi. In the past, they were included among protozoa, but after genetic studies, it was discovered that they were related to fungi, so it is common to find older works identifying *P. locustae* as protozoa.

Entomopathogenic fungi have a broad spectrum of action, being able to colonize several species of insects, and the main advantage in relation to other groups of microorganisms is their infection capacity at all stages of host's development. These fungi invade the body of insects, mainly through the cuticle or skin (integument), rapidly multiplying throughout the body and causing their death by the destruction of tissues and, occasionally, by toxins produced. After infection, the insects stop feeding and become slower, leading to their death. Entomopathogenic fungi have the characteristic of emerging from the body of insects, producing spores that can be dispersed by wind, rain and contact with other insects or by ejecting their spores in

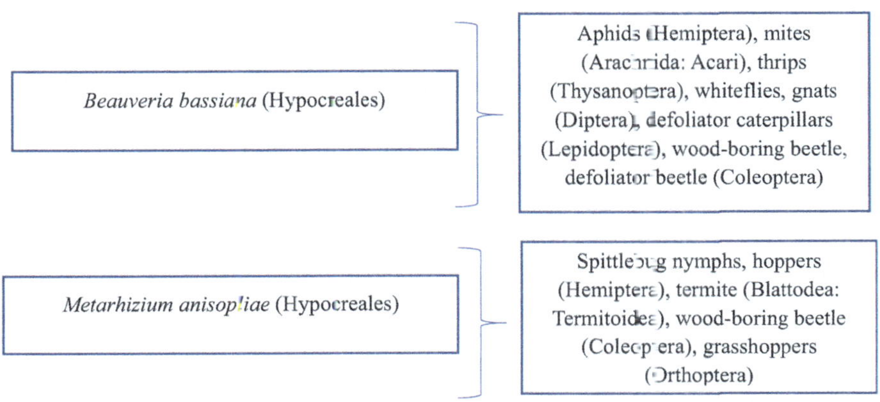

Fig. 4.2 Main groups of insect pests controlled with *B. bassiana* and *M. anisopliae* (Silva 2000; AGROFIT 2020b, c; ABCBio 2020)

the air (Hajeck and Leger 1994; Alves et al. 2008; Valadares-Inglis et al. 2020). To be effective in insect control, entomopathogenic fungi require moisture to enable infection, so their efficiency increases considerably in hot and humid periods. Thus, the microclimate conditions found on the surface of plants and in the bodies of insect pests are fundamental for the establishment of infection. Regions of the insect's body, such as the mouth, anus and respiratory orifices, have a moist microenvironment, favourable to the entry and beginning of infection, positively influencing the efficiency of the pathogen (Valadares-Inglis et al. 2020).

When applied to the environment, fungi can suffer abiotic damage that compromises their efficiency, such as exposure to ultraviolet (UV) radiation, which has a negative effect on spore germination, as well as on survival, reproduction, dispersion and virulence of the fungi (Alves 1998). Thus, the abiotic conditions at the time of application of products based on fungi must be carefully monitored, giving priority to applications at the end of the day or during the night, when solar radiation is less or absent and the relative humidity of the air is favourable to the action of the entomopathogen (Valadares-Inglis et al. 2020).

Currently, commercial products with entomopathogenic fungi as an active compound are found in a significant number of countries, in different commercial formulations such as granules, emulsifiable oil, baits, dry and wettable powder, concentrated suspension and emulsifiable concentrate. In Brazil, the wettable powder is the most common formulation, and there are currently 60 registered mycoinsecticides, with *B. bassiana* (22) and *M. anisopliae* (34) species with the highest number of registered products (ABCBio 2020; AGROFIT 2020b, c). Brazil stands out in the sustainable management of some insect pests with mycoinsecticides, such as leafhoppers (Hemiptera: Cercopidae) from pastures and sugar cane (*Saccharum officinarum* L.) where the area treated with the fungus *M. anisopliae* exceeds one million hectares. The use of *B. bassiana* in coffee plantations (*Coffea* sp. L.) to control the fruit borer (*Hypothenemus hampei* Ferrari (1867), Coleoptera) and in banana cultivation (*Musa* sp. L.) for the rhizome borer (*Cosmopolites sordidus* (Germar) Coleoptera) also deserves mention with significant results for the management of these pest species (Batista Filho et al. 2013).

Endophytic fungi live inside plants, propagating in inter- or intracellular spaces without harming or causing apparent disease symptoms (Li et al. 2012; Park et al. 2018). Several studies have shown that these fungi improve plant resistance to insects and pathogens, because they colonize all parts of the host, to activate primary plant defences and producing several metabolites with antimicrobial and antiviral activities (Gunatilaka 2006; Macia-Vicente et al. 2008; Maciá-Vicente et al. 2009; O'Hanlon et al. 2012; Segaran and Sathiavelu 2019). They also increase plant tolerance to abiotic stresses such as drought, extreme temperature and salinity (Ownley et al. 2010; Park et al. 2018). Thus, due to their beneficial effects on the development, growth and fitness of host plants, these fungi have been considered valuable natural resources that can be exploited as BCAs against microbial pathogens (Zhang et al. 2014; Gholami et al. 2019).

Trichoderma spp. is undoubtedly one of the most studied and used fungi in the biological control of phytopathogens, including *Fusarium* spp., *Rhizoctonia solani*

Kuhn, *Phytophthora* spp., *Sclerotium rolfsii* Sacc, *Pythium* spp., *Botrytis cinerea* and *Alternaria* spp. (Zeilinger and Omann 2007; Bettiol et al. 2008; Yao et al. 2016; Zeilinger et al. 2016). It is estimated that this microorganism could suppress at least 29 species from 18 genera of pathogenic fungi (Ahmed et al. 2010; Zhu et al. 2020), and this is one of the reasons that several species of *Trichoderma* were developed as commercial products, which are now used to control leaf pathogens and soil-borne diseases in several crops (Zeilinger et al. 2016; Zhu et al. 2020). In Latin America, the species *Trichoderma asperellum*, *Trichoderma harzianum*, *Trichoderma stromaticum* and *Trichoderma viride* have been the most commercialized (Morandi and Bettiol 2009). Some endophytic fungi, like *Trichoderma citrinoviride*, can produce large amounts of enzymes like 1,4-β-D-glucanase and cellulases that degrade the cell wall of important phytopathogenic fungi such as *Botrytis cinerea*, *Cylindrocarpon destructans*, *Phytophthora cactorum* and *Pythium* sp. (McQuilken and Gemmell 2004; Shi et al. 2013; Lin et al. 2016). Also, the use of *Trichoderma* can result in increased plant growth and soil remediation; therefore, these fungi provide more advantages as BCAs than viruses, bacteria and protozoa (Cavalcante et al. 2008).

4.2.3 Viruses

Viruses are infectious agents, whose genome contains only one type of nucleic acid (RNA or DNA), which is surrounded by a protein layer and can be encompassed by a membrane containing lipids. They are inert in the extracellular environment and can replicate only in living cells, being mandatory intracellular parasites of any living being (Reis and Santos 2016). Thus, they are also important agents for microbial control of insects, with *Baculoviridae*, *Poxviridae* and *Reoviridae* being the viral families already reported effective against insects.

The *Baculoviridae* family represents the main group of entomopathogenic viruses studied, and the literature reports more than 700 species of arthropods naturally infected with baculovirus. These viruses were first reported during the development of the silk industry in China in the sixteenth century, when they were responsible for causing a disease in the silkworm (*Bombyx mori* L. (Lepidoptera: Bombycidae)). Baculoviruses have a simple circular double DNA strand and comprise the genera *Nucleopolyhedrovirus* (NPV) and *Granulovirus* (GV). NPVs have polyhedral inclusion bodies, containing multiple viral particles, while GVs generally contain single particles, occluded in ovoid protein bodies (Castro et al. 1999; Usta 2013; Das et al. 2019). In most cases, baculoviruses are quite efficient because they are highly virulent and specific to hosts, in addition to being safe for human health and the environment, as they do not negatively impact plants, mammals, birds, fish or even non-target insects. (Castro et al. 2020). Thus, they are the main viruses used in the biological control of insects, and in Fig. 4.3, the baculovirus genera and the respective groups of insects that they control are presented.

The infection of insects by viruses occurs by ingesting viral polyhedra, and the penetration occurs in epithelial cells of the midgut. After entering the intestinal cells,

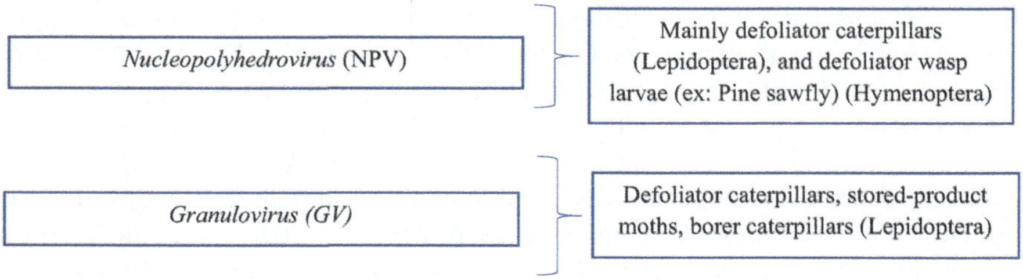

Fig. 4.3 Groups of insect pests controlled by *Nucleopolyhedrovirus* and *Granulovirus* (Silva 2000; Castro et al. 2020)

the spread of the infection through the insect's body occurs with the arrival of the virus in the circulatory (hemolymph) and/or respiratory (trachea) system, with or without previous viral replication in the intestinal cells (Gelernter and Lomer 2000; Castro et al. 2020). During infection, a series of behavioural and morphological changes occur in the insect, starting with loss of appetite, followed by decreased growth, lethargy, discoloration and loss of the natural brightness of the integument due to tissue dissolution and accumulation of viral particles, until death of the insect. In the final stage of the infection process, a very common characteristic infection is observed, which is the migration of the infected insect to a higher position of the branch of the tree or plant, in order to facilitate the dispersion of the occlusion bodies. This behaviour of the insect (negative geotropism), caused by the viral infection, is called the "zombie effect" (Castro et al. 2020).

Insects of the order Lepidoptera are, for the most part, important pests of cultivated systems; thus, most pathology studies of viral diseases are concentrated in this order. There are reports that more than 700 species of insects were naturally infected with baculovirus and more than 90% of baculoviruses were isolated from Lepidoptera species, although they can be found in other orders, such as Diptera and Hymenoptera (Das et al. 2019; Castro et al. 2020).

The largest baculovirus biological control program was established in Brazil to control the caterpillar *Anticarsia gemmatalis* Hübner, (1818) (Lepidoptera: Noctuidae), an important pest species in soybean plantations (*G. max*) in an area comprising 2,000,000 ha (Usta 2013). Currently, in the country, 17 commercial baculovirus-based products are available to control caterpillars, mainly in soybeans (ABCBio 2020).

4.2.4 Nematoda

The phylum Nematoda is considered the most numerous and varied group among multicellular animals. They are pseudocoelomate beings, with bilateral symmetry and filiform in at least one stage of their life cycle, occurring in the most different environments, and in certain ecosystems, they constitute the largest part of the soil microfauna. In the last three decades, they represent one of the most studied groups of biological insect control agents, since several species of nematodes were experimentally proven to associate with different species of insect pests. They are currently considered an effective tool in integrated pest management programs, representing an important part of BCAs (Dolinski et al. 2017; Voss et al. 2019; Dolinski 2020).

Entomopathogenic nematodes belong to the order Rhabditida, in which the families Steinernematidae and Heterorhabditidae can be found, and where the genera *Steinernema* Travassos (1927) and *Heterorhabditis* Poinar (1976) are the main nematodes used in biological control, generally for soil insects. Both genera have a wide geographical distribution, with reports of occurrence in soils from six different biomes worldwide (Van Driesche and Bellows 1996; Voss et al. 2019). Figure 4.4 shows the main groups of insect pests with potential for biocontrol by the genera *Steinernema* and *Heterorhabditis*. Currently, about 89 species of the genus *Steinernema* and 22 of the genus *Heterorhabditis* have been described, more than 70% of which have been described in the last 20 years. These nematodes act in symbiotic association with g-proteobacteria of the *Enterobacteriaceae* family. These bacteria are mobile, Gram-negative and produce toxins that cause the death of the insect, and antibiotics that prevent the growth of other opportunistic microorganisms. Two genera of bacteria were found associated with these nematodes: *Xenorhabdus* (Thomas and Poinar 1979) associated with the genus *Steinernema* and *Photorhabdus* (Boemare et al. 1993) associated to *Heterorhabditis* (Dolinski 2020). These nematodes invade hosts through natural openings (mouth, spiracles, anus) or wounds and penetrate the hemocoel. When introduced into the insect's hemocoel by the nematode, the bacterial cells are released into the hemolymph, where they excrete toxins, which kill the insect in 24 to 48 h. The nematode digests substances of the insect, in addition to feeding the bacteria, and multiplies for several cycles presenting four larval (juvenile) phases, an adult and egg phase inside the insect. When the insect's nutritional reserves are exhausted, the third larval phase ingests some bacteria and goes in search of another insect to infect. They are the so-called infective juveniles, the only phase adapted to survive outside the insect (Van Driesche and Bellows 1996; Voss et al. 2019).

Several studies indicate that entomopathogenic nematodes are more efficient in controlling insects that are distributed in closed habitats with high humidity, low incidence of UV light and mild temperatures, and that is why they have been most effective for soil insect pests or borers. According to Dolinski et al. (2017), several studies tried to indicate the most favourable conditions and formulations for storage and application of nematodes in the field in Brazil, since several factors, such as

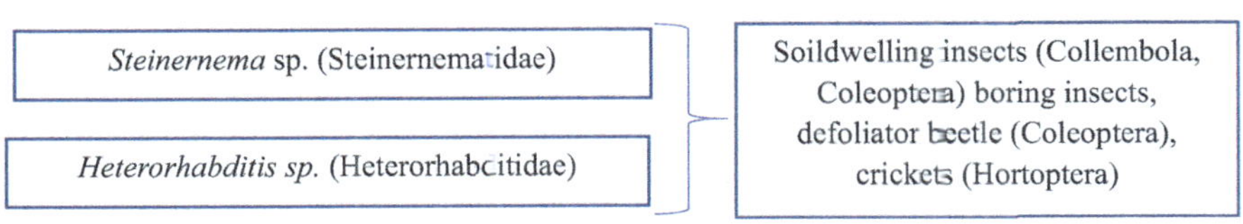

Fig. 4.4 Groups of insect pests with potential for biocontrol by the genera *Steinernema* and *Heterorhabditis* (Silva 2000; Dolinski 2020)

temperature and oxygen, directly affect the survival and infectivity of organisms stored in aqueous suspension. Almenara et al. (2012) highlight that the field application of pathogenic nematodes can be done with sprays containing different carriers, with dead bodies of insects or baits. In the case of using dead insects, infected larvae of *Galleria mellonella* L. (Lepidoptera: Pyralidae) are wrapped in clay and distributed throughout the field. This type of application is often more efficient than spraying with aqueous or gelatinous suspensions. Baits may contain nematodes complexed with different inert materials. However, the main difficulty in disseminating the use of this group of insect pathogens is the small number of companies that study and produce these nematodes, which compromises their large-scale application in the field (Dolinski et al. 2017).

In addition to the genera *Steinernema* and *Heterorhabditis* already mentioned as potential BCAs, species *Deladenus siricidicola* Bedding (1968) (Neotylenchidae) also deserves mention, since it has been widely used to control the wood wasp (*Sirex noctilio* Fabricius (1773) (Hymenoptera: Siricidae)), an important pest species in *Pinus* sp. L. plantations. The nematode *D. siricidicola* has a bicyclic life cycle, parasitizing wasp larvae and remaining in the host until pupation and, when it moves to the reproductive organs, infesting and sterilizing the female's eggs. The female wasp then oviposits sterile eggs to host trees, which helps spread the nematode. Then, in a second part of its life cycle, *D. siricidicola* feeds on *Amylostereum areolatum* (Chaillet ex Fr.) Boidin, a symbiotic wood wasp fungus. The life cycle of the nematode in the fungus lasts only about 2 weeks, which means that several nematodes can be created in laboratory conditions in a short period. Mass-created nematodes are shipped to application areas in breathable plastic containers with water, and the nematode-water mixture is inoculated into holes made in trees with inoculation hammers; then the nematode goes in search of the fungus for food and is consumed by the wasp (Kenis et al. 2019).

It is noteworthy that the only nematicide registered in Brazil is a product formulated based on *D. siricidicola* (ABCBio 2020), and this nematode has been used in several countries with pine cultivations in the southern hemisphere to control *S. noctilio*.

4.3 Advantages, Disadvantages and Challenges in Microbial Control

The advantages of using microbial insecticides are mainly related to the ecological benefit, because these agents are less harmful to the environment, mainly acting on a single species or on a specific group of insects and preventing the occurrence of secondary pests. Usta (2013) highlights that these products are often effective in small quantities and decompose rapidly, resulting in lower exposures and avoiding environmental problems. Another important point concerns the compatibility of this type of biocontrol with other methods and the smaller possibility of the insect to acquire resistance to pathogens.

As a main disadvantage, Carrano-Moreira (2014) mentions that specificity can represent an economic downside, since the cost of developing a bioinsecticide to combat a single pest species may not be competitive, since these products already have high production and registration costs. Another issue is related to the time of action, because consumers are used to chemical products that deliver results almost immediately after application, and microbial agents may require a few days or even weeks to be effective and achieve the expected results. One of the main limitations of microbial control is the inability of these microorganisms to persist in habitats exposed to UV radiation or desiccation. Thus, obtaining commercial formulations that allow these pathogens to increase their action time when applied is essential. In this context, Mandal (2019) considers that the development of nanoscale insecticides called nanopesticides may be an alternative to increase the efficiency of microbial insecticides. Das et al. (2019) report that studies involving the use of nanotechnology to increase the potential of baculovirus as an insecticide have been carried out, since this virus has a sensitivity to UV radiation, thus limiting its large-scale application as an insecticide.

According to Sampaio (2018), among the challenges of controlling insects with microorganisms, it is possible to highlight the necessary expansion of the availability of commercial products, currently concentrated in a few pathogens. Therefore, constant research involving the identification of new agents is necessary, increasing the number of insects controlled by microbial products. Finally, another important challenge in the use of microbial insecticides is cultural, since consumers need to think about this control in the context of integrated pest management, with constant assessment and monitoring of cultivated ecosystems. They also need to know all the factors that involve the population dynamics of insect pests, thus applying BCAs before the pest reaches the level of economic damage and using this control in an integrated way with other regulation tactics. Thus, Sampaio (2018) suggests that in addition to generating knowledge, actions aiming at disseminating, interacting, training and providing technical assistance to consumers about the use of entomopathogens may result in a favourable environment for the innovation of production systems, contributing to overcoming the issues imposed by the agrochemicals.

4.4 Mechanisms of Microorganism-Associated Biological Control

In the last four decades, there has been a considerable advance in the screening of antagonistic microorganisms capable of acting as BCAs and in the understanding of the mechanisms behind their action (Alabouvette et al. 2006; De Corato 2020).

The antagonistic effects of BCAs responsible for suppressing a disease promoted by a phytopathogenic organism can result from two main strategies: (1) through microbial interactions directly against the pathogen, mainly during its saprophytic phase, known as direct antagonism (parasitism, antibiosis and nutrient competition), or (2) through an indirect action, inducing the development of systemic resistance in the host plant (Alabouvette et al. 2006; Mhatre et al. 2019).

4.4.1 Parasitism

Different microorganisms directly inhibit the development of plant pathogens, thus acting as their natural parasites. Parasitism begins with the release of compounds by the pathogen, which act as chemotrophic molecules, guiding the growth of the parasite towards the pathogen. Then, the parasite adheres to the pathogen's surface and penetrates its cells through the release of lytic enzymes, causing damage to its internal organs and resulting in their death (Kiss 2003; Heydari and Pessarakli 2010; Bardin and Pugliese 2020). This type of antagonism, which causes the death of the target organism through parasitism, results in decreased density of the pathogens, thus reducing or eliminating the negative effects they have on plants (Milgroom and Cortesi 2004). Some strains of *Trichoderma* spp. are known to be parasite of a variety of phytopathogens, such as *B. cinerea*, *Pythium* spp., *S. rolfsii*, *Sclerotinia sclerotiorum* and *Fusarium* spp. (Waghunde et al. 2016). In addition, some studies show the use of *Trichoderma* spp. in the control of pathogens in agriculture, like the pathogenic fungus *R. solani*, which causes various plant diseases, such as burning rice and black spots on potatoes (Jia et al. 2013; Rahman et al. 2014). Other mycoparasites, such as *Coniothyrium minitans* and *Sporidesmium sclerotivorum*, were also shown to be effective in controlling diseases caused phytopathogens like *Sclerotinia* spp. and others (Adams and Fravel 1993; Jones et al. 2004).

4.4.2 Antibiosis

Antibiosis is a common phenomenon, being one of the main mechanisms of action among BCAs and is defined as the antagonism resulting from the microorganism production of secondary metabolites that are toxic to other microorganisms. These metabolites are generally low molecular weight organic compounds and play an important role in the biocontrol of several pathogens (Raguchander et al. 2011; Köhl et al. 2019). Several bacteria and fungi have been reported to produce antimicrobial metabolites, most with broad-spectrum activity, such as *Agrobacterium*, *Bacillus*, *Pantoea*, *Pseudomonas*, *Serratia*, *Stenotrophomonas*, *Streptomyces*, *Trichoderma* and *Clonostachys* (former *Gliocladium*) (Ongena and Jacques 2008; Ghorbanpour et al. 2018; Prakash and Arora 2021).

A great diversity of molecules secreted by these microorganisms have been well studied for their role in suppressing several types of plant pathogens (Fravel 1988; Loper and Lindow 1993; Weller and Thomashow 1993). They include not only *stricto sensu* antibiotics but also bacteriocins, enzymes, alkaloids, flavonoids, peptides, phenols, quinones, steroids, terpenoids, polyketides and volatile organic compounds (VOCs) like terpenoids and polyketides.

Among the antibiotics produced, 2,4-diacetylphloroglucinol (2,4-DAPG) is widely studied and has antifungal, antibacterial, anthelmintic and also phytotoxic activities (Weller et al. 2007), being considered the main antimicrobial metabolite involved in the biocontrol of phytopathogens (Sonnleitner and Haas 2011). The efficiency of phytopathogens inhibition by the action of 2,4-DAPG has been demonstrated in several plant-pathogen interactions such as *Xanthomonas oryzae* pv. *oryzae* on *Oryza sativa* (Velusamy and Gnanamanickam 2003), *C. michiganensis* subsp. *michiganensis* on *Solanum lycopersicum* (Lanteigne et al. 2012) and *Fusarium oxysporum* f. sp. *cubense FOC* on *Musa* ssp. (Ayyadurai et al. 2006).

Several agents are known to produce more than one antibiotic that can act in biocontrol (Pal and Gardener 2006). For instance, syringomycin, syringopeptin and syringotoxin are antibiotics produced by *Pseudomonas syringae* pv. *syringae* and are important for biocontrol, because their main mechanism of action is the induction of ion channels in the membrane of the pathogen, which can lead to cytolysis and death of these organisms (Raaijmakers et al. 2006). However, Raaijmakers and Mazzola (2012) revealed that the action of antibiotics occurs in a concentration-dependent manner, since in high concentrations they can act as inhibitors and in low concentrations they mediate the intercellular signalling.

Although the knowledge about antibiotics and their application in biological control is broad, an analysis of the microbial genome revealed many antibiotic gene clusters that code for unknown antibiotics (Köhl et al. 2019). In addition, most of these substances are identified in in vitro techniques, which may not truly reflect field production and can generate false-positive results for inhibiting pathogens (Koch et al. 2018). This highlights the importance of conducting in vivo tests, analysis of concentration and research about the lifetime of antibiotics in the rhizosphere (Raaijmakers and Mazzola 2012; Arseneault and Filion 2017). A single strain of a BCA can produce several types of secondary metabolites, with different and effective functions against diverse species of pathogens, whether they are fungi, bacteria or even nematodes (Yu et al. 2010; Mousa and Raizada 2013; Lugtenberg et al. 2016). For instance, the CHAO strain of *Pseudomonas fluorescens* produces siderophores, phenazine, DAPG and cyanide, and the combination of these metabolites is responsible for the antagonism against *Gaeumannomyces graminis* var. *tritici* and *Chalara elegans* (Défago and Haas 1990). It is important to emphasize that a single metabolite does not usually account for all the antagonistic activity of a BCA. Different metabolites and their combinations can be responsible for the efficiency of an antagonistic action against a pathogen, and the exclusion of one of them may shorten or even cancel the entire activity (Köhl et al. 2019).

4.4.3 Lytic Enzymes

One of the compounds most produced by microorganisms and used in the biocontrol of phytopathogens are lytic enzymes. There are several kinds of microbial biocontrol enzymes like chitinases, cellulases, proteases and β-1,3-glucanases (Mishra et al. 2020).

The suppression of plant phytopathogens activity can occur directly, in which the lytic enzymes will stress or lyse the cell wall of living organisms trough hydrolysis of a variety of polymeric compounds like chitin, proteins, cellulose, hemicellulose and DNA (Pal and Gardener 2006). This mechanism has been reported in the BCA *Trichoderma* sp., which produces cell wall-degrading enzymes such as β-1,3-glucanase, N-acetylglucosaminidases and chitinase that degrade the mycelium of phytopathogenic fungi (Cheong et al. 2017).

The mode of action of lytic enzymes or their activity can be modified in order to increase the efficiency of this mechanism in the BCAs, as demonstrated in the control of *R. solani* by *T. harzianum*, which was attributed to the secretion of chitinase and cellulase, and a gamma mutation increased the efficiency and amount of enzymes in *T. harzianum*, improving its ability as a BCA (Ghasemi et al. 2019). Also, postharvest inhibition of pathogens using the native or recombinant exo-β-1,3-glucanase secreted by *Pichia guilliermondii*, showing the involvement of natural and modified enzymes in the biocontrol activity has been demonstrated (Zhang et al. 2011). The activity of lytic enzymes can also contribute to indirectly suppress the diseases cause by phytopathogens trough the activation of plant defences. Some products of lytic enzyme activity like oligosaccharides or glycoproteins can elicit an induced systemic resistance (ISR) in plants before infection and effectively enhance the plant resistance against various pathogens (Xu et al. 2017; Savitha and Sriram 2017). Also, several defence-related enzymes can be activated by the BCA upon phytopathogen attack, like peroxidases (POD), polyphenol oxidases, phenylalanine ammonia-lyase (PAL), glucanases and chitinases (Zhu and Ma 2015; Doley et al. 2019).

Zhao et al. (2012) showed that *Streptomyces bikiniensis* HD-087 can effectively induce the resistance in cucumber by increasing activities of enzymes such as POD, PAL and β-1,3-glucanase. Moreover, arbuscular mycorrhizal fungi have been reported to regulate catalase, superoxide dismutase, ascorbate peroxidase and glutathione peroxidase content in potato plants infected with the bacteria *Pectobacterium carotovorum*, which causes blackleg disease (Bagy et al. 2019). However, the mechanism by which antagonistic microorganisms induces this enzymatic resistance has not been clearly elucidated due to the complex interactions among BCAs, hosts, pathogens and their nutritional and metabolic environments (Droby et al. 2009; Parafati et al. 2017a).

4.4.4 Volatile Compounds

Volatile organic compounds (VOCs) are small (typically less than 300 Da), carbon-based molecules, characterized by low water solubility and high vapour pressure, which makes them available in a gaseous state under normal environmental conditions (Pagans et al. 2006; Morath et al. 2012). They belong to different molecular classes, including hydrocarbons, alcohols, thioalcohols, aldehydes, ketones, thioesters, cyclohexanes, heterocyclic compounds, phenols and benzene derivatives (Wheatley et al. 1997; Chiron and Micherlot 2005; Morath et al. 2012). These volatile metabolites are produced by virtually all living microorganisms, being identified as active in the control of oomycetes, fungi and phytopathogenic bacteria through antimicrobial activity and other cross-interactions. Due to their antimicrobial effects and because they present a reduced risk to the environment and to humans, as well as their possible application without the need for a spray, they are a promising and sustainable tool to replace artificial fungicides in the control of phytopathogens (Mercier and Jiménez 2004; Fialho et al. 2010; Parafati et al. 2017b). In agriculture, the use of VOCs of microbial origin in the biocontrol of plant pathogens has not yet been implemented, mainly because this area is only being studied recently. However, recent progress has highlighted the potential benefits of microbial VOCs in this field (Tilocca et al. 2020).

VOCs are generally effective in low concentrations, and due to their volatility, they diffuse between soil particles and spread in the atmosphere over great distances from the point of application, where they can exert their inhibitory activity, without requiring direct contact between the producing microorganism and the target pathogen (Minerdi et al. 2009; Mitchell et al. 2010; Heydari and Pessarakli 2010; Raza et al. 2016). Among the activities already related to VOCs, we can highlight inhibition of pathogens, inhibition of fungal spores germination, biocontrol of plant parasitic nematodes and induction of resistance mechanisms in plants, preventing them from being colonized by pathogens and also promoting their growth (Van Loon et al. 1998; Ryu et al. 2003; Compant et al. 2005; Mercier and Manker 2005; Farag et al. 2006; Gu et al. 2007; McKee and Robinson 2009; Fialho et al. 2011; Minerdi et al. 2011; Morath et al. 2012; Yuan et al. 2012; Xu et al. 2015).

Among yeasts, alcohols (ethyl alcohol, 3-methyl-1-butanol and phenylethyl alcohol) and esters (ethyl acetate and isoamyl acetate) are the main VOCs produced. It was demonstrated that the application of yeast strains in the soil during the postharvest period suppressed the growth of some pathogenic fungi in a very efficient way and a production of high percentages of ethyl acetate and carbon dioxide was observed (Contarino et al. 2019). Strains of beneficial bacteria also produce VOCs capable of inhibiting the growth of certain soil pathogens, being effective against both fungi and bacteria. Two bacteria commonly associated with plants, *P. fluorescens* and *Serratia plymuthica*, have been reported to emit volatiles that strongly suppress the growth of *Agrobacterium tumefaciens* and *Agrobacterium vitis*, and it was identified that the release of dimethyl disulfide was the cause of this bacteriostatic effect (Dandurishvili et al. 2011). It has also been reported that *P. fluorescens* effectively control the growth of the phytopathogen *Ralstonia solanacearum* in tomato plants, with the volatiles released by these bacteria having inhibited the growth of the pathogen in both agar and infested soil. In addition, these volatiles seemed to supress the expression of virulence characteristics involved in host colonization, such as motility, root colonization and biofilm formation, effectively controlling the tomato wilt (Raza et al. 2016).

More studies are still needed for VOCs to be used in agriculture in a practical way; nevertheless, their implementation is seen as an excellent alternative to the use of artificial pesticides, mainly due to their easy applicability and their effectiveness against several pathogens, both in soil and plants.

4.4.5 Nutrient Competition

Competitive exclusion is a determining factor in the composition of the soil microbiome, being a probable mechanism by which beneficial microorganisms can prevent the colonization of plants by phytopathogens (Zabalgogeazcoa 2008; Martinuz et al. 2012). Nutrient competition is a general phenomenon that regulates the population dynamics of microorganisms that share the same ecological niche and have the same physiological requirements when trophic resources are limited. The competition for nutrients occurs, for example, on the surface of a leaf, preventing the germination of fungal spores and reducing a possible infection (Fokkema 1996). Competition for nutrients, especially carbon, is also quite common in the soil and is responsible for the well-known phenomenon of fungistase (Lockwood 1977), which describes the inhibition of fungal spores' germination in the soil. Although it is difficult to clearly demonstrate, competition for nutrients is certainly one of the mechanisms of many BCAs, and one of the best documented examples is the competition for carbon between pathogenic and non-pathogenic *F. oxysporum*. According to Couteaudier and Alabouvette (1990), non-pathogenic *F. oxysporum* is more aggressive in nutrient uptake, multiplying more quickly, thus having an advantage over pathogenic strains, partially explaining why they suppress the growth of these microorganisms. Recently, a study conducted by Solanki et al. (2019) demonstrated that competition for nutrients in the rhizosphere zone stimulates BCAs to produce compounds that act in the control of pathogenic microorganisms, such as lytic enzymes, indicating that the suppression mechanisms used by BCAs are not only correlated but also occur simultaneously.

Competition for smaller elements often occurs in the soil; for instance, competition for iron is one of the modes by which *P. fluorescens* limit the growth of phytopathogenic fungi and reduce the incidence or severity of the disease caused by these organisms (Lemanceau and Alabouvette 1993; Loper and Henkels 1997). Several strains of soil *Pseudomonas*, which are PGPR, have been shown to be effective in suppressing soil-borne diseases (Haas and Defago 2005; Weller 2007), and among its biocontrol activities, the release of siderophores can be highlighted, since they act as competitors, reducing the availability of iron for plant pathogens, including *F. oxysporum* and *Pythium ultimum* (Van Wees et al. 2008; Ahmed and Holmström 2014). It is important to note that siderophores produced by other bacterial species, such as *Bacillus subtilis*, also exhibit biocontrol activities (Yu et al. 2011). In addition to their ability to compete with soil pathogens by iron, the siderophores produced by PGPR have also been shown to protect different plant species by indirectly activating the mechanisms of ISR (Trapet et al. 2016). Despite being common and well-studied, competition for nutrients between BCAs and pathogens is still a difficult phenomenon to explore from a practical point of view, especially for biological control in agriculture.

4.4.6 Induction of Systemic Response in Plants

Plants have several defence mechanisms against phytopathogens, which can be activated by biotic or abiotic elicitors (Vallad and Goodman 2004) and can be induced by the varied groups of microorganisms, both by intact living cells or by their extracts (Morandi and Bettiol 2009). The induction of plant resistance can be defined as the activation of latent mechanisms that are expressed after exposure of the plant to suitable elicitors (Sticher et al. 1997). Thus, the exposure of plants to certain pathogenic microorganisms or to physical and/or chemical stimuli caused by them can result in the expression of several locally and systemic defence responses (Agrios 2005; Dodds and Rathjen 2010). These responses are typically divided into two groups: systemic acquired resistance (SAR) that occurs when there is an increased plant defence capacity against a wide range of pathogens, after local induction by a pathogen or its molecules, and induced systemic resistance (ISR) that occurs when there is an increased defence capacity of the entire plant against a wide range of pathogens, after local induction by beneficial microorganisms. These different types of resistance help the plant to contain the attacker and are characterized, for example, by the direct induction of antimicrobial proteins (Mauch-Mani et al. 2017). However, even though ISR and SAR are effective against several pathogens, their effectiveness can be variable, like the infection of the pathogens *Alternaria brassicicola* and *B. cinereal* in *Arabidopsis thaliana*, in which only ISR was effective against these pathogens (Bisen et al. 2016).

Adenylate cyclase and cyclic adenosine monophosphate (cAMP) are key components of signal transduction processes involved in pathogenicity, morphogenesis and differentiation in many fungi. As reported for the necrotrophic pathogen *B. cinerea*, the cAMP signalling pathway is linked to the regulatory subunits of the *PKA1* gene, which plays an important role in its vegetative growth and pathogenicity (Klimpel et al. 2002). Several studies have demonstrated the ability of microbial BCAs to regulate the formation of cAMP, thus supressing the pathogenicity of fungi (Lengeler et al. 2000). In the work of Park et al. (2018), the expression of six genes involved in the growth and virulence of *B. cinerea* (*BAC*, *BCG1*, *BOT*, *CHS3*, *PKA1* and *TRE1*) was significantly reduced in ginseng plants inoculated with the endophytic fungus *T. citrinoviride*, while the expression of genes related to plant defence (*PR2*, *PR4*, *PR5* and *PR10*) were significantly upregulated in samples of pre-inoculated plant leaves (Park et al. 2018).

In a study by Schumacher et al. (2008), mutants of *B. cinerea* for the *BCG1* gene lost their pathogenicity and failed to produce the botrydial phytotoxin, which confirm the importance of *BCG1* for signal transduction due to its involvement in the regulation of pathogenicity and vegetative growth (Schulze et al. 2004). Also, the *BOT1* gene is controlled by signal transduction to develop the symptoms of grey mould disease (Siewers et al. 2005), and during spore germination, *TRE1* regulates carbohydrate metabolism by decomposing intercellular trehalose (Argüelles 2000). Some studies have demonstrated that all genes mentioned above are involved with the virulence and growth of phytopathogenic fungi and may have their expression reduced when the host plants are pre-inoculated with the BCA *Trichoderma* sp. (Perazzolli et al. 2008; Martínez-Medina et al. 2013; Park et al. 2018).

The fact that beneficial microorganisms can promote ISR in plants of different genera demonstrates their ability to successfully protect plants against a wide variety of pathogens (Kloepper et al. 2004; Van Loon and Bakker 2006; Van Wees et al. 2008). Among beneficial microorganisms, *Pseudomonas* spp. and *Bacillus* spp. are the main documented PGPRs that induce ISR in plants during phytopathogen attack (Kloepper et al. 2004; Van Loon and Bakker 2006). The BCA *P. fluorescens*, for example, was able to trigger ISR against *P. syringae* pv. *tomato* in *A. thaliana* (Van Loon et al. 1998). Also, the PGPR *Bacillus cereus* was able to induce systemic resistance in *A. thaliana* while simultaneously activating the salicylate and jasmonate-/ethylene-dependent signalling routes (Niu et al. 2011).

One of the most common mechanisms related to ISR and modulated by the colonization of plants by beneficial microorganisms is the induction of phytohormones (Harman et al. 2004) like salicylic acid (SA), jasmonic acid (JA) and ethylene (ET) that act in signalling and stimulating the host's defence response against a variety of phytopathogens (Beneduzi et al. 2012; Pieterse et al. 2014). Studies on the ability to control diseases and promote plant growth using *Bacillus* spp. identified the production of SA, JA and ET by these bacteria as the main mechanism of biocontrol, reducing the incidence of root pathogens in inoculated plants (García et al. 2004; Wang et al. 2020). Inoculation of ginseng (*Panax ginseng*) plants with *T. citrinoviride* also resulted in the induction of phytohormones (zeatin, abscisic acid and JA) that contribute to root development, in response to injuries caused by herbivorous attacks, and, also, induced a SAR against the six most relevant ginseng pathogens (Garay-Arroyo et al. 2012; Pieterse et al. 2012; Park et al. 2018).

In addition to phytohormones, pathogenesis-related proteins (PR-proteins), such as β-1,3-glucanases, chitinases and ribosome-inactivating proteins, have antimicrobial activity and contribute to plant resistance against infection by pathogens (Pulla et al. 2010; Park et al. 2018). The PR-proteins cause a "priming" effect, which corresponds to a physiological state of the plant, in which the cells respond to very low levels of a stimulus; thus the plant previously exposed to the inducing agent becomes sensitized (Saikkonen et al. 2010; Dallagnol 2018). The biocontrol of plant diseases through resistance induction has been verified in several studies involving BCAs, confirming its potential use as a tool for the biological control of diseases, since it has two highly desirable characteristics from the market point of view: simultaneous protection against multiple pathogens and systemicity (Morandi and Bettiol 2009).

4.5 In Vitro Screening of Potential Biological Control Agents

Screening of new microorganisms for use in biocontrol of plant pathogens is a complex process, in which many criteria are evaluated besides the antagonistic efficacy. The use of refined methodologies is crucial to explore the diversity of the plant's microbiome and to identify new potential BCAs. Köhl et al. (2011) demonstrated that the commercial use of BCAs is defined in nine steps, with step 2 of this process aiming at the origin and isolation of potential candidates. Different strategies can be used in the search for potential BCAs, and the selection of a more suitable screening method is dependent on several factors. Most of the adopted methodologies are carried out in two distinct stages: the first is conducted using in vitro techniques to identify potential agents, which will be used in the second phase of the study that is developed in vivo (Lecomte et al. 2016). Thus, in vitro screening techniques play a fundamental role in screening, isolating and identifying potential BCAs (Lecomte et al. 2016; Besset-Manzoni et al. 2019).

In vitro techniques for screening potential agents have been used to identify thousands of BCAs for more than 50 years (Johnson et al. 1960; Dennis and Webster 1971a, b). These techniques are used for the selection of microorganisms that exhibit direct pathogen antagonism using plate assays, with only one or two different microorganisms (Pliego et al. 2011). Raymaekers et al. (2020) reported that there are three main techniques used for in vitro evaluation of the direct antagonistic effect in relation to growth and infection of the pathogen: (I) dual culture assay; (II) dual culture-volatile compounds; and (III) liquid medium. The dual culture consists in the simultaneous cultivation of the agent with the pathogen, and the evaluation is carried out by means of the pathogen growth inhibition index in the solid medium (Fig. 4.5a). This technique is widely used, and more than 2500 potential agents have been identified (Raymaekers et al. 2020), in which the growth of the pathogen colonies is inhibited in the proximity of the outgrowing agents' colonies (Comby et al. 2017). Using dual culture, Ramesh and Phadke (2012) screened agents for their antibacterial activity against *R. solanacearum* and found 22 isolates with an inhibition radius higher than 4 cm. Similar positive results were obtained by Adesina et al. (2007) that evaluated the antagonistic activity between *R. solani* and *F. oxysporum* and found 327 isolates with action potential. Although widely used and showing positive results, this type of culture excludes agent-pathogen interaction and cannot select agents that provide disease control by other mechanisms, such as root colonization and induction of systemic resistance (Pliego et al. 2011).

An adaptation to the dual culture technique is the cultivation of the pathogen and BCA in a solid medium, but physically separated (Fig. 4.5b). This strategy has been adopted to identify the exclusive action of volatile compounds produced by the BCA and allows the exclusion of other inhibition fonts (Raymaekers et al. 2020). Volatile compounds produced by microorganisms have been shown to be efficient in inhibiting the growth of plant pathogens (Ezra et al. 2004). Fialho et al. (2010) analysing VOCs produced by *Saccharomyces cerevisiae* observed an 87% inhibition of the growth of the pathogen *Guignardia citricarpa*, which causes black spot disease. Fernando et al. (2005) identified 14 potential BCAs and observed a high rate of pathogen growth inhibition, demonstrating the efficiency of the method for the screening of agents. However, as observed in dual culture assay, this type of culture also excludes agent-pathogen interaction and cannot select agents that provide disease control by other mechanisms.

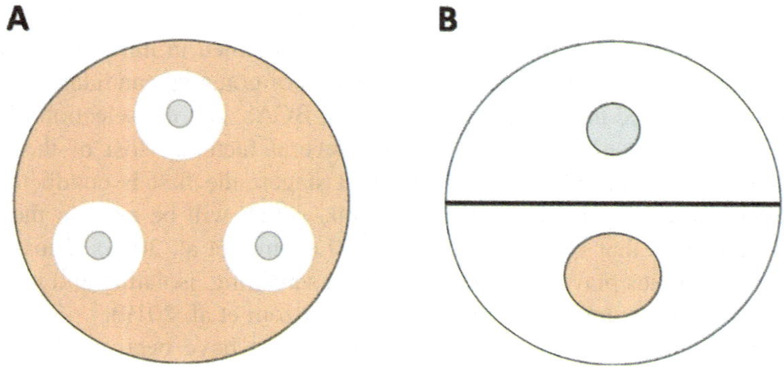

Fig. 4.5 In vitro screening techniques for direct biocontrol mechanisms. (**a**) Dual culture assay in which BCAs (grey) and the pathogen (pink) are spotted on the solid agar medium exhibiting the pathogen inhibition halo. (**b**) Dual culture assay-volatile compounds in which BCAs (grey) and pathogen (pink) are placed in solid agar medium with separate compartments. (adapted from Raymaekers et al. 2020)

Different to the methodologies that use solid media, screening for new BCAs can also be performed in liquid culture media, in which the pathogen and agent are inoculated together, and the inhibition of pathogen growth is assessed by reading the optical density of the culture (Raymaekers et al. 2020). Since the 1990s (Broekaert et al. 1990), this technique has shown promising results in the identification of new agents as demonstrated by Omar and Abd-Alla (1998) and Oh and Lee (2000). Even though it is used in many medical studies to identify agents that inhibit the growth of pathogens (Xie et al. 2018), there are still some studies that demonstrate the use of this methodology for screening BCAs of plant diseases. However, the application of this technique is difficult when living BCAs are used, because during the inhibition evaluation process, a differentiation between pathogen and agent is necessary (Raymaekers et al. 2020).

Aiming to evaluate the direct pathogen antagonism and the production of secondary metabolites (Raaijmakers et al. 2002), in vitro techniques are considered an important strategy for the initial screening, in order to save time and financial resources when compared to in vivo screening techniques (Köhl et al. 2011). Therefore, it is from the results of in vitro assays that agents will be selected according to the growth inhibition of the pathogen (Askew and Laing 1994; Abdalla et al. 2014). However, the efficiency of these assays has been questioned, since the results obtained in vitro are not always the same as those observed in the field (Besset-Manzoni et al. 2019). One of the main factors associated with in vitro assays that can result in a false positive is the nutritional concentration of the medium, which is generally richer in nutrients compared to the rhizosphere and can lead to an exacerbated production of secondary metabolites (Köhl et al. 2019). Different studies have demonstrated the nutritional effect of the medium on the production of secondary metabolites of BCAs (Bode et al. 2002; Raaijmakers et al. 2002; Bosmans et al. 2016); therefore, the development of new strategies for the screening of new agents is crucial. According to Pliego et al. (2011), molecular approaches combined with screening methodologies can be important to optimize the process of identifying new agents. In this context, transcriptomics has emerged as a powerful tool to improve the screening techniques of BCAs. Atanasova et al. (2013) and Hennessy et al. (2017) revealed through transcriptomic analysis that secondary metabolites detected in vitro may not be expressed in vivo or have a secondary action in inhibiting the growth of the pathogen. Other studies have used transcriptomic analysis to investigate the expression of important genes associated to secondary metabolites production during the in vitro co-cultivation of agent and pathogen (Lysøe et al. 2017) and observed a combination of different types of mechanisms during this interaction.

Even with methodological advances in recent years, there are still several difficulties to obtain a high success rate in in vitro screening techniques. Some studies discuss the possibility of using high-throughput *in planta* assays for optimization of the screening (Barnett et al. 2017); however, they also showed disadvantages due to their artificial character (Schreiter et al. 2014; Rybakova et al. 2016). Although the effectiveness of conducting in vitro screening is discussed, these methodologies are widely used, especially because potential BCAs are detected in an easy, fast and reproducible manner, and it is possible to perform screening on a relatively large collection of organisms (Köhl et al. 2019). In addition to in vitro assays, it is important to extend this line of research to plants grown on natural substrates, increasing the selection efficiency of the most promising BCAs (Pliego et al. 2011). Finally, the constant improvement of in vitro methodologies, as well as the combination of resources in the initial isolation process, can be fundamental for the advancement in the scientific and commercial field of efficient BCAs production for hundreds of plant diseases.

4.6 Biological Control Agents from a Commercial Point of View

Biological control can occur in four different ways: natural, classic, augmentative and by conservation (Van Lenteren et al. 2018; Baker et al. 2020). Natural biological control occurs in all ecosystems, through the action of beneficial microorganisms that naturally occur in the environment, without human interference (Van Lenteren et al. 2018). The classic biological control was the first to be widely practiced and consists in the implantation and release of natural enemies, which are previously collected in areas of pest origin and then released in areas where the pest is invasive (Van Lenteren et al. 2018; Baker et al. 2020). Although this type of control is not always able to reduce the levels of invasive species, it can successfully supply the level of general equilibrium of the population, bringing enormous economic benefits (DiTomaso et al. 2017). In the augmentative biological control, natural enemies are produced in large scale to be released in large numbers, aiming at immediate control or control over several generations of crops with long production cycle (seasonal control). Finally, biological control by conservation involves implementing better survival conditions for existing natural species, by reducing the use of pesticides or using selective pesticides to conserve natural microorganisms, maintaining weed borders or providing cover with flowers to improve the performance of natural enemies (Van Lenteren et al. 2018; Baker et al. 2020). However, one or more biocontrol strategies must be considered for the implementation of biological control programs. The selection of the type of biological agent must be closely related to the cost, convenience, efficacy and reliability of the products, in addition to the adoption of cultural practices, such as good sanitation, soil preparation and water management, since the highest successes of biological control are achieved in situations where environmental conditions are more controlled or predictable (Pal and Gardener 2006).

Most of the world's production of BCAs is concentrated in relatively small companies such as BioWorks, AgraQuest, Novozymes, Prophyta and Kemira Agro (Pal and Gardener 2006). However, there is a growing interest from large agrochemical companies, such as Monsanto, Bayer and Syngenta, as well as microbial industries, such as yeast producers for baking, beer and wine manufacturing, to acquire technologies for the development of BCAs. This is reflected in the number of subsidiaries and acquisitions made by large multinational companies that are concerned with offering their customers a broad portfolio to protect their cultures, with options for conventional products and more environmentally friendly products (Pal and Gardener 2006; Droby et al. 2016).

In 2015, the global pesticide market generated around U$ 58.46 billion, with the market of BCAs accounting for around 1.7 billion, which corresponds to less than 2% of the market. However, there has been a percentage increase in sales since 2005, with growth projections of more than 15% per year (Van Lenteren et al. 2018). The rapid annual growth of the BCAs market is the result of a combination of factors. The increase in the oil price, the increased demand for organic food, the concern for the health of workers and farmers living in rural areas and the increased demand for "safer" and more sustainable pesticides that do not cause damage to plantations and are able to provide a higher yield and healthier crops are the main factors for reducing dependence on chemical inputs (Pal and Gardener 2006; Van Lenteren et al. 2018). In addition, the use of BCAs facilitates trade between exporting countries of agricultural products, which must comply with increasingly strict food safety laws, and can only commercialize products with low levels of chemical residues (Praneetvatakul et al. 2013; Van Lenteren et al. 2018).

All forms of biological control have been extensively explored in the past 50 years, and the science behind it has gone from a trial and error-based method to more predictive, with several economic and environmental benefits. However, to ensure that its implementation is a viable and sustainable strategy is still a major challenge (Messing and Brodeur 2018). The main problems to further dissemination are the lack of economic investments and incentives for new research, mainly by government entities, the political and marketing power of the dominant chemical industries in the market and the national and international laws that end up hampering a greater worldwide dissemination of the use of BCAs, such as the Convention on Biological Diversity (CBD 1993; Messing and Brodeur 2018). The terms established in that treaty give signatory countries sovereign rights over genetic resources and agreements over access and sharing the benefits resulting from their use. This means that the permission to collect potential BCAs, which are accidentally exported between countries, can only be granted by the country that has sovereign rights over the genetic resources, making research mass production and the commercialization of products even more difficult (Van Lenteren et al. 2018). These obstacles that still exist in relation to the use of BCAs need to be resolved immediately, so that trade and products commercialization could be widely encouraged and implemented in agricultural practices. The recognition of benefits conferred by the use of BCAs in total or partial substitution to agrochemicals by the main agricultural economies in the world, and the application of a "real cost" for the commercialization of these products, would be a good alternative to highlight and boost the market of BCAs (Praneetvatakul et al. 2013; Van Lenteren et al. 2018). This trend may further stimulate the development and registration of new BCAs, and projections of marketing strategies may help to increase sales and guarantee greater market share.

4.7 Conclusion and Future Prospects

BCAs of pest and plant diseases play an important role in restoring natural enemies and ecological balance in agricultural systems whose sustainability has been compromised by environmental degradation. In this context, the use of microbial agents to control pests and diseases should focus on agroforestry systems that guarantee, through management techniques, the maintenance of these microorganisms in agroecosystems and/or in the use of commercial products that can assist in recomposing their populations.

The implications of using microorganisms for biological control of insect pests and plant diseases may be limited by the difficulty in identifying many promising groups of microorganisms, since their morphological determination becomes very difficult due to the homoplasia of the characters used to differentiate species, even for specialists. Thus, in vitro techniques associated with molecular analysis are a fundamental tool for screening and identification of new potential BCAs, since they limit false-positive results. These types of techniques also help to develop research in a more refined way, ensuring specific results that allow the identification of potential BCAs together with their mode of action in the environment.

Regardless of the species of microorganisms used as a BCA, one thing is common: biotic and abiotic factors, environmental conditions such as temperature and relative humidity, soil type, resistant host genotypes (Barretti et al. 2012) and disease pressure (Gerbore et al. 2014) can affect the performance of the BCA. The use of fungi of the genus *Trichoderma* in combination with PGPR could be a viable solution for sustainable agriculture; however, some gaps regarding the double inoculation of these two biocontrol agents need to be filled. It is necessary to research the mechanisms involved in this process in different agricultural systems, and also the impact of the concomitant use of these BCAs on the biodiversity of other microorganisms in the field, so that the maximum benefits can be obtained in terms of better plant growth, especially under stress conditions.

Despite recent advances on the importance of BCAs for plant health, there are still doubts about the effectiveness of their use in agriculture, mainly due to the lack of knowledge of how they would behave in complex environments. Therefore, a more comprehensive selection of beneficial microorganisms, adapted to different environmental conditions, will be fundamental for their applicability in agriculture. Based on advances in this area, it will be possible to identify which beneficial microorganism will be best adapted for the environmental conditions, as well as which mechanism will be the most suitable to protect plants from possible pathogens. Nevertheless, even with several challenges mentioned above, BCAs emerge as an important tool for combating pests and diseases in agriculture, as they are a sustainable and more economic alternative to the use of pesticides.

References

ABCBio- Brazilian Association of Biological Control Companies (2020) Biological control products registered in Brazil since 2005. https://www.abcbio.org.br/biodefensivos-registrados/

Abdalla SA, Algam SAA, Ibrahim EA, El Naim AM (2014) In vitro screening of *Bacillus* isolates for biological control of early Blight disease of tomato in Shambat soil. World J Agric Res 2:47–50

Adams PB, Fravel DR (1993) Dynamics of *Sporidesmium*, a naturally occurring fungal mycoparasite. In: Lumsden RD, Vaughn JL (eds) Pest Management: biologically based technologies. American Chemical Society, Washington, DC, pp 189–195

Adesina MF, Lembke A, Costa R, Speksnijder A, Smalla K (2007) Screening of bacterial isolates from various European soils for in vitro antagonistic activity towards *Rhizoctonia solani* and *Fusarium oxysporum*: site-dependent composition and diversity revealed. Soil Biol Biochem 39:2818–2828

Agrios GN (2005) Plant pathology, 5th edn. Elsevier, Academic Press, Amsterdam

AGROFIT (2020a) *Bacillus thuringiensis*. http://agrofit.agricultura.gov.br/agrofit_cons/principal_agrofit_cons. Accessed 6 May 2020

AGROFIT (2020b) *Beauveria bassiana*. http://bi.agricultura.gov.br/reports/rwservlet?agrofit_cons&ingredientes_ativos.rdf&p_id_ingrediente_ativo=&p_nm_comum_port=Beauveria%20bassiana&p_id_grupo_quimico=&p_id_classe=&p_id_cultura=&p_tipo_relatorio=DETALHADO&paramform=no. Accessed 5 May 2020

AGROFIT (2020c) *Metarhizium*. http://bi.agricultura.gov.br/reports/rwservlet?agrofit_cons&ingredientes_ativos.rdf&p_id_ingrediente_ativo=&p_nm_comum_port=Metarhizium&p_id_grupo_quimico=&p_id_classe=&p_id_cultura=&p_tipo_relatorio=DETALHADO&paramform=no. Accessed 7 May 2020

Ahmed E, Holmström SJ (2014) Siderophores in environmental research: roles and applications. J Microbial Biotechnol 7(3):196–208

Ahmed AS, Sanchez CP, Candela ME (2010) Evaluation of induction of systemic resistance in pepper plants (*Capsicum annuum*) to *Phytopthora capsici* using *Trichoderma harzianum* and its relation with capsidiol accumulation. Eur J Plant Pathol 106:817–824

Alabouvette C, Olivain C, Steinberg C (2006) Biological control of plant diseases: the European situation. Eur J Plant Pathol 114(3):329–341

Almassi F, Ghisalberti EL, Narbey MJ, Sivasithamparam K (1991) New antibiotics from strains of *Trichoderma harzianum*. J Nat Prod 54:396–402

Almenara DP, Rossi C, Camargo MR, Carlos Edurdo Winter CE (2012) Entomopathogenic nematodes. INCT (orgs) Advanced topics in molecular entomology. National Institute of Science and Technology in Molecular Entomology, Rio de Janeiro, pp 1–40

Alves SB (1998) Entomopathogenic fungi. In: Alves SB (ed) Microbial control of insects. FEALQ, Piracicaba, pp 289–382

Alves SB, Lopes RB, Vieira AS, Tamai MA (2008) Entomopathogenic fungi used in pest control in Latin America. In: Alves SB, Lopes RB (eds) Microbial pest control in Latin America: advances and challenges. FEALQ, Piracicaba, pp 69–110

Angeli D, Saharan KR, Mauchofer M, Gessler C, Pertot I (2013) Increasing efficacy of *Ampelomyces quisqualis* against powdery mildew pathogen. IOBC-WPRS Bull 86:195–196

Anisha C, Jishma P, Bilzamol VS, Radhakrishnan EK (2018) Effect of ginger endophyte *Rhizopycnis vagum* on rhizome bud formation and protection from phytopathogens. Biocatal Agric Biotechnol 14:116–119

Argüelles JC (2000) Physiological roles of trehalose in bacteria and yeasts: a comparative analysis. Arch Microbiol 174:217–224

Arora NK (2018) Agricultural sustainability and food security. Environ Sustain 1(3):1–3

Arseneault T, Filion M (2017) Biocontrol through antibiosis: exploring the role played by subinhibitory concentrations of antibiotics in soil and their impact on plant pathogens. Can J Plant Pathol 39(3):267–274

Askew DJ, Laing MD (1994) The in vitro screening of 118 *Trichoderma* isolates for antagonism to *Rhizoctonia solarti* and an evaluation of different environmental sites of *Trichoderma* as sources of aggressive strains. Plant and Soil 159:277–281

Atanasova L, Le Crom S, Gruber S, Coulpier F, Seidl-Seiboth V, Kubicek CP et al (2013) Comparative transcriptomics reveals different strategies of *Trichoderma* mycoparasitism. BMC Genomics 14:121–136

Audenaert K, Pattery T, Cornelis P, Hofte M (2002) Induction of systemic resistance to *Botrytis cinerea* in tomato by *Pseudomonas aeruginosa* 7NSK2: role of salicylic acid, pyochelin, and pyocyanin. Mol Plant Microbe Interact 15:1147–1156

Awad NE, Kassem HA, Hamed MA, El-Feky AM, Elnaggar MAA, Mahmoud K et al (2018) Isolation and characterization of the bioactive metabolites from the soil derived fungus *Trichoderma viride*. Mycology 9(1):70–80

Ayukea FO, Lagerlöfa J, Jorge G, Söderlunda S, Muturid JJ, Saroshe BE et al (2017) Effects of biocontrol bacteria and earthworms on the severity of *Alternaria brassicae* disease and the growth of oilseed rape plants (*Brassica napus*). Appl Soil Ecol 117:63–69

Ayyadurai N, Naik PR, Rao SM, Kumar SR, Samrat SK, Manohar M et al (2006) Isolation and characterization of a novel banana rhizosphere bacterium as fungal antagonist and microbial adjuvant in micropropagation of banana. J Appl Microbiol 100:926–937

Bagy HMM, Hassan KEA, Nafady NA, Dawoodb MFA (2019) Efficacy of arbuscular mycorrhizal fungi and endophytic strain *Epicoccum nigrum* ASU11 as biocontrol agents against blackleg disease of potato caused by bacterial strain *Pectobacterium carotovora subsp. atrosepticum* PHY7. Biol Control 134:103–113

Baker BP, Green TA, Loker AJ (2020) Biological control and integrated pest management in organic and conventional systems. Biol Control 140:1–9

Bardin M, Pugliese M (2020) Biocontrol agents against diseases. In: Gullino M, Albajes R, Nicot P (eds) Integrated pest and disease management in Greenhouse crops. Plant pathology in the 21st century, vol 9. Springer, Cham, pp 385–407

Barnett S, Zhao S, Ballard R, Franco C (2017) Selection of microbes for control of Rhizoctonia root rot on wheat using a high throughput pathosystem. Biol Control 113:45–57

Barretti PB, de Souza RM, Pozza EA, de Souza JT (2012) Combination of endophytic bacteria and resistant cultivars improves control of *Ralstonia* wilt of tomato. Aust Plant Pathol 41:189–195

Batista Filho JEM, Almeida LG, Leite H (2013) Retrospective of microbial control of agricultural pests in Brazil. Biol Inst 9:31–33

Bedding RA (1968) *Deladenus wilsoni* n. sp. and *D. siricidicola* n. sp.(Neotylenchidae), entomophagous-mycetophagous nematodes parasitic in siricid woodwasps. Nematologica 14(4):515–525

Bencheqroun SK, Bajji M, Massart B, Labhilili M, Jaafari SE, Jijakli MH (2007) In vitro and in situ study of postharvest apple blue mold biocontrol by *Aureobasidium pullulans*: evidence for the involvement of competition for nutrients. Postharvest Biol Technol 46:128–135

Beneduzi A, Ambrosini A, Passaglia LMP (2012) Plant growth-promoting rhizobacteria (PGPR): their potential as antagonists and biocontrol agents. Genet Mol Biol 35:1044–1051

Berliner E (1915) Über die Schlaffsucht der Mehlmottenraupe (Ephestia kühniella Zell.) und ihren Erreger *Bacillus thuringiensis* n. sp. Zeitschrift für angewandte Entomologie 2(1):29–56

Besset-Manzoni Y, Joly P, Brutel A, Gerin F, Soudière O, Langin T, Prigent-Combaret C (2019) Does in vitro selection of biocontrol agents guarantee success in planta? A study case of wheat protection against Fusarium seedling blight by soil bacteria. PLoS One 14(12):e0225655

Bettiol W, Ghini R, Morandi MAB, Stadnik MJ, Kraus U, Stefanova M, Prado AMC (2008) Biological control of plant diseases in Latin America. In: Alves SB, Lopes RB (eds) Microbial pest control in Latin America—advances and challenges. Piracicaba, FEALQ, pp 303–331

Bisen K, Keswani C, Patel JS, Sarma BK, Singh HB (2016) *Trichoderma* spp.: efficient inducers of systemic resistance in plants. In: Choudhary D, Varma A (eds) Microbial-mediated induced systemic resistance in plants. Springer, Singapore, pp 185–195

Bode HB, Bethe B, Höfs R, Zeeck A (2002) Big effects from small changes: possible ways to explore nature's chemical diversity. ChemBioChem 3:619–627

Boemare NE, Akhurst RJ, Mourant RG (1993) DNA relatedness between *Xenorhabdus* spp. (Enterobacteriaceae), symbiotic bacteria of entomopathogenic nematodes, and a proposal to transfer *Xenorhabdus luminescens* to a new genus, Photorhabdus gen. nov. Int J Syst Evol Microbiol 43(2):249–255

Bosmans L, De Bruijn I, De Mot R, Rediers H, Lievens B (2016) Agar composition affects in vitro screening of biocontrol activity of antagonistic microorganisms. J Microbiol Methods 127:7–9

Boukerma L, Benchabane M, Charif A, Khélifi L (2017) Activity of plant growth promoting rhizobacteria (PGPRs) in the biocontrol of tomato *Fusarium* wilt. Plant Protect Sci 53:78–84

Brazil. Ministry of Agriculture Livestock and Supply (2019) Biodefensive snow market grows more than 70% in Brazil in one year. https://ciorganicos.com.br/sustentabilidade/mercado-de-biodefensivos-cresce-mais-de-70-no-brasil-em-um-ano/. Accessed 6 May 2020

Broekaert WF, Terras FRG, Cammue BPA, Vanderleyden J (1990) An automated quantitative assay for fungal growth inhibition. FEMS Microbiol Lett 69:55–59

Brum MCP, Araújo WL, Maki CS, Azevedo JL (2012) Endophytic fungi from *Vitis labrusca* L. ('Niagara Rosada') and its potential for the biological control of *Fusarium oxysporum*. Genet Mol Res 11:4187–4197

Cao R, Liu X, Gao K, Mendgen K, Kang Z, Gao J et al (2009) Mycoparasitism of endophytic fungi isolated from reed on soilborne phytopathogenic fungi and production of cell wall-degrading enzymes in vitro. Curr Microbiol 59:584–592

Carrano-Moreira AF (2014) Integrated management of forest pests: ecological foundations, concepts and control tactics. Technical Books, Rio de Janeiro

Castro MEB, Souza ML, Sihler W, Rodrigues JCM, Ribeiro BM (1999) Molecular biology of baculovirus and its use in biological pest control in Brazil. Braz Agric Res 34:1733–1761

Castro MEB, Bergmann MRS, Rocha C, Ward Inglis P, Valicente FH (2020) Control of pest arthropods with entomopathogenic viruses. In: EMG Sources, Valadares-Inglis MC (eds) Biological control of agricultural pests. Embrapa, Brasília, pp 237–273

Cavalcante RS, Lima HLS, Pinto GAS, Gava CAT, Rodrigues S (2008) Effect of moisture on *Trichoderma conidia* production on corn and wheat bran by solid state fermentation. Food Bioproc Tech 1:100–104

Cavalcanti VP, Araújo NAF, Machado NB, Costa Júnior PSP, Pasquala M, Alves E et al (2020) Yeasts and Bacillus spp. as potential biocontrol agents of Sclerotinia sclerotiorum in garlic. Sci Hortic 261

CBD (1993) Convention on biological diversity (with annexes). Concluded at Rio de Janeiro on 5 June 1992. United Nations Treaty Ser 1760(30619):142–382

Chen S, Dickinson DW (2004) Biological control of nematodes with bacterial antagonists. In: Chen Z, Chen S, Dickinson DW (eds) Nematology—advances and perspectives. Nematode management and utilization, vol 2. Tsinghua University Press & CABI, Beijing and Wallingford, pp 1041–1082

Chen J, Sun S, Miao C, Wu K, Chen Y, Xu L et al (2016) Endophytic *Trichoderma gamsii* YIM PH30019: a promising biocontrol agent with hyperosmolar, mycoparasitism, and antagonistic activities of induced volatile organic compounds on root-rot pathogenic fungi of *Panax notoginseng*. J Ginseng Res 40:315–324

Cheong SL, Cheow YL, Ting ASY (2017) Characterizing antagonistic activities and host compatibility (via simple endophyte-calli test) of endophytes as biocontrol agents of *Ganoderma boninense*. Biol Control 105:86–92

Chernin L, Ismailov Z, Haran S, Chet I (1995) Chitinolytic *Enterobacter agglomerans* antagonistic to fungal plant pathogens. Appl Environ Microbiol 61(5):1720–1726

Chiron N, Micherlot D (2005) Odeurs des champignons: chimie et rôle dans les interactions biotiques - une revue. Cryptogam Mycol 26:299–364

Cohen-Kupiec R, Chet I (1998) The molecular biology of chitin digestion. Curr Opin Biotechnol 9: 270–277

Cohn F (1872) Untersuchungen über bakterien. Beitrage zur Biologie der Pflanzen 1:127–224

Comby M, Gacoin M, Robineau M, Rabenoelina F, Ptas S, Dupont J et al (2017) Screening of wheat endophytes as biological control agents against *Fusarium* head blight using two different in vitro tests. Microbiol Res 202:11–20

Compant S, Duffy B, Nowak J, Clément C, Barka EA (2005) Use of plant growth promoting bacteria for biocontrol of plant diseases: principles, mechanisms of action, and future prospects. Appl Environ Microbiol 71:4951–4959

Contarino R, Brighina S, Fallico B, Cirvilleri G, Parafati L, Restuccia C (2019) Volatile organic compounds (VOCs) produced by biocontrol yeasts. Food Microbiol 82:70–74

Couteaudier Y, Alabouvette C (1990) Quantitative comparison of *Fusarium oxysporum* competitiveness in relation with carbon utilization. FEMS Microbiol Ecol 74:261–268

Cui L, Yang C, Wei L, Li T, Chen X (2019) Isolation and identification of an endophytic bacteria *Bacillus velezensis* 8-4 exhibiting biocontrol activity against potato scab. Biol Control 141: 104156

Dallagnol LJ (2018) Genetic Resistance: From Plants to Pathogens. UFPel Publisher, Pelotas

Dandurishvili N, Toklikishvili N, Ovadis M, Eliashvili P, Giorgobiani N, Keshelava R et al (2011) Broad-range antagonistic rhizobacteria *Pseudomonas fluorescens* and *Serratia plymuthica* suppress *Agrobacterium* crown gall tumours on tomato plants. J Appl Microbiol 110:341–352

Das S, Goswami A, Debnath N (2019) Application of baculoviruses as biopesticides and the possibilities of nanoparticle mediated delivery. In: Koul O (ed) Nano-biopesticides today and future perspectives. Academic Press, New York, pp 261–280

De Corato U (2020) Disease-suppressive compost enhances natural soil suppressiveness against soil-borne plant pathogens: a critical review. Rhizosphere 13:100–192

Défago G, Haas D (1990) Pseudomonads as antagonists of soilborne plant pathogens: modes of action and genetic analysis. In: Bollag JM, Stotsky G (eds) Soil biochemistry. Marcel Dekker, New York, pp 249–291

Dennis C, Webster J (1971a) Antagonistic properties of species-groups of *Trichoderma*. I. Production of non-volatile antibiotics. Trans Br Mycol Soc 57:25–39

Dennis C, Webster J (1971b) Antagonistic properties of species-groups of *Trichoderma* II. production of volatile antibiotics. Trans Br Mycol Soc 57:41–48

Dickinson JM, Hanson JR, Hitchcock PB, Claydon N (1989) Structure and biosynthesis of harzianopyridone, an antifungal metabolite of *Trichoderma harzianum*. J Chem Soc Perkin Trans 1:1885–1887

Dickson DW, Oostendorp M, Mitchell DJ (1992) Development of *Pasteuria penetrans* on *Meloidogyne arenaria* race 1 in the field. In: Gommers FJ, Maas PW (eds) Nematology from molecule to ecosystem. European Society of Nematologists, Dundee, pp 213–218

DiTomaso JM, Steenwyk RAV, Nowierskic RM, Meyerson LA, Doeringe OC, Lane E et al (2017) Addressing the needs for improving classical biological control programs in the USA. Biol Control 106:35–39

Djaya L, Hersantia IN, Hartatia S, Joni IM, Bagya HMMK et al (2019) In vitro study of plant growth promoting rhizobacteria (PGPR) and endophytic bacteria antagonistic to *Ralstonia solanacearum* formulated with graphite and silica nanoparticles as a biocontrol delivery system (BDS). Biocatal Agric Biotechnol 19:101153

Dodds PN, Rathjen JP (2010) Plant immunity: towards an integrated view of plant pathogen interactions. Nat Rev Genet 11:539–548

Doley K, Borde M, Kulkarni M (2019) AM Fungi and *Trichoderma* interaction for biological control of soilborne plant pathogen *Fusarium oxysporum*. Plant Microbe Interface 4:95–128

Dolinski CM (2020) Control of pest arthropods with entomopathogenic nematodes. In: EMG Sources, Valadares-Inglis MC (eds) Biological control of agricultural pests. Embrapa, Brasflia, pp 275–288

Dolinski C, Monteiro C, Andaló V, Leite LG (2017) Studies on entomopathogenic nematodes in Brazil: past and future. Nematoda 4:102–117

Droby S, Wisniewski M, Macarisin D, Wilson C (2009) Twenty years of postharvest biocontrol research: is it time for a new paradigm? Postharvest Biol Technol 52:137–145

Droby S, Wisniewski M, Teixidó N, Spadaro D, Jijakli MH (2016) The science, development, and commercialization of postharvest biocontrol products. Postharvest Biol Technol 122:22–29

Dukare AS, Paul S, Nambi VE, Gupta RK, Singh R, Sharma K (2018) Exploitation of microbial antagonists for the control of postharvest diseases of fruits: a review. Crit Rev Food Sci 2:1–16

Ezra D, Hess WM, Strobel GA (2004) New endophytic isolates of Muscodor albus, a volatile-antiobiotic-producing fungus. Microbiology 150:4023–4031

Farag MA, Ryu CM, Sumner LW, Paré PW (2006) GC-MS SPME profiling of rhizobacterial volatiles reveals prospective inducers of growth promotion and induced systemic resistance in plants. Phytochemistry 67:2262–2268

Favaro LCL, de Sebastianes SFL, Araújo WL (2012) *Epicoccum nigrum* P16, a sugarcane endophyte, produces antifungal compounds and induces root growth. PLoS One 7:1–10

Feng YJ, Ge Y, Tan SQ, Zhang KQ, Ji R, Shi WP (2015) Effect of *Paranosema locustae* (Microsporidia) on the behavioural phases of *Locusta migratoria* (Orthoptera: Acrididae) in the laboratory. Biocontrol Sci Technol 25:48–55

Fernando WGD, Ramarathnam R, Krishnamoorthy AS, Savchuk SC (2005) Identification and use of potential bacterial organic antifungal volatiles in biocontrol. Soil Biol Biochem 37:955–964

Fialho MB, Toffano L, Pedroso MP, Augusto F, Pascholati SF (2010) Volatile organic compounds produced by *Saccharomyces cerevisiae* inhibit the *in vitro* development of *Guignardia citricarpa*, the causal agent of citrus black spot. World J Microbiol Biotechnol 26:925–932

Fialho MB, Ferreira LFR, Monteiro RTR, Pascholati SF (2011) Antimicrobial volatile organic compounds affect morphogenesis-related enzymes in *Guignardia citricarpa*, causal agent of citrus black spot. Biocontrol Sci Technol 21:797–807

Fokkema NJ (1996) Biological control on aerial plant surfaces. In: Food and Fertilizer Technology Center for the ASPAC Region Book Series, vol 42, pp 41–48

Fontes EMG, Pires CSS, Sujii ER (2020) Usage strategies and history. In: EMG Sources, Valadares-Inglis MC (eds) Biological control of agricultural pests. Embrapa, Brasflia, pp 21–45

Fravel D (1988) Role of antibiosis in the biocontrol of plant diseases. Annu Rev Phytopathol 26:75–91

Gao P, Qin J, Li D, Zhou S (2018) Inhibitory effect and possible mechanism of a *Pseudomonas* strain QBA5 against gray mold on tomato leaves and fruits caused by *Botrytis cinerea*. PLoS One 13:e0190932

Garay-Arroyo A, De La Paz SM, García-Ponce B, Azpeitia E, Alvarez-Buylla ER (2012) Hormone symphony during root growth and development. Dev Dyn 241:1867–1885

García JA, Probanza A, Ramos B, Palomino MR, Mañero FJ (2004) Effect of inoculation of *Bacillus licheniformis* on tomato and pepper. Agronomie:169–176

Gardener BB, Fravel DR (2002) Biological control of plant pathogens: research, commercialization, and application in the USA. Plant Health Prog 3(1):17

Gautama S, Chauhanb A, Sharmac R, Sehgala R, Shirkota CK (2019) Potential of *Bacillus amyloliquefaciens* for biocontrol of bacterial canker of tomato incited by *Clavibacter michiganensis* ssp. Michiganensis. Microb Pathog 130:196–203

Gelernter WD, Lomer CJ (2000) Success in biological control of above-ground insects by pathogens. In: Gurr G, Wratten S (eds) Biological control: measures of success. Springer, Dordrecht, pp 297–322

Gerbore J, Benhamou N, Vallance J, Le Floch G, Grizard D, Regnault-Roger C et al (2014) Biological control of plant pathogens: advantages and limitations seen through the case study of *Pythium oligandrum*. Environ Sci Pollut Res 21:4847–4860

Ghasemi S, Safaie N, Shabazi S, Shams-Bakhsh M, Askari H (2019) Enhancement of lytic enzymes activity and antagonistic traits of *Trichoderma harzianum* using γ-radiation induced mutation. J Agric Sci Technol 21:1035–1048

Gholami M, Amini J, Abdollahzadeh J, Ashengroph M (2019) Basidiomycetes fungi as biocontrol agents against take-all disease of wheat. Biol Control 130:34–43

Ghorbanpour M, Omidvari M, Abbaszadeh-Dahaji P, Omidvar R, Kariman K (2018) Mechanisms underlying the protective effects of beneficial fungi against plant diseases. Biol Control 117: 147–157

Gouda S, Kerry RG, Das G, Paramithiotis S, Shin HS, Patra JK (2018) Revitalization of plant growth promoting rhizobacteria for sustainable development in agriculture. Microbiol Res 206: 131–140

Gu YQ, Mo MH, Zhou JP, Zou CS, Zhang KQ (2007) Evaluation and identification of potential organic nematicidal volatiles from soil bacteria. Soil Biol Biochem 39:2567–2575

Guardado-Valdivia L, Tovar P, Chacón LA, López GU, Gutíerrez MP, Selene A (2018) Identification and characterization of a new *Bacillus atrophaeus* strain E5 as biocontrol agent of postharvest anthracnose disease in soursop (*Annona muricata*) and avocado (*Persea americana*). Microbiol Res 210:26–32

Gunatilaka AAL (2006) Natural products from plant-associated microorganisms: distribution, structural diversity, bioactivity, and implications of their occurrence. J Nat Prod 69(3):509–526

Guo D, Yuan C, Luo Y et al (2020) Biocontrol of tobacco black shank disease (*Phytophthora nicotianae*) by *Bacillus velezensis* Ba168. Pestic Biochem Phys 165:104523

Haas D, Defago G (2005) Biological control of soil-borne pathogens by fluorescent pseudomonads. Nat Rev Microbiol 3:307–319

Hajeck AE, Leger ST (1994) Interactions between fungal pathogens and insect hosts. Annu Rev Entomol 39:293–322

Han SH, Anderson AJ, Yang KY, Cho BH, Kim KY, Lee MC et al (2006) Multiple determinants influence root colonization and induction of induced systemic resistance by *Pseudomonas chlororaphis* O6. Mol Plant Pathol 7:463–472

Harman GE, Howell CR, Viterbo A, Chet I, Lorito M (2004) *Trichoderma* species—opportunistic, avirulent plant symbionts. Nat Rev Microbiol 2:43–56

Havey MJ (2018) Onion breeding. Plant Breed Rev 42:39–85

Hennessy RC, Glaring MA, Olsson S, Stougaard P (2017) Transcriptomic profiling of microbe–microbe interactions reveals the specific response of the biocontrol strain *P. fluorescens* In5 to the phytopathogen *Rhizoctonia solani*. BMC Res Notes 10:376

Herrero S, Oppert B, Ferré J (2001) Different mechanisms of resistance to *Bacillus thuringiensis* toxins in the Indianmeal moth. Appl Environ Microbiol 67:1085–1089

Heydari A, Pessarakli M (2010) A review on biological control of fungal plant pathogens using microbial antagonists. Aust J Biol Sci 10:273–290

ISAAA-International Service for the Acquisition of Agri-Biotech Applications (2020) Bt Insect Resistance Technology. https://www.isaaa.org/resources/publications/pocketk/foldable/Pocket%20K6%20(English).pdf. Accessed 7 May 2020

Jia Y, Liu G, Park D, Yang Y (2013) Inoculation and scoring methods for rice sheath blight disease. In: Yang Y (ed) Rice protocols. Methods in molecular biology (methods and protocols), vol 956. Humana Press, Totowa, pp 257–268

Jiang C-H, Liao M-J, Wang H-K, Zheng M-Z, Xud J-J, Guo J-H (2018) *Bacillus velezensis*, a potential and efficient biocontrol agent in control of pepper gray mold caused by *Botrytis cinerea*. Biol Control 126:147–157

Jiang CH, Yao XF, Mi DD, Li ZJ, Yang BY, Zheng Y et al (2019) Comparative transcriptome analysis reveals the biocontrol mechanism of *Bacillus velezensis* F21 against *Fusarium* wilt on watermelon. Front Microbiol 10:652

Johnson LF, Curl EA, Bond JH, Fribourg HA (1960) Methods for studying soil microflora-plant disease relationships. Burgess, Minneapolis, p 178

Jones EE, Mead A, Whipps JM (2004) Effect of inoculum type and timing of application of *Coniothyrium minitans* on *Sclerotinia sclerotiorum*: control of sclerotinia disease in glasshouse lettuce. Plant Pathol 53:611–620

Jones EE, Rabeendran N, Stewart A (2014) Biocontrol of *Sclerotinia sclerotiorum* infection of cabbage by *Coniothyrium minitans* and *Trichoderma* spp. Biocontrol Sci Technol 24:1363–1382

Kakemboa D, Leea YH (2019) Analysis of traits for biocontrol performance of *Pseudomonas parafulva* JBCS1880 against bacterial pustule in soybean plants. Biol Control 134:72–81

Kandula D, Jones E, Stewart A, McLean K, Hampton J (2015) *Trichoderma* species for biocontrol of soil-borne plant pathogens of pasture species. Biocontrol Sci Technol 25:1052–1069

Kanini GS, Katsifas EA, Savvides AL, Karagouni AD (2013) *Streptomyces rochei* ACTA1551, an Indigenous Greek isolate studied as a potential biocontrol agent against *Fusarium oxysporum f. sp. lycopersici*. Biomed Res Int 2013:387230

Kenis M, Hurley BP, Colombari F, Lawson S, Sun J, Wilcken C et al (2019) Guide to the classical biological control of insect pests in planted and natural forests, N. 182. FAO Forestry Paper, Rome, p 182

Khalili E, Javed MA, Huyop F, Rayatpanah S, Jamshidi S, Wahab RA (2016) Evaluation of *Trichoderma* isolates as potential biological control agent against soybean charcoal rot disease caused by *Macrophomina phaseolina*. Biotechnol Biotec Eq 30:479–488

Kiss L (2003) A review of fungal antagonists of powdery mildews and their potential as biocontrol agents. Pest Manag Sci 59:475–483

Klimpel A, Gronover CS, Williamson B, Stewart JA, Tudzynski B (2002) The adenylate cyclase (BAC) in *Botrytis cinerea* is required for full pathogenicity. Mol Plant Pathol 3:439–450

Kloepper JW, Ryu CM, Zhang SA (2004) Induced systemic resistance and promotion of plant growth by *Bacillus* spp. Phytopathology 94:1259–1266

Koch E, Becker JO, Berg G, Hauschild R, Jehle J, Köhl J et al (2018) Biocontrol of plant diseases is not an unsafe technology. J Plant Dis Prot 125:121–125

Köhl J, Postma J, Nicot P, Ruocco M, Blum B (2011) Stepwise screening of microorganisms for commercial use in biological control of plant-pathogenic fungi and bacteria. Biol Control 57:1–12

Köhl J, Kolnaar R, Ravensberg WJ (2019) Mode of action of microbial biological control agents against plant diseases: relevance beyond efficacy. Front Plant Sci 10:845

Landum MC, Félix R, Alho J, Garcia R, João M, Rei F et al (2016) Antagonistic activity of fungi of *Olea europaea* L. against *Colletotrichum acutatum*. Microbiol Res 183:100–108

Lanteigne C, Gadkar VJ, Wallon T, Novinscak A, Filion M (2012) Production of DAPG and HCN by Pseudomonas sp. lbum300 contributes to the biological control of bacterial canker of tomato. Am Phytopathol Soc 102:967–973

Lecomte C, Alabouvette C, Edel-Hermann V, Robert F, Steinberg C (2016) Biological control of ornamental plant diseases caused by *Fusarium oxysporum*. A review. Biol Control 101:17–30

Lemanceau P, Alabouvette C (1993) Suppression of Fusarium-wilts by fluorescent pseudomonads: mechanisms and applications. Biocontrol Sci Technol 3:219–234

Lengeler KB, Davidson RC, D'souza C, Harashima T, Shen WC, Wang P et al (2000) Signal transduction cascades regulating fungal development and virulence. Microbiol Mol Biol Rev 64:746–785

Léon-Kloosterziel KM, Verhagen BWM, Keurentjes JJB et al (2005) Colonization of the *Arabidopsis rhizosphere* by fluorescent *Pseudomonas spp.* activates a root-specific, ethylene-responsive PR-5 gene in the vascular bundle. Plant Mol Biol 57:731–748

Li HY, Wei DQ, Shen M, Zhou ZP (2012) Endophytes and their role in phytoremediation. Fungal Divers 54:11–18

Li Y, Feng X, Wang X, Zheng L, Liu H (2020) Inhibitory effects of *Bacillus licheniformis* BL06 on *Phytophthora capsici* in pepper by multiple modes of action. Biol Control 144:104210

Lin H, Travisano M, Kazlauskas RJ (2016) Experimental evolution of *Trichoderma citrinoviride* for faster deconstruction of cellulose. PLoS One 11:1–17

Liu X, Fang W, Liu L, Yu T, Lou B, Zheng X (2010) Biological control of postharvest sour rot of citrus by two antagonistic yeasts. Lett Appl Mcrobiol 51:30–35

Lockwood JL (1977) Fungistasis in soils. Biol Rev 52:1–43

Loper JE, Henkels MD (1997) Availability of iron to *Pseudomonas fluorescens* in rhizosphere and bulk soil evaluated with an ice nucleation reporter gene. Appl Environ Microbiol 63(1):99–105

Loper JE, Lindow SE (1993) Roles of competition and antibiosis in suppression of plant diseases by bacterial biological control agents. In: Lumsden D, Vaugh JL (eds) Pest management: biologically bases technologies. American Chemical Society, Washington, DC, pp 144–155

Lopes RBA (2009) Industry in biological control: production and marketing of microorganisms in Brazil. In: Bettiol W, Morandi MAB (eds) Biocontrol of plant diseases: use and perspectives. Embrapa Meio Ambiente, Jaguariúna, pp 15–28

Lugtenberg BJJ, Caradus JR, Johnson LJ (2016) Fungal endophytes for sustainable crop production. FEMS Microbiol Ecol 92(12):1–17

Lysøe E, Dees MW, Brurberg MB (2017) A three-way transcriptomic interaction study of a biocontrol agent (*Clonostachys rosea*), a fungal pathogen (*Helminthosporium solani*), and a potato host (*Solanum tuberosum*). MPMI 30:646–655

Macia-Vicente JG, Jansson HB, Lopez-Liorca LV (2008) Colonization of barley root by endophytic fungi and their reducing of take-all caused by *Gaeumannomyces graminis* var. *tritici*. Can J Microbiol 54:600–609

Maciá-Vicente JG, Jansson HB, Talbot NJ, Lopez-Llorca LV (2009) Real-time PCR quantification and live-cell imaging of endophytic colonization of barley (*Hordeum vulgare*) roots by *Fusarium equiseti* and *Pochonia chlamydosporia*. New Phytol 182:213–218

Mandal BK (2019) Silver nanoparticles: potential as insecticidal and microbial biopesticides. In: Koul O (ed) Nano-Biopesticides today and future perspectives. Academic Press, San Diego, CA, pp 281–302

Marco JLD, Valadares-Inglis MC, Felix CR (2003) Production of hydrolytic enzymes by *Trichoderma* isolates with antagonistic activity against *Crinipellis perniciosa*, the causal agent of witches' broom of cocoa. Braz J Microbiol 34:33–38

Marian M, Ohno T, Suzuki H, Kitamura H, Kuroda K, Shimizu M (2020) A novel strain of endophytic *Streptomyces* for the biocontrol of strawberry anthracnose caused by *Glomerella cingulate*. Microbiol Res 234:126428

Martínez-Medina A, Fernández I, Sánchez-Guzmán MJ, Jung SC, Pascual JA, Pozo MJ (2013) Deciphering the hormonal signalling network behind the systemic resistance induced by *Trichoderma harzianum* in tomato. Front Plant Sci 4:206

Martinuz A, Schouten A, Sikora RA (2012) Systemically induced resistance and microbial competitive exclusion: plant ecology & diversity implications on biological control. Phytopathology 102(3):260–266

Mauch-Mani B, Baccelli I, Luna E, Flors V (2017) Defense priming: an adaptive part of induced resistance. Annu Rev Plant Biol 68:485–512

Mbarga JB, Begoude BAD, Ambang Z, Meboma M, Kuate J, Ewbank W et al (2020) Field testing an oil-based *Trichoderma asperellum* formulation for the biological control of cacao black pod disease, caused by *Phytophthora megakarya*. Crop Prot 132:105134

McKee ND, Robinson PM (2009) Production of volatile inhibitors of germination and hyphal extension by *Geotrichum candidum*. Trans Br Mycol Soc 91:157–160

McQuilken MP, Gemmell J (2004) Enzyme production by the mycoparasite *Verticillium biguttatum* against *Rhizoctonia solani*. Mycopathologia 157:201–205

Mercier J, Jiménez JI (2004) Control of fungal decay of apples and peaches by the biofumigant fungus *Muscodor albus*. Postharvest Biol Technol 31:1–8

Mercier J, Manker DC (2005) Biocontrol of soil-borne diseases and plant growth enhancement in greenhouse soilless mix by the volatile-producing fungus *Muscodor albus*. Crop Prot 24:355–362

Messiha NAS, van Diepeningen AD, Farag NS, Abdallah SA, Janse JD, van Bruggen AHC (2007). *Stenotrophomonas maltophilia*: a new potential biocontrol agent of *Ralstonia solanacearum*, causal agent of potato brown rot. Eur J Plant Pathol 118:211–225

Messing R, Brodeur J (2018) Current challenges to the implementation of classical biological control. BioControl 63:1–9

Meziane H, Gavriel S, Ismailov Z, Chet I, Chernin L, Hofte M (2006) Control of green and blue mould on orange fruit by *Serratia plymuthica* strains IC14 and IC1270 and putative modes of action. Postharvest Biol Technol 39:125–133

Mhatre PH, Karthik C, Kadirvelu K, Divya KL, Venkatasalam EP, Srinivasan S et al (2019) Plant growth promoting rhizobacteria (PGPR): a potential alternative tool for nematodes bio-control. Biocatal Agric Biotechnol 17:119–128

Migheli Q, González-Candelas L, Dealessi L, Camponogara A (1998) Transformants of *Trichoderma longibrachiatum* overexpressing the β-1,4-Endoglucanase Gene egl1 show enhanced biocontrol of *Pythium ultimum* on cucumber. Phytopathology 88(7):673–677

Milgroom MG, Cortesi P (2004) Biological control of chestnut blight with hypovirulense: a critical analysis. Annu Rev Phytopathol 42:331–638

Minerdi D, Bossi S, Gullino ML, Garibaldi A (2009) Volatile organic compounds: a potential direct long-distance mechanism for antagonistic action of *Fusarium oxysporum* strain MSA 35. Environ Microbial 11(4):844–854

Minerdi D, Bossi S, Maffei ME, Gullino ML, Garibaldi A (2011) *Fusarium oxysporum* and its bacterial consortium promote lettuce growth and expansin A5 gene expression through microbial volatile organic compound (MVOC) emission. FEMS Microbiol Ecol 76:342–351

Mishra P, Mishra J, Dwivedi SK, Arora NK (2020) Microbial enzymes in biocontrol of phytopathogens. In: Arora NK, Mishra J, Mishra V (eds) Microbial enzymes: roles and applications in industries. Microorganisms for sustainability. Springer Nature, Singapore, pp 259–285

Mitchell AM, Strobel GA, Moore E, Robison R, Sears J (2010) Volatile antimicrobials from *Muscodor crispans*, a novel endophytic fungus. Microbiology 156:270–277

Moraes IO, Capalbo DMF, Arruda ROM (1998) Production of entomopathogenic bacteria. In: Alves AS (ed) Microbial control of insects. FEALQ, pp 815–843

Morandi MAB, Bettiol W (2009) Biological control of plant diseases in Brazil. In: Bettiol W, Morandi MAB (eds) Biocontrol of plant diseases: use and perspectives. Embrapa Meio Ambiente, Jaguariúna, pp 07–14

Morath SU, Hung R, Bennett JW (2012) Fungal volatile organic compounds: a review with emphasis on their biotechnological potential. Fungal Biol Rev 26:73–83

Mousa WK, Raizada MN (2013) The diversity of antimicrobial secondary metabolites produced by fungal endophytes: an interdisciplinary perspective. Front Microbiol 4:65

Nees von Esenbeck CG (1816) Dan System der Pilze and Schwamme. Stahelschen Buchhandlung, Wurzburg

Niu D, Liu H, Jiang C, Wang Y, Wang Q, Jin H et al (2011) The plant growth promoting rhizobacterium *Bacillus cereus* AR156 induces systemic resistance in *Arabidopsis thaliana* by simultaneously activating salicylate- and jasmonate/ethylene dependent signaling pathways. Mol Plant Microbe Interact 24:533–542

O'Hanlon KA, Knorr K, Jorgensen LN, Nicolaisen M, Boelt B (2012) Exploring the potential of symbiotic fungal endophytes in cereal disease suppression. Biol Control 63:69–78

Oh HS, Lee YH (2000) A target-site-specific screening system for antifungal compounds on appressorium formation in *Magnaporthe grisea*. Phytopathology 90:1162–1168

Omar SA, Abd-Alla MH (1998) Biocontrol of fungal root rot diseases of crop plants by the use of rhizobia and bradyrhizobia. Folia Microbiol 43:431–437

Ongena M, Jacques P (2008) *Bacillus* lipopeptides: versatile weapons for plant disease biocontrol. Trends Microbiol 16(3):115–125

Ongena M, Jourdan E, Schafer M, Kech C, Budzikiewicz H, Luxen A, Thonart P (2005) Isolation of an N-alkylated benzylamine derivative from *Pseudomonas putida* BTP1 as elicitor of induced systemic resistance in bean. Mol Plant Microbe Interact 18:562–569

ONU (2012) United Nations, Department of Economic and Social Affairs, The United Nations, Population Division, Population Estimates and Projections Section

Ownley BH, Gwinn KD, Vega FE (2010) Endophytic fungal entomopathogens with activity against plant pathogens: ecology and evolution. BioControl 55:113–128

Pagans E, Font X, Sánchez A (2006) Emission of volatile organic compounds from composting of different solid wastes: abatement by biofiltration. J Hazard Mater 131:179–186

Pal KK, Gardener MB (2006) Biological control of plant pathogens. The plant health instructor. https://doi.org/10.1094/PHI-A-2006-1117-02

Parafati L, Cirvilleri G, Restuccia C, Wisniewsk M (2017a) Potential role of exoglucanase genes (WaEXG1 and WaEXG2) in the biocontrol activity of *Wickerhamomyces anomalus*. Microb Ecol 73:876–884

Parafati L, Vitale A, Restuccia C, Cirvilleri G (2017b) Performance evaluation of volatile organic compounds by antagonistic yeasts immobilized on hydrogel spheres against gray, green and blue postharvest decays. Food Microbiol 63:191–198

Park YH, Mishra RC, Yoon S, Kim H, Park C, Seo ST et al (2018) Endophytic *Trichoderma citrinoviride* isolated from mountain-cultivated ginseng (*Panax ginseng*) has great potential as a biocontrol agent against ginseng pathogens. J Ginseng Res 43(3):408–420

Paulina M, Viviana B, Josefina C, Carolina B, Lucila K, Andrea T et al (2019) *Trichoderma* as biocontrol and plant growth-promoting agent in the pathosystem *Pyrenophora teres*-barley in Argentina. Biol Control

Peferöen M (1997) Progress and prospects for field use of Bt genes in crops. Trends Biotechnol 15: 173–177

Perazzolli M, Dagostin S, Ferrari A, Elad Y, Pertot I (2008) Induction of systemic resistance against *Plasmopara viticola* in grapevine by *Trichoderma harzianum* T39 and benzothiadiazole. Biol Control 47:228–234

Pieterse CMJ, Van der Does D, Zamioudis C, Leon-Reyes A, Van Wees SCM (2012) Hormonal modulation of plant immunity. Annu Rev Cell Dev Biol 28:489–521

Pieterse CM, Zamioudis C, Berendsen RL, Weller DM, Van Wees SC, Bakker PA (2014) Induced systemic resistance by beneficial microbes. Annu Rev Phytopathol 52:347–375

Pliego C, Ramos C, de Vicente A, Cazorla FM (2011) Screening for candidate bacterial biocontrol agents against soilborne fungal plant pathogens. Plant and Soil 340:505–520

Poinar GO Jr (1976) Description and biology of a new insect parasitic rhabitoid, *Heterorhabditis bacteriophora* n. gen. n. sp. (Rhabditida; Heterorhabditidae n. family). Nematologica 21:463–470

Prakash J, Arora NK (2021) Novel metabolites from *Bacillus safensis* and their antifungal property against *Alternaria alternata*. Antonie Van Leeuwenhoek. https://doi.org/10.1007/s10482-021-01598-4

Praneetvatakul S, Schreinemachers P, Pananurak P, Tipraqsa P (2013) Pesticides, external costs and policy options for Thai agriculture. Environ Sci Policy 27:103–113

Pulla RK, Lee OR, In JG, Kim YJ, Senthil K, Yang DC (2010) Expression and functional characterization of pathogenesis-related protein family 10 gene, PgPR10-2, from *Panax ginseng* C.A. Meyer. Physiol Mol Plant Pathol 74:323–329

Raaijmakers JM, Mazzola M (2012) Diversity and natural functions of antibiotics produced by beneficial and plant pathogenic bacteria. Annu Rev Phytopathol 50:403–424

Raaijmakers JM, Vlami M, de Souza JT (2002) Antibiotic production by bacterial biocontrol agents. Antonie Van Leeuwenhoek 81:537–547

Raaijmakers JM, de Bruijn I, de Kock MJD (2006) Cyclic lipopeptide production by plant-associated *Pseudomonas* spp.: diversity, activity, biosynthesis, and regulation. Mol Plant Microbe Interact 19:699–710

Raguchander T, Saravanakumar D, Balasurramanian P (2011) Molecular approaches to improvement of biocontrol agents of plant diseases. J Biol Control 25:71–84

Rahman M, Ali AM, Dey TK et al (2014) Evolution of disease and potential biocontrol activity of *Trichoderma* sp. against *Rhizoctonia solani* on potato. Biosci J 30:1108–1117

Rai A, Nabti E (2017) Plant growth-promoting bacteria: importance in vegetable production. In: Zaidi A, Khan MS (eds) Microbial strategies for vegetable production. Springer, Cham, pp 23–48

Ramadasappa S, Rai AK, Jaat RS, Singh A, Rai R (2012) Isolation and screening of phldþ plant growth promoting rhizobacteria antagonistic to *Ralstonia solanacearum*. World J Microbiol Biotechnol 28:1681–1690

Ramesh R, Phadke GS (2012) Rhizosphere and endophytic bacteria for the suppression of eggplant wilt caused by *Ralstonia solanacearum*. Crop Prot 37:35–41

Raymaekers K, Poneta L, Holtappels D, Berckmans B, Cammue BPA (2020) Screening for novel biocontrol agents applicable in plant disease management—a review. Biol Control 144

Raza W, Ling N, Yang L, Huang Q, Shen Q (2016) Response of tomato wilt pathogen *Ralstonia solanacearum* to the volatile organic compounds produced by a biocontrol strain *Bacillus amyloliquefaciens* SQR-9. Sci Rep 6:24856

Reis AAS, Santos RS (2016) Basic microbiology. Faculdade Alfredo Nasser, Aparecida de Goiânia

Rivera-Méndez W, Obregón M, Morán-Diez ME, Hermosa R, Monte E (2020) *Trichoderma asperellum* biocontrol activity and induction of systemic defenses against *Sclerotium cepivorum* in onion plants under tropical climate conditions. Biol Control 141:104145

Rojo FG, Reynoso MM, Ferez S, Chulze SN, Torres AM (2007) Biological control by *Trichoderma* species of *Fusarium solani* causing peanut brown root rot under field conditions. Crop Prot 26(4):549–555

Ruanpanun P, Nimnoi P (2020) Evaluation on the efficiency and persistence of *Streptomyces jietaisiensis* strain A034 in controlling root knot disease and promoting plant growth in the plant-parasitic nematode infested soils. Biol Control 144:104221

Rybakova D, Schmuck M, Wetzlinger U, Varo-Suarez A, Murgu O, Müller H et al (2016) Kill or cure? The interaction between endophytic *Paenibacillus* and *Serratia* strains and the host plant is shaped by plant growth conditions. Plant and Soil 405:65–79

Ryu CM, Farag MA, Hu CH, Reddy MS, Pare PW, Kloepper JW (2003) Volatiles produced by PGPR elicit plant growth promotion and induced resistance in Arabidopsis. In: Proceedings of the Sixth International Workshop on Plant Growth Promoting Rhizobacteria, 6th International PGPR Workshop, Calicut, India, pp 436–443

Saddler GS (2005) Management of bacterial wilt disease. In: Allen C, Prior P, Hayward AC (eds) Bacterial wilt disease and *Ralstonia Solanacearum* species complex. APS Press, St. Paul, MN, pp 121–132

Saikkonen K, Saari S, Helander M (2010) Defensive mutualism between plants and endophytic fungi? Fungal Divers 41:101–113

Sampaio RM (2018) Controle Biológico: tecnologias na construção de oportunidades no Brasil. Análises e Indicadores do Agronegócio 13:1–4

Saravanakumar D, Ciavorella A, Spadaro D, Garibaldi A, Gullino ML (2008) *Metschnikowia pulcherrima* strain MACH1 out competes *Botrytis cinerea, Alternaria alternata* and *Penicillium expansum* in apples through iron depletion. Postharvest Biol Technol 49:121–128

Sathiyabama M, Balasubramanian R (2017) Protection of groundnut plants from rust disease by application of glucan isolated from a biocontrol agent *Acremonium obclavatum*. Int J Biol Macromol 116:316–319

Savitha MJ, Sriram S (2017) Induced systemic resistance (ISR) in hot pepper against *Phytophthora capsici* infection triggered by cell wall oligosaccharide elicitors from *Trichoderma* species. Indian J Hortic 74:233–239

Schreiter S, Ding GC, Heuer H, Neumann G, Sandmann M, Grosch R et al (2014) Effect of the soil type on the microbiome in the rhizosphere of field-grown lettuce. Front Microbiol 5:144

Schulze GC, Schorn C, Tudzynski B, Gronover CS, Schorn C, Tudzynski B et al (2004) Identification of *Botrytis cinerea* genes up-regulated during infection and controlled by the Galpha subunit BCG1 using suppression subtractive hybridization (SSH). Mol Plant Microbe Interact 17:537–546

Schumacher J, Kokkelink L, Huesmann C, Jimenez-Teja D, Collado IG, Barakat R et al (2008) The cAMP-dependent signaling pathway and its role in conidial germination, growth, and virulence of the gray mold *Botrytis cinerea*. Mol Plant Microbe Interact 21:1443–1459

Segaran G, Sathiavelu M (2019) Fungal endophytes: a potent biocontrol agent and a bioactive metabolites reservoir. Biocatal Agric Biotechnol 21:101284

Selim S, Roisin-Fichter C, Andry JB, Bogdanow B, Sambou R (2014) Real-time PCR to study the effect of timing and persistence of fungicide application and wheat varietal resistance on *Mycosphaerella graminicola* and its sterol 14α-demethylation inhibitor-resistant genotypes. Pest Manag Sci 70(1):60–90

Shanmugaiah V, Balasubramanian N, Gomathinayagam S, Manoharan P, Rajendran A (2009) Effect of single application of *Trichoderma viride* and *Pseudomonas fluorescens* on growth promotion in cotton plants. Afr J Agr Res 4:1220–1225

Sharon E, Orion D, Spiegel Y (1993) Binding of soil microorganisms and red blood cells by the gelatinous matrix and eggs of *Meloidogyne javanica* and *Rotylenchulus reniformis*. Fundam Appl Nematol 16:5–9

Shi J, Liu A, Li X, Chen W (2013) Control of *Phytophthora incotianae* disease, induction of defense responses and genes expression of papaya fruits treated with *Pseudomonas putida* MGP1. J Sci Food Agric 53:568–574

Siewers V, Viaud M, Jimenez-Teja D, Collado IG, Gronover CS, Pradier J-M et al (2005) Functional analysis of the cytochrome P450 monooxygenase gene bcbot1 of *Botrytis cinerea* indicates that botrydial is a strain-specific virulence factor. Mol Plant Microbe Interact 602:602–612

Silva CAD (2000) Entomopathogenic microorganisms associated with insects and mites of the odoper. National Cotton Research Center, Campina Grande, Embrapa, p 77

Smith KP, Handelsman J, Goodman RM (1999) Genetic basis in plants for interactions with disease-suppressive bacteria. Proc Natl Acad Sci, Agric Sci 96(9):4786–4790

Solanki MK, Yandigeri MS, Kumar S, Singh RK, Srivastava AK (2019) Co-inoculation of different antagonists can enhance the biocontrol activity against *Rhizoctonia solani* in tomato. Anton Leeuw Int J G 112(11):1633–1644

Sonnleitner E, Haas D (2011) Small RNAs as regulators of primary and secondary metabolism in Pseudomonas species. Appl Microbiol Biotechnol 91:63–79

Sticher L, Mauch Mani B, Metraux JP (1997) Systemic acquired resistance. Annu Rev Phytopathol 35:235–270

Subedi N, Taylor CG, Paul PA, Miller SA (2020) Combining partial host resistance with bacterial biocontrol agents improves outcomes for tomatoes infected with *Ralstonia pseudosolanacearum*. Crop Prot 135:104776

Sun D, Zhuo T, Hu X, Fan X, Zou H (2017) Identification of a *Pseudomonas putida* as biocontrol agent for tomato bacterial wilt disease. Biol Control 114:45–50

Taqarort N, Echairi A, Chaussod R, Nouaim R, Boubaker H, Benaoumar A (2008) Screening and identification of epiphytic yeasts with potential for biological control of green mold of citrus fruits. World J Microbiol Biotechnol 24:3031–3038

Thomas GM, Poinar GO Jr (1979) Xenorhabdus gen. nov., a genus of entomopathogenic, nematophilic bacteria of the family Enterobacteriaceae. Int J Syst Evol Microbiol 29(4): 352–360

Tilocca B, Cao A, Migheli Q (2020) Scent of a Killer: microbial volatilome and its role in the biological control of plant pathogens. Front Microbiol 11:41

Trapet P, Avoscan L, Klinguer A, Pateyron S, Citerne S, Chervin C et al (2016) The *Pseudomonas fluorescens* siderophore pyoverdine weakens *Arabidopsis thaliana* defense in favor of growth in iron-deficient conditions. Plant Physiol 171(1):675–693

Travassos L (1927) Sobre o Genera Oxysomatium. Bol Biol 5:20–21

Usta C (2013) Microorganisms in biological pest control—a review (bacterial toxin application and effect of environmental factors). In: Silva-Opps M (ed) Current progress in biological research. IntechOpen, London, pp 287–317

Valadares-Inglis MC, Lopes RB, Faria MR (2020) Controle de artrópodes-praga com fungos entomopatogênicos. In: Fontes EMG, Valadares-Inglis MC (eds) Controle biológico de pragas da agricultura. Embrapa, Brasília, pp 201–236

Vallad GE, Goodman RM (2004) Systemic acquired resistance and induced systemic resistance in conventional agriculture. Crop Sci 44:1920–1934

Van Driesche RG, Bellows TS (1996) Biological control. Chapman & Hall, London

Van Lenteren JC, Bolckmans K, Kohl J, Ravensberg WL, Urbaneja A (2018) Biological control using invertebrates and microorganisms: plenty of new opportunities. BioControl 63:39–59

Van Loon LC, Bakker PAHM (2006) Root-associated bacteria inducing systemic resistance. In: Gnanamanickam SS (ed) Plant-associated Bacteria. Springer, Dordrecht, pp 269–316

Van Loon LC, Bakker PAHM, Pieterse CMJ (1998) Systemic resistance induced by rhizosphere bacteria. Annu Rev Phytopathol 36:453–483

Van Wees SC, Van der Ent S, Pieterse CM (2008) Plant immune responses triggered by beneficial microbes. Curr Opin Plant Biol 11(4):443–448

Velusamy P, Gnanamanickam SS (2003) Identification of 2,4-diacetylphloroglucinol production by plant-associated bacteria and its role in suppression of rice bacterial blight in India. Curr Sci 85: 1270–1273

Verma M, Mishra J, Arora NK (2019) Plant growth-promoting rhizobacteria: diversity and applications. In: Sobti RC, Arora NK, Kothari R (eds) Environmental biotechnology: for sustainable future. Springer, Singapore, pp 129–173

Vinale F, Marra R, Scala F, Ghisalberti E, Lorito M, Sivasithamparam K (2006) Major secondary metabolites produced by two commercial *Trichoderma* strains active against different phytopathogens. Lett Appl Microbiol 43:143–148

Vinale F, Flematti G, Sivasithamparam K, Lorito M, Marra R, Skelton BW et al (2009) Harzianic acid, an antifungal and plant growth promoting metabolite from *Trichoderma harzianum*. J Nat Prod 72:2032–2035

Vinale F, Sivasithamparam K, Ghisalberti EL, Woo SL, Nigro M, Marra R et al (2014) *Trichoderma* secondary metabolites active on plants and fungal pathogens. Open Mycol J 8: 127–139

Voss M, Andaló V, Negrisoli Júnior AS, Barbosa-Negrisoli CR (2019) Manual of laboratory techniques for obtaining, maintaining and characterizing entomopathogenic nematodes. Embrapa Trigo, Passo Fundo

Waghunde RR, Shelake RM, Sabalpara AN (2016) *Trichoderma*: a significant fungus for agriculture and environment. Afr J Agric Res 11:1952–1965

Wang M, Geng L, Sun X, Shu C, Song F, Zhang J (2020) Screening of *Bacillus thuringiensis* strains to identify new potential biocontrol agents against *Sclerotinia sclerotiorum* and *Plutella xylostella* in *Brassica campestris* L. Biol Control 145:104262

Wei Z, Huang J, Tan S, Mei X, Shen Q, Xu Y (2013) The congeneric strain *Ralstonia pickettii* QL-A6 of *Ralstonia solanacearum* as an effective biocontrol agent for bacterial wilt of tomato. Biol Control 65:278–285

Weller DM (2007) *Pseudomonas* biocontrol agents of soilborne pathogens: looking back over 30 years. Phytopathology 97(2):250–256

Weller DM, Thomashow LS (1993) Microbial metabolites with biological activity. In: Lumsden D, Vaugh JL (eds) Pest management: biologically based technologies. American Chemical Society, Washington, DC, pp 173–180

Weller DM, Landa BB, Mavrodi OV, Schroeder KL, De La Fuente L, Bankhead SB et al (2007) Role of 2,4 diacetylphloroglucinol-producing *fluorescent Pseudomonas* spp. in the defense of plant roots. Plant Biol 9:4–20

Wheatley R, Hackett C, Bruce A, Kundzewicz A (1997) Effect of substrate composition on production of volatile organic compounds from *Trichoderma* spp. inhibitory to wood decay fungi. Int Biodeter Biodegr 39:199–205

Xie Y, Wahab L, Gill JJ (2018) Development and validation of a microtiter plate-based assay for determination of bacteriophage host range and virulence. Viruses 10:189

Xu YY, Lu H, Wang X, Zhang KQ, Li GH (2015) Effect of volatile organic compounds from bacteria on nematodes. Chem Biodivers 12:1415–1421

Xu H, Li J, Wang S, Xu L, Cheng R, Zhang J (2017) Oligosaccharide elicitor prepared from Salecan triggers the defense responses of *Arabidopsis thaliana* Col0 against *Botrytis cinerea* infection. World J Microbiol Biotechnol 33:165

Xue QI, Chen Y, Li SM, Chen LF, Ding GC, Guo DW et al (2009) Evaluation of strains of *Acinetobacter* and *Enterobacter* as potential biocontrol agents against *Ralstonia* wilt of tomato. Biol Control 48:252–258

Yao YP, Li Y, Chen ZQ, Zheng BQ, Zhang LT, Wang Q (2016) Biological control of potato late blight using isolates of *Trichoderma*. Am J Potato Res 93:33–42

Yu H, Zhang L, Li L, Zheng C, Guo L, Li W et al (2010) Recent developments and future prospects of antimicrobial metabolites produced by endophytes. Microbiol Res 165(6):437–449

Yu X, Ai C, Xin L, Zhou G (2011) The siderophore-producing bacterium, *Bacillus subtilis* CAS15, has a biocontrol effect on *Fusarium* wilt and promotes the growth of pepper. Eur J Soil Biol 47(2):138–145

Yuan J, Raza W, Shen Q, Huang Q (2012) Antifungal activity of *Bacillus amyloliquefaciens* NJN-6 volatile compounds against *Fusarium oxysporum* f. sp. *cubense*. Appl Environ Microbiol 78: 5942–5944

Zabalgogeazcoa I (2008) Fungal endophytes and their interaction with plant pathogens. Spanish J Agric Res 6(S1):138–146

Zeilinger S, Omann M (2007) Trichoderma biocontrol: signal transduction pathways involved in host sensing and mycoparasitism. Gene Regul Syst Biol 1:227–234

Zeilinger S, Gruber S, Bansal R, Mukherjee PK (2016) Secondary metabolism in Trichoderma—chemistry meets genomics. Fungal Biol Rev 30:74–90

Zhang D, Spadaro D, Valente S, Garibaldi A, Gullino ML (2011) Cloning, characterization and expression of an exo-1, 3-beta-glucanase gene from the antagonistic yeast, *Pichia guilliermondii* strain M8 against grey mold on apples. Biol Control 59:284–293

Zhang QH, Zhang J, Yang L, Zhang L, Jiang DH, Chen WD et al (2014) Diversity and biocontrol potential of endophytic fungi in *Brassica napus*. Biol Control 72:98–108

Zhao S, Du CM, Tian CY (2012) Suppression of *Fusarium oxysporum* and induced resistance of plants involved in the biocontrol of *Cucumber Fusarium* wilt by *Streptomyces bikiniensis* HD-087. World J Microbiol Biotechnol 28:2919–2927

Zhu S, Ma B (2015) Benzothiadiazole- or methyl jasmonate-induced resistance to *Colletotrichum musae* in harvested banana fruit is related to elevated defense enzyme activities. J Hortic Sci Biotech 82:500–506

Zhu H, Ma Y, Guo Q, Xu B (2020) Biological weed control using *Trichoderma polysporum* strain HZ-31. Crop Prot 134:105161

Zou C-S, Mo M-H, Gu Y-Q, Zhou J-P, Zhang K-Q (2007) Possible contributions of volatile-producing bacteria to soil fungistasis. Soil Biol Biochem 39:2371–2379

An Insight into Role of Microorganisms in Composting and Its Applications in Agriculture

5

Bouchra El Hayany, Loubna El Fels, Lamfeddal Kouisni, Abdelaziz Yasri, and Mohamed Hafidi

Abstract

Several important improvements in the food security situation in the world were realized during the last decade. However, according to the report on 'The State of Food Security and Nutrition in the World', jointly published by Food and Agriculture Organisation (FAO), PAM (United Nations World Food Programme), IFAD (International Fund for Agricultural Development), UNICEF (United Nation's Children Fund) and WHO (World Health Organisation), the number of malnourished people increased by 36 million between 2015 and 2018, rising to 820 million in 2018. Climate change and soil degradation, leading to a decrease in soil fertility, have worsened the situation. In this alarming context, the development of agriculture in emerging and developing countries is one of the conditions for coming out of this systemic crisis. Many developing countries are

B. El Hayany · L. El Fels
Microbial Biotechnologies, Agrosciences and Environment, Faculty of Sciences Semlalia, Cadi Ayyad University (UCA), Marrakech, Morocco

L. Kouisni
African Sustainable Agriculture Research Institute (ASARI), Mohammed VI Polytechnic University (UM6P), Laâyoune, Morocco

A. Yasri
INRA, Rabat, Morocco

AgroBioSciences Program, Mohammed VI Polytechnic University (UM6P), Benguerir, Morocco

M. Hafidi (✉)
Microbial Biotechnologies, Agrosciences and Environment, Faculty of Sciences Semlalia, Cadi Ayyad University (UCA), Marrakech, Morocco

AgroBioSciences Program, Mohammed VI Polytechnic University (UM6P), Benguerir, Morocco
e-mail: hafidi@uca.ac.ma

N. K. Arora, B. Bouizgarne (eds.), *Microbial BioTechnology for Sustainable Agriculture Volume 1*, Microorganisms for Sustainability 33,
https://doi.org/10.1007/978-981-16-4843-4_5

relying on plant nutrients through the improvement of soil fertility to achieve their agricultural development goals. Integrated management of plant nutrition can be achieved through the judicious use of both mineral and organic fertilizers for a sustainable agricultural development. Use of biofertilizers was declared as a safe, healthy and environment-friendly solution for soil improvement. Biofertilizers ameliorate the soil content in nutrients and enhance the diversity of microbial population, which is a major process to increase the productivity of agro-ecosystems.

Compost as organic biofertilizer is rich in organic matter and has an important role in soil fertility and in sustainable agriculture. During composting, a complex bacterial process is applied with many interacting factors, which degrade and stabilize the organic biodegradable waste. Microorganism's dynamics, influenced by the substrate composition and compost conditions, are the essential factors for a successful composting. This chapter describes the diversity and microbial succession during the composting process and highlights the behaviour of two groups of microorganisms in compost: (1) beneficial microorganisms which consist in bacteria and fungi degrading organic matter with positive effects on the soil and plants after compost utilization and (2) harmful microorganisms which can be introduced in compost through the original substrate by which the composting process started (yeast, fungi, protozoa, bacteria). These non-beneficial microorganisms are almost inactivated by high temperatures during composting or other mechanisms such as pH, ammoniac and antagonistic interaction.

Compost applications have also a beneficial impact on microbiological activity of the soil by increasing the microbial biomass such as phosphate-solubilizing and nitrogen-fixing bacteria, which consequently improve the ammonification rate and fluorescein diacetate hydrolysis activity. Additionally, compost spreading introduces various beneficial fungi in soil, which increase soil fertility and protect plant from diseases.

Keywords

Soil fertility · Biofertilizers · Composting · Organic fertilizers

5.1 Introduction

The massive utilization of chemical fertilizers to multiply and accelerate the agricultural production can have negative environmental impacts such as soil erosion and loss of its physicochemical and microbiological equilibria (Sánchez et al. 2017), hence the need for alternative environmentally friendly agricultural supplies. To avoid these environmental pollutions, the use of composting technique can be the alternative method to improve agricultural production. Composting is a cost-effective and suitable technique to produce a biofertilizer that can enrich the soil with essential elements and contributes to improve the physicochemical and microbiological proprieties of soil. It is a process requiring a complex ecosystem with many interacting factors (Yu et al. 2007). It is used as a major process of organic wastes stabilizing, through the degradation of biodegradable components by microbial communities.

Microbial populations are instrumental and part of fundamental process that degrades and stabilizes organic matter and drive the productivity of agro-ecosystems. Microorganisms are the essential factors in the successful operation of composting and for perfect soil amendment. Their metabolic pathways lead to significant changes in the physicochemical parameters of the composting substrate, and that, in turn, contribute to changes in the structure of the microbial community (Chroni et al. 2009). Mesophilic, thermophilic and thermotolerant microorganisms, including bacteria, actinomycetes, yeasts and fungi, have been widely reported in composts (Moreno et al. 2021). Compost produced from biosolid wastes can provide a portion of the nutrient requirements for growth of crops and increase the microbial activity of soils. The organic matter in compost is notably beneficial as a soil conditioner by improving its quality, increasing water holding of soils and aeration and drainage of clay soils (Shammas and Wang 2009).

Improving the diversity, dynamics and importance of compost microbial communities and their beneficial effect in agricultural productivity is important to determine the degree of compost quality. This chapter describes the diversity and microbial succession during composting and highlights the role of beneficial microorganisms degrading organic matter and having a beneficial effect on the soil and plant after compost spreading, and presents the conditions of destruction of harmful microorganisms, which can be introduced in compost with original substrate.

5.2 Composting Process Technologies

Composting is a biological and sustainable way of organic waste recycling by transformation of organic material into a soil-like product called humus. Compost can be made from most organic by-products (Cooperband 2002). The common feedstocks used are: poultry, hog and cattle manures, food processing wastes, municipal leaves, brush and grass clippings, sawdust, green waste (GW); biowaste; municipal solid waste and sewage sludge (Rudnik 2019). Several technologies exist for composting such as windrows, aerated static piles, bunkers and in-vessel systems (Shammas and Wang 2009). In the windrow composting (conventional system), the oxygen supply and porosity should be improved through windrows turning and redistribution of cooler and hotter portions of the pile; furthermore, composting needs special conditions, mainly of temperature, moisture, aeration, pH and carbon to nitrogen (C/N) ratio, linked to highest biological activity in the various stages of the process (Rudnik 2019). Composting-based microbial communities have an important role in the biodegradation of organic materials and compost stability. In general, microorganisms are present in the initial stage of the composting process and are introduced with the primal substrate of composting (Fuchs 2010).

5.3 Source and Role of Microorganism Community During Composting

Microorganisms (bacteria, fungi, yeast, etc.) present during composting can have an urban or industrial source, and they can arise from human or animal faecal matter (human waste, sewage sludge, animal dung), as well as vegetable waste (food and plant waste).

5.3.1 Yeasts and Fungi

These microorganisms are almost more abundant in food waste due to vegetable residues with low pH, which favours the proliferation of fungi and yeasts, because of the ability of yeasts to grow at a lower pH (Ryckeboer et al. 2003; Fuchs 2010). In sewage sludge waste, fungi are relatively less abundant; their roles are considered to be minor because of actions in a limited range of physicochemical conditions (Zhao et al. 2019). In sewage sludge, the fungal community is generally characterized by the dominance of the order *Saccharomycetales* and *Agaricales*, belonging to phyla *Ascomycota* and *Basidiomycota*, respectively, and by the existence of phylum *Chytridiomycota* (Robledo-Mahón Gómez-silván et al. 2019). *Saccharomycetales* is an order of spore forming, and it is more tolerant to high temperature (Gu et al. 2017). Some fungi have the ability to attack lignin and its derivatives through releasing ligninolytic enzymes. For instance, the white rot fungus (*Phanerochaete chrysosporium*) has been extensively studied due to its capacity to produce cellulose- and lignin-degrading enzymes at 36–40 °C with maximum temperatures of 46–40 °C (Mouchacca 1997; El Fels et al. 2014a).

In general the role of fungal communities on composting is not underestimated, and its performance in mesophilic and maturation phase could increase in the degree of polycondensation and the cross-linking of organic compounds which improves the degree of stability and maturity during sludge composting (Amir et al. 2008; Zhao et al. 2019). Fungi community is known to be associated with the latter stage of composting and the degradation of complex molecules, such as lignocellulose (Robledo-Mahón Gómez-silván et al. 2019). On the other hand, fungal consortium can influence compost quality.

Fungi can produce a diversity of extracellular enzymes that degrade organic matter; some of them can tolerate high temperatures within the composting environment so they can still be in the maturation phase to degrade complex organic matter (Gu et al. 2017). Fungi could be correlated with bacterial community evolution during composting of sludge, which indicate that the interactions between fungal and bacterial communities contribute largely to organic matter stability and compost safety. Zhao et al. (2019) showed that fungal community could participate in the transformation of dissolved organic matter components and the formation of humic-like substances through interacting with bacterial community.

5.3.2 Bacteria

Bacteria are known as the dominant microorganisms during composting; they are responsible for degradation of easily degradable organic material by using carbonaceous and nitrogenous organic matter as their main sources of energy. During composting of sewage sludge, bacteria are more important than fungi in the initial stages of the process due to more richness of sewage sludge with bacteria (Robledo-Mahón Gómez-silván et al. 2019); some studies indicate that order of *Rhizobiales* and *Actinomycetales* are the more abundant bacteria in sewage sludge followed by *Clostridiales* and *Burkholderiales* (Zhang et al. 2012; Robledo-Mahón Gómez-silván et al. 2019). In food waste and especially during spoilage, the *Lactobacillus* and *Enterobacteriaceae* are the most dominant species (Wu et al. 2018). According to Anjali and Anitha (1997), the organic waste sample from agro-based industries and from the agriculture fields contains a large diversity of bacteria, such as *Escherichia coli*, *Klebsiella*, *Pseudomonas*, *Bacillus*, *Micrococcus*, *Serratia*, *Enterobacteriaceae*, *Shigella*, *Proteus* spp., *Staphylococcus* spp. and *Salmonella* spp. Saha and Santra (2014) showed that the bacteria present in municipal solid waste are metabolically active which leads to the production of various enzymes and bioactive compounds stimulating the biodegradation of waste.

Actinobacteria (also known as actinomycetes) are part of a large phylum of filamentous Gram-positive bacteria. These microorganisms are widely present throughout the world in different climatic zones both in soil and in the water. Besides being responsible for the characteristic smell of soil (Sánchez et al. 2017), they are important to the process of converting organic waste to stable and humified organic matter and degrading the relatively complex substrates such as lignocellulosic materials due to their decomposition ability. Actinobacteria have a special interest in composting application. They prefer neutral or alkaline pH; several are thermotolerant, in a temperature range from 50 to 60 °C (Moreno et al. 2021). Most actinobacteria grow well when the substrate is moist and the oxygen supply is good and the most easily degradable substrates have already been consumed by bacteria (Insam and De Bertoldi 2007).

The degradation mechanism is through the action of synthesized enzymes such as α-amylase, proteases, lignin peroxidase and glucoamylase (Sánchez et al. 2017). *Streptomyces thermovulgaris*, *Actinobifida chromogena*, *Thermoactinomyces vulgaris* (Sánchez et al. 2017) and *Micromonospora carbonacea* (Kausar et al. 2011) are among the actinobacteria mostly found in compost.

5.4 Microorganism's Succession During Composting Phases and Microbial Diversity at Different Stages of Composting

Bacterial diversity profile during the composting depends on composting phases, on initial substrate and on composting conditions. In household biowaste composting, bacteria are less dominating in mesophilic phase than fungi, streptomycetes and yeasts. However, in thermophilic phase, fungi and yeasts decrease or disappear due to higher temperature during this phase, which gives the opportunity for the dominance of thermophilic bacteria (Moreno et al. 2021). Thus, bacteria have a greater capacity to persist during composting than fungi (Suárez-Estrella et al. 2007).

For the process of sewage sludge composting, Robledo-Mahón Gómez-silván et al. (2019) and Nakasaki et al. (2019) showed a predominance of *Bacillales* and *Actinomycetales*; during thermophilic phase, *Ascomycota* phylum was noticed as the predominant fungi, with a shift to *Basidiomycota* at the end of the process. Indeed, some group of *Bacillales* increased in the final stage of composting when organic matter was thoroughly degraded. In general, the microbial community compositions can be influenced by several physicochemical parameters during composting processes. Some studies declared that water-soluble carbon and temperature may influence microorganisms more than other parameters (Ishii and Takii 2003; Tang et al. 2007). While in another study (Liang et al. 2003), moisture content even had an impact on the microbial activity greater than the temperature (Zhang et al. 2011). The temperature phase can be determined during composting evolution: mesophilic, thermophilic, cooling and maturation (Ishii et al. 2000; Moreno et al. 2021). Changes in compost temperature and chemical conditions affect the evolution of the bacterial population in the compost. The composting process consists of three phases: mesophilic (<45 °C), thermophilic (≥ 45 °C) and maturation. It is a highly dynamic process where bacteria and fungi have a major role in the decomposition and mineralization of organic wastes, with protozoa, worms and insects which have a lower or minor function in decomposition (Storey et al. 2015).

5.4.1 Mesophilic Phase

The mesophilic phase starts with breakdown of the soluble and easily degradable compounds of organic matter performed by the mesophilic microorganisms such as bacteria from families *Enterobacteriaceae*, *Pseudomonadaceae*, *Erythrobacteraceae*, *Streptomycetaceae*, *Comamonadaceae* and genera *Caulobacteraceae*, *Staphylococcus* and *Bacillus* (Chang et al. 2021), some fungi and actinobacteria. These microorganisms degrade the easily assimilated compounds such as amino, lipids and sugars present in the feedstocks used for making compost (Insam and De Bertoldi 2007; Sánchez et al. 2017). The exothermic reactions, generated by microbial metabolic activity, increase the composting temperature; these conditions reduce the tolerance of mesophilic organisms which are then replaced by thermophilic microorganisms (Insam and De Bertoldi 2007; El Fels et al. 2014a, b, 2015).

5.4.2 Thermophilic Phase

The compost reaches the thermophilic phase when the temperature increase to 40 °C. Thermophilic microorganisms take over, and the degradation rate of organic matter increases (Rudnik 2019). During thermophilic phase, the microbial populations are dominated by members of the genus *Bacillus*, and then microbial activity and diversity decrease dramatically; the proliferation of actinobacteria and other thermophilic bacteria occurs. These thermophilic microorganisms have the capacity to degrade complex molecules through enzymes excretion (Bernal et al. 2009; Sánchez et al. 2017). Among the *Bacillus* species that appear during thermophilic phase are *Bacillus coagulans* which can grow up to 65 °C and *Bacillus thermoamylovorans* which can grow in a temperature of about 50 °C and a pH from 5.4 to 8.5 (Combet-Blanc et al. 1995). Furthermore, Zhang et al. (2020) declared the bacterium *Streptosporangiaceae* and fungus *Chaetomiaceae* as biomarkers in the thermophilic and cooling stage of composting, respectively.

The heat reached during thermophilic phases depends on composting conditions, initial substrate and composting environment, for example, in the windrow process; 65 °C is the maximum temperature reached during composting of activated sludge with palm waste (El Fels et al. 2014b). Although during composting of activated sludge with green plant waste the temperature reaches 72 °C (Jouraiphy et al. 2005), during this stage the elimination of pathogenic agents and seeds ability to germinate occurs. After thermophilic period, the pH stabilizes to a neutral level. This phase can last from a few days to several months (Rudnik 2019).

5.4.3 Cooling and Maturation Phase

After the high microbial activity during thermophilic phase, the carbon sources are reduced, and the compost starts to become stable after the cooling phase. Mesophilic bacteria and fungi reappear, and the maturation phase follows. This stage is characterized by the appearance of mesophilic bacteria, actinobacteria, some protists and a wide range of macro-organisms (including mites, springtails, ants, millipedes, centipedes, spiders, beetles and worms). During this stage, the precursors of the humic substances are formed (El Fels et al. 2014a; El Glauoi et al. 2020); its duration depends on the composition of the organic matter and the efficiency of the process (Rudnik 2019). Several authors have indicated the role of maturation phase in biodegradation of complex and recalcitrant compounds (Ait Baddi et al. 2004; El Fels et al. 2015) through the cellulolytic and hemicellulolytic activities of microorganisms which are primarily achieved in thermophilic stage and continued during maturation (Vargas-García et al. 2010; El Hayany et al. 2021a). Fungi are decomposers in this stage; the most representative genera are *Acremonium*, *Chrysosporium*, *Aspergillus*, *Penicillium*, *Mortierella* and *Trichoderma* (Sánchez et al. 2017).

Due to the microbial degradation and nitrifying bacteria activity, various changes in physicochemical property of compost have been shown during the maturation phase; as decreasing of C/N ratio to values around 10 and increasing of nitrates which cause the reduction in NH_4^+/NO_3^- ratio up to values less than 1 (El Fels et al. 2019). Then, biodegradation leads to obtaining a mature and stable product available for agricultural spreading without impact on the environment. In general, the role of microorganisms during composting is summarized by transforming the recyclable waste to final product "compost" that respects the international standards of agricultural use of compost (NFU 44–095 and NFU 44–051). These standards investigate the compost quality in terms of the degree of biodegradation of organic matter, heavy metal-tolerated levels; pathogenic bacteria, maturity and plant growth performance (Rudnik 2019). Therefore, final product can enrich the soil quality by enhancing absorbable nutrients and increase productivity of agricultural crops.

5.5 Composting as Source of Beneficial Microorganisms for Agriculture and Potential Tool for Pathogen Suppression

Compost application in agriculture has a positive influence on plant growth, health and disease suppressiveness (Lutz et al. 2020). These effects are attributed to beneficial microorganism's existence in compost. Several microorganisms which are harmful for humans, animals or plants can be introduced into the compost with the initial substrates (Fuchs 2010). High temperature is the most important factor for the destruction and the low persistence of pathogenic microorganisms during composting. The pathogens have different degree of sensibility to high temperature and to the duration of the heat period (Fuchs 2010); for instance, the deadly temperature for *E. coli* is 60 °C for 20 min, although for *Salmonella* spp. and *Shigella* spp. it is about 55 °C for 1 h (Golueke 1991). The pathogenic bacteria elimination has been confirmed through different studies which confirm the hygienic aspects of compost against the persistence of *Pythium* spp., *Rhizoctonia* spp., *Erwinia* spp. or *Pseudomonas* spp. (Christensen et al. 2002; Suárez-Estrella et al. 2007).

High temperature is not the only factor of the hygienization. Even when the composting process does not reach the temperature level required for thermal kill, pathogens can be inactivated in compost (Fuchs 2010). Other factors can influence pathogens removal and compost hygienization such as phenolic compounds (Elorrieta et al. 2003), ammonia (Lazarovits 2001) and ethyl esters of hydroxyoleic acids as inhibitors of the pathogen *Phytophthora* spp. (Hoitink and Fahy 1986). Some microorganisms produce volatile substances, often secondary metabolites, which can have toxic effect on other microorganisms, including pathogens (Wheatley 2002; El Fels et al. 2015). Furthermore, there are indirect and direct factors affecting the pathogens survival like the solar exposition time, as well as the orientation of the composting pile (Suárez-Estrella et al. 2007) and the complex interactions among microorganisms, compost substrate and environmental parameters (e.g. pH, temperature and moisture content) (Zhao et al. 2019). Several research studies have revealed the strong effect of factors on the persistence of pathogens, such as annual period, season, location and exposition time (Suárez-Estrella et al. 2004); organic acids, aldehydes and alcohols (El Fels et al. 2018); and antagonism and antibiotics produced by compost microorganisms, especially actinobacteria (El Fels et al. 2015). According to El Hayany et al. (2018), the combination of humidity, temperature, ammonium and pH can lead to helminth eggs destruction during composting. However, the removal of bacteria belonging to coliform groups can be linked to the high temperature during thermophilic phase (El Fels et al. 2015).

The composting process leads to a product containing mainly beneficial microorganisms; these are supposed to play an important role in disease suppression; they are more competitive in colonizing organic residues during the composting process than pathogens (Fuchs 2010). Thornton (2004) work indicates that antagonistic *Trichoderma* species have the ability to outcompete the pathogenic fungus *Rhizoctonia solani* for nutrients, which prevents its growth (Speiser et al. 2020). Nelson et al. (1983) also used *Trichoderma* strains to increase the suppressive potential of compost. Kwok et al. (1987) showed that bacterial antagonists could protect cucumber against *R. solani* (Fuchs 2010). The interactions between organisms can continue in the soil after the application of the compost. Fuchs (2010) showed that the microbiological activity of soils increased by compost amendments.

Microorganisms have an essential role in the transformation and migration of nutrients in the soil and compost; there exist different bacteria that are instrumental in the nitrogen transformation as well as fungi and bacteria favouring the solubilization of phosphorus and potassium. Among the decomposers, bacteria present in compost are *Pseudomonas*, *Burkholderia*, *Zymomonas* and *Xanthomonas* genera, and nitrogen-fixing aerobic bacteria could be found (Sánchez et al. 2017). Nitrogen-transforming microorganisms that convert the ammonium into nitrites and nitrates during composting process are mostly decomposing bacteria, bacteria fixing the atmospheric nitrogen, nitrifying bacteria and archaea oxidizing the ammonium to nitrites and nitrates (Vargas-García et al. 2010). Among the nitrogen-fixing bacterial genera present in compost are *Stenotrophomonas*, *Xanthomonas*, *Pseudomonas*, *Klebsiella*, *Alcaligenes*, *Achromobacter* and *Caulobacter* (Pepe et al. 2013) which are in turn involved in processes for organic matter degradation during composting (Insam and De Bertoldi 2007; Sánchez et al. 2017). The potassium-solubilizing bacteria can solubilize potassium, such as *Bacillus mucilaginosus*, *Enterobacter hormaechei* and *Bacillus edaphicus*. The genus *Bacillus* is related to the solubilization of potassium (Sugumaran and Janarthanam 2007) and phosphorus, and at the same time, it is also involved with waste degradation in the composting process (Insam and De Bertoldi 2007; Sánchez et al. 2017). Phosphate-solubilizing microorganisms are type of microorganisms which are widely distributed in nature. In the soil, although their number varies with the soil and compost type, they assimilate phosphate and solubilize it (Restrepo and Marulanda 2015). Phosphate is solubilized by bacteria and fungi by lowering the pH of the medium through the synthesis of organic acids that act on the insoluble phosphorus present in materials such as tricalcium phosphate, dicalcium phosphate and phosphoric rock (Sánchez et al. 2017). Rhizobia have the ability to solubilize phosphate, which is associated to the synthesis of 2-ketogluconic acid (Khan et al. 2014; Khan et al. 2007). Other bacteria as cyanobacteria may contribute to enhance agriculture crops; they represent beneficial microorganisms and renewable biomass source. Cyanobacteria play a major role in different chemical transformations of soils and thus influence the bioavailability of major nutrients like P to plants and increasing of crop production (Singh et al. 2011).

The beneficial microorganisms which are introduced to soil through compost spreading can breakdown and ferment organic fraction of the soil system and convert it into humus comprising nutrients while releasing hormones that improve plant growth, and they enhance the nutrients absorption by root system also retaining nutrients and moisture in the soil (Joshi et al. 2019). In addition, microorganisms can consume contaminants present in soils, by their transformation into non-toxic products, their volatilization and the formation of humus and inert by-products (Bagyaraj 2014).

5.6 Monitoring and Characterization of Microbial Community Evolution During Composting

5.6.1 Composting Effect on Harmful Microorganisms

The composting windrow method based on lagooning sludge (LS) was proposed to assess the evolution of harmful microorganisms. Variation of faecal and total coliforms population in the compost was examined at different stages of composting. Physicochemical parameters and temperature were also measured during composting process of two mixtures with different proportions of LS and GW on a semi-industrial scale—M1 (1/2 LS + 1/2 GW) and M2 (1/3 LS + 1/3 GW)—for 200 days. The temperature measurement indicates that in thermophilic phase, the temperature increased to 45 °C and 50 °C, respectively; for mixtures M1 and M2 (Fig. 5.1), high temperature can contribute directly in destruction of pathogenic bacteria which leads to the safety and the hygienic quality of end products (El Fels

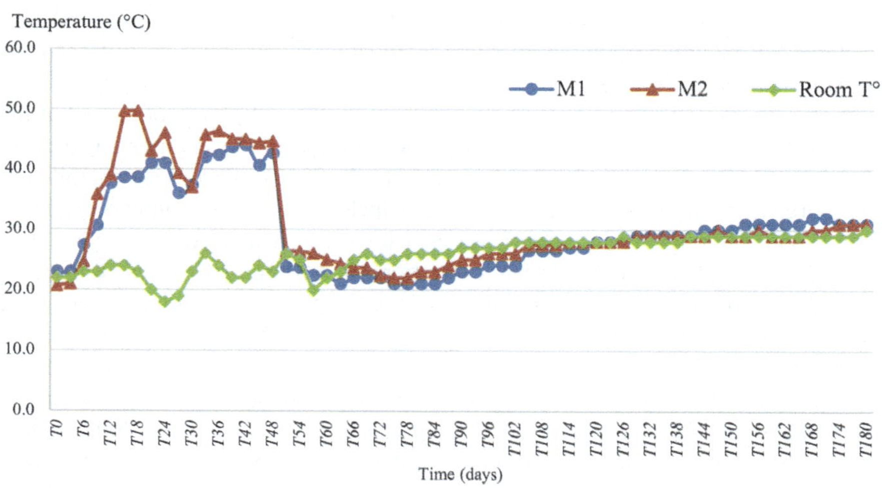

Fig. 5.1 Temperature evolution during composting of M1 and M2

Table 5.1 Physicochemical parameters evolution during composting of M1 and M2

	M1			M2		
Time (days)	0	45	200	0	45	200
C/N	29.00	16.26	8.24	25.49	14.57	9.40
Decomposition rate (%)	–	19.29	47.62	–	25.30	36.41
TKN (%)	0.84	1.03	1.42	0.88	1.15	1.36
TOC (%)	24.36	16.75	11.71	23.31	16.76	12.79

et al. 2019; El Hayany et al. 2021b). Thereafter, it decreases to reach room temperature; the second phase is called maturation phase during which only compounds resistant to degradation are remaining (El Fels et al. 2014b; El Hayany et al. 2018). El Fels et al. (2014b) found that the temperature increased to reach 65 °C at the 15th day of composting of sewage sludge with palm waste in a windrow, which indicates the higher degree of biodegradation of organic compounds.

Total organic matter decomposition increased during composting to reach a decomposition rate of 47.62% and 36.41% after 200 days, for M1 and M2 respectively (Table 5.1). Carbon/nitrogen ratio (C/N), a good indicator to evaluate compost maturity (El Glauoi et al. 2020), increased during composting for the two mixtures and reached final values 8.24 and 9.40, respectively, for M1 and M2 (Table 5.1). According to compost quality standards (NFU 44–051 and NFU 44–095) for agricultural use, a final C/N value around 8–12 is an indicator of mature compost. Faecal and total coliforms were used as faecal contamination indicator organisms (Redlinger et al. 2001; El Hayany et al. 2021b). The standards indicate that the recommended faecal coliforms (FC) and faecal streptococci (FS) densities are 10^3 CFU/g and the final compost should be stable and suitable for use as soil amendment composts and should not contain pathogen microorganisms or viable seeds (Tønner-Klank et al. 2007; El Fels et al. 2019).

The evolution of faecal and total coliforms during LS composting with GW showed a prominent decreasing (Fig. 5.2) for the three mixtures; the decreasing rates are about 99.95% and 99.97%, respectively, for M1; M2 for FC and 99.88% and 98.70%, respectively, for M1; M2 for total coliforms. This decrease is due to the high temperatures reached during the thermophilic phase, the moisture reached during the maturation phase (27.8% and 30.09%, for M1; M2 respectively), emission of ammonia gas during composting in windrows along with treatment period (6 months) and non-availability of nutrients. The results indicate that the final values obtained are lower than the international standards recommended for the agricultural use. The results also indicate the good hygienic quality of the final products of composted sewage sludge and GW, which confirm the usefulness of composting processes for LS treatment. In another study based on windrows composting of activated sludge conducted by El Fels et al. (2015), the decreasing rate of FC reached 99.95% and 95.50%, respectively, for mixture A (1/3 sludge +2/3 date palm tree waste) and mixture B (1/2 sludge +1/2 date palm tree waste) and 99.96% for total

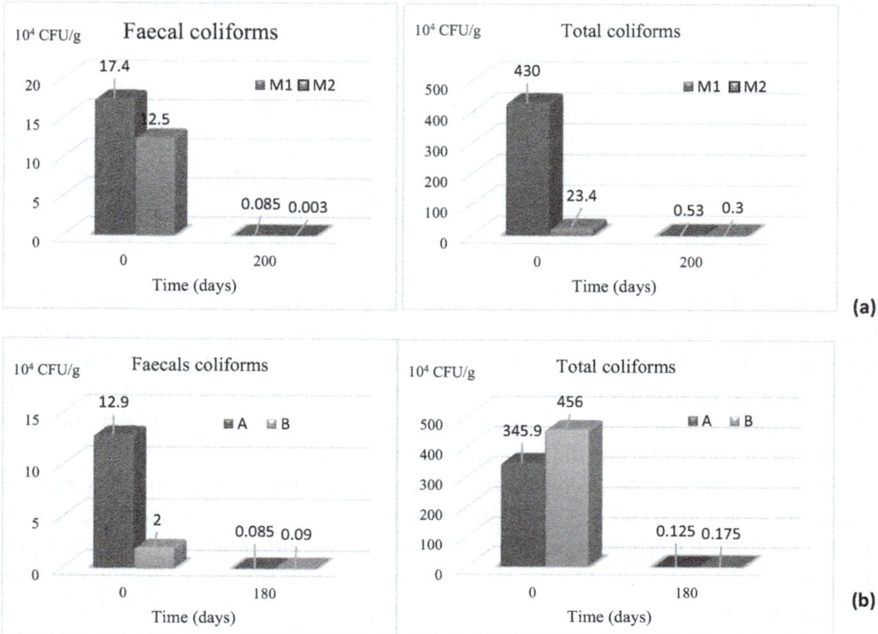

Fig. 5.2 Evolution of total and faecal coliforms during composting of LS (**a**) and activated sludge (**b**) (El Fels et al. 2015)

coliforms in the both mixtures, after 180 days of composting (Fig. 5.2). That indicates the safety of final product and decrease of agricultural risks.

Compost sanitization process and antimicrobial activities of actinobacteria showed that 12 active actinobacteria isolated from several composting phases were screened for their antagonistic activity spectrum against several wastewater-linked microorganisms known as human and plant potential pathogens (El Fels et al. 2015; 2016). Results indicate that the intensity of inhibition is greater among isolates from the composting of 3 and 6 months, which indicates that development of the compost actinobacteria is connected to sanitization in the later stages of composting. El Fels et al.'s (2015) work indicates that according to antagonistic interactions screening out of the 12 active actinobacteria, 9 were active against *Bacillus cereus*, 7 against *Staphylococcus aureus* CCMB3, 6 against *Micrococcus luteus*, 3 against *E. coli* K12 and 2 against *Pseudomonas aeruginosa* and *Enterobacter cloacae*. In the antifungal activities, 7 out 12 were active against *Rhodotorula rubra*, 2 against *Botrytis cinerea* and 6 against *Saccharomyces cerevisiae*, which means that 3 strains were active against both Gram-positive (*B. cereus*; *S. aureus*; *M. luteus*) and Gram-negative bacteria (*E. coli*; *P. aeruginosa*; *E. cloacae*), 9 against Gram-positive bacteria and only 3 against Gram-negative bacteria. This study revealed the antagonistic activity of actinobacteria from composts and showed that heat is not the only cause of compost sanitization. It is a combination of various factors, including antagonism and antibiotics produced by compost microorganisms.

5.6.2 Dynamics of Mesophilic and Thermophilic Microorganisms During Composting

El Fels et al. (2015) studied the dynamic of three main aerobic microorganisms during composting of sewage sludge mixed with tree palm waste; fungi, actinobacteria and bacteria, which were already presented as dominating taxa present during composting (Fig. 5.3). In this study, mesophilic and thermophilic fungi represent a peak of 40% and 50%, respectively, after 15 days of composting; at thermophilic phase, the total fungal mesophilic microflora decreased after 180 days of composting to reach 20% and 18%, respectively, for mixtures A and B. Chroni et al. (2009) showed that thermophilic fungi exhibited a small increase when temperature increases in the first month of composting; furthermore, cooling and

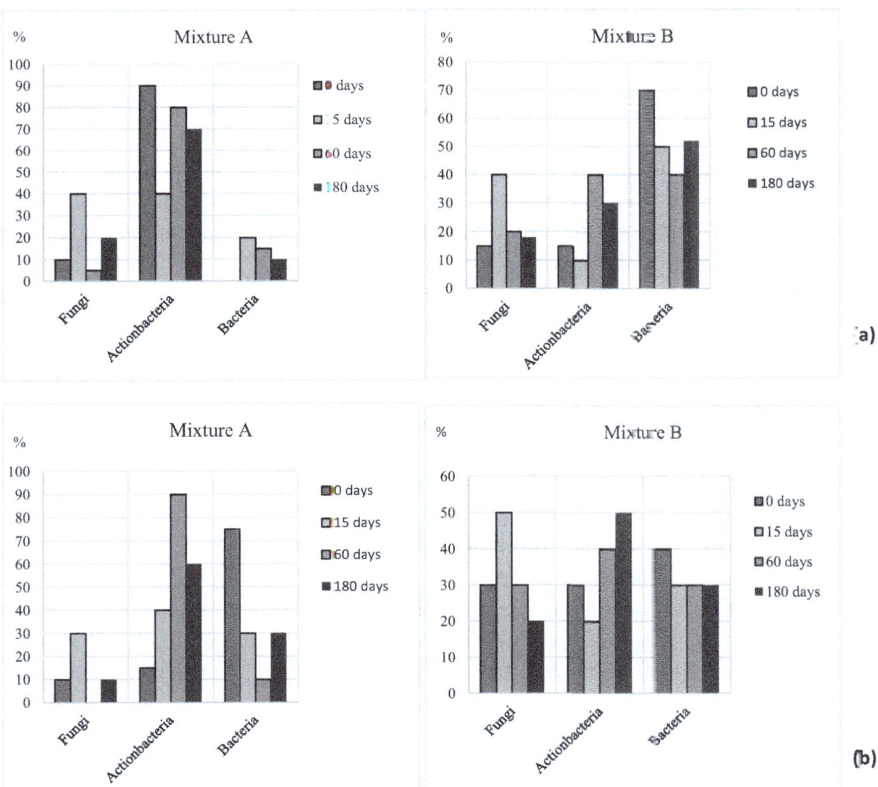

Fig. 5.3 (a) Mesophilic indigenous microflora groups during composting process: fungi = moulds and yeasts. (b) Thermophilic microflora groups versus time of composting (El Fels et al. 2015)

maturation stages are characterized by the development of a new mesophilic fungal community (Ishii et al. 2000). The development of fungi in lignocellulosic substrates at the end of composting promotes their degradation and stabilization by their lignolytic enzymatic capacity (El Fels et al. 2015). Actinobacteria showed a peak in the order of 80% and 40% and 90% and 40%, for the mesophilic and thermophilic microflora of mixtures A and B after 2 months of composting respectively (Fig. 5.3); the dominance of actinobacteria indicate good composting conditions (Kurola et al. 2011; Zhang et al. 2020). The proliferation of actinobacteria in maturation phase is linked to their enzymatic capacity to break down recalcitrant molecules; this could confirm the increase of actinobacteria groups during the composting of lignocellulose-based waste. The mesophilic and thermophilic bacteria showed strong decrease during composting of the two mixtures which is due to the inhibition and destruction of pathogenic type after thermophilic phase and to more proliferation of actinobacteria and fungi responsible for biodegradation process.

In general, indigenous microflora such as bacteria, actinobacteria and fungi, which can break down organic matter in composted sewage sludge, showed a rise–fall–rise trend during the composting time according to temperature, oxygen, water and nutrient availability. The results from recent literature and presented in this chapter give a global information on the microorganisms' dynamic and evolution during composting by studying the behaviour of pathogenic (FC) and beneficial microorganisms during sewage sludge composting. This data justifies the usefulness of composting as green biotechnology, which can destroy pathogenic bacteria and fungi thorough the antagonistic activity of actinobacteria and temperature. Furthermore, the proliferation of actinobacteria in the final compost will have a beneficial effect on soil; its antagonistic activity has a positive influence on sanitary quality of soil and on plaint growth.

The plant germination is a process involving complex mechanisms, which can be affected by the presence of pollutants; therefore, the percentage of germination could indicate the quality of biofertilizers and its capacity to promote plants' germination and growth. Composts of LS showed a higher value in germination index: 124.67% and 158.31% for turnip, 126.51% and 108.79% for radish, 108.37% and 109.30% for cress and 139.44% and 143.48% for alfalfa, respectively, for M1 and M2 (El Glauoi et al. 2020) which indicates the perfect reduction of phytotoxic compounds and the agricultural quality of compost.

The compost application contributes certainly to improving the quality and quantity of agricultural crops through the disease suppression (Zhao et al. 2019; Lutz et al. 2020) and enhancing germination and radical growth. Several studies have indicated the positive effect of compost on crops. Hadar et al. (1985) showed that mature compost promotes the growth of tomato. Furthermore compost amendments can contribute to improvement and control of soil-borne diseases and in increasing sustainability and productivity in potato production systems (Larkin and Tavantzis 2013; Lutz et al. 2020); in improving leafy crops of spinach and turnip (Ansari 2008). Goswami et al. (2017) indicated through a field experiment that application of drum compost and vermicompost enhances soil fertility and crop yield in an agro-ecosystem involving tomato and cabbage; they showed an amelioration in soil health, nutrient availability, microbial diversity and physical stability and concluded that the combined use of drum compost and vermicompost with the recommended chemical fertilization enhances crop growth, yield and product quality of tomato and cabbage. In general, the productivity of vegetable crops is enhanced by compost and vermicompost application leading to sustainable and productive agriculture.

5.7 Future Prospects

The understanding and in-depth knowledge of composting mechanisms and the development of the methods for evaluating compost maturity lead to quality compost that is not only mature and high yielding but can be produced in a limited time, in order to satisfy the agricultural needs (Estrella-González et al. 2020; El Hayany et al. 2021a). Abdellah et al. (2021) revealed that psychrotrophic strains accelerate the composting process by reducing the maturity period and improving compost quality. Tran et al. (2015) have studied the effect of lactic acid bacterium on the organic matter degradation and composting time. Other studies have suggested temperature control as the main factor to enhance composting period and quality. Xiao et al. (2009) have proposed thermophilic compost for rapid biodegradation and maturation of organic municipal waste. This application may be limited in comparison to hyperthermophilic composting which can accelerate composting period, humic acid formation (Liu et al. 2020), eliminate pathogenic microorganisms and enhance biodegradation of some organic pollutants (Chen et al. 2020). Faster and large-scale techniques can be used for production of organic amendments such as industrial-scale and hyperthermophilic composting. The use of compost can be improved by raising awareness among farmers of the benefits and beneficial effects of compost on soil.

5.8 Conclusion

The different phases of composting process contain various communities of microorganisms. The first decomposition is done by mesophilic microorganisms, which rapidly break down the readily degradable compounds. The biodegradation process promotes the temperature increasing; reaching 50 °C and 65 °C, for the compost of LS with GW and activated sludge with date palm tree waste respectively. The heat contributes to suppression of microorganisms that may be human or plant pathogens; the examples presented in this study is the almost total destruction of faecal and total coliforms, which are an indicator of faecal contamination, after 200 days and 180 days, for composting of LS and activated sludge respectively. Compost sanitization is important, and there is clear indication of the crucial role of actinobacteria as antagonists for the inhibition of pathogenic microorganisms. Bacteria in compost are mostly heterotrophic, responsible for biodegradation, but denitrifying, nitrogen-fixing, hydrogen-oxidizing bacteria and sulphur-oxidizing bacteria can also be present, and their beneficial effects can be transported to soil and plants by proper application in agro-ecosystems.

References

Abdellah YAY, Li T, Chen X, Cheng Y, Sun S, Wang Y et al (2021) Role of psychrotrophic fungal strains in accelerating and enhancing the maturity of pig manure composting under low-temperature conditions. Bioresour Technol 320:124402

Ait Baddi G, Alburquerque JA, González J, Cegarra J, Hafidi M (2004) Chemical and spectroscopic analyses of organic matter transformations during composting of olive mill wastes. Int Biodeter Biodegr 54:39–44

Amir S, Merlina G, Pinelli E, Winterton P, Revel JC, Hafidi M (2008) Microbial community dynamics during composting of sewage sludge and straw studied through phospholipid and neutral lipid analysis. J Hazard Mater 159:593–601

Anjali KU, Anitha K (1997) Isolation, characterization and identification of bacteria from organic waste. Philipp Agric Sci 80:57–64

Ansari AA (2008) Effect of vermicompost and vermiwash on the productivity of spinach (*Spinacia oleracea*), onion (*Allium cepa*) and potato (*Solanum tuberosum*). World J Agric Sci 4:554–557

Bagyaraj DJ (2014) Microorganisms in sustainable agriculture. Proc Indian Natl Sci Acad 80:357

Bernal MP, Alburquerque JA, Moral R (2009) Composting of animal manures and chemical criteria for compost maturity assessment: a review. Bioresour Technol 100:5444–5453

Chang H, Zhu X, Wu J, Guo D, Zhang L, Feng Y (2021) Dynamics of microbial diversity during the composting of agricultural straw. J Integr Agric 20(5):1121–1136

Chen Z, Zhao W, Xing R, Xie S, Yang X, Cui P et al (2020) Enhanced in situ biodegradation of microplastics in sewage sludge using hyperthermophilic composting technology. J Hazard Mater 384:121271

Christensen KK, Carlsbaek M, Kron E (2002) Strategies for evaluating the sanitary quality of composting. J Appl Microbiol 92:1143–1158

Chroni C, Kyriacou A, Georgaki I, Manios T, Kotsou M, Lasaridi K (2009) Microbial characterization during composting of biowaste. Waste Manag 29:1520–1525

Combet-Blanc Y, Ollivier B, Streicher C, Patel BKC, Dwivedi PP, Pot B et al (1995) *Bacillus themoamylovorans* sp. nov., a moderately thermophilic and Amylolytic bacterium. Int Union Microbiol Soc 45:9–16

Cooperband L (2002) The art and science of composting. A resource for farmers and compost producers. University of Wisconsin-Madison, Center for Integrated Agricultural Systems, pp 1–14

El Fels L, Lemee L, Ambles A, Hafidi M (2014a) Identification and biotransformation of lignin compounds during co-composting of sewage sludge-palm tree waste using pyrolysis-GC/MS. Int Biodeter Biodegr 92:26–35

El Fels L, Zamama M, El Asli A, Hafidi M (2014b) Assessment of biotransformation of organic matter during co-composting of sewage sludge-lignocelullosic waste by chemical, FTIR analyses, and phytotoxicity tests. Int Biodeterior Biodegradation 87:128–137

El Fels L, Ouhdouch Y, Hafidi M (2015) Use of the co-composting time extract agar to evaluate the microbial community changes during the co-composting of activated sludge and date palm waste. Int J Recycl Org Waste Agric 4:95–103

El Fels L, Hafidi M, Ouhdouch Y (2016) Date palm and the activated sludge co-composting actinobacteria sanitization potential. Environ Technol 37(1):129–135

El Fels L, El Hayany B, El Faiz A, Saadani M, Houari M, Hafidi M (2018) Sludge nematodes, cestodes, and trematodes eggs variation from lagooning, activated sludge and infiltration-percolation wastewater treatment system under semi-arid climate. Environ Sci Pollut Res 26: 570–576

El Fels L, El Hayany B, Aguelmous A, Boutafda A, Zegzouti Y, Kouisni L et al (2019) The use of microorganisms for the biodegradation of sewage sludge and the production of biocompost for sustainable agriculture. In: Giri B, Prasad R, Wu QS, Varma A (eds) Biofertilizers for sustainable agriculture and environment. Soil biology, vol 55. Springer, Cham, pp 301–316

El Glauoi GE, El Hayany B, El Fels L, El Faiz A, Ezzariai A, Rihani M et al (2020) Physico-chemical and spectroscopy assessment of sludge biodegradation during semi-industrial composting under semi-arid climate. Waste Biomass Valor 11(4):1217–1228

El Hayany B, El Glaoui GEM, Rihanni M, Ezzariai A, El Faiz A, El Gharous M et al (2018) Effect of dewatering and composting on helminth eggs removal from lagooning sludge under semi-arid climate. Environ Sci Pollut Res 25:10988–10996

El Hayany B, El Fels L, Dignac MF, Quenea K, Rumpel C, Hafidi M (2021a) Pyrolysis-GCMS as a tool for maturity evaluation of compost from sewage sludge and green waste. Waste Biomass Valori 12(5):2639–2652

El Hayany B, El Fels L, Ouhdouch Y, Hafidi M (2021b) Fate of pathogenic microorganisms during lagooning sludge composting and exploration of bacteriophages as indicator of hygienization. Environ Technol Innov 21:101268

Elorrieta MA, Suárez-Estrella F, López MJ, Vargas-García MC, Moreno J (2003) Survival of phytopathogenic bacteria during waste composting. Agric Ecosyst Environ 96:141–146

Estrella-González MJ, Suárez-Estrella F, Jurado MM, López MJ, López-González JA, Siles-Castellano AB et al (2020) Uncovering new indicators to predict stability, maturity and biodiversity of compost on an industrial scale. Bioresour Technol 313:123557

Fuchs JC (2010) Interactions between beneficial and harmful microorganisms: from the composting process to compost application. In: Insam H, Franke-Whittle I, Goberna M (eds) Microbes at work: from wastes to resources. Springer, Berlin, Heidelberg, pp 1–329

Golueke CG (1991) Principles of composting. In: The staff of biocycle journal of waste recycling the art and science of composting. The JG Press, Emmaus, PA, pp 14–27

Goswami L, Nath A, Sutradhar S, Bhattacharya SS, Kalamdhad A, Vellingiri K et al (2017) Application of drum compost and vermicompost to improve soil health, growth, and yield parameters for tomato and cabbage plants. J Environ Manage 200:243–252

Gu W, Lu Y, Tan Z, Xu P, Xie K, Li X et al (2017) Fungi diversity from different depths and times in chicken manure waste static aerobic composting. Bioresour Technol 239:447–453

Hadar Y, Inbar Y, Chen Y (1985) Effect of compost maturity on tomato seedling growth. Sci Hortic 27(3–4):199–208

Hoitink HAJ, Fahy PC (1986) Basis for the control of Soilborne plant pathogens with composts. Annu Rev Phytopathol 24:93–114

Insam H, De Bertoldi M (2007) Microbiology of the composting process Waste Manag Ser 8:25–48

Ishii K, Takii S (2003) Comparison of microbial communities in four different composting processes as evaluated by denaturing gradient gel electrophoresis analysis. J Appl Microbiol 95:109–119

Ishii K, Fukui M, Takii S (2000) Microbial succession during a composting process as evaluated by denaturing gradient gel electrophoresis analysis. J Appl Microbiol 89:768–777

Joshi H, Duttand S, Choudhary P, Mundra SL (2019) Role of effective microorganisms (EM) in sustainable agriculture. Int J Curr Microbiol App Sci 8:172–181

Jouraiphy A, Amir S, El Gharous M, Revel JC, Hafidi M (2005) Chemical and spectroscopic analysis of organic matter transformation during composting of sewage sludge and green plant waste. Int Biodeter Biodegr 56:101–108

Kausar H, Sariah M, Saud M, Alam Z, Ismail R (2011) Isolation and screening of potential actinobacteria for rapid composting of rice straw. Biodegradation 22:367–375

Khan MS, Zaidi A, Wani PA (2007) Role of phosphate-solubilizing microorganisms in sustainable agriculture—a review. Agron Sustain Dev 27(1):29–43

Khan MS, Zaidi A, Ahmad E (2014) Mechanism of phosphate solubilization and physiological functions of phosphate-solubilizing microorganisms. In: Khan M, Zaidi A, Musarrat J (eds) Phosphate solubilizing microorganisms. Springer, Cham, pp 31–62

Kurola JM, Arnold M, Kontro MH, Talves M, Romantschuk M (2011) Wood ash for application in municipal biowaste composting. Bioresour Technol 102:5214–5220

Kwok OCH, Fahy PC, Hoitink AJ, Kuter GA (1987) Interactions between Bacteria and *Trichoderma hamatum* in suppression of Rhizoctonia damping-off in bark compost media. Phytopathology 77(8):1206–1212

Larkin RP, Tavantzis S (2013) Use of biocontrol organisms and compost amendments for improved control of Soilborne diseases and increased potato production. Am J Potato Res 90:261–270

Lazarovits G (2001) Management of soil-borne plant pathogens with organic soil amendments: a disease control. Agriculture 7:1–7

Liang C, Das KC, McClendon RW (2003) The influence of temperature and moisture contents regimes on the aerobic microbial activity of a biosolids composting blend. Bioresour Technol 86:131–137

Liu X, Hou Y, Li Z, Yu Z, Tang J, Wang Y et al (2020) Hyperthermophilic composting of sewage sludge accelerates humic acid formation: elemental and spectroscopic evidence. Waste Manag 103:342–351

Lutz S, Thuerig B, Oberhaensli T, Mayerhofer J, Fuchs JG, Widmer F et al (2020) Harnessing the microbiomes of suppressive composts for plant protection: from metagenomes to beneficial microorganisms and reliable diagnostics. Front Microbiol 11:1810

Moreno J, López-González JA, Arcos-Nievas MA, Suárez-Estrella F, Jurado MM, Estrella-González MJ et al (2021) Revisiting the succession of microbial populations throughout composting: a matter of thermotolerance. Sci Total Environ 773:145587

Mouchacca J (1997) Thermophilic fungi: biodiversity and taxonomic status. Cryptogam Mycol 18: 19–69

Nakasaki K, Hirai H, Mimoto H, Quyen TNM, Koyama M, Takeda K (2019) Succession of microbial community during vigorous organic matter degradation in the primary fermentation stage of food waste composting. Sci Total Environ 671:1237–1244

Nelson EB, Kuter GA, Hoitink HAJ (1983) Effects of fungal antagonist and compost age on suppression of damping-off in compost bark media. Phytopathology 73:1457–1462

Pepe O, Ventorino V, Blaiotta G (2013) Dynamic of functional microbial groups during mesophilic composting of agro-industrial wastes and free-living (N2)-fixing bacteria application. Waste Manag 33:1616–1625

Redlinger T, Graham J, Corella-Barud V, Raquel A (2001) Survival of fecal coliforms in dry-composting toilets. Appl Environ Microbiol 67:38

Restrepo G, Marulanda S (2015) Bacterias solubilizadoras de fosfato y sus potencialidades de uso en la promoción del crecimiento de cultivos de importancia económica. Rev CENIC Ciencias Biológicas 46:63–76

Robledo-Mahón Gómez-silván C, Andersen GL, Calvo C, Aranda E (2019) Assessment of bacterial and fungal communities in a full-scale thermophilic sewage sludge composting pile under a semipermeable cove. Bioresour Technol 298:122550

Rudnik E (2019) Composting methods and legislation. In: Compostable polymer materials, pp 127–161

Ryckeboer J, Mergaert J, Coosemans J, Deprins K, Swings J (2003) Microbiological aspects of biowaste during composting in a monitored compost bin. J Appl Microbiol 94:127–137

Saha A, Santra SC (2014) Isolation and characterization of bacteria isolated from municipal solid waste for production of industrial enzymes and waste degradation. J Microbiol Exp 1:12–19

Sánchez ÓJ, Ospina DA, Montoya S (2017) Compost supplementation with nutrients and microorganisms in composting process. Waste Manag 69:136–153

Shammas NK, Wang LK (2009) Biosolids composting. Biol Treat Process 8:669–714

Singh JS, Pandey VC, Singh DP (2011) Efficient soil microorganisms: a new dimension for sustainable agriculture and environmental development. Agric Ecosyst Environ 140:339–353

Speiser B, Tamm L, Roggli M et al (2020) Betriebsmittelliste 2020. Research Institute of Organic Agriculture, Frick

Storey S, Chualain DN, Doyle O, Clipson N, Doyle E (2015) Comparison of bacterial succession in green waste composts amended with inorganic fertiliser and wastewater treatment plant sludge. Bioresour Technol 179:71–77

Suárez-Estrella F, Vargas-García MC, López MJ, Moreno J (2004) Survival of *Fusarium oxysporum* f.sp. melonis on plant waste. Crop Prot 23:127–133

Suárez-Estrella F, Vargas-García MC, López MJ, Moreno J (2007) Effect of horticultural waste composting on infected plant residues with pathogenic bacteria and fungi: integrated and localized sanitation. Waste Manag 27:886–892

Sugumaran P, Janarthanam B (2007) Solubilization of potassium-containing minerals by bacteria and their effect on plant growth. World J Agric Sci 3:350–355

Tang JC, Shibata A, Zhou Q, Katayama A (2007) Effect of temperature on reaction rate and microbial community in composting of cattle manure with rice straw. J Biosci Bioeng 104:321–328

Thornton CR (2004) An immunological approach to quantifying the saprotrophic growth dynamics of *Trichoderma* species during antagonistic interactions with *Rhizoctonia solani* in a soil-less mix. Environ Microbiol 6:323–334

Tønner-Klank L, Møller J, Forslund A, Dalsgaard A (2007) Microbiological assessments of compost toilets: in situ measurements and laboratory studies on the survival of fecal microbial indicators using sentinel chambers. Waste Manag 27:1144–1154

Tran QNM, Mimoto H, Nakasaki K (2015) Inoculation of lactic acid bacterium accelerates organic matter degradation during composting. Int Biodeter Biodegr 104:377–383

Vargas-García MC, Suárez-Estrella F, López MJ, Moreno J (2010) Microbial population dynamics and enzyme activities in composting processes with different starting materials. Waste Manag 30:771–778

Wheatley RE (2002) The consequences of volatile organic compound mediated bacterial and fungal interactions. Antonie van Leeuwenhoek Int J Gen Mol Microbiol 81:357–364

Wu S, Xu S, Chen X, Sun H, Hu M, Bai Z et al (2018) Bacterial communities changes during food waste spoilage. Sci Rep 8:1–9

Xiao Y, Zeng G-M, Yang Z-H, Shi W-J, Huang C, Fan C-Z, Xu Z-Y (2009) Continuous thermophilic composting (CTC) for rapid biodegradation and maturation of organic municipal solid waste. Bioresour Technol 100(20):4807–4813

Yu H, Zeng G, Huang H, Xi X, Wang R, Huang D et al (2007) Microbial community succession and lignocellulose degradation during agricultural waste composting. Biodegradation 18:793–802

Zhang J, Zeng G, Chen Y, Yu M, Yu Z, Li H (2011) Effects of physico-chemical parameters on the bacterial and fungal communities during agricultural waste composting. Bioresour Technol 102: 2950–2956

Zhang T, Shao MF, Ye L (2012) 454 pyrosequencing reveals bacterial diversity of activated sludge from 14 sewage treatment plants. ISME J 6:1137–1147

Zhang C, Gao Z, Shi W, Li L, Tian R, Huang J et al (2020) Material conversion, microbial community composition and metabolic functional succession during green soybean hull composting. Bioresour Technol 316:123823

Zhao X, Wei Y, Zhang F, Tan W, Fan Y, Xi B (2019) How do fungal communities and their interaction with bacterial communities influence dissolved organic matter on the stability and safety of sludge compost? Environ Sci Pollut Res 26:4141–4146

Plant Growth-Promoting Bacteria as Biostimulants of Crops in Saline Agroecosystems

6

Priya Mishra, Ankita Bhattacharya, Priyanka Verma, Chanda Bharti, and Naveen Kumar Arora

Abstract

Economy of many countries in the world hinges on the agriculture and thus directly or indirectly on soil quality. Soil salinity is one of the major abiotic constraints negatively impacting the agricultural productivity in arid and semiarid regions. Physical and chemical amendments (irrigation and leaching with chemicals) used to reclaim soil salinization pollute groundwater, waterbodies, and aquatic life. Use of plant growth-promoting bacteria (PGPB) for amelioration of salt stress in soil is a sustainable and ecofriendly measure, which not only reconciles saline soil but also augments nutritional values in soil. PGPB use array of mechanisms to promote growth and endurance in plants by minimizing the adverse effect of salinity. PGPB can enhance the expression of various salt stress-related genes in plants that supplement growth and yield under salinity. This chapter focuses on the effects of salinity on agroecosystems and role of PGPB in alleviation of salt stress in plants. The chapter also emphasizes on the utilization of PGPB-based biostimulants in improvising soil health and strengthen plant productivity.

Keywords

Salinity · PGPB · Nutrients uptake · Phytohormones · Antioxidants · Reclamation

P. Mishra · A. Bhattacharya · P. Verma · C. Bharti · N. K. Arora (✉)
Department of Environmental Science, School of Earth and Environmental Sciences, Babasaheb Bhimrao Ambedkar University, Lucknow, Uttar Pradesh, India

N. K. Arora, B. Bouizgarne (eds.), *Microbial BioTechnology for Sustainable Agriculture Volume 1*, Microorganisms for Sustainability 33,
https://doi.org/10.1007/978-981-16-4843-4_6

6.1 Introduction

A rapid increase in the global human population has posed the challenge of food security. According to the Global Agricultural Productivity Report (GAP) (2019), intensification in agricultural productivity is necessary to fulfill the demand for food of continuously increasing global population. Currently, global agricultural productivity is growing at an annual rate of 1.63% which is lesser than the required, i.e., 1.73% to sustainably produce food for ten billion population in 2050.[1] There are several abiotic constraints such as temperature anomaly, drought, salinity, nutrient deficiency, heavy metals contamination, and surplus use of synthetic chemicals, responsible for decreased productivity of the agroecosystems (Zhu 2016). Among them, soil salinity is a well-known factor adversely hampering global agricultural production. Worldwide the total productive agricultural land is 13.4 billion hectares of which 11% (5.4 billion hectare) land surface is used in crop production.[2] According to Food and Agriculture Organisation (FAO) (2017), approximately 954.8 million ha of land in the world is covered with saline soil, and almost every continent of world is affected with soil salinization.

In high saline conditions, accumulation of excess sodium (Na^+) and chloride (Cl^-) ions in various plant organs creates cytotoxicity that counterfires plant growth and development process. Moreover, salinity induces oxidative stress by the generation of reactive oxygen species (ROS) that causes irreversible changes in cellular and physiological processes (Isayenkov and Maathuis 2019). Most of the crops including important cereals like rice, millet, and leguminous crops (beans, pigeon pea) are sensitive to salt stress (Grieve et al. 2012). Some plants may tolerate salt stress by the means of ion homeostasis, modification in cellular membrane structure, and production of osmoregulatory compounds (Egamberdieva et al. 2019). In contrary to salt-tolerant varieties, salt-sensitive plants have lesser ability to cope salt toxicity which results in huge loss in yield and productivity. However, there are several approaches that can be used to abate the salt stress in the saline agroecosystems. Among them physical methods that include leaching, scraping, and flushing with irrigation and artificial drainage are being used for decades (Ayyam et al. 2019). Chemicals such as gypsum and $CaCl_2$ are also being used for improving soil water infiltration in arid and semiarid regions (Oster 1982; Osman 2018). As gypsum also contains phosphoric acid and chlorides, it leaches into groundwater and other water bodies through irrigation that negatively impacts aquatic ecosystems (Naveen et al. 2018). Besides this, physical methods used for saline soil reclamation are also time-consuming, labor intensive, and economically not feasible.

Development of salt-resistant crop varieties through efficient technologies such as plant breeding and genetic engineering has also been practiced since long for

[1]Global agricultural productivity is growing at an annual rate of 1.63% which is lesser than the required, i.e., 1.73% to sustainably produce food for ten billion population in 2050.

[2]http://www.fao.org/3/y4252e/y4252e06.htm

enhancing productivity in saline agroecosystems. Even after undeniable success in the area of genetic engineering and plant breeding, there are few drawbacks; they are highly expensive and time taking and require government approvals which may result in unnecessary hurdles. All such drawbacks reduce the dependency on such techniques (Coleman-Derr and Tringe 2014). Therefore, development of sustainable approaches that not only remediate saline soil but also increase crop yield is the need of the hour. Salt-tolerant plant growth-promoting rhizobacteria (ST-PGPR) emerged as ecofriendly and greener tools for programming crops for salt stress. ST-PGPR form mutualistic association with plants, and via salt stress responsive mechanisms, they help in plant's survival under saline conditions (Arora et al. 2012; Arora et al. 2021). Their application in the form of biostimulants is proven to be beneficial in terms of enhancing productivity and providing resistance against salt stress conditions (Mishra et al. 2018). This chapter targets to define the various mechanisms by which ST-PGPB enhance plant growth and help in survival in such harsh conditions. Apart from this, current use of ST-PGPB-based biostimulants for the better management of saline soil in comparison to other conventional approaches is also deliberated upon.

6.2 Effect of Salinity on Agroecosystems

Salinity adversely impacts agroecosystems by counteracting with the key metabolic processes in plants and properties of soil. Increased salt concentration in soil affects seed germination, photosynthesis, respiration, reproduction, solute transportation, defense mechanisms, gene expression, and hormonal responses in plants (Khan and Abdullah 2003; Parihar et al. 2015; Fallah et al. 2017). Salt-induced stress causes oxidative damage to the plants by producing ROS such as hydrogen peroxides (H_2O_2), hydroxyl radicals (OH), and superoxides ($O2^{\cdot -}$). Production of ROS causes oxidative damage in cell membranes, lipids, proteins, and chloroplast (Gong et al. 2018; Goharrizi et al. 2019). The process of nutrient uptake is also affected under salinity stress. As the concentration of soluble salts increases, it reduces formation of organic matter in soil and bioavailability of nutrients in different growth phases of plants (Walpola and Arunakumara 2011; Moradi et al. 2019). Under extreme saline conditions, germination of seed is either delayed or completely inhibited. High salt concentration results in the low availability of water molecules which slows down activation of key enzymes involved in initial germination process (Khan and Weber 2008). Under higher salt concentration, seed vigor, shoot length, and root length were found to be decreased in *Perilla frutescens* and *Helianthus annuus* L. (Zhang et al. 2011; Tewari and Arora 2016, 2018). In a study by Jafarzadeh and Aliasgarhzad (2007), high salt (NaCl) concentration in the soil decreased seed germination and seedling root length in sugar beet.

In the saline conditions, reduction in ribulose-1,5-bisphosphate carboxylase/oxygenase (RuBisCO, EC 4.1.1.39) enzyme is found to be linked with low yield in agroecosystems. Apart from this, under high salt concentration, carotenoid and chlorophyll contents are also decreased which inhibit the process of photosynthesis

(Gong et al. 2018). Salinity affects crop yield by the accumulation of Na^+ ions in the reproductive tissues which subsequently inhibit process of fertilization and embryogenesis in many plants. Salinity causes harmful effects to stigma receptivity, pollen viability, and starch synthase activity in *Oryza sativa* (Khan and Abdullah 2003), while accumulation of Na^+ ions in high concentration in the laminae of young leaves is found to be involved in delayed flowering, which in turn affects reproductive success (Pushpavalli et al. 2016). In cotton, developmental stages like germination, fruiting, flowering and activities of metabolic enzymes are impacted due to high salt accumulation in vital tissues (Sharif et al. 2019). According to Zhang et al. (2013), toxicity caused by Na^+ accumulation is the primary reason of early leaf senescence and decrease in the crop yield in cotton. Fernandes et al. (2004) reported that lesser deposition of cellulose due to impairment of sucrose metabolism under saline conditions results in deterioration of fiber quality.

Soil salinization also affects various physical and chemical properties of soil. A large amount of soluble salts extract the water molecules from soil and affect moisture content, while introduction of carbonates and bicarbonates of sodium, potassium, calcium, and magnesium in the soil decreases H^+ ions and increases OH^- ions that also results in high pH value of the soil (Warrence et al. 2002; Pearson and Bauder 2006; Cemek et al. 2007). Under high salt concentration, calcium-dominated soil may be converted into sodium-dominated soil which affects exchangeable sodium percentage (ESP) and makes soil dry and infertile (Chen and Banin 1975; Warrence et al. 2002; Cemek et al. 2007). In saline conditions, aggregation of fine particles decreases porosity and aeration of soil, and affects soil structure and texture (Oster and Shainberg 2001). Waterlogging increases in saline soils due to reduced hydraulic conductivity which also affects aeration and slows down decomposition of organic matter (Oster and Shainberg 2001; Pearson and Bauder 2006; Cemek et al. 2007).

Apart from interfering with physiological process in plants, salinity also affects microbial processes that help in marinating soil fertility (Rath et al. 2019). Salinity can alter microbial composition, diversity, and key biochemical processes such as soil respiration and enzymatic activities that collectively take part in making soil living and fertile (Hasbullah and Marschner 2014; Zhang et al. 2019; Zhang et al. 2021). Hence, it can be clearly stated that salinity is a major factor that is responsible for loss of fertility and productivity in saline agroecosystems.

6.3 Salt-Tolerant Plant Growth-Promoting Bacteria

Microorganisms are ubiquitously present living entities, and some can live and tolerate extreme surroundings with the help of special mechanisms that are the result of adaptation and evolution. Plants growing in harsh and unfavorable environment are surrounded by array of these minute creatures that help in tolerance to various biotic and abiotic stresses, in addition to providing plants with many growth-promoting factors in stressed conditions. A rich population of PGPB is found to be concentrated in the rhizospheric zone of plants as this region is quite rich in many nutrients which attract soil-borne microbes. Hence this rhizospheric region can be designated as a mutually benefiting cite for both plant and microbes where exchange of active substances takes place. These PGPB can also thrive inside various plant tissues (endophytes) (Khare et al. 2018) as well as above the plant surface (epiphytes) (Arora 2013). PGPB may help plants by providing resistance to unfavorable environmental conditions like drought salinity, high and low pH, heavy metals, etc. Stress-tolerant PGPB provide beneficiary effects to plants by a number of direct and indirect mechanisms even under abiotic stress conditions like salinity. These mechanisms include hormone production, nutrient acquisition, improving soil structure and composition by enhancing binding, cation exchange capacity, water holding capacity, bulk density, production of exopolysaccharides, antioxidants, osmolytes, and many others (Arora 2015; Vaishnav et al. 2016; Egamberdieva et al. 2019). Common salt-tolerant PGPB genera include *Arthrobacter*, *Azotobacter*, *Azospirillum*, *Bacillus*, *Burkholderia*, *Klebsiella*, *Pantoea*, *Pseudomonas*, *Paenibacillus*, *Rhizobium*, *Rhodococcus*, *Serratia*, *Sphingomonas*, *Thiobacillus*, and *Xanthomonas* (Arora et al. 2012; Arora et al. 2020).

Gram-negative *Pseudomonas* and Gram-positive *Bacillus* are among the most widely studied and best known PGPB genera (Etesami and Glick 2020). Among *Pseudomonas* spp., *P. fluorescens*, *P. aeruginosa*, *P. putida*, *P. cepacia*, *P. entomophila*, *P. alcaligenes*, and *P. pseudoalcaligenes* are most commonly studied PGPB (Deshwal and Kumar 2013). Recently, salt-tolerant strain *P. entomophila* PE3 was reported to provide salt stress tolerance to plants by producing exopolysaccharides (EPS) along with other growth-stimulating substances like phytohormones, siderophores, and mineral solubilization under saline conditions (Fatima and Arora 2021). In a study, flucrescent *Pseudomonas* sp. PF17 and its EPS-based formulation was documented to promote growth of sunflower as well as show biocontrol activity by solubilizing phosphates and producing metabolites like phytohormones, hydrogen cyanide, pyocyanin, and siderophores in high salt concentration even in vivo conditions (Tewari and Arora 2014; Tewari and Arora 2016). Likewise, *P. putida* has been found to upregulate transcriptional expression of five early salt stress-responsive genes, namely, ROS scavenging (*APX2*), detoxification (*GLYI7*), ABA signaling (*RD29A and RD29B*), and JA synthesis (*LOX2*), causing enhancement of germination rate and survival of *Arabidopsis thaliana* under salt stress (Chu et al. 2019).

Bacillus strains like *B. cereus*, *B. amyloliquefaciens*, *B. megaterium*, *B. toyonensis*, *B. licheniformis*, *B. pumilus*, *B. subtilis*, and *B. safensis* are also well reported to promote growth of plants under salinity stress (Wang et al. 2018; Goswami et al. 2014; Upadhyay et al. 2012, 2019; Prakash and Arora 2021). In a study, *B. toyonensis* COPE52 has been reported to enhance growth of tomato under salinity stress by modifying lipid and fatty acid composition of plant cells. The bacterium stimulated growth of tomato by producing indole-3-acetic acid (IAA), biofilm, and protease and also showed biocontrol against the fungi *Botrytis cinerea* (Rojas-Solis et al. 2020). Apart from rhizospheric PGPB, an endophytic strain *B. safensis* ZY16 isolated from the root of halotolerant plant *Chloris virgata* Sw. was also reported to be engaged with phytoremediation of petroleum-contaminated soil, production of biofilm, IAA, siderophore, and phosphate solubilization under induced hypersaline conditions (Wu et al. 2019).

Other than *Pseudomonas* and *Bacillus*, several other PGPB genera have shown magnificent role in plant growth promotion and alleviation of salt stress. For instance, PGPR strain *Serratia liquefaciens* KM4 was investigated and found to reduce salinity stress and induced plant growth in maize by regulating various salt stress-related genes, ion homeostasis, leaf gas exchange, and redox potential (El-Esawi et al. 2018a). Salt and heavy metal-tolerant endophytic strain of *Serratia rubidaea* isolated from roots of quinoa plants solubilized complex form of phosphate and zinc and efficiently produced IAA, siderophore, and ammonia in quinoa (Mahdi et al. 2021). Similarly, *Azospirillum lipoferum* and *Azotobacter chroococcum* inoculated maize plants showed significantly enhanced K^+/Na^+ ratio, antioxidative enzymes and osmolytes under saline condition (Latef et al. 2020). In a similar study, inoculation of *A. lipoferum* and *A. chroococcum* in hopbush shrub (*Dodonaea viscosa* L.) elevated the germination rate as well as shoot and root length to many folds under saline conditions (Yousefi et al. 2017). In another study, endophytic bacterium *Burkholderia phytofirmans* PsJN was found to be able to scavenge ROS and improved osmotic balances in halophyte quinoa (*Chenopodium quinoa* Willd.) (Yang et al. 2020).

Rhizobia-legume symbiosis is known to be sensitive to salt stress, but some of the salt-tolerant *Rhizobium* spp. have been reported which promote plant growth and fix nitrogen (N_2) under salt stress. For example, when *Sinorhizobium meliloti* RM1521 was inoculated in alfalfa, it showed higher number of nodulations even under salt stress (Bertrand et al. 2020). Rhizobia are also being explored for their plant growth-promoting attributes in non-legume plants like rice, maize, wheat, and canola under stressed environment including salinity (Hussain et al. 2009; Saghafi et al. 2018). Saghafi et al. (2018) have studied that *S. meliloti* and *Rhizobium leguminosarum* bv. *phaseoli* inoculation in non-legume canola (*Brassica napus* L.) may ameliorate salt stress by production of 1-aminocyclopropane-1-carboxylic acid (ACC) deaminase, IAA, and phosphate solubilization. Additionally, co-inoculation of *Azospirillum brasilense* and *Rhizobium tropici* in maize helped in production of ROS-scavenging enzymes and osmolytes conferring salinity tolerance (Fukami et al. 2018b). Therefore, a wide array of PGPB have been reported and studied for their significant potential to ameliorate salt stress along with plant growth promotion, which in turn enhances productivity of crops grown in saline agroecosystems.

6.4 PGPB-Based Biostimulants in Saline Agroecosystems

Plant growth-promoting bacterial biostimulants can play magnificent role in stimulation of plant growth under salt stress conditions. These bacteria have a variety of mechanisms to withstand in saline conditions and enhance plant growth via increasing nutrient uptake, phytohormone production, antioxidant production, and many more (Fig. 6.1). These mechanisms have been briefly described below:

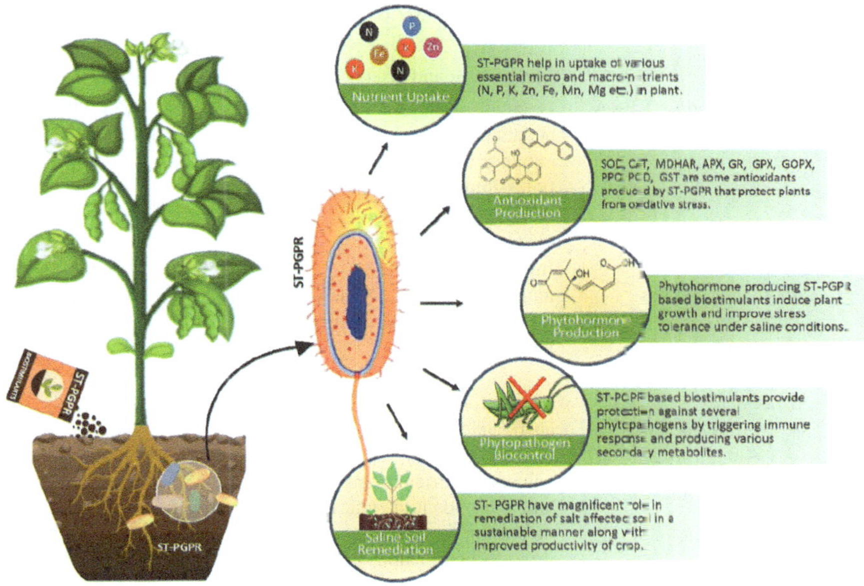

Fig. 6.1 Mechanisms used by PGPB to combat salt stress

6.4.1 Biostimulants for Nutrient Uptake

Plants require 17 essential elemental nutrients for their overall growth and development, which are categorized into macronutrients and micronutrients according to the amount required. Macronutrients include carbon (C), hydrogen (H), oxygen (O), nitrogen (N), phosphorus (P), potassium (K), calcium (Ca), magnesium (Mg), and sulfur (S), and micronutrients include zinc (Zn), copper (Cu), iron (Fe), manganese (Mn), boron (B), molybdenum (Mo), chlorine (Cl), and nickel (Ni). Although required in minute quantity, micronutrients are also essentially required, and their deficiency may give rise to various deformity symptoms in plants (Fageria et al. 2008). Effect of salinity on the nutrient composition was observed in twelve *B. napus* L. cultivars and it was concluded to have negative impact on plant growth along with reduction in water and nutrient composition. K^+, K^+/Na^+, and Ca^+ content decreased, and stress ions (Na^+ and Cl^-) concentration increased in all cultivars (Tunçtürk et al. 2011). Several studies are available where PGPB are reported to improvise nutrient uptake in plants under saline conditions (Table 6.1).

Nitrogen (N) is an element that is required by plants in large amounts, as most of the cell components require N as building block, that is why deficiency of N rapidly inhibits plant growth. Under salinity conditions N uptake by plant shoots decreases (Hu and Schmidhalter 2005). Biological nitrogen fixation makes global contribution of about 180×10^6 metric tons/year (Mayak et al. 2004) of which 80% contribution is from symbiotic nitrogen fixers that are *Rhizobium* in legume plants and *Frankia*

Table 6.1 Nutrient uptake mechanisms used by PGPB to alleviate salt stress in plants

Sr. no.	PGPB	Mechanism and effects	Host/plant	References
1	*Bacillus subtilis* (BERA 71)	Decrease sodium accumulation and enhance the nitrogen (N), potassium (K), calcium (Ca), and magnesium (Mg) content	Chickpea (*Cicer arietinum* L.)	Abd-Allah et al. (2018)
2	*Bradyrhizobium japonicum* and the endophyte *Stenotrophomonas rhizophila*	Improved root, shoot length, dry weight, N and P uptake, and number of nodules	Soybean (*Glycine max* L.)	Egamberdieva et al. (2018)
3	*Azotobacter* sp. (Az1-Az6)	Production of indole-3-acetic acid (IAA) and nitrogen fixation, enhances nutrient uptake and antioxidant enzymes, increasing fresh weight, dry weight, root and shoot length	Wheat (*Triticum aestivum* L.)	El-Nahrawy and Yassin (2020)
4	*Streptomyces laurentii*, *Sinorhizobium* sp., and *Bacillus safensis*	Showed nitrogen fixation, solubilization of inorganic phosphate, production of phytohormones, ammonia, and acetoin	Mint (*Ocimum basilicum* L.)	AlAli et al. (2021)
5	*Bacillus* strains, *Acinetobacter* strains, *Pseudomonas* strains, *Brevibacillus*, *Gordonia terrae*, *Chryseobacterium lathyri*, *Ensifer sesbaniae*, and *Paenibacillus illinoisensis*	Showed high potential for dissolving calcium phosphate [$Ca_3(PO_4)_2$]	Peanut (*Arachis hypogaea* L.)	Jiang et al. (2019)
6	*Bacillus halotolerans* and *Lelliottia amnigena*	Significantly increased K^+/Na^+ in root-shoot and reduced Na^+ in root-shoot, fix elemental nitrogen and solubilize phosphate and reduce proline accumulation	Wheat (*Triticum aestivum* L.)	El-Akhdar et al. (2020)
7	*Azotobacter chroococcum*	Boosted N, P, and K concentration, exopolysaccharide (EPS) production, and melatonin production under salinity stress	Faba bean (*Vicia faba* L.)	El-Ghany et al. (2020)

(continued)

Table 6.1 (continued)

Sr. no.	PGPB	Mechanism and effects	Host/plant	References
8	*Pseudomonas koreensis* (MU2)	Significantly increase N, P, and K contents in plant leaf and upsurge K^+ uptake by 46%	Soybean (*Glycine max* L.)	Adhikari et al. (2020)
9	*Pseudomonas argentinensis* and *P. azotoformans*	Induce production of aminolaevulinic acid, indole-3-acetic acid (IAA), 1-aminocyclopropane-1-carboxylate deaminase, and solubilization of potassium and phosphorus	Mustard (*Brassica juncea* L.)	Phour and Sindhu (2020)
10	*Enterobacter* sp.	Enhance phosphate (P) solubilization, IAA production, siderophore production, HCN production, and possess ACC deaminase activity	Rice (*Oryza sativa* L.)	Sarkar et al. (2018a)
11	*Alcaligenes* sp.	Produces siderophores, zinc (Zn), IAA, GA, EPS, comprehend proline, antioxidants, and hydroxyl ion scavenging activity	Rice (*Oryza sativa* L.)	Fatima et al (2020)
12	*Achromobacter denitrificans* and *Ochrobactrum intermedium*	Showed higher amount of atmospheric nitrogen fixation, phosphate solubilization, and indole-3-acetic acid production, an exopolysaccharide layer outside the cell surface	Rice (*Oryza sativa* L.)	Sultana et al. (2020)
13	*Pseudomonas entomophila* (PE3)	Showed potential to solubilize potassium (K), zinc (Zn), and phosphorus (P) along with production of indole-3-acetic acid (IAA), gibberellic acid (GA), exopolysaccharides (EPS), and siderophore	Sunflower (*Helianthus annuus* L.)	Fatima and Arora (2021)
14	*Pseudomonas* sp. (PF17)	Solubilize phosphate, produce siderophores, IAA, pyocyanin, HCN, chitinase and β-1,3-glucanase	Sunflower (*Helianthus annuus* L.)	Tewari and Arora (2016)

(continued)

Table 6.1 (continued)

Sr. no.	PGPB	Mechanism and effects	Host/plant	References
15	*Rhizobium radiobacter* LB2	Solubilize zinc and phosphate, fix atmospheric nitrogen, produce EPS, IAA, and siderophores	Lettuce (*Lactuca sativa* L.)	Verma et al. (2020)

colonizing in non-legume plants. Other 20% belong to free-living nitrogen fixers such as *Acetobacter*, *Alcaligenes*, *Arthrobacter*, *Azospirillum*, *Azotobacter*, *Bacillus*, *Cyanobacteria*, *Clostridium*, *Derxia*, *Enterobacter*, *Klebsiella*, *Pseudomonas*, *Rhodospirillum*, and *Xanthobacter* (Young and Haukka 1996; Sivasakthi et al. 2014; Ghorai et al. 2015; Di Benedetto et al. 2017). Earlier, symbiotic biological nitrogen fixation was believed to occur only in legume plants with the help of legume-*Rhizobium* symbiosis, and many attempts were made to imitate this symbiosis in non-legume plants. In the late1980s, nitrogen-fixing microbes were discovered from root tissues of non-leguminous crops like sugarcane, maize, sorghum, and rice, which transformed the whole approach (Baldani et al. 1986; Cavalcante and Dobereiner 1988). These N-fixing, non-nodule forming endophytes are termed as endophytic diazotrophic bacteria (Döbereiner 1992). Apart from these, free-living nitrogen fixers also play very important role in agroecosystems. For instance, *Azospirillum* sp. which is known to be an excellent free-living nitrogen fixer is also found to produce phytohormones like IAA, jasmonic acid, and salicylic acid, and the phytohormone-mediated signaling is also believed to be contributing in abiotic stress tolerance in plants (Fukami et al. 2018a). Similarly, *Rhizobium* sp. was also reported to induce salinity stress tolerance both in legumes and non-legumes like canola (*B. napus* L.) (Saghafi et al. 2018). A salt-tolerant strain *Rhizobium radiobacter* LB2 isolated from root nodules of *Leucaena leucocephala* was found to improve growth of lettuce by showing PGP traits such as zinc and phosphate solubilization and production of IAA, EPS, and siderophore (Verma et al. 2020). Many studies have shown that PGPB can improve the *Rhizobium*-legume symbiosis that can fulfill need of N in plants under saline conditions (Saharan and Nehra 2011; Bhattacharyya and Jha 2012). Egamberdieva et al. (2013) reported that the inoculation with *Rhizobium galegae* along with PGPB strains *Pseudomonas extremorientalis* TSAU20 or *Pseudomonas trivialis* 3Re27 gave better nodulation, growth, and N assimilation in *Galega officinalis* L. (goat's rue) in salt stressed environment. Application of magnetite nanoparticle-based formulation including *Rhizobium* sp., endophyte *Stenotrophomonas maltophilia*, and mycorrhizal fungi induced the nodulation, nitrogen fixation, and growth of chickpea (*Cicer arietinum* L.) under saline conditions (Abd-Alla et al. 2019).

Phosphorus (P) is another important macronutrient required by plants as it is constituent of genetic material, proteins, and lipids required in storage and transfer of energy, photosynthesis, enzyme regulation processes in plants, etc. Plants may take only primary ($H_2PO_4^-$) orthophosphate and secondary (HPO_4^{2-}) orthophosphate ions. The insoluble forms present in soil such as tricalcium phosphate $(Ca_3PO_4)_2$, aluminum phosphate (Al_3PO_4), iron phosphate (Fe_3PO_4), and complex organic phosphates can be solubilized by the help of variety of soil microorganisms either by solubilization or by mineralization. They do so by producing low molecular weight organic acids which can chelate mineral ions or can lower the pH to convert P into soluble form. On the other hand, mineralization process involves enzymatic degradation by enzymes like phosphatases and phytases which release phosphates from the complex P compounds (Maougal et al. 2014; Khan et al. 2013a; Alaylar et al. 2020). Salt-tolerant PGPB have been reported to display significant role in solubilization of P and make it available to plants grown under saline conditions. In a study, Srivastava and Srivastava (2020) have reported that phosphate-solubilizing bacteria (PSB) *P. putida* MTCC 5279 can contribute in alleviation of P deficiency and promote growth of *A. thaliana* under salinity-induced phosphate-starved conditions. In another study, *Arthrobacter* sp. and *Bacillus* sp. have been investigated to solubilize and enhance P uptake in maize grown in P-deficient saline soil (Vanissa et al. 2020). Wang et al. (2018) reported some excellent phosphate-solubilizing strains of *Bacillus* spp. with potential to produce ACC deaminase, phytohormones, and fix nitrogen in *Capsicum annuum* L. under saline conditions. In a lime-induced salinity stress, consortia of PSB including strains from genera like *Bacillus*, *Burkholderia*, *Erwinia*, *Pseudomonas*, *Rhizobia*, and *Achromobacter* with organic P manure improved P uptake efficiency as well as growth of maize plant (Adnan et al. 2020). Mukhtar et al. (2020) used consortia of phosphate-solubilizing halophilic PGPB belonging to genera *Bacillus*, *Halobacillus*, and *Pseudomonas*, which promoted plant growth by producing IAA, ACC deaminase, siderophore, HCN, and fixing N along with P solubilization. Furthermore, the synergistic interaction between phosphate-solubilizing fungi and PSB is also being used to alleviate salt stress (Etesami and Alikhani 2019). For instance, recently, Zai et al. (2021) reported the synergistic effect of arbuscular mycorrhizal fungi *Funneliformis mosseae* and PSB *Apophysomyces spartima* on nutrient uptake (NPK) and photosynthesis in beach plum (*Prunus maritima*). Inoculated plant was found with increased P uptake by 68.9% under salt stress, in comparison to control.

Potassium (K) is an important macronutrient as 98% of it is found in earth crust in insoluble form as rocks and silicates. In the saline environment, it becomes necessary to increase the concentration of K^+ ions in the cytoplasm of plant cell to maintain osmotic pressure. However, high salinity tends to displace K^+ ions causing high ratios of Na^+/K^+ inside root and shoot systems disturbing osmotic balance and metabolic processes (Botella et al. 1997). Sodic and saline soils are rich in K^+ minerals but in unavailable form. In such conditions, PGPB could work as a magnificent tool for enhancing salt tolerance and K-solubilizing activity in plants. K^+-solubilizing bacteria (KSB) belonging to *Bacillus* spp. and *Pseudomonas* spp. are among the most dominant genera under salinity range of 2.3–3.5 dSm^{-1} (Jha and Subramanian 2016). Salt-tolerant *Enterobacter hormaechei* isolated from water have been found to solubilize insoluble K and Ca minerals along with growth promotion in tomato under salinity stress (Ranawat et al. 2021). In a study, Ashfaq et al. (2020) evaluated phosphate-solubilizing ability of PGPRs isolated from paddy rhizosphere at different level of NaCl stress and also found KSB strains *Acinetobacter pittii*, *Rhizobium pusense*, *Cupriavidus oxalaticus*, and *Ochrobactrum ciceri* with K^+ solubilization potential under salinity stress.

Iron (Fe), an essential micronutrient is required for metalloenzymes involved in processes like photosynthesis and respiration. Fe is available in soil in two interconvertible oxidation states Fe^{2+} and Fe^{3+}, and among them Fe^{3+} is biologically unavailable to plants. Abiotic stresses like salinity and other environmental factors affect availability of Fe^{2+} ion to the plant as it is converted to Fe^{3+}. Also, under alkaline pH conditions, solubility of Fe reduces, causing chlorosis in plants. Under such conditions, siderophore-producing PGPB can act as a promising tool to make Fe available to plants (Ferreira et al. 2019). Fe-chelating siderophores produced by many PGPB help to overcome lower Fe deficiency under saline conditions (Deshwal and Kumar 2013; Numan et al. 2018). *Bacillus aryabhattai*, *Ochrobactrum intermedium*, and *Achromobacter denitrificans* were documented to upregulate the *entD* gene expression responsible for siderophore production under salt stress (Sultana et al. 2021). Recently, Fe deficiency and salinity stress in crops are found to be resolved by the application of siderophore-producing PGPB belonging to genera *Streptomyces* (Nozari et al. 2021). In another study, PGPB *P. fluorescens*, *B. megaterium*, and *Variovorax paradoxus* were reported to be efficient siderophore producers under salt stress of 2% and 5% (Nadeem et al. 2016).

It has been studied that zinc-solubilizing bacteria (ZSB) can help plant to overcome Zn deficiency under saline conditions. Fluorescent pseudomonads when applied along with Zn in pistachio (*Pistacia vera* L.) were able to reduce salinity stress and enhance uptake of Zn in plants under saline conditions (Azarmi et al. 2016). In another example, halotolerant endophytic bacteria of genera *Enterobacter*, *Bacillus*, and *Pseudomonas* isolated from internal tissues of sugarcane were found to solubilize Zn and K along with showing antifungal activity against *Fusarium* sp. (Pirhadi et al. 2016). *R. radiobacter* LB2 strain isolated from the root nodules of salt-tolerant plant *L. leucocephala* was reported to promote growth of lettuce by solubilizing minerals like Zn and phosphate, and producing siderophore and IAA under salinity stress (Verma et al. 2020). Hence, application of PGPB-based biostimulants enhances nutrient uptake in plants which indirectly helps plant to withstand salt stressed conditions. It can be concluded that salt-tolerant PGPB can play a very important role in acquisition of both macro- and micronutrients in saline soils and maintaining the health and vigor of plant along with enhancing the productivity.

6.4.2 Biostimulants for Phytohormones

Phytohormones have significant role in plant growth and developmental processes, and are produced during different life stages of plants at varied concentrations. Major phytohormones are auxins IAA, gibberellins, cytokinins (CKs), abscisic acid (ABA), ethylene, salicylic acid (SA), and jasmonates that regulate plant growth and development. Salinity hampers the ability of plant to produce phytohormones which negatively impact plant growth and metabolic reactions (Arora et al. 2020). PGPB are reported to produce various phytohormones even under abiotic stresses such as salinity (Table 6.2).

IAA was the first phytohormone identified; it is an important molecule needed in vascular differentiation process (Wang et al. 2001). It has key role in regulating plant growth and controls cell elongation, vascular tissue development, and apical dominance (Wang et al. 2001; Fahad et al. 2015). Level of IAA production in plants is found to decrease under salinity stress. Application of IAA producing bacteria can avail IAA to plant under saline condition. This exogenous IAA produced by PGPB may help plant in changing root architecture in a way that can absorb more water and minerals under drought-like conditions caused by salinity (Etesami and Beattie 2018). In a study, *Bacillus* spp. were found to produce significant amount of IAA along with providing salinity stress tolerance by producing EPS and ACC deaminase (Din et al. 2019). In a study, *Leclercia adecarboxylata* was reported to significantly augment IAA (9.815 μg/mL) levels in tomato under saline conditions (Kang et al. 2019). A halotolerant rhizobacteria, *Rheinheimera* sp. isolated from the rhizosphere of *Cuminum cyminum* was identified to produce huge amount of IAA (250 μg/mL) under salt stress conditions (Raval and Saraf 2020). Ul Hassan and Bano (2019) also documented the positive effect of IAA-producing PGPB *Pseudomonas moraviensis*, on the growth of wheat grown in saline and sodic soil; IAA-deficient mutant strain of *P. moraviensis* showed poor growth of wheat under saline conditions.

Ethylene hampers the growth and development of plant tissues under abiotic stresses; PGPB have potential to produce enzyme called ACC deaminase that cleaves the precursor ACC molecule of ethylene, thus saving plant from ethylene accumulation due to salinity (Glick et al. 1998). A number of studies are available which documented the ameliorative effect of ACC deaminase activity by PGPB under salinity. For example, *Enterobacter* sp. was found to show ACC deaminase activity along with some other growth promotion activities like phosphate solubilization, IAA, siderophore, and HCN production, in rice seedlings growing under salinity stress (Sarkar et al. 2018a). *Aneurinibacillus aneurinilyticus* and *Paenibacillus* sp. isolated from rhizosphere of garlic when applied in consortia were found to produce ACC deaminase and significantly lower stress ethylene levels (nearly by 60%) in French bean (Gupta and Pandey 2019). In another study, *Bacillus* sp. were found to be potent modulators of ACC deaminase, ACC oxidase (ACO), and ACC synthase (ACS), thus alleviating salt stress in *Zea mays* (Misra and Chauhan 2020). In another study, application of ACC deaminase-producing PGPB *Enterobacter* sp. reported to promote growth of rice and millets under salinity stress (Sagar et al. 2020).

ABA plays an important role in saving plants during environmental stress conditions by enabling various signaling pathways in plants; it is an important part of plant life cycle essentially required from seed development, dormancy, stomatal opening and closure, maintaining water status during stress by acting as messenger and upregulating genes responsible for cell dehydration tolerance, etc. (Zhu 2002; Mahajan and Tuteja 2005; Fahad et al. 2015). PGPB have caliber to maintain the level of ABA in plant suffering from salt stressed environment. In a study,

Table 6.2 Phytohormone production mechanisms used by PGPB to alleviate salt stress in plants

Sr. no.	PGPB	Mechanism/effect	Host/plant	References
1	*LBEndo1* as *Bacillus* sp. and *KBEcto4* as *Pseudomonas lini*	Induced IAA and siderophore production, as well as phosphate solubilization. Additionally, the KBEcto4 strain expresses the ACC deaminase enzyme (*acdS* gene)	Saltgrass (*Distichlis spicata*)	Palacio-Rodríguez et al. (2017)
2	*Enterobacter cloacae* (HSNJ4)	Increase endogenous IAA content and reduced ethylene emission, produce both IAA and ACC deaminase, led to increase in root length, shoot length, number of lateral roots, and chlorophyll content	Canola (*Brassica napus* L.)	Li et al. (2017)
3	*Rhizobium* sp. LSMR-32 and *Enterococcus mundtii* LSMRS-3	Production of -3-acetic acid (IAA), exopolysaccharides (EPS), and siderophores, solubilization of phosphate and formation of biofilm, ACC deaminase activities, and salt tolerance were all found highly significant in dual inoculant (LSMR-32 + LSMRS-3)	Mungbean (*Vigna radiata* L.)	Kumawat et al. (2020)
4	*Hartmannibacter diazotrophicus*	Reduced ethylene emission by ACC deaminase production	Barley (*Hordeum vulgare* L.)	Suarez et al. (2015)
5	*Bacillus subtilis* (NBRI 28B), *B. subtilis* (NBRI 33 N), and *B. safensis* (NBRI 12 M)	Extenuate the adverse effects of ethylene by modulating ACC accumulation, ACC oxidase (ACO), and ACC synthase (ACS) activities; induced plant response for defense enzymes, chlorophyll, proline, and soluble sugar	Maize (*Zea mays* L.)	Misra and Chauhan (2020)
6	*Glutamicibacter* sp. (YD01)	Showed ACC deaminase enzyme activity and IAA production; regulate the plant ethylene production, ACC content, and ACC	Rice (*Oryza sativa* L.)	Ji et al. (2020)

(continued)

Table 6.2 (continued)

Sr. no.	PGPB	Mechanism/effect	Host plant	References
		oxidase activity; improved K^+ acquisition; reduce ROS accumulation, thereby increasing photosynthetic capacity		
7	*Kocuria rhizophila* (Y1)	IAA and ABA genes encoding antioxidants (ZmGR1 and ZmAPX1) and genes involved in salt tolerance (ZmNHX1, ZmNHX2, ZmNHX3, ZmWRKY58, and ZmDREB2A), improved nutrient acquisition, biomass production, seed germination rate, photosynthetic capacity, antioxidant levels, relative water content, and chlorophyll accumulation	Maize (*Zea mays* L.)	Li et al. (2020)
8	*Arthrobacter protophormiae* (SA3) and *Dietzia natronolimnaea* (STR1)	Increase in IAA and abscisic acid (ABA) content, production of ACC deaminase enzyme, enhanced photosynthetic efficiency of wheat	Wheat (*Triticum aestivum* L.)	Barnawal et al. (2017)
9	*Bacillus pumilus* (EU927414)	Produces IAA, GA, and some antioxidants like SOD, APX, CAT, proline, siderophore, and solubilize phosphate	Wheat (*Triticum aestivum* L.)	Upadhyay et al. (2019)
10	*Streptomyces venezuelae ATCC 10712*	Reduce ethylene concentration, ROS, and Na^+ ion by expressing *acds* gene	Rice (*Oryza sativa* L. *cv. KDML105*)	Yoolong et al. (2019)
11	*Bacillus strains*	Produced IAA, exopolysaccharides of PGPR sequester Na^+ ion and degrade ACC into ammonia and α-ketobutyrate, improve vegetative shoot growth and root elongation	Wheat (*Triticum aestivum* L.)	Din et al. (2019)
12	*Alcaligenes faecalis JBCS1294*	Produced IAA, bacterial VOCs induced salt tolerance of plant	*Arabidopsis thaliana*	Bhattacharyya and Lee (2017)

(continued)

Table 6.2 (continued)

Sr. no.	PGPB	Mechanism/effect	Host/plant	References
13	*Paraburkholderia phytofirmans*	Activate specific genes such as ACC deaminase, auxin catabolism, *N*-acyl-homoserine-lactone, along with VOCs that reduce salinity stress	*Arabidopsis thaliana*	Ledger et al. (2016)

inoculation of soybean crop with PGPB isolated from rhizosphere of weeds provided salinity tolerance by production of phytohormones like IAA, gibberellic acid (GA), trans-zeatin riboside, and ABA (Naz et al. 2009). GA is required by plant for seed dormancy, lateral shoot growth, and floral organ development. It helps wheat crop under salinity stress by improving seed yield, replacing Na^+ ions by K^+ and Ca^{2+}, and maintaining hormone homeostasis (Iqbal and Ashraf 2013). Kumar et al. (2018) found that endophytic bacteria belonging to genus *Bacillus* and *Pseudomonas* significantly reduced the amount of stress-induced ABA production and enhanced production of GA4 and SA in drought and salinity-stressed tomato. Endophytic bacteria *B. amyloliquefaciens* RWL-1 isolated from rice seeds was reported to produce good quantity of gibberellins and promoted rice growth through modulating endogenous hormone level (Shahzad et al. 2016). Later the same endophyte was found to reduce the amount of endogenous ABA by 40.77% and 29.31% in 120 and 250 mM NaCl-stressed rice plants. PGPB also enhanced some of the amino acid levels like that of aspartic acid, glutamic acid, proline, alanine, threonine, glycine, serine, isoleucine, and lysine (Shahzad et al. 2017).

SA has important role in tolerance against various abiotic stresses and is recently gaining attention. SA is required by plant at an optimal concentration, and too much or too less may make plant more susceptible to abiotic stresses by disturbing plant redox status (Hara et al. 2012). Under abiotic stress, higher production of endogenous SA is not good for plant health. However, exogenous sources and application of SA and production of SA by endophytic bacteria can help in plant growth under abiotic stresses (Kim et al. 2017). Inoculation of *B. subtilis* is reported to produce SA reduces the endogenous level (of SA) in wheat under saline conditions. On the other hand, in another study application of PGPB *Bacillus* sp. is found to reduce the synthesis of stress-induced SA, thus helping plant from oxidative stress caused by higher levels of the hormone under salinity (Lastochkina et al. 2017). Furthermore, Tewari and Arora (2018) reported the application of EPS- and SA-producing strain *P. aeruginosa* PF23 in growth promotion of sunflower under salinity. The extracted SA also showed biocontrol against phytopathogen *Macrophomina phaseolina* under salinity. Hence it can be concluded that phytohormones produced by PGPB could help plant to better survive under salt stress conditions.

6.4.3 Biostimulants for Antioxidants

Extreme environmental conditions such as salinity, drought, and waterlogging induce oxidative stress by generating ROS and reactive nitrogen species (RNS) in plants. Due to high salinity, concentration of oxidative stress markers such as malondialdehyde (MDA), superoxide (O_2^-), and H_2O_2 also increases, which cause high membrane permeability, electrolytes leakage, lipid peroxidation, and other cell injuries (Mudgal et al. 2010; Khan et al. 2013b). To overcome salinity-induced oxidative stress, plants produce antioxidative enzymes such as superoxide dismutase (SOD), catalase (CAT), monodehydroascorbate reductase (MDHAR), ascorbate peroxidase (APX), dehydroascorbate reductase (DHAR), glutathione reductase (GR), glutathione peroxidase (GPX), guaiacol peroxidase (GOPX), polyphenol oxidase (PPO), peroxidase (POD), and glutathione S-transferase (GST). Non-enzymatic antioxidants such as ascorbic acid, glutathione, alkaloids, phenolic compounds, non-protein amino acids, and a-tocopherols have also been reported in confronting oxidative stress in many plants (Kapoor et al. 2019).

Presence of ST-PGPB may augment the antioxidant defense system in plant under saline conditions (Santos et al. 2018; Fatima et al. 2020) (Table 6.3). PGPB reduce salt stress by enhancing the production of antioxidative enzymes and reducing the concentration of oxidative stress markers in plants. In a study, sweet pepper plant inoculated with PGPB *Bacillus thuringiensis* enhanced plant growth by reducing lipid peroxidation, electrolyte leakage, O_2^-, and H_2O_2 content and regulating the proline content under salt stress (Al Kahtani et al. 2020). Similarly, PGPB *Serratia marcescens* isolated from *Capparis decidua* was found to decrease MDA, O_2^-, and H_2O_2 content and improve the level of SOD, CAT, and POD enzymes in wheat (Singh and Jha 2016). In another study, *A. chroococcum* and/or *Alcaligenes faecalis* were reported to alleviate salt stress, augmenting the SOD, APX, and POD in canola (Abdel Latef et al. 2021). ST-PGPR are also reported to modulate the expression of plant's genes involved in antioxidant defense system under salinity. For instance, in a study, Sapre et al. (2018) found that application of *Klebsiella* sp. decreased the MDA and H_2O_2 content and significantly increased SOD and POD enzymes in oat seedlings. Application of *Klebsiella* sp. also upregulated the expression of salt stress-related genes *rbcL* and *WRKY1* in oat seedling. Application of *Dietzia natronolimnaea* as biostimulant in wheat enhanced the expression of antioxidative genes including *APX*, *MnSOD*, *CAT*, *POD*, *GPX*, and *GR* (Bharti et al. 2016). Use of *B. megaterium* and *Enterobacter* sp. as biostimulants promoted growth of okra plant by the augmentation of ROS-scavenging enzymes, CAT, SOD, and APX and upregulation of antioxidative genes *CAT*, *APX*, *GR*, and *DHAR* (Habib et al. 2016). Li et al. (2020) reported that use of a novel salt stress mediating bacterium *Kocuria rhizophila* promoted growth in maize by enhancing antioxidant levels and removing deleterious effect of salt via regulating antioxidant encoding genes *ZmGR1* and *ZmAPX1* under salt stress.

Apart from the antioxidative enzymes, ST-PGPR could also alleviate negative impact of salt stress in plant via modulating the production of certain non-enzymatic antioxidant compounds in plants. For instance, ST-PGPR *Bacillus halotolerans*-

Table 6.3 Antioxidant production mechanisms used by PGPB to alleviate salt stress in plants

Sr. no.	PGPB	Mechanism/effect	Host plant	References
1	*Serratia liquefaciens* (KM4)	Upregulation of stress-related genes (APX, CAT, SOD, RBCS, RBCL, H^+-PPase, HKT1, and NHX1), improved plant growth by regulating ion homeostasis, redox potential, leaf gas exchange, and stress-related genes expression	Maize (*Zea mays* L.)	El-Esawi et al. (2018b)
2	*Bacillus firmus* (SW5)	(APX, CAT, POD, Fe-SOD) promoted the growth and biomass yield, chlorophyll synthesis, nutrient uptake, gas exchange parameters, osmolytes levels, total phenolic, and flavonoid contents	Soybean (*Glycine max* L.)	El-Esawi et al. (2018a)
3	*Azospirillum lipoferum* (FK1)	Expressions of three antioxidants genes (CAT, APX, and SOD), significantly improved nutrient acquisition, growth, biomass, photosynthetic pigment synthesis, osmolytes level, gas exchange attributes, phenols and flavonoids content, and enzymatic and non-enzymatic antioxidant levels	Chickpea (*Cicer arietinum* L.)	El-Esawi et al. (2019)
4	*Azospirillum lipoferum* or *Azotobacter chroococcum*	Enhanced growth parameters, pigments, K^+, osmolytes, K^+/Na^+ ratio, and the activity of CAT, POD, and APX of the salt-affected maize, lower MDA and Na^+	Maize (*Zea mays* L.)	Latef et al. (2020)
5	*Pseudomonas pseudoalcaligenes* (SRM-16) and *Bacillus subtilis* (SRM-3)	Showed ACC deaminase activity, siderophore, and indole-3-acetic acid (IAA) production, enhancing plant biomass, increased activities of defense-related system such as ion transport,	Soybean (*Glycine max* L.)	Yasmin et al. (2020)

(continued)

Table 6.3 (continued)

Sr. no.	PGPB	Mechanism/effect	Host plant	References
		antioxidant enzymes, proline, and MDA content in shoots and roots		
6	*Azospirillum Brasilense*	Encompass higher aerial biomass, chlorophyll and ascorbic acid content, better overall visual quality, hue, Chroma, and antioxidant capacity	Lettuce (*Lactuca sativa* L.)	Fasciglione et al. (2015)
7	*Pseudomonas stutzeri* (ISE12) strain	Decreased the accumulation of non-enzymatic antioxidants, activating their antioxidant defense system and triggering the rearrangement of cell walls	Canola (*Brassica napus* L.)	Szymańska et al. (2019)
8	*Bacillus flexus* (KLBMP)	Promoted the accumulation of chlorophyll to enhance photosynthesis, increased osmotic regulator contents, enhanced flavonoid and antioxidant enzymes, and regulated Na^+/K^+ homeostasis	Sea lavender (*Limonium sinense* L.)	Xiong et al. (2020)
9	*Kocuria rhizophila* KF875448 (14ASP) and *Cronobacter sakazakii* KM042090 (OF115)	Increase the antioxidant activity by reducing Na^+ contents, significantly increase root shoot length, fresh dry weight, proline, and chlorophyll contents	Wheat (*Triticum aestivum* L.)	Afridi et al. (2019)
10	*Pseudomonas fluorescence*, *Bacillus pumilus*, and *Exiguobacterium aurantiacum*	Showed maximum free proline, total soluble proteins, dry biomass, CAT activity, and shoot K uptake, after inoculation	Wheat (*Triticum aestivum* L.)	Nawaz et al. (2020)
11	*Staphylococcus sciuri* (ET101)	Increased quantum efficiency (ΦPSII) with photosynthetic electron transport rate (ETR), decrease electrolyte level with ROS scavenging and antioxidant enzymatic activities	Rice (*Oryza sativa* L.) and tomato (*Lycopersicon esculentum*)	Taj and Challabathula (2020)

(continued)

Table 6.3 (continued)

Sr. no.	PGPB	Mechanism/effect	Host plant	References
12	*Burkholderia* sp. *MTCC 12259*	Enhanced IAA, EPS, proline content, and antioxidant activities of plants	Rice (*Oryza sativa* L.)	Sarkar et al. (2018b)
13	*Sphingobacterium* BHU-AV3	Showed reduction in ROS level with increase in peroxidase (POD), polyphenol oxidase (PPO), SOD, CAT, ATP synthase, and carbohydrate metabolism	Tomato (*Lycopersicon esculentum* L.)	Vaishnav et al. (2020)
14	*Bacillus licheniformis* and *PB5* with *Pseudomonas plecoglossicida*	Contract oxidative stress through upregulation of CAT, GPX, and SOD, increase nutrient use efficiency, root length, shoot length, fresh and dry biomass of plant	Sunflower (*Helianthus annuus* L.)	Yasmin et al. (2020)

based biostimulant was reported to enhance development and growth in *Coriandrum sativum* L. by increasing the level of phenolic compounds 5-O-caffeoylquinic acid, cinnamic acid, 4-methoxy-cinnamic acid hexoside, K-3-O rutinoside, Q-3-O-rutinoside, Q-3-O-glucoside, Q-3-O-glucuronide and flavonoid compound quercetin 3-O-glucuronide under saline conditions (Jiménez-Gómez et al. 2020). In another study, ST-PGPR *Rhizobium laguerreae* was documented to increase certain phenolic and flavonoid derivatives including caffeoyl acid and quercetin in lettuce under salt stressed conditions (Ayuso-Calles et al. 2020). Likewise, El-Esawi et al. (2018b) also reported an increase in phenols, total free amino acids, soluble proteins, and flavonoid compounds and upregulation of antioxidative enzyme-related genes (*APX*, *CAT*, *SOD*) in *S. liquefaciens* inoculated maize plant under salinity stress. These findings disclose that application of antioxidant modulating ST-PGPR-based biostimulants could be useful to mitigate the negative impacts of salt-induced oxidative stress in plants.

6.4.4 Biostimulants for Soil Health and Reclamation

Saline soils contain large amount of soluble salts accumulated in root zones. Many methods are being applied for improving soil health and reclamation of salt from saline soils. Application of gypsum is the most common method used for reclamation of saline soil at global level (Hanay et al. 2004; Ezeaku et al. 2015). There are several other physical and chemical methods that are applied for restoring the quality of salt-affected soil. However, in relation to their effect on environment, economics, and ease of availability, these methods have both pros and cons. In the present scenario, there is increasing demand for opting for long-term effective and sustainable approaches for reclamation of salt-affected soil. PGPB can be the potent option for improving soil health and ameliorating salt from saline soil. Numerous mechanisms of PGPB are found to be involved in remediation of saline soils. For instance, in a study by Grobelak et al. (2018), it has been reported that bacterial strains amended with sewage sludge converted the phosphorus into bioavailable form and increased the amount of organic nitrogen which helped in restoration of properties of saline soil. In a study by Tirry et al. (2021), it has been found that halotolerant PGPR strains *P. putida*, *Alcaligenes* sp., *Klebsiella* sp., and *Pseudomonas cedrina* were found to enhance the soil functioning and enzymatic activities by making soil biota more diversified. Application of these isolates in alfalfa (*Medicago sativa*) also improved the colonization of root arbuscular mycorrhiza (AM) which subsequently increased content of organic matter in soil. In a study done by Chaudhary et al. (2020), it has been observed that PGPR isolated from roots of *Arthrocnemum indicum* improved soil quality by enhancing phosphatase activity and bioavailability of nutrients (N, P, K, and Fe) in peanut grown in halophilic conditions. In a pot experiment performed under stressed conditions by Fazal and Bano (2016), the role of biochar incorporated with *Pseudomonas* sp. in improving soil moisture content, proline content, and POD activity was reported.

Bioremediation of soil affected with stresses imposed by pollutants, heavy metals, and pesticides can be done by the application of halotolerant PGPB (Arora et al. 2020). Biofilm formation by halophilic bacterial strains *Kocuria flava* AB402 and *Bacillus vietnamensis* AB403 remediated contamination of arsenic (As) from hypersaline soils by the process of bioadsorption. In a study, Mallick et al. (2018) showed that EPS production by PGPR in saline soil helped in root colonization of beneficial microbes and limits the uptake of As by plants. *P. aeruginosa* strain FZ-2 isolated from leachates of non-active sanitary landfill has been reported to remove the toxicity of mercury (Hg) with more than 90% efficiency at different salt concentrations (Imron et al. 2021). Saline soil with cadmium (Cd) contamination was reported to be ameliorated by inoculation of wheat seeds with *Pseudomonas* sp., which increased the proline expression in plants and lowered the proline degrading genes along with concentration of Cd in plant tissues, making it tolerant towards the salinity stress (Shavalikohshori et al. 2020). Sequestration of carbon by PGP is also an important feature that helps in abatement of environmental stresses. Although plants and soil are the significant sink of carbon (Kumar et al. 2006), mutualistic association of PGPB accelerates this process. For instance, mutualistic association of *Bacillus* spp. and mycorrhizal fungi in cassava crop (*Manihot esculenta crantz*) not only enhanced crop yield and but also helped in mineralization and nutrient cycling through carbon sequestration activity (Otaiku et al. 2019). It has also been suggested that constant application of salt-tolerant PGPB-based biostimulants in soil can help in improvising the organic matter, nutrient level, microbial load, as well as improving microbial diversity in saline soil via rhizoremediation (Arora et al. 2021). Hence, studies available on PGPB prove that these microbes could be utilised as a sustainable approach to remediate saline soil and enhance plant growth and production.

6.5 Conclusion

Enhancing the agricultural production in saline agroecosystems by sustainable means is an intimidating task. Soil salinity is one among the abiotic stresses that is responsible for reduced agricultural productivity in agroecosystems around the globe. Physical and chemical means for combating soil salinity are found to be less effective due to their environmental impact, while genetic engineering and plant breeding techniques for developing salt-tolerant crop varieties are also of limited use. On the other hand, application of salt-tolerant PGPR-based biostimulants has proven to be much effective in alleviating the salt stress than other conventional approaches. Application of ST-PGPR may intensify the capacity of plant to withstand such harsh conditions, ultimately increasing productivity of saline agroecosystems. Although there are a large number of ST-PGPR with capability to overcome ill effects of salinity, exploration of their mechanisms of salt-tolerance and interaction with host plant needs to be further investigated. Furthermore, translation of this knowledge for the development of biostimulants could extend their pertinence in restoration of fertility of saline agroecosystems.

References

Abd-Alla MH, Nafady NA, Bashandy SR, Hassan AA (2019) Mitigation of effect of salt stress on the nodulation, nitrogen fixation and growth of chickpea (*Cicer arietinum* L.) by triple microbial inoculation. Rhizosphere 10:100148

Abd-Allah EF, Alqarawi AA, Hashem A, Radhakrishnan R, Al-Huqail AA, Al-Otibi FON, Egamberdieva D (2018) Endophytic bacterium *Bacillus subtilis* (BERA 71) improves salt tolerance in chickpea plants by regulating the plant defense mechanisms. J Plant Interact 13(1):37–44

Abdel Latef AAH, Omer AM, Badawy AA, Osman MS, Ragaey MM (2021) Strategy of salt tolerance and interactive impact of *Azotobacter chroococcum* and/or *Alcaligenes faecalis* inoculation on canola (*Brassica napus* L.) plants grown in saline soil. Plants 10:110

Adhikari A, Khan MA, Lee KE, Kang SM, Dhungana SK, Bhusal N, Lee IJ (2020) The halotolerant Rhizobacterium—*Pseudomonas koreensis* MU2 enhances inorganic silicon and phosphorus use efficiency and augments salt stress tolerance in soybean (*Glycine max* L.). Microorganisms 8(9):1256

Adnan M, Fahad S, Zamin M, Shah S, Mian IA, Danish S et al (2020) Coupling phosphate-solubilizing bacteria with phosphorus supplements improve maize phosphorus acquisition and growth under lime induced salinity stress. Plants J 9(7):900

Afridi MS, Mahmood T, Salam A, Mukhtar T, Mehmood S, Ali J, Chaudhary HJ (2019) Induction of tolerance to salinity in wheat genotypes by plant growth promoting endophytes: involvement of ACC deaminase and antioxidant enzymes. Plant Physiol Biochem 139:569–577

Al Kahtani MD, Attia KA, Hafez YM, Khan N, Eid AM, Ali MA, Abdelaal KA (2020) Chlorophyll fluorescence parameters and antioxidant defense system can display salt tolerance of salt acclimated sweet pepper plants treated with chitosan and plant growth promoting rhizobacteria. Agronomy 10:1180

AlAli HA, Khalifa A, Al-Malki M (2021) Plant growth-promoting rhizobacteria from *Ocimum basilicum* improve growth of *Phaseolus vulgaris* and *Abelmoschus esculentus*. S Afr J Bot 139: 200–209

Alaylar B, Egamberdieva D, Gulluce M, Karadayi M, Arora NK (2020) Integration of molecular tools in microbial phosphate solubilization research in agriculture perspective. World J Microbiol Biotechnol 36(7):93
Arora NK (2013) Plant microbe symbiosis: fundamentals and advances. Springer, New Delhi
Arora NK (2015) Plant microbes symbiosis: applied facets. Springer, New Delhi
Arora NK, Tewari S, Singh S, Lal N, Maheshwari DK (2012) PGPR for protection of plant health under saline conditions. In: Maheshwari DK (ed) Bacteria in agrobiology: stress management. Springer Verlag, Berlin, Heidelberg, pp 239–258
Arora NK, Fatima T, Mishra J, Mishra I, Verma S, Verma R et al (2020) Halo-tolerant plant growth promoting rhizobacteria for improving productivity and remediation of saline soils. J Adv Res 26:69–82
Arora NK, Egamberdieva D, Mehnaz S, Li W-J, Mishra I (2021) Editorial: salt tolerant Rhizobacteria: for better productivity and remediation of saline soils. Front Microbiol 12: 660075
Ashfaq M, Hassan HM, Ghazali AHA, Ahmad M (2020) Halotolerant potassium solubilizing plant growth promoting rhizobacteria may improve potassium availability under saline conditions. Environ Monit Assess 192(11):1–20
Ayuso-Calles M, García-Estévez I, Jiménez-Gómez A, Flores-Félix JD, Escribano-Bailón MT, Rivas R (2020) *Rhizobium laguerreae* improves productivity and phenolic compound content of lettuce (*Lactuca sativa* L.) under saline stress conditions. Foods 9:1166
Ayyam V, Palanivel S, Chandrakasan S (2019) Coastal ecosystems of the tropics-adaptive management. Springer, Singapore
Azarmi F, Mozaffari V, Hamidpour M, Abbaszadeh-Dahaji P (2016) Interactive effect of fluorescent pseudomonads rhizobacteria and Zn on the growth, chemical composition, and water relations of pistachio (*Pistacia vera* L.) seedlings under NaCl stress. Commun Soil Sci Plant Anal 47(8):955–972
Baldani JI, Baldani VLD, Seldin L, Döbereiner J (1986) Characterization of *Herbaspirillum seropedicae* gen. Nov., sp. nov., a root-associated nitrogen-fixing bacterium. Int J Syst Evol 36(1):86–93
Barnawal D, Bharti N, Pandey SS, Pandey A, Chanotiya CS, Kalra A (2017) Plant growth-promoting rhizobacteria enhance wheat salt and drought stress tolerance by altering endogenous phytohormone levels and TaCTR1/TaDREB2 expression. Physiol Plant 161(4):502–514
Bertrand A, Gatzke C, Bipfubusa M, Lévesque V, Chalifour FP, Claessens A et al (2020) Physiological and biochemical responses to salt stress of alfalfa populations selected for salinity tolerance and grown in symbiosis with salt-tolerant *rhizobium*. Agron J 10(4):569
Bharti N, Pandey SS, Barnawal D, Patel VK, Kalra A (2016) Plant growth promoting rhizobacteria *Dietzia natronolimnaea* modulates the expression of stress responsive genes providing protection of wheat from salinity stress. Sci Rep 6:1–16
Bhattacharyya PN, Jha DK (2012) Plant growth-promoting rhizobacteria (PGPR): emergence in agriculture. World J Microbiol Biotechnol 28(4):1327–1350
Bhattacharyya D, Lee YH (2017) A cocktail of volatile compounds emitted from Alcaligenes faecalis JBCS1294 induces salt tolerance in *Arabidopsis thaliana* by modulating hormonal pathways and ion transporters. J Plant Physiol 214:64–73
Botella MA, Martinez V, Pardines J, Cerda A (1997) Salinity induced potassium deficiency in maize plants. J Plant Physiol 150(1–2):200–205
Cavalcante VA, Dobereiner J (1988) A new acid-tolerant nitrogen-fixing bacterium associated with sugarcane. Plant Soil 108(1):23–31
Cemek B, Güler M, Kiliç K, Demir Y, Arslan H (2007) Assessment of spatial variability in some soil properties as related to soil salinity and alkalinity in Bafra plain in northern Turkey. Environ Monit Assess 124(1):223–234
Chaudhary DR, Rathore AP, Sharma S (2020) Effect of halotolerant plant growth promoting rhizobacteria inoculation on soil microbial community structure and nutrients. Appl Soil Ecol 150:103461

Chen Y, Banin A (1975) Scanning electron microscope (SEM) observations of soil structure changes induced by sodium calcium exchange in relation to hydraulic conductivity. Soil Sci Soc Am J 120:428–436

Chu TN, Tran BTH, Hoang MTT (2019) Plant growth-promoting rhizobacterium *Pseudomonas* PS01 induces salt tolerance in *Arabidopsis thaliana*. BMC Res Notes 12(1):1–7

Coleman-Derr D, Tringe SG (2014) Building the crops of tomorrow: advantages of symbiont-based approaches to improving abiotic stress tolerance. Front Microbiol 5:283

Deshwal VK, Kumar P (2013) Effect of salinity on growth and PGPR activity of pseudomonads. J Acad Ind Res 2(6):353–356

Di Benedetto NA, Corbo MR, Campaniello D, Cataldi MP, Bevilacqua A, Sinigaglia M et al (2017) The role of plant growth promoting bacteria in improving nitrogen use efficiency for sustainable crop production: a focus on wheat. AIMS Microbiol 3(3):413

Din BU, Sarfraz S, Xia Y, Kamran MA, Javed MT, Sultan T et al (2019) Mechanistic elucidation of germination potential and growth of wheat inoculated with exopolysaccharide and ACC-deaminase producing *Bacillus* strains under induced salinity stress. Ecotoxicol Environ Saf 183:109466

Döbereiner J (1992) Recent changes in concepts of plant bacteria interactions: endophytic N2 fixing bacteria. Ciênc Cult 44(5):310–313

Egamberdieva D, Berg G, Lindström K, Räsänen LA (2013) Alleviation of salt stress of symbiotic *Galega officinalis* L.(goat's rue) by co-inoculation of *rhizobium* with root-colonizing *Pseudomonas*. Plant Soil 369(1):453–465

Egamberdieva D, Jabborova D, Wirth SJ, Alam P, Alyemeni MN, Ahmad P (2018) Interactive effects of nutrients and *Bradyrhizobium japonicum* on the growth and root architecture of soybean (*Glycine max* L.). Front Microbiol 9:1000

Egamberdieva D, Wirth S, Bellingrath-Kimura SD, Mishra J, Arora NK (2019) Salt-tolerant plant growth promoting Rhizobacteria for enhancing crop productivity of saline soils. Front Microbiol 10:2791

El-Akhdar I, Elsakhawy T, Abo-Koura HA (2020) Alleviation of salt stress on wheat *(Triticum aestivum* L.*)* by plant growth promoting Bacteria strains *Bacillus* halotolerans MSR-H4 and *Lelliottia amnigena* MSR-M49. J Adv Microbiol:44–58

El-Esawi MA, Alaraidh IA, Alsahli AA, Alamri SA, Ali HM, Alayafi AA (2018a) *Bacillus firmus* (SW5) augments salt tolerance in soybean *(Glycine max* L.*)* by modulating root system architecture, antioxidant defense systems and stress-responsive genes expression. Plant Physiol Biochem 132:375–384

El-Esawi MA, Alaraidh IA, Alsahli AA, Alzahrani SM, Ali HM, Alayafi AA, Ahmad M (2018b) *Serratia liquefaciens* KM4 improves salt stress tolerance in maize by regulating redox potential, ion homeostasis, leaf gas exchange and stress-related gene expression. Int J Mol Sci 19(11): 3310

El-Esawi MA, Al-Ghamdi AA, Ali HM, Alayafi AA (2019) *Azospirillum lipoferum* FK1 confers improved salt tolerance in chickpea *(Cicer arietinum* L.*)* by modulating osmolytes, antioxidant machinery and stress-related genes expression. Environ Exp Bot 159:55–65

El-Ghany A, Mona F, Attia M (2020) Effect of exopolysaccharide-producing bacteria and melatonin on faba bean production in saline and non-saline soil. Agron 10(3):316

El-Nahrawy S, Yassin M (2020) Response of different cultivars of wheat plants *(Triticum aestivum* L.*)* to inoculation by *Azotobacter sp.* under salinity stress conditions. J Adv Microbiol 20:59–79

Etesami H, Alikhani HA (2019) Halotolerant plant growth-promoting Fungi and Bacteria as an alternative strategy for improving nutrient availability to salinity-stressed crop plants. In: Kumar M, Etesami H, Kumar V (eds) Saline soil-based agriculture by halotolerant microorganisms. Springer, Singapore, pp 103–146

Etesami H, Beattie GA (2018) Mining halophytes for plant growth-promoting halotolerant bacteria to enhance the salinity tolerance of non-halophytic crops. Front Microbiol 9:148

Etesami H, Glick BR (2020) Halotolerant plant growth–promoting bacteria: prospects for alleviating salinity stress in plants. Environ Exp Bot 178:104124

Ezeaku PI, Ene J, Shehu JA (2015) Application of different reclamation methods on salt affected soils for crop production. J Exp Agric Int:1–11

Fageria NK, Baligar VC, Li YC (2008) The role of nutrient efficient plants in improving crop yields in the twenty first century. J Plant Nutr 31(6):1121–1157

Fahad S, Hussain S, Matloob A, Khan FA, Khaliq A, Saud S et al (2015) Phytohormones and plant responses to salinity stress: a review. Plant Growth Regul 75(2):391–404

Fallah F, Nokhasi F, Ghaheri M, Kahrizi D, Beheshti Ale AA, Ghorbani T et al (2017) Effect of salinity on gene expression, morphological and biochemical characteristics of *Stevia rebaudiana* Bertoni under in vitro conditions. Cell Mol Biol 63(7):102–106

FAO (2017) Global network on integrated soil management for sustainable use of salt affected soils. Natural Resources and Environment, FAO, Rome. www.fao.org/ag/AGL/agII/spush/intro.htm

Fasciglione G, Casanovas EM, Quillehauquy V, Yommi AK, Goñi MG, Roura SI, Barassi CA (2015) *Azospirillum* inoculation effects on growth, product quality and storage life of lettuce plants grown under salt stress. Sci Hortic 195:154–162

Fatima T, Arora NK (2021) *Pseudomonas entomophila* PE3 and its exopolysaccharides as biostimulants for enhancing growth, yield and tolerance responses of sunflower under saline conditions. Microbiol Res 244: 126671

Fatima T, Mishra I, Verma R, Arora NK (2020) Mechanisms of halotolerant plant growth promoting *Alcaligenes* sp. involved in salt tolerance and enhancement of the growth of rice under salinity stress. 3 Biotech 10:361

Fazal A, Bano A (2016) Role of plant growth-promoting rhizobacteria (pgpr), biochar, and chemical fertilizer under salinity stress. Commun Soil Sci Plant Anal 47(17):1985–1993

Fernandes F, Arrabaca M, Carvalho L (2004) Sucrose metabolism in *Lupinus albus* L. under salt stress. Biol Plantarum 48:317

Ferreira MJ, Silva H, Cunha A (2019) Siderophore-producing rhizobacteria as a promising tool for empowering plants to cope with iron limitation in saline soils: a review. Pedosphere 29(4): 409–420

Fukami J, Cerezini P, Hungria M (2018a) *Azospirillum*: benefits that go far beyond biological nitrogen fixation. AMB Express 8(1):1–12

Fukami J, de la Osa C, Ollero FJ, Megías M, Hungria M (2018b) Co-inoculation of maize with *Azospirillum brasilense* and *rhizobium tropici* as a strategy to mitigate salinity stress. Funct Plant Biol 45(3):328–339

Ghorai S, Pal KK, Dey R (2015) Alleviation of salinity stress in groundnut by application of PGPR. Int J Res Eng Technol 2:742–750

Glick BR, Penrose DM, Li J (1998) A model for the lowering of plant ethylene concentrations by plant growth-promoting bacteria. J Theor Biol 190(1):63–68

Global Agricultural Productivity Report (2019) Productivity growth for sustainable diets and more. https://globalagriculturalproductivity.org/2019-gap-report/

Goharrizi KJ, Riahi-Madvar A, Rezaee F, Pakzad R, Bonyad FJ, Ahsaei MG (2019) Effect of salinity stress on enzymes' activity, ions concentration, oxidative stress parameters, biochemical traits, content of Sulforaphane, and CYP79F1 gene expression level in *Lepidium draba* plant. J Plant Growth Regul:1–20

Gong DH, Wang GZ, Si WT, Zhou Y, Liu Z, Jia J (2018) Effects of salt stress on photosynthetic pigments and activity of ribulose-1, 5-bisphosphate carboxylase/oxygenase in *Kalidium foliatum*. Russ J Plant Physiol 65(1):98–103

Goswami D, Dhandhukia P, Patel P, Thakker JN (2014) Screening of PGPR from saline desert of Kutch: growth promotion in *Arachis hypogea* by *Bacillus licheniformis* A2. Microbiol Res 169(1):66–75

Grieve CM, Grattan SR, Maas EV (2012) Plant salt tolerance. In: ASCE Manual and Reports on Engineering Practice. ASCE, Reston, pp 405–459

Grobelak A, Kokot P, Hutchison D, Grosser A, Kacprzak M (2018) Plant growth-promoting rhizobacteria as an alternative to mineral fertilizers in assisted bioremediation-sustainable land and waste management. J Environ Manag 227:1–9

Gupta S, Pandey S (2019) ACC deaminase producing bacteria with multifarious plant growth promoting traits alleviates salinity stress in French bean (*Phaseolus vulgaris*) plants. Front Microbiol 10:1506

Habib SH, Kausar H, Saud HM (2016) Plant growth-promoting rhizobacteria enhance salinity stress tolerance in okra through ROS-scavenging enzymes. BioMed Res. Int 2016:6284547

Hanay A, Büyüksönmez F, Kiziloglu FM, Canbolat MY (2004) Reclamation of saline-sodic soils with gypsum and MSW compost. Compost sci util 12(2):175–179

Hara M, Furukawa J, Sato A, Mizoguchi T, Miura K (2012) Abiotic stress and role of salicylic acid in plants. In: Ahmad P, Prasad MNV (eds) Abiotic stress responses in plants: metabolism productivity and sustainability. Springer Verlag, New York, pp 235–251

Hasbullah H, Marschner P (2014) Residue properties influence the impact of salinity on soil respiration. Biol Fertil Soils 51:99–111

Hu Y, Schmidhalter U (2005) Drought and salinity: a comparison of their effects on mineral nutrition of plants. J Plant Nutr Soil Sci 168(4):541–549

Hussain MB, Mehboob I, Zahir ZA, Naveed M, Asghar HN (2009) Potential of *rhizobium* spp. for improving growth and yield of rice (*Oryza sativa* L.). Soil Environ 28(1):49–55

Imron MF, Kurniawan SB, Abdullah SR (2021) Resistance of bacteria isolated from leachate to heavy metals and the removal of hg by *Pseudomonas aeruginosa* strain FZ-2 at different salinity levels in a batch biosorption system. Sust Environ Res 31(1):1–3

Iqbal M, Ashraf M (2013) Gibberellic acid mediated induction of salt tolerance in wheat plants: growth, ionic partitioning, photosynthesis, yield and hormonal homeostasis. Environ Exp Bot 86:76–85

Isayenkov SV, Maathuis FJ (2019) Plant salinity stress: many unanswered questions remain. Front Plant Sci 10:80

Jafarzadeh AA, Aliasgharzad N (2007) Salinity and salt composition effects on seed germination and root length of four sugar beet cultivars. Biologia 62(5):562–564

Jha Y, Subramanian RB (2016) Regulation of plant physiology and antioxidant enzymes for alleviating salinity stress by potassium-mobilizing Bacteria. In: Meena V, Maurya B, Verma J, Meena R (eds) Potassium solubilizing microorganisms for sustainable agriculture. Springer, New Delhi, pp 149–162

Ji J, Yuan D, Jin C, Wang G, Li X, Guan C (2020) Enhancement of growth and salt tolerance of rice seedlings *(Oryza sativa* L.*)* by regulating ethylene production with a novel halotolerant PGPR strain *Glutamicibacter sp.* YD01 containing ACC deaminase activity. Acta Physiol Plant 42(4): 1–17

Jiang H, Qi P, Wang T, Chi X, Wang M, Chen M, Pan L (2019) Role of halotolerant phosphate-solubilising bacteria on growth promotion of peanut (*Arachis hypogaea* L.*)* under saline soil. Ann Appl Biol 174(1):20–30

Jiménez-Gómez A, García-Estévez I, García-Fraile P, Escribano-Bailón MT, Rivas R (2020) Increase in phenolic compounds of Coriandrum sativum L. after the application of a Bacillus halotolerans biofertilizer. J Sci Food Agric 100:2742–2749

Kang SM, Shahzad R, Bilal S, Khan AL, Park YG, Lee KE et al (2019) Indole-3-acetic-acid and ACC deaminase producing *Leclerciaade carboxylata* MO1 improves *Solanum lycopersicum* L. growth and salinity stress tolerance by endogenous secondary metabolites regulation. BMC Microbiol 19(1):1–14

Kapoor D, Singh S, Kumar V, Romero R, Prasad R, Singh J (2019) Antioxidant enzymes regulation in plants in reference to reactive oxygen species (ROS) and reactive nitrogen species (RNS). Plant Gene 19:100182

Khan MA, Abdullah Z (2003) Salinity–sodicity induced changes in reproductive physiology of rice (*Oryza sativa*) under dense soil conditions. Environ Exp Bot 49(2):145–157

Khan MA, Weber DJ (2008) Ecophysiology of high salinity tolerant plants: tasks for vegetation science, 1st edn. Springer, Dordrecht

Khan MM, Al-Mas'oudi RS, Al-Said F, Khan I (2013a) Salinity effects on growth, electrolyte leakage, chlorophyll content and lipid peroxidation in cucumber (*Cucumis sativus* L.). In: International conference on food and agricultural sciences. IACSIT Press, Malaysia, pp 28–32

Khan MS, Ahmad E, Zaidi A, Oves M (2013b) Functional aspect of phosphate-solubilizing bacteria: importance in crop production. In: Maheshwari D, Saraf M, Aeron A (eds) Bacteria in agrobiology: crop productivity springer. Heidelberg, Berlin, pp 237–263

Khare E, Mishra J, Arora NK (2018) Multifaceted interactions between endophytes and plant: developments and prospects. Front Microbiol 9:2732

Kim Y, Kim S, Shim IS (2017) Exogenous salicylic acid alleviates salt-stress damage in cucumber under moderate nitrogen conditions by controlling endogenous salicylic acid levels. Hortic Environ Biotechnol 58(3):247–253

Kumar R, Pandey S, Pandey A (2006) Plant roots and carbon sequestration. Curr Sci 91:885–890

Kumar M, Sharma S, Gupta S, Kumar V (2018) Mitigation of abiotic stresses in *Lycopersicon esculentum* by endophytic bacteria. Environ Sust 1(1):71–80

Kumawat KC, Sharma P, Nagpal S, Gupta RK, Sirari A, Nair RM, Singh S (2020) Dual microbial inoculation, a game changer?—bacterial biostimulants with multifunctional growth promoting traits to mitigate salinity stress in spring Mungbean. Front Microbiol 11:600576

Lastochkina O, Pusenkova L, Yuldashev R, Babaev M, Garipova S, Blagova DY et al (2017) Effects of *Bacillus subtilis* on some physiological and biochemical parameters of *Triticum aestivum* L. (wheat) under salinity. Plant Physiol Biochem 121:80–88

Latef AAHA, Alhmad MFA, Kordrostami M, Abo-Baker ABAE, Zakir A (2020) Inoculation with *Azospirillum lipoferum* or *Azotobacter chroococcum* reinforces maize growth by improving physiological activities under saline conditions. J Plant Growth Regul 39(3):1293–1306

Ledger T, Rojas S, Timmermann T, Pinedo I, Poupin MJ, Garrido T, Donoso R (2016) Volatile-mediated effects predominate in *Paraburkholderia phytofirmans* growth promotion and salt stress tolerance of *Arabidopsis thaliana*. Front Microbiol 7:18–38

Li H, Lei P, Pang X, Li S, Xu H, Xu Z, Feng X (2017) Enhanced tolerance to salt stress in canola (*Brassica napus* L.) seedlings inoculated with the halotolerant *Enterobacter cloacae* HSNJ4. Appl Soil Ecol 119:26–34

Li X, Sun P, Zhang Y, Jin C, Guan C (2020) A novel PGPR strain Kocuria rhizophila Y1 enhances salt stress tolerance in maize by regulating phytohormone levels nutrient acquisition, redox potential, ion homeostasis, photosynthetic capacity and stress-responsive genes expression. Environ Exp Botany 174:104023

Mahajan S, Tuteja N (2005) Cold, salinity and drought stresses: an overview. Arch Biochem Biophys 444(2):139–158

Mahdi I, Hafidi M, Allaoui A, Biskri L (2021) Halotolerant endophytic bacterium *Serratia rubidaea* ED1 enhances phosphate solubilization and promotes seed germination. Agriculture 11(3):224

Mallick I, Bhattacharyya C, Mukherji S, Dey D, Sarkar SC, Mukhopadhyay UK et al (2018) Effective rhizoinoculation and biofilm formation by arsenic immobilizing halophilic plant growth promoting bacteria (PGPB) isolated from mangrove rhizosphere: a step towards arsenic rhizoremediation. Sci Total Environ 610:1239–1250

Maougal RT, Brauman A, Plassard C, Abadie J, Djekoun A, Drevon JJ (2014) Bacterial capacities to mineralize phytate increase in the rhizosphere of nodulated common bean (*Phaseolus vulgaris*) under P deficiency. Eur J Soil Biol 62:8–14

Mayak S, Tirosh T, Glick BR (2004) Plant growth-promoting bacteria that confer resistance in tomato and pepper to salt stress. Plant Physiol Biochem 167:650–656

Mishra J, Fatima T, Arora NK (2018) Role of secondary metabolites from plant growth-promoting rhizobacteria in combating salinity stress. In: Plant microbiome: stress response. Springer, Berlin, pp 127–163

Misra S, Chauhan PS (2020) ACC deaminase-producing rhizosphere competent *Bacillus* spp. mitigate salt stress and promote *Zea mays* growth by modulating ethylene metabolism. 3 Biotech 10(3):1–14

Moradi S, Rasouli-Sadaghiani M, Sepehr E, Khodaverdiloo H, Barin M (2019) The role of organic carbon in the mineralization of nitrogen, carbon and some of nutrient concentrations in soil salinity conditions. J Soil Manage Sustain Product 9(3):153–169

Mudgal V, Madaan N, Mudgal A (2010) Biochemical mechanisms of salt tolerance in plants: a review. Int J Botany 6:136–143

Mukhtar S, Zareen M, Khaliq Z, Mehnaz S, Malik KA (2020) Phylogenetic analysis of halophyte-associated rhizobacteria and effect of halotolerant and halophilic phosphate-solubilizing biofertilizers on maize growth under salinity stress conditions. J Appl Microbiol 128(2): 556–573

Nadeem SM, Ahmad M, Naveed M, Imran M, Zahir ZA, Crowley DE (2016) Relationship between in vitro characterization and comparative efficacy of plant growth-promoting rhizobacteria for improving cucumber salt tolerance. Arch Microbiol 198(4):379–387

Naveen B, Sumalatha J, Malik R (2018) A study on contamination of ground and surface water bodies by leachate leakage from a landfill in Bangalore, India. International Journal of Geo-Engineering 9:1–20

Nawaz A, Shahbaz M, Asadullah AI, Marghoob MU, Imtiaz M, Mubeen F (2020) Potential of salt tolerant PGPR in growth and yield augmentation of wheat *(Triticum aestivum* L.*)* under saline conditions. Front Microbiol 11:2019

Naz I, Bano A, Ul-Hassan T (2009) Isolation of phytohormones producing plant growth promoting rhizobacteria from weeds growing in Khewra salt range, Pakistan and their implication in providing salt tolerance to *Glycine max* L. Afr J Biotechnol 8(21)

Nozari RM, Ortolan F, Astarita LV, Santarém ER (2021) *Streptomyces* spp. enhance vegetative growth of maize plants under saline stress. Braz J Microbiol 52:1–13

Numan M, Bashir S, Khan Y, Mumtaz R, Shinwari ZK, Khan AL et al (2018) Plant growth promoting bacteria as an alternative strategy for salt tolerance in plants: a review. Microbiol Res 209:21–32

Osman KT (2018) Saline and sodic soils. In: Management of soil problems. Springer, Berlin, pp 255–298

Oster J (1982) Gypsum usage in irrigated agriculture: a review. Fertilizer research 3:73–89

Oster JD, Shainberg I (2001) Soil responses to sodicity and salinity: challenges and opportunities. Soil Res 39(6):1219–1224

Otaiku AA, Mmom PC, Ano AO (2019) Biofertilizer impacts on cassava (*Manihot Esculenta Crantz*) cultivation: improved soil health and quality. Igbariam, Nigeria

Palacio-Rodríguez R, Coria-Arellano JL, López-Bucio J, Sánchez-Salas J, Muro-Pérez G, Castañeda-Gaytán G, Sáenz-Mata J (2017) Halophilic rhizobacteria from *Distichlis spicata* promote growth and improve salt tolerance in heterologous plant hosts. Symbiosis 73(3): 179–189

Parihar P, Singh S, Singh R, Singh VP, Prasad SM (2015) Effect of salinity stress on plants and its tolerance strategies: a review. Environ Sci Pollut Res 22(6):4056–4075

Pearson KE, Bauder JW (2006) The basics of salinity and sodicity effects on soil physical properties. MSU Ext Water Qual:1–11

Phour M, Sindhu SS (2020) Amelioration of salinity stress and growth stimulation of mustard *(Brassica juncea* L.*)* by salt-tolerant Pseudomonas species. Appl Soil Ecol 149:103518

Pirhadi M, Enayatizamir N, Motamedi H, Sorkheh K (2016) Screening of salt tolerant sugarcane endophytic bacteria with potassium and zinc for their solubilizing and antifungal activity. Biosci Biotechnol Res Commun 9(3):530–538

Prakash J, Arora NK (2021) Novel metabolites from *Bacillus safensis* and their antifungal property against *Alternaria alternata*. Antonie Van Leeuwenhoek. https://doi.org/10.1007/s10482-021-01598-4

Pushpavalli R, Quealy J, Colmer TD, Turner NC, Siddique KH, Rao MV et al (2016) Salt stress delayed flowering and reduced reproductive success of chickpea (*Cicer arietinum* L.), a response associated with Na+ accumulation in leaves. J Agron Crop Sci 202(2):125–138

Ranawat B, Mishra S, Singh A (2021) *Enterobacter hormaechei* (MF957335) enhanced yield, disease and salinity tolerance in tomato. Arch Microbiol 203(5):2659–2667

Rath KM, Murphy DN, Rousk J (2019) The microbial community size, structure, and process rates along natural gradients of soil salinity. Soil Biol Biochem 138:107607

Raval VH, Saraf M (2020) Biosynthesis and purification of indole-3-acetic acid by halotolerant rhizobacteria isolated from little Runn of Kachchh. Biocatal Agric Biotechnol 23:101435

Rojas-Solis D, Vences-Guzmán MÁ, Sohlenkamp C, Santoyo G (2020) Antifungal and plant growth–promoting bacillus under saline stress modify their membrane composition. J Soil Sci Plant Nutr 20(3):1549–1559

Sagar A, Sayyed RZ, Ramteke PW, Sharma S, Marraiki N, Elgorban AM et al (2020) ACC deaminase and antioxidant enzymes producing halophilic *Enterobacter* sp. PR14 promotes the growth of rice and millets under salinity stress. Physiol Mol Biol Plants 26(9):1847–1854

Saghafi D, Ghorbanpour M, Lajayer BA (2018) Efficiency of rhizobium strains as plant growth promoting rhizobacteria on morpho-physiological properties of *Brassica napus* L. under salinity stress. J Soil Sci Plant Nutr 18(1):253–268

Saharan BS, Nehra V (2011) Plant growth promoting rhizobacteria: a critical review. Life Sci Med Res 21(1):30

Santos AA, Silveira JAG, Bonifacio A, Rodrigues AC, Figueiredo MVB (2018) Antioxidant response of cowpea co-inoculated with plant growth-promoting bacteria under salt stress. Braz J Microbiol 49:513–521

Sapre S, Gontia-Mishra I, Tiwari S (2018) *Klebsiella* sp. confers enhanced tolerance to salinity and plant growth promotion in oat seedlings (*Avena sativa*). Microbiol Res 206:25–32

Sarkar A, Ghosh PK, Pramanik K, Mitra S, Soren T, Pandey S et al (2018a) A halotolerant *Enterobacter* sp. displaying ACC deaminase activity promotes rice seedling growth under salt stress. Res Microbiol 169(1):20–32

Sarkar A, Pramanik K, Mitra S, Soren T, Maiti TK (2018b) Enhancement of growth and salt tolerance of rice seedlings by ACC deaminase-producing *Burkholderia sp.* MTCC 12259. J Plant Physiol 231:434–442

Shahzad R, Waqas M, Khan AL, Asaf S, Khan MA, Kang SM et al (2016) Seed-borne endophytic *Bacillus amyloliquefaciens* RWL-1 produces gibberellins and regulates endogenous phytohormones of *Oryza sativa*. Plant Physiol Biochem 106:236–243

Shahzad R, Khan AL, Bilal S, Waqas M, Kang SM, Lee IJ (2017) Inoculation of abscisic acid-producing endophytic bacteria enhances salinity stress tolerance in *Oryza sativa*. Environ Exp Bot 136:68–77

Sharif I, Aleem S, Farooq J, Rizwan M, Younas A, Sarwar G et al (2019) Salinity stress in cotton: effects, mechanism of tolerance and its management strategies. Physiol Mol Biol Plants:1–14

Shavalikohshori O, Zalaghi R, Sorkheh K, Enaytizamir N (2020) The expression of proline production/degradation genes under salinity and cadmium stresses in *Triticum aestivum* inoculated with *Pseudomonas* sp. Int J Environ Sci Technol 17(4):2233–2242

Singh RP, Jha PN (2016) The multifarious PGPR Serratia marcescens CDP-13 augments induced systemic resistance and enhanced salinity tolerance of wheat (*Triticum aestivum* L.). PLoS One 11:e0155026

Sivasakthi S, Usharani G, Saranraj P (2014) Biocontrol potentiality of plant growth promoting bacteria (PGPR)-*Pseudomonas fluorescens* and *Bacillus subtilis*: a review. AfrJ Agric Res 9(16):1265–1277

Srivastava S, Srivastava S (2020) Prescience of endogenous regulation in *Arabidopsis thaliana* by *Pseudomonas putida* MTCC 5279 under phosphate starved salinity stress condition. Sci Rep 10(1):1–15

Suarez C, Cardinale M, Ratering S, Steffens D, Jung S, Montoya AMZ, Schnell S (2015) Plant growth-promoting effects of *Hartmannibacter diazotrophicus* on summer barley (*Hordeum vulgare* L.) under salt stress. Appl Soil Ecol 95:23–30

Sultana S, Paul SC, Parveen S, Alam S, Rahman N, Jannat B, Karim MM (2020) Isolation and identification of salt-tolerant plant-growth-promoting rhizobacteria and their application for rice cultivation under salt stress. Can J Microbiol 66(2):144–160

Sultana S, Alam S, Karim MM (2021) Screening of siderophore-producing salt-tolerant rhizobacteria suitable for supporting plant growth in saline soils with iron limitation. J Food Agric Environ 4:100150

Szymańska S, Dąbrowska GB, Tyburski J, Niedojadło K, Piernik A, Hrynkiewicz K (2019) Boosting the Brassica napus L. tolerance to salinity by the halotolerant strain *Pseudomonas stutzeri* ISE12. Environ Exp Bot 163:55–68

Taj Z, Challabathula D (2020) Protection of photosynthesis by halotolerant *Staphylococcus sciuri* ET101 in tomato *(Lycopersicon esculentum)* and rice (*Oryza sativa* L.) plants during salinity stress: possible interplay between carboxylation and oxygenation in stress mitigation. Front Microbiol 11:547750

Tewari S, Arora NK (2014) Talc based exopolysaccharides formulation enhancing growth and production of *Helianthus annuus* under saline conditions. Cell Mol Biol 60(5):73–81

Tewari S, Arora NK (2016) *Fluorescent Pseudomonas* sp. PF17 as an efficient plant growth regulator and biocontrol agent for sunflower crop under saline conditions. Symbiosis 68(1–3): 99–108

Tewari S, Arora NK (2018) Role of salicylic acid from *Pseudomonas aeruginosa* PF23 EPS+ in growth promotion of sunflower in saline soils infested with phytopathogen *Macrophomina phaseolina*. Environ Sustain 1(1):49–59

Tirry N, Kouchou A, Laghmari G, Lemjereb M, Hnadi H, Amrani K et al (2021) Improved salinity tolerance of *Medicago sativa* and soil enzyme activities by PGPR. Biocatal Agric Biotechnol 31:101914

Tunçtürk M, Tunçtürk R, Yildirim B, Çiftçi V (2011) Changes of micronutrients, dry weight and plant development in canola (*Brassica napus* L.) cultivars under salt stress. Afr J Biotechnol 10(19):3726–3730

Ul Hassan T, Bano A (2019) Construction of IAA-deficient mutants of *Pseudomonas moraviensis* and their comparative effects with wild type strains as bio-inoculant on wheat in saline sodic soil. Geomicrobiol J 36(4):376–384

Upadhyay SK, Singh JS, Saxena AK, Singh DP (2012) Impact of PGPR inoculation on growth and antioxidant status of wheat under saline conditions. Plant Biol 14(4):605–611

Upadhyay SK, Saxena AK, Singh JS, Singh DP (2019) Impact of native ST-PGPR (*bacillus pumilus*; EU927414) on PGP traits, antioxidants activities, wheat plant growth and yield under salinity. Clim Change Environ Sustain 7(2):157–168

Vaishnav A, Varma A, Tuteja N, Choudhary DK (2016) PGPR-mediated amelioration of crops under salt stress. In: Choudhary D, Varma A, Tuteja N (eds) Plant-microbe interaction: an approach to sustainable agriculture. Springer, Singapore, pp 205–226

Vaishnav A, Singh J, Singh P, Rajput RS, Singh HB, Sarma BK (2020) *Sphingobacterium* sp. BHU-AV3 induces salt tolerance in tomato by enhancing antioxidant activities and energy metabolism. Front Microbiol 11:443

Vanissa TTG, Berger B, Patz S, Becker M, Turečková V, Novák O et al (2020) The response of maize to inoculation with *Arthrobacter* sp. and *Bacillus* sp. in phosphorus-deficient, salinity-affected soil. Microorganisms 8(7):1005

Verma M, Singh A, Dwivedi DH, Arora NK (2020) Zinc and phosphate solubilizing *rhizobium radiobacter* (LB2) for enhancing quality and yield of looseleaf lettuce in saline soil. Environ Sustain 3:209–218

Walpola BC, Arunakumara KK (2011) Carbon and nitrogen mineralization of a plant residue amended soil: the effect of salinity stress. Bangladesh J Sci Ind Res 46(4):565–572

Wang Y, Mopper S, Hasenstein KH (2001) Effects of salinity on endogenous ABA, IAA, JA, and SA in *Iris hexagona*. J Chem Ecol 27(2):327–342

Wang W, Wu Z, He Y, Huang Y, Li X, Ye BC (2018) Plant growth promotion and alleviation of salinity stress in *Capsicum annuum* L. by *Bacillus* isolated from saline soil in Xinjiang. Ecotoxicol Environ Saf 164:520–529

Warrence NJ, Bauder JW, Pearson KE (2002) Basics of salinity and sodicity effects on soil physical properties. Department of Land Resources and Environmental Sciences, Montana State University-Bozeman, MT 129

Wu T, Xu J, Liu J, Guo WH, Li XB, Xia JB et al (2019) Characterization and initial application of endophytic *Bacillus safensis* strain ZY16 for improving phytoremediation of oil-contaminated saline soils. Front Microbiol 10:991

Xiong YW, Li XW, Wag TT, Gong Y, Zhang CM, Xing K, Qin S (2020) Root exudates-driven rhizosphere recruitment of the plant growth-promoting rhizobacterium *Bacillus flexus* KLBMP 4941 and its growth-promoting effect on the coastal halophyte *Limonium sinense* under salt stress. Ecotoxicol Environ Saf 194:110374

Yang A, Akhtar SS, Fu Q, Naveed M, Iqbal S, Roitsch T et al. (2020) *Burkholderia phytofirmans* PsJN stimulate growth and yield of quinoa under salinity stress. Plan Theory 9(6):672

Yasmin H, Naeem S, Bakhtawar M, Jabeen Z, Nosheen A, Naz R, Hassan MN (2020) Halotolerant rhizobacteria *Pseudomonas pseudoalcaligenes* and *Bacillus subtilis* mediate systemic tolerance in hydroponically grown soybean (*Glycine max* L.) against salinity stress. PLoS One 15(4): e0231348

Yoolong S, Kruasuwan W, Phạm HTT, Jaemsaeng R, Jantasuriyarat C, Thamchaipenet A (2019) Modulation of salt tolerance in Thai jasmine rice (*Oryza sativa* L. cv. KDML105) by *Streptomyces venezuelae* ATCC 10712 expressing ACC deaminase. Sci Rep 9(1):1–10

Young JPW, Haukka KE (1996) Diversity and phylogeny of rhizobia. New Phytol 133(1):87–94

Yousefi S, Kartoolinejad D, Bahmani M, Naghdi R (2017) Effect of *Azospirillum lipoferum* and *Azotobacter chroococcum* on germination and early growth of hopbush shrub (*Dodonaea viscosa* L.) under salinity stress. J Sustain Forestry 36(2):107–120

Zai XM, Fan JJ, Hao ZP, Liu XM, Zhang WX (2021) Effect of co-inoculation with arbuscular mycorrhizal fungi and phosphate solubilizing fungi on nutrient uptake and photosynthesis of beach palm under salt stress environment. Sci Rep 11(1):1–11

Zhang Z, Li H, Qiao S, Zhang X, Liu P, Liu X (2011) Effect of salinity on seed germination, seedling growth, and physiological characteristics of *Perilla frutescens*. Plant Biosyst 146:1–7

Zhang L, Zhang G, Wang Y, Zhou Z, Meng Y, Chen B (2013) Effect of soil salinity on physiological characteristics of functional leaves of cotton plants. J Plant Res 126:293–304

Zhang J, Yang H, Wang J, Tian D, Li Y, He N et al (2019) Soil and climate determine differential responses of soil respiration to nitrogen and acid deposition along a forest transect. Eur J Soil Biol 93:103097

Zhang G, Bai J, Tebbe CC, Zhao Q, Jia J, Wang W (2021) Salinity controls soil microbial community structure and function in coastal estuarine wetlands. Environ Microbiol 23(2): 1020–1037

Zhu JK (2002) Salt and drought stress signal transduction in plants. Annu Rev Plant Biol 53(1): 247–273

Zhu J-K (2016) Abiotic stress signaling and responses in plants. Cell 167:313–324

Signalling of Rhizosphere Microbiomes: Benign and Malign Borders

7

Subhoshmita Mondal and Sibashish Baksi

Abstract

In the course of evolution, symbiotic relationship is the most prominent and practical depiction of the term 'survival of the fittest'. Symbiosis is a long-term biological interaction between two distinct organisms and can be classified as mutualistic, parasitic and commensalism. Plant-microbiome and microbe-microbe interactions are the two most crucial symbiotic relationships in the biosphere that play an important regulatory role in crop production, plant development and growth, resistance against plant pathogens, value-added metabolites production, etc. The rhizosphere is the hotspot of such symbiotic interactions where soil microbes influence the physiological and biochemical expressions of the host plants by secreting metabolites, and, simultaneously, plant roots also discharge several different kinds of chemical molecules that affect the cellular expression of the neighbouring microbes. Metabolic expressions of both microbes and plants are critically regulated by each other and induced by different kinds of biotic and abiotic stresses. However, the rhizosphere of different plants houses the same or different kinds of microbial communities, and they have been found to respond uniformly or differently to the same or different types of stresses. Concisely, the existence of the living entities in the planet depends on the synchronized interkingdom symbiosis where plant and rhizomicrobiome play a key regulatory role. Therefore, detailed insight into the rhizospheric symbiosis is essentially needed to understand this interplay at the molecular level.

Keywords

Symbiosis · Rhizosphere · Endosphere · Rhizomicrobiome · Root exudates

S. Mondal · S. Baksi (✉)
Department of Chemical Engineering, Jadavpur University, Kolkata, West Bengal, India

N. K. Arora, B. Bouizgarne (eds.), *Microbial BioTechnology for Sustainable Agriculture Volume 1*, Microorganisms for Sustainability 33,
https://doi.org/10.1007/978-981-16-4843-4_7

7.1 Introduction to Microbiome

Signal transduction is a very popular area of research associated with almost every focus of biological sciences. In plants and microbes, signalling or communication from the cellular level to the level of the whole organism requires the involvement of vast diversity of functional organic compounds that are considered as secondary metabolites or bioactive compounds. There are immense demands for these bioactive compounds in the pharmaceutical, agro- and food-based industries, and plant-associated microbes are an attractive and promising source of these compounds (Müller et al. 2016; Gupta et al. 2020).

Before introducing microbiome-plant relation, it is important to first understand the concept of the microbiome. Fundamentally, a microbiome is a community of microorganisms that either can be beneficial to the host plants or can be harmful. Plants are crowded with microbial organisms that colonize internal tissues as well as adhere to the external surfaces. The wide diversity of microorganisms in the soil, surrounding the root system, is collectively called rhizomicrobiome, and the surrounding soil area is known as the rhizosphere. Rhizomicrobiome critically regulates the physiological conditions of the host plant and also serves as a tank of additional genes that plants can access when needed (Compant et al. 2019; Rodriguez et al. 2019; Qu et al. 2020).

7.2 Who Is There? What Are they Doing?—The Rhizosphere Microbiome

Conceptually rhizosphere portrays the volume of soil influenced by root exudates and root microbiome including fungi, bacteria, protists, nematodes and invertebrates. Rhizosphere is complex and is a highly dynamic environment where the biogeochemical cycling of elements is steered by an enormous number of interactions between roots, minerals, organic compounds, solutes, gases and microorganisms (Paterson 2013; Verma et al. 2019). The core plant microbiome defines the habitat, which is extended from the whole plant, as an individual, to particular regions (roots, leaves, shoots, flowers and seeds), as well as interactions between the roots and adhering or surrounding soil—the rhizosphere (Compant et al. 2019; Rodriguez et al. 2019; Qu et al. 2020).

7.2.1 Biodiversity of the Rhizomicrobiome: Classification and Significance

7.2.1.1 Diazotrophs: The Nitrogen-Fixing Rhizomicrobiome

The gamma diversity of the rhizomicrobiome community is greatly affected by different biotic and abiotic stresses which eventually affect the mode and intensity of the interactions, around the rhizosphere, and also affect the functionality of the associated ecosystem to a great extent. Nitrogen-fixing bacteria like *Azotobacter* and

Rhizobium species are the most predominant candidates in the rhizosphere ecosystem that assimilate aerial nitrogen molecules into various nitrogenous compounds. Although the genus *Rhizobium* is primarily referred to by the term 'rhizobia', including several other genera like *Bradyrhizobium*, *Sinorhizobium*, *Azorhizobium*, *Mesorhizobium*. More than 70 species, disseminated over 13 different genera, are included under the designation rhizobia (Zaidi et al. 2017). Some other putative nitrogen-fixing actinobacterial species are *Arthrobacter*, *Microbacterium*, *Georgenia*, *Mycobacterium* and *Leifsonia*. Firmicutes like *Bacillus* and *Psychrobacillus* and proteobacteria like *Bosea* and *Roseomonas* are other nitrogen-fixing rhizosphere microbes exhibiting promising symbiotic effect (Rilling et al. 2018). An example of different nitrogen-fixing bacteria and their taxonomic classification from wheat rhizosphere and endosphere has been given in Table 7.1.

7.2.1.2 Phosphate Assimilating Rhizomicrobiome

Phosphate assimilation is another outstanding activity of rhizobacterium that supports the development and growth of the plant. Plant growth-promoting fungi, *Aspergillus*, *Sclerotium*, *Fusarium*, *Penicillium* and mycorrhiza are probably the most prominent rhizomicrobiome that carry out solubilization and mineralization of phosphorus. Additionally, *Bacillus*, *Rhizobium* and *Pseudomonas* have been found as the most efficient phosphate-solubilizing bacteria that support the growth of plants

Table 7.1 Taxonomic identity of nitrogen-fixing bacteria from wheat rhizosphere and endosphere (adapted from Rilling et al. 2018)

Taxonomy group	Closest relatives
Assembly compartment: Rhizosphere	
Firmicutes; *Bacilli*; *Bacillales*; *Bacillaceae*; *Bacillus*	*Bacillus megaterium* strain from maize rhizosphere
Firmicutes; *Bacilli*; *Bacillales*; *Bacillaceae*; *Bacillus*	Nitrogen-fixing *Bacillus megaterium* strain from date palm soil
Firmicutes; *Bacilli*; *Bacillales*; *Bacillaceae*; *Bacillus*	*Bacillus megaterium* from acid soil
Firmicutes; *Bacilli*; *Bacillales*; *Bacillaceae*; *Bacillus*	*Bacillus* sp. strain from straw decomposition
Firmicutes; *Bacilli*; *Bacillales*; *Bacillaceae*; *Bacillus*	*Bacillus megaterium* strain from *Juglans regia* rhizosphere
Actinobacteria; *Micrococcales*; *Micrococcaceae*; *Arthrobacter*	*Arthrobacter* sp. strain from phthalate-contaminated soil
Firmicutes; *Bacilli*; *Bacillales*; *Bacillaceae*; *Bacillus*	*Bacillus megaterium* strain from acid soil
Assembly compartment: root endosphere	
Firmicutes; *Bacilli*; *Bacillales*; *Bacillaceae*; *Bacillus*	*Bacillus aryabhattai* strain from arsenic-contaminated soil
Bacteroidetes; *Chitinophagia*; *Chitinophagales*; *Chitinophagaceae*; *Chitinophaga*	*Chitinophaga* sp. from tomato rhizosphere
Actinobacteria; *Corynebacteriales*; *Mycobacteriaceae*; *Mycobacterium*	*Mycobacterium* sp. from manure-fertilized grassland
Actinobacteria; *Micrococcales*; *Microbacteriaceae*; *Leifsonia*	*Leifsonia* sp. from grassland rhizosphere

by releasing organic acids, like gluconic and ketogluconic acids, that affect the mobility of phosphorous through ionic interactions or by releasing phosphatases that unbound phosphate groups from organic matter (Sindhu et al. 2014; Kundan et al. 2015; Alaylar et al. 2020). Additionally, bacteria like *Serratia*, *Achromobacter*, *Brevibacterium*, *Alcaligenes* and *Xanthomonas* have also been identified as efficient phosphate solubilises (Sindhu et al. 2014). Several different phosphate solubilising rhizobacteria and their host plants are presented in Table 7.2.

7.2.1.3 Microbial Iron Assimilation

Iron is another crucial element that is required to carry out different physiological activities in living organisms. However, total iron content in soil ranges between 2% and 4%, and iron deficiency stimulates the secretion of chelators like phytosiderophores and microbial siderophores and plants also release protons which eventually acidify the rhizosphere. Phytosiderophores, non-proteinogenic amino acids, usually bind to Fe^{3+} and transport back to the plant root tissue. Amongst microbes, fluorescent pseudomonads are the typical examples that secrete siderophores to promote iron intake in plants. Fungal siderophores have also been found carrying ferric ions with equal efficiency. Rhizoferrin is a very good example of siderophore that is produced by *Rhizopus arrhizus*, a fungus of the family *Mucoraceae*. Rhizobacteria *Bacillus subtilis* GB03 was found efficient for iron intake mechanisms for the plants (Yanes and Bajsa 2016; Ab Rahman et al. 2018; Oleńska et al. 2020). Dicots and non-graminaceous monocots have their indigenous iron up-taking mechanisms that include (1) reduction of soil pH by releasing of protons, (2) modification of iron solubility and (3) morphological modifications of root network. However, plants of the Poaceae family depend on phytosiderophores that bind to different trace elements like iron, zinc and copper and transport them into the root cells in chelated form. The rhizomicrobiome community secretes siderophores or organic acid anions that chelate the trace elements and eventually transport them into the root cells. Enterobactin, pyoverdine and ferrioxamines are some well-known bacterial siderophores, whereas ferrichromes are produced by fungi. Bacterial siderophores are not as effective source of iron for both monocotyledons and dicotyledons (Marschner et al. 2011; Sah et al. 2017).

Table 7.2 Examples of phosphate-solubilizing bacteria with their host plants (Kundan et al. 2015)

Plant growth-promoting rhizobacteria	Host plant
Azotobacter chroococcum	Wheat
Bacillus circulans *Cladosporium herbarum*	Mung beans
Bradyrhizobium japonicum	Soybeans
Enterobacter agglomerans	Tomato
Pseudomonas chlororaphis *Pseudomonas putida*	Soybeans
Rhizobium leguminosarum	Beans
Bacillus megaterium	Tea

7.2.1.4 Rhizomicrobiome Biodiversity: Communities and Habitats

Genetic manipulation of plants significantly regulates the assembly and nature of the immediate rhizomicrobiome community. Tian and co-workers investigated the rhizomicrobiome community of the wild and cultivated Afro-Asian rice species and found that the wild species possess an improved nucleotide and carbon metabolism capacity whereas cultivated species contain rhizobacteria engaged in xenobiotic degradation and metabolism of lipid, nitrogen and vitamins. Moreover, rhizomicrobiome in wild rice species overproduce and utilize methane in a balanced manner in comparison to the cultivated rice species (Tian et al. 2020a). Moreover, several instances have been found by researchers where the rhizomicrobiome community, associated with wild type strains, plays a crucial role in disease resistance. Tian et al. (2017) suggested that restructuring the rhizomicrobiome community can be a solution to rejuvenate the soil quality and productivity. A striking result was provided by Tian et al. (2020b) where wild and commercial soybean were simultaneously cultivated under similar conditions and found a continuous change in the rhizobacterial community, even after four continuous planting seasons, whereas the fungal community maintained a stable assembly. *Spizellomycetaceae* is the predominant fungal group of cultivated soybeans, whereas *Chaetomiaceae* and *Orbiliaceae* are more highly enriched in wild soybean (Tian et al. 2020b). Medicinal plants like *Mentha haplocalyx*, *Perilla frutescens*, *Astragalus membranaceus* and *Glycyrrhiza uralensis* also harbour different kinds of rhizobacteria like *Pseudomonas koreensis*, *Sulfitobacter* sp., *Mesorhizobium* sp., *Pseudoxanthomonas* sp. and *Methylocystis* sp., along with some other rhizoids like *Coprinellus radians*, *Tilletiopsis pallescens*, *Paraglomus* sp. Medicinal plants *G. uralensis* and *Astragalus membranaceus* have rhizomicrobiome community with a higher degree of diversity than the rhizomicrobiome present in *M. haplocalyx* and *P. frutescens* suggesting that plants growing under the same conditions may have different rhizomicrobiome preferences (Shi et al. 2018).

7.2.1.5 The Impact of Agricultural Practices on Rhizomicrobiome Diversity

Shifting cultivation practices in the hilly regions are also responsible for altering the soil characteristics of the farming area which, in turn, affects the rhizomicrobiome community. *Acidobacteria*, *Proteobacteria* and *Verrucomicrobia* are the predominant bacterial phyla in the hilly region where shifting cultivation is practised. Among them, *Acidobacteria* and *Proteobacteria* are the two most abundant phyla present in the natural forests, whereas *Verrucomicrobia* is the principal rhizobacterial phyla in fallow land. On the contrary, *Actinobacteria* phyla are the most predominant in the oil plant plantation soil. *Fibrobacteres*, *Elusimicrobia*, *Lentisphaerae*, *Parvarchaeota*, *Armatimonadetes*, *Planctomycetes* and *Bacteroidetes* are some other low abundant phyla found in the cultivation land of the hilly regions (Shi et al. 2019; Manpoong et al. 2020). Surprisingly, continuous cultivation of watermelon, potatoes and ginseng is often linked to the reduction of beneficial microorganisms, destruction of soil ecosystem and growth of pathogenic rhizomicrobiome. For example, the monoculture of peanuts (*Arachis hypogaea*) is found responsible for the destruction of *Glomeromycota* which is a beneficial rhizomicrobiome phylum, and the same also supports the growth of pathogenic rhizoids like *Fusarium* spp. and *Alternaria* spp. (Cardarelli et al. 2020; Tian et al. 2020b).

7.2.2 Mode of Action: Perception and Attribution

Rhizomicrobiome is often considered as biological control system that employs several direct and indirect mechanisms to protect the plants and support their growth. Secretion of phytohormones, assimilation of trace elements and soil minerals are some of the most important direct mechanisms, whereas indirect mechanisms include quorum quenching among pathogenic bacteria, imposition of resistance against pathogens by rhizomicrobiome-producing metabolites, eco-niche competition, etc. (Chen et al. 2018; Rahman et al. 2018). *Enterobacter* spp., *Bacillus* spp., *Streptomyces* spp., *Aspergillus* spp., *Pseudomonas* spp., *Gliocladium* spp., *Fusarium* spp., *Petriella* spp., *Lysobacter* spp. and *Pantoea* spp. are significantly used as potential biocontrol agents. *Agrobacterium radiobacter* is another potential biocontrol agent that prevents crown gall bacterium infection (De Silva et al. 2019). Suppression of autogenetic diseases of the plants is another crucial phenomenon regulated by the rhizomicrobiome communities. General suppression of autogenetic diseases is quite common and can be cured by maintaining balanced microbial consortia in the rhizosphere, whereas suppression of specific diseases requires proliferation of specific microbes which can be according to the root exudates (Tian et al. 2020c). Primary metabolites, like amino and organic acids, along with the secondary metabolites, like phenolics, terpenoids and alkaloids, are secreted by plant roots, and the same has been found responsible for captivating and promoting beneficial microorganisms while confronting the growth of pathogenic microbiome in the rhizosphere. Niche colonization in the rhizosphere is often found essential as it helps the plants to evolve and adapt to a dynamic environment (Berendsen et al. 2012). Antibiotics like iturin A and surfactin are produced by *Bacillus* sp., whereas *Agrobacterium* sp. produces agrocin 84 which is another potential antibiotic. Pyoluteorin, pyrrolnitrin and herbicolin A, which is a lipopeptide antibiotic in nature, are secreted by rhizomicrobiome. Secretion of two important enzymes, chitinases and glucanases, produced by rhizobacteria is induced by phytopathogenic fungi. *Pseudomonas* spp. are responsible for the production of antifungal metabolites like pyoluteorin, phenazines, 2,4-diacetylphloroglucinol (DAPG) and pyrrolnitrin (Ab Rahman et al. 2018). Plants release organic carbon compounds from their roots which are eventually decomposed to CO_2 using a mechanism named as rhizosphere priming effect (Haldar and Sengupta 2015). The large amounts of organic carbon released by the plant roots amplify the microbial numbers and activities in the rhizosphere (Jones et al. 2009; Lugtenberg and Kamilova 2009; Mendes et al. 2013). Plant roots are also equally responsible for rhizomicrobiome sustainability and maintaining a healthy rhizosphere environment. Plant roots secrete a group of substances which are collectively known as root exudates or rhizodeposits

Table 7.3 Root-associated rhizodeposits (Tian et al. 2020c)

Component	Effect
Epidermal cells	Quickly degraded to serve as nutrients
Border cells	Produce metabolites as nutrients
Soluble lysates	Serve as potential carbon and nitrogen sources
Root exudates	Rich in sugar components, phenolics, amino acids; provide defence against pathogenic microbes and function as chemoattractants
Mucilage	Function as chelator

that can regulate the assembly of rhizomicrobiome and are essential for their sustainability (Marschner 2012). When the soil of the rhizosphere is subjected to qualitative comparison with bulk soil, increased microbial activities have been observed in the rhizosphere, and the increased microbial activity is associated with the secretion of organic carbon compounds by roots of the plant (Haldar and Sengupta 2015).

Rhizodeposits are the group of substances that are released from plant roots and include the following:

1. Soluble ions, monosaccharides of high and low molecular weight, and organic and amino acids.
2. Biomacromolecules like proteins, carbohydrates and lipids.
3. Unsolvable mucilage comprised of polygalacturonic acid and polysaccharides.
4. Secondary metabolites composed of flavonoids, antimicrobial compounds and nematicides (Haldar and Sengupta 2015).

Several different rhizodeposits, secreted from plant roots, are mentioned in Table 7.3.

7.3 Role of the Rhizosphere Microbiome in Signal Conversation

The rhizosphere is not only the essential component in plant-plant and microbe-microbe interactions but also takes part in communicating between plant and microbill communities (Fig. 7.1).

However, microbe-assisted transformations of plant metabolism always remains in the focus of plant-microbe communications. Quorum sensing (QS) is another category of communication that takes place among bacterial families at both intra-species and interspecies levels. It endorses bacterial populations to orchestrate their behaviour and cell density by exchanging of numerous signalling molecules (Whiteley et al. 2017; Stephens and Bentley 2020). Additionally, several biophysical and physiochemical phenomena of microbial population, like biofilm development, degree of virulence of human pathogens and antibiotic production by plant growth-promoting rhizobacteria (PGPR), have been found critically regulated by QS. Quorum sensing also plays a vital role in interkingdom communication where some specific microbial signalling molecules induce systemic resistance in plant and also improve their immune system against plant pathogens (Venturi and Keel 2016; Raymaekers et al. 2020). Some antibiotics, when present in low concentration, are involved in microbial communication and trigger gene expression regulation in the surrounding bacterial cells. Primary and secondary metabolites like organic and amino acids, hormones, and phenolic or terpenoid compounds are the fundamental elements of rhizospheric communication that convey signals within a microbial population (van Dam and Bouwmeester 2016; Shastri and Kumar 2019). Additionally, the high nutritional value of primary and secondary metabolites modifies the physiochemical assembly of soil microbes and also induces microbial cell division and colonization. Phytohormones like auxins, cytokinins and gibberellins are also an essential part of the interkingdom signal transduction that are essentially associated with plant growth and development. For example, when the plants are exposed to biotic and/or abiotic stress, the production of phytohormone ethylene is upregulated which eventually slows down or ceases the growth of plant roots. A specific compound, 1-aminocyclopropane-1-carboxylic acid (ACC), is used as the precursor for synthesizing ethylene. However, some bacteria produce ACC deaminase that hydrolyses ACC which eventually terminate the ACC to ethylene conversion. Finally, the decreased ethylene content triggers root development and maturation. Matsuoka et al. (2016) found that common fruits and vegetables were a promising

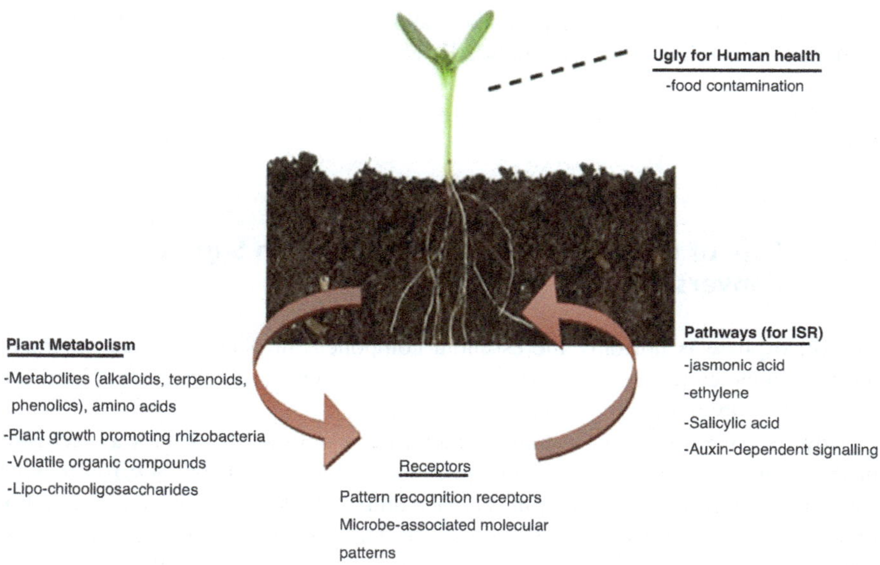

Fig. 7.1 Plant-microbe interactions taking place in the rhizosphere involving various chemical and signalling pathways

source of ACC deaminase producing endophytes. For example, endophytes, isolated from carrot, produce ACC deaminase that modifies stress-induced volatile metabolic profiles of carrots (flavour of carrots) (Venturi and Keel 2016; Foo et al. 2019; Basso et al. 2020).

An alternative theory of rhizomicrobiome recruitment in the rhizosphere proposed that both the plants and rhizomicrobiome respond to the low molecular weight compounds, secreted by either member of the rhizosphere. Activated cellular molecules subsequently take part in a cascade of cellular signal transductions which eventually leads to transcription of various genetic loci. Signalling mechanisms of the rhizosphere can be divided into three categories (Mommer et al. 2016):

(a) Quorum sensing-based signal transduction among microbial intraspecies and interspecies.
(b) Plant metabolite-based signalling from plant to rhizomicrobiome as symbiotic relationships.
(c) As a result of signalling from rhizomicrobiome to plants, extracellular metabolites of the microbiome alter plant gene expression, defence system and root structure (Mommer et al. 2016).

7.3.1 Microbe-Microbe Communication

Microbial cell density or population in the rhizosphere triggers several microorganisms to synthesize signalling compounds that synchronize their genetic expressions, and the same defines QS. Rhizosphere is comprised of diverse kinds of microbiomes with different population density, and QS signals play very important roles in shaping the rhizomicrobiome community. QS has been found responsible for developing and stabilizing microbial community and their communication among themselves as well as with the plants, which eventually affects the development of plant development (Padder et al. 2018; Mukherjee and Bassler 2019). For example, several proteobacterial rhizosphere isolates like *P. chlororaphis*, *P. putida*, *P. syringae*, *Burkholderia*, *Serratia*, *Erwinia* and *Ralstonia* act in response to N-acyl homoserine lactone (AHL) QS signals. The evolution of AHL has been found tightly associated with the modifications of plant gene expressions which, in turn, affect the growth and development of plants (Calatrava-Morales et al. 2018; Mhlongo et al. 2018; Balasundararajan and Dananjeyan 2019). It has been found that rhizosphere-residing species such as *Burkholderia* spp. and *S. maltophilia*, along with other Gram-negative bacteria, produce QS signals under diffusible signal factor family (cis-2-unsaturated fatty acids) [DSF]. Surprisingly, DSF is also found involved in ascribing innate immunity in plants and therefore are considered as interkingdom signals. Additionally, peptides like pheromones also act as QS signalling molecules in Gram-positive inhabitants that regulate scores of cellular processes at both intra- and interspecies level. Several types of microorganisms are found to release a group of volatile organic compounds

(VOCs) which play a significant role in interactions among microbial and plant communities. The volatile exudates from roots show a strong antifungal effect if supplemented with amino acids. On the contrary, soil-borne volatiles have an antibacterial effect when boosted with artificial root exudates but without any amino acids. Apart from exhibiting antimicrobial activity, VOCs act as signalling molecules that coordinate among intra- as well as interspecies gene expressions and interfere in other rhizospheric QS systems also. Involvement of VOCs in gene expression often makes them regulatory molecules for microbial biofilm generation, stress resistance, plant growth, immunity development and different genetic expressions, involved in hormone signalling pathways (Mommer et al. 2016; Venturi and Keel 2016; Huedo et al. 2018; Ye et al. 2020).

7.3.2 Plants -Microorganisms Communication

Irrespective of the parasitic or mutualist nature of microorganisms, plants play the role of host, and therefore interaction between plant and microbe substantially relies on phytochemical secretions. For example, plants cannot intake atmospheric nitrogen directly, and for this reason, they have to depend on nitrogen fixing bacteria such as rhizobia. Upon signal exchange and host-pathogen agreement, plants allow rhizobacteria to colonize root. Eventually, the nitrogen fixing rhizobacteria convert atmospheric nitrogen into ammonia (to proteins) via nitrogen fixation. The flavonoid compounds (2-phenyl-1,4-benzopyrone derivatives) were the first signal molecules discovered that induce bacterial nod genes during the signal exchange in between the host plant and its rhizobial symbiont (Mhlongo et al. 2018; Sasse et al. 2018). Nod genes are responsible for the production of nodulation factors (lipo-chitooligosaccharides (LCOs)) which are the first signal molecules that initiate nodule formation. Following identification of nod secreted LCOs, by epidermal kinase receptors of root nodules, a cascade of signal transduction is activated which eventually leads to nodule formation. Besides, LCO-assisted mutualism in between arbuscular mycorrhizal fungi (AMF or AM) and plants established the fact that these fungi substantially increase the capability of roots to access immobile phosphates from soil, whereas the fungi receive carbohydrates from plants. Despite the ecological importance, signalling-mediated symbiotic relationship is not clearly understood at the molecular level (Feng et al. 2019; Ratet 2019; Sakamoto et al. 2019).

Strigolactone plant hormones secreted from roots are considered one of the primary signals to initiate AMF-plant symbiosis. Surprisingly, arresting of strigolactone by parasitic plants, like *Striga* spp. and *Orobanche* spp., helps them to sense their host and persuade seed germination at the proximate area of the roots. It has been recently drawn out that two loci RAM1 and RAM2, which encode GRAS domain transcription factor and an acyltransferase respectively, are badly affected in *Medicago truncatula* mutants (Castillejo et al. 2019; Clear and Hom 2019; Dhanker et al. 2020). GRAS domain transcription factor and acyltransferase are solely responsible for the synthesis of cutin monomers which, in turn, play a crucial role in AM stimulation. Therefore, the affected GRAS domain transcription factor and acyltransferase make *M. truncatula* mutants unfit for AM symbiosis (Jiang et al. 2018; Vigneron et al. 2018). Like primary metabolites, plant secondary metabolites hold equal importance in signal transductions. Being very low molecular weight molecules, plant secondary metabolites, like phenylpropanoids, fatty acid and amino acid derivatives, are believed to cross cell membranes and cell wall freely and secreted into the rhizosphere soil. Advantages of secondary metabolites are often associated with reproductive and defense mechanisms of plants against several kinds of herbivores and pathogens. In the roots, secreted volatiles are often composed of substances that attract the enemies of root-consuming herbivorous, or they often comprise antimicrobial materials that protect roots from harmful microorganisms. Attractant properties of the root secreted volatiles can be well understood with the interaction between (E)-β-caryophyllene and Western corn rootworm (WCR) (Bécard 2017; Lin et al. 2017; Dixit et al. 2019). WCR is a herbivorous that feeds on the root of maize. However, during consumption, a volatile metabolite (E)-β-caryophyllene is secreted from roots that attracts entomopathogenic nematodes which eventually paralyse WCR, and within a few days, larvae of WCR are dead. This phenomenon illustrates how the volatiles in roots act as a signalling molecules (Bécard 2017; Lin et al. 2017; Dixit et al. 2019). An additional example, along with the signalling mechanism, of symbiotic relationship has been investigated and found that plants use some precise pattern recognition receptors (PRRs) to recognize microbe-associated molecular patterns (MAMPs) which are well-preserved molecules (like peptidoglycans, chitin, lipopolysaccharides and flagellin) in microbes. Following the identification of MAMPs, rhizomicrobiome, like mycorrhizae and rhizobia, and other plant growth-promoting fungi are recognized by the plant (Quiza et al. 2015; Venturi and Keel 2016; Saijo et al. 2018). Upon identification of non-self MAMPs, plant triggers basal immunological defence in the root system, commonly known as induced systemic resistance (ISR), which ultimately protects plants from leaf-feeding herbivorous and foliar pathogens. ISR is substantially distinct from systemic acquired resistance (SAR) (Kamle et al. 2020; Pršić and Ongena 2020). Ethylene and jasmonate signalling pathways are the main regulatory signalling pathways of ISR, whereas SAR is triggered by plant pathogens, and the salicylic acid signalling pathway is a perfect example of SAR. Priming is another symbiotic mechanism where beneficial microorganisms tempt the plants to respond efficiently to different kinds of environmental stresses. Other than MAMP-mediated plant response, there are numerous other mechanisms through which rhizosphere bacteria induce plant response through symbiotic understanding (Kouzai et al. 2016; Mhlongo et al. 2016; Nguvo and Gao 2019; Yuan et al. 2019; Gupta and Bar 2020). Symbiosis with the aid of interkingdom signalling molecules has also been investigated and found that upon root contact, *Laccaria bicolor*, an ectomycorrhizal fungus, secretes a 7 kDa MiSSP7 protein from hyphae. Afterwards, the protein of low molecular weight enters into the host cell nucleus and interacts with hormone co-receptors and eventually influences the jasmonic acid signalling pathway negatively. Salicylic acid and ethylene-dependent ISR have been identified in *Arabidopsis* where the sub-inhibitory concentration of 2,4-diacetylphloroglucinol

(DAPG) in the root or rhizosphere soil induces a defensive response against microbial leaf pathogens. Additionally, DAPG has been found to be a crucial molecular compound that takes part in root development via an auxin-mediated signalling pathway. In recent researches it has been proved that VOCs have gained considerable attention for being priming agents. 2,3-Butanediol (2,3-BD), a VOC secreted by *Bacillus* strains, identified as the first microbial VOC influencing the growth of *Arabidopsis* and provokes ethylene signalling-based ISR against phytopathogen *Erwinia*. *Enterobacter*, a root endophyte, secretes 2,3-BD that elevates the fungal resistance of maize and disturbs tritrophic interactions with herbivorous. Another important PGPR-producing VOC, indole, involves maturing the *Arabidopsis* root system using auxin signalling pathway. One interesting nature of indole is that it acts as a volatile signal to neighbouring plants against herbivorous and insect invasions. Phytohormone, including auxins, gibberellins and cytokinins, are identified with impacts on plant growth and organ development, host immune responses and hormonal signalling cascades. Although the symbiosis, at the rhizosphere, represents the interaction between rhizomicrobiome and plant roots, it ultimately influences the whole plant body implying the contribution of root-shoot signalling networks (Newcombe et al. 2010; Mabood et al. 2014; Quiza et al. 2015; Venturi and Keel 2016; Chagas et al. 2018; Mhlongo et al. 2018; Olowe et al. 2020; Vishwakarma et al. 2020).

As shown in Fig. 7.2, before infection, the microbes compete among themselves for supplements and space. Beneficial microbes restrict the development and growth of plant pathogens through some physiological and physiochemical actions like the production of biostatic compounds, ingesting micronutrients or stimulating the immune response of the host plant. However, most of the microbes are symbiont, and they settle in different ecological niches. Therefore, they neither engage in affecting the host plants nor the competing pathogens but form a compound network of interactions that eventually affect plants and every other organism through infection or production of phytotoxic compounds (Berendsen et al. 2012).

7.4 The Benefice and Bull Effects: Plant and Human Health

Due to high intrinsic sensitivity, rhizomicrobiome can critically sense small changes in soil atmosphere as well as in the environmental conditions and respond accordingly in an immediate manner which labels them as bioindicators for assessing soil condition and quality. It is also important to describe the facts which show the impact of rhizosphere microbes on plant growth and human health. The following sections describe the good and ugly faces of the rhizomicrobiome (Mendes et al. 2013).

7.4.1 Conducive Behaviour

The association of rhizobiome with plants eventually supports overall plant growth and protects against plant pathogens. Additionally, nutrient acquisition of plants is

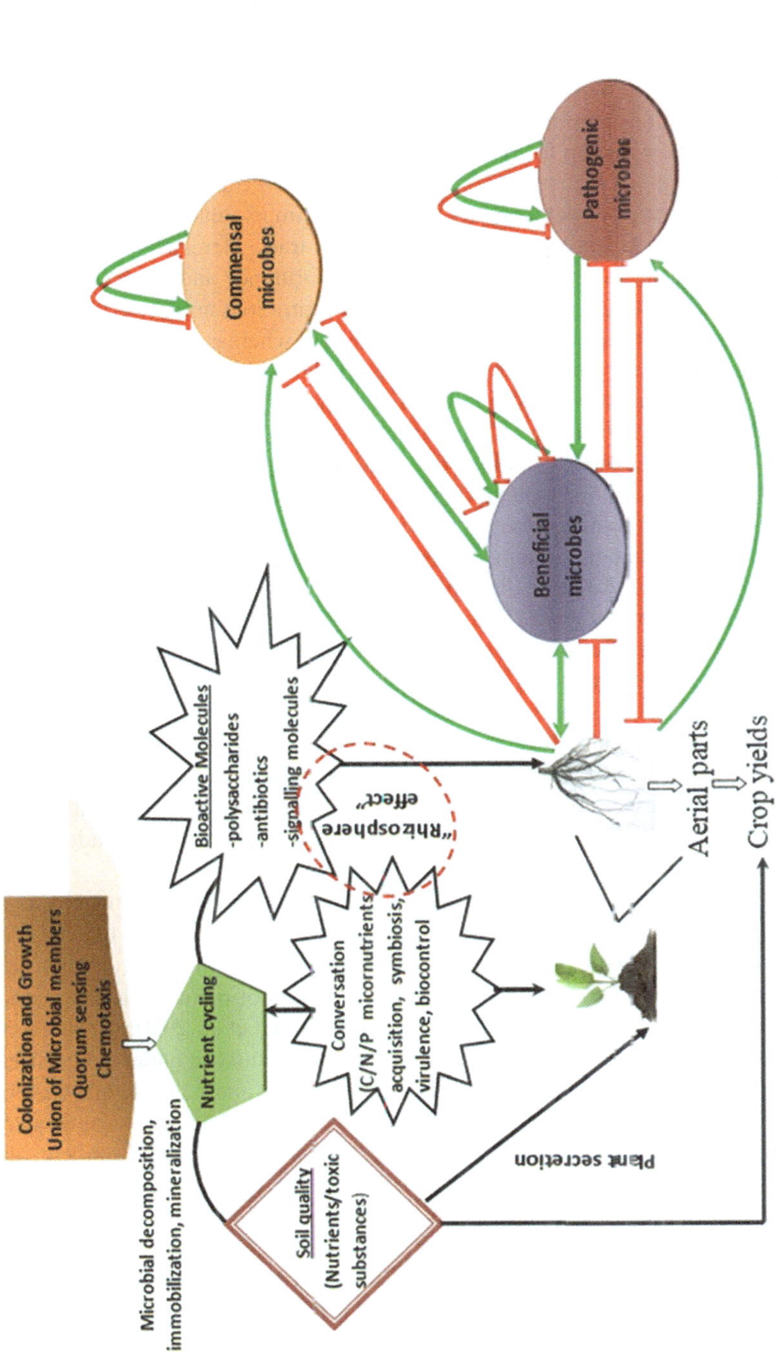

Fig. 7.2 Flow chart of a combined rhizobiome communication where the influence of plants for activation of microbes via exudation of compounds stimulates (green arrows) or inhibits (red blocked arrows)

also significantly influenced by the rhizosphere microbiome. It is well documented that nitrogen fixation in plants is essential for synthesizing amino and nucleic acids. Some well-known symbiotic nitrogen-fixing microbes also facilitate phosphorus intake and soil to plant nutrient translocation. However, the development of stable nutrient-rich soil depends on the presence of rhizomicrobiome which also suppress the soil-borne plant pathogens (Cocking 2003; Turner et al. 2013). Rhizosphere regulates several principal pathways that attribute primary immunity to the plants and also takes part in antibiosis, competition for trace elements, nutrients and microsites, parasitism, intrusion with QS affecting virulence and induced systemic resistance. An interesting aspect is that at low concentration, antibiotic compounds act as mediators of intracellular signalling pathway, whereas paradoxical effects, such as growth inhibition, have been found associated with a high concentration of antibiotics. On the other hand, additional benefits of bacterial volatiles are associated with regulated mycorrhizal network establishment and tripartite interactions between bacteria, fungi and nematodes (Schouteden et al. 2015; Maurya et al. 2018; Sarkar et al. 2019). Concerning this context, it has been found that *Paenibacillus polymyxa* and *Paenibacillus lentimorbus* arrest substantial antifungal properties and eventually restrict the communications among nematode *Meloidogyne incognita* and fungus *Fusarium oxysporum*. Additionally, the rhizobacterial QS mechanism has been found to activate various genes like MPK3, MPK6, WRKY22, WRKY29 and Pdf1.2 that are responsible for plant defence. Following a combination of metabolic pathways two separate strains of *P. fluorescens* minimised the carbon intake of host plants, whereas the same strains were found beneficial as well as they protected the host plant against pathogen *P. syringae*. These findings indicate that the rhizobacteria have diverse and reflective effects on the metabolic pathways and immunogenic responsees of their host plants, thus regulating the biosynthesis of unfamiliar metabolites, along with known secondary metabolites (Mendes et al. 2013; Gkizi et al. 2016; Kurokawa et al. 2021).

It has been observed that the rhizomicrobiome promotes the development of some plants under severe conditions which can be illustrated by an example of *Achromobacter piechaudii* ARV8. Soil-borne bacterium, isolated from the arid and salty environment, increase the growth and biomass of plants following exposure to either drought stress or flooding conditions (Sunita et al. 2020; Mishra et al. 2021). It is well known; halotolerant bacteria succeed to survive under saline stress conditions. However, the alliance between the host-plant and halotolerant bacteria eventually promotes plant growth. For example, 130 rhizobacteria were isolated from the rhizosphere of wheat plants which were grown in a saline environment, and 24 of the isolates had a high salt tolerance level ([NaCl] = 8%). Additionally, all the isolates, having high salt tolerance level, produced indole-3-acetic acid, whereas only six isolates of them produced gibberellin, and ten of them could solubilise phosphate. However, only two of the isolates had the *nif*H gene, which makes them potential nitrogen fixers. *Salicornia brachiata*, *Brachybacterium saurashtrense* sp. nov., *Pseudomonas fluorescens*, *Halomonas* sp., *Zhihengliuella* sp., *Haererehalobacter* sp., *Vibrio* sp., *Streptomyces* sp. *GMKU 336*, *Cronobacter sakazakii*, *Pseudomonas* spp., *Brevibacterium casei*, *Pseudomonas* spp., *Rhizobium* *radiobacter* and *Mesorhizobium* sp. are some diazotrophic bacteria, newly isolated from roots, that produce 1-aminocyclopropane-1-carboxylic acid (ACC) and also have the phosphate dissolution and 1-aminocyclopropane-1-carboxylic acid (ACC) deaminase property (Jha et al. 2012; Alexander et al. 2019; Jha 2019; Jha 2020). Low temperature has serious effects on nodulation and nitrogen fixation. However, the legumes from the high arctic and temperate region can efficiently nodulate and fix nitrogen at relatively same rates in cold conditions. Another area of interest, in horticulture and agriculture, is the enhancement of plant growth by using microbial inoculants under cold conditions. For example, enhanced physiochemical activities and root growth of grapevine plants can be achieved by introducing *Burkholderia phytofirmans* PsJN microbes at 4 °C. The same phenomenon has been encountered during the growth and nodulation of the soybean plant. It has been found that at 15 °C, nodule development in soybean plants is inhibited which eventually arrests the nitrogen fixation. However, when *Serratia proteamaculans* is co-inoculated with *Bradyrhizobium japonicum*, soybean growth and nodulation at 15 °C have been observed. To identify the cellular mechanisms involved in the growth and development of plants at low temperatures, cold-tolerant, like *P. fluorescens* (at 10 °C) *and Pseudomonas corrugate* (at 4 °C to 28 °C), were selected due to their capacity of solubilizing phosphate, which eventually promoted the growth and development of plants. pH and high toxic compound concentrations are the two other main factors that regulate the plant growth. When corn plants are harvested in low pH soil, increased incitation of foliar lesions have been observed on corn but radically reduced when the plants are contaminated with 1-(3-Acetyl-2,4,6-trihydroxyphenyl)ethanone (DAPG)-producing bacteria *P. fluorescens*. This was possibly the first instance where the introduction of DAPG was found essential for restructuring the plants to fight against abiotic stress, whereas previously, DAPG was used as a pathogen controlling agent (Bordiec et al. 2011 Berendsen et al. 2012; Mommer et al. 2016).

Nowadays, bioremediation is being used as a potential substitute for removal of soil pollution. The idea of rhizoremediation is a combination of phytoremediation and bioaugmentation. Degradation of pollutants with rhizobacteria is a promising strategy to clean polluted sites. An evidential study showed an increased population of *Actinobacteria*, *Verrucomicrobia* and *Denitratisoma* in the polycyclic aromatic hydrocarbon-contaminated rhizospheres, signifying that these microbes may have a vital role in the rhizoremediation. Additionally, rhizoremediation of hydrocarbons has been achieved by using fungi *Lewia* sp. (Berendsen et al. 2012; Mendes et al. 2013).

However, successful application of microbial inoculants remains in its initial stages because of several limitations like shelf life, effective in different plant species, and irregular effectiveness across environments, etc Finally, more insight is required on how rhizomicrobiome communicates with the host plant, and alters the metabolic pathways, to resolve the research gaps (Mendes et al. 2013).

7.4.2 Wicked Behaviour: Opportunistic Microbes

Apart from the conventional rhizomicrobiome community, a separate microbial community has been identified which is generally non-pathogenic but acts as a pathogen under certain circumstances and termed 'opportunistic pathogens'. These opportunistic pathogens stay dormant in the host cell for a long period and attack immunocompromised hosts. The fanciful nature of climate reduces the quantity and regularity of annual rainfall in several geographical areas, which eventually forces the plants to experience drought stress. This stress induces several physiological changes in plants which also weakens the immune system of the plants and gives opportunity to the pathogens to flourish and attack the host plants (Overstreet and Lotz 2016; Milici et al. 2020). Recent studies revealed the opportunistic and virulent nature of plant symbiont *Trichoderma reesei* which is parasitic to other fungal strains also (Harman et al. 2004). On the other hand, *Pseudomonas aeruginosa* is an opportunistic multidrug-resistant pathogen that is responsible for various diseases in plants and humans. They also secrete autoinducers, an external metabolite, that interrupts the signalling cascades of other microbial communities. Additionally, *P. aeruginosa* secretes several distinct signalling molecules: rhl, pqs, iqs, las and pch. Iqs, rhl and pqs responsible for intercellular communications, QS and quinolone signalling. With the collective effects of these signalling molecules, *P. aeruginosa* grows as a biofilm within the lungs of the patients, suffering from cystic fibrosis (Allesen-Holm et al. 2006; Dekimpe and Deziel 2009; Lee and Zhang 2015). Apart from human pathogenicity, *P. aeruginosa* is a promising endophyte having the capability of phosphate and zinc solubilization and also secretes siderophores, vitamins and antibiotics which eventually support plant growth and provide resistance against phytopathogens (Kumawat et al. 2019). *Burkholderia cepacia* complex refers to a group of 17 closely related species that were believed pathogenic to the host plants. Unfortunately, recent studies revealed that the same microorganisms are also potential opportunistic human pathogen, responsible for human morbidity. *B. cepacia* complex is responsible for several diseases like urinary tract infections, chronic granulomatous disease, septic arthritis and meningitis. Most serious characteristic of the *B. cepacia* complex is that they can survive for a long period of time and multiply in aqueous environments even in presence of disinfectant agents and intravenous fluids (Baylan 2012). The symbiotic nature of *B. cepacia* gives them the advantages of being present as endophytes or by forming nodules in the leguminous hosts like rice, coffee and agave. However, despite being a potential human pathogen, *B. cepacia* is extensively used as phytopathogens and applied in agriculture for promoting plant growth at a significant level. The nitrogen and phosphorus fixation ability of *B. cepacia* makes them advantageous in agricultural practices. Moreover, *B. cepacia* can grow on crystal violet, methyl parathion and β-endosulfan, following degradation of these compounds, and the same characteristic promotes *B. cepacia* to be used for bioremediation (Rojas-Rojas et al. 2018). Another potential multidrug-resistant opportunistic pathogen is *Alcaligenes faecalis*. *A. faecalis*-induced infections are almost incurable as this pathogen is multidrug-resistant at a significant level. Postoperative endophthalmitis, diabetic foot infection, acute pyelonephritis and severe pneumonia are some of the very common fatal diseases caused by the opportunistic pathogen *A. faecalis*. Bloodstream, middle ear and soft tissues are the most common parts of the human body vulnerable to *A. faecalis* infection (Kaliaperumal et al. 2006; Huang 2020). The most striking advantage of using *A. faecalis* is that this pathogen promotes the growth of the plants in saline soil. For example, enhanced growth of canola plants has been achieved when *A. faecalis* was applied alone or in combination with *Azotobacter chroococcum*. This pathogen stimulates the growth of the plants following the synthesis of photosynthetic pigments and solubilization of proteins and sugars. *A. faecalis* also stimulates the production of antioxidant enzymes like ascorbate peroxidase, superoxide dismutase and peroxidases which are responsible for cellular damage repair and plant growth promotion. Additionally, *A. faecalis* can degrade 48% of the pyrene compound that makes this pathogen useful for bioremediation (Singha et al. 2018; Abdel Latef et al. 2021). *Enterococcus faecium* has been found in the gastrointestinal tract of the healthy human beings but can be life-threatening if they are exposed to the hospital environment where they acquire resistance against different types of generic drugs. On the other hand, *Enterococcus faecium* is a potential source of gibberellin that regulates the growth of various parts of the plants (Singh 2019). *Rhodococcus equi*, has no direct plant growth-promoting property, however, is used for bioremediation as it can withstand high mercury and arsenic-contaminated soil and successfully degrade the toxic xenobiotics. On the other hand, *R. equi* has been found responsible for cavitary pulmonary disease in AIDS patients and also cause pyogranulomatous pneumonia in young foals (Prescott 1991; Franchi et al. 2017).

Several soil-borne plant pathogens, other than bacteria and fungi, are also responsible for reduced crop production, and nematode and fungi are two of them. In extreme temperatures, nematodes, along with plant pathogenic fungi, become agronomically more vibrant than plant pathogenic bacteria. Bacterial pathogens like *Dickeya dadantii*, *Dickeya solani*, *Agrobacterium tumefaciens*, *Ralstonia solanacearum*, *Pectobacterium carotovorum* and *Pectobacterium atrosepticum* are the most well-known plant pathogens that infect plants through roots. Apart from pathogenic bacteria, fungi, nematodes and viruses are also responsible in the same manner causing plant damage. Viruses generally invade host plants using vector assisted transfer mechanisms. Nematodes or zoosporic fungi are often used as a vector by viruses to penetrate through the root tissue and infect the host plant (Berendsen et al. 2012; Mendes et al. 2013). Moreover several opportunistic human pathogens like *B. cepacia*, *P. aeruginosa*, *Stenotrophomonas maltophilia*, *Bacillus cereus* and *Proteus vulgaris* have been found in rhizosphere environments, hosted by many wild and cultivated plants. Unfortunately, studies have been conducted focussing only on the existence of opportunistic pathogens. However, their relative virulence in clinical counterparts has been carelessly overlooked (Kumar et al. 2013).

7.5 Prospects and Outlook

The relationship among human, plant and soil microbiome exhibits a significant impact on the environment and also can influence the microbial communities which, in turn, makes it an interesting subject for further investigation and study. Detailed insight of methodologies, like environmental effect on genetic expression, identification and quantification of the plant and microbial exudates, will provide a clear picture about how one and all factors are interconnected and regulated by the other factor(s). Cutting-edge techniques associated with genomics, transcriptomics and proteomics can provide a detailed picture of the specific expression of genetic loci in the rhizomicrobiome (Phour et al. 2020; Tian et al. 2020c).

Modification of rhizomicrobiome signalling, through genetic recombination, has been found responsible for extensive remodelling of cell signalling, and the same also discloses new avenues to develop genetically modified crops with better productivity with minimum agrochemical supplementation. Making use of selective beneficial rhizomicrobiome, plant metabolic pathways can be significantly modified, and the same also escalates the plant resistance against varying stresses. Depletion of human pathogens in the plant environment is another strategy to arrest the spreading of human diseases, and for this reason, it is essential to understand, in detail, how different factors make plant surfaces a suitable niche for human pathogens. To maintain the limitations of plant and human pathogens, varying and balanced trickeries must be developed that transmit the rhizobiome signals in favour of indigenous microorganisms only and hinder the biophysical activities of human pathogens. The introduction of plant breeding programs untying the molecular level of communications among plant lines and beneficial rhizomicrobiome could be a potential approach. The rhizosphere is important field of research for ecologists, molecular biologists and plant biologists to further study new outlooks of interactions stirring in this complex zone and rhizobiome (Mendes et al. 2013; Huang et al. 2014; Venturi and Keel 2016; Balasubramanian et al. 2020).

7.6 Conclusion

A complex symbiotic relationship, orchestrated between plants and microbial consortia, is responsible for morphogenesis, host-pathogen interactions, nutrients assimilation and QS. The relationship between plants and microbes is carried out by several inter and intraspecies signalling cascades where metabolites, hormones, volatile organic carbon, sugar molecules and phytochemicals act as signalling molecules. However, the composition of the phytochemicals and root exudates is often found revised under a modified environment that ultimately endorses the growth of some specific rhizomicrobiome communities, who, in turn, look after the host. Types of crops, local weather and soil conditions and rotational cultivation are some important independent variables that regulate the types and characteristics of the associated rhizomicrobiome community. Thus, deciphering the chemical language of interkingdom communication has become a difficult challenge for researchers. Therefore, in-depth investigation and implementation of different 'omics approaches' are strongly suggested to better understand the communication at the molecular level.

Acknowledgements None. No funding to declare.

Conflict of Interest: The authors declare that there is no conflict of interests regarding the publication of this article.

References

Ab Rahman SFS, Singh E, Pieterse CM et al (2018) Emerging microbial biocontrol strategies for plant pathogens. Plant Sci 267:102–111

Abdel Latef AAH, Omer AM, Badawy AA et al (2021) Strategy of salt tolerance and interactive impact of *Azotobacter chroococcum* and/or *Alcaligenes faecalis* inoculation on canola (*Brassica napus L.*) plants grown in saline soil. Plants 10:110

Alaylar B, Egamberdieva D, Gulluce M, Karadayi M, Arora NK (2020) Integration of molecular tools in microbial phosphate solubilization research in agriculture perspective. World J Microbiol Biotechnol 36(7):93

Alexander A, Mishra A, Jha B (2019) Halotolerant rhizobacteria: a promising probiotic for saline soil-based agriculture. In: Kumar M, Etesami H, Kumar V (eds) Saline soil-based agriculture by halotolerant microorganisms. Springer, Singapore, pp 53–73

Allesen-Holm M, Barken KB, Yang L et al (2006) A characterization of DNA release in *Pseudomonas aeruginosa* cultures and biofilms. Mol Microbiol 59:1114–1128

Balasubramanian V, Sur A, Nayak KK et al (2020) Plant root exudates as determinant of Rhizomicrobiome. In: Sharma SK, Singh UB, Sahu PK, Singh HV, Sharma PK (eds) Rhizosphere microbes. Microorganisms for sustainability, vol 23. Springer, Singapore, pp 105–126

Balasundararajan V, Dananjeyan B (2019) Occurrence of diversified N-acyl homoserine lactone mediated biofilm-forming bacteria in rice rhizoplane. J Basic Microbiol 59:1031–1039

Basso V, Kohler A, Miyauchi S et al (2020) An ectomycorrhizal fungus alters sensitivity to jasmonate, salicylate, gibberellin, and ethylene in host roots. Plant Cell Environ 43:1047–1068

Baylan O (2012) An opportunistic pathogen frequently isolated from immunocompromised patients: *Burkholderia cepacia* complex. Mikrobiyol Bul 46:304–318

Bécard G (2017) How plants communicate with their biotic environment, vol 82, 1st edn. Academic Press, New York, p 404

Berendsen RL, Pieterse CM, Bakker PA (2012) The rhizosphere microbiome and plant health. Trends Plant Sci 17:478–486

Bordiec S, Paquis S, Lacroix H et al (2011) Comparative analysis of defence responses induced by the endophytic plant growth-promoting rhizobacterium *Burkholderia phytofirmans* strain PsJN and the non-host bacterium *Pseudomonas syringae* pv. Pisi in grapevine cell suspensions. J Exp Bot 62:595–603

Calatrava-Morales N, McIntosh M, Soto MJ (2018) Regulation mediated by N-acyl homoserine lactone quorum sensing signals in the rhizobium-legume symbiosis. Gene 9:263

Cardarelli M, Rouphael Y, Kyriacou MC et al (2020) Augmenting the sustainability of vegetable cropping systems by configuring rootstock-dependent rhizomicrobiomes that support plant protection. Agronomy 10:1185

Castillejo MÁ, Fernández-Aparicio M, Rubiales D (2019) Characterization of defense mechanisms to parasitic plants in the model *Medicago truncatula*. In: Bruijn FD (ed) The model legume *Medicago truncatula*. Wiley, Hoboken, NJ, pp 378–383

Chagas FO, de Cassia PR, Caraballo-Rodríguez AM et al (2018) Chemical signaling involved in plant–microbe interactions. Chem Soc Rev 47:1652–1704

Chen Y, Wang J, Yang N et al (2018) Wheat microbiome bacteria can reduce virulence of a plant pathogenic fungus by altering histone acetylation. Nat Commun 9:1–14

Clear MR, Hom EF (2019) The evolution of symbiotic plant–microbe signalling. Annu Plant Rev Online:785–836

Cocking EC (2003) Endophytic colonization of plant roots by nitrogen-fixing bacteria. Plant Soil 252:169–175

Compant S, Samad A, Faist H et al (2019) A review on the plant microbiome: ecology, functions, and emerging trends in microbial application. J Adv Res 19:29–37

De Silva NI, Brooks S, Lumyong S et al (2019) Use of endophytes as biocontrol agents. Fungal Biol Rev 33:133–148

Dekimpe V, Deziel E (2009) Revisiting the quorum-sensing hierarchy in *Pseudomonas aeruginosa*: the transcriptional regulator RhlR regulates LasR-specific factors. Microbiology 155:712–723

Dhanker R, Chaudhary S, Kumari A et al (2020) Symbiotic signaling: insights from *arbuscular mycorrhizal* symbiosis. In: Varma A, Tripathi S, Prasad R (eds) Plant microbe symbiosis. Springer, Cham, pp 75–103

Dixit S, Shukla A, Upadhyay SK et al (2019) Mode of communication between plants during environmental stress. In: Singh S, Upadhyy S, Pandey A, Kumar S (eds) Molecular approaches in plant biology and environmental challenges. Energy, environment, and sustainability. Springer, Singapore, pp 127–147

Feng F, Sun J, Radhakrishnan GV et al (2019) A combination of chitooligosaccharide and lipochitooligosaccharide recognition promotes arbuscular mycorrhizal associations in *Medicago truncatula*. Nat Commun 10:1–12

Foo E, Plett JM, Lopez-Raez JA et al (2019) The role of plant hormones in plant-microbe symbioses. Front Plant Sci 10:1391

Franchi E, Rolli E, Marasco R et al (2017) Phytoremediation of a multi contaminated soil: mercury and arsenic phytoextraction assisted by mobilizing agent and plant growth promoting bacteria. J Soil Sediment 17:1224–1236

Gkizi D, Lehmann S, L'Haridon F et al (2016) The innate immune signaling system as a regulator of disease resistance and induced systemic resistance activity against *Verticillium dahliae*. Mol Plant Microbe Interact 29:313–323

Gupta R, Bar M (2020) Plant immunity, priming, and systemic resistance as mechanisms for *Trichoderma* spp. biocontrol. In: Sharma A, Sharma P (eds) Trichoderma. Rhizosphere biology. Springer, Singapore, pp 81–110

Gupta S, Chaturvedi P, Kulkarni MG et al (2020) A critical review on exploiting the pharmaceutical potential of plant endophytic fungi. Biotechnol Adv 39:107462

Haldar S, Sengupta S (2015) Plant-microbe cross-talk in the rhizosphere: insight and biotechnological potential. Open Microbiol J 9:1

Harman GE, Howell CR, Viterbo A et al (2004) *Trichoderma* species—opportunistic, avirulent plant symbionts. Nat Rev Microbiol 2:43–56

Huang C (2020) Extensively drug-resistant *Alcaligenes faecalis* infection. BMC Infect Dis 20:1–11

Huang X-F, Chaparro JM, Reardon KF et al (2014) Rhizosphere interactions: root exudates, microbes, and microbial communities. Botany 92:267–275

Huedo P, Coves X, Daura X et al (2018) Quorum sensing signaling and quenching in the multidrug-resistant pathogen *Stenotrophomonas maltophilia*. Front Cell Infect Microbiol 8:122

Jha Y (2019) Endophytic bacteria as a modern tool for sustainable crop management under stress. In: Giri B, Prasad R, Wu QS, Varma A (eds) Biofertilizers for sustainable agriculture and environment. Soil biology, vol 55. Springer, Cham, pp 203–223

Jha Y (2020) Plant microbiomes with phytohormones: attribute for plant growth and adaptation under the stress conditions. In: Yadav A, Rastegari A, Yadav N, Kour D (eds) Advances in plant microbiome and sustainable agriculture. Microorganisms for sustainability, vol 20. Springer, Singapore, pp 85–103

Jha B, Gontia I, Hartmann A (2012) The roots of the halophyte *Salicornia brachiata* are a source of new halotolerant diazotrophic bacteria with plant growth-promoting potential. Plant Soil 356: 265–277

Jiang Y, Xie Q, Wang W et al (2018) Medicago AP2-domain transcription factor WRI5a is a master regulator of lipid biosynthesis and transfer during mycorrhizal symbiosis. Mol Plant 11:1344–1359

Jones DL, Nguyen C, Finlay RD (2009) Carbon flow in the rhizosphere: carbon trading at the soil–root interface. Plant Soil 321:5–33

Kaliaperumal S, Srinivasan R, Gupta A et al (2006) Postoperative endophthalmitis due to an unusual pathogen: *Alcaligenes faecalis*. Eye 20:968–969

Kamle M, Borah R, Bora H et al (2020) Systemic acquired resistance (SAR) and induced systemic resistance (ISR): role and mechanism of action against phytopathogens. In: Hesham AL, Upadhyay R, Sharma G, Manoharachary C, Gupta V (eds) Fungal biotechnology and bioengineering. Fungal biology. Springer, Cham, pp 457–470

Kouzai Y, Kimura M, Yamanaka Y et al (2016) Expression profiling of marker genes responsive to the defence-associated phytohormones salicylic acid, jasmonic acid and ethylene in *Brachypodium distachyon*. BMC Plant Biol 16:1–11

Kumar A, Munder A, Aravind R et al (2013) Friend or foe: genetic and functional characterization of plant endophytic *Pseudomonas aeruginosa*. Environ Microbiol 15:764–779

Kumawat K, Sharma P, Sirari A et al (2019) Synergism of *Pseudomonas aeruginosa* (LSE-2) nodule endophyte with *Bradyrhizobium* sp.(LSBR-3) for improving plant growth, nutrient acquisition and soil health in soybean. World J Microbiol Biotechnol 35:1–17

Kundan R, Pant G, Jadon N et al (2015) Plant growth promoting rhizobacteria: mechanism and current prospective. J Fertil Pestic 6:9

Kurokawa M, Nakano M, Kitahata N et al (2021) An efficient direct screening system for microorganisms that activate plant immune responses based on plant–microbe interactions using cultured plant cells. Sci Rep 11:1–14

Lee J, Zhang L (2015) The hierarchy quorum sensing network in *Pseudomonas aeruginosa*. Protein Cell 6:26–41

Lin Y, Qasim M, Hussain M et al (2017) The herbivore-induced plant volatiles methyl salicylate and menthol positively affect growth and pathogenicity of entomopathogenic fungi. Sci Rep 7: 1–11

Lugtenberg B, Kamilova F (2009) Plant-growth-promoting rhizobacteria. Annu Rev Microbiol 63: 541–556

Mabood F, Zhou X, Smith DL (2014) Microbial signaling and plant growth promotion. Can J Plant Sci 94:1051–1063

Manpoong C, De Mandal S, Bangaruswamy DK et al (2020) Linking rhizosphere soil biochemical and microbial community characteristics across different land use systems in mountainous region in Northeast India. Meta Gene 23:100625

Marschner P (2012) Rhizosphere biology. In: Mukerji KG, Manoharachary C, Singh J (eds) Microbial activity in the rhizosphere. Soil biology, vol 7. Springer, Berlin, Heidelberg, pp 369–388

Marschner P, Crowley D, Rengel Z (2011) Rhizosphere interactions between microorganisms and plants govern iron and phosphorus acquisition along the root axis–model and research methods. Soil Biol Biochem 43:883–894

Matsuoka H, Ohwaki Y, Terakado-Tonooka J et al (2016) Changes in volatiles in carrots inoculated with ACC deaminase-producing bacteria isolated from organic crops. Plant Soil 407:173–186

Maurya AK, Kelly MP, Mahaney SM et al (2018) Arbuscular mycorrhizal symbiosis alters plant gene expression and aphid weight in a tripartite interaction. J Plant Interact 13:294–305

Mendes R, Garbeva P, Raaijmakers JM (2013) The rhizosphere microbiome: significance of plant beneficial, plant pathogenic, and human pathogenic microorganisms. FEMS Microbiol Rev 37: 634–663

Mhlongo MI, Piater LA, Madala NE et al (2016) Phenylpropanoid defences in *Nicotiana tabacum* cells: overlapping metabolomes indicate common aspects to priming responses induced by lipopolysaccharides, chitosan and flagellin-22. PLoS One 11:e0151350

Mhlongo MI, Piater LA, Madala NE et al (2018) The chemistry of plant–microbe interactions in the rhizosphere and the potential for metabolomics to reveal signaling related to defense priming and induced systemic resistance. Front Plant Sci 9:112

Milici VR, Dalui D, Mickley JG et al (2020) Responses of plant–pathogen interactions to precipitation: implications for tropical tree richness in a changing world. J Ecol 108:1800–1809

Mishra P, Mishra J, Arora NK (2021) Plant growth promoting bacteria for combating salinity stress in plants – recent developments and prospects: a review. Microbiol Res 252:126861

Mommer L, Hinsinger P, Prigent-Combaret C et al (2016) Advances in the rhizosphere: stretching the interface of life. Plant Soil 407:1–8

Mukherjee S, Bassler BL (2019) Bacterial quorum sensing in complex and dynamically changing environments. Nat Rev Microbiol 17:371–382

Müller CA, Obermeier MM, Berg G (2016) Bioprospecting plant-associated microbiomes. J Biotechnol 235:171–180

Newcombe G, Martin F, Kohler A (2010) Defense and nutrient mutualisms in Populus. In: Jansson S, Bhalerao R, Groover A (eds) Genetics and genomics of Populus. Plant genetics and genomics: crops and models, vol 8. Springer, New York, NY, pp 247–277

Nguvo KJ, Gao X (2019) Weapons hidden underneath: bio-control agents and their potentials to activate plant induced systemic resistance in controlling crop *Fusarium* diseases. J Plant Dis Prot 126:177–190

Oleńska E, Małek W, Wójcik M et al (2020) Beneficial features of plant growth-promoting rhizobacteria for improving plant growth and health in challenging conditions: a methodical review. Sci Total Environ:140682

Olowe OM, Akanmu AO, Asemoloye MD (2020) Exploration of microbial stimulants for induction of systemic resistance in plant disease management. Ann Appl Biol 177:282–293

Overstreet RM, Lotz JM (2016) Host–symbiont relationships: understanding the change from guest to pest. In: Hurst C (ed) The Rasputin effect: when commensals and symbionts become parasitic. Advances in environmental microbiology, vol 3. Springer, Cham, pp 27–64

Padder SA, Prasad R, Shah AH (2018) Quorum sensing: a less known mode of communication among fungi. Microbiol Res 210:51–58

Paterson A (2013) Genomes of herbaceous land plants, vol 69, 1st edn. Academic Press, Elsevier, London

Phour M, Sehrawat A, Sindhu SS et al (2020) Interkingdom signaling in plant-rhizomicrobiome interactions for sustainable agriculture. Microbiol Res:126589

Prescott JF (1991) *Rhodococcus equi*: an animal and human pathogen. Clin Microbiol Rev 4:20–34

Pršić J, Ongena M (2020) Elicitors of plant immunity triggered by beneficial bacteria. Front Plant Sci 11:594530

Qu Q, Zhang Z, Peijnenburg W et al (2020) Rhizosphere microbiome assembly and its impact on plant growth. J Agric Food Chem 68:5024–5038

Quiza L, St-Arnaud M, Yergeau E (2015) Harnessing phytomicrobiome signaling for rhizosphere microbiome engineering. Front Plant Sci 6:507

Rahman M, Sabir AA, Mukta JA et al (2018) Plant probiotic bacteria *Bacillus* and *Paraburkholderia* improve growth, yield and content of antioxidants in strawberry fruit. Sci Rep 8:1–11

Ratet P (2019) Symbiosis signaling: Solanaceae symbiotic LCO receptors are functional for rhizobium perception in legumes. Curr Biol 29:R1312–R1314

Raymaekers K, Ponet L, Holtappels D et al (2020) Screening for novel biocontrol agents applicable in plant disease management—a review. Biol Control 144:104240

Rilling JI, Acuña JJ, Sadowsky MJ et al (2018) Putative nitrogen-fixing bacteria associated with the rhizosphere and root endosphere of wheat plants grown in an andisol from southern Chile. Front Microbiol 9:2710

Rodriguez PA, Rothballer M, Chowdhury SP et al (2019) Systems biology of plant-microbiome interactions. Mol Plant 12:804–821
Rojas-Rojas FU, López-Sánchez D, Meza-Radilla G et al (2018) The controversial *Burkholderia cepacia* complex, a group of plant growth promoting species and plant, animals and human pathogens. Rev Argent Microbiol 51:84–92
Sah S, Singh N, Singh R (2017) Iron acquisition in maize (*Zea mays* L.) using *Pseudomonas* siderophore. 3 Biotech 7:1–7
Saijo Y, Loo EP, Yasuda S (2018) Pattern recognition receptors and signaling in plant–microbe interactions. Plant J 93:592–613
Sakamoto K, Ogiwara N, Kaji T et al (2019) Transcriptome analysis of soybean (*Glycine max*) root genes differentially expressed in rhizobial, arbuscular mycorrhizal, and dual symbiosis. J Plant Res 132:541–568
Sarkar D, Rovenich H, Jeena G et al (2019) The inconspicuous gatekeeper: endophytic *Serendipita vermifera* acts as extended plant protection barrier in the rhizosphere. New Phytol 224:886–901
Sasse J, Martinoia E, Northen T (2018) Feed your friends: do plant exudates shape the root microbiome? Trends Plant Sci 23:25–41
Schouteden N, De Waele D, Panis B et al (2015) Arbuscular mycorrhizal fungi for the biocontrol of plant-parasitic nematodes: a review of the mechanisms involved. Front Microbiol 6:1280
Shastri B, Kumar R (2019) Microbial secondary metabolites and plant–microbe communications in the rhizosphere. In: Singh JS (ed) New and future developments in microbial biotechnology and bioengineering. Elsevier, Amsterdam, pp 93–111
Shi S, Tian L, Ma L et al (2018) Community structure of rhizomicrobiomes in four medicinal herbs and its implication on growth management. Microbiology 87:425–436
Shi S, Chang J, Tian L et al (2019) Comparative analysis of the rhizomicrobiome of the wild versus cultivated crop: insights from rice and soybean. Arch Microbiol 201:879–888
Sindhu SS, Phour M, Choudhary SR et al (2014) Phosphorus cycling: prospects of using rhizosphere microorganisms for improving phosphorus nutrition of plants. In: Parmar N (ed) Geomicrobiology and biogeochemistry. Soil biology, vol 39. Springer, Berlin, Heidelberg, pp 199–237
Singh JS (2019) New and future developments in microbial biotechnology and bioengineering: microbes in soil, crop and environmental sustainability. Elsevier, Amsterdam
Singha LP, Sinha N, Pandey P (2018) Rhizoremediation prospects of polyaromatic hydrocarbon degrading rhizobacteria, that facilitate glutathione and glutathione-S-transferase mediated stress response, and enhance growth of rice plants in pyrene contaminated soil. Ecotoxicol Environ Saf 164:579–588
Stephens K, Bentley WE (2020) Synthetic biology for manipulating quorum sensing in microbial consortia. Trends Microbiol 28(8):633–643
Sunita K, Mishra I, Mishra J, Prakash J, Arora NK (2020) Secondary metabolites from halotolerant plant growth promoting rhizobacteria for ameliorating salinity stress in plants. Front Microbiol 11:567768
Tian L, Zhou X, Ma L et al (2017) Root-associated bacterial diversities of *Oryza rufipogon* and *Oryza sativa* and their influencing environmental factors. Arch Microbiol 199:563–571
Tian L, Chang J, Shi S et al (2020a) Comparison between wild and related cultivated rice species reveals strong impacts of crop domestication on methane metabolism of the rhizomicrobiome. Res Squ. https://doi.org/10.21203/rs.3.rs-28865/v1
Tian L, Shi S, Ma L et al (2020b) Community structures of the rhizomicrobiomes of cultivated and wild soybeans in their continuous cropping. Microbiol Res 232:126390
Tian T, Reverdy A, She Q et al (2020c) The role of rhizodeposits in shaping rhizomicrobiome. Environ Microbiol Rep 12:160–172
Turner TR, James EK, Poole PS (2013) The plant microbiome. Genome Biol 14:1–10
van Dam NM, Bouwmeester HJ (2016) Metabolomics in the rhizosphere: tapping into belowground chemical communication. Trends Plant Sci 21:256–265
Venturi V, Keel C (2016) Signaling in the rhizosphere. Trends Plant Sci 21:187–198

Verma M, Mishra J, Arora NK (2019) Plant growth-promoting rhizobacteria: diversity and applications. In: Sobti RC, Arora NK, Kothari R (eds) Environmental biotechnology: for sustinable future. Springer, Singapore, pp 129–173

Vigneron N, Radhakrishnan GV, Delaux P-M (2018) What have we learnt from studying the evolution of the arbuscular mycorrhizal symbiosis? Curr Opin Plant Biol 44:49–56

Vishwakarma K, Kumar N, Shandilya C et al (2020) Revisiting plant–microbe interactions and microbial consortia application for enhancing sustainable agriculture: a review. Front Microbiol 11:3195

Whiteley M, Diggle SP, Greenberg EP (2017) Progress in and promise of bacterial quorum sensing research. Nature 551:313–320

Yanes ML, Bajsa N (2016) Fluorescent *Pseudomonas*: a natural resource from soil to enhance crop growth and health. In: Castro-Sowinski S (ed) Microbial models: from environmental to industrial sustainability. Microorganisms for sustainability, vol 1. Springer, Singapore, pp 323–349

Ye T, Zhou T, Li Q et al (2020) *Cupriavidus* sp. HN-2, a novel quorum quenching bacterial isolate, is a potent biocontrol agent against *Xanthomonas campestris* pv. *Campestris*. Microorganisms 8:45

Yuan M, Huang Y, Ge W et al (2019) Involvement of jasmonic acid, ethylene and salicylic acid signaling pathways behind the systemic resistance induced by *Trichoderma longibrachiatum* H9 in cucumber. BMC Genomics 20:1–13

Zaidi A, Khan MS, Saif S et al (2017) Role of nitrogen-fixing plant growth-promoting rhizobacteria in sustainable production of vegetables: current perspective. In: Zaidi A (ed) Microbial strategies for vegetable production. Springer, Cham, pp 49–79

Ecology of Arbuscular Mycorrhizae and Influence on Drought Tolerance in Crop Plants

8

Kamlesh K. Meena, Praveen Kumar, Ajay M. Sorty, U. Bitla, and H. Pathak

Abstract

Drought stress critically affects plant growth and productivity. Alleviation of drought-induced detrimental effects in plants is urgently required to achieve sustainable crop production. Consequently, there is constant demand of controlling strategies that can sustainably promote growth, development, and productivity of plants under limited moisture conditions. Among the proposed strategies, use of arbuscular mycorrhizal fungi (AMF) has gained significant attention due to their multifaceted capabilities. AMF-induced tolerance in plants against abiotic stresses such as heat, salinity, drought, and extreme temperatures is well-known. AMF symbiosis significantly strengthens the host plant against multiple abiotic stresses including drought and improves productivity. This chapter will mainly explore the ecology of AMF, their interaction with host plant and soil microbial communities, and influence of AMF-plant-soil microbiome interactions on plant-drought tolerance.

Keywords

Arbuscular mycorrhizal fungi · Drought · AMF interactions · Soil microbiology · Triangular response

K. K. Meena (✉) · P. Kumar
Division of Integrated Farming System, ICAR-Central Arid Zone Research Institute, Jodhpur, Rajasthan, India
e-mail: kk.meena@icar.gov.in

A. M. Sorty · U. Bitla · H. Pathak
Agricultural Microbiology Laboratory, School of Edaphic Stress Management, ICAR-National Institute of Abiotic Stress Management, Baramati, Maharashtra, India

N. K. Arora, B. Bouizgarne (eds.), *Microbial BioTechnology for Sustainable Agriculture Volume 1*, Microorganisms for Sustainability 33,
https://doi.org/10.1007/978-981-16-4843-4_8

8.1 Introduction

Drought is an increasingly detrimental abiotic stress that negatively affects the growth and development of plants. To restrict the ill effects of drought on crop productivity, availability of diverse adaptation and mitigation measures are needed (Nilsen and Orcutt 1996). Drought stress induces cellular damage through increased production of reactive oxygen species (ROS). These lethal molecules (ROS) can cause oxidative destruction of proteins, DNA, and lipids and critically affect vital cellular processes (Miller et al. 2010; Impa et al. 2012). To sustain crop yield under limited water supply, an effective rooting system is essential for plants to take up nutrients and water (Fohse et al. 1988). Arbuscular mycorrhizal fungi (AMF) are a unique class of beneficial soil microorganisms that can form a symbiotic association with plant roots forming arbuscular mycorrhizae (AMs), which play a vital role in regulating the plant growth even under variety of unfavorable abiotic circumstances including drought (Philippot et al. 2013). Most ancient, well-known mutualistic association is the endomycorrhizal symbiosis, in which specific soil fungi—arbuscular mycorrhizal (AM) fungi—actively colonize the roots of up to 74–90% of the land plants on earth (Bonfante and Genre 2010; van der Heijden et al. 2015). Generally, AMF exhibit a broad-spectrum symbiosis; Klironomos (2000) confirmed that AMF were not host specific but strong functional variants of fungal species. Owing to the multifaceted role of AMF, their symbiotic association with plant roots is a well-known approach to improve plant tolerance to environmental stress such as drought and phosphorus (P) limitation (Brachmann and Parniske 2006). AMF provide nutrients to their host plants by producing hyphae that grow out from plant roots, effectively increasing the soil volume from which immobile nutrients can be acquired. This fact is corroborated by the phenomenon that most agricultural crops perform better and are more productive when they are colonized by AMF compared with non-mycorrhizal plants. Apart from beneficial effects on plant biomass, AMF also frequently increase plant-nutrient concentrations by facilitating the nutrient uptake and improve the amounts of secondary metabolites in plants (Mohamed et al. 2014).

Microbial communities native to the soil also play key role in the germination and hyphal growth of AMF. Microbes promote the establishment of AMF and also participate in the process of plant growth promotion. More importantly AMF bring modification of the root architecture for improving access of water and nutrients (Hameed et al. 2014). AMF symbiosis-mediated growth elevation is not only due to enhanced uptake of water and mineral nutrients from the adjoining soil but also due to protection of the plants from fungal pathogens (Smith and Read 2008).

Despite having significant knowledge on influence of bacteria and AMF on plant growth and nutrition, there is a general lack of information on their performance under natural drought conditions. More studies elucidating the effect of these microorganisms are urgently needed to mitigate the influence of changing agro-climate conditions on crop productivity. Further, there is a dearth of knowledge on plant-AMF-soil microbe interactions under drought conditions. In this chapter we focus on importance of collective studies on plant-AMF-soil microbe interactions and their key role in alleviation of drought stress in plants.

8.2 Plant-AMF Interactions

Arbuscular mycorrhizal (AM) symbiosis is among the hoariest and most prevalent plant-fungal symbioses (Redecker et al. 2000; Smith and Read 2008). AMF can form symbiotic associations with plants and affect interspecific interactions by enlarging nutrient-absorbing part of plant root systems and further promote plant nutrient and water uptake and also induce tolerance to abiotic stress conditions (Smith and Read 2008). AMF are microscopic filamentous fungi that colonize the plant roots and rhizosphere and spread several centimeters in the form of ramified filaments (Bonfante and Genre 2010) that perform an array of biological interactions in the soil habitat.

8.2.1 AMF, their Types, and Diversity

AMF are omnipresent root symbionts of above 90% of vascular plants and over 80% of the existing terrestrial plants (Wang and Qui 2006). Taxonomy of AMF is described in 214 known species in 4 orders, 13 families, and 19 genera, in the class *Glomeromycetes* of the phylum *Glomeromycota* (Muthukumar et al. 2009). According to the fossil archives and molecular data, AM symbiosis is evolutionarily primitive and could have had occurred with the first terrestrial plants (Lee et al. 2012). AMF or *Glomeromycota* are obligate symbiotic fungi that penetrate plant roots and form the arbuscules—a highly specialized hyphal structure that develops inside cortex cells and represents the main site of nutrient exchange between partners (Smith and Read 2008; Schußler and Walker 2010). AMF represent elements which are important to agricultural productivity and biogeochemical processes (van der Heijden et al. 2015). These fungi are known to play an important ecological role in management practices in low-input agricultural systems (Ma et al. 2005). There are two distinct kinds of AMF, categorized by intraradical hyphal adaptations which are also termed as the *Paris*-type where hyphal development is exclusively intracellular, forming the characteristic coil-like structures in host plants' cortical cells, while the other one is *Arum*-type, which is characterized by intraradical hyphal development that mostly remains intercellular and forms arbuscules in root cortical cells. AMF are mostly of asexual type in reproduction; however, exchange genetic material between genetically distinct strains takes place when their hyphae anastomose (Hijri and Sanders 2005).

8.2.2 Establishment of Symbiotic Association

The contribution of plant-derived and fungal signaling molecules that promote AM association encompasses a high degree of adaptation and genetic/metabolic coordination between mycorrhizal partners (Foo et al. 2013). Establishment of AM symbiosis is an intricate process involving a dedicated signaling pathway initiated by root-borne signal strigolactones, which are known to stimulate AM fungal activity (Van de Velde et al. 2017). AMF respond to the plant signals through consequent secretion of lipochito-oligosaccharide moieties, which induce a signal transduction pathway in plants upon perception. As this signal transduction pathway exhibits significant similarity with the one involved in root-nodule symbiosis, it is also known as the common symbiosis signaling pathway (CSSP), which has been elucidated in great detail in recent years (Gutjahr and Parniske 2013). Recent studies showed that formation of an infection structure called prepenetration apparatus (PPA) promotes cellular invasion (Genre et al. 2008) and development of the intracellular arbuscules that serve as nutritional edge between the associates (Gutjahr and Parniske 2013). Although the molecular mechanism involved in PPA formation is elusive, PPAs are thought to be a requirement for AMF contagion of host roots and to necessitate signaling through the CSSP (Genre et al. 2005). Establishment of AM is concomitant with an intricate process of reprogramming of the host cells including the stimulation of hundreds of genes (Calabrese et al. 2017) of which some are expressed absolutely in cells with arbuscules. Majority of these genes are thought to be indispensable for intracellular establishment, functioning of the fungus and for correlation of symbiotic activity.

8.3 Significance of AMF Symbiosis

AMF promote many aspects of plant life. In particular, AMF improve plant nutrition and enhance growth, stress tolerance, and disease resistance. In addition, the hyphal networks of AMF improve soil characters such as soil particle aggregation, thereby improving the resistance of soil towards erosion due to wind and water. Finally, AMF decrease nutrient leaching from the soil and contribute to retention of nutrients in the soil and also decrease the risks of groundwater contamination.

8.3.1 Nutritional Enhancement

Importance of AMF symbiosis in nutrient transport has been extensively studied. Capability of nutrient transport varies with AMF species and is highly dependent on their root colonization ability. For instance, members of *Gigasporaceae* are sluggish root colonizers; however, they are superior soil colonizers, where they form denser extraradical mycelium network than the members of the *Glomeraceae* (Hart and Reader 2002a). Therefore, it should be clearly noted that *Gigasporaceae* are more essential in soil structure establishment and maintenance than *Glomeraceae*.

Gigasporaceae seem to be less efficient in transferring P to the host plant when compared with *Glomeraceae* (Hart and Reader 2002b). AMF possess high-affinity inorganic phosphate (Pi) transporters; the expression of this Pi transporter was localized to the extraradical hyphae of *Glomus versiforme*, which is the main site of phosphate uptake from the soil. Accumulated as polyphosphate, Pi is then rapidly translocated along the aseptate mycelium to the host plant (Hijikata 2010). In mycorrhizal plants, nearly 70% of the whole phosphate uptake can be acquired via the AM pathway (Yang et al. 2012). Recent studies revealed nitrogen uptake in the AM symbiosis, with a significant role played equally for plant nutrition and for the regulation of the symbiosis functioning itself. In the soil, inorganic nitrogen is present as nitrate (NO_3) and ammonium ($NH4^+$), and AMF possess specific transporters for both the N forms. In *Rhizophagus irregularis*, three sequences refer to ammonium transporters, and one nitrate transporter has been identified (Tisserant et al. 2013). Role of sulfur (S) in AM symbiosis is poorly understood. However, some studies have reported AMF-mediated S transfer in plants. Allen and Shachar-Hill (2009) reported that AMF can take up both organic and inorganic S and transfer it to the host plant. *Lotus japonicas* and *Medicago truncatula* have sulfate transporters *LjSultr1;2* and *MtSultr1;2* that respond to mycorrhizal symbiosis (Guether et al. 2009). Macronutrient like potassium (K) is not well studied with respect to AMF-plant symbiosis. A few studies have been performed in this regard where a number of plant K transporters were identified like *Trk* (transporter of K), *HAK* (high-affinity towards K uptake), and *SKC* (Shaker-like channels), but their role in AMF symbiosis is still not demonstrated (Garcia and Zimmermann 2014).

8.3.2 Abiotic Stress Tolerance

Importance of AMF-plant symbiosis in mitigation of abiotic stresses like drought, salinity, high temperature, and heavy metal in different agriculture crop is well-known (Moradtalab et al. 2019; El-Nashar 2017). AMF symbiosis protects plants against a variety of abiotic stresses through various processes such as improved photosynthetic rate, uptake and accumulation of mineral nutrients, cellular accumulation of osmoprotectants, upregulation of antioxidant enzyme activity, and change in the rhizosphere microecosystem (Table 8.1) (Bárzana et al. 2015; Calvo-Polanco et al. 2016; Yin et al. 2016). Exploiting AMF symbiosis may boost P and N uptake and improve plant growth and productivity. The mycorrhizosphere formed by roots and AMF mycelium in soil effectively enhances water and nutrient uptake because of its larger expanse (Smith and Smith 2012). AMF symbiosis decreases oxidative stress through declining membrane lipid peroxidation in plant under salinity stress (Abdel Latef and Chaoxing 2014). Further, it also improves integrity of the plasma membrane by lowering lipid peroxidation and alleviating electrolyte leakage due to membrane damage (Evelin et al. 2012). AMF defend plants against detrimental effects of ROS by increasing antioxidant enzymes like superoxide dismutase (SOD), catalase (CAT), peroxidase (POD), ascorbate peroxidase (APX), and glutathione reductase (GR) (Table 8.1) (Evelin and Kapoor 2014). Improvement in cold

Table 8.1 Alleviation of abiotic stresses through AMF inoculation

Sr. no.	Types of AMF	Crop	Stress	Function	Reference
1	*Rhizophagus irregularis*, *Funneliformis mosseae*, *Funneliformis geosporum*, *Claroideoglomus claroideum*	*Triticum aestivum* L.	Heat	Increased grain quantity, nutrient distribution, and nutrient composition in root	Cabral et al. (2016)
2	*Glomus* sp.	Tomato	Salt stress	Amplified leaf area, and nutrient content, reflecting superior plant growth	Balliu et al. (2015)
3	*Glomus iranicum* var. *tenuihypharum* sp.	*Euonymus japonica*	Sewage water irrigation and salinity stress	Enhanced levels of P, Ca, and K	Gómez-Bellot et al. (2015)
4	*Glomus mosseae* and *Glomus intraradices*	*Pistachio* plants	Drought stress	Increased levels of P, K, Zn, and Mn	Bagheri et al. (2012)
5	*Funneliformis mosseae* and *Diversispora versiformis*	*Chrysanthemum morifolium (Hangbaiju)*	Salt stress	Increased root length, shoot and root dry weight, total dry weight, and root N concentration	Wang et al. (2018)
6	*Glomus intraradices*	*Medicago sativa* L.	Cd stress (heavy metal stress)	Plant growth enhancement, increased roots and shoot biomass	Wang et al. (2012)
7	*Rhizophagus intraradices*, *Funneliformis mosseae*, *Funneliformis geosporum*	*Zea mays*	High temperature	Increased leaf length, plant height, leaf number, chlorophyll a, photosynthetic rate, stomatal conductance, and transpiration rate	Mathur et al. (2018)
8	*Glomus monosporum*, *Glomus clarum*, *Gigaspora nigra*, and *Acaulospora laevis*	*Trigonella foenum-graecum* L.	Cadmium	Increased antioxidant enzymes activities and malondialdehyde content	Abdelhameed and Rabab (2019)
	Rhizophagus irregularis	*Solanum lycopersicum* L.	Salinity	Improved shoot FW, leaf area, leaf number, root FW, and levels of growth hormones	Khalloufi et al. (2017)

9	*Glomus intraradices*	*Solanum lycopersicum* L.	Salinity	Enhanced dry matter, ion uptake, growth parameters, and chlorophyll content	Hajiboland et al. (2010)
10	*Glomus mosseae*	*Leymus chinensis*	Salinity-alkali	Improved colonization level, seedling mass, water contents, and both P and N	Lin et al. (2017)

stress tolerance in plants due to inoculation of AMF was reported by Gamalero et al. (2009), where the authors linked AMF-induced mitigation through enhancement of photosynthesis. Cucumber root colonization by AMF might efficiently enhance the accumulation of phenolics, flavonoids, and lignin in the leaves and significantly reduce H_2O_2 accumulation under low-temperature conditions (Chen et al. 2013). Elevated temperature, however, negatively impacts the process of mycorrhizae development, which can also hamper AMF colonization, spore germination, and hyphal growth of the fungi (Jahromi et al. 2008). AMF-mediated improvements in photosynthetic rate, stomatal conductance, and transpiration rate indicate that AM symbiosis can deliver high gas exchange ability by reducing stomatal resistances and increasing CO_2 assimilation and transpiration fluxes (Zhu et al. 2012). These outcomes evidently demonstrate that AM symbiosis can efficiently protect the PSII reaction center and structural and functional disruption of photosynthetic apparatus from stress-induced damages. AMF symbiosis plays a key role and also improves the plant health in heavy metal-contaminated soils (Nadeem et al. 2014). AMF-induced enhanced plant tolerance towards heavy metals can be attributed to the mechanisms like restriction of heavy metals by AMF-secreted biomolecules, precipitation in polyphosphate granules in the soil, metal adsorption to chitin in the cell wall, chelation of metals inside the fungus, changes in rhizosphere pH, and regulation of variety of stress-responsive genes under stress conditions (Malekzadeh et al. 2011).

8.3.2.1 Rhizosphere Microbiology and Biochemistry

AM colonization exerts significant influence on the species composition of the soil microbial community by aggregating some groups and declining the others (Krishnaraj and Sreenivasa 1992). Participation of AM symbiosis in numerous physiological and biochemical processes including direct uptake and transfer of water and nutrients by AMF improves osmotic regulation, enhanced gas exchange and water use efficiency, and superior protection against oxidative damage has also been reported (Rapparini and Penuelas 2014). Sustenance of AMF hyphae improves physical defense of SOM by stimulating soil aggregation, which is important for sandy soils and reclaimed mine soil where other binding agents are scarce (Birgitte and Leif 2000). AM symbiosis can enhance decomposition and increase N capture from organic material; mycorrhizal hyphae can also produce phosphatase to hydrolyze the organic P compounds (Hodge and Fitter 2010). Available nutrient content in the rhizosphere soil directly affects soil enzyme activity (He et al. 2010). Soil enzymes activity is related to the nutrients such as mineralization of N, P, and C and can be used as indicator to detect changes in the fertility and microbial functioning of soil. Saccharase enzyme imitates the law of accumulation and decomposition of soil organic carbon and indicates soil C cycling and important biochemical activities. The saccharase activity in the AMF-inoculated plants appeared significantly higher than that in the AMF-non-inoculated treatments; also enhanced activity of dehydrogenase, urease, and phosphatases in AMF-inoculated treatments remained higher than those in AMF-non-inoculated plants (Qian et al. 2012). Esterase, trehalase, phosphatase, and chitinase activities were higher in the rhizosphere of mycorrhiza-inoculated maize (Vázquez et al. 2000). Enhanced invertase, catalase, urease, and alkaline phosphatase activities were observed in rhizospheric soil of AMF-inoculated *Amorpha fruticosa* (Qin et al. 2018) and also showed higher root mycorrhizal colonization, glomalin-related soil protein concentrations, as well as an increase in soil organic carbon (SOC), total N, Olsen P, and available K.

8.4 AMF Interactions with Indigenous Soil Microbes

Rhizospheric microbial communities contribute a vital role in plant health and development. Bacteria appear to represent the third component of AM symbiosis because there exists an influence by bacteria associated with the mycorrhizosphere on establishment and function of AMF (Frey-Klett et al. 2007). Diverse types of bacterial communities thrive within the mycorrhizosphere, where they form association with mycorrhizal roots, spores, sporocarps, and extraradical hyphae (Rambelli 1973). From *Rhizoglomus irregulare* spores, as many as 374 bacterial strains were isolated (Battini et al. 2016).

In general, three ways of interaction of between AMF and other microbial communities could be possible—synergistic, antagonistic, and neutral. Literature is available to generate an overview of these interactions. Mycorrhizospheric microbiota performs characteristic functional activities such as contributing as mycorrhiza helpers, in addition to direct plant growth promotion. Mycorrhizae helper bacteria may accelerate the germination of spores and establishment of mycorrhizal symbiosis. As an example, *Streptomyces* spp., *Pseudomonas* sp., and *Corynebacterium* sp. improved the germination of *Funneliformis mosseae*, *G. versiforme*, and *Glomus margarita* spores (Frey-klett et al. 2007). The enhancement of spore germination was credited to *Actinobacteria*—a group of bacteria that regularly interact with AMF spores, able to hydrolyze chitin which is a key component of the spore walls (Agnolucci et al. 2015). *Klebsiella pneumoniae*, *Trichoderma* sp., and *Paenibacillus validus* increased germlings hyphal growth (Calvet et al. 1992; Hildebrandt et al. 2006), while a bacterial strain belonging to *Oxalobacteraceae* not only improved spore germination and germling growth but also enhanced root colonization to establish symbiosis (Pivato et al. 2009). In addition, the development of AMF extraradical mycelia may be promoted by strains of *Paenibacillus rhizosphaerae*, *Azospirillum* sp., *Rhizobium etli*, *Pseudomonas* spp., *Burkholderia cepacia*, and *Ensifer meliloti* (Bidondo et al. 2011; Ordoñez et al. 2016; Battini et al. 2017).

Microbial biocontrol agents are capable to produce antibiotic effect, while some antifungal components originating from non-AMF microbes can affect the process of spore production and mycelial growth of AMF (Bitla et al. 2017). Germination rate of *G. mosseae* was accelerated, and expansions of mycelia from germinated spores were improved due to the presence of *Trichoderma* spp.; however, the presence of *Penicillium decumbens* and *Aspergillus fumigatus* inhibited spore germination of *G. mosseae* (Calvet et al. 1992). Some of the plant growth-promoting microbial agents could produce volatile, hormones and other secondary metabolites (Sorty et al. 2016; Meena et al. 2019) that can be detrimental to growth of AMF hypha and spores; *Bacillus subtilis* produces some unknown volatiles that inhibit the spore germination and hyphal growth of *Glomus etunicatum* (Xiao et al. 2008). Filion et al. (1999) observed that conidial germination of *Trichoderma harzianum* was enhanced by the fungal extract secreted by AMF while the *Fusarium oxysporum* f. sp. *chrysanthemi* were decreased; these effects were directly correlated with extract concentration. AMF- and phosphate-solubilizing bacteria (PSB) could interact synergistically because PSB solubilize thriftily accessible phosphorous compounds into orthophosphate that AMF can absorb and transport to the host plant (Nacoon et al. 2020).

8.4.1 Influence on Soil Microbial Community

The colonization of plants by AMF affects the quality and the quantity of the host plant root exudates and hence the structure of microbial communities in the rhizosphere and chemotactic responses of specific bacteria (Tahat and Sijam 2012). Presence of AMF also exerts an influence on soil microorganisms concomitant with their extraradical mycelium, principal to the formation of a specific zone of soil which is generally called the mycorrhizosphere (Artursson et al. 2006). Denaturing gradient gel electrophoresis (DGGE) profiles of 16S rRNA amplicons from total DNA extracts of pea rhizosphere appeared relatively similar between AMF-inoculated and AMF-non-inoculated rhizosphere. However, AMF (*Glomus intraradices*)-inoculated plants showed suppression of four to five specific bright bands (Wamberg et al. 2003). Direct root colonization with either *G. mosseae* or *G. intraradices* significantly modified the DGGE profiles of bacterial community from tomato rhizosphere. Both the AMF species had similar bacterial communities after 4 weeks. The bacterial taxa associated with the rhizosphere of tomato plants inoculated with *G. mosseae* were identified as *Pseudomonas*, *Herbaspirillum*, and *Acidobacterium*, while *Bacillus simplex* (clone TR03) was found to be affiliated only with *G. intraradices* (Lioussanne et al. 2010).

8.4.2 Long-Term Impact on Soil Microbiology

An AMF-rhizosphere microbial interaction significantly influences the soil microbiology and also affects the soil physiological and chemical characteristics. AMF influence the soil structure by binding and enmeshing soil particles into macro-aggregates and by producing glomalin (Rillig and Mummey 2006; Treseder and Turner 2007). Carbon (C) fraction of glomalin can range from 9 to 22%, habitation time in soil from 6 to 42 years and may represent >5% of total soil C (Rillig et al. 2001). Soil aggregates are important for storage of C and nutrients. Soil aggregates are important for flourishing microbial communities, where they provide shelter in the form of micropores. Stable soil aggregates are long-lasting source of C because

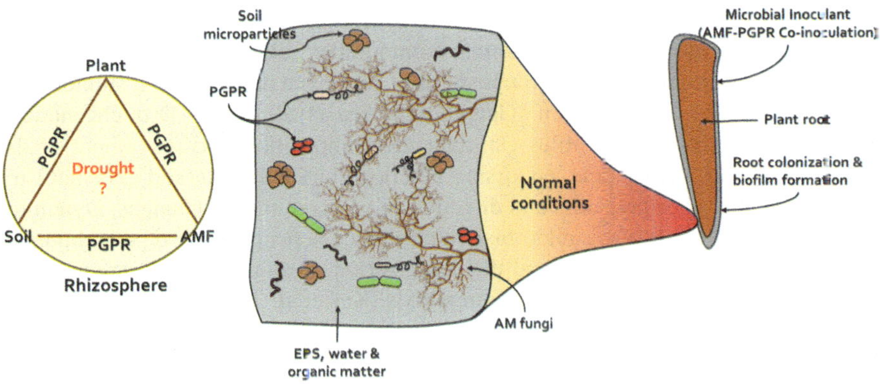

Fig. 8.1 An overview of AMF associations within the plant rhizosphere under normal conditions

they significantly enhance the microbial activity and the process of C sequestration in soil (Fig. 8.1). Soil C is important for all microbial activities that are connected to the phenomenon of microbial plant growth promotion. AMF-rhizosphere microbial interaction maintains the C:N ratio through symbiotic exchange of nutrients and C. Long-term mesocosm exploration with single AMF strain showed that there is significant influence on the diversity of associative microbial communities and on alterations in aggregate formation. In a diverse AM community, this may specify a highly composite biotic heterogeneity in mycorrhizosphere soils (Rillig et al. 2005). Fascinatingly, investigators constantly demonstrate enhanced concentration of available P in the smallest aggregates, in ultisols (Thao et al. 2008). AMF symbiosis affects the diversity of the microbial community as well as their function within the rhizospheric soil (Schreiner et al. 1997; Olsson et al. 1997). AMF can improve the quantity and movement of beneficial soil microorganisms including nitrogen fixers and phosphate solubilizers, which are important for growth, development, and productivity of plants (Linderman 1992).

8.5 Impact of Drought Stress on Plant-AMF-Rhizosphere Microbial Interactions

8.5.1 Recognition of Drought Conditions and Consequences

Plant-microbe interactions at biochemical, physiological, and molecular levels established that microbial associations largely direct plant responses towards environmental stress(es) (Meena et al. 2017, 2020). Extended moisture stress decreases leaf water potential and stomatal opening; reduces leaf size; suppresses root growth; decreases seed number, size, and sustainability; delays flowering and fruiting; and restricts plant growth and productivity (Osakabe et al. 2014). Drought condition also affects the plant root exudation (Sorty et al. 2018; Bitla et al. 2017). Root exudates act as direct communication pathway between plant and rhizosphere microbial communities; deviation in the composition of root exudates governs the diversity and function of the colonizing microbial communities (Sasse et al. 2018). Drought-induced changes in amount or composition of root exudates may be a situation- or species-dependent phenomenon (Gargallo-Garriga et al. 2018). Drought-induced changes in root exudate profiles can alter the composition and movement of the surrounding soil microbiome, endorsing additional alterations to soil geochemistry that in turn alter enormousness and direction of soil community changes. Dynamics and structure of rhizosphere microbial communities depend on the type and quality of the decaying root materials and the composition of root exudates, ranging from simple sugars to complex aromatic compounds (Kos et al. 2015). Drought condition has a substantial influence on the biomass and structure of soil microbial communities and their involvement in nutrient cycling (Schimel 2018). Water stress not only directly affects microbial community but also modifies the relative significance of the impact on plant AMF interaction (Monokrousos et al. 2020). Although the available literature signifies beneficial interaction of AMF under drought conditions, the knowledge relating to the underlying mechanisms of AMF-mediated plant growth enhancement under drought conditions and the potential beneficial effects on soil structure and indigenous microbial community is very limited. Thus, rigorous initiatives are critically needed to address the major knowledge gaps in the area of AMF-plant-microbe interactions (Fig. 8.1).

8.5.2 Counter Responses of AMF to Drought Conditions

AMF symbiosis might play a significant role in alleviating the impact of drought on crop yield. AMF symbiosis could considerably enhance plant resistance to drought via increased dehydration avoidance and increased tolerance (Ruiz-Lozano and Azcon 2000). The root colonization by the AMF is represented by active absorptive surface zone and enhanced water and nutrient uptake even under limited moisture conditions. AMF symbiosis could increase drought tolerance via the increased soil water drive to the plant roots (Ruiz-Lozano et al. 2016). Impact of AMF on drought-plant relation also extends to the soil environment in terms of higher rate of soil aggregate formation and improved soil water conditions to the rhizosphere (Rillig et al. 2002). AMF also regulate plant physiological performance to alleviate drought stresses by the upregulation of antioxidant enzyme activity and jasmonate synthesis to reduce peroxidative damage and also improve stomatal conductance (Pedranzani et al. 2016). AMF-mediated enhancement of drought tolerance in C3 (*Leymus chinensis*) and C4 (*Hemarthria altissima*) plant species through upregulation of antioxidant system has been reported (Li et al. 2019). Colonization of AMF can modulate morphological adaptation to improve drought tolerance of the host plant (Table 8.2). Previous studies showed that less epicuticular wax and lower cuticle weight were observed in leaves of AMF-inoculated rose plants than non-AMF-inoculated plants during drought acclimation (Henderson and Davies 1990). Fluctuations are observed in root structure of trifoliate orange inoculated with or without *F. mosseae*. Considerably enhanced root total length, projected area, surface

Table 8.2 Alleviation of drought stress through AMF inoculation

Sr. no.	AMF strain	Host plant	Abiotic stress	Influence on host	Reference
1	*Funneliformis mosseae*	Rose geranium	Water stress	Increased the concentrations of N, P, and Fe	Amiri et al. (2017)
2	*Glomus mosseae* and *Glomus intraradices*	*Pistachio* plants	Drought stress	Nutrient content is improved	Bagheri et al. (2012)
3	*Glomus deserticola*	*Antirrhinum majus* cv. *Butterfly*	Water stress	Increase the shoot and root dry mass, rise the water use efficiency, flower yield, nutrient and chlorophyll content	Asrar et al. (2012)
4	*Septoglomus constrictum*	*Tomato*	Drought and heat stress	Enhanced stomatal conductance, leaf water potential, and relative water content	Duc et al. (2018)
5	*Rhizophagus irregularis*	Tomato and lettuce	Drought stress	Improved growth rate and efficiency of photosystem II, increase strigolactone production	Ruiz-Lozano et al. (2016)
6	*Glomus mosseae*	Sunflower	Drought	Increase dry matter, heavier seeds and greater seed and oil yield, increase the N percentage in leaves and seeds	Gholamhoseini et al. (2013)
7	*Rhizophagus irregularis*	Hybrid poplar	Drought	A higher net photosynthetic rate and Chl fluorescence, better water use efficiency and water uptake, enhanced growth, and reduced loss of biomass	Liu et al. (2015)
8	*Glomus intraradices*	Rice	Drought	Enhanced the photosynthetic efficiency and the antioxidative enzyme activity	Ruiz-Sánchez et al. (2010)
9	*Glomus versiforme*	Citrus	Drought	Improving reactive oxygen metabolism	Wu and Zou (2009)
10	*Funneliformis mosseae* and *Paraglomus occultum*	Trifoliate orange	Drought	Escalated the leaf sucrose and fructose concentration, significantly increase the leaf sucrose	Wu et al. (2017)

(continued)

Table 8.2 (continued)

Sr. no.	AMF strain	Host plant	Abiotic stress	Influence on host	Reference
				phosphate synthase and neutral invertase, regulating the sucrose- and proline-metabolized enzyme activities	
11	*Glomus mosseae* or *Glomus etunicatum*	Wheat	Drought	Improved growth, yield, and nutrient uptake	Al-Karaki et al. (2004)
12	*G. mosseae*	*Wheat*	Drought	Enhancement of P, Zn, Cu, Mn, and Fe uptake	Al-Karaki and Al-Raddad (1997)
13	*G. mosseae*	(*Zea mays* L.)	Drought	Increased P uptake	Abdelmoneim et al. (2014)
14	*Glomus* spp.	*Ipomoea batatas*	Drought	Proline and soluble sugars adjust osmotic potential	Yooyongwech et al. (2016)

area, average diameter, volume, and number of first-, second-, and third-order lateral roots are observed in AMF trifoliate orange seedlings under well-watered and drought situations compared with non-AMF seedlings (Liu et al. 2015, 2016). Comas et al. (2013) observed that alteration in root structure caused by AMF can deliver more exploration of soil volume to absorb water and nutrients from the soil, therefore potentially enhancing drought tolerance of the host plant.

8.5.3 Integrated Response Triangle of Plant-AMF-Rhizosphere Microbes

Under drought condition, root colonization by AMF generally decreases (Augé 2001). Still AMF inoculation in crop is one of the best biological methods to alleviate drought stress. AMF can promote plant drought fitness either via enhanced nutrition or more direct effects on stomatal conductance and enhanced water use efficiency (Augé 2001). Mycorrhizal helper bacteria play a vital role in establishment of AMF symbiosis under drought stress. The symbiotic relationship between legumes and nitrogen-fixing rhizobia is susceptible to drought which induces failure of the infection and nodulation developments (Bouhmouch et al. 2005); however, presence of AMF improves the performance of legume-rhizobial symbiosis under drought. Bacterial strains isolated from AMF spores also exhibit the ability of plant growth promotion; 17 actinobacterial strains were able to produce siderophores and IAA, mineralize phytate and solubilize inorganic phosphate and 10 putative N fixers to produce siderophore and solubilize P (Battini et al. 2016). Nitrogen and P are transported towards the host through AMF extraradical hypha. Inoculation of drought-adapted AMF and *Bacillus thuringiensis* consortium increased growth a of maize under drought stress; it was observed that the co-inoculation increased plant nutrition and plant drought tolerance including regulation of plant aquaporins with several putative physiological functions (Armada et al. 2015).

8.5.4 Influence of the Triangular Response on Plant Performance under Drought Conditions

Under drought conditions, plants undergo variety of morphological and metabolic changes that affect root exudations and soil microbial communities. Plant also alters the root-shoot ratio and stimulates root growth to gain access areas to the moisture from additional part of the adjacent soil. Drought-induced changes in root metabolites may also explain observed changes in root-associated microbiome (Xu et al. 2018). Most microbes are susceptible to limited C. A significant amount of C is provided through root exudates which are thus important governing factors of microbial processes and shaping root-associated microbial communities (Peterson 2003). AMF-induced improved soil aggregation and SOC are known to increase the abundance of plant growth promoting (PGP) bacteria that accelerate the processes like nitrogen fixation, phosphate solubilization, hormone production, and nutrient mobilization. Involvement of phytohormones such as strigolactones, gibberellins, auxin, abscisic acid, ethylene, jasmonic acid, salicylic acid, cytokinins, and brassinosteroids in AMF symbiosis and plant growth is extensively reviewed by Liao et al. (2018). These hormones improve extraradical hyphal growth of AMF (Bidondo et al. 2011) and also improve plant performance under varying abiotic conditions (Meena et al. 2011, 2017; Sorty et al. 2016, 2018). The hyphal growth regulates transport of nutrient and water to host plant under drought conditions. Thus the plant-AMF-soil microbe triangular interactions improve drought tolerance in plants and promote growth, development, and productivity.

Additional to nutrients mobilization, and phytohormones production, PGP microbes are involved in different activities of plant growth promotion through the production of wide types of metabolites and volatile compounds that directly or indirectly enhance plant growth and sustenance (Barea et al. 2002; Rouphael et al. 2015; Sorty et al. 2016, 2018; Meena et al. 2017). These activities also influence AMF-plant symbiosis, thus making it important to consider the activities performed by associative microbiota before selecting the AMF and bacterial combination for co-inoculation. P is quickly immobilized in the soil, forming insoluble compounds with aluminum/iron and with calcium in acid and alkaline soil therefore becoming inaccessible to plants; P-solubilizing bacteria could work in interaction with AMF to enhance P availability and subsequent uptake by the plant. P-mobilizing bacteria, such as *Streptomyces* spp., *Leifsonia* sp., *Bacillus pumilus*, *Lysinibacillus fusiformis*, and *E. meliloti*, isolated from AMF spores of *Rhizoglomus irregulare*, showed synergistic action with AMF, promoting mineralization of soil phytate and facilitating P uptake by mycorrhizal plants (Battini et al. 2017).

Co-inoculation of PGP bacteria and AMF is known to significantly enhance plant growth than inoculation with the PGP bacteria or AMF alone. Co-inoculation also improves crop yield and shows improvement in uptake of N and P than single inoculated crop (Kim et al. 1998; Singh and Kapoor 1998). Most of the researchers have already shown the importance of microbial co-inoculation in the mitigation of drought stress in crops under in vitro and field conditions (Meena et al. 2010). Drought-resistant *B. thuringiensis* inoculated with indigenous AMF isolate could decrease water requirement by 42% in *Retama sphaerocarpa*, a positive AMF-plant growth promoting rhizobacteria (PGPR) consequence on plants grown under drought stress (Marulanda et al. 2006).

8.6 Recent Trends and Advancements on AMF Inoculants in Modern Agriculture

A variety of strategies for application of AMF have been developed during the last few decades. The prominent ones include combination of different AMF sp., AMF-PGPR co-inoculation, AMF inoculation with organic compost, and liquid inoculum. Marketable inocula are prepared with *R. irregulare* (syn. *Rhizophagus irregularis*, formerly *G. intraradices*) and *F. mosseae* (formerly *G. mosseae*) that are generalist symbionts and are common all over the world in nearly all soils and climatic zones (Smith and Read 2008). However, recently a very common practice of using AMF with other commercial biofertilizers of PGP microbes has gained significant popularity. It is a proven fact that co-inoculation of AMF and PGPR under drought stress can induce several beneficial changes in plants. Some studies have reported that co-inoculation of more than one AMF species with bacterial inoculants can significantly promote plant growth in addition to alleviating the drought stress. For instance, co-inoculation of *G. mosseae* and *G. intraradices* along with *Bacillus* sp. alleviates drought responses in *Lactuca sativa* (Vivas et al. 2003). Consortium of *Claroideoglomus etunicatum*, *R. intraradices*, and *F. mosseae* inoculated with *B. subtilis* alleviates salt stress impact on *Acacia gerrardii* (Hashem et al. 2016), while a consortium of *B. thuringiensis*, *Bacillus megaterium*, and *Pseudomonas putida* inoculated with single *R. intraradices* sp. also enhances plant growth under drought stress (Ortiz et al. 2015). Compost-amended substrate inoculated with AMF *Glomus iranicum* in micropropagated date palm plantlets (cv. *Feggous*) revealed increased biomass, chlorophyll, and mineral nutrient contents than plantlets transplanted into compost-amended substrate or without compost amendment (El Kinany et al. 2019). Combination of biochar and AMF *C. etunicatum* (syn. *G. etunicatum*), *Rhizophagus irregularis* (syn. *G. intraradices*), and *F. mosseae* (syn. *G. mosseae*) enhanced the growth of chickpea by improving synthesis of chlorophylls, rate of photosynthesis, membrane stability index (MSI), and relative water content (RWC); it also enhanced the nitrogen fixation ability of *Cicer arietinum*. Plants may have considerably contributed to growth promotion and the superior drought tolerance by retaining the optimal concentration of amino acids and other stress-responsive metabolites (Hashem et al. 2019). Organic compost applied in combination with the inoculation of a native AMF synergistically promoted the growth of *Anthyllis cytisoides* in a neutral mine tailing under field conditions (Kohler et al. 2015). Organic compost amendment enhances soil nutrient status, stimulates mycorrhiza formation, enhances soil microbial function and soil enzyme activity, thereby inducing the nutrient turnover and plant performance. *Glomus hoi* applied as liquid inoculum form in *Oryza sativa* L promoted growth and development and improved the production of rice plants under saline conditions. Liquid inoculum also reduced the detrimental effect of salinity on AMF-like mycorrhizal colonization, and decreased the ectophyte growth in soil (Fernández et al. 2011). Liquid inoculum of AMF also increased the size and number of fruits, especially under deficit irrigation or low nitrogen input condition (Robinson Boyer et al. 2016).

8.7 Future Perspectives on AMF-Based Strategies for Agricultural Sustainability

Owing to their multifaceted function in agricultural systems, AMF are gaining significant popularity at grassroots level. Consequently, there is constantly increasing demand for AMF-based formulations. However, there is still a dearth of knowledge particularly with respect to the appropriate dosage, propagule density, optimization-related issues for varying agro-climatic conditions, post-inoculation survival, establishment and function of AMF under different biogeochemical and environmental regimes, etc. Further, a regulatory framework to determine quality standards of AMF-based products is yet to be formulated. Finally, there remains a huge scope for exploitation of new AMF strains from diverse natural habitats for promoting plant performances. Chronically stressed habitats can serve as excellent resources of resilient AMF diversity. Extensive use of mycorrhizal inocula in agriculture remains challenging due to their cost, variability in terms of quality, responses on plants, and incompatibility with high available P levels in soils under certain circumstances. Further, there is a need to develop the AMF products that are compatible with rhizospheric microbial communities and are tolerant to biotic and abiotic stresses. Although different types of combinations and amendments are already studied, there is a huge scope for optimization of AMF, crop, associated soil microbial communities, and soil physicochemical properties. Thus, there is a critical need to evaluate the AMF-plant-soil microbial response triangle and its importance in alleviation of abiotic and biotic stresses.

Acknowledgments Authors are grateful to the Indian Council of Agricultural Research (ICAR) for their financial support through the network program—Application of Microorganisms in Agriculture and Allied Sectors (AMAAS).

References

Abdel Latef AA, Chaoxing H (2014) Does the inoculation with *Glomus mosseae* improve salt tolerance in pepper plants? J Plant Growth Regul 33:644–653

Abdelhameed RE, Rabab AM (2019) Alleviation of cadmium stress by arbuscular mycorrhizal symbiosis. Int J Phytoremediation 21(7):663–671

Abdelmoneim TS, Tarek AA, Almaghrabi OA (2014) Increasing plant tolerance to drought stress by inoculation with arbuscular mycorrhizal fungi. Life Sci J 11:10–17

Agnolucci M, Battini F, Cristani C, Giovannetti M (2015) Diverse bacterial communities are recruited on spores of different arbuscular mycorrhizal fungal isolates. Biol Fertil Soils 51: 379–389

Al-Karaki GN, Al-Raddad A (1997) Effects of arbuscular mycorrhizal fungi and drought stress on growth and nutrient uptake of two wheat genotypes differing in drought resistance. Mycorrhiza 7:83–88

Al-Karaki G, McMichael B, Zak J (2004) Field response of wheat to arbuscular mycorrhizal fungi and drought stress. Mycorrhiza 14:263–269

Allen JW, Shachar-Hill Y (2009) Sulfur transfer through an arbuscular mycorrhiza. Plant Physiol 149:549–560

Amiri R, Nikbakht A, Etemadi N, Sabzalian MR (2017) Nutritional status, essential oil changes and water-use efficiency of rose geranium in response to arbuscular mycorrhizal fungi and water deficiency stress. Symbiosis 73:15–25

Armada E, Azcón R, López-Castillo OM, Calvo-Polanco M, Ruiz-Lozano JM (2015) Autochthonous arbuscular mycorrhizal fungi and *Bacillus thuringiensis* from a degraded Mediterranean area can be used to improve physiological traits and performance of a plant of agronomic interest under drought conditions. Plant Physiol Biochem 90:64–74

Artursson V, Finlay RD, Jansson JK (2006) Interactions between arbuscular mycorrhizal fungi and bacteria and their potential for stimulating plant growth. Environ Microbiol 8:1–10

Asrar AA, Abdel-Fattah GM, Elhindi KM (2012) Improving growth, flower yield, and water relations of snapdragon *Antirrhinum majus* L. plants grown under well-watered and water-stress conditions using arbuscular mycorrhizal fungi. Photosynthetica 50:305–316

Augé RM (2001) Water relations, drought and vesicular-arbuscular mycorrhizal symbiosis. Mycorrhiza 11:3–42

Bagheri V, Shamshiri MH, Shirani H, Roosta HR (2012) Nutrient uptake and distribution in mycorrhizal pistachio seedlings under drought stress. J Agric Sci Technol 14:1591–1604

Balliu A, Sallaku G, Rewald B (2015) AMF inoculation enhances growth and improves the nutrient uptake rates of transplanted, salt-stressed tomato seedlings. Sustainability 7:15967–15981

Barea JM, Azcón R, Azcón-Aguilar C (2002) Mycorrhizosphere interactions to improve plant fitness and soil quality. Antonie Van Leeuwenhoek 81:343–351

Bárzana G, Aroca R, Ruiz-Lozano JM (2015) Localized and nonlocalized effects of arbuscular mycorrhizal symbiosis on accumulation of osmolytes and aquaporins and on antioxidant systems in maize plants subjected to total or partial root drying. Plant Cell Environ 38:1613–1627

Battini F, Cristani C, Giovannetti M, Agnolucci M (2016) Multifunctionality and diversity of culturable bacterial communities strictly associated with spores of the plant beneficial symbiont *Rhizophagus intraradices*. Microbiol Res 183:68–79

Battini F, Grønlund M, Agnolucci M, Giovannetti M, Jakobsen I (2017) Facilitation of phosphorus uptake in maize plants by mycorrhizosphere bacteria. Sci Rep 7:4686

Bidondo LF, Silvani V, Colombo R, Pérgola M, Bompadre J, Godeas A (2011) Pre-symbiotic and symbiotic interactions between *Glomus intraradices* and two *Paenibacillus* species isolated from AM propagules. In vitro and in vivo assays with soybean (AG043RG) as plant host. Soil Biol Biochem 43:1866–1872

Birgitte NB, Leif P (2000) Influence of arbuscular mycorrhizal fungi on soil structure and aggregate stability of a vertisol. Plant Soil 218:173–183

Bitla UM, Sorty AM, Meena KK, Singh NP (2017) Rhizosphere signaling cascades: fundamentals and determinants. In: Singh DP, Singh HB, Prabha R (eds) Plant-microbe interactions in agro-ecological perspectives. Springer Nature, Singapore, pp 211–226

Bonfante P, Genre A (2010) Mechanisms underlying beneficial plant fungus interactions in mycorrhizal symbiosis. Nat Commun 1:48

Bouhmouch I, Souad-Mouhsine B, Brhada F, Aurag J (2005) Influence of host cultivars and *rhizobium* species on the growth and symbiotic performance of *Phaseolus vulgaris* under salt stress. J Plant Physiol 162:1103–1113

Brachmann A, Parniske M (2006) The most widespread symbiosis on earth. PLoS Biol 4:1111–1112

Cabral C, Ravnskov S, Tringovska I, Wollenweber B (2016) Arbuscular mycorrhizal fungi modify nutrient allocation and composition in wheat (*Triticum aestivum* L.) subjected to heat-stress. Plant Soil 408:385–399

Calabrese S, Kohler A, Niehl A, Veneault-Fourrey C, Boller T, Courty PE (2017) Transcriptome analysis of the *Populus trichocarpa-Rhizophagus irregularis* mycorrhizal symbiosis: regulation of plant and fungal transportomes under nitrogen starvation. Plant Cell Physiol 58:1003–1017

Calvet C, Barea JM, Pera J (1992) In vitro interactions between the ves cular-arbuscular mycorrhizal fungus *Glomus mosseae* and some saprophytic fungi isolated from organic substrates. Soil Biol Biochem 24:775–780

Calvo-Polanco M, Sánchez-Romera B, Aroca R, Asins MJ, Declerck S, Dodd IC, Martínez-Andújar C, Albacete A, Ruiz-Lozano JM (2016) Exploring the use of recombinant inbred lines in combination with beneficial microbial inoculants (AM fungus and PGPR) to improve drought stress tolerance in tomato. Environ Exp Bot 131:47–57

Chen S, Jin W, Liu A, Zhang S, Liu D, Wang F, Lin X, He C (2013) Arbuscular mycorrhizal fungi (AMF) increase growth and secondary metabolism in cucumber subjected to low temperature stress. Sci Hortic 160:222–229

Comas L, Becker S, Cruz VM, Byrne PF, Dierig DA (2013) Root traits contributing to plant productivity under drought. Front Plant Sci 4:442

Duc NH, Csintalan Z, Posta K (2018) Arbuscular mycorrhizal fungi mitigate negative effects of combined drought and heat stress on tomato plants. Plant Physiol Biochem 132:297–307

El Kinany S, Achbani E, Faggroud M, Ouahmane L, El Hilali R, Haggoud A, Bouamri R (2019) Effect of organic fertilizer and commercial arbuscular mycorrhizal fungi on the growth of micropropagated date palm cv. Feggouss J Sau Soc Agri Sci 18:411–417

El-Nashar YI (2017) Response of snapdragon *Antirrhinum majus* L. to blended water irrigation and arbuscular mycorrhizal fungi inoculation: uptake of minerals and leaf water relations. Photosynthetica 55:201–209

Evelin H, Kapoor R (2014) Arbuscular mycorrhizal symbiosis modulates antioxidant response in salt-stressed *Trigonella foenum-graecum* plants. Mycorrhiza 24(3):197–208

Evelin H, Giri B, Kapoor R (2012) Contribution of *Glomus intraradices* inoculation to nutrient acquisition and mitigation of ionic imbalance in NaCl-stressed *Trigonella foenum-graecum*. Mycorrhiza 22:203–217

Fernández F, Dellrsquo JM, Angoa MV, de la Providencia IE (2011) Use of a liquid inoculum of the arbuscular mycorrhizal fungi *Glomus hoi* in rice plants cultivated in a saline Gleysol: a new alternative to inoculate. J Plant Breed Crop Sci 3:24–33

Filion M, St Arnaud M, Fortin JA (1999) Direct interaction between the arbuscular mycorrhizal fungus *Glomus intraradices* and different rhizosphere microorganisms. New Phytol 141:525–533

Fohse D, Claassen N, Jungk A (1988) Phosphorus efficiency of plants. Plant Soil 110:101–109

Foo E, Ross JJ, Jones WT, Reid JB (2013) Plant hormones in arbuscular mycorrhizal symbioses: an emerging role for gibberellins. Ann Bot 111:769–779

Frey-Klett P, Garbaye J, Tarkka M (2007) The mycorrhiza helper bacteria revisited. New Phytol 176:22–36

Gamalero E, Lingua G, Berta G, Glick BR (2009) Beneficial role of plant growth promoting bacteria and arbuscular mycorrhizal fungi on plant responses to heavy metal stress. Can J Microbiol 55:501–514

Garcia K, Zimmermann SD (2014) The role of mycorrhizal associations in plant potassium nutrition. Front Plant Sci 5:337

Gargallo-Garriga A, Preece C, Sardans J, Oravec M, Urban O, Peñuelas J (2018) Root exudate metabolomes change under drought and show limited capacity for recovery. Sci Rep 8:1–15

Genre A, Chabaud M, Timmers T, Bonfante P, Barker DG (2005) Arbuscular mycorrhizal fungi elicit a novel intracellular apparatus in *Medicago truncatula* root epidermal cells before infection. Plant Cell 17:3489–3499

Genre A, Chabaud M, Faccio A, Barker DG, Bonfante P (2008) Prepenetration apparatus assembly precedes and predicts the colonization patterns of arbuscular mycorrhizal fungi within the root cortex of both *Medicago truncatula* and *Daucus carota*. Plant Cell 20:1407–1420

Gholamhoseini M, Ghalavand A, Dolatabadian A (2013) Effects of arbuscular mycorrhizal inoculation on growth, yield, nutrient uptake and irrigation water productivity of sunflowers grown under drought stress. Agric Water Manag 117:106–114

Gómez-Bellot MJ, Ortuño MF, Nortes PA, Vicente-Sánchez J, Bañón S, Sánchez-Blanco MJ (2015) Mycorrhizal euonymus plants and reclaimed water: biomass, water status and nutritional responses. Sci Hortic 186:61–69

Guether M, Balestrini R, Hannah M, He J, Udvardi MK, Bonfante P (2009) Genome-wide reprogramming of regulatory networks, transport, cell wall and membrane biogenesis during arbuscular mycorrhizal symbiosis in *Lotus japonicus*. New Phytol 182:200–212

Gutjahr C, Parniske M (2013) Cell and developmental biology of arbuscular mycorrhiza symbiosis. Annu Rev Cell Dev Biol 29:593–617

Hajiboland R, Aliasgharzadeh N, Laiegh SF, Poschenrieder C (2010) Colonization with arbuscular mycorrhizal fungi improves salinity tolerance of tomato *Solanum lycopersicum* L. plants. Plant and Soil 331:313–327

Hameed A, Wu QS, Abd-Allah EF, Hashem A, Kumar A, Lone HA, Ahmad P (2014) Role of AM fungi in alleviating drought stress in plants. In: Miransari M (ed) Use of microbes for the alleviation of soil stresses. Springer, New York, pp 55–75

Hart MM, Reader RJ (2002a) Host plant benefit from association with arbuscular mycorrhizal fungi: variation due to differences in size of mycelium. Biol Fertil Soils 36:357–366

Hart MM, Reader RJ (2002b) Taxonomic basis for variation in the colonization strategy of arbuscular mycorrhizal fungi. New Phytol 153:335–344

Hashem A, Abd-Allah E, Alqarawi A, Al-Huqail A, Shah M (2016) Induction of osmoregulation and modulation of salt stress in *Acacia gerrardii* Benth. by arbuscular mycorrhizal fungi and *Bacillus subtilis* (BERA 71). Biomed Res Int:6294098

Hashem A, Kumar A, Al-Dbass AM, Alqarawi AA, Al-Arjani AB, Singh G, Farooq M, Abd-Allah EF (2019) Arbuscular mycorrhizal fungi and biochar improves drought tolerance in chickpea. Saudi J Biol Sci 26:614–6024

He XL, Li YP, Zhao LL (2010) Dynamics of arbuscular mycorrhizal fungi and glomalin in the rhizosphere of *Artemisia ordosica* Krasch. In Mu us sandland, China. Soil Biol Biochem 42: 1313–1319

Henderson JC, Davies FT (1990) Drought acclimation and the morphology of mycorrhizal *Rosa hybrida* L cv Ferdy is independent of leaf elemental content. New Phytol 115:503–510

Hijikata N (2010) Polyphosphate has a central role in the rapid and massive accumulation of phosphorus in extraradical mycelium of an arbuscular mycorrhizal fungus. New Phytol 186: 285–289

Hijri M, Sanders IR (2005) Low gene copy number shows that arbuscular mycorrhizal fungi inherit genetically different nuclei. Nature 433:161–163

Hildebrandt U, Ouziad F, Marner FJ, Bothe H (2006) The bacterium *Paenibacillus validus* stimulates growth of the arbuscular mycorrhizal fungus *Glomus intraradices* up to the formation of fertile spores. FEMS Microbiol Lett 254:258–267

Hodge A, Fitter AH (2010) Substantial nitrogen acquisition by arbuscular mycorrhizal fungi from organic material has implications for N cycling. Proc Natl Acad Sci 107:13754–13759

Impa SM, Nadaradjan S, Jagadish SVK (2012) Drought stress induced reactive oxygen species and anti-oxidants in plants. In Ahmad P, Prasad MNV (eds) Abiotic stress responses in plants. Springer, Berlin, pp 131–147

Jahromi F, Aroca R, Porcel R, Ruiz-Lozano JM (2008) Influence of salinity on the in vitro development of *Glomus intraradices* and on the in vivo physiological and molecular responses of mycorrhizal lettuce plants. Microb Ecol 55:45–53

Khalloufi M, Martínez-Andújar C, Lachaâl M, Karray-Bouraoui N, Pérez-Alfocea F, Albacete A (2017) The interaction between foliar GA3 application and arbuscular mycorrhizal fungi inoculation improves growth in salinized tomato (*Solanum lycopersicum* L.) plants by modifying the hormonal balance. J Plant Physiol 214:134–144

Kim Y, Jordan D, McDonald GA (1998) Effect of phosphate-solubilizing bacteria and vesicular-arbuscular mycorrhizae on tomato growth and soil microbial activity. Biol Fertil Soils 26:79–87

Klironomos JJ (2000) Host-specificity and functional diversity among arbuscular mycorrhizal fungi. Microb Biosyst New Front:845–851

Kohler J, Caravaca F, Azcón R, Díaz G, Roldán A (2015) The combination of compost addition and arbuscular mycorrhizal inoculation produced positive and synergistic effects on the phytomanagement of a semiarid mine tailing. Sci Total Environ 514:42–48

Kos M, Tuijl MA, de Roo J, Mulder PP, Bezemer TM (2015) Plant–soil feedback effects on plant quality and performance of an aboveground herbivore interact with fertilisation. Oikos 124: 658–667

Krishnaraj PU, Sreenivasa MN (1992) Increased root colonization by bacteria due to inoculation of vesicular– arbuscular mycorrhizal fungus in chilli (*Capsicum annuum*). Z Allg Mikrobiol 147: 131–133

Lee B-R, Muneer S, Avice J-C, Jung W-J, Kim T-H (2012) Mycorrhizal colonization and P-supplement effects on N uptake and N assimilation in perennial ryegrass under well-watered and drought-stressed conditions. Mycorrhiza 22:525–534

Li J, Meng B, Chai H, Yang X, Song W, Li S, Lu A, Zhang T, Sun W (2019) Arbuscular mycorrhizal fungi alleviate drought stress in C3 (Leymus chinensis) and C4 (Hemarthria altissima) grasses via altering antioxidant enzyme activities and photosynthesis. Front Plant Sci 10:499. https://doi.org/10.3389/fpls.2019.00499

Liao D, Wang S, Cui M, Liu J, Chen A, Xu G (2018) Phytohormones regulate the development of arbuscular mycorrhizal symbiosis. Int J Mol Sci 19:3146

Lin J, Wang Y, Sun S, Mu C, Yan X (2017) Effects of arbuscular mycorrhizal fungi on the growth, photosynthesis and photosynthetic pigments of *Leymus chinensis* seedlings under salt-alkali stress and nitrogen deposition. Sci Total Environ 576:234–241

Linderman RG (1992) Vesicular-arbuscular mycorrhizae and soil microbial interactions. Myco Sustain Agric 54:45–70

Lioussanne L, Perreault F, Jolicoeur M, St-Arnaud M (2010) The bacterial community of tomato rhizosphere is modified by inoculation with arbuscular mycorrhizal fungi but unaffected by soil enrichment with mycorrhizal root exudates or inoculation with *Phytophthora nicotianae*. Soil Biol Biochem 42(3):473–483

Liu T, Sheng M, Wang CY, Chen H, Li Z, Tang M (2015) Impact of arbuscular mycorrhizal fungi on the growth, water status, and photosynthesis of hybrid poplar under drought stress and recovery. Photosynthetica 53:250–258

Liu J, Guo C, Chen ZL, He JD, Zou YN (2016) Mycorrhizal inoculation modulates root morphology and root phytohormone responses in trifoliate orange under drought stress. Emir J Food Agric 28:251–256

Ma WK, Siciliano SD, Germida J (2005) A PCR-DGGE method for detecting arbuscular mycorrhizal fungi in cultivated soils. Soil Biol Biochem 37:1589–1597

Malekzadeh E, Alikhni HA, Savaghebi-Firoozabadi GR, Zarei M (2011) Influence of arbuscular mycorrhizal fungi and an improving growth bacterium on cd uptake and maize growth in cd-polluted soils. Spanish J Agric Res 9:1213–1223

Marulanda A, Barea JM, Azcón R (2006) An indigenous drought-tolerant strain of *Glomus intraradices* associated with a native bacterium improves water transport and root development in *Retama sphaerocarpa*. Microb Ecol 52:670–678

Mathur S, Sharma MP, Jajoo A (2018) Improved photosynthetic efficacy of maize (*Zea mays*) plants with arbuscular mycorrhizal fungi (AMF) under high temperature stress. J Photochem Photobiol B 180:149–154

Meena KK, Mesapogu S, Kumar M, Yandigeri MS, Singh G, Saxena AK (2010) Co-inoculation of the endophytic fungus *Piriformospora indica* with the phosphate-solubilising bacterium *Pseudomonas striata* affects population dynamics and plant growth in chickpea. Biol Fertil Soils 46: 169–174

Meena KK, Kumar M, Kalyuzhnaya M, Yandigeri MS, Singh DP, Saxena AK, Arora DK (2011) Epiphytic pink pigmented methylotrophic bacteria enhance germination and early seedling growth of wheat (*Triticum aestivum*) by producing phytohormone. Antonie Leeuwenhoek 101(4):777–786

Meena KK, Sorty AM, Bitla UM, Choudhary K, Gupta P, Pareek A, Singh DP, Prabha R, Sahu PK, Gupta VK, Singh HB, Krishnani KK, Minhas PS (2017) Abiotic stress responses and microbe-mediated mitigation in plants: the omics strategies. Front Plant Sci 8(172)

Meena KK, Shinde AL, Sorty AM, Bitla UM, Meena H, Singh NP (2019) Microbe- mediated biotic and abiotic stress tolerance in crop plants. In: Singh DP, Prabha R (eds) Microbial interventions in agriculture and environment. Springer Nature, Singapore, pp 315–329

Meena KK, Bitla UM, Sorty AM, Singh DP, Gupta VK, Wakchaure GC, Kumar S (2020) Mitigation of salinity stress in wheat seedlings due to application of phytohormone rich culture filtrate extract of methylotrophic actinobacterium *Nocardioides* sp. NIMe6. Front Microbiol 11: 2099

Miller G, Suzuki N, Ciftci-Yilmaz S, Mittler R (2010) Reactive oxygen species homeostasis and signalling during drought and salinity stresses. Plant Cell Environ 33:453–467

Mohamed AA, Eweda WE, Heggo AM, Hassan EA (2014) Effect of dual inoculation with arbuscular mycorrhizal fungi and Sulphur-oxidising bacteria on onion (*Allium cepa* L.) and maize (*Zea mays* L.) grown in sandy soil under greenhouse conditions. Ann Agric Sci 59:109–118

Monokrousos N, Papatheodorou EM, Orfanoudakis M, Jones DG, Scullion J, Stamou GP (2020) The effects of plant type, AMF inoculation and water regime on rhizosphere microbial communities. Eur J Soil Sci 71:265–278

Moradtalab N, Hajiboland R, Aliasgharzad N, Hartmann TE, Neumann G (2019) Silicon and the association with an arbuscular-mycorrhizal fungus (*Rhizophagus clarus*) mitigate the adverse effects of drought stress on strawberry. Agronomy 9(1):41

Muthukumar T, Radhika KP, Vaingankar J, D'Souza J, Dessai S, Rodrigues BF (2009) Taxonomy of AM fungi—an update. In: Rodrigues BF, Muthukumar T (eds) Arbuscular mycorrhizae of Goa—a manual of identification protocols. Goa University, Goa

Nacoon S, Jogloy S, Riddech N, Mongkolthanaruk W, Kuyper TW, Boonlue S (2020) Interaction between phosphate solubilizing bacteria and arbuscular mycorrhizal Fungi on growth promotion and tuber inulin content of *Helianthus tuberosus* L. Sci Rep 10:4916

Nadeem SM, Ahmad M, Zahir ZA, Javaid A, Ashraf M (2014) The role of mycorrhizae and plant growth promoting rhizobacteria (PGPR) in improving crop productivity under stressful environments. Biotechnol Adv 32:429–448

Nilsen ET, Orcutt DM (1996) Physiology of plants under stress. In: Abiotic factors. Wiley, New York, p 689

Olsson PA, Baath E, Jakobsen I (1997) Phosphorus effects on the mycelium and storage structures of an arbuscular mycorrhizal fungus as studied in the soil and roots by analysis of fatty acid signatures. Appl Environ Microbiol 63:3531–3538

Ordoñez YM, Fernandez BR, Lara LS, Rodriguez A, Uribe-Velez D, Sanders IR (2016) Bacteria with phosphate solubilizing capacity alter mycorrhizal fungal growth both inside and outside the root and in the presence of native microbial communities. PLoS One 11:e0154438

Ortiz N, Armada E, Duque E, Roldán A, Azcón R (2015) Contribution of arbuscular mycorrhizal fungi and/or bacteria to enhancing plant drought tolerance under natural soil conditions: effectiveness of autochthonous or allochthonous strains. J Plant Physiol 1:87–96

Osakabe Y, Osakabe K, Shinozaki K, Tran LS (2014) Response of plants to water stress. Front Plant Sci 5:86

Pedranzani H, Rodríguezrivera M, Gutiérrez M, Porcel R, Hause B, Ruizlozano JM (2016) Arbuscular mycorrhizal symbiosis regulates physiology and performance of Digitaria eriantha plants subjected to abiotic stresses by modulating antioxidant and jasmonate levels. Mycorrhiza 26:141–152. https://doi.org/10.1007/s00572-015-0653-4

Peterson E (2003) Importance of rhizodeposition in the coupling of plant and microbial productivity. Eur J Soil Sci 54:741–750

Philippot L, Raaijmakers JM, Lemanceau P, Van Der Putten WH (2013) Going back to the roots: the microbial ecology of the rhizosphere. Nat Rev Microbiol 11:789–799

Pivato B, Offre P, Marchelli S, Barbonaglia B, Mougel C, Lemanceau P, Berta G (2009) Bacterial effects on arbuscular mycorrhizal fungi and mycorrhiza development as influenced by the bacteria, fungi, and host plant. Mycorrhiza 19:81–90

Qian K, Wang L, Yin N (2012) Effects of AMF on soil enzyme activity and carbon sequestration capacity in reclaimed mine soil. Int J Min Sci Technol 22:553–557

Qin Z, Xie JF, Quan GM, Zhang JE, Mao DJ, Wang JX (2018) Changes in the soil meso- and micro-fauna community under the impacts of exotic *Ambrosia artemisiifolia*. Ecol Res 34:265–276

Rambelli A (1973) The rhizosphere of mycorrhizae. In: Marks GC, Kozlowski TT (eds) Ectomycorrhizae: their ecology and physiology. Academic Press, New York, USA pp, pp 299–343

Rapparini F, Penuelas J (2014) Mycorrhizal fungi to alleviate drought stress on plant growth. In: Miransari M (ed) Use of microbes for the alleviation of soil stresses. Springer Science + Business Media New York, 21–42

Redecker D, Kodner R, Graham LE (2000) Glomalean fungi from the Ordovician. Science 289: 1920–1921

Rillig MC, Mummey DL (2006) Mycorrhizas and soil structure. New Phytol 171:41–53

Rillig MC, Wright SF, Kimball BA, Pinter PJ, Wall GW, Ottman MJ, Leavitt SW (2001) Elevated carbon dioxide and irrigation effects on water stable aggregates in a *Sorghum* field: a possible role for arbuscular mycorrhizal fungi. Glob Chang Biol 7:333–337

Rillig MC, Treseder KK, Allen MF (2002) Global change and mycorrhizal Fungi. In: van der Heijden MGA, Sanders IR (eds) Mycorrhizal ecology. Ecological studies, vol 157. Springer, Berlin, pp 135–160

Rillig MC, Lutgen ER, Ramsey PW, Klironomos JN, Gannon JE (2005) Microbiota accompanying different arbuscular mycorrhizal fungal isolates influence soil aggregation. Pedobiologia 49: 251–259

Robinson Boyer L, Feng W, Gulbis N, Hajdu K, Harrison RJ, Jeffries P, Xu X (2016) The use of arbuscular mycorrhizal Fungi to improve strawberry production in coir substrate. Front Plant Sci 7:1237

Rouphael Y, Franken P, Schneider C (2015) Arbuscular mycorrhizal fungi act as biostimulants in horticultural crops. Sci Hortic 196:91–108

Ruiz-Lozano JM, Azcon R (2000) Symbiotic efficiency and infectivity of an autochthonous arbuscular mycorrhizal *Glomus* sp. from saline soils and *Glomus deserticola* under salinity. Mycorrhiza 10:137–143

Ruiz-Lozano JM, Aroca R, Zamarreño ÁM, Molina S, Andreo-Jiménez B, Porcel R, García-Mina JM, Ruyter-Spira C, López-Ráez JA (2016) Arbuscular mycorrhizal symbiosis induces strigolactone biosynthesis under drought and improves drought tolerance in lettuce and tomato. Plant Cell Environ 39:441–452

Ruiz-Sánchez M, Aroca R, Muñoz Y, Polón R, Ruiz-Lozano JM (2010) The arbuscular mycorrhizal symbiosis enhances the photosynthetic efficiency and the antioxidative response of rice plants subjected to drought stress. J Plant Physiol 167:862–869

Sasse J, Martinoia E, Northen T (2018) Feed your friends: do plant exudates shape the root microbiome? Trends Plant Sci 23:25–41

Schimel JP (2018) Life in dry soils: effects of drought on soil microbial communities and processes. Annu Rev Ecol Evol Syst 49:409–432

Schreiner R, Mihara K, McDaniel H, Bethlenfalvay G (1997) Mycorrhizal fungi influence plant and soil functions and interactions. Plant Soil 188:199–209

Schußler A, Walker C (2010) The Glomeromycota. A species list with new families and new genera. Published in libraries at the Royal Botanic Garden Edinburgh, the Royal Botanic Garden Kew, Botanische Staatssammlung Munich, and Oregon State University electronic version freely available online at www.amf-phylogeny.com

Singh S, Kapoor KK (1998) Effects of inoculation of phosphate-solubilizing microorganisms and an arbuscular mycorrhizal fungus on mungbean grown under natural soil conditions. Mycorrhiza 7:249–253

Smith SE, Read DJ (2008) Mycorrhizal symbiosis, 3rd edn. Academic Press, London

Smith SE, Smith FA (2012) Fresh perspectives on the roles of arbuscular mycorrhizal fungi in plant nutrition and growth. Mycologia 104:1–13

Sorty AM, Meena KK, Choudhary K, Bitla UM, Minhas PS, Krishnani KK (2016) Effect of plant growth promoting Bacteria associated with halophytic weed (*Psoralea corylifolia* L) on germination and seedling growth of wheat under saline. Appl Biochem Biotechnol 180(5):872–882

Sorty AM, Bitla UM, Meena KK, Singh NP (2018) Role of microorganisms in alleviating abiotic stresses. In: Panpatte DG, Jhala YK, Shelat HN, Vyas R (eds) Microorganisms for green revolution. Microorganisms for sustainability, vol 7. Springer Nature, Singapore, pp 115–128

Tahat MM, Sijam K (2012) Arbuscular mycorrhizal fungi and plant root exudates bio-communications in the rhizosphere. Afr J Microbiol Res 6:7295–7301

Thao HT, George T, Yamakawa T, Widowati LR (2008) Effects of soil aggregate size on phosphorus extractability and uptake by rice (*Oryza sativa* L.) and corn (*Zea mays* L.) in two Ultisols from the Philippines. Soil Sci Plant Nutr 54:148–158

Tisserant E, Malbreil M, Kuo A, Kohler A, Symeonidi A, Balestrini R, Charron P, Duensing N, Dit Frey NF, Gianinazzi-Pearson V, Gilbert LB (2013) Genome of an arbuscular mycorrhizal fungus provides insight into the oldest plant symbiosis. Proc Natl Acad Sci 110:20117–20122

Treseder KK, Turner KM (2007) Glomalin in ecosystems. Soil Sci Soc Am J 71:1257–1266

Vázquez MM, César S, Azcón R, Barea JM (2000) Interactions between arbuscular mycorrhizal fungi and other microbial inoculants (Azospirillum, Pseudomonas, Trichoderma) and their effects on microbial population and enzyme activities in the rhizosphere of maize plants. Appl Soil Ecol 15(3):0–272. https://doi.org/10.1016/s0929-1393(00)00075-5

Van De Velde K, Ruelens P, Geuten K, Rohde A, Van Der Straeten D (2017) Exploiting DELLA signaling in cereals. Trends Plant Sci 22:880–893

van der Heijden MG, Martin FM, Selosse MA, Sanders IR (2015) Mycorrhizal ecology and evolution: the past, the present, and the future. New Phytol 205:1406–1423

Vivas A, Marulanda A, Ruiz-Lozano JM, Barea JM, Azcón R (2003) Influence of a *Bacillus* sp. on physiological activities of two arbuscular mycorrhizal fungi and on plant responses to PEG-induced drought stress. Mycorrhiza 13:249–256

Wamberg C, Christensen S, Jakobsen I, Müller AK, Sørensen SJ (2003) The mycorrhizal fungus (*Glomus intraradices*) affects microbial activity in the rhizosphere of pea plants (*Pisum sativum*). Soil Biol Biochem 35:1349–1357

Wang B, Qui YL (2006) Phylogenetic distribution and evolution of mycorrhizae in land plants. Mycorrhiza 16:299–363

Wang Y, Jing H, Gao Y (2012) Arbuscular mycorrhizal colonization alters subcellular distribution and chemical forms of cadmium in *Medicago sativa* L. and resists cadmium toxicity. PLoS One 7:3161–3164

Wang Y, Wang M, Li Y, Wu A, Huang J (2018) Effects of arbuscular mycorrhizal fungi on growth and nitrogen uptake of *Chrysanthemum morifolium* under salt stress. PLoS One 13(4):e0196408

Wu QS, Zou YN (2009) Mycorrhiza has a direct effect on reactive oxygen metabolism of drought-stressed citrus. Plant Soil Environ 55:436–442

Wu HH, Zou YN, Rahman MM, Ni QD, Wu QS (2017) Mycorrhizas alter sucrose and proline metabolism in trifoliate orange exposed to drought stress. Sci Rep 7(1):1–10

Xiao X, Chen H, Chen H, Wang J, Ren C, Wu L (2008) Impact of *Bacillus subtilis* JA, a biocontrol strain of fungal plant pathogens, on arbuscular mycorrhiza formation in *Zea mays*. World J Microbiol Biotechnol 24:1133–1137

Xu L, Naylor D, Dong Z, Simmons T, Pierroz G, Hixson KK, Kim YM, Zink EM, Engbrecht KM, Wang Y, Gao C (2018) Drought delays development of the sorghum root microbiome and enriches for monoderm bacteria. Proc Natl Acad Sci U S A 115:e4284–E4293

Yang SY, Grønlund M, Jakobsen I, Grotemeyer MS, Rentsch D, Miyao A, Hirochika H, Kumar CS, Sundaresan V, Salamin N, Catausan S (2012) Nonredundant regulation of rice arbuscular mycorrhizal symbiosis by two members of the phosphate transporter1 gene family. Plant Cell 24:4236–4251

Yin N, Zhang Z, Wang L, Qian K (2016) Variations in organic carbon, aggregation, and enzyme activities of gangue-fly ash-reconstructed soils with sludge and arbuscular mycorrhizal fungi during 6-year reclamation. Environ Sci Pollut Res 23:17840–17849

Yooyongwech S, Samphumphuang T, Tisarum R (2016) Arbuscular mycorrhizal fungi (AMF) improved water deficit tolerance in two different sweet potato genotypes involves osmotic adjustments *via* soluble sugar and free proline. Sci Hortic 198:107–117

Zhu XC, Song FB, Liu SQ, Liu TD, Zhou X (2012) Arbuscular mycorrhizae improves photosynthesis and water status of *Zea mays* L. under drought stress. Plant Soil Environ 58:186–191

Role of Metabolites Produced by Plant Growth-Promoting Bacteria in Biocontrol of Phytopathogens Under Saline Conditions

9

Tahmish Fatima, Priyanka Verma, Sushma Verma, Burak Alaylar, and Naveen Kumar Arora

Abstract

Soil salinity has continuously degraded the quality and quantity of crops and has challenged their health and defense mechanisms. Although relationship between salinity and biotic stress is poorly understood, yet due to negative impact of salts on plants, they become susceptible to diseases. Limiting the use of agro-chemicals, salt-tolerant plant growth-promoting rhizobacteria (ST-PGPR) are emerging as potential replacements for use against combinatorial stresses such as salinity and pathogens. Under combinatorial stress, i.e., salt and biotic, ST-PGPR produce an array of metabolites which directly act upon the pathogens and also induce a number of molecular and signaling pathways activating a two-way defensive response. Siderophores, antibiotics, volatile organic compounds, and biosurfactants are among the metabolites produced by PGPR for suppressing the phytopathogens. Apart from these, PGPR initiate induced systemic resistance and resultantly protect and prepare the plants against the diseases. With dual role against salinity and pathogens, PGPR produce various metabolites such as exopolysaccharides, osmoprotectants, and antioxidants and initiate nutrient chelation and biofilming activity, thereby improving the cropping system. Quorum sensing and quorum quenching are also initiated by PGPR to

T. Fatima · S. Verma
Department of Environmental Microbiology, School of Earth and Environmental Sciences, Babasaheb Bhimrao Ambedkar University, Lucknow, Uttar Pradesh, India

P. Verma · N. K. Arora (✉)
Department of Environmental Science, School of Earth and Environmental Sciences, Babasaheb Bhimrao Ambedkar University, Lucknow, Uttar Pradesh, India

B. Alaylar
Department of Molecular Biology and Genetics, Faculty of Arts and Sciences, Agri Ibrahim Cecen University, Agri, Turkey

N. K. Arora, B. Bouizgarne (eds.), *Microbial BioTechnology for Sustainable Agriculture Volume 1*, Microorganisms for Sustainability 33,
https://doi.org/10.1007/978-981-16-4843-4_9

limit the negative impact of salinity and pathogens and promote plant growth. Therefore, ST-PGPR exhibiting combinatorial defense mechanisms can be applied as stress relievers, antagonists, and biofertilizers to saline pathogen-infested soils and improve the cropping system by maintaining agricultural sustainability.

Keywords

Salt-tolerant plant growth-promoting rhizobacteria · Biocontrol · Antibiotics · Exopolysaccharides · Biosurfactants · Nutrient chelation

9.1 Introduction

The increasing human population along with the declining soil fertility and agricultural productivity has affected the global food security. Salinity stress has been a hassling issue, and about 20% of total agricultural lands are affected by the stress which is continuously increasing (Gupta and Huang 2014). Currently, almost 1 billion ha of land across the globe is known to be degraded by salinity stress (Arora et al. 2018). Saline soils are characterized by EC (electrical conductivity) > 4.0 dS/m along with low water holding capacity (WHC), structural instability, poor filtration rate, reduced nutrient content, and loss of beneficial microbes (Tanji 2002; Fatima and Arora 2019; Arora et al. 2020; Sunita et al. 2020). Among the accumulated salts, NaCl causes the highest level of toxicity and growth retardation in plants. Under saline conditions, the plants are exposed to physical and biochemical challenges including osmotic imbalance, impaired nitrogen fixing ability, reduced uptake of water and nutrients, inhibition of seed germination, cell flaccidity and plasmolysis, stomatal closure, and ethylene stress (Arora et al. 2016). The increasing risks of climate change and the resultant abiotic stresses also expose plants to biotic stresses. The complex interaction between the adverse environmental factors, phytopathogens, and plants is often referred as "disease triangle" (Pandey et al. 2017). Salinity is reported to aggravate incidence of powdery mildew (by *Botrytis cinerea*) in *Solanum habrochaites* (Kissoudis et al. 2015, 2016), wilt (by *Fusarium oxysporum*) in *Solanum lycopersicum* (Siddique et al. 2019), charcoal rot (by *Macrophomina phaseolina*) in chick pea and sunflower (Khare et al. 2011; Tewari and Arora 2014), crown rot disease (by *Fusarium culmorum*) in wheat (Albdaiwi et al. 2019), and *Fusarium* sp., *Verticillium*, and *Phytophthora* in potato (Nachmias et al. 1993; Grossi et al. 2020). Strange and Scott (2005) reported that about 10–20% of global agricultural productivity is affected by phytopathogens. The scenario of disease incidence is worse in arid and semi-arid areas. Egamberdieva et al. (2011) reported that 17% of the cucumber grown in saline soils of Uzbekistan were infected with *Fusarium solani*. The disease incidence was noticed to enhance proportionally from 40 to 93% when salinity increased from 0.01 to 5 dS m^{-1} (Al-Sadi et al. 2010).

To overcome the increasing pathogenicity in plants along with climatic adversity, farmers are forced to make use of chemicals pesticides. But this, in turn, further aggravates the problem of soil salinity and degradation, challenging agricultural sustainability. Thus, there arises the urgent need of biological solutions to combat the multiple stresses faced by the plants. Plant growth-promoting bacteria/ rhizobacteria (PGPB/PGPR) association in the soil and rhizoplane are reported as efficient biostimulants which can reduce the pathogenicity in plants and improve the growth parameters in a hyperosmotic surrounding (Arora et al. 2018). Biological control under salinity is dependent on the secondary metabolites produced by salt-tolerant PGPR (ST-PGPR) which are responsible for plant growth promotion and reduction in severity against phytopathogens under biotic and abiotic stresses (Mishra and Arora 2018). ST-PGPR can act as biocontrol agents and can even mitigate salt stress through various mechanisms (Table 9.1, Fig. 9.1). These include direct action against the phytopathogenic virulence factors by synthesis of secondary metabolites such as antibiotics, phenolic compounds, siderophores, lytic enzymes, volatile organic compounds (VOCs), exopolysaccharides (EPS), and biosurfactants (Haas and Défago 2005; Suryadi et al. 2019). The indirect mechanism of action involves increasing the competition for nutrients and space by chelating the necessary micro- and macronutrients such as nitrogen (N), phosphorous (P), potassium (K), iron (Fe), and zinc (Zn) and establishing a healthy PGPR-plant rhizospheric association (Kumar et al. 2012). PGPR also enhance the innate and acquired immunity of plants (induced systemic resistance/systemic acquired resistance (ISR/SAR), by stimulating various distal and local immunity responses through activation of jasmonic acid and ethylene pathways (Nurnberger et al. 2004). At present various antifungal metabolites of ST-PGPR including siderophores, hydrogen cyanide (HCN), antibiotics, and VOCs are being utilized to control disease incidences in economically significant crops (Timmusk et al. 2019). The action of ST-PGPR involves two-way response mechanisms in multi-stressed plants: firstly, by limiting the infection in plants and secondly by promoting the growth of plants against the odds of salinity. Performing the latter function, these beneficial microbes enhance synthesis of phytohormones such as indole acetic acid (IAA), gibberellic acid (GA), chelation of nutrients, accumulation of osmoprotectants, antioxidants, and synthesis of ACC deaminase to support plant growth under salinity stress (Fatima and Arora 2019).

In the current scenario of climate change and global warming, the microbial technology can be used to overcome the problems of agro-productivity to diseases and salinity in an ecofriendly manner. However, to avail all the beneficial factors of ST-PGPR, a deeper knowledge about the spectrum of antifungal metabolites and their functionality under salt stress is required to be elucidated.

Table 9.1 Role and action of different metabolites produced by salt-tolerant PGPR in biocontrol of phytopathogens under salinity stress

Mechanisms	Metabolites	Action against phytopathogens	Name of ST-PGPR	References
Nutrient acquisition	Nitrogen fixation	Direct chelation of nitrogen by rhizobia and other diazotrophic plant growth-promoting rhizobacteria (PGPR) reduces the pathogenicity and survival chances of pathogens. Lipochitooligosaccharides (LCOs) involved in rhizobia-legume association induce synthesis of antioxidant enzymes and PR proteins against phytopathogens	*Bacillus safensis*, *Bacillus pumilus*, *Kocuria rosea*, *Enterobacter aerogenes*, and *Aeromonas veronii*	Mukhtar et al. (2020)
	Siderophore production	Iron is necessary for survival of plants, PGPR, as well as phytopathogens. Chelation of iron by siderophore-producing ST-PGPR reduces the chances of survival of phytopathogens. Siderophore also elicit induced systemic resistance (ISR) in plants and even promote their growth under salt stress	*Pseudomonas chlororaphis*-66, *Mesorhizobium ciceri*-4, *Stenotrophomonas maltophilia* SBP-9, *Streptomyces* sp., *Bacillus amyloliquefaciens*	Sadeghi et al. (2012), Singh and Jha (2017), Shurigin et al. (2018) and Masum et al. (2018)
	Phosphate (P) solubilization	Chelation of P by ST-PGPR hinders essential metabolic processes in pathogens, and in return the solubilized nutrient benefits the plant growth. The organic acids produced by phosphate-solubilizing bacteria (PSM) also extend the	*B. pumilus*, *Bacillus subtilis*, *Planococcus rifietoensis*, *Bacillus licheniformis*, *Enterobacter cloacae R13*, *E. cloacae R33*, *Paenibacillus lactis*, *Pseudomonas* sp., *Methylobacterium* sp., *Pseudomonas argentinensis*,	Damodaran et al. (2013), Rajput et al. (2013), Goswami et al. (2014), Lamizadeh et al. (2016), Phour and Sindhu (2020) and Grossi et al. (2020)

		antimicrobial and antifungal activities excluding the phytopathogens	and *Pseudomonas azotoformans*	
	Solubilization of potassium and zinc (Zn)	K solubilization improves the thickening of plant cell wall and also helps in maintaining osmotic balance under salt stress Zn exclusion by ST-PGPR inhibits many catabolic and biochemical processes in pathogens which involve Zn as co-factor or catalyst. Zn assimilation enhances antioxidant activity protecting against oxidative stress due to salinity and pathogen attack and increases disease resistive (by expression of Znf genes and MT) in plants	*E. cloacae*, *B. pumilus*, *Pseudomonas* sp.	Pirhadi et al. (2016) and Ćalić et al. (2017)
Antibiosis	Pyocyanin	The phenazine derivative toxin is thought to be involved in induction of ISR and thereby helps in pathogen biocontrol	Fluorescent *Pseudomonas* strain EKi	Khare et al. (2011)
	Bacitracin	Bactericidal activity of this cyclic polypeptide antibiotic includes cell wall inhibition	*Bacillus* strain GU057	Amin et al. (2012)
	Bacylisin, surfactin, and fengycin	Antimicrobial peptide (AMP) biosynthetic metabolites have strong direct antifungal effect	*B. subtilis* RH5	Jamali et al. (2020)

(continued)

Table 9.1 (continued)

Mechanisms	Metabolites	Action against phytopathogens	Name of ST-PGPR	References
		and/or also induce resistance in plants		
	2,4-diacetylphloroglucinol (DAPG)	They are low-molecular-weight polyketide and encourages systemic resistance in plants to pathogens	*Pseudomonas protegens*, *P. chlororaphis* subsp. aureofaciens, *Lysobacter capsici*, *Pseudomonas* spp.	Ashwitha et al. (2013) and Wang et al. (2015)
Lytic enzymes	Chitinase, dehydrogenase, β-1,3 glucanase, lipase, phosphatase, and proteases	Hydrolytic enzymes break the glyosidic bonds of fungal cell wall and henceforth directly contribute in the biocontrol of phytopathogens	*Pseudomonas fluorescens*, *Streptomyces rimosus*, *Streptomyces monomycini*, *Bacillus cereus*, *Bacillus* sp., *Enterobacter* sp., *Streptomyces aureofaciens*, *Bacillus xiamenensis*, *Bacillus toyonensis*	Kamaly et al. (2020), Karthika et al. (2020), Abdelshafy-Mohamad et al. (2020), Rojas-Solis et al. (2020) and Xia et al. (2020)
Induced systemic resistance (ISR) and systemic acquired resistance (SAR)		Triggering of ISR and SAR in plants by ST-PGPR initiates the upregulation of PR proteins, antioxidants, phytohormones, osmolytes, enhancement of plant cell wall rigidity, deposition of callose, and production of antibiotics	*Serratia marcescens*, *B. licheniformis*, *Bacillus* sp.	Singh and Jha (2016), Amaresan et al. (2016) and Sukkasem et al. (2018)
Other mechanisms	Antioxidant enzymes	Increase in activities of peroxidase (POD), catalase (CAT), and superoxide dismutase (SOD) improves the accumulation of proline, nitrogen, potassium, and phosphorus and also decreases	*B. cereus* Pb25, *Pseudomonas extremorientalis* TSAU20, *P. chlororaphis* TSAU13	Islam et al. (2016) and Egamberdieva et al. (2017)

		the sodium accumulation in plants under saline conditions. These enzymes also help plants to cope up with the oxidative stresses generated in salinity and pathogenic stresses. The antioxidant enzymes also initiate ISR in plants		
	Volatile organic compounds (VOCs) (acetoin and 2,3-butanediol)	VOCs paly an important role in mitigating salt stress in plants through increment in plant growth parameters, biomass, photosynthetic activity, regulation of ionic balance (K^+/Na^+; HKT transporters), and eliciting IST. Their role in biocontrol mainly involves in escalating plant resistance to infections, i.e., ISR	*Bacillus velezensis* XT1	Torres et al. (2020)
	HCN production (hydrogen cyanide)	The cyanide ion is a potent inhibitor of various metalloenzymes, specifically copper-containing cytochrome c oxidases. In addition, cyanide may also contribute in amplifying the nutrient availability to plants because of their ability to sequestrate metal ions	*B. cereus*, Fluorescent *Pseudomonas* sp. PF17, *Azotobacter species*, viz., *A. armeniacus*, *A. chroococcum*, *A. salinestris*, *A. tropicalis*, and *A. vinelandii*	Chakraborty et al. (2011), Tewari and Arora (2016) and Chennappa et al. (2016)

(continued)

Table 9.1 (continued)

Mechanisms	Metabolites	Action against phytopathogens	Name of ST-PGPR	References
	Exopolysaccharides (EPS)	One of the main components of the polysaccharide, i.e., uronic acids, is of critical importance in antagonism and play significant role in the sequestration of cations and in formation of biofilms. Biocontrol agent EPS matrix with high water-holding capacity (WHC) also serves as important factor in protecting the plants from various pathogens	*Pseudomonas aeruginosa* PF23, *Pseudomonas entomophila* PE3, *Alcaligenes faecalis* AF7	Tewari and Arora (2014); Fatima et al. (2020); Fatima and Arora (2021)
	Root exudates	Although root exudates are directly not involved in biocontrol, prior inoculation of biocontrol agents (ST-PGPR) changes the pattern and composition of root exudates which further excludes phytopathogens and acts as attractants toward beneficial microbes	*Pseudomonas* strain AKC-O11	Khan et al. (2018)
	N-acyl homoserine lactones (AHLs)	AHLs are responsible for establishment of PGPR-plant association through quorum sensing. AHLs are also involved in signaling the induction of systemic resistivity (ISR) against	*Staphylococcus equorum EN21*	Vega et al. (2020)

		phytopathogens and systemic tolerance (IST) against salinity in plants. Additionally, quorum quenching (QQ) activity in PGPR, involving enzymatic degradation of phytopathogen-generated AHLs, serves as an important mechanism of biocontrol		
	Biosurfactants	Biosurfactants directly protect the plant by biofilm formation, which leads to cell wall destabilization of pathogens, and are also associated with ISR, antibiosis, parasitism, virulence, and competition for space and nutrients	*Idiomarina zobelli* FMH6, *Nesterenkonia halotolerans* FMH10, *Halomonas janggokensis* FMH54, *B. velezensis* FMH2, *B. subtilis* subsp. spizizenii FMH4, *Bacillus altitudinis* MS16	Masmoudi et al. (2019) and Goswami and Deka (2019)

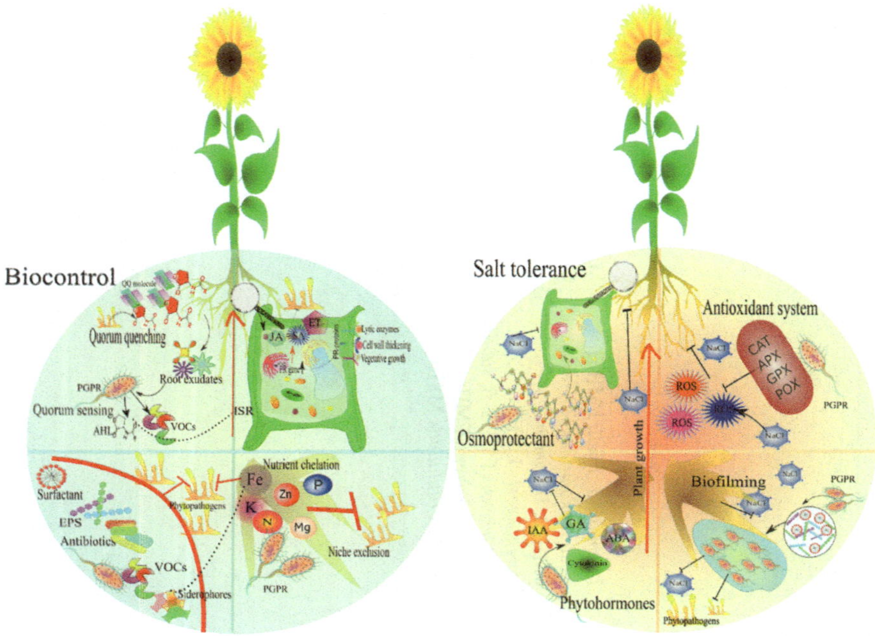

Fig. 9.1 The dual role of ST-PGPR in inducing salinity tolerance and biocontrol of phytopathogens in plants under combinatorial stress and the associated metabolites involved

9.2 Role of PGPR Metabolites Involved in Suppression of Phytopathogens

The increasing risk of disease incidence in plants under saline conditions also induces various responses by rhizospheric microbes which lead to the exclusion of phytopathogens. The exogenous application of microbial biostimulants protects the plants through various mechanisms involved in synthesizing broad spectrum of secondary metabolites. Table 9.2 illustrates various ST-PGPR which are reported in biocontrol of pathogens in different plants and the associated mechanisms involved. The following section describes the metabolites and mechanisms of PGPR involved in suppression of phytopathogens specifically under saline conditions.

9.2.1 Nutrient Competence

The severity and growth of phytopathogens are primarily dependent upon nutrient uptake. Majority of the pathogens spend a larger span of their life cycle in soil and thereby survive independent of the host. In this significant life stage, pathogens feed

Table 9.2 Different salt-tolerant PGPR as biocontrol agents against phytopathogens under salinity stress and the mechanisms involved

S. no	Biocontrol agent	Plant	Disease	Phytopathogens	Mechanism of action	Salt concentration	Reference
1	*Streptomyces*	Sugar beet	Root rot on sugar beet	*Rhizoctonia solani* AG-2, *Fusarium solani*, and *Phytophthora drechsleri*	Protease, chitinase, β-glucanase, cellulase, lipase, and α-amylase activity Siderophore and salicylic acid (SA) production	300 mM NaCl	Karimi et al. (2012)
2	*Bacillus halotolerans* and *B. pumilus*	*Prosopis strombulifera* (isolated from)	Leaf spots, rots, and blights	*Alternaria* sp.	Presence of both lipopeptide genes (ItuD and SrfC) and iturin A responsible for synthesis of surfactin and iturin antifungal compounds	0.9% NaCl	Sgroy et al. (2009)
3	*Bacillus* sp. and *Ochrobactrum* sp.	Maize	Downy mildews, *Fusarium* rots, powdery mildews, rusts, sclerotium rots	*Sclerotinia* and *Fusarium*	Secretion of bioactive compounds, cellulase and protease	150 mM NaCl	Príncipe et al. (2007)
4	*P. chlororaphis* TSAU13 and *P. extremorientalis* TSAU20	Tomato	Foot and root rot disease	*F. solani*	Improved N and P uptake, activity of antioxidant enzymes	5.6 dSm	Egamberdieva et al. (2017)
5	*Bacillus*, *Oceanobacillus*, and *Halomonas*	Durum wheat	Crown rot	*Fusarium culmorum*	Multiple plant growth-promoting traits including nutrient uptake, phosphate solubilization, IAA production	80 mM and 160 mM NaCl	Albdaiwi et al. (2019)

(continued)

Table 9.2 (continued)

S. no	Biocontrol agent	Plant	Disease	Phytopathogens	Mechanism of action	Salt concentration	Reference
6	*S. maltophilia* SBP-9	Wheat	*Fusarium* head blight	*Fusarium graminearum*	Antioxidative enzymatic activities such as SOD, CAT, and POX. Accumulation of Na^+ in both shoot and roots, increase in K^+ uptake	150 mM and 200 mM NaCl	Singh and Jha (2017)
7	*P. aeruginosa* GS-33	Soybean	Charcoal rot	*Macrophomina phaseolina*	Production of phenazine-1-carboxylic acid, siderophores, HCN and protease	100 mM NaCl	Patil et al. (2016)
8	*Pseudomonas*	Cucumber	*Fusarium* wilt	*Fusarium oxysporum f.* sp. conglutinans and *F. oxysporum f.* sp. cucumerinum	Action of antifungal and hydrolytic enzymes such as chitinase, β-1,3-glucanase and proteases		Teng et al. (2010)
9	Rhizobacteria	Asparagus	*Fusarium* crown and root rot	*F. oxysporum* and *Fusarium proliferatum*	Increase in rhizospheric bacterial densities especially fluorescent pseudomonads and Mn-reducing bacteria	0.5 or 1.0% NaCl	Elmer (2003)
10	*Pseudomonas* (tolerant to 1.0 M NaCl)	Rice	Bacterial leaf blight (BB) and sheath blight (ShB)	*Xanthomonas oryzae* pv. oryzae and *R. solani*	Antibiosis	25% artificial sea water	Rangarajan et al. (2003)

11	*Pseudomonas* and *Bacillus* spp.	–	–	*Botrytis ricini*, *F. oxysporum f.* sp. ricini, *M. phaseolina*, *R. solani* and *Sclerotium rolfsii*	Production of hydrolytic enzymes, HCN, antibiotics, and siderophores	1.2 M	Kumar et al. (2014)
12	*P. aeruginosa* PF23	Sunflower	Charcoal rot disease	*M. phaseolina*	Salicylic acid production and secretion of exopolysaccharides	125 mM NaCl (pot trial) 10 dS/m (field trial)	Tewari and Arora (2018)
13	*Pseudomonas* sp. PF17	Sunflower	Charcoal rot	*M. phaseolina*	Enhanced root colonization, increased RAS/RT ratio, production of IAA	10 dS/m	Tewari and Arora (2016)
14	Fluorescent *Pseudomonas* strain EKi	Chickpea	Charcoal rot	*M. phaseolina*	Production of pyocyanin, siderophore, and HCN	100 mM NaCl	Khare et al. (2011)
15	*P. aeruginosa* TO3	Chickpea	Charcoal rot	*M. phaseolina*	Secretion of biocontrol metabolites	100 mM NaCl	Khare and Arora (2010)
16	*Pseudomonas putida* R4 and *P. chlororaphis* R5	Cotton	Root rot	*F. solani*	IAA production	6.8 dS/m	Egamberdieva et al. (2015)

on soil nutrients and plant exudates to extend their virulence factors and ensure their survival, and then they multiply to a threshold population sufficient enough to cause infection in plants (Köhl et al. 2019). Pal and Gardener (2006) reported that phytopathogenic fungal species including *Fusarium* and *Pythium* mainly infect the plant through mycelial contact instead of sporulating and germinating on plant necrotic tissues. Therefore, these phytopathogens are susceptible to nutrient chelation by beneficial microbes reducing their chances of infection. In saline soils although the nutrient-deficient conditions impose some threat to the pathogens, their efficient salt-tolerance mechanisms in a way support their growth and proliferation. However, increasing nutrient competence by using microbial biocontrol agents could be a significant call in reducing the population of pathogens as well as lowering their communication with plants. Negating involvement of virulence-mitigating factors (such as antibiotics, volatiles), the nutrient competence mode of action has to be further explored at epidemiology level to illustrate how each nutrient's deficiency impacts the stages of pathogen development. Potential ST-PGPR with remarkable nutrient chelation property stimulate distal and local substrate competition in soil excluding the phytopathogens (Kessel et al. 2005). The competition for carbon, Fe along with limited N content is the key action of biocontrol agents to initiate resource capturing (Spadaro and Droby 2016). PGPR stimulate production of various secondary metabolites, enzymes, and organic/inorganic acids to ensure chelation of micro- and macronutrients. Primarily, these microbes very efficiently colonize the hotspots of nutrient cycling in plants such as exit points of secondary roots, damaged epidermal cells, and nectaries and directly take up water- and carbon-containing nutrients also feeding upon root mucilage (Pal and Gardener 2006).

Nitrogen, the most limiting nutrient in saline soil, is highly essential for phytopathogens for their survival and pathogenicity. As a universal fact, N improves plant structure and development even under environmental stresses such as salinity. It has been suggested that due to improvement in plant canopy structure, the disease susceptibility and survival chances of phytopathogens under salinity (or other abiotic stresses) increase (Tavernier et al. 2007; Mur et al. 2017). Correlatively, Robert et al. (2002) found that wheat plants with low nutrient content showed 70% lesser spore production by rust fungus *Puccinia triticina*. These stats suggest that plant has to compromise its nutrient status in order to protect from pathogens. However, this cannot be the ultimate solution, and therefore there is the need of an intermediate link which could provide nutrients to the plant through intimate association while still excluding the chances of colonization by these phytopathogens. The symbiotic nitrogen fixation between legume and rhizobia can be considered as the most potential association to exclude infection of pathogens. This particular interaction is governed by a specific "entry molecule," i.e., lipochitooligosaccharides (LCOs) (Nod factors) which are produced by rhizobia after receiving root exudate signals (flavonoids) (Checcucci et al. 2017). Therefore, the application of such symbionts would not only improve the nutrient status of plants under salt stress but would also help in excluding the colonization of phytopathogens (which lack these LCOs) and even deprive them of N content. LCOs have also been reported as elicitors of systemic resistance in plants (Limpens et al. 2015). Apart from these symbionts, free-living diazotrophic nitrogen fixers have also been reported in saline soil which efficiently mitigate salt toxicity in plants and protect them from disease incidence. Rodríguez et al. (2020) declared salt-tolerant *Pseudomonas segetis* strain P6 as a potential biocontrol agent on the basis of their ability to fix nitrogen, assimilate Fe and P and possession of quorum quenching (QQ) activity. The biopriming of tomato plants with the isolate showed protection against *Pseudomonas syringae* pv. tomato. Príncipe et al. (2007) isolated biocontrol and PGP native strains from saline soil of Argentina. The isolates identified as *Bacillus* sp. were able to grow up to 1M NaCl and possessed nitrogen-fixing, IAA-producing, and phosphate-solubilizing abilities along with the antifungal activity. The strains successfully inhibited *Sclerotinia sclerotiorum*, *Sclerotinia minor*, *Sclerotium rolfsii*, *Fusarium verticillioides* RC 2000, *F. solani*, *Fusarium proliferatum* RC 479, and *Fusarium graminearum* RC 664. Explaining the role of LCOs in salinity tolerance and biocontrol, Subramanian et al. (2016a) reported that soybean seeds treated with LCO and exposed to 80 mM and 100 mM NaCl stimulated signals of carbon, N, and energy metabolic pathways and also showed upregulation of genes including production of antioxidants and pathogenesis-related PR5 protein expression. Additionally, in non-legumes LCO induced *enod* genes expression which stimulated synthesis of chitinase, PR proteins, and antioxidant enzymes such as peroxidase and L-phenylalanine ammonia-lyase (PAL) (Schultze and Kondorosi 1996, 1998; Inui et al. 1997; Wang et al. 2012). Subramanian et al. (2016b) also proved the concept by reporting the expression of the above genes and metabolites in salt stressed LCO inoculated *Arabidopsis thaliana* plants, confirming the role of LCO in biocontrol.

Another limiting factor for the growth of phytopathogens includes the availability of Fe. Several phytopathogens have higher requirement of Fe than plants, and the chelation decides their severity of disease incidence and development of symptoms. Therefore, Fe is the prime requirement of pathogens as compared to other micronutrients such as manganese (Mn), copper (Cu), and boron (B), and while the former can be either beneficial or cause of infection in the host, the latter only benefit the host (Dordas 2008). In saline soil, high concentration of bicarbonates immobilizes Fe, and alkaline pH leads to oxidation of soluble ferrous to insoluble ferric (Arora et al. 2020), and the element concentration may be as low as 10^{-13} M (Lindsay 1979). Under this Fe-deficient environment, no microbes can survive, and thus salt-tolerant PGPR initiate the production of low molecular weight secondary metabolites, i.e., siderophores to approach the minimum levels of 10^{-6} M supporting their growth (Pal and Gardener 2006). Domiciling these high affinity Fe chelating siderophore-producing ST-PGPR helps to sequester the element from saline environment, and this renders the pathogens devoid of the necessary micronutrient. The production of siderophores also enhances the growth of plants under nutrient scant conditions of salinity, and these low molecular weight metabolites are even reported as elicitors of ISR in plants (Rosier et al. 2018). Tortora et al. (2011, 2012) demonstrated that *Azospirillum brasilense* strain REC2 and REC3 colonizing the rhizosphere of strawberry produced siderophore under iron-limiting conditions. The chemical composition revealed category of catechol-type siderophore which triggered ISR in plants and showed antifungal activity against *Colletotrichum acutatum* M11. Albdaiwi et al. (2019) found that halotolerant strains belonging to genera *Bacillus*, *Oceanobacillus*, and *Halomonas*, isolated from extremely saline environment of the Dead Sea region, showed significant PGP properties including siderophore production. The isolates were also able to inhibit the growth of phytopathogen *F. culmorum*.

In addition to primary role of N and Fe deficiency in exclusion of phytopathogens, other micro- and macronutrients are also chelated by PGPR; such as P, K, and Zn. However, their exact role in suppressing pathogens is not clear. May be the improved growth of plants by uptake of these elements could play important role in exaggerating the defense responses and maintain the health of the plant (Khare and Arora 2021). Phosphorous, being an important constituent of nucleic acids (DNA and RNA), phospholipids, phosphoproteins, enzymes, and energy-rich compounds (ATP, ADP), is required for survival of plants, PGPR, as well as pathogens (Prakash and Arora 2019). Therefore its direct chelation by potential phosphate-solubilizing bacteria (PSB) would restrict the essential metabolic processes in pathogens. Another mechanism by Rasul et al. (2019) proposes that PSB mainly produce organic acids for nutrient assimilation, and these acids also induce antimicrobial and antifungal activity protecting the plant from unwanted associations. The application of P was able to suppress bacterial leaf blight in rice, pod, and stem blight in soybean, downy mildew, blue mold, brown stripe disease in sugarcane, leaf curl virus disease in tobacco, blast disease in rice, and yellow dwarf virus disease in barley (Potash and Phosphate Institute (PPI) 1988; Reuveni et al. 1998, 2000; Huber et al. 1999; Kirkegaard et al. 1999). Potassium is another very important element to mitigate the salt toxicity in plants, establishing osmotic balance and regulating various biochemical processes such as stomatal closure (Arora et al. 2020). With respect to biocontrol, K induces the thickening of plant cell wall protecting against the entry of phytopathogens (Dordas 2008). Therefore, the combined activity of K chelation helps the plant to respond efficiently in presence of phytopathogens and salt stress. Gao et al. (2018) found that K application suppressed *Heterodera glycines* in soybean by the release of cinnamic, ferulic, and salicylic acids and antioxidant enzymes. However, Davis et al. (2018) report that the exact relationship between K accumulation in plants and disease incidence is more or less species dependent. Among the micronutrients, Zn is also required to extend the plant defense strategies against diseases. Along with its well-established role as catalyst and protein co-factor, Zn also initiates production of antioxidant enzymes (Khederi et al. 2018; Cabot et al. 2019), activation of Znf gene (required to provide resistance against pathogens) (Wang et al. 2017), and metallothionein expression (MT; develops disease-resistant plant varieties) (Ćalić et al. 2017) in plants. Wadhwa et al. (2014) showed that exogenous application of Zn to cluster bean reduced the disease incidence by *Rhizoctonia* through upregulation of PAL and tyrosine ammonia-lyase (TAL) synthesis, which further regulated the phenylpropanoid metabolism and performed defense-related functions. However, in the absence of Zn solubilizing bacteria, the availability of nutrients can benefit the pathogenic fungi in establishing and enhancing their pathogenicity, which is also the case for P and

K. Henceforth, beneficial salt-tolerant biocontrol agents are domiciled in the rhizosphere to capture the essential nutrients before the pathogens take advantage. There are several reports addressing potent nutrient-solubilizing ST-PGPR candidates, but their direct or indirect role in reducing pathogenicity is lesser explored. Paul and Nair (2008) demonstrated that salt-tolerant biocontrol strain *Pseudomonas fluorescens* MSP-393, isolated from saline coastal ecosystems, showed various mechanisms of stress mitigation along with strategies for exclusion of pathogens. The isolate produced toxins and even had efficient nutrient chelation property to increase resource competition under saline conditions. Similarly, Sadfi-Zouaoui et al. (2008) suggested that the biocontrol activity of halotolerant *Bacillus* sp. against gray mold disease on tomato fruits was mainly due to the production of antifungal compounds and nutrient competence. Abdelshafy-Mohamad et al. (2020) studied the endophytic microbial population of medicinal plant *Thymus vulgaris* growing in naturally arid regions of Egypt. The study found that salt-tolerant *Bacillus* and *Enterobacter* strains showed biocontrol activity against *F. oxysporum* in tomato plants under salt stress. The biocontrol mechanism reported by the strains included production of antifungal compounds and chelation of Fe, K, Zn, and P, thereby minimizing the chances of pathogen survival under extreme conditions of salinity.

9.2.2 Antibiosis

Antibiosis is the most popular mechanism of root-colonizing rhizobacteria to fight against phytopathogens (Haas and Défago 2005; Arseneault and Filion 2017). Antibiosis is governed by production of low molecular weight compounds called antibiotics involved in biocontrol against various bacterial and fungal phytopathogens and plant parasitic nematodes (Beneduzi et al. 2012). Pseudomonads and *Bacillus* are well-known and extensively studied biocontrol agents because of production of wide range of antibiotics including pseudomonic acid, phenazine, pyoluteorin, pyrrolnitrin, pyoverdine, pyocyanin, 2,4-diacetylphloroglucinol (DAPG), cyclic lipopeptides, mycobacillin, surfactin, and subtilin (Gupta et al. 2015; Singh et al. 2017; Suryadi et al. 2019). Among these the most widely reported antibiotic compounds for disease management include phenazines, DAPG, pyrrole compounds (e.g., pyoluteorin and pyrrolnitrin), and cyclic (lipo-) peptides (Shameer and Prasad 2018). In a study, Egamberdieva et al. (2017) found that *Pseudomonas chlororaphis* TSAU13 and *Pseudomonas extremorientalis* TSAU2 efficiently colonized the roots of tomato, promoted plant growth, and inhibited the growth of *F. solani* (causing foot and root rot disease in plants) under saline conditions. *Pseudomonas* sp. has been reported to produce antibiosis-related compound pyocyanin along with other PGP traits such as IAA and siderophores, which reduced the infection of *M. phaseolina* causing charcoal rot disease in sunflower under salt stress (Khare and Arora 2011; Tewari and Arora 2016). *Pseudomonas* spp. MCC 3145 isolated from rhizospheric zone of rice was found to be very effective in biocontrol of major phytopathogens in mung bean (*Vigna radiata*). The coated seeds showed significant production of secondary metabolites responsible for antibiosis including phenazine (phenazine-1-carboxylic acid) and biosurfactant under salt and heavy metal stress conditions (Patil et al. 2017). *P. extremorientalis* TSAU20 showed significant results in control of damping off disease caused by *Rhizoctonia solani* in the seedlings of cotton plants growing under saline conditions (Egamberdieva and Jabborova 2013). Wang et al. (2015) reported salt-tolerant PGP strains including *Pseudomonas protegens*, *P. chlororaphis* subsp. *aureofaciens*, and *Serratia plymuthica* containing DAPG-producing genes (*phlD*, *prnD*, *pltC*, or *phzF)* capable of inhibiting growth of *R. solani* and *S. sclerotiorum* under saline conditions. Similarly, Rangarajan et al. (2003) elucidated antibiosis activity by *Pseudomonas* sp. strains against rice pathogens including *Xanthomonas oryzae* pv. *oryzae* and *R. solani* causing leaf and sheath blight, respectively, under saline conditions. Barriuso et al. (2008) isolated salt-tolerant strains of *Bacillus* sp. strain L81 and *Arthrobacter oxidans* strain BB1 and elucidated their role in biocontrol against foliar pathogen *P. syringae* as well as enhanced salt tolerance in *A. thaliana*.

9.2.3 Lytic Enzyme Production

Another mechanism involving biocontrol against phytopathogens by PGPR includes production of lytic/hydrolytic enzymes. These enzymes are capable of degrading organic components present in cell wall of phytopathogens and use them as carbon source. Chitinase, dehydrogenase, β-1,3 glucanase, lipase, phosphatase, and proteases are some of the important enzymes secreted by PGPR in response to the onset of biotic stress (Hayat et al. 2010; Kumar et al. 2020). Mechanism of action involves breakdown of glyosidic bonds present in polysaccharides of pathogen cell wall structure, inhibiting the growth of fungal hyphal tips and germ tubes and causing partial swelling in the hyphae (Jadhav et al. 2017). Mohamed and Gomaa (2012) found that *Bacillus subtilis* and *P. fluorescens* could be very effective in growth promotion of radish (*Raphanus sativus*) and biocontrol through the production of lytic enzymes even under saline conditions. According to Tewari and Arora (2016), production of HCN, IAA, siderophore along with chitinase and β-1,3-glucanase activities by *Pseudomonas* sp. PF17 was helpful in degrading cell wall and inhibiting *M. phaseolina* mycelial growth, thereby stimulating the productivity of sunflower under saline conditions. Salt-tolerant strains of *Streptomyces* (C and S2) were found to be effective in protecting sugar beet from root rot-causing phytopathogens, viz., *R. solani* AG-2, *F. solani*, and *Phytophthora drechsleri* by production of cell wall-degrading lytic enzymes both under saline and non-saline conditions (Karimi et al. 2012). *P. chlororaphis-66* and *Mesorhizobium ciceri-4* have been reported for antifungal activity against several phytopathogenic fungi (*F. oxysporum f.* sp. ciceris, *F. verticillioides*, *R. solani*, *F. solani*, *Alternaria alternata*), protecting chickpea by producing hydrolytic enzymes under saline conditions (Shurigin et al. 2018). *Ochrobactrum pseudogrignonense* isolated from the rhizospheric zone of *Imperata cylindrical* (halophyte) showed upregulation of genes associated with the synthesis of lytic and antioxidant enzymes against salinity. These enzymes played significant role in protection of plants by inducing resistivity and improving ability of plants to survive under salt-stressed conditions (Chakraborty et al. 2019).

9.2.4 Induced Systemic Resistance (ISR) and Systemic Acquired Resistance (SAR)

Defense mechanisms play a very significant role in disease management through activation of various reactions which improves the immunity and growth of plants. The internal defense mechanisms in plants including induced systemic resistance (ISR) and systemic acquired resistance (SAR) are stimulated by PGPR in response to biotic and abiotic stresses. ISR is mainly governed by ethylene (ET) and jasmonic acid (JA) signaling pathways (Sharma et al. 2019). Application of PGPR has been significant in triggering numerous defense reactions including stimulation of phytohormones, expressing the genes encoding defense-related proteins, enhancement in the rigidity of cell wall, production of antioxidants, phosphorylation mechanisms, and deposition of callose for the protection of plant against various phytopathogens (Nurnberger et al. 2004; Van Loon and Bakker 2005). Interestingly, due to triggering of ISR the ET level in plants shoots up which could hinder their growth and productivity (Glick et al. 2007). Therefore, under such stressed conditions, beneficial microbes produce ACC deaminase (ACCD) enzyme to reduce the level of ethylene while still initiating ISR (Romera et al. 2019). This complex signaling and maintenance of ET balance in plants by PGPR is a fascinating area which could unravel many interrelated pathways and molecular ambiguities involved in stress mitigation. PGPR produce a number of signaling molecules to communicate with biotic/abiotic or combinatorial stressed plants and resultantly initiate the ISR responses (Arora et al. 2020). These signaling molecules include *N*-acyl homoserine lactones (AHLs), VOCs, and cyclodipeptides (CDPs) (Rosier et al. 2018; Hartman et al. 2019). AHLs play prominent role in the expression of genes related to the elicitation of ISR through quorum sensing (Viswanath et al. 2016). The complex signaling by AHL not only induces ISR against pathogens but is also responsible for triggering induced systemic tolerance (IST) in salt-stressed plants (Yang et al. 2009). Pang et al. (2009) observed that regulation of AHL signaling mediated by *S. plymuthica* HRO-C48 increased the resistivity in plants against *Pythium aphanidermatum* and *B. cinerea*. Additionally, categorizing the ISR elicitors, many molecules often referred as MAMPs (microbial-associated molecular patterns) include lipopolysaccharides (LPS), EPS, siderophores, biosurfactants, antibiotics such as DAPG, lipopeptides, pyocyanin, and pseudobactins are also responsible for imparting disease resistivity (Han et al. 2016). Therefore, most of these metabolites which are produced by PGPR under salt stress can serve the dual role of ISR as well as IST in plants. Rhizospheric bacterial strain of *Stenotrophomonas maltophilia* SBP-9 isolated from *Sorghum bicolor* showed PGP traits such as IAA, ACCD, GA, siderophore production, and phosphate solubilization. These traits were found helpful in growth promotion of plant and also induced systemic resistivity and salt tolerance in wheat crop (*Triticum aestivum* L.) (Singh and Jha 2017). Many studies have reported the role of *Pseudomonas*, *Bacillus*, and *Rhizobium* strains in elicitation of ISR which is found efficient in the protection of plants from phytopathogens prior to infection (Elbadry et al. 2006; Choudhary and Johri 2009; Kumar et al. 2018). Singh and Jha (2016) observed that PGPR *Serratia marcescens* CDP-13 decreased the infectivity of various fungal pathogens by inducing systemic resistance along with improvement of salt tolerance in wheat (*T. aestivum* L.). A halotolerant PGPR, *Klebsiella* spp. strain MBE02 was found responsible for expression of genes associated with ISR through activation of JA and ET signaling pathways. The strain was effective in inducing disease resistance against the fungal pathogens *Aspergillus niger* and *Aspergillus flavus* in peanut (Sharma et al. 2019). *Bacillus vallismortis* BS07M strain was found effective against biotic stress by triggering ISR in cabbage against soft rot pathogen *Pectobacterium carotovorum* along with the upregulated expression of salinity response gene P5CS (Park et al. 2016).

SAR is related to signaling of phytohormones and other protective compounds at the site of infection, and these signals induce resistance in plants to cope with biotic and abiotic stresses. These signals are generated at the primary site of infection and then are transported to the whole plant by phloem, hence improving the defense system of the plant (Shine et al. 2019). Salicylic acid (SA) is the natural phenolic compound and phytohormone which improves resistivity of plants against phytopathogens and also supports growth of plant in salinity stress (Van Loon 2007). SA also plays vital role in regulation of *NPR1*, *NPR3*, and *NPR4* proteins inducing the signaling of SAR in stressed plants (Ali et al. 2018). Notably, Sticher et al. (1997) found loss of SAR in *npr1* mutant transgenic *NahG* plants which were unable to accumulate SA. The increasing level of SAR induction in plants is correlated with higher accumulation of pathogenesis-related proteins (PRs) and expression of gene encoding PR-1 proteins (De Vleesschauwer and Höfte 2009). Although SAR is mainly initiated by pathogens, PGPR can indirectly regulate the pathways through upregulation of SA synthesis in plants under abiotic and biotic stresses (Vallad and Goodman 2004; Kang et al. 2014). In *A. thaliana*, two PGPR, i.e., *Bacillus* sp. strain L81 and *A. oxidans* strain BB1 isolated from *Pinus* spp., were found to be responsible for inducing SAR through signaling of SA-dependent pathway. The resultant PR activation genes helped in defending against the foliar pathogen *P. syringae* DC3000 under salinity (Barriuso et al. 2008). An endophytic strain of *Pseudomonas aeruginosa*, when inoculated in combination with neem cake (*Azadirachta indica*), increased the SA content and thereby mediated SAR mechanisms, inducing tolerance against *M. phaseolina* and salinity in cotton (Rahman et al. 2016). SA along with EPS-producing isolate *P. aeruginosa* PF23EPS$^+$ also remarkably reduced the pathogenicity of *M. phaseolina* under high salt concentrations in sunflower crop (Tewari and Arora 2018).

9.2.5 Other Mechanisms

Biological control of plant diseases by PGPR has been a notable alternative for chemical inputs in agriculture. Apart from the above mentioned secondary metabolites, PGPR can also involve other mechanisms and metabolites to combat phytopathogens in presence of salt stress. The dual activity of salt tolerance and biocontrol by ST-PGPR tremendously helps in increasing plant growth and productivity. This section discusses other metabolites and their possible mechanisms of action that are reported to combat phytopathogens in presence of salt stress.

9.2.5.1 Antioxidant System

In salt stress conditions, the generation of reactive oxygen species (ROS) peaks due to high buildup of sodium ions, and thereby it causes alterations in redox state, membrane-bounded protein denaturation, DNA deterioration, reduction in membrane fluidity, variations in protein formation, and impairment in cell homeostasis devastating cellular system and lastly results in cell death (Halo et al. 2015). A number of enzymatic as well as non-enzymatic antioxidants are able to degrade ROS and are reported to favorably augment the endurance level of oxidative stresses in plants. Inoculation of *Enterobacter* spp. in tomato seedling under salt stress conditions showed high ascorbate peroxidase (APX) activities (Sandhya et al. 2010). Catalase (CAT) and superoxide dismutase (SOD) activities were also reported to be amplified in gladiolus plant by application of PGPR inoculants (Damodaran et al. 2013; Kim et al. 2014). Apart from salt stress tolerance, PGPR also suppress the oxidative damage due to plant diseases with the aid of antioxidant enzymes, e.g., SOD, CAT, glutathione peroxidase (GPX), glutathione S transferase (GST), peroxidase, ascorbate oxidase, and polyphenol oxidase (PPO) (Starlin and Gopalakrishnan 2013). Egamberdieva et al. (2017) found that salt-tolerant *P. chlororaphis* TSAU13 and *P. extremorientalis* TSAU20 strains enhanced the antioxidant activity along with increased accumulation of proline in tomato under saline conditions. The strains also acted as biocontrol agents against the foot and root rot of tomatoes caused by *F. solani* extending their dual mechanism in plant growth promotion. The study correlated the fact that antioxidant systems help in extending the plant defense machinery. A study by Meena et al. (2000) summarizes that various defense enzymes including PAL, PPO, and peroxidase (PO) are associated with induction of systemic resistance against phytopathogens. PAL and PO are related to biosynthesis of phytoalexins and phenolics (Daayf et al. 1997); PO and PPO are directly related to oxidative damage and lignification of host plant cells leading to restriction of disease incidence (Silva et al. 2004). Singh and Jha (2017) confirmed the concept by reporting that salt-tolerant *S. maltophilia* SBP-9 strain promoted growth of *S. bicolor* under salinity, and the ability of the bacteria to produce β-1,3 glucanase, PAL, PPO, and PO inhibited the fungal pathogen *F. graminearum*.

9.2.5.2 Volatile Organic Compounds (VOCs)

Microorganisms living in the rhizosphere also produce VOCs for signaling and establishing plant-microbe associations. VOCs are organic lipophilic fluids possessing high vapor pressure and are known to be responsible for communication between two or more organisms via signal transduction (Schulz-Bohm et al. 2017). The wide spectrum of VOCs produced by PGPR include short-chain aliphatic aldehydes, esters, alcohols, organic acids, ethers, ketones, sulfur compounds, and hydrocarbons (Vaishnav et al. 2017). The level, composition, and nature of VOCs produced by microbes varies according to the rhizospheric conditions, environmental stimuli, and even species of microbes. VOCs are actually multifarious molecules with strong ability to sustain salt tolerance in plants through increasing plant biomass, leaf area, chlorophyll content, flowering/fruiting, auxin, ABA, ethylene synthesis, nutrient uptake, regulation of HKT transporters (important for regulating K^+/Na^+ ratio), and sugar assimilation (Sharifi and Ryu 2018). VOCs have also been directly related to activation of ISR in plants against phytopathogens (Raza et al. 2016). *Bacillus amyloliquefaciens* GB03, a model organism reported for the production of over 25 VOCs, promoted growth of various plants under different ecological conditions (Ryu et al. 2004; Banchio et al. 2009; Choudhary and Johri 2009). Cappellari and Banchio (2019) found that salt-stressed *Mentha piperita* plants when exposed to VOCs of *B. amyloliquefaciens* GB03 showed higher plant growth parameters and nutrient status, reduced Na^+ uptake, induced SA pathway, enhanced accumulation of proline, and thereby imparted better resilience against salinity. The study declared the major VOC involved was acetoin (3-hydroxy-2-butanone). Correlating this study with previous reports, it was found that acetoin has been regarded as systemic resistance elicitor (Ann et al. 2013), with biocontrol activity against *P. carotovorum* in tobacco, and triggering of PR (pathogenesis related) proteins in *Arabidopsis* (Rudrappa et al. 2010). Ledger et al. (2016) found production of organic compounds such as 2-undecanone, 7-hexanol, 3-methylbutanol, and dimethyl disulfide by *Paraburkholderia phytofirmans* promoted growth of *A. thaliana* in the presence of salinity. Therefore, it can be concluded that these bioactive molecules can serve more than one function depending upon the stresses faced. To date there are almost no reports defining the combined action of VOCs against salinity as well as biocontrol (Fincheira and Quiroz 2018), and wide study is required about these metabolites.

Among the various metabolites produced by biocontrol agents, hydrogen cyanide (HCN) is highly acclaimed due to its remarkable toxic properties (Cucu et al. 2019). The primary mechanism behind HCN toxicity involves inhibition of cytochrome c oxidase and metallo-enzyme production in phytopathogens (Nandi et al. 2017). Interestingly, HCN-producing microbial biocontrol agents are also found to be producing other antibiotics and cell wall-degrading enzymes (Ramette et al. 2006; Olanrewaju et al. 2017), suggesting some sort of common signaling. Production of HCN by PGPR also helps in increasing P availability and sequestration of the geochemical entities (Rijavec and Lapanje 2016). In a study, salt-tolerant isolates of *Bacillus* sp. were screened and were found to be positive for HCN production under salinity stress conditions (Sharma et al. 2015). Patil et al. (2016) showed that

HCN-producing *P. aeruginosa* GS-33 mitigated the infestation of *M. phaseolina* in soybean plants under salt stress conditions.

9.2.5.3 Exopolysaccharides

Bacteria also protect plants from the constraints of high levels of salinity by producing biopolymers such as EPS, capsular polysaccharides (CPS), LPS, teichoic acids (TAs), and β-1,2-glycans (Ormeño-Orrillo et al. 2012). EPS have been recognized as the most promising components produced under saline conditions that bind to the available cations by forming a protective barrier known as rhizosheath around roots (Fatima et al. 2020). Apart from direct protection against environmental stresses such as salinity, EPS also have role in competence and defense against phytopathogens. High WHC of this polymer, which is more than 15 times its weight (Costa et al. 2018), plays a dual role in mitigating salinity stress and also exclusion of pathogens. Further, the presence of uronic acids in EPS is already well recognized as the property and structure defining moiety of the macro-molecule, which aids in chelation of cations, formation of biofilms, and modification of sugar backbones (Liang and Wang 2015; Mukherjee et al. 2019). Timmusk et al. (2019) through kernel assay showed that biofilming ability of *Paenibacillus polymyxa* successfully inhibited *Fusarium* head blight (FHB) caused by *F. graminearum* in wheat. Interestingly, the study confirmed that uronates present in polysaccharides are the critical components for antagonism. Mukherjee et al. (2019) showed that EPS-producing halotolerant *Halomonas* sp. Exo1 strain showed biocontrol against *F. oxysporum* both under salinity and heavy metal toxicity stress, and the possible mechanisms hypothesized were siderophore production, P chelation, and EPS production. Tewari and Arora (2014) illustrated the role of EPS produced by *P. aeruginosa* PF23 in antagonizing phytopathogen *M. phaseolina* in sunflower under salt stress conditions. In the same study, EPS not only functioned as osmoprotectant but also acted as an important biocontrol metabolite. The role of EPS in plant growth and salt stress mitigation has also been reported by Fatima and Arora (2021), concluding that at optimized concentration, the metabolite enhanced the production and yield of sunflower and elevated the salt tolerance efficiency under field conditions. The study also showed the presence of uronates in *Pseudomonas entomophila* PE3 extracted EPS, which are reported in antagonism, therefore signifying the potential of EPS in biocontrol specifically under salt stress. Fatima et al. (2020) also reported growth enhancement of rice in saline soil upon inoculation with *Alcaligenes faecalis* extracted EPS.

9.2.5.4 Root Exudation

The phytochemicals discharged by the roots of plants as a result of secretion, diffusion, and cell lysis are called root exudates, and they play dynamic roles in governing plant-microbe interactions (Ortíz-Castro et al. 2009). A number of organic and inorganic compounds along with various signals including sugars, amino acids, vitamins, mucilage, phenols, flavonoids, and organic acids attract the microbial population through chemotaxis (Schreiter et al. 2018). About 30–50% of carbon fixed by photosynthesis are released as root exudates which serve as signals for

PGPR to initiate colonization (Kumar et al. 2006). The composition of root exudates depends upon the species of plant, growth substrate, developmental stages, as well as the stress involved (Dakora and Phillips 2002; Bertin et al. 2003). According to Schreiter et al. (2014), the root exudation and bacterial colonization are also dependent upon plant growth. Therefore, under saline conditions prior inoculation of plants with ST-biocontrol agents could be helpful in redesigning the root exudation by maintaining the growth of plants. Interestingly, Wang et al. (2019a) reported that pretreated tomato plants with a model biocontrol agent, *Bacillus cereus* AR156, upregulated the concentration of lactic acid and hexanoic acid in root exudates. These components in the root exudates further extended the biofilming ability of *B. cereus* AR156 and even reduced the efficacy of *Ralstonia solanacearum*. However, under saline conditions this complex signaling is not explored yet, and there are just hypotheses to illustrate the role and pattern of exudation modification by ST-PGPR. Tsegaye et al. (2019) characterized and identified strains of PGPR that have exploited carbohydrates, carboxylic acids, and amino acids as root exudates, and the strains also upheld multiple PGP traits, biocontrol properties, and ability to endure salt stress.

9.2.5.5 Quorum Sensing and Quorum Quenching

Another such signaling biocontrol mechanism involves AHL-induced quorum sensing. Although AHLs are widely reported as beneficial molecules for establishment of plant-microbe associations, mechanism of biocontrol by degrading AHLs through quorum quenching (QQ) is gaining attention (Chane et al. 2019). There are a number of phytopathogens which enhance their virulence factors through quorum sensing, and thus QQ could be an effective mechanism of controlling the disease incidence while still reducing the chances of pesticide resistance in pathogens (Vesuna and Nerurkar 2020). Ma et al. (2013) reported various strains inhabiting tobacco leaves, belonging to genera *Bacillus* sp., *Acinetobacter* sp., *Lysinibacillus* sp., *Serratia* sp., *Pseudomonas* sp., and *Myroides* sp. which showed QQ activity and potential as biocontrol agents. Surprisingly, although these strains quenched the signaling molecule (AHL), they were still successfully able to establish themselves in plants. Wang et al. (2019b) isolated strains of *Pseudoalteromonas*, *Pontibacillus*, and *Altererythrobacter* showing QQ activity and were also capable of growing under high salinity stress (58 g/L NaCl), pH 7.8, and temperature up to 50° C. The study concluded that the enzymes produced by these QQ microbes could be used to solve pathogenesis-related problems in saline and high temperature stressed environments.

9.2.5.6 Biosurfactants

Biosurfactants are another structurally diverse group of amphiphilic compounds produced by microorganisms that confer the ability to pile up between fluid phases. Surfactants include fatty acids, phospholipids, polysaccharides, proteins, as well as some of their combinations (e.g., lipoproteins, glycoproteins, and glycolipids) (Yañez-Ocampo and Wong-Villarreal 2013). The primary mechanism of biocontrol by these surfactants include cell wall destabilization, inhibition of phytopathogens' attachment to plant surfaces, directly safeguarding the plants by formation of biofilms, replacing as a toxin during recognition, and binding of the pathogen receptors on plant cell membranes (Nihorimbere et al. 2011; Lahkar et al. 2018). Role of biosurfactants as promising biocontrol compounds and the associated mechanisms including parasitism, antibiosis, competition, ISR, and hypovirulence have been confirmed by several researchers (Thein and Prathuangwong 2010; Velho et al. 2011; Krzyzanowska et al. 2012; Akladious et al. 2019). Rivas-Garcia et al. (2019) found that the combination of yeast *Debaryomyces hansenii* and bacteria *Stenotrophomonas rhizophila* isolated from marine environment, upon application to muskmelon, caused inhibition of *Fusarium proliferum*. The mechanism of biocontrol reported was production of VOCs and biosurfactants which possibly had reduced the attachment of pathogen to plants and reduced their virulence factors. Hassen et al. (2018) isolated novel *Pseudomonas rhizophila* S211 strain which was able to tolerate salt concentration up to 300 mM NaCl pH range 6–10, and temperature up to 90 °C. The strain showed various PGP traits and even production of rhamnolipid biosurfactant. The study reported the potentiality of the strain in biocontrol, biofertilization, and rhizoremediation. A study by Elazzazy et al. (2015) showed biosurfactant-producing *Virgibacillus salarius* to effectively reduce surface tension under high salt stress conditions, wide temperature range (30 and 100 °C), and alkaline pH. Zhang and Tang (2019) illustrated that halotolerant rhamnolipid-producing strain upon inoculation to strawberry (pot assay) not only treated the symptoms of powdery mildew of strawberry caused by *Fragaria ananassa* but also improved root and shoot growth. Khare and Arora (2021) reported that biosurfactant extracted from *Pseudomonas guariconensis* LE3 had multifarious functions including biocontrol against charcoal rot in sunflower crop, growth enhancement of plants, improvement in shelf life of LE3-based bioformulation, enhanced plant-microbe interaction, and better soil properties. Mishra et al. (2020) found enhanced growth of mustard plant upon inoculation with various concentrations of *Pseudomonas putida* biosurfactant under field conditions. Therefore, the metabolite can be regarded as multifunctional with the potential of biocontrol as well as biofertilization and remediation. The addition of biosurfactants in bioformulation could be helpful in enhancing nutrient content of soil and microbial colonization which are otherwise highly depressed in saline soil. Henceforth, the metabolite could be a hallmark in designing novel bioformulations for saline soil infested with pathogens.

9.3 Future Prospects

Soil pollution is a major concern across the globe which is directly hitting the food basket and the sustainability of agro-ecosystems. The increasing salinization of agro-ecosystems has led to the creation of marginal lands and has steeply brought down the productivity of crops. In the current scenario of climate change and global warming, the issues faced by agriculture are not only limited to abiotic stresses but have also increased the incidence of plant diseases. Salinity stress along with prevailing plant pathogens is an excellent example of abiotic and biotic stresses co-occurring in fields and declining the food production. Although there are many evidences about the synergistic effect of salinity and phytopathogens in agriculture, there are few reports signifying the combined impact of stresses on crops and their response (Nejat and Mantri 2017; Zhang and Sonnewald 2017). The increasing degradation of soil and plant health demands the most sustainable and cost-effective biological approaches which can act to defend the odds of salinity as well as protect the plants from diseases. The application of ST-PGPR hold the potential to mitigate the salinity stress and also enhance the immunity responses in plants against phytopathogens. With the co-occurrence of abiotic and biotic stresses, the cross-tolerance initiated by PGPR shares some common signals, responsive genes, metabolic products, and physiological responses which need to be investigated in detail (Foyer et al. 2016). For example, Rejeb et al. (2014) reported that plants which were primarily affected by abiotic stress showed effective response against the subsequent biotic stress in comparison to the non-primed plant that lacked activation of defense signaling. Thus, the prior application of ST-PGPR can safeguard the plant against salinity and can also make plants immune form phytopathogens.

The complex stress signaling in plants upon inoculation with PGPR constitutes a number of directly and indirectly related responses which synergistically provide protection against multi-stresses (Chojak-Koźniewska et al. 2018). Some responses of PGPR including improvement of root system architecture (RSA), phytohormonal signaling, antioxidant activity, EPS production, and nutrient chelation act as common recovery methods both against salinity and pathogens (Tang et al. 2015). Yet there are some antagonistic interactions taking place at molecular and physiological levels which although provide resilience against one stress but increase the risk of the other. For example, the increasing concentration of abscisic acid (ABA) in plants under inoculation with ST-PGPR increases the defense against salinity, but ABA accumulation suppresses the disease resistance mechanism of SA and JA/ET signaling (Sivakumaran et al. 2016). Similarly, Sunita et al. (2020) summarized that production of VOCs by ST-PGPR, apart from regulating plant-microbe interaction, and virulence factors also are responsible for biosynthesis of osmoprotectants, phytohormones, and siderophores. Therefore, these complex interactions need to be further studied via the future techniques of "interactomics" (Yadav and Mani 2019). Apart from intracellular interactions, there are various chemical kinetics and signaling taking place in the rhizosphere. The establishment of plant-microbe association requires a number of plant metabolites such as root exudates (flavonoids, tryptophan, ACC) and microbial metabolites including LCOs, VOCs, AHLs, and EPS (Sasse et al. 2018). In combinatorial stress (i.e., salinity and biotic stress), these interactions taking place in the rhizosphere can provide a better illustration of how ST-PGPR exclude the soil pathogens from colonizing the plants and also the metabolic signals generating from such novel associations. These metabolite "blends" present in the symbiotic associations and the dynamics involved can be studied using the advanced field of "signalomics" (Rothballer et al. 2018).

The wide role of different microbial metabolites involved in mitigating biotic and abiotic stresses generates the idea of reclamation of saline lands excluding the soil pathogens, thereby increasing the fertility of soil and upgrading the health of plants. The injudicious increase in degraded lands casts another limitation for microbial technology, i.e., its application. Although there are many bioinoculants available in the market, due to soil dynamics and climatic variations the field results are less efficient. Therefore, if the exact mechanism of action of ST-PGPR is reported, the next-generation application technology can be applied to "mimic" the mechanism in soil either by directly adding metabolites and chemical attractants or by exposing the plants to stresses before plantation. Thereby, this would help in preparing the plants for the forthcoming environmental and biological challenges (Rosier et al. 2018). Arora et al. (2020) discussed that addition of osmoprotectants or cell protectants in bioformulation aids in preparing the microbial cells and plants against initial environmental stresses and then acclimatizing to the unhospitable condition. Microbial technology thus holds the future potential of land reclamation and enhanced agricultural productivity.

9.4 Conclusion

Soil salinization has been a hassling issue ever since the beginning of industrial era. The loss of agricultural productivity due to increasing salt toxicity is increasing without any viable large scale biological solutions. Climate change and environmental factors have also increased the incidence of diseases in immuno-suppressed plants which are already facing stress due to salinity. Therefore, popularization of microbial techniques to mitigate the combinatorial biotic and abiotic stresses and ensure agricultural sustainability has become important. ST-PGPR produce an array of secondary metabolites which provide resilience against phytopathogens. Apart from indirectly excluding the soil pathogens by chelating essential ions and competing for space, these beneficial microbes also directly act upon the phytopathogens by secretion of several metabolites. ST-PGPR also induce the immune system of plant through ISR and SAR and help to defend against the pathogens. Thus, the dual role of ST-PGPR in plant growth promotion and protection under salt stress can be harnessed to achieve the goals of agriculture sustainability and food security by increasing the quality and quantity of produce.

References

Abdelshafy-Mohamad OA, Ma JB, Liu YH, Zhang D, Hua S, Bhute S et al (2020) Beneficial endophytic bacterial populations associated with medicinal plant *Thymus vulgaris* alleviate salt stress and confer resistance to *Fusarium oxysporum*. Front Plant Sci 11:47

Akladious SA, Gomaa EZ, El-Mahdy OM (2019) Efficiency of bacterial biosurfactant for biocontrol of *Rhizoctonia solani* (AG-4) causing root rot in faba bean (*Vicia faba*) plants. Eur J Plant Pathol 153(1):15–35

Albdaiwi RN, Khyami-Horani H, Ayad JY, Alananbeh KM, Al-Sayaydeh R (2019) Isolation and characterization of halotolerant plant growth promoting rhizobacteria from durum wheat (*Triticum turgidum* subsp. *durum*) cultivated in saline areas of the dead sea region. Front Microbiol 10:1639

Ali A, Shah L, Rahman S, Riaz MW, Yahya M, Xu YJ et al (2018) Plant defense mechanism and current understanding of salicylic acid and NPRs in activating SAR. Physiol Mol Plant Pathol 104:15–22

Al-Sadi AM, Al-Masoudi RS, Al-Habsi N, Al-Said FA, Al-Rawahy SA, Ahmed M (2010) Effect of salinity on pythium damping-off of cucumber and on the tolerance of *Pythium aphanidermatum*. Plant Pathol 59(1):112–120

Amaresan N, Kumar K, Madhuri K, Kannamma GU (2016) Isolation and characterization of salt tolerant plant growth promoting rhizobacteria from plants grown in tsunami affected regions of Andaman and Nicobar Islands. Geomicrobiol J 36(20):942–947

Amin A, Khan MA, Ehsanullah M, Haroon U, Azam SM, Hameed A (2012) Production of peptide antibiotics by *Bacillus* sp. GU 057 indigenously isolated from saline soil. Braz J Microbiol 43(4):1340–1346

Ann MN, Cho YE, Ryu HJ, Kim HT, Park K (2013) Growth promotion of tobacco plant by 3-hydroxy-2-butanone from *Bacillus vallismortis* EXTN-1. Korean J Pestic Sci 17(4):388–393

Arora S, Singh YP, Vanza M, Sahni D (2016) Bio-remediation of saline and sodic soils through halophilic bacteria to enhance agricultural production. J Soil Water Conserv 15(4):302–305

Arora NK, Fatima T, Mishra I, Verma M, Mishra J, Mishra V (2018) Environmental sustainability: challenges and viable solutions. Environ Sustain 1(4):309–340

Arora NK, Fatima T, Mishra J, Mishra I, Verma S, Verma R et al (2020) Halo-tolerant plant growth promoting rhizobacteria for improving productivity and remediation of saline soils. J Adv Res 26:69–82

Arseneault T, Filion M (2017) Biocontrol through antibiosis: exploring the role played by sub-inhibitory concentrations of antibiotics in soil and their impact on plant pathogens. Can J Plant Pathol 39(3):267–274

Ashwitha K, Rangeshwaran R, Vajid NV, Sivakumar G, Jalali SK, Rajalaxmi K, Manjunath H (2013) Characterization of abiotic stress tolerant *Pseudomonas* spp. occurring in Indian soils. Biol Control 27(4):319–328

Banchio E, Xie X, Zhang H, Pare PW (2009) Soil bacteria elevate essential oil accumulation and emissions in sweet basil. J Agric Food Chem 57(2):653–657

Barriuso J, Solano BR, Gutiérrez Mañero FJ (2008) Protection against pathogen and salt stress by four plant growth-promoting rhizobacteria isolated from *Pinus* sp. on *Arabidopsis thaliana*. Phytopathology 98(6):666–672

Beneduzi A, Ambrosini A, Passaglia LM (2012) Plant growth-promoting rhizobacteria (PGPR): their potential as antagonists and biocontrol agents. Genet Mol Biol 35(4):1044–1051

Bertin C, Yang X, Weston LA (2003) The role of root exudates and allelochemicals in the rhizosphere. Plant Soil 256(1):67–83

Cabot C, Martos S, Llugany M, Gallego B, Tolrà R, Poschenrieder C (2019) A role for zinc in plant defense against pathogens and herbivores. Front Plant Sci 10:1171

Ćalić I, Koch J, Carey D, Addo-Quaye C, Carlson JE, Neale DB (2017) Genome-wide association study identifies a major gene for beech bark disease resistance in American beech (*Fagus grandifolia Ehrh.*). BMC Genom 18(1):1–14

Cappellari LDR, Banchio E (2019) Microbial volatile organic compounds produced by *Bacillus amyloliquefaciens* GB03 ameliorate the effects of salt stress in *Mentha piperita* principally through acetoin emission. J Plant Growth Regul 39:764–775

Chakraborty U, Roy S, Chakraborty AP, Dey P, Chakraborty B (2011) Plant growth promotion and amelioration of salinity stress in crop plants by a salt-tolerant bacterium. Res Sci Technol 3(11): 61–70

Chakraborty U, Chakraborty BN, Dey PL, Chakraborty AP, Sarkar J (2019) Biochemical responses of wheat plants primed with *Ochrobactrum pseudogrignonense* and subjected to salinity stress. Agri Res 8(4):427–440

Chane A, Barbey C, Bourigault Y, Maillot O, Rodrigues S, Bouteiller M et al (2019) A flavor lactone mimicking AHL quorum-sensing signals exploits the broad affinity of the QsdR

regulator to stimulate transcription of the rhodococcal qsd operon involved in quorum-quenching and biocontrol activities. Front Microbiol 10:786
Checcucci A, DiCenzo GC, Bazzicalupo M, Mengoni A (2017) Trade, diplomacy, and warfare: the quest for elite rhizobia inoculant strains. Front Microbiol 8:2207
Chennappa G, Naik MK, Adkar-Purushothama CR, Amaresh YS, Sreenivasa MY (2016) PGP potential, abiotic stress tolerance and antifungal activity of *Azotobacter* strains isolated from paddy soils. Indian J Exp Biol 54(5):322–331
Chojak-Koźniewska J, Kuźniak E, Zimny J (2018) The effects of combined abiotic and pathogen stress in plants: Insights from salinity and *Pseudomonas syringae* pv *lachrymans* interaction in cucumber. Front Plant Sci 9:1591
Choudhary DK, Johri BN (2009) Interactions of *Bacillus* spp. and plants–with special reference to induced systemic resistance (ISR). Microbiol Res 164(5):493–513
Costa OY, Raaijmakers JM, Kuramae EE (2018) Microbial extracellular polymeric substances: ecological function and impact on soil aggregation. Front Microbiol 9:1636
Cucu MA, Gilardi G, Pugliese M, Matić S, Gisi U, Gullino ML et al (2019) Influence of different biological control agents and compost on total and nitrification-driven microbial communities at rhizosphere and soil level in a lettuce-*Fusarium oxysporum* f. sp. *lactucae* pathosystem. J Appl Microbiol 126(3):905–918
Daayf F, Schmitt A, Bélanger RR (1997) Evidence of phytoalexins in cucumber leaves infected with powdery mildew following treatment with leaf extracts of *Reynoutria sachalinensis*. Plant Physiol 113(3):719–727
Dakora FD, Phillips DA (2002) Root exudates as mediators of mineral acquisition in low-nutrient environments. Plant Soil 245:35–47
Damodaran T, Sah V, Rai R, Sharma D, Mishra V, Jha S, Kannan R (2013) Isolation of salt-tolerant endophytic and rhizospheric bacteria by natural selection and screening for promising plant growth-promoting rhizobacteria (PGPR) and growth vigor in tomato under sodic environment. Afr J Microbiol Res 7:5082–5089
Davis JL, Armengaud P, Larson TR, Graham IA, White PJ, Newton AC (2018) Contrasting nutrient–disease relationships: potassium gradients in barley leaves have opposite effects on two fungal pathogens with different sensitivities to jasmonic acid. Plant Cell Environ 41(10): 2357–2372
De Vleesschauwer D, Höfte M (2009) Rhizobacteria-induced systemic resistance. Adv Bot Res 51: 223–281
Dordas C (2008) Role of nutrients in controlling plant diseases in sustainable agriculture. A review. Agron Sustain Dev 28(1):33–46
Egamberdieva D, Jabborova D (2013) Biocontrol of cotton damping-off caused by *Rhizoctonia solani* in salinated soil with rhizosphere bacteria. Asian Australas J Plant Sci Biotechnol 7(2): 31–38
Egamberdieva D, Jabborova D, Hashem A (2015) Pseudomonas induces salinity tolerance in cotton (*Gossypium hirsutum*) and resistance to Fusarium root rot through the modulation of indole-3-acetic acid. Saudi J Biol Sci 22(6):773–779
Egamberdieva D, Kucharova Z, Davranov K, Berg G, Makarova N, Azarova T et al (2011) Bacteria able to control foot and root rot and to promote growth of cucumber in salinated soils. Biol Fertile Soils 47(2):197–205
Egamberdieva D, Davranov K, Wirth S, Hashem A, Abd Allah EF (2017) Impact of soil salinity on the plant-growth—promoting and biological control abilities of root associated bacteria. Saudi J Biol Sci 24(7):1601–1608
Elazzazy AM, Abdelmoneim TS, Almaghrabi OA (2015) Isolation and characterization of biosurfactant production under extreme environmental conditions by alkali-halo-thermophilic bacteria from Saudi Arabia. Saudi J Biol Sci 22(4):466–475
Elbadry M, Taha RM, Eldougdoug KA, Gamal-Eldin H (2006) Induction of systemic resistance in faba bean (*Vicia faba* L.) to bean yellow mosaic potyvirus (BYMV) via seed bacterization with plant growth promoting rhizobacteria. J Plant Dis Prot 113(6):247–251

Elmer WH (2003) Local and systemic effects of NaCl on root composition, rhizobacteria, and Fusarium Crown and root rot of Asparagus. Phytopathology 93(2):186–192

Fatima T, Arora NK (2019) Plant growth-promoting rhizospheric microbes for remediation of saline soils. In: Arora NK, Kumar N (eds) Phyto and rhizo remediation. Microorganisms for sustainability, vol 9. Springer, Singapore, pp 121–146

Fatima T, Arora NK (2021) *Pseudomonas entomophila* PE3 and its exopolysaccharides as biostimulants for enhancing growth, yield and tolerance responses of sunflower under saline conditions. Microbiol Res 244:126671

Fatima T, Mishra I, Verma R, Arora NK (2020) Mechanisms of halotolerant plant growth promoting *Alcaligenes* sp. involved in salt tolerance and enhancement of the growth of rice under salinity stress. 3 Biotech 10(8):1–12

Fincheira P, Quiroz A (2018) Microbial volatiles as plant growth inducers. Microbiol Res 208:63–75

Foyer CH, Rasool B, Davey JW, Hancock RD (2016) Cross-tolerance to biotic and abiotic stresses in plants: a focus on resistance to aphid infestation. J Exp B 67(7):2025–2037

Gao X, Zhang S, Zhao X, Wu Q (2018) Potassium-induced plant resistance against soybean cyst nematode via root exudation of phenolic acids and plant pathogen-related genes. PLoS One 13(7):e0200903

Glick BR, Cheng Z, Czarny J, Duan J (2007) Promotion of plant growth by ACC deaminase-producing soil bacteria. Eur J Plant Pathol 119:329–339

Goswami M, Deka S (2019) Biosurfactant production by a rhizosphere bacteria *Bacillus altitudinis* MS16 and its promising emulsification and antifungal activity. Colloids Surf B Biointerfaces 178:285–296

Goswami D, Dhandhukia P, Patel P, Thakker JN (2014) Screening of PGPR from saline desert of Kutch: growth promotion in *Arachis hypogea* by *Bacillus licheniformis* A2. Microbiol Res 169(1):66–75

Grossi CEM, Fantino E, Serral F, Zawoznik MS, Fernandez Do Porto DA, Ulloa RM (2020) *Methylobacterium* sp. 2A is a plant growth-promoting rhizobacteria that has the potential to improve potato crop yield under adverse conditions. Front Plant Sci 11:71

Gupta B, Huang B (2014) Mechanism of salinity tolerance in plants: physiological, biochemical, and molecular characterization. Int J Genomics 2014:701596. https://doi.org/10.1155/2014/701596

Gupta G, Parihar SS, Ahirwar NK, Snehi SK, Singh V (2015) Plant growth promoting rhizobacteria (PGPR): current and future prospects for development of sustainable agriculture. J Microb Biochem Technol 7(2):096–102

Haas D, Défago G (2005) Biological control of soil-borne pathogens by fluorescent Pseudomonads. Nat Rev Microbiol 3:307–319

Halo BA, Khan AL, Waqas M, Al-Harrasi A, Hussain J, Ali L et al (2015) Endophytic bacteria (*Sphingomonas* sp. LK11) and gibberellin can improve *Solanum lycopersicum* growth and oxidative stress under salinity. J Plant Interact 10(1):117–125

Han S, Li D, Trost E, Mayer KF, Vlot AC, Heller W et al (2016) Systemic responses of barley to the 3-hydroxy-decanoyl-homoserine lactone producing plant beneficial endophyte *Acidovorax radicis* N35. Front Plant Sci 7:1868

Hartman S, Liu Z, Van Veen H, Vicente J, Reinen E, Martopawiro S et al (2019) Ethylene-mediated nitric oxide depletion pre-adapts plants to hypoxia stress. Nat Commun 10(1):1–9

Hassen W, Neifar M, Cherif H, Najjari A, Chouchane H, Driouich RC et al. (2018) *Pseudomonas rhizophila* S211, a new plant growth-promoting rhizobacterium with potential in pesticide-bioremediation. Front Microbiol 9:34

Hayat R, Ali S, Amara U, Khalid R, Ahmed I (2010) Soil beneficial bacteria and their role in plant growth promotion: a review. Ann Microbiol 60(4):579–598

Huber DM, Graham RD, Rengel Z (1999) Mineral nutrition of crops: fundamental mechanisms and implication. The role of nutrition in crop resistance and tolerance to disease. Food Products Press, New York, pp 169–206

Inui H, Yamaguchi Y, Hirano S (1997) Elicitor actions of N-acetylchitooligosaccharides and laminarioligosaccharides for chitinase and L-phenylalanine ammonia-lyase induction in rice suspension culture. Biosci Biotechnol Biochem 61(6):975–978

Islam F, Yasmeen T, Arif MS, Ali S, Ali B, Hameed S, Zhou W (2016) Plant growth promoting bacteria confer salt tolerance in *Vigna radiata* by up-regulating antioxidant defense and biological soil fertility. Plant Growth Regul 80:23–36

Jadhav HP, Shaikh SS, Sayyed RZ (2017) Role of hydrolytic enzymes of rhizoflora in biocontrol of fungal phytopathogens: an overview. In: Mehnaz S (ed) Rhizotrophs: plant growth promotion to bioremediation. Microorganisms for sustainability, vol 2. Springer Singapore, pp 183–203

Jamali H, Sharma A, Roohi SAK (2020) Biocontrol potential of *Bacillus subtilis* RH5 against sheath blight of rice caused by *Rhizoctonia solani*. J Basic Microbiol 60:268–280

Kamaly A, Sadeghi A, Arghand Z, Ahmadzadeh M (2020) Ectoines mitigate the reduction of antagonistic activity of bacteria against *Phytophthora drechsleri* Tucker in saline conditions. J Agric Sci Technol 22(4):1083–1098

Kang SM, Khan AL, Waqas M, You YH, Kim JH, Kim JG, Lee IJ (2014) Plant growth-promoting rhizobacteria reduce adverse effects of salinity and osmotic stress by regulating phytohormones and antioxidants in *Cucumis sativus*. J Plant Interact 9(1):673–682

Karimi E, Sadeghi A, Abbaszadeh Dahaji P, Dalvand Y, Omidvari M, Kakuei Nezhad M (2012) Biocontrol activity of salt tolerant *Streptomyces* isolates against phytopathogens causing root rot of sugar beet. Biocontrol Sci Technol 22(3):333–349

Karthika S, Varghese S, Jisha MS (2020) Exploring the efficacy of antagonistic rhizobacteria as native biocontrol agents against tomato plant diseases. 3 Biotech 10(7):320

Kessel GJT, Kohl J, Powell JA, Rabbinge R, Werf VDW (2005) Modelling spatial characteristics in the biological control of fungi at the leaf scale: competitive substrate colonization by *Botrytis cinerea* and the saprophytic antagonist *Ulocladium atrum*. Phytopathology 95:439–448

Khan AI, Bhandari RR, Pokhrel A, Yadav RN (2018) A study on root exudation pattern and effect of plant growth promoting fungi during biotic and abiotic stress in pigeon pea. World J Agric Res 6(4):122–131

Khare E, Arora NK (2010) Effect of indole-3-acetic acid (IAA) produced by *Pseudomonas aeruginosa* in suppression of charcoal rot disease of chickpea. Curr Microbiol 61(1):64–68

Khare E, Arora NK (2011) Physiologically stressed cells of fluorescent *Pseudomonas* EKi as better option for bioformulation development for management of charcoal rot caused by *Macrophomina phaseolina* in field conditions. Curr microbiol 62(5):1789

Khare E, Arora NK (2021) Biosurfactant based formulation of *Pseudomonas guariconensis* LE3 with multifarious plant growth promoting traits controls charcoal rot disease in *Helianthus annuus*. World J Microbiol Biotechnol 37(4):1–14

Khare E, Singh S, Maheshwari DK, Arora NK (2011) Suppression of charcoal rot of chickpea by *fluorescent Pseudomonas* under saline stress condition. Curr Microbiol 62(5):1548–1553

Khederi SJ, Khanjani M, Gholami M, Panzarino O, de Lillo E (2018) Influence of the erineum strain of *Colomerus vitis* (Acari: Eriophyidae) on grape (*Vitis vinifera*) defense mechanisms. Exp Appl Acarol 75(1):1–24

Kim K, Jang YJ, Lee SM, Oh BT, Chae JC, Lee KJ (2014) Alleviation of salt stress by *Enterobacter* sp. EJ01 in tomato and *Arabidopsis* is accompanied by up-regulation of conserved salinity responsive factors in plants. Mol Cells 37(2):109

Kirkegaard JA, Munns R, James RA, Neate SM (1999) Does water and phosphorus uptake limit leaf growth of rhizoctonia-infected wheat seedlings? Plant Soil 209:157–166

Kissoudis C, Chowdhury R, van Heusden S, van de Wiel C, Finkers R, Visser RG et al (2015) Combined biotic and abiotic stress resistance in tomato. Euphytica 202(2):317–332

Kissoudis C, Sunarti S, van de Wiel C, Visser RG, van der Linden CG, Bai Y (2016) Responses to combined abiotic and biotic stress in tomato are governed by stress intensity and resistance mechanism. J Exp Bot 67(17):5119–5132

Köhl J, Kolnaar R, Ravensberg WJ (2019) Mode of action of microbial biological control agents against plant diseases: relevance beyond efficacy. Front Plant Sci 10:345

Krzyzanowska DM, Potrykus M, Golanowska M, Polonis K, Gwizdek-Wisniewska A, Lojkowska E et al (2012) Rhizosphere bacteria as potential biocontrol agents against soft rot caused by various *Pectobacterium* and *Dickeya* spp. strains. J Plant Pathol 94:367–378

Kumar R, Pandey S, Pandey A (2006) Plant roots and carbon sequestration. Curr Sci 91(7):885–890

Kumar P, Dubey RC, Maheshwari DK (2012) *Bacillus* strains isolated from rhizosphere showed plant growth promoting and antagonistic activity against phytopathogens. Microbiol Res 167(8):493–499

Kumar GP, Ahmed SKMH, Desai S, Amalraj ELD, Rasul A (2014) In vitro screening for abiotic stress tolerance in potent biocontrol and plant growth promoting strains of *Pseudomonas* and *Bacillus* spp. Int J Bacteriol 2014:195946

Kumar A, Singh VK, Tripathi V, Singh PP, Singh AK (2018) Plant growth-promoting rhizobacteria (PGPR): perspective in agriculture under biotic and abiotic stress. In: Prasad R, Gill SS, Tuteja N (eds) Crop improvement through microbial biotechnology. Elsevier, Amsterdam, pp 333–342

Kumar M, Kumar P, Das P, Solanki R, Kapur MK (2020) Potential applications of extracellular enzymes from *Streptomyces* spp. in various industries. Arch Microbiol 202:1597–1615

Lahkar J, Goswami D, Deka S, Ahmed G (2018) Novel approaches for application of biosurfactant produced by *Pseudomonas aeruginosa* for biocontrol of *Colletotrichum capsici* responsible for anthracnose disease in chilli. Eur J Plant Pathol 150(1):57–71

Lamizadeh E, Enayatizamir N, Motamedi H (2016) Isolation and identification of plant growth-promoting rhizobacteria (PGPR) from the rhizosphere of sugarcane in saline and non-saline soil. Int J Curr Microbiol Appl Sci 5(10):1072–1083

Ledger T, Rojas S, Timmermann T, Pinedo I, Poupin MJ, Garrido T, Donoso R (2016) Volatile-mediated effects predominate in *Paraburkholderia phytofirmans* growth promotion and salt stress tolerance of *Arabidopsis thaliana*. Front Microbiol 7:1838

Liang TW, Wang SL (2015) Recent advances in exopolysaccharides from *Paenibacillus* spp.: production, isolation, structure, and bioactivities. Mar drugs 13(4):1847–1863

Limpens E, van Zeijl A, Geurts R (2015) Lipochitooligosaccharides modulate plant host immunity to enable endosymbioses. Annu Rev Phytopathol 53:311–334

Lindsay WL (1979) Chemical equilibria in soils. Wiley, Hoboken, NJ

Ma A, Lv D, Zhuang X, Zhuang G (2013) Quorum quenching in culturable phyllosphere bacteria from tobacco. Int J Mol Sci 14(7):14607–14619

Masmoudi F, Abdelmalek N, Tounsi S, Dunlap CA, Trigui M (2019) Abiotic stress resistance, plant growth promotion and antifungal potential of halotolerant bacteria from a Tunisian solar saltern. Microbiol Res 229:126331

Masum MM, Liu L, Yang M, Hossain MM, Siddiqa MM, Supty ME et al (2018) Halotolerant bacteria belonging to operational group *Bacillus amyloliquefaciens* in biocontrol of the rice brown stripe pathogen *Acidovorax oryzae*. J Appl Microbiol 125(6):1852–1867

Meena B, Radhajeyalakshmi R, Marimuthu T, Vidhyasekaran P, Doraiswamy S, Velazhahan R (2000) Induction of pathogenesis-related proteins, phenolics and phenylalanine ammonia-lyase in groundnut by *Pseudomonas fluorescens*. J Plant Dis Prot 107:514–527

Mishra J, Arora NK (2018) Secondary metabolites of fluorescent pseudomonads in biocontrol of phytopathogens for sustainable agriculture. Appl Soil Ecol 125:35–45

Mishra I, Fatima T, Egamberdieva D, Arora NK (2020) Novel bioformulations developed from *Pseudomonas putida* BSP9 and its biosurfactant for growth promotion of *Brassica juncea* (L.). Plants 9(10):1349

Mohamed HI, Gomaa EZ (2012) Effect of plant growth promoting *Bacillus subtilis* and *Pseudomonas fluorescens* on growth and pigment composition of radish plants (*Raphanus sativus*) under NaCl stress. Photosynthetica 50(2):263–272

Mukherjee P, Mitra A, Roy M (2019) *Halomonas* rhizobacteria of *Avicennia marina* of Indian Sundarbans promote rice growth under saline and heavy metal stresses through exopolysaccharide production. Front Microbiol 10:1207

Mukhtar S, Zareen M, Khaliq Z, Mehnaz S, Malik K (2020) Phylogenetic analysis of halophyte-associated rhizobacteria and effect of halotolerant and halophilic phosphate-solubilizing biofertilizers on maize growth under salinity stress conditions. J Appl Microbiol 128:556–573

Mur LA, Simpson C, Kumari A, Gupta AK, Gupta KJ (2017) Moving nitrogen to the centre of plant defence against pathogens. Ann Bot 119(5):703–709

Nachmias A, Kaufman Z, Livescu L, Tsror L, Meiri A, Caligari PDS (1993) Effects of salinity and its interactions with disease incidence on potatoes grown in hot climates. Phytoparasitica 21(3): 245

Nandi M, Selin C, Brawerman G, Fernando WD, de Kievit T (2017) Hydrogen cyanide, which contributes to *Pseudomonas chlororaphis* strain PA23 biocontrol, is upregulated in the presence of glycine. Biol Control 108:47–54

Nejat N, Mantri N (2017) Plant immune system: crosstalk between responses to biotic and abiotic stresses the missing link in understanding plant defence. Signal 2:02

Nihorimbere V, Marc Ongena M, Smargiassi M, Thonart P (2011) Beneficial effect of the rhizosphere microbial community for plant growth and health. Biotechnol Agron Soc Environ 15:327–337

Nurnberger T, Brunner F, Kemmerling B, Piater L (2004) Innate immunity in plants and animals: striking similarities and obvious differences. Immunol Rev 198(1):249–266

Olanrewaju OS, Glick BR, Babalola OO (2017) Mechanisms of action of plant growth promoting bacteria. World J Microbiol Biotechnol 33(11):1–16

Ormeño-Orrillo E, Menna P, Almeida LGP, Ollero FJ, Nicolás MF, Rodrigues EP et al (2012) Genomic basis of broad host range and environmental adaptability of *Rhizobium tropici* CIAT 899 and *Rhizobium* sp. PRF 81 which are used in inoculants for common bean (*Phaseolus vulgaris* L.). BMC Genom 13:735

Ortíz-Castro R, Contreras-Cornejo HA, Macías-Rodríguez L, López-Bucio J (2009) The role of microbial signals in plant growth and development. Plant Signal Behav 4(8):701–712

Pal KK, Gardener BM (2006) Biological control of plant pathogens. The Plant Health Instructor. https://doi.org/10.1094/PHI-A-2006-1117-02

Pandey P, Irulappan V, Bagavathiannan MV, Senthil-Kumar M (2017) Impact of combined abiotic and biotic stresses on plant growth and avenues for crop improvement by exploiting physio-morphological traits. Front Plant Sci 8:537

Pang Y, Liu X, Ma Y, Chernin L, Berg G, Gao K (2009) Induction of systemic resistance, root colonisation and biocontrol activities of the rhizospheric strain of *Serratia plymuthica* are dependent on N-acyl homoserine lactones. Eur J Plant Pathol 124(2):261–268

Park K, Dutta S, Park YS, Sang MK, Moon SS (2016) Induction of systemic resistance and tolerance against biotic and abiotic stress in Chinese cabbage by cyclic peptides producing *Bacillus vallismortis* strain BS07M. In: Proceeding of Recent trends in PGPR research for sustainable crop productivity, 4th Asian PGPR Conference, Hanoi, Vietnam, 3-6, pp 200–205

Patil S, Paradeshi J, Chaudhari B (2016) Suppression of charcoal rot in soybean by moderately halotolerant *Pseudomonas aeruginosa* GS-33 under saline conditions. J Basic Microbiol 56(8): 889–899

Patil S, Nikam M, Anokhina T, Kochetkov V, Chaudhari A (2017) Multi-stress tolerant plant growth promoting *Pseudomonas* spp. MCC 3145 producing cytostatic and fungicidal pigment. Biocatal Agric Biotechnol 10:53–63

Paul D, Nair S (2008) Stress adaptations in a plant growth promoting rhizobacterium (PGPR) with increasing salinity in the coastal agricultural soils. J Basic Microbiol 48(5):378–384

Phour M, Sindhu SS (2020) Amelioration of salinity stress and growth stimulation of mustard (*Brassica juncea* L.) by salt-tolerant *Pseudomonas* species. Appl Soil Ecol 149:103518

Pirhadi M, Enayatizamir N, Motamedi H, Sorkheh K (2016) Screening of salt tolerant sugarcane endophytic bacteria with potassium and zinc for their solubilizing and antifungal activity. Biosc Biotech Res Comm 9(3):530–538

Potash and Phosphate Institute (PPI) (1988) Phosphorus nutrition improves plant disease resistance. In: PPI (ed) Better crops with plant food, Atlanta, GA, pp 22–23

Prakash J, Arora NK (2019) Phosphate-solubilizing *Bacillus* sp. enhances growth, phosphorus uptake and oil yield of *Mentha arvensis* L. 3 Biotech 9(4):1–9

Príncipe A, Alvarez F, Castro MG, Zacchi LF, Fischer SE, Mori GB, Jofré E (2007) Biocontrol and PGPR features in native strains isolated from saline soils of Argentina. Curr Microbiol 55(4): 314–322

Rahman A, Sultana V, Ara J, Ehteshamul-Haque S (2016) Induction of systemic resistance in cotton by the neem cake and *Pseudomonas aeruginosa* under salinity stress and *Macrophomina phaseolina* infection. Pak J Bot 48(4):1681–1689

Rajput L, Imran A, Mubeen F, Hafeez FY (2013) Salt-tolerant PGPR strain *Planococcus rifietoensis* promotes the growth and yield of wheat (*Triticum aestivum* L.) cultivated in saline soil. Pak J Bot 45:1955–1962

Ramette A, Moënne-Loccoz Y, Défago G (2006) Genetic diversity and biocontrol potential of fluorescent pseudomonads producing phloroglucinols and hydrogen cyanide from Swiss soils naturally suppressive or conducive to *Thielaviopsis basicola*-mediated black root rot of tobacco. FEMS Microbiol Ecol 55(3):369–381

Rangarajan S, Saleena L, Vasudevan P, Nair S (2003) Biological suppression of rice diseases by *Pseudomonas* spp. under saline soil conditions. Plant Soil 251(1):73–82

Rasul M, Yasmin S, Zubair M, Mahreen N, Yousaf S, Arif M (2019) Phosphate solubilizers as antagonists for bacterial leaf blight with improved rice growth in phosphorus deficit soil. Biol Control 136:103997

Raza W, Ling N, Yang L, Huang Q, Shen Q (2016) Response of tomato wilt pathogen *Ralstonia solanacearum* to the volatile organic compounds produced by a biocontrol strain *Bacillus amyloliquefaciens* SQR-9. Sci Rep 6(1):1–13

Rejeb IB, Pastor V, Mauch-Mani B (2014) Plant responses to simultaneous biotic and abiotic stress: molecular mechanisms. Plants 3(4):458–475

Reuveni R, Dor G, Reuveni M (1998) Local and systemic control of powdery mildew (*Leveillula taurica*) on pepper plants by foliar spray of mono-potassium phosphate. Crop Prot 17(9): 703–709

Reuveni R, Dor G, Raviv M, Reuveni M, Tuzun S (2000) Systemic resistance against *Sphaerotheca fuliginea* in cucumber plants exposed to phosphate in hydroponics system, and its control by foliar spray of mono-potassium phosphate. Crop Prot 19(5):355–361

Rijavec T, Lapanje A (2016) Hydrogen cyanide in the rhizosphere: not suppressing plant pathogens, but rather regulating availability of phosphate. Front Microbiol 7:1785

Rivas-Garcia T, Murillo-Amador B, Nieto-Garibay A, Rincon-Enriquez G, Chiquito-Contreras RG, Hernandez-Montiel LG (2019) Enhanced biocontrol of fruit rot on muskmelon by combination treatment with marine *Debaryomyces hansenii* and *Stenotrophomonas rhizophila* and their potential modes of action. Postharvest Biol Tec 151:61–67

Robert C, Bancal MO, Lannou C (2002) Wheat leaf rust uredospore production and carbon and nitrogen export in relation to lesion size and density. Phytopathology 92(7):762–768

Rodríguez M, Torres M, Blanco L, Béjar V, Sampedro I, Llamas I (2020) Plant growth-promoting activity and quorum quenching-mediated biocontrol of bacterial phytopathogens by *Pseudomonas segetis* strain P6. Sci Rep 10(1):1–12

Rojas-Solis D, Vences-Guzmán MA, Sohlenkamp C, Santoyo G (2020) *Bacillus toyonensis* COPE52 modifies lipid and fatty acid composition, exhibits antifungal activity, and stimulates growth of tomato plants under saline conditions. Curr Microbiol 77:2735–2744

Romera FJ, García MJ, Lucena C, Martínez-Medina A, Aparicio MA, Ramos J et al (2019) Induced systemic resistance (ISR) and Fe deficiency responses in dicot plants. Front Plant Sci 10:287

Rosier A, Medeiros FH, Bais HP (2018) Defining plant growth promoting rhizobacteria molecular and biochemical networks in beneficial plant-microbe interactions. Plant soil 428(1):35–55

Rothballer M, Uhl J, Kunze J, Schmitt-Kopplin P, Hartmann A (2018) Detection of the bacterial quorum-sensing signaling molecules N-acylhomoserine lactones (HSL) and N-acyl-homoserine (HS) with an enzyme-linked immunosorbent assay (ELISA) and via ultrahigh-performance liquid chromatography coupled to mass spectrometry (UHPLC-MS). Methods Mol Biol 1673: 61–72

Rudrappa T, Biedrzycki ML, Kunjeti SG, Donofrio NM, Czymmek KJ, Paul WP, Bais HP (2010) The rhizobacterial elicitor acetoin induces systemic resistance in *Arabidopsis thaliana*. Commun Integr Biol 3(2):130–138

Ryu CM, Farag MA, Hu CH, Reddy MS, Kloepper JW, Paré PW (2004) Bacterial volatiles induce systemic resistance in *Arabidopsis*. Plant Physiol 134(3):1017–1026

Sadeghi A, Karimi E, Dahaji PA, Javid MG, Dalvand Y, Askari H (2012) Plant growth promoting activity of an auxin and siderophore producing isolate of *Streptomyces* under saline soil conditions. World J Microbiol Biotechnol 28(4):1503–1509

Sadfi-Zouaoui N, Essghaier B, Hajlaoui MR, Fardeau ML, Cayaol JL, Ollivier B et al (2008) Ability of moderately halophilic bacteria to control grey mould disease on tomato fruits. J Phytopathol 156(1):42–52

Sandhya V, Ali SZ, Grover M, Reddy G, Venkateswarlu B (2010) Effect of plant growth promoting *Pseudomonas* spp. on compatible solutes, antioxidant status and plant growth of maize under drought stress. Plant Growth Regul 62:21–30

Sasse J, Martinoia E, Northen T (2018) Feed your friends: do plant exudates shape the root microbiome. Trends Plant Sci 23(1):25–41

Schreiter S, Sandmann M, Smalla K, Grosch R (2014) Soil type dependent rhizosphere competence and biocontrol of two bacterial inoculant strains and their effects on the rhizosphere microbial community of field-grown lettuce. PLoS One 9(8):e103726

Schreiter S, Babin D, Smalla K, Grosch R (2018) Rhizosphere competence and biocontrol effect of *Pseudomonas* sp. RU47 independent from plant species and soil type at the field scale. Front Microbiol 9:97

Schultze M, Kondorosi Á (1996) The role of lipochitooligosaccharides in root nodule organogenesis and plant cell growth. Curr Opin Genet Dev 6(5):631–638

Schultze M, Kondorosi A (1998) Regulation of symbiotic root nodule development. Annu Rev Genet 32(1):33–57

Schulz-Bohm K, Martín-Sánchez L, Garbeva P (2017) Microbial volatiles: small molecules with an important role in intra-and inter-kingdom interactions. Front Microbiol 8:2484

Sgroy V, Cassán F, Masciarelli O, Del Papa MF, Lagares A, Luna V (2009) Isolation and characterization of endophytic plant growth-promoting (PGPB) or stress homeostasis-regulating (PSHB) bacteria associated to the halophyte *Prosopis strombulifera*. Appl Microbiol Biotechnol 85:371–381

Shameer S, Prasad TNVKV (2018) Plant growth promoting rhizobacteria for sustainable agricultural practices with special reference to biotic and abiotic stresses. Plant Growth Regul 84(3): 603–615

Sharifi R, Ryu CM (2018) Revisiting bacterial volatile-mediated plant growth promotion: lessons from the past and objectives for the future. Ann Bot 122(3):349–358

Sharma A, Singh P, Kumar S, Kashyap PL, Srivastava AK, Chakdar H et al (2015) Deciphering diversity of salt-tolerant bacilli from saline soils of eastern Indo-gangetic plains of India. Geomicrobiol J 32(2):170–180

Sharma S, Chen C, Navathe S, Chand R, Pandey SP (2019) A halotolerant growth promoting rhizobacteria triggers induced systemic resistance in plants and defends against fungal infection. Sci Rep 9(1):1–17

Shine MB, Xiao X, Kachroo P, Kachroo A (2019) Signaling mechanisms underlying systemic acquired resistance to microbial pathogens. Plant Sci 279:81–86

Shurigin VV, Ruzimova KK, Davranov K (2018) Biological control of phytopathogenic fungi causing chickpea root diseases by means of PGPR in the saline soil conditions. Int J Adv Biotechnol Res 9(4):36–47

Siddique NS, Hiremath V, Abhiram P, Kumar Y, Khedikar A, Kunghatkar A et al (2019) Biological control of fusarium wilt of tomato (*Solanum lycopersicum* L.) by antagonistic fungi. J Pharmacogn Phytochem 8(4):2252–2259

Silva HS, Aromeiro RS, Macagnan D, Halfeld-Vierira BA, Pereira MCB, Mounteer A (2004) Rhizobacterial induction of systemic resistance in tomato plants: non-specific protection and increase in enzyme activities. Biol Control 29(2):288–295
Singh RP, Jha PN (2016) The multifarious PGPR *Serratia marcescens* CDP-13 augments induced systemic resistance and enhanced salinity tolerance of wheat (*Triticum aestivum* L.). PLos One 11(6):e0155026
Singh RP, Jha PN (2017) The PGPR *Stenotrophomonas maltophilia* SBP-9 augments resistance against biotic and abiotic stress in wheat plants. Front Microbiol 8:1945
Singh SR, Joshi D, Tripathi N, Singh P, Srivastava TK (2017) Plant growth-promoting bacteria: an emerging tool for sustainable crop production under salt stress. In: Bioremediation of salt affected soils: an Indian perspective. Springer, Cham, pp 101–131
Sivakumaran A, Akinyemi A, Mandon J, Cristescu SM, Hall MA, Harren FJ et al (2016) ABA suppresses *Botrytis cinerea* elicited NO production in tomato to influence H_2O_2 generation and increase host susceptibility. Front Plant Sci 7:709
Spadaro D, Droby S (2016) Development of biocontrol products for postharvest diseases of fruit: the importance of elucidating the mechanisms of action of yeast antagonists. Trends Food Sci Technol 47:39–49
Starlin T, Gopalakrishnan VK (2013) Enzymatic and non-enzymatic antioxidant properties of *Tylophora pauciflora* Wight and Arn.—an in vitro study. Asian J Pharm Clin Res 6(4):68–71
Sticher L, Mauch-Mani B, Métraux AJ (1997) Systemic acquired resistance. Annu Rev Phytopathol 35(1):235–270
Strange RN, Scott PR (2005) Plant disease: a threat to global food security. Annu Rev Phytopathol 43:83–116
Subramanian S, Ricci E, Souleimanov A, Smith DL (2016a) A proteomic approach to lipo-chitooligosaccharide and thuricin 17 effects on soybean germination unstressed and salt stress. PLoS One 11(8):e0160660
Subramanian S, Souleimanov A, Smith DL (2016b) Proteomic studies on the effects of lipo-chitooligosaccharide and thuricin 17 under unstressed and salt stressed conditions in *Arabidopsis thaliana*. Front Plant Sci 7:1314
Sukkasem P, Kurniawan A, Kao TC, Chuang HW (2018) A multifaceted rhizobacterium *Bacillus licheniformis* functions as a fungal antagonist and a promoter of plant growth and abiotic stress tolerance. Environ Exp Bot 155:541–551
Sunita K, Mishra I, Mishra J, Prakash J, Arora NK (2020) Secondary metabolites from halotolerant plant growth promoting rhizobacteria for ameliorating salinity stress in plants. Front Microbiol 11:2619
Suryadi Y, Susilowati DN, Fauziah F (2019) Management of plant diseases by PGPR-mediated induced resistance with special reference to tea and rice crops. In: Sayyed R (ed) Plant growth promoting rhizobacteria for sustainable stress management. microorganisms for sustainability, vol 13. Springer, Singapore, pp 65–110
Tang X, Mu X, Shao H, Wang H, Brestic M (2015) Global plant-responding mechanisms to salt stress: physiological and molecular levels and implications in biotechnology. Crit Rev Biotechnol 35(4):425–437
Tanji KK (2002) Salinity in the soil environment. In: Läuchli A, Lüttge U (eds) Salinity: environment-plants—molecules. Springer, Dordrecht
Tavernier V, Cadiou S, Pageau K, Laugé R, Reisdorf-Cren M, Langin T (2007) The plant nitrogen mobilization promoted by *Colletotrichum lindemuthianum* in *Phaseolus* leaves depends on fungus pathogenicity. J Exp Bot 58(12):3351–3360
Teng S, Liu Y, Zhao L (2010) Isolation, identification and characterization of ACC deaminase-containing endophytic bacteria from halophyte *Suaeda salsa*. Wei Sheng Wu Xue Bao 50: 1503–1509
Tewari S, Arora NK (2014) Multifunctional exopolysaccharides from *Pseudomonas aeruginosa* PF23 involved in plant growth stimulation, biocontrol and stress amelioration in sunflower under saline conditions. Curr Microbiol 69(4):484–494

Tewari S, Arora NK (2016) *Fluorescent Pseudomonas* sp. PF17 as an efficient plant growth regulator and biocontrol agent for sunflower crop under saline conditions. Symbiosis 68(1–3): 99–108

Tewari S, Arora NK (2018) Role of salicylic acid from *Pseudomonas aeruginosa* PF23 EPS+ in growth promotion of sunflower in saline soils infested with phytopathogen *Macrophomina phaseolina*. Environ Sustain 1(1):49–59

Thein A, Prathuangwong S (2010) Novel strains of *Xanthomonas oryzae* pv. *oryzae* UV mutated induce systemic resistance in rice against bacterial leaf blight disease. Kasersart J Nat Sci 44 1026–1043

Timmusk S, Copolovici D, Copolovici L, Teder T, Nevo E, Behers L (2019) *Paenibacillus polymyxa* biofilm polysaccharides antagonise *Fusarium graminearum*. Sci Rep 9(1):1–11

Torres M, Llamas I, Torres B, Toral L, Sampedro I, Bejar V (2020) Growth promotion on horticultural crops and antifungal activity of *Bacillus velezensis* XT1. Appl Soil Ecol 150: 103453

Tortora ML, Díaz-Ricci JC, Pedraza RO (2011) *Azospirillum brasilense* siderophores with antifungal activity against *Colletotrichum acutatum*. Arch Microbiol 193(4):275–286

Tortora ML, Díaz-Ricci JC, Pedraza RO (2012) Protection of strawberry plants (*Fragaria ananassa* Duch.) against anthracnose disease induced by *Azospirillum brasilense*. Plant soil 356(1): 279–290

Tsegaye Z, Gizaw B, Tefera G, Feleke A, Chaniyalew S, Alemu T et al (2019) Isolation and biochemical characterization of plant growth promoting (pgp) bacteria colonizing the rhizosphere of Tef crop during the seedling stage. J Plant Sci Phytopathol 14:13–27

Vaishnav A, Varma A, Tuteja N, Choudhary DK (2017) Characterization of bacterial volatiles and their impact on plant health under abiotic stress. In: Choudhary D, Sharma A, Agarwal P, Varma A, Tuteja N (eds) Volatiles and food security. Springer, Singapore, pp 15–24

Vallad GE, Goodman RM (2004) Systemic acquired resistance and induced systemic resistance in conventional agriculture. Crop Sci 44(6):1920–1934

Van Loon LC (2007) Plant responses to plant growth-promoting rhizobacteria. Eur J Plant Pathol 119:243–254

Van Loon LC, Bakker PAHM (2005) Induced systemic resistance as a mechanism of disease suppression by rhizobacteria. In: Siddiqui ZA (ed) pgpr: biocontrol and biofertilization. Springer, Dordrecht, pp 39–66

Vega C, Rodríguez M, Llamas I, Béjar V, Sampedro I (2020) Silencing of phytopathogen communication by the halotolerant PGPR *Staphylococcus Equorum* strain EN21. Microorganisms 8(1):42

Velho RV, Medina LF, Segalin J, Brandelli A (2011) Production of lipopeptides among *Bacillus* strains showing growth inhibition of phytopathogenic fungi. Folia Microbiol 56(4):297

Vesuna AP, Nerurkar AS (2020) Biocontrol impact of AHL degrading actinobacteria on quorum sensing regulated virulence of phytopathogen *Pectobacterium carotovorum* subsp. *carotovorum* BR1. Plant Soil 453(1):371–388

Viswanath G, Sekar J, Prabavathy VR (2016) Acyl homoserine lactone-producing rhizobacteria elicit systemic resistance in plants. In: Choudhary D, Varma A (eds) Microbial-mediated induced systemic resistance in plants. Springer, Singapore, pp 135–146

Wadhwa N, Joshi UN, Mehta N (2014) Zinc induced enzymatic defense mechanisms in *Rhizoctonia* root rot infected clusterbean seedlings. J Bot:735760

Wang N, Khan W, Smith DL (2012) Changes in soybean global gene expression after application of lipo-chitooligosaccharide from *Bradyrhizobium japonicum* under sub-optimal temperature. PLoS One 7(2):e31571

Wang X, Mavrodi DV, Ke L, Mavrodi OV, Yang M, Thomashow LS et al (2015) Biocontrol and plant growth-promoting activity of rhizobacteria from Chinese fields with contaminated soils. Micro Biotechnol 8(3):404–418

Wang P, Karki S, Biswal AK, Lin HC, Dionora MJ, Rizal G et al (2017) Candidate regulators of early leaf development in maize perturb hormone signalling and secondary cell wall formation when constitutively expressed in rice. Sci Rep 7(1):1–5

Wang N, Wang L, Zhu K, Hou S, Chen L, Mi D et al (2019a) Plant root exudates are involved in *Bacillus cereus* AR156 mediated biocontrol against *Ralstonia solanacearum*. Front Microbiol 10:98

Wang TN, Guan QT, Pain A, Kaksonen AH, Hong PY (2019b) Discovering, characterizing, and applying acyl homoserine lactone-quenching enzymes to mitigate microbe-associated problems under saline conditions. Front Microbiol 10:823

Xia Y, Farooq MA, Javed MT, Kamran MA, Mukhtar T, Ali J (2020) Multi-stress tolerant PGPR *Bacillus xiamenensis* PM14 activating sugarcane (*Saccharum officinarum* L.) red rot disease resistance. Plant Physiol Biochem 151:640–649

Yadav BS, Mani A (2019) Functional genomics and systems biology approach for understanding agroecosystems. In: Tripathi V, Kumar P, Tripathi P, Kishore A (eds) Microbial genomics in sustainable agroecosystems. Springer, Singapore, pp 67–86

Yañez-Ocampo G, Wong-Villarreal A (2013) Biosurfactantes microbianos, producción potencial con residuos agroindustriales de Chiapas. Bio Tecnología 17(3):12–28

Yang J, Kloepper JW, Ryu CM (2009) Rhizosphere bacteria help plants tolerate abiotic stress. Trends Plant Sci 14(1):1–4

Zhang H, Sonnewald U (2017) Differences and commonalities of plant responses to single and combined stresses. Plant J 90(5):839–855

Zhang X, Tang B (2019) Production of Rhamnolipids by a halotolerant bacterial strain with seawater and its application in the treatment of powdery mildew of Strawberry (*Fragaria ananassa*). In: Kumar M, Etesami H, Kumar V (eds) Saline soil-based agriculture by halotolerant microorganisms. Springer, Singapore, pp 175–185

Endophytic Fungi: Important Source of Biologically Active Molecules

10

Marwa Jardak, Mani Jayaprakashvel, Sami Aifa, and Sami Mnif

Abstract

Fungal endophytes are considered plant mutualists that protect their hosts against pathogens. Therefore, the search for such microorganisms from different ecosystems and the investigation of their biological activities could open up new promising possibilities for the control of infectious diseases and their vectors. Endophytic fungi can provide their macroscopic hosts with a kind of protection against surface pathogens and the predation of herbivorous insects. They excrete secondary antifungal, antibacterial metabolites, and/or cytotoxic agents. Accordingly, this article provides updated knowledge on natural compounds with antimicrobial (sessile and adherent form) and/or anticarcinogenic activities produced by endophytic fungi isolated from different plants in different regions of the world.

Keywords

Endophytic fungi · Bioactive secondary metabolites · Pathogens · Infectious diseases · Biofilm · Cancer

M. Jardak · S. Aifa · S. Mnif (✉)
Laboratory of Molecular and Cellular Screening Processes, Centre of Biotechnology of Sfax, P O Box 1177, Sfax, Tunisia

M. Jayaprakashvel
Department of Marine Biotechnology, AMET University, Chennai, Tamilnadu, India

N. K. Arora, B. Bouizgarne (eds.), *Microbial BioTechnology for Sustainable Agriculture Volume 1*, Microorganisms for Sustainability 33,
https://doi.org/10.1007/978-981-16-4843-4_10

10.1 Introduction

The emergence of diseases and deadly viruses, the development of drug resistance, the failure of cancer chemotherapies, the complications associated with organ transplantation, and the increasing incidence of microbial infections are serious public health burdens. In addition, the risk of toxicity associated with the excessive use of synthetic drugs has increased. Many researchers around the world are trying to explore natural resources to obtain new drugs that are efficient and have low toxicity and low environmental impact. Important sources of bioactive molecules include fungi, which were the origin of the world's first discovered antibiotic, penicillin, produced by *Penicillium notatum* (Fleming 1929). Other novel fungal molecules were discovered, including griseofulvin from *Penicillium griseofulvum* and cyclosporin from *Tolypocladium niveum* (Oxford et al. 1939; Borel et al. 1994; Valente et al. 2020). In this context, endophytic fungi that colonize living plant tissue without causing apparent symptoms could be an important source of bioactive compounds (Powthong et al. 2018; Vasundhara et al. 2019; Khalil et al. 2020; Manganyi and Ateba 2020).

Bioactive compounds from endophytic fungi are becoming an interesting field of research particularly in cancer therapy (Uzma et al. 2018). Taxol, an anti-cancer molecule that was originally isolated from the pacific yew (*Taxus brevifolia*), can be found as a cytotoxic drug that kills cancer cells on the World Health Organization website model list of Essential Medicines. It is used to treat breast cancer, ovarian cancer, non-small cell lung cancer, pancreatic cancer, and AIDS-related Kaposi sarcoma. However, Taxol can only be extracted from yew trees in very low concentrations, which has led researchers to look for alternative sources of natural Taxol. Interestingly, they succeeded in identifying endophytic fungal species from yew trees which have been repeatedly reported to produce Taxol (Wani et al. 1971; Stierle et al. 1993; Mirjalili et al. 2012; Isah 2015). These important discoveries have fueled hopes of finding novel bioactive metabolites from endophytic fungi in various host plants that can be used for the treatment of infectious diseases and cancer.

10.2 Definition of an Endophyte

The etymological origin of the word "endophyte" comes from the Greek "endo" or "endon" which means "inside" and "phyton" which means "plant." In contrast to epiphytes, which are located outside the plant surfaces, endophytes are microorganisms that colonize the inner tissue of the plant below the epidermal cell layers without causing negative effects on the host plants (Hardoim et al. 2015; Vasundhara et al. 2019; Fadiji and Babalola 2020). Endophytic microorganisms colonize the various parts of the plant essentially leaves, stems, and roots (dos Santos et al. 2018). Different groups of microorganisms such as fungi, bacteria, actinomycetes, and mycoplasmas are called endophytes, but fungi are the most commonly isolated (Rodriguez et al. 2009; Padhi et al. 2013; Kharkwal and Kulkarni 2020). Endophytes have an intimate relationship with host plants by producing components that promote vegetative growth and protect the host against pathogens (Jia et al. 2016). According to many studies, endophytic fungi can produce the same active components that are produced by the host plant (Zhao et al. 2010; Wang et al. 2015; Venieraki et al. 2017). This is explained by the presence of fungal biosynthetic pathways similar to those of host plants (Stierle et al. 1993; Jennewein and Croteau 2001; Naik et al. 2019).

10.3 Plant-Endophyte Interaction Advantages

Each plant can host one or more species of endophytes (Gamboa and Bayman 2001; Strobel and Daisy 2003). Depending on the target species, there is a broad spectrum of endophytic and plant interactions ranging from simple commensalism to mutualism or symbiosis and then from commensalism to parasitism (Redman et al. 2001; Kogel et al. 2006; Aly et al. 2011; Wani et al. 2015). Environmental conditions could be responsible for the definition of the type of interaction (Schulz and Boyle 2005; Kogel et al. 2006).

Colonization by endophytic microorganisms gives the plant many benefits, including the promotion of plant growth (Khare et al. 2018). In fact, the plant can benefit from increased growth while acquiring resistance to adverse abiotic stresses such as drought, heat, metal toxicity, and high salinity via several molecular mechanisms (Lata et al. 2018; Rho et al. 2018; Eid et al. 2019; Sampangi-Ramaiah et al. 2020). Endophytic microorganisms are regarded as inhibitors of the growth of phytopathogenic microorganisms (Bruisson et al. 2019) through the production of antibiotics or the induction of host defense mechanisms against phytopathogens (Gimenez et al. 2007; Martinez-Klimova et al. 2017). Besides the endophyte-host interaction, there are also endophyte-endophyte or endophyte-pathogen interactions. These types of interactions lead to the acquisition of several functions by endophytes such as the production of bioactive molecules and the control of phytopathogens through quorum quenching mechanisms (Kusari et al. 2013, 2014).

10.4 Endophytic Fungi as Source of Bioactive Natural Molecules

Infection of plants by endophytes could cause a change in the level of gene expression of the host plant by stimulating the production of bioactive molecules (Deepika et al. 2016). In addition, environmental factors, including biotic and abiotic conditions, balance of carbon-nutrition, genotype, and ontogeny of plant control, usually regulate the biosynthesis of plant secondary metabolites (Isah 2019). Moreover, endophytes are considered chemical synthesizers within plant tissues and a source of secondary metabolites and could therefore be used as potential sources for pharmaceutical products (Owen and Hundley 2004; Nisa et al. 2015; Gouda et al. 2016; Vasundhara et al. 2019; Rustamova et al. 2020). Indeed, endophytic fungi have been found to be useful in the discovery of new drugs such as antibiotics, antimicrobials, antioxidants, anti-tumor, antiviral, and anti-inflammatory substances

(Bedi et al. 2018; Uzma et al. 2018; Abdel-Azeem et al. 2019; Gupta and Sharma 2020; Mishra et al. 2020; Pal et al. 2020). These biomolecules can be alkaloids, steroids, flavonoids, and other types of structures (Lugtenberg et al. 2016; Vasundhara et al. 2019).

10.5 Antimicrobial and Antiviral Molecules from Endophytic Fungi

Currently, several studies report on a large number of antimicrobial compounds isolated from endophytic fungi and belonging to different structural classes such as alkaloids, peptides, steroids, terpenoids, phenols, isocoumarins, lactones, flavonoids, quinones, and others (Lugtenberg et al. 2016; Bedi et al. 2018; Vasundhara et al. 2019). Taking the example of hypericin, a derivative of naphthodianthrone, and its precursor emodin produced by an endophytic fungus isolated from an Indian medicinal plant *Hypericum perforatum*, showed antimicrobial activities against several bacteria including *Staphylococcus aureus*, *Klebsiella pneumoniae* subsp. ozaenae, *Pseudomonas aeruginosa*, *Salmonella enterica* subsp. enterica, and *Escherichia coli* and against fungal pathogens such as *Aspergillus niger* and *Candida albicans* (Kusari et al. 2008) (Table 10.1). Helvolic acid with antibacterial and antimycotic activities was isolated for the first time from an endophytic fungus *Pichia guilliermondii* Ppf9, which was collected from the rhizomes of the medicinal plant *Paris polyphylla* var. yunnanensis. Helvolic acid exhibits important antibacterial activity against all the bacteria tested, i.e., *Agrobacterium tumefaciens*, *E. coli*, *Pseudomonas lachrymans*, *Ralstonia solanacearum*, *Xanthomonas vesicatoria*, *Bacillus subtilis*, *S. aureus*, and *Staphylococcus haemolyticus*, with minimum inhibitory concentrations MICs in the range of 1.56 to 6.25 μg/mL and IC_{50} values from 0.98 μg/mL to 33.19 μg/mL comparable to the positive control streptomycin sulfate. Moreover, the helvolic acid inhibits the spore germination of *Magnaporthe oryzae* with an IC_{50} value of 7.20 μg/mL (Zhao et al. 2010). The same compound was obtained along with hydrohelvolic acid and a new derivative called helvolic acid methyl ester from *Fusarium* sp., an endophytic fungus isolated from the leaves of *Ficus carica*. These compounds showed effectiveness against bacterial strains of *B. subtilis*, *S. aureus*, *E. coli*, and *P. aeruginosa* with MIC values ranging from 3.13 to 12.5 μg/mL and against some plant pathogenic fungi *Botrytis cinerea*, *Colletotrichum gloeosporioides*, *Fusarium oxysporum*, *Fusarium graminearum*, and *P. capsici* with MIC values between 12.5 and 25 μg/mL (Liang et al. 2016).

An endophytic fungal strain belonging to the genus *Periconia* was isolated from the plant *Piper longum*. This strain is capable of producing piperine, a compound characterized by a good anti-mycobacterial activity against *Mycobacterium tuberculosis* and *Mycobacterium smegmatis* with MICs of 1.74 and 2.62 μg/mL, respectively (Table 10.1). Piperine also has anti-inflammatory, antimutagenic, and hepatoprotective properties (Verma et al. 2011). Piperine is also produced by the endophytic fungus *C. gloeosporioides*, which is isolated from the stems of *Piper*

Table 10.1 Structures of some antimicrobial compounds isolated from endophytic fungi

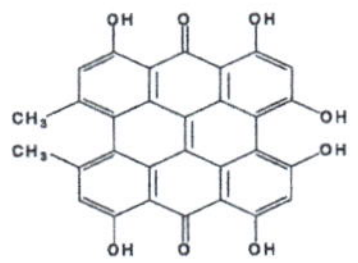 Hypericin	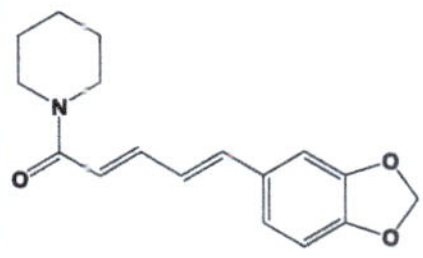 Piperine	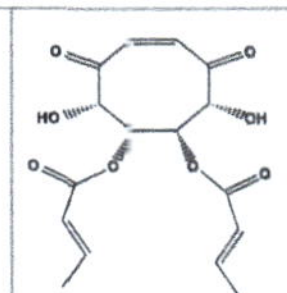 6,7-(20 E) dibutenyl-5,8-dihydroxy-(Z)-cyclooct-2-ene-1,4-dione
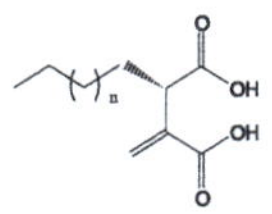 2S-Butylitaconic acid: $n = 1$ 2S-hexylitaconic acid: $n = 3$	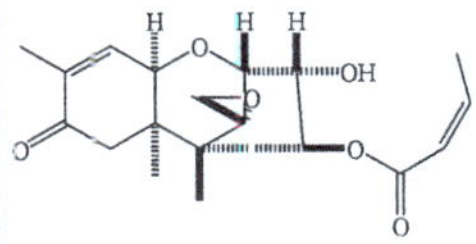 Tricothecinol A	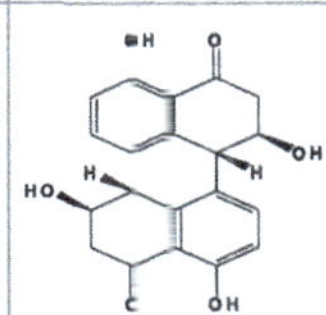 Stemphyperylenol
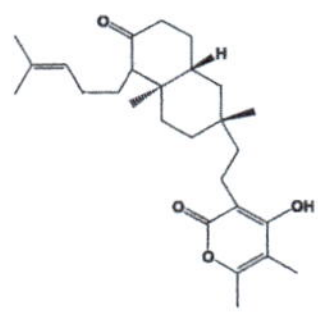 Pycnophorin	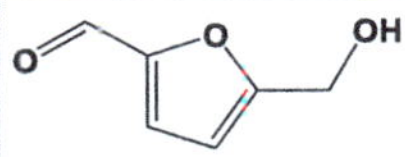 5-hydroxymethylfurfural	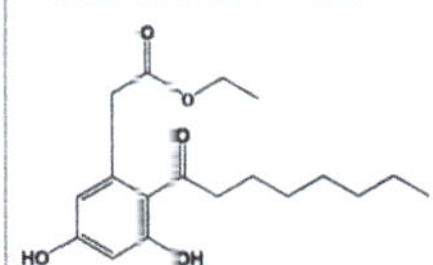 Cytosporone B
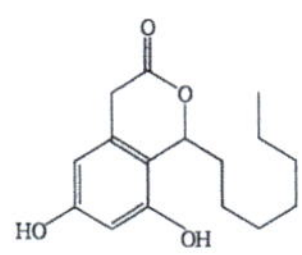 Cytosporone C	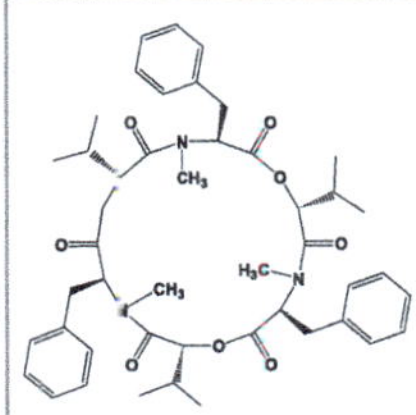 Beauvericin	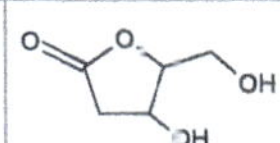 2′-deoxyribolactone
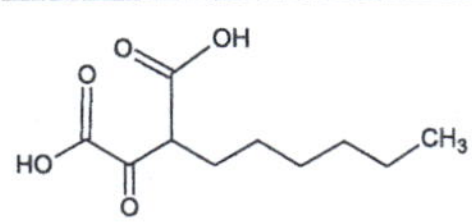 Hexylitaconic acid	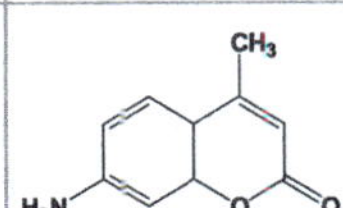 7-amino-4-methylcoumarin	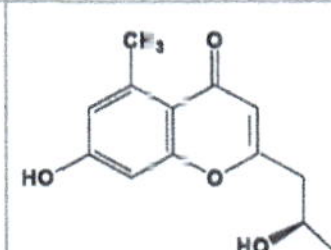 (2′*S**)-2-(2′-hydroxypropyl)-5-methyl-7,8 dihydroxy-chromone
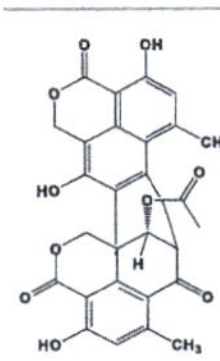 **(20)** Bacillisporin A	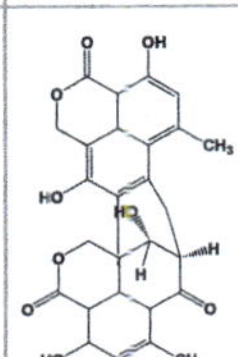 Bacillisporin B	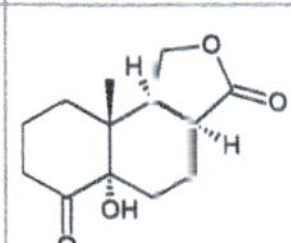 Phomanolide

(continued)

Table 10.1 (continued)

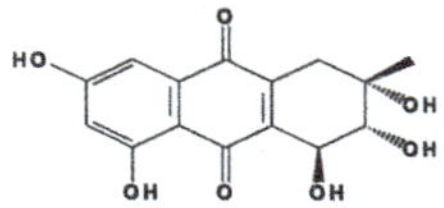 6-O-demethyl-4-dehydroxyaltersolanol A	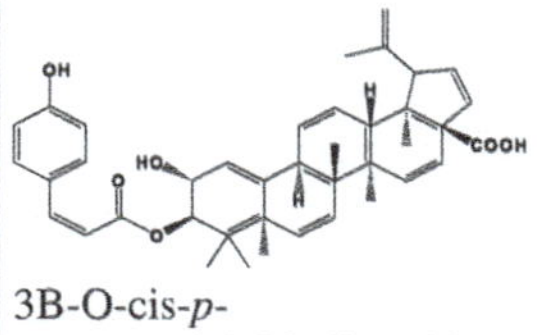 3B-O-cis-*p*-coumaroylalphitolic acid	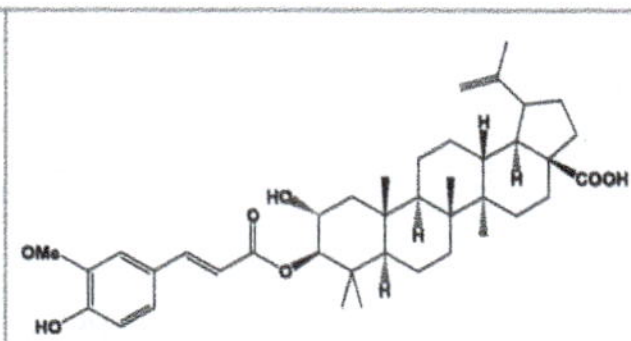 Eucalyptic acid
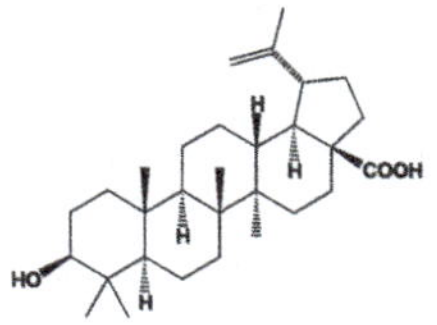 Betulinic acid	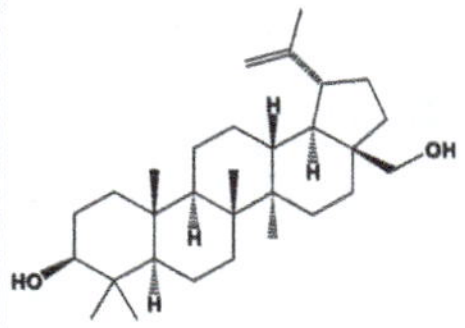 Betulin	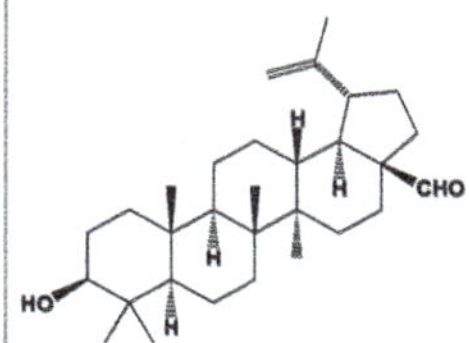 Betulinicaldehyde

nigrum (Chithra et al. 2014). Yu et al. reported for the first time the endophytic fungus BS002 *Penicillium* sp. M-01, which is isolated from the seeds of the medicinal plant *Sophora flavescens* to produce 6,7-(2′E) dibutenyl-5,8-dihydroxy-(Z)-cyclooct-2-ene-1,4-dione (Table 10.1). This compound inhibits the growth of various microorganisms such as *Botryosphaeria berengeriana* f. sp. piricola, *Physalospora piricola*, *Cladosporium cucumerinum* Ell. *Arthur*, *F. oxysporum*. sp. cucumerinum, and *Fusarium moniliforme* having a strong activity against *P. piricola* demonstrated by an antibacterial diameter of 45 mm (Yu et al. 2014).

Another endophytic fungus GDGJ-91 identified as *Penicillium vulpinum* isolated from the root of *Sophora tonkinensis*, a Chinese medicinal plant, yielded two compounds (−)-3-carboxypropyl-7-hydroxyphthalide and (−)-3-carboxypropyl-7-hydroxyphthalide methyl ester, new phthalide derivatives. The compound (−)-3-carboxypropyl-7-hydroxyphthalide showed antibacterial activities against *Enterobacter aerogenes*, *Shigella dysenteriae*, *Bacillus megaterium*, *B. subtilis*, and *Micrococcus lysodeikticus* with MIC values ranging from 12.5 to 50 μg/mL. The second compound (−)-3-carboxypropyl-7-hydroxyphthalide methyl ester exhibited antibacterial effect against *E. aerogenes*, *B. subtilis*, *B. megaterium*, and *M. lysodeikticus* with MIC values between 12.5 and 100 μg/mL (Qin et al. 2019). Another study conducted by Qin et al. reported the production of 10-formyl andrastone, 10- demethylated andrastone A, and andrastin G, new andrastin derivatives, along with four known andrastin analogues from the same endophytic fungus *P. vulpinum*. These compounds showed antibacterial activities against several bacterial strains. Taking the example of 10- demethylated andrastone A, it inhibited the growth of *B. megaterium* with MIC value of 6.25 μg/mL (Qin et al. 2020).

Two bioactive metabolites (2S)-butylitaconic acid and (2S)-hexylitaconic acid were isolated from an endophytic fungus, *Eupenicillium* sp. LG41, from the roots of the Chinese medicinal plant *Xanthium sibiricum* (Table 10.1). These compounds showed strong antibacterial activity against *Acinetobacter* sp. BD4 (DSM 586) with an MIC of 1 μg/mL being more important than the positive controls streptomycin and gentamicin (MICs = 5 μg/mL) (Li et al. 2014). The same microorganism *Eupenicillium* sp. LG41 was able to produce eupenicinicol B, a new compound effective against *S. aureus* subsp. aureus (DSM 799) with a MIC value of 1 μg/mL with almost the same efficacy of gentamicin (Li et al. 2014). Ebada and Ebrahim (2020) reported the production of two epimers, aniduquinolone A and a new quinolone derivative by the endophytic fungus *Aspergillus versicolor* strain Eich.5.2.2 isolated from the petals of *Eichhornia crassipes*, an Egyptian plant. These epimers altogether inhibited the growth of *S. aureus* (ATCC700699) with a MIC value of 0.4 mg/mL.

The endophytic fungus *Trichothecium* sp., which was isolated from the medicinal herb *Phyllanthus amarus*, produces Trichothecinol-A and has an antimycotic activity against a number of filamentous fungi and yeast and a moderate activity against the yeast *Cryptococcus albidusvar diffluens* (NCIM 3372) with up to 20 μg/mL (Taware et al. 2014) (Table 10.1). Stemphyperylenol, from *Botryosphaeria dothidea* KJ-1 of the stems of white cedar (*Melia azedarach* L) (Table 10.1) showed a strong antifungal activity against the plant pathogen *Alternaria solani* with a MIC of 1.57 μM comparable to the commonly used fungicide carbendazim. The same endophytic fungus KJ-1 also produced other compounds, namely, pycnophorin, β-sitosterol glucoside, and 5-hydroxymethylfurfural (Table 10.1), which showed moderate antifungal activity against *A. solani* with the same efficacy as the positive control hymexazol (Xiao et al. 2014).

The two compounds cytosporone B and C (Table 10.1) were isolated from *Phomopsis* sp. ZSU-H76, an endophytic fungus from the stem of *Excoecaria agallocha* collected in South China Sea. Cytosporone B and C showed antifungal activity against *C. albicans* and *F. oxysporum* with MIC values in the range of 32–64 μg/mL (Huang et al. 2008). Beauvericin (Table 10.1), a depsipeptide, was obtained from an endophytic fungus *Epicoccum nigrum* isolated from the leaves of *Entada abyssinica* Steud, Cameroon. Beauvericin showed important activity against *Bacillus cereus* and *Salmonella typhimurium* with MIC of 3.12 and 6.25 μg/mL, respectively (Dzoyem et al. 2017).

Antibacterial compounds, 2′-deoxyribolactone and hexylitaconic acid (Table 10.1), were obtained from endophytic fungi *Curvularia* sp. T12 isolated from the medicinal plant *Rauwolfia macrophylla* from Cameroon. 2′-deoxyribolactone showed antibacterial activities against four bacterial strains *E. coli*, *Micrococcus luteus*, *Pseudomonas agarici*, and *Staphylococcus warneri* with MIC values between 0.17 and 0.32 μg/mL. The compound hexylitaconic also exhibited antibacterial activities against the same tested strains with MIC values ranging from 0.29 to 0.58 μg/mL (Kaaniche et al. 2019). Kheiric acid, a polyhydroxy acid, is produced by the endophytic fungi *Curvularia papendorfii* isolated from the Sudanese medicinal plant *Vernonia amygdalina*. This compound showed antibacterial effect against methicillin-resistant *S. aureus* (MRSA) with MIC value of 62.5 μg/mL (Khiralla et al. 2020). (+)- and (−)-preuisolactone A, two rare naturally obtained sesquiterpenoidal enantiomers, were produced by the endophytic fungus *Preussia isomera* XL-1326 isolated from *Panax notoginseng* plant. (+)-

preuisolactone A displayed a strong antibacterial effect against *M. luteus* with an MIC value of 10.2 μM (Xu et al. 2019).

Antimicrobial compounds can be used as preservatives in the control of food spoilage and foodborne infections such as the molecule 7-amino-4-methylcoumarin (Table 10.1) isolated from the fungal endophytic *Xylaria* sp. from *Ginkgo biloba*. This compound showed activity against several foodborne and food-spoilage microorganisms, including *S. aureus*, *E. coli*, *Salmonella typhi*, *Salmonella enteritidis*, *Aeromonas hydrophila*, *Vibrio anguillarum*, *C. albicans*, *Penicillium expansum*, and *A. niger* (Liu et al. 2008). Another strain *Muscodor albus* (cz-620) isolated from a small vine in Indonesia produced a mixture of volatile organic compounds (VOCs) that showed antifungal activity against *Stachybotrys chartarum*. This strain of mold was found in the cellulose of building construction materials damaged by water (Samson et al. 2010). The VOCs of *Muscodor albus* can be also used in the treatment of interior surfaces of buildings (Atmosukarto et al. 2005).

A new chromone derivative, (2′S*)-2-(2′-hydroxypropyl)-5-methyl-7, 8-dihydroxy-chromone (Table 10.1), was reported to be produced by an endophytic fungus *Penicillium aculeatum* (No. 9 EB), isolated from the leaves of *Kandelia candel*. This compound showed antibacterial activity against *Salmonella* strains with a MIC of 2 μM. Other compounds (bacillisporin A and bacillisporin B) isolated from the same microorganism (Table 10.1) showed strong antibacterial activity against *B. subtilis* with MIC values of 0.13 μM (Huang et al. 2017). Pereira et al. (2016) reported two molecules: 2-amino-3,4-dihydroxy-2-25-(hydroxymethyl)-14-oxo-6,12-eicosenoic acid and myriocin from an endophytic fungus *Mycosphaerella* sp. isolated from plant *Eugenia bimarginata* DC. (*Myrtaceae*). 2-amino-3,4-dihydroxy-2-25-(hydroxymethyl)-14-oxo-6,12-eicosenoic acid and myriocin are two eicosanoic acids which showed antifungal activities against *Cryptococcus neoformans* and *Cryptococcus gattii* with MIC values of 0.49 to 7.82 μM and 0.48 to 1.95 μM, respectively (Pereira et al. 2016). Wu et al. (2019) evaluated the antifungal potential of three isocoumarin derivatives, botryospyrones A, B, and C, and a new natural tryptamine, (3aS, 8aS)-1-acetyl-1, 2, 3, 3a, 8, 8a-hexahydropyrrolo [2,3b] indol-3a-ol. These compounds were produced by *Botryosphaeria ramosa* L29, a marine mangrove endophytic fungus, isolated from the leaf of *Myoporum bontioides* and showed antifungal activities against *F. oxysporum*, *P. italicum*, and *F. graminearum*. All compounds are more active against *F. oxysporum* than the commonly used agricultural fungicide triadimefon (Wu et al. 2019).

On the other hand, metabolites produced by endophytic microorganisms can play the role of antiviral agents. Taking the example of 6-*O*-demethyl-4-dehydroxyaltersolanol A (Table 10.1), a hydroanthraquinone derivative, obtained from the endophytic fungus *Nigrospora* sp. YE3033, which was isolated from the plant *Aconitum carmichaelii*. 6-*O*-demethyl-4-dehydroxyaltersolanol A inhibited the influenza virus strain of H1N1 (A/Puerto Rico/8/34) with IC_{50} of 2.59 μg/mL (Zhang et al. 2016). Liu et al. (2019) discovered phomanolide, a rare 14-nordrimane-type sesquiterpenoid, produced by the endophytic fungus YE3135 identified as *Phoma* sp. isolated from the roots of *Aconitum vilmorinianum*. This new compound exhibited an important antiviral inhibitory effect against influenza A virus (A/Puerto Rico/8/34, H1N1) with IC50 value of 2.96 ± 0.64 μg/mL (Liu et al. 2019).

An endophytic fungus *Phomopsis* sp. from the plant *Diospyros carbonaria*, collected in French Guiana (Saint Elie), produces 3b-*O*-cis-p-coumaroylalphitolic acid, 3b-*O*-trans-p-coumaroylalphitolic acid, eucalyptic acid, betulinic acid, betulin, and betulinic aldehyde (Table 10.1), which inhibited the replication of the dengue virus with IC_{50} values of 3.0, 2.2, 2.3, 4.3, 4.1, and 7.5 μM, respectively (Peyrat et al. 2017).

The structures of some antimicrobial molecules produced by endophytic fungi are grouped in Table 10.1. Other bioactive molecules from endophytic fungi with their antimicrobial activities are grouped in Table 10.2 and with their respective MICs on representative strains.

10.6 Antibiofilm and Quorum Quenching Molecules

In nature, bacteria can adopt two radically different lifestyles, either in a planktonic state or in sessile form known as biofilm mode. Biofilm is probably the predominant lifestyle of bacteria that show cooperative behavior (Wong and O'Toole 2011; Penesyan et al. 2019). Within the biofilm, bacteria either interact physically with each other to maintain "intimate" relationships, or they use the quorum sensing communication system by secreting specific signal molecules (Schuster et al. 2013; Li and Tian 2016; Subramani and Jayaprakashvel 2019).

Quorum sensing is a bacterial communication system from cell to cell that is responsible for the regulation of genes responsible for virulence, infection, invasion, colonization, biofilm formation, and pathogenesis in prokaryotic and eukaryotic organisms (LaSarre and Federle 2013; Ćirić et al. 2019). Quorum sensing is facilitated by specific signaling molecules known as autoinducers (AIs), L-homoserine N-acylated lactones (AHL) for Gram-negative bacteria, and oligopeptides (AIPs) for Gram-positive bacteria (Miller and Bassler 2001; Haque et al. 2019; Zhao et al. 2020). There is another family of common autoinducers for Gram-positive and Gram-negative bacteria known as autoinducers 2 (AI-2) (Miller and Bassler 2001). While quorum quenching is a process that disrupts quorum sensing (Murugayah and Gerth 2019), endophytes can present both quorum sensing and quorum quenching. Taking the example of the bacterial genera *Acinetobacter* and *Burkholderia*, which colonize the rhizosphere of the ginger plant (*Zingiber officinale*), they showed the coexistence of quorum sensing and quorum quenching systems (Chan et al. 2011) (Fig. 10.1).

AHL degradation, which is a quorum quenching mechanism, is generally performed by the two main enzymes lactonase and acylase (Kusada et al. 2019). AHL-lactonase hydrolyzes the homoserine lactone ring of AHL independently of acyl chain length while decreasing its efficiency to bind to the transcriptional regulator (Dong et al. 2000, 2001). AHL-acylase irreversibly hydrolyzes the amide ring, producing HSL and fatty acid, as shown in Fig. 10.2, which shows no signaling activity (Lin et al. 2003). Another class of enzyme, known as oxidoreductase,

Table 10.2 Antimicrobial compounds reported from endophytic fungi along with their activities

Bioactive compounds	Name of the endophytic fungi	Source of endophytes	Specific antimicrobial activities	References
Tribacopin AV	*Trichoderma lixii* (IIIM-B4)	*Bacopa monnieri* L.	*Candida albicans* (MIC 25 μg/mL)	Katoch et al. (2019)
Trypethelone	*Coniothyrium cereale*	Marine green alga *Enteromorpha* sp.	*Mycobacterium phlei* (zone of inhibition of 8 mm at 20 μg/disk) *S. aureus* (zone of inhibition of 14 mm at 20 μg/disk) *E. coli* (zone of inhibition of 12 mm at 20 μg/disk)	Elsebai et al. (2011)
Microsporaline B, Microsporaline C	*Pestalotiopsis microspora*	Leaves of *Scaevola taccada* (Gaertn.) Roxb	*C. albicans* ATCC 10231 (MIC 25.0 μg/mL)	Liao et al. (2020)
Gamahorin	*Pestalotiopsis microspora*	Leaves of *Scaevola taccada* (Gaertn.) Roxb	*C. albicans* ATCC 10231 (MIC 12.5 μg/mL)	Liao et al. (2020)
Halobacillin	*Trichoderma asperellum*	Chinese medicinal plant *Panax notoginseng*	*E. faecium* (CGMCC 1.2025) (IC50 5.24 μM) *S. aureus* COL (CGMCC1.2465) (IC50 14.00 μM)	Ding et al. (2012)
6-Formamide-chetomin	*Chaetomium* sp. M336	*Huperzia serrata* (Thunb. ex Murray) Trev	*Escherichia coli* (MIC 0.78 μg/mL) *Staphylococcus aureus* (MIC 0.78 μg/mL) *Salmonella typhimurium* ATCC 6539 (MIC 0.78 μg/mL)	Yu et al. (2018)
Cyclopeptides PF1022F	*Trichoderma asperellum*	Chinese medicinal plant *Panax notoginseng*	*E. faecium* (CGMCC 1.2025) (IC50 7.30 μM) *S. aureus* COL (CGMCC1.2465) (IC50 19.02 μM)	Ding et al. (2012)
Dicerandrol C	*Phomopsis longicolla* C81	Red seaweed *Bostrychia radicans*	*S. aureus* (ATCC 6538) (MIC 1 μg/mL) *S. saprophyticus* (ATCC 15305) (MIC 2 μg/mL)	Erbert et al. (2012)

(continued)

Table 10.2 (continued)

Bioactive compounds	Name of the endophytic fungi	Source of endophytes	Specific antimicrobial activities	References
Xanalteric acids I	*Alternaria* sp.	Mangrove plant *Sonneratia alba*	MRSA (MIC 125 μg/mL)	Kjer et al. (2009)
Ophiobolins P	*Ulocladium* sp.	Endolichenic	*B. subtilis* (MIC 62.5 μM) MRSA (MIC 31.3 μM)	Wang et al. (2013)
Alterporriol N	*Stemphylium globuliferum*	*Mentha pulegium*	MRSA (MIC 62.5 μg/mL) *E. faecalis* (MIC 15.63 μg/mL)	Debbab et al. (2009)
Chermesin A	*Penicillium chermesinum* EN-480	Marine red alga *Pterocladiella tenuis*	*Micrococcus luteus* (MIC 8 μg/mL) *Candida albicans* (MIC 32 μg/mL) *Escherichia coli* (MIC 32 μg/mL) *Vibrio alginolyticus* (MIC 32 μg/mL)	Liu et al. (2016)
Flavipesins A	*Aspergillus flavipes*	Mangrove plant *Acanthus ilicifolius*	*Staphylococcus aureus* (MIC 8.0 μg/mL) *Bacillus subtilis* (MIC 0.25 μg/mL)	Bai et al. (2014)

catalyzes the reduction of carbonyl to hydroxyl of the acyl side chain (Uroz et al. 2005) (Fig. 10.2).

The autoinducers (AIPs) of Gram-positive bacteria are oligopeptides typically consisting of 5 to 17 amino acids. Post-transcriptional modifications of the side chain by the incorporation of a thiolactone, lanthionine, lactone, or isoprenyl group lead to the formation of various structures and binding selectivity for Gram-positive bacteria (Lazazzera and Grossman 1998; Federle and Bassler 2003; Williams et al. 2007; Bouyahya et al. 2017). Gram-positive bacteria synthesize and secrete AIPs in the extracellular environment by surface oligopeptide transporters since the wall of these bacteria is impermeable to AIPs (Federle and Bassler 2003). Some studies reported on the ability of endophytes to degrade or modify AHL. Examples would be the endophytic fungus *Meliniomyces variabilis* JH2, which showed the ability to degrade C6- and 3O and C6-HSL, and the disappearance of AHL signal molecules using the *Chromobacterium violaceum* assay (Uroz and Heinonsalo 2008).

The endophytic fungus *Penicillium restrictum*, isolated from the stems of a milk thistle plant (*Silybum marianum*), produced a number of polyhydroxyanthraquinones, five new compounds with ω-hydroxyemodin, emodic

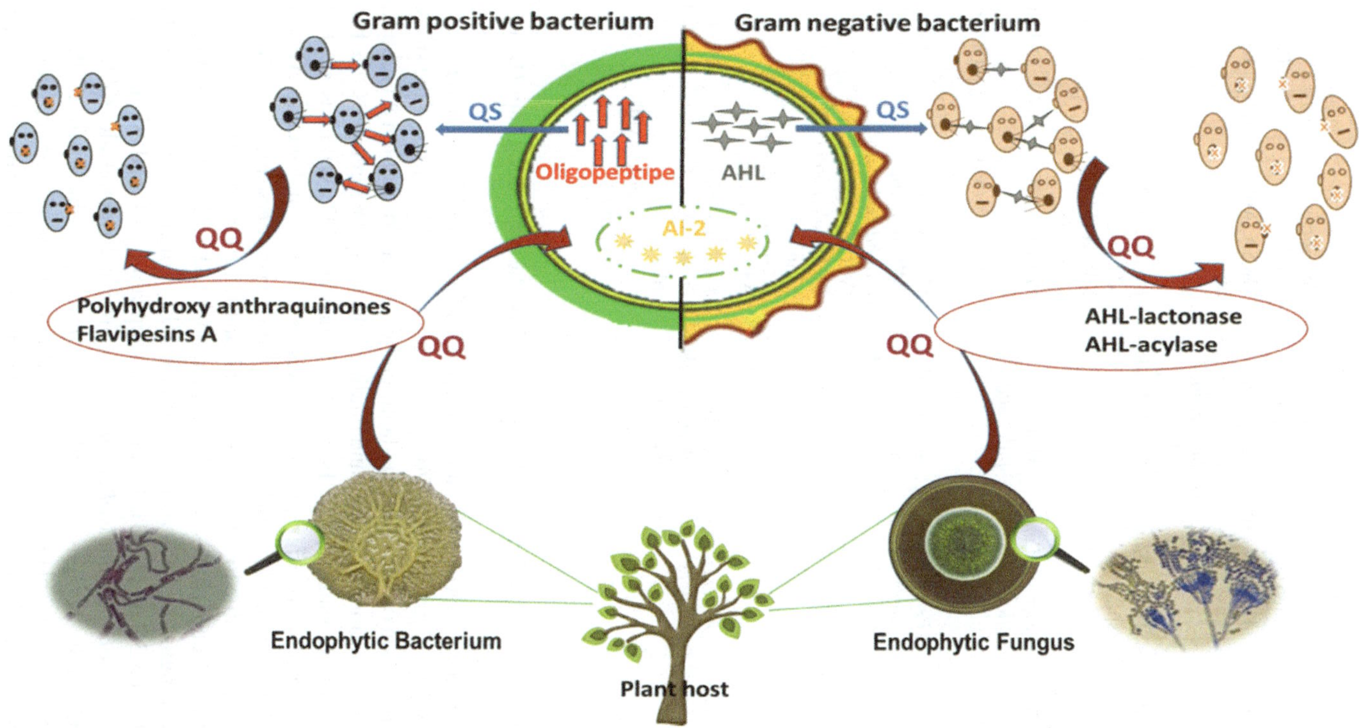

Fig. 10.1 Schematic representation of quorum quenching responses of endophytic microorganisms; *QQ* quorum quenching, *QS* quorum sensing, *AHL N*-acyl-L-homoserine lactone

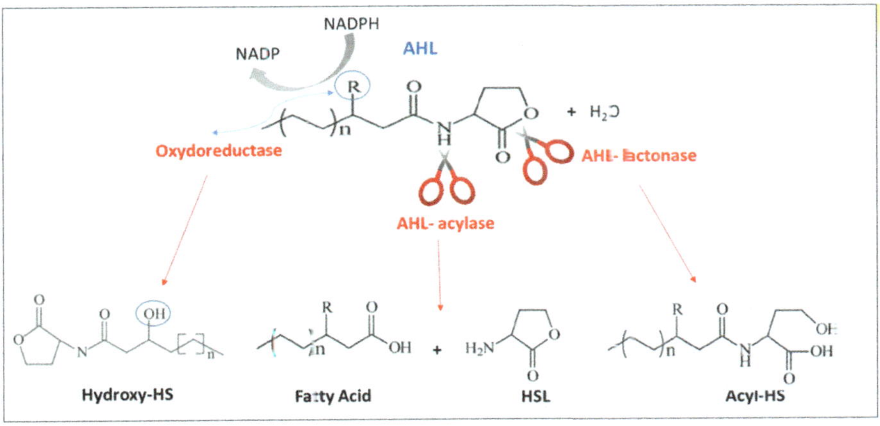

Fig. 10.2 The degradation or modification of quorum sensing molecule AHL by three different enzymes of quorum quenching: AHL-lactonase, AHL-acylase, and oxidoreductase. Mechanism of quorum quenching enzymes: AHL-acylase, AHL-lactonase, and oxidoreductase

acid, (+)-2′S-isorhodoptilometrin, and emodin. The evaluation of the ability of polyhydroxyanthraquinones to quench the quorum sensing system of the gene regulator (agr), responsible for the production of toxins and the pathogenesis of methicillin-resistant *S. aureus* (MRSA) (Thoendel et al. 2011), showed the ability of these compounds to suppress quorum sensing with IC_{50} in the range of 8–120 μM (Figueroa et al. 2014). The evaluation of the quorum sensing inhibition by 100 extracts of several species of *Penicillium* showed that 33% of the tested extracts produce QS-inhibiting molecules. Two QSI, patuline and penicillic acid, were produced by *Penicillium coprobium* and *Penicillium radicicola*, respectively. In addition, patulin could eliminate *P. aeruginosa* biofilms. Biofilm treated with patulin is more sensitive to the treatment of tobramycin with an increase of cellular mortality within the biofilm (Rasmussen et al. 2005).

The fungus *Aspergillus flavipes* AIL8, which was isolated from the mangrove plant *Acanthus ilicifolius* in China, produced a new aromatic butyrolactone flavipesins A. This compound exhibited an antibiofilm activity against *S. aureus* by interfering in the biofilm matrix and causing the decrease of living cells using concentrations from 390.6 to 97.7 μg/mL (Bai et al. 2014). Phukhamsakda and collaborators discovered a new abscisic acid derivative, roussoellenic acid, which is produced by the endophytic fungus *Roussoella* sp. (MFLUCC 17–2059) isolated from *Clematis subumbellata* Kurz plant from Thailand. The antibiofilm assay revealed that roussoellenic acid is able to inhibit 34% and 33% of *S. aureus* DSM 1104 biofilm at a concentration of 16 and 128 μg/mL, respectively (Phukhamsakda et al. 2018). Parasuraman et al. reported the ability of *Blastobotrys parvus* PPR3 metabolites, an endophytic fungus isolated from the wood of *Avicennia marina*, to inhibit QS-controlled virulence factor in *P. aeruginosa* PAO1 by reducing the production of exopolymeric substance (EPS), pyocyanin, protease, chitinase, elastase, alginate, motility, and biofilm formation (Parasuraman et al. 2020).

10.7 Anticancer Molecules

Cancer is one of the most serious diseases worldwide, causing death of 9.6 million and 18.1 million new cancer cases in 2018 (Bray et al. 2018). A wide number of fungal endophytes present a potential source of new anticancer drugs (Uzma et al. 2018) including the famous drug, generating a turnover of more than three billion dollars a year, known as Taxol or paclitaxel (Zhu and Chen 2019). Taxol is a diterpenoid produced for the first time by the endophytic fungus *Taxomyces andreanae*, isolated from the plant *T. brevifolia* (Stierle et al. 1993). This compound was characterized by its unique mode of action compared with other anticancer agents since it blocks the mechanism of growth of cancer cells (Yan-Hua et al. 2020). Taxol is a cytoskeletal drug that targets microtubules which are fundamental structural components in cells formed by the association of protofilaments of tubulin, α- and β-subunits. Microtubules are very dynamic and undergo rapid cycles of polymerization and depolymerization (Kerssemakers et al. 2006; Yan-Hua et al. 2020). They play a role in cellular organization including the constitution of mitotic fibers necessary for cell division in M phase (Wilson 1975). Taxol could penetrate the cell and bind to microtubule β-tubulin subunits (Löwe et al. 2001). Taxol is a microtubule stabilizer that inhibits depolymerization of microtubes and subsequently promotes the increase of free tubulin. Thereafter the microtubules thus formed are rigid and nonfunctional (Yang and Horwitz 2017). The cell cycle is stopped, and cells remain blocked in G2-M phase of mitosis preventing any multiplication and, thereby, any tumor proliferation and lead finally to apoptosis (Miller et al. 2013) (Fig. 10.3a). In addition, the expression of androgen receptors (AR) can be inhibited by the action of Taxol, at the transcriptional level, blocking the AR nuclear translocation and FOXO1-mediated repression of transcriptional activity (Gan et al. 2009) (Fig. 10.3b). Finally, Taxol could inhibit anti-apoptotic Bcl-2 expression by inducing the release of BAX-mediated cytochrome c and subsequent apoptotic cell death (Miller et al. 2013) (Fig. 10.3c). The FDA (Food and Drug Administration) approved Taxol for the treatment of advanced breast, lung, and refractory ovarian cancers (Cremasco et al. 2009). Zhou et al. (2010) reported that more than 19 genera of endophytic fungi produced Taxol and its analogues (Zhou et al. 2010).

Raj et al. (2015) isolated Taxol from endophytic fungal *Cladosporium oxysporum* from the *Moringa oleifera* plant. This Taxol inhibits human colon cancer cell HTC 15 growth with an IC_{50} value of 3.5 μM (Raj et al. 2015). Another anticancer compound, Trichothecinol-A, was produced from endophyte *Trichothecium sp.* isolated from the medicinal herb *P. amarus*. Trichothecinol-A was confirmed to have a potent cytotoxicity against HeLa and B16F10 cells with an IC_{50} value of 500 nM (Taware et al. 2014). This compound is considered as an anti-metastatic drug by the inhibition of wound migration of MDA-MB-231 cells (Taware et al. 2014) (Table 10.3).

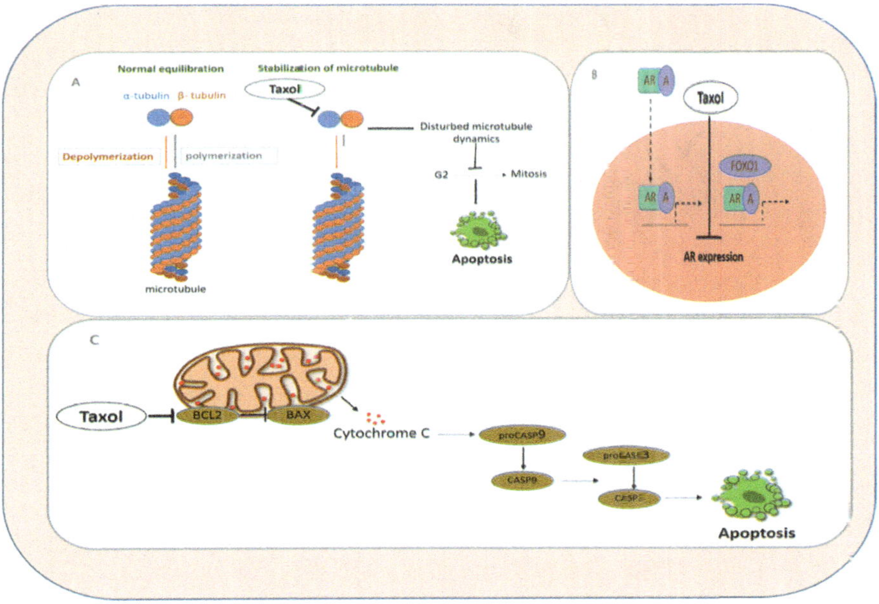

Fig. 10.3 Mechanism of action of anticancer compound Taxol

Another compound, digoxin, was used in recent years as anticancer agent alone or in combination with Taxol (Huang et al. 2018; Pereira et al. 2019; Wang et al. 2020). El-Sayed et al. reported the production of digoxin by the endophytic fungi *E. nigrum* isolated from the bark of Egyptian *Terminalia arjuna* plant. Digoxin showed anticancer activity against CHO-K1, A-549, HepG-2, MCF-7, and HEp-2 cancer cell lines with IC_{50} values of 35.14, 10.76, 33.42, 34.66, and 12.99 μg/mL, respectively (El-Sayed et al. 2020).

Multirostratin A, a new cytochalasin, was isolated from the endophytic fungi *Phoma multirostrata* EA-12 from *Eupatorium adenophorum*. Multirostratin A showed anticancer activities against HL-60, A-549, SMMC-7721, SW-480, and MCF-7 cell lines with IC_{50} of 7.8, 10.8, 9.8, 15.8, and 11.6 μM, respectively (Chen et al. 2015). Recently, the isolation of five new cytochalasans, named diaporthichalasins D–H, was reported from the endophytic fungus *Diaporthe* sp. SC-J0138 isolated from the leaves of *Cyclosorus parasiticus*. Diaporthichalasins D and H exhibited cytotoxic effect against human carcinoma A549, HeLa, and HepG2 cells with IC50 ranging from 8.8 to 20 μM (Yang et al. 2020) (Table 10.3). Moreover, four isoprenylated epoxy derivatives pestalofones F–H and pestalodiols C were obtained from the endophytic fungi *Pestalotiopsis fici* isolated from *Camellia sinensis*, a Chinese plant. These compounds showed cytotoxic effects against HeLa and MCF-7 cells with IC_{50} of 11.9–36.4 μmol/L (Shu-Chun et al. 2011). In 2013, another active molecule identified as siccayne was isolated from the same endophytic fungus *P. fici*. This compound displayed

Table 10.3 Structures of some anticancer metabolites from endophytic fungi

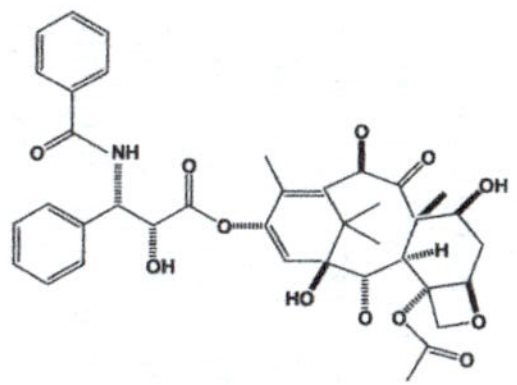 Taxol	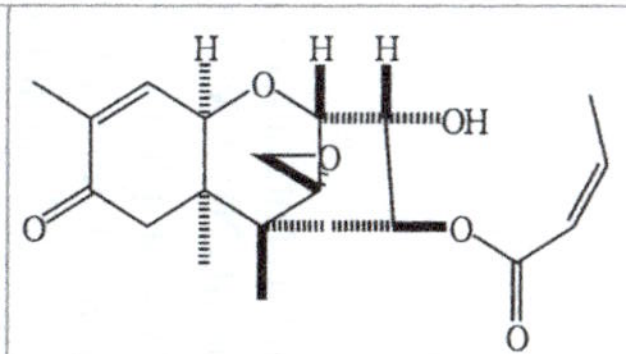 Trichothecinol A	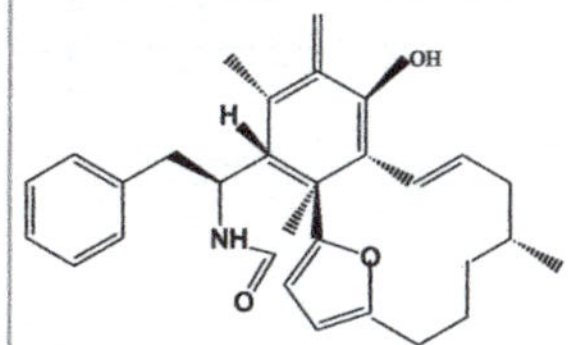 Multirostratin A
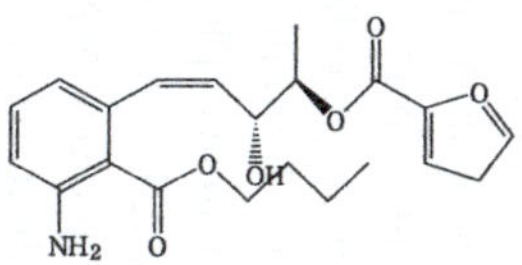 Pestalamine A	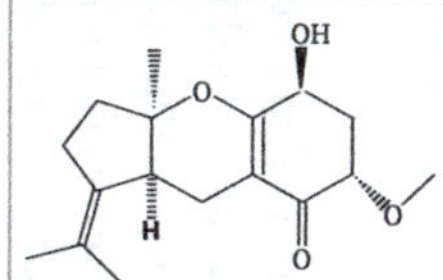 Guignardone S	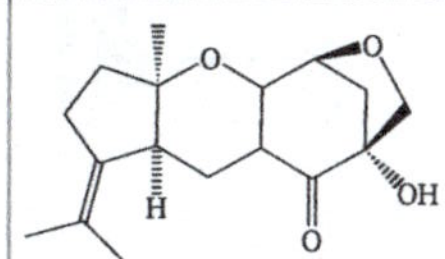 Guignardone Q
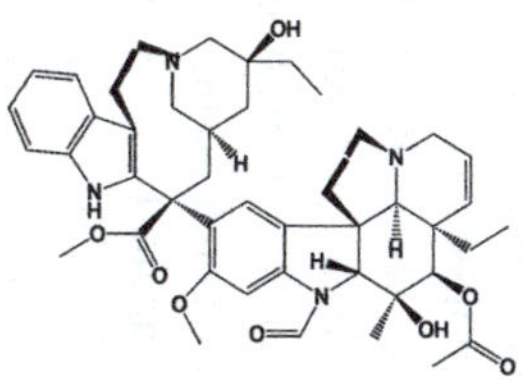 Vincristine	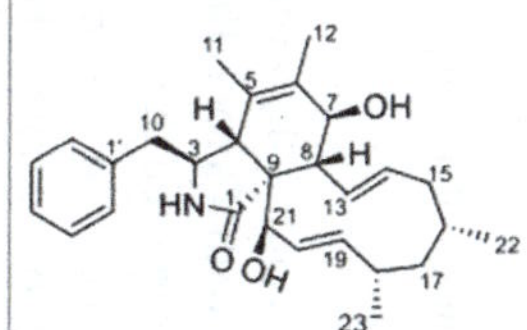 Diaporthichalasin D	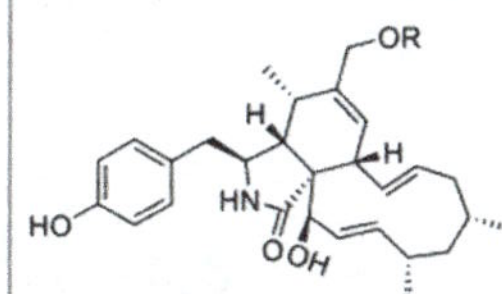 R = Ac Diaporthichalasin H

anticancer activity against HeLa and HT29, with IC_{50} values of 48.2 and 33.9 μM, respectively (Liu et al. 2013). *Pestalotiopsis vaccinii*, an endophytic fungus isolated from the mangrove plant *K. candel* (L.) Druce, in China, was reported to produce pestalamine A. This compound exhibited cytotoxic activity against HeLa, HepG2, and MCF-7 cell lines with IC_{50} values of 22.0, 32.8, and 40.3 μM, respectively (Zhou et al. 2014). Another endophytic fungus from the genus *Pestalotiopsis*, *Pestalotiopsis microspora* HF 12440 isolated from the stem of *Artocarpus heterophyllus*, was reported for the first time to produce (+)-sydowic acid, a bisabolane-type sesquiterpene. (+)-sydowic acid displayed a cytotoxic effect against murine leukemia P-388 cells with IC_{50} of 2.56 μg/mL (Riga et al. 2020). Liu et al. (2020a) reported the cytotoxic activity of four new cyclodepsipeptides, trichodestruxins A – D (1–4), produced by the *Trichoderma harzianum*, an endophytic fungus from the fruit of *Physalis angulata* L. in China. These compounds exhibited cytotoxic effect against human cancer cell lines P388, HT-29, and A549, with IC_{50} values ranging from 0.7 to 19.1 μM (Liu et al. 2020a).

Guignardones Q and S with anticancer activities were isolated from an endophytic fungus *Guignardia mangiferae* A348 collected from the plant *Smilax glabra*. They showed cytotoxicity against MCF-7 cell line with IC_{50} values of 83.7 and 92.1 μM, respectively (Sun et al. 2015). Vincristine, a terpenoid indole alkaloid, was isolated from the endophytic fungus *Talaromyces radicus* from *Catharanthus roseus*. This compound showed selective anticancer activities against HeLa, MCF7, A549, U251, and A431 with IC_{50} values of 4.2, 4.5, 5.5, 5.5, and 5.8 μg/mL, respectively, but no effect was registered against HEK293 (normal cells). Furthermore, Vincristine allows the apoptotic induction of Hela cell line (Palem et al. 2015). Kuriakose et al. isolated Vincristine from another endophytic fungus *Eutypella* spp. - CrP14 from the same plant. They reported the antiproliferative effect of this compound against HeLa, A459, and A431 cells with IC_{50} of 14, 10, and 4.8 μg/mL, respectively. Vincristine was able to induce apoptotic death in A431 cell lines (Kuriakose et al. 2016) (Table 10.3).

Cerrenin D, a new triquinane-type sesquiterpenoid, with cytotoxic effect was obtained from the endophytic fungus *Cerrena* sp. A593 isolated from *Pogostemon cablin*. Cerrenin D showed cytotoxicity against human cancer cell lines SF-268 (IC_{50} 41.01 μM), MCF-7 (14.43 μM), NCI-H460 (29.67 μM), and HepG-2 (44.32 μM) (Liu et al. 2020b).

The structures of some important anticancer molecules produced by endophytic fungi are grouped in Table 10.3. A list of some anticancer compounds found as endophytic fungi metabolites is shown in Table 10.4.

10.8 Molecules with Antimicrobial and Anticancer Activities

Recently, molecules produced by endophytic fungi have showed both antimicrobial and anticancer activities. Terrein (4,5-dihydroxy-3-(1-propenyl)-2-cyclopenten-1-one) was obtained from *Aspergillus terreus* (JAS-2), isolated from *Achyranthes aspera*, and showed antifungal, antibacterial, and cytotoxic activities (Goutam

Table 10.4 Anticancer compounds from endophytic fungi

Chemical compound	Name of the endophytic fungi	Name of the host plants	Activities	References
(6S,7S,8R)-hydroxypestalotin	*Pestalotiopsis microspora* HF 12440	*Artocarpus heterophyllus*	Murine leukemia P388 cell (IC_{50} 3.34 μg/mL)	Riga et al. (2019)
Incarxanthone B	*Peniophora incarnata* Z4	*Bruguiera gymnorrhiza*	Melanoma cell line A375 (IC_{50} of 8.6 μM) Breast cancer cell line MCF-7 (6.5 μM) Leukemia cell line HL-60 (4.9 μM)	Li et al. (2020)
Oblongolide X	*Phomopsis* sp. BCC 9789	Leaf of *Musa acuminata*	BC cell (IC_{50} 48 μM)	Bunyapaiboonsri et al. (2010)
Oblongolide Y	*Phomopsis* sp. BCC 9789	Leaf of *Musa acuminata*	KB cell (IC_{50} 37 μM) BC cell (IC_{50} 26 μM) NCI-H187 cell (IC_{50} 32 μM)	Bunyapaiboonsri et al. (2010)
Sclerotiorin	*Cephalotheca faveolata*	Leaves of *Eugenia jambolana*	ACHN, Panc-1, Calu-1, HCT-116, and H460 cancer cell lines (IC_{50} from 0.63 to 2.1 μM) Normal breast epithelium cells (MCF10A) ($IC_{50} > 10$ μM)	Giridharan et al. (2012)
9-Deacetoxyfumigaclavine C	*Aspergillus fumigatus*	Stem of *Cynodon dactylon*	Human leukemia cells (K562) (IC_{50} 3.1 μM)	Ge et al. (2009)
Citriquinochroman	*Penicillium citrinum*	Stem of the Moroccan plant *Ceratonia siliqua*	Murine lymphoma L5178Y (IC_{50} 6.1 μM)	El-Neketi et al. (2013)
Polonidine A	*Penicillium polonicum* TY12	*Aconitum vilmorinianum* Kom	Hepatoma cell line MHCC97H (IC_{50} 7.1 μg/mL) Breast cancer cell line BT549 (IC_{50} 6.1 μg/mL) Lung cancer cell line H1299 (IC_{50} 20 μg/mL) Colon cancer cell line SW620 (IC_{50} 7.6 μg/mL Human glioma cell line T98G (IC_{50} 18 μg/mL) Human lung cancer cell line A549 (IC_{50} 15 μg/mL)	Cai et al. (2020)
Chaetoglobosin X	*Chaetomium globosum* L18	*Curcuma wenyujin*	MFC cells (IC_{50} 6.25 μg/mL) H22 cells (IC_{50} 3.125 μg/mL)	Wang et al. (2012)

6-formamide-chetomin	*Chaetomium* sp. M336	*Huperzia serrata* (Thunb. ex Murray) Trev	Human cervical cancer cell line HeLa (IC_{50} 21.6 nM) Human gastric cancer cell line SGC-7901 (IC_{50} 523.0 nM) Lung adenocarcinoma cell line A-549 (IC_{50} 27.1 nM)	Yu et al. (2018)

et al. 2017). In fact, it was able to inhibit the growth of phytopathogenic fungi *B. sorokiniana* (57.14%), *A. flavus* (52.5%), and *Alternaria alternata* at 10 μg/μL. At a concentration of 20 μg/mL, Terrein exhibited 50% inhibition against *Enterococcus faecalis* and 38.3% and 48% inhibition against *Aeromonas hydrophila* and *S. aureus*, respectively. Moreover, lung cancer cell lines A-549 were sensible to this compound with an IC_{50} of 121.9 μg/mL (Goutam et al. 2017). Perveen et al. (2017) purified three metabolites from the endophyte fungi *E. nigrum* (2-methyl-3-nonyl prodiginine, Bis (2-ethylhexyl) phthalate and a meroterpenoid, preaustinoid A, having antimicrobial and cytotoxic activities. They concluded that these compounds have great potential, and these may be applied in textile (dyeing), pharmaceutical (drug, infectious agents), and food (preservatives) industries. *Monarda citriodora*, a plant that harbors diverse endophytic fungi, among them, the isolated endophyte MC-24L (*Cladosporium tenuissimum*), has the potential to be a source of novel cytotoxic/antimicrobial compounds (Katoch et al. 2017).

10.9 Future Prospects

For drug discovery, natural products have the most significant pharmaceutical potential. A study was carried out years ago on molecules from medicinal plants. Nowadays, endophytic fungi are considered as the most precious natural source of bioactive molecules. Many bioactive endophyte-generated molecules have already been characterized. Researchers should try to set up databases of all molecules generated by endophytes and their tested activities to help perform in silico molecular docking for recognizing new targets of these molecules and thus find out further applications. Although we already know a number of molecules from endophytes, research has to continue to find more molecules since there is a huge diversity of fungal endophytes. One of the most recent approaches in drug discovery from endophytes is co-cultivation. In the co-cultivation strategy, a culture of two or more organisms from different species can be conducted. Thus, their physiology could be modified to synthesize cryptic compounds that cannot be produced in single-strain cultivation strategy.

10.10 Conclusion

Endophytic fungi represent an important group of microorganisms which are capable of synthesizing biologically active compounds combating against various pathogens and destroying cancer cell lines. This achievement became possible by the molecules produced by these endophytes and which could be synthesized inside and even outside the plants in the lab and at industrial scales. In this context, this review highlighted the previous discoveries of molecules from endophytes and their use against prokaryotic and eukaryotic cells. This is to attract the attention of researchers all over the world toward this field and the importance of exploitation of endophytic microbes for medical and pharmaceutical purposes.

Acknowledgments This work was supported by a grant provided by the Tunisian Ministry of Higher Education and Scientific Research.

Disclosure Statement: No potential conflict of interest was reported by the authors.

References

Abdel-Azeem AM, Abdel-Azeem MA, Khalil WF (2019) Endophytic fungi as a new source of antirheumatoid metabolites. In: Bioactive food as dietary interventions for arthritis and related inflammatory diseases. Elsevier, Amsterdam, pp 355–384

Aly AH, Debbab A, Proksch P (2011) Fungal endophytes: unique plant inhabitants with great promises. Appl Microbiol Biotechnol 90:1829–1845

Atmosukarto I, Castillo U, Hess WM, Sears J, Strobel G (2005) Isolation and characterization of *Muscodor albus* I-413 s, a volatile antibiotic producing fungus. Plant Sci 169:854–861

Bai Z-Q, Lin X, Wang Y, Wang J, Zhou X, Yang B, Liu J, Yang X, Wang Y, Liu Y (2014) New phenyl derivatives from endophytic fungus *Aspergillus flavipes* AIL8 derived of mangrove plant *Acanthus ilicifolius*. Fitoterapia 95:194–202

Bedi A, Adholeya A, Deshmukh SK (2018) Novel anticancer compounds from endophytic fungi. Curr Biotechnol 7:168–184

Borel JF, Feurer C, Gubler HU, Stähelin H (1994) Biological effects of cyclosporin a: a new antilymphocytic agent. Agents Actions 43:179–186

Bouyahya A, Dakka N, Et-Touys A, Abrini J, Bakri Y (2017) Medicinal plant products targeting quorum sensing for combating bacterial infections. Asian Pac J Trop Med 10:729–743

Bray F, Ferlay J, Soerjomataram I, Siegel RL, Torre LA, Jemal A (2018) Global cancer statistics 2018: GLOBOCAN estimates of incidence and mortality worldwide for 36 cancers in 185 countries. CA Cancer J Clin 68:394–424

Bruisson S, Zufferey M, L'Haridon F, Trutmann E, Anand A, Dutartre A, De Vrieze M, Weisskopf L (2019) Endophytes and epiphytes from the grapevine leaf microbiome as potential biocontrol agents against phytopathogens. Front Microbiol 10:2726

Bunyapaiboonsri T, Yoiprommarat S, Srikitikulchai P, Srichomthong K, Lumyong S (2010) Oblongolides from the endophytic fungus *Phomopsis* sp. BCC 9789. J Nat Prod 73:55–59

Cai X-Y, Wang J-P, Shu Y, Hu J-T, Sun C-T, Cai L, Ding Z-T (2020) A new cytotoxic indole alkaloid from the fungus. *Penicillium polonicum* TY12. Nat Prod Res:1–7

Chan K-G, Atkinson S, Mathee K, Sam C-K, Chhabra SR, Camara M, Koh C-L, Williams P (2011) Characterization of N-acylhomoserine lactone-degrading bacteria associated with the *Zingiber officinale* (ginger) rhizosphere: co-existence of quorum quenching and quorum sensing in *Acinetobacter* and *Burkholderia*. BMC Microbiol 11:51

Chen Z, Chen H-P, Li Y, Feng T, Liu J-K (2015) Cytochalasins from cultures of endophytic fungus *Phoma multirostrata* EA-12. J Antibiot (Tokyo) 68:23–26

Chithra S, Jasim B, Sachidanandan P, Jyothis M, Radhakrishnan EK (2014) Piperine production by endophytic fungus *Colletotrichum gloeosporioides* isolated from *Piper nigrum*. Phytomedicine 21:534–540

Ćirić AD, Petrović JD, Glamočlija JM, Smiljković MS, Nikolić MM, Stojković DS, Soković MD (2019) Natural products as biofilm formation antagonists and regulators of quorum sensing functions: a comprehensive review update and future trends. South Afr J Bot 120:65–80

Cremasco MA, Hritzko BJ, Linda Wang N-H (2009) Experimental purification of paclitaxel from a complex mixture of taxanes using a simulated moving bed. Braz J Chem Eng 26:207–218

Debbab A, Aly AH, Edrada-Ebel R, Wray V, Müller WE, Totzke F, Zirrgiebel U, Schachtele C, Kubbutat MH, Lin WH (2009) Bioactive metabolites from the endophytic fungus *Stemphylium globuliferum* isolated from *Mentha pulegium*. J Nat Prod 72:626–631

Deepika VB, Murali TS, Satyamoorthy K (2016) Modulation of genetic clusters for synthesis of bioactive molecules in fungal endophytes: a review. Microbiol Res 182:125–140

Ding G, Chen AJ, Lan J, Zhang H, Chen X, Liu X, Zou Z (2012) Sesquiterpenes and Cyclopeptides from the endophytic fungus *Trichoderma asperellum* Samuels, Lieckf & Nirenberg. Chem Biodivers 9:1205–1212

Dong Y-H, Xu J-L, Li X-Z, Zhang L-H (2000) AiiA, an enzyme that inactivates the acylhomoserine lactone quorum-sensing signal and attenuates the virulence of *Erwinia carotovora*. Proc Natl Acad Sci 97:3526–3531

Dong Y-H, Wang L-H, Xu J-L, Zhang H-B, Zhang X-F, Zhang L-H (2001) Quenching quorum-sensing-dependent bacterial infection by an N-acyl homoserine lactonase. Nature 411:813–817

dos Santos ML, Berlitz DL, Wiest SLF, Schünemann R, Knaak N, Fiuza LM (2018) Benefits associated with the interaction of endophytic bacteria and plants. Braz Arch Biol Technol:61

Dzoyem JP, Melong R, Tsamo AT, Maffo T, Kapche DG, Ngadjui BT, McGaw LJ, Eloff JN (2017) Cytotoxicity, antioxidant and antibacterial activity of four compounds produced by an endophytic fungus *Epicoccum nigrum* associated with *Entada abyssinica*. Rev Bras Farm 27:251–253

Ebada SS, Ebrahim W (2020) A new antibacterial quinolone derivative from the endophytic fungus *Aspergillus versicolor* strain Eich522. South Afr J Bot 134:151–155

Eid AM, Salim SS, Hassan SE-D, Ismail MA, Fouda A (2019) Role of endophytes in plant health and abiotic stress management. In: Kumar V, Prasad R, Kumar M, Choudhary D (eds) Microbiome in plant health and disease. Springer, Singapore, pp 119–144

El-Neketi M, Ebrahim W, Lin W, Gedara S, Badria F, Saad H-EA, Lai D, Proksch P (2013) Alkaloids and polyketides from *Penicillium citrinum*, an endophyte isolated from the Moroccan plant *Ceratonia siliqua*. J Nat Prod 76:1099–1104

El-Sayed ER, Ahmed AS, Abdelhakim HK (2020) A novel source of the cardiac glycoside digoxin from the endophytic fungus *Epicoccum nigrum*: isolation, characterization, production enhancement by gamma irradiation mutagenesis and anticancer activity evaluation. J Appl Microbiol 128:747–762

Elsebai M, Natesan L, Kehraus S, Mohamed I, Schnakenburg G, Sasse F, Shaaban S, Gütschow M, König G (2011) HLE-inhibitory alkaloids with a polyketide skeleton from the marine-derived fungus *Coniothyrium cereale*. J Nat Prod 74:2282–2285

Erbert C, Lopes AA, Yokoya NS, Furtado NA, Conti R, Pupo MT, Lopes JLC, Debonsi HM (2012) Antibacterial compound from the endophytic fungus *Phomopsis longicolla* isolated from the tropical red seaweed *Bostrychia radicans*. Bot Mar 55:435–440

Fadiji AE, Babalola OO (2020) Elucidating mechanisms of endophytes used in plant protection and other bioactivities with multifunctional prospects. Front Bioeng Biotechnol 8:467

Federle MJ, Bassler BL (2003) Interspecies communication in bacteria. J Clin Invest 112:1291–1299

Figueroa M, Jarmusch AK, Raja HA, El-Elimat T, Kavanaugh JS, Horswill AR, Cooks RG, Cech NB, Oberlies NH (2014) Polyhydroxyanthraquinones as quorum sensing inhibitors from the guttates of *Penicillium restrictum* and their analysis by desorption electrospray ionization mass spectrometry. J Nat Prod 77:1351–1358

Fleming A (1929) On the antibacterial action of cultures of a *Penicillium*, with special reference to their use in the isolation of *B. influenzæ*. Br J Exp Pathol 10:226–236

Gamboa MA, Bayman P (2001) Communities of endophytic fungi in leaves of a tropical timber tree (*Guarea guidonia*: Meliaceae) 1. Biotropica 33:352–360

Gan L, Chen S, Wang Y, Watahiki A, Bohrer L, Sun Z, Wang Y, Huang H (2009) Inhibition of the androgen receptor as a novel mechanism of taxol chemotherapy in prostate cancer. Cancer Res 69:8386–8394

Ge HM, Yu ZG, Zhang J, Wu JH, Tan RX (2009) Bioactive alkaloids from endophytic *Aspergillus fumigatus*. J Nat Prod 72:753–755

Gimenez C, Cabrera R, Reina M, Gonzalez-Coloma A (2007) Fungal endophytes and their role in plant protection. Curr Org Chem 11:707–720

Giridharan P, Verekar SA, Kharna A, Mishra PD, Deshmukh SK (2012) Anticancer activity of sclerotiorin, isolated from an endophytic fungus *Cephalotheca faveolata*. Indian J Exp Biol 50(7):464–468

Gouda S, Das G, Sen SK, Shin H-S, Patra JK (2016) Endophytes: a treasure house of bioactive compounds of medicinal importance. Front Microbiol 7:1538

Goutam J, Sharma G, Tiwari VK, Mishra A, Kharwar RN, Ramaraj V, Koch B (2017) Isolation and characterization of "Terrein" an antimicrobial and antitumor compound from endophytic fungus *Aspergillus terreus* (JAS-2) associated from *Achyranthus aspera* Varanasi, India. Front Microbiol 8:1334

Gupta J, Sharma S (2020) Endophytic fungi: a new hope for drug discovery. In: Singh J, Gehlot P (eds) New and future developments in microbial biotechnology and bioengineering. Elsevier, Amsterdam, pp 39–49

Haque S, Yadav DK, Bisht SC, Yadav N, Singh V, Dubey KK, Jawed A, Wahid M, Dar SA (2019) Quorum sensing pathways in gram-positive and-negative bacteria: potential of their interruption in abating drug resistance. J Chemother 31:161–187

Hardoim PR, van Overbeek LS, Berg G, Pirttilä AM, Compant S, Campisano A, Döring M, Sessitsch A (2015) The hidden world within plants: ecological and evolutionary considerations for defining functioning of microbial endophytes. Microbiol Mol Biol Rev 79:293–320

Huang Z, Cai X, Shao C, She Z, Xia X, Chen Y, Yang J, Zhou S, Lin Y (2008) Chemistry and weak antimicrobial activities of phomopsins produced by mangrove endophytic fungus *Phomopsis* sp. ZSU-H76. Phytochemistry 69:1604–1608

Huang H, Liu T, Wu X, Guo J, Lan X, Zhu Q, Zheng X, Zhang K (2017) A new antibacterial chromone derivative from mangrove-derived fungus *Penicillium aculeatum* (no 9EB). Nat Prod Res 31:2593–2598

Huang L, Injac SG, Cui K, Braun F, Lin Q, Du Y, Zhang H, Kogiso M, Lindsay H, Zhao S (2018) Systems biology–based drug repositioning identifies digoxin as a potential therapy for groups 3 and 4 medulloblastoma. Sci Transl Med 10(464):eaat0150

Isah T (2015) Natural sources of Taxol. Br J Pharm Res 6:214–227

Isah T (2019) Stress and defense responses in plant secondary metabolites production. Biol Res 52: 39

Jennewein S, Croteau R (2001) Taxol: biosynthesis, molecular genetics, and biotechnological applications. Appl Microbiol Biotechnol 57:13–19

Jia M, Chen L, Xin H-L, Zheng C-J, Rahman K, Han T, Qin L-P (2016) A friendly relationship between endophytic fungi and medicinal plants: a systematic review. Front Microbiol 7:906

Kaaniche F, Hamed A, Abdel-Razek AS, Wibberg D, Abdissa N, El Euch IZ, Allouche N, Mellouli L, Shaaban M, Sewald N (2019) Bioactive secondary metabolites from new endophytic fungus *Curvularia* sp. isolated from *Rauwolfia macrophylla*. PLoS One 14:e0217627

Katoch M, Phull S, Vaid S, Singh S (2017) Diversity, phylogeny, anticancer and antimicrobial potential of fungal endophytes associated with *Monarda citriodora* L. BMC Microbiol 17:44

Katoch M, Singh D, Kapoor KK, Vishwakarma RA (2019) Trichoderma lixii (IIIM-B4), an endophyte of *Bacopa monnieri* L. producing peptaibols. BMC Microbiol 19:1–10

Kerssemakers JW, Munteanu EL, Laan L, Noetzel TL, Janson ME, Dogterom M (2006) Assembly dynamics of microtubules at molecular resolution. Nature 442:709–712

Khalil AMA, Abdelaziz AM, Khaleil MM, Hashem AH (2020) Fungal endophytes from leaves of *Avicennia marina* growing in semi-arid environment as a promising source for bioactive compounds. Lett Appl Microbiol 72:263–274

Khare E, Mishra J, Arora NK (2018) Multifaceted interactions between endophytes and plant: developments and prospects. Front Microbiol 9:2732

Kharkwal ASBMH, Kulkarni GT (2020) Therapeutic agents from endophytes harbored in Asian medicinal plants. Phytochem Rev 19:691–720

Khiralla A, Spina R, Varbanov M, Philippot S, Lemiere P, Slezack-Deschaumes S, André P, Mohamed I, Yagi SM, Laurain-Mattar D (2020) Evaluation of antiviral, antibacterial and

Antiproliferative activities of the endophytic fungus *Curvularia papendorfii*, and isolation of a new Polyhydroxyacid. Microorganisms 8:1353

Kjer J, Wray V, Edrada-Ebel R, Ebel R, Pretsch A, Lin W, Proksch P (2009) Xanalteric acids I and II and related phenolic compounds from an endophytic *Alternaria* sp. isolated from the mangrove plant *Sonneratia alba*. J Nat Prod 72:2053–2057

Kogel K-H, Franken P, Hückelhoven R (2006) Endophyte or parasite–what decides? Curr Opin Plant Biol 9:358–363

Kuriakose GC, Palem PP, Jayabaskaran C (2016) Fungal vincristine from Eutypella spp-CrP14 isolated from *Catharanthus roseus* induces apoptosis in human squamous carcinoma cell line-A431. BMC Complement Altern Med 16:1–8

Kusada H, Zhang Y, Tamaki H, Kimura N, Kamagata Y (2019) Novel N-acyl homoserine lactone-degrading bacteria isolated from penicillin-contaminated environments and their quorum-quenching activities. Front Microbiol 10:455

Kusari S, Lamshöft M, Zühlke S, Spiteller M (2008) An endophytic fungus from *Hypericum perforatum* that produces hypericin. J Nat Prod 71:159–162

Kusari S, Pandey SP, Spiteller M (2013) Untapped mutualistic paradigms linking host plant and endophytic fungal production of similar bioactive secondary metabolites. Phytochemistry 91: 81–87

Kusari S, Singh S, Jayabaskaran C (2014) Biotechnological potential of plant-associated endophytic fungi: hope versus hype. Trends Biotechnol 32:297–303

LaSarre B, Federle MJ (2013) Exploiting quorum sensing to confuse bacterial pathogens. Microbiol Mol Biol Rev 77:73–111

Lata R, Chowdhury S, Gond SK, White JF Jr (2018) Induction of abiotic stress tolerance in plants by endophytic microbes. Lett Appl Microbiol 66:268–276

Lazazzera BA, Grossman AD (1998) The ins and outs of peptide signaling. Trends Microbiol 6: 288–294

Li Y-H, Tian X-L (2016) Microbial interactions in biofilms: impacts on homeostasis and pathogenesis. IntechOpen, London

Li G, Kusari S, Lamshöft M, Schueffler A, Laatsch H, Spiteller M (2014) Antibacterial secondary metabolites from an endophytic fungus, *Eupenicillium* sp. LG41. J Nat Prod 77(11):2335–2341

Li SJ, Jiao FW, Li W, Zhang X, Yan W, Jiao RH (2020) Cytotoxic Xanthone derivatives from the mangrove-derived endophytic fungus *Peniophora incarnata* Z4. J Nat Prod 83:2976–2982

Liang X-A, Ma Y-M, Zhang H-C, Liu R (2016) A new helvolic acid derivative from an endophytic *Fusarium* sp. of *Ficus carica*. Nat Prod Res 30:2407–2412

Liao G, Wu P, Liu Z, Xue J, Li H, Wei X (2020) 2 H-pyranone and isocoumarin derivatives from the endophytic fungus *Pestalotiopsis microspora* SC3082 derived from *Scaevola taccada* (Gaertn). Roxb. Nat Prod Res:1–8

Lin Y-H, Xu J-L, Hu J, Wang L-H, Ong SL, Leadbetter JR, Zhang L-H (2003) Acyl-homoserine lactone acylase from *Ralstonia* strain XJ12B represents a novel and potent class of quorum-quenching enzymes. Mol Microbiol 47:849–860

Liu X, Dong M, Chen X, Jiang M, Lv X, Zhou J (2008) Antimicrobial activity of an endophytic *Xylaria* sp. YX-28 and identification of its antimicrobial compound 7-amino-4-methylcoumarin. Appl Microbiol Biotechnol 78:241–247

Liu S, Guo L, Che Y, Liu L (2013) Pestaloficiols Q–S from the plant endophytic fungus *Pestalotiopsis fici*. Fitoterapia 85:114–118

Liu H, Li X-M, Liu Y, Zhang P, Wang J-N, Wang B-G (2016) Chermesins A–D: meroterpenoids with a drimane-type spirosesquiterpene skeleton from the marine algal-derived endophytic fungus *Penicillium chermesinum* EN-480. J Nat Prod 79:806–811

Liu S-S, Jiang J-X, Huang R, Wang Y-T, Jiang B-G, Zheng K-X, Wu S-H (2019) A new antiviral 14-nordrimane sesquiterpenoid from an endophytic fungus *Phoma* sp. Phytochem Lett 29:75–78

Liu Z, Sun Y, Tang M, Sun P, Wang A, Hao Y, Wang Y, Pei Y (2020a) Trichodestruxins A–D cytotoxic Cyclodepsipeptides from the endophytic fungus *Trichoderma harzianum*. J Nat Prod 83(12):3635–3641

Liu H-X, Tan H-B, Chen Y-C, Li S-N, Li H-H, Zhang W-M (2020b) Cytotoxic triquinane-type sesquiterpenoids from the endophytic fungus *Cerrena* sp. A593. Nat Prod Res 34:2430–2436

Löwe J, Li H, Downing KH, Nogales E (2001) Refined structure of αβ-tubulin at 35 Å resolution 1 1Edited by I a Wilson. J Mol Biol 313:1045–1057

Lugtenberg BJ, Caradus JR, Johnson LJ (2016) Fungal endophytes for sustainable crop production. FEMS Microbiol Ecol 92(12)

Manganyi MC, Ateba CN (2020) Untapped potentials of endophytic Fungi: a review of novel bioactive compounds with biological applications. Microorganisms 8:1934

Martinez-Klimova E, Rodríguez-Peña K, Sánchez S (2017) Endophytes as sources of antibiotics. Biochem Pharmacol 134:1–17

Miller MB, Bassler BL (2001) Quorum sensing in bacteria. Annu Rev Microbiol 55:165–199

Miller AV, Hicks MA, Nakajima W, Richardson AC, Windle JJ, Harada H (2013) Paclitaxel-induced apoptosis is BAK-dependent, but BAX and BIM-independent in breast tumor. PLoS One 8:e60685

Mirjalili MH, Farzaneh M, Bonfill M, Rezadoost H, Ghassempour A (2012) Isolation and characterization of *Stemphylium sedicola* SBU-16 as a new endophytic taxol-producing fungus from *Taxus baccata* grown in Iran. FEMS Microbiol Lett 328:122–129

Mishra RC, Goel M, Barrow CJ, Deshmukh SK (2020) Endophytic Fungi - an untapped source of potential antioxidants. Curr Bioact Compd 16:944–964

Murugayah SA, Gerth ML (2019) Engineering quorum quenching enzymes: progress and perspectives. Biochem Soc Trans 47:793–800

Naik S, Shaanker RU, Ravikanth G, Dayanandan S (2019) How and why do endophytes produce plant secondary metabolites? Symbiosis 78:193–201

Nisa H, Kamili AN, Nawchoo IA, Shafi S, Shameem N, Bandh SA (2015) Fungal endophytes as prolific source of phytochemicals and other bioactive natural products: a review. Microb Pathog 82:50–59

Owen NL, Hundley N (2004) Endophytes—the chemical synthesizers inside plants. Sci Prog 87: 79–99

Oxford AE, Raistrick H, Simonart P (1939) Studies in the biochemistry of micro-organisms: griseofulvin, C17H17O6Cl, a metabolic product of *Penicillium griseo-fulvum* Dierckx. Biochem J 33:240–248

Padhi L, Mohanta YK, Panda SK (2013) Endophytic fungi with great promises: a review. J Adv Pharm Educ Res 3(3)

Pal PP, Shaik AB, Begum AS (2020) Prospective leads from endophytic Fungi for anti-inflammatory drug discovery. Planta Med 86(13/14):941–959

Palem PP, Kuriakose GC, Jayabaskaran C (2015) An endophytic fungus, *Talaromyces radicus*, isolated from *Catharanthus roseus*, produces vincristine and vinblastine, which induce apoptotic cell death. PLoS One 10:e0144476

Parasuraman P, Devadatha B, Sarma VV, Ranganathan S, Ampasala DR, Siddhardha B (2020) Anti-quorum sensing and antibiofilm activities of *Blastobotrys parvus* PPR3 against *Pseudomonas aeruginosa* PAO1. Microb Pathog 138:103811

Penesyan A, Nagy SS, Kjelleberg S, Gillings MR, Paulsen IT (2019) Rapid microevolution of biofilm cells in response to antibiotics. NPJ Biofilms Microbiomes 5:1–14

Pereira CB, de Sá NP, Borelli BM, Rosa CA, Barbeira PJS, Cota BB, Johann S (2016) Antifungal activity of eicosanoic acids isolated from the endophytic fungus *Mycosphaerella* sp. against *Cryptococcus neoformans* and *C. gattii*. Microb Pathog 100:205–212

Pereira DG, Rendeiro MM, Cortes VF, Barbosa LA, Quintas LEM (2019) Antagonistic anticancer effect of paclitaxel and digoxin combination. J Cell Biochem 120:13107–13114

Perveen I, Raza MA, Iqbal T, Naz I, Sehar S, Ahmed S (2017) Isolation of anticancer and antimicrobial metabolites from *Epicoccum nigrum*; endophyte of *Ferula sumbul*. Microb Pathog 110:214–224

Peyrat L-A, Eparvier V, Eydoux C, Guillemot J-C, Litaudon M, Stien D (2017) Betulinic acid, the first Lupane-type triterpenoid isolated from both a *Phomopsis* sp. and its host plant *Diospyros carbonaria*. Benoist. Chem Biodivers 14:e1600171

Phukhamsakda C, Macabeo APG, Yuyama KT, Hyde KD, Stadler M (2018) Biofilm inhibitory abscisic acid derivatives from the plant-associated *Dothideomycete* fungus, *Roussoella* sp. Molecules 23:2190

Powthong P, Jantrapanukorn B, Thongmee A, Suntornthiticharoen P (2018) Screening of antimicrobial activities of the endophytic Fungi isolated from *Sesbania grandiflora* (L.) Pers. J Agric Sci Technol 15:1513–1522

Qin Y, Liu X, Lin J, Huang J, Jiang X, Mo T, Xu Z, Li J, Yang R (2019) Two new phthalide derivatives from the endophytic fungus *Penicillium vulpinum* isolated from *Sophora tonkinensis*. Nat Prod Res 35(3):1–7

Qin Y-Y, Huang X-S, Liu X-B, Mo T-X, Xu Z-L, Li B-C, Qin X-Y, Li J, Schäberle TF, Yang R-Y (2020) Three new andrastin derivatives from the endophytic fungus *Penicillium vulpinum*. Nat Prod Res:1–9

Raj KG, Manikandan R, Arulvasu C, Pandi M (2015) Anti-proliferative effect of fungal taxol extracted from *Cladosporium oxysporum* against human pathogenic bacteria and human colon cancer cell line HCT 15. Spectrochim Acta A Mol Biomol Spectrosc 138:667–674

Rasmussen TB, Skindersoe ME, Bjarnsholt T, Phipps RK, Christensen KB, Jensen PO, Andersen JB, Koch B, Larsen TO, Hentzer M (2005) Identity and effects of quorum-sensing inhibitors produced by *Penicillium* species. Microbiology 151:1325–1340

Redman RS, Dunigan DD, Rodriguez RJ (2001) Fungal symbiosis from mutualism to parasitism: who controls the outcome, host or invader? New Phytol 151:705–716

Rho H, Hsieh M, Kandel SL, Cantillo J, Doty SL, Kim S-H (2018) Do endophytes promote growth of host plants under stress? A meta-analysis on plant stress mitigation by endophytes. Microb Ecol 75:407–418

Riga R, Happyana N, Hakim EH (2019) Chemical constituents of HF 12440 *Pestalotiopsis microspore*. J Appl Pharm Sci 9:108–124

Riga R, Happyana N, Hakim EH (2020) Sesquiterpenes produced by *Pestalotiopsis microspora* HF 12440 isolated from *Artocarpus heterophyllus*. Nat Prod Res 34:2229–2231

Rodriguez RJ, White JF, Arnold AE, Redman RS (2009) Fungal endophytes: diversity and functional roles. New Phytol 182:314–330

Rustamova N, Bozorov K, Efferth T, Yili A (2020) Novel secondary metabolites from endophytic fungi: synthesis and biological properties. Phytochem Rev:1–24

Sampangi-Ramaiah MH, Dey P, Jambagi S, Kumari MV, Oelmüller R, Nataraja KN, Ravishankar KV, Ravikanth G, Shaanker RU (2020) An endophyte from salt-adapted Pokkali rice confers salt-tolerance to a salt-sensitive rice variety and targets a unique pattern of genes in its new host. Sci Rep 10:1–14

Samson R, Houbraken J, Thrane U, Frisvard J, Andersen B (2010) Food and indoor fungi. CBS-KNAW Fungal Biodiversity Centre, Utrecht

Schulz B, Boyle C (2005) The endophytic continuum. Mycol Res 109:661–686

Schuster M, Joseph Sexton D, Diggle SP, Peter Greenberg E (2013) Acyl-homoserine lactone quorum sensing: from evolution to application. Annu Rev Microbiol 67:43–63

Shu-Chun LIU, Xin YE, Liang-Dong GUO, Ling LIU (2011) Cytotoxic isoprenylated epoxycyclohexanediols from the plant endophyte *Pestalotiopsis fici*. Chin J Nat Med 9:374–379

Stierle A, Strobel G, Stierle D (1993) Taxol and taxane production by *Taxomyces andreanae*, an endophytic fungus of Pacific yew. Science 260:214–216

Strobel G, Daisy B (2003) Bioprospecting for microbial endophytes and their natural products. Microbiol Mol Biol Rev 67:491–502

Subramani R, Jayaprakashvel M (2019) Bacterial quorum sensing: biofilm formation, survival behaviour and antibiotic resistance. In: Bramhachari PV (ed) Implication of quorum sensing and biofilm formation in medicine, agriculture and food industry. Springer, Singapore, pp 21–37

Sun Z-H, Liang F-L, Wu W, Chen Y-C, Pan Q-L, Li H-H, Ye W, Liu H-X, Li S-N, Tan G-H (2015) Guignardones P–S, new meroterpenoids from the endophytic fungus *Guignardia mangiferae* A348 derived from the medicinal plant smilax glabra. Molecules 20:22900–22907

Taware R, Abnave P, Patil D, Rajamohananan PR, Raja R, Soundarajan G, Kundu GC, Ahmad A (2014) Isolation, purification and characterization of Trichothecinol-A produced by endophytic fungus *Trichothecium* sp and its antifungal, anticancer and antimetastatic activities. Sustain Chem Process 2:8

Thoendel M, Kavanaugh JS, Flack CE, Horswill AR (2011) Peptide signaling in the *staphylococci*. Chem Rev 111:117–151

Uroz S, Heinonsalo J (2008) Degradation of N-acyl homoserine lactone quorum sensing signal molecules by forest root-associated fungi. FEMS Microbiol Ecol 65:271–278

Uroz S, Chhabra SR, Camara M, Williams P, Oger P, Dessaux Y (2005) N-Acylhomoserine lactone quorum-sensing molecules are modified and degraded by *Rhodococcus erythropolis* W2 by both amidolytic and novel oxidoreductase activities. Microbiology 151:3313–3322

Uzma F, Mohan CD, Hashem A, Konappa NM, Rangappa S, Kamath PV, Singh BP, Mudili V, Gupta VK, Siddaiah CN (2018) Endophytic fungi—alternative sources of cytotoxic compounds: a review. Front Pharmacol 9:309

Valente S, Cometto A, Piombo E, Meloni GR, Ballester A-R, González-Candelas L, Spadaro D (2020) Elaborated regulation of griseofulvin biosynthesis in *Penicillium griseofulvum* and its role on conidiation and virulence. Int J Food Microbiol 328:108687

Vasundhara M, Reddy MS, Kumar A (2019) Secondary metabolites from endophytic Fungi and their biological activities. In: Gupta VK (ed) New and future developments in microbial biotechnology and bioengineering: microbial secondary metabolites biochemistry and applications. Elsevier, Amsterdam, pp 237–258

Venieraki A, Dimou M, Katinakis P (2017) Endophytic fungi residing in medicinal plants have the ability to produce the same or similar pharmacologically active secondary metabolites as their hosts. Hell Plant Prot J 10:51–66

Verma VC, Lobkovsky E, Gange AC, Singh SK, Prakash S (2011) Piperine production by endophytic fungus *Periconia* sp. isolated from *Piper longum* L. J Antibiot (Tokyo) 64:427–431

Wang Y, Xu L, Ren W, Zhao D, Zhu Y, Wu X (2012) Bioactive metabolites from Chaetomium globosum L18, an endophytic fungus in the medicinal plant *Curcuma wenyujin*. Phytomedicine 19:364–368

Wang Q-X, Bao L, Yang X-L, Liu D-L, Guo H, Dai H-Q, Song F-H, Zhang L-X, Guo L-D, Li S-J, Liu H-W (2013) Ophiobolins P–T, five new cytotoxic and antibacterial sesterterpenes from the endolichenic fungus *Ulocladium* sp. Fitoterapia 90:220–227

Wang Z, Ma Z, Wang L, Tang C, Hu Z, Chou G, Li W (2015) Active anti-acetylcholinesterase component of secondary metabolites produced by the endophytic fungi of *Huperzia serrata*. Electron J Biotechnol 18:399–405

Wang Y, Ma Q, Zhang S, Liu H, Zhao B, Du B, Wang W, Lin P, Zhang Z, Zhong Y, Kong D (2020) Digoxin enhances the anticancer effect on non-small cell lung Cancer while reducing the cardiotoxicity of Adriamycin. Front Pharmacol 11:186

Wani MC, Taylor HL, Wall ME, Coggon P, McPhail AT (1971) Plant antitumor agents VI isolation and structure of taxol, a novel antileukemic and antitumor agent from *Taxus brevifolia*. J Am Chem Soc 93:2325–2327

Wani ZA, Ashraf N, Mohiuddin T, Riyaz-Ul-Hassan S (2015) Plant-endophyte symbiosis, an ecological perspective. Appl Microbiol Biotechnol 99:2955–2965

Williams P, Winzer K, Chan WC, Cámara M (2007) Look who's talking: communication and quorum sensing in the bacterial world. Philos Trans R Soc B Biol Sci 362:1119–1134

Wilson L (1975) Microtubules as drug receptors: pharmacological properties of microtubule protein. Ann N Y Acad Sci 253:213–231

Wong GC, O'Toole GA (2011) All together now: integrating biofilm research across disciplines. MRS Bull 36:339–342

Wu Z, Chen J, Zhang X, Chen Z, Li T, She Z, Ding W, Li C (2019) Four new Isocoumarins and a new natural tryptamine with antifungal activities from a mangrove endophytic fungus *Botryosphaeria ramosa* L29. Mar Drugs 17:88

Xiao J, Zhang Q, Gao Y-Q, Tang J-J, Zhang A-L, Gao J-M (2014) Secondary metabolites from the endophytic *Botryosphaeria dothidea* of *Melia azedarach* and their antifungal, antibacterial, antioxidant, and cytotoxic activities. J Agric Food Chem 62:3584–3590

Xu L-L, Chen H-L, Hai P, Gao Y, Xie C-D, Yang X-L, Abe I (2019) (+)-and (−)-Preuisolactone a: a pair of caged Norsesquiterpenoidal enantiomers with a Tricyclo [44 01, 602, 8] decane carbon skeleton from the endophytic fungus *Preussia isomera*. Org Lett 21:1078–1081

Yang C-PH, Horwitz SB (2017) Taxol®: the first microtubule stabilizing agent. Int J Mol Sci 18: 1733

Yang X, Wu P, Xue J, Li H, Wei X (2020) Cytochalasans from endophytic fungus *Diaporthe* sp. SC-J0138. Fitoterapia 145:104611

Yan-Hua Y, Jia-Wang M, Xiao-Li T (2020) Research Progress on the source, production, and anticancer mechanisms of paclitaxel. Chin J Nat Med 18:10–17

Yu N, He L, Liu N, Wang Y, Xu H, Liu D (2014) Antimicrobial action of an endophytic fungi from *Sophor flavescens* and structure identification of its active constituent. Biotechnol Biotechnol Equip 28:327–332

Yu F-X, Chen Y, Yang Y-H, Li G-H, Zhao P-J (2018) A new epipolythiodioxopiperazine with antibacterial and cytotoxic activities from the endophytic fungus *Chaetomium* sp. M336. Nat Prod Res 32:689–694

Zhang S-P, Huang R, Li F-F, Wei H-X, Fang X-W, Xie X-S, Lin D-G, Wu S-H, He J (2016) Antiviral anthraquinones and azaphilones produced by an endophytic fungus *Nigrospora* sp from *Aconitum carmichaeli*. Fitoterapia 112:85–89

Zhao J, Zhou L, Wang J, Shan T, Zhong L, Liu X, Gao X (2010) Endophytic fungi for producing bioactive compounds originally from their host plants. Curr Res Technol Educ Trop Appl Microbiol Microb Biotechnol 1:567–576

Zhao X, Yu Z, Ding T (2020) Quorum-sensing regulation of antimicrobial resistance in bacteria. Microorganisms 8:425

Zhou X, Zhu H, Liu L, Lin J, Tang K (2010) A review: recent advances and future prospects of taxol-producing endophytic fungi. Appl Microbiol Biotechnol 86:1707–1717

Zhou X, Lin X, Ma W, Fang W, Chen Z, Yang B, Liu Y (2014) A new aromatic amine from fungus *Pestalotiopsis vaccinii*. Phytochem Lett 7:35–37

Zhu L, Chen L (2019) Progress in research on paclitaxel and tumor immunotherapy. Cell Mol Biol Lett 24(1):1–11

Drought Stress and Sustainable Sugarcane Production

11

Varucha Misra, S. Solomon, A. K. Mall, Murtaza Abid, M. M. Abid Ali Khan, and Mohammad Israil Ansari

Abstract

Drought, limiting crop production and productivity factor, is known to hinder cane growth and photosynthetic ability along with a decrease in biomass production and yield. The occurrence of drought is a common problem around the world. Several cane-growing countries are facing this problem causing losses in cane yield by up to 60%. Recently, 40% production loss in cane due to drought conditions has been reported in India in the states of Maharashtra and Karnataka. The decline in the formation of tillers, rolling, folding, shredding and discoloration in leaves, decrease in the area of leaves with narrow width, etc., are some of the prominent effects of drought seen in sugarcane. In the tolerant cane varieties, there are some other chief characteristics that have been observed such as enhancement in malondialdehyde (MDA) and proline content and drop in photochemical efficiency. Identification of selective traits for developing drought-tolerant varieties through breeding is a way to face such conditions without leading to much loss in yield and sugar production. Other management practices, including the use of plant growth-promoting rhizobacteria can also help in

V. Misra · A. K. Mall
ICAR-Indian Institute of Sugarcane Research, Lucknow, Uttar Pradesh, India

S. Solomon
Chandra Shekhar Azad University of Agriculture and Technology, Kanpur, Uttar Pradesh, India

M. Abid
Department of Biochemistry, King George's Medical University, Lucknow, Uttar Pradesh, India

M. M. Abid Ali Khan (✉)
Department of Botany, Shia P. G. College, Lucknow, Uttar Pradesh, India

M. I. Ansari
Department of Botany, University of Lucknow, Lucknow, Uttar Pradesh, India

N. K. Arora, B. Bouizgarne (eds.), *Microbial BioTechnology for Sustainable Agriculture Volume 1*, Microorganisms for Sustainability 33,
https://doi.org/10.1007/978-981-16-4343-4_11

lessening the loss in cane productivity in agro-ecosystems prone to drought conditions.

Keywords

Sugarcane · Drought stress · Climate change · Sugarcane production · Yield loss

11.1 Introduction

Drought is one of the most devastating and constraining abiotic stresses which causes hindrance to the development and growth of plant and affects its productivity by causing severe losses (Saranga et al. 2002; Apel and Hirt 2004). These losses depend on genotype, duration of the drought, the intensity of the drought, and the growth phase at which drought strikes (Shao et al. 2009; Anjum et al. 2011). Drought is defined as a condition when there is a deficiency in the plant water status wherein alteration in plant growth and its development occurs, which induces changes in the physiological as well as biochemical processes of plant (Anjum et al. 2017; Zargar et al. 2011). In general, plant experience such stress only when water does not reach the roots for its uptake or when rate of transpiration is very high implying more water loss from the plant (Anjum et al. 2011). A number of factors determine drought in plants, viz., reduction in water status of the cells, reduction in turgor pressure, reduction in growth and development of cells, enhancement in stomatal closure, and wilting (Lisar et al. 2012). When drought conditions are severe, photosynthesis may be either reduced or stopped leading to aggravation of other metabolic processes (Cornic and Massacci 1996; Pinheiro and Chaves 2011). In such a condition, loss of turgidity in cells is seen that causes an increase in the concentration of solutes in the cytosol and extracellular matrix of the plant cells resulting in a reduction of enlargement process of cells.

Sugarcane, like all other crops also shows a prominent drought effect. It is well known that a fully ripened sugarcane crop undergoes four stages: germination, tillering, grand growth, and ripening wherein larger water uptake has been reported in tillering and grand growth phase (Ramesh 2000), and these two critical stages are known to be used for identifying and distinguishing tolerant cane varieties for drought conditions (Endres et al. 2010). Taiz and Zeiger (2006) had revealed that for uptake of carbon dioxide by plants, the opening of stomata is a necessary condition; however, this needs a high transpiration rate which might be hindered due to less water content (Molina 2002). Sugarcane belonging to C_4 family may have the capability to facilitate its growth and development in such a situation with the help of capability to reduce the photorespiration and water loss. Different physiological changes have been observed in sugarcane crop when exposed to drought or water-deficient conditions (Fig. 11.1). It is well known that roots are the first part of the plant which detect the water-deficient conditions. Under such a situation, sugarcane crop senses the hydraulic perturbation which activates roots to send signals to plant for changing the stomatal processes under drought conditions (Buckley 2005).

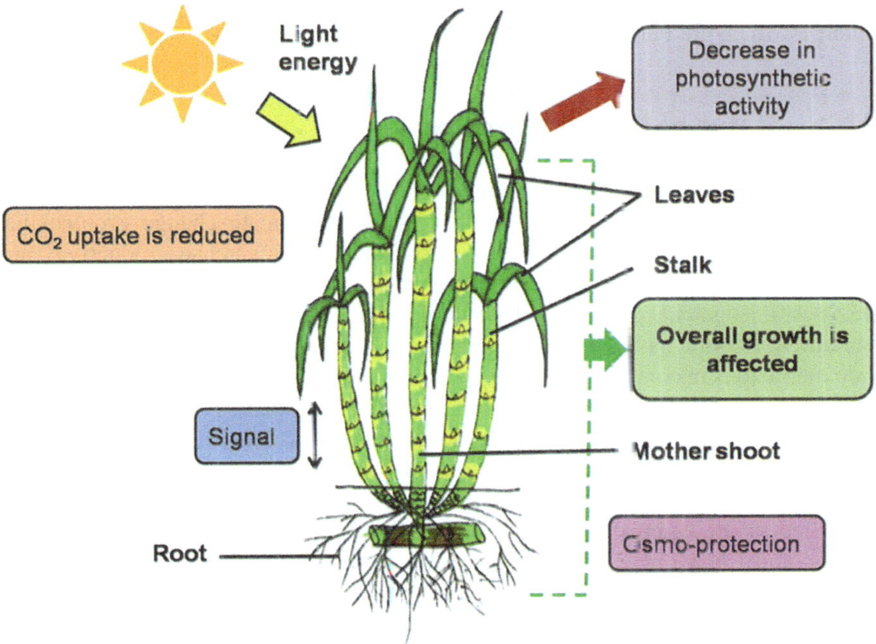

Fig. 11.1 Physiological response of sugarcane under drought stress conditions

But this crop has the capability to tolerate such a condition by causing accumulation of osmoprotectants, such as glycine betaine, proline, etc. (Wahid 2007).

Several studies have illustrated the changes that sugarcane undergoes when faced with drought stress (Misra et al. 2020a, b). Activation of a large number of perturbations starting from initial phases of growth to its maturity status, disturbances in the water status of the crop, variations in transport, and uptake of ions have been reported under drought conditions (Lutts et al. 1996; Bajji et al. 2000). Reduction in tillering, rolling of leaves, folding, shredding and leaf reflectance, leaf coloration, coating of wax on sugarcane stalks, reduction in leaf area, and narrower leaves are the alterations which have been reported in leaf morphology under such conditions (Shrivastava and Srivastava 2006). Besides, decrease in membrane lipid per oxidation; increase in plasma permeability, mainly in drought-sensitive cultivars; reduced transpiration and respiration rates; increase in ABA and ethylene; and increase in proline (de Oliveira et al. 2018) and malondialdehyde (MDA) content are also being observed in drought stress (Shrivastava and Srivastava 2006). Sanji et al. (2002) observed a decrease in photochemical efficiency (F_v/F_m) as well as potential activity (F_v/F_o) of Photosystem II in canes grown under drought conditions, while higher utilization rate of soil water has also been reported in such a case in this crop (Li-huan et al. 2010). Furthermore, less production of fresh and dry matter, delay in tillering, and short internodal length are some more morphological changes which are being observed due to drought stress (Zlatev and Lidon 2012).

11.2 Status of Drought Occurrence

Drought is an important crop-limiting factor around the globe. The occurrence of drought is a common feature nowadays since the climate change problem has emerged (Shrivastava et al. 2016). Drought influences sugarcane productivity and sugar production in many cane-growing parts of the world. Brazil, the highest producer of sugarcane, faces the problem of drought stress due to seasonal variations in rainfall. Reports have shown that there was a fall in cane production to 545.9 million metric tons in comparison to the previous years having an average production of 597.1 million tons (Lewis 2014). In the year 2012, sugarcane production of 52 million tons had been reported from the northeast part of Brazil illustrating a fall of 10–15% as compared to the previous year. This had resulted in the downfall of sugar output (10–15%) from the previous year. In 2009, due to the dry weather that prevailed in the country, there was a sharp decline in productivity to 18.17% in the month of November. In 2010, a drop in productivity was also reported due to water stress. This fall has been reported to be 7.26% from growth to cutting phase in comparison to the year 2009. In Cuba, although the country faces adequate rainfall throughout the year, the plant faces water stress conditions in some or other parts of the year. In Myanmar, sugarcane production is also reported being limited by this abiotic stress that generally occurs during the months of March and April. Another sugar-producing country, Pakistan, has also shown a decrease in sugarcane production due to less rainfall and longer tenure of drought stress (Anonymous 2002). An uneven pattern of rainfall in the Philippines also causes low cane production in this country (Smith 1978). Seasonal pattern of rainfall has also been seen in China due to which lower cane yield was reported (Smith 1978); however, southern regions of China, particularly Guangxi province, had more prominent problems of drought for cane growth and production (Li and Solomon 2003). About 30 provinces of Thailand have been regarded as disaster zones on basis of the occurrence of severe drought causing huge damage in the province of Khon Kaen (National News Agency of Thailand 2014). In India, 23 droughts have been faced by this crop from 1891 to 2004, and recently in 2015 and 2016, thus a frequent occurrence of drought has been reported (Anand 2004; Shrivastava and Srivastava 2006; Shrivastava et al. 2016). Zarabi (2004) had shown that drought causes losses in productivity and yield that was expected to affect productivity by 10%. In 2015, just 88% of monsoon rain occurred that caused drought in half of the world, while the other half showed excess rainfall leading to waterlogging conditions. Maharashtra, the highest cane-producing state of India, faced rigorous drought over the past 3 years, especially Marathwada and Vidarbha regions. Kelkar-Khambete (2014) had revealed that due to deficiency in rainfall and fall of groundwater level, the productivity and growth of agricultural crops were hampered to a great extent. Ram (2017) had shown that in 2014–2015, sugarcane covers an area of 5.06 million ha, but in 2015–2016, it decreased to

4.297 mha, while in 2016–2017, it further decreased to 4.503 mha due to continuous drought in Tamil Nadu, Maharashtra, Karnataka, and other states. Shrivastava and Srivastava (2006) had shown the tenure of drought occurring in various states of cane-growing area of India. Assam showed recurrence of deficient rainfall once in 15 years, while states like West Bengal, M.P., Coastal A.P., Maharashtra, Konkan, Kerala, Bihar, Orissa, South interior Mysore, Uttar Pradesh, Vidarbha Gujarat, Eastern Rajasthan, Tamil Nadu, Kashmir, Rayalaseema, and Telangana of India showed once in 5 years. However, western Rajasthan of India showed once in 2.5 years.

11.3 Sugarcane Adaptations in Drought Stress

Sugarcane crop, just like any other crop, needs to survive drought conditions by certain adaptive mechanisms such as escape or avoidance or tolerance (Levitt 1980; Chaves et al. 2003) or a combination of these (Pimentel 2004; Bueno et al. 2006). In the escape mechanism, success will only be attained when the plant escapes the critical stages of growth before the onset of stress (Campos et al. 2004). In avoidance mechanism, the plant needs to avoid dehydration process by maintaining high water status during the stress period (Blum 2005; Kooyers 2015). In such a strategy, water loss is either minimized or maximized. The minimization of water loss involves closure of stomata, reduction in leaf area, and senescence of old leaves, while maximization involves the proliferation of roots to reach the level of water wherever available (Lopes and Reynolds 2011). Such an adaptive strategy is generally taken up by the plant in the case when severe water stress conditions occur (Ferreira et al. 2017). In the tolerance adaptive mechanism, the plant needs to maintain its proper functioning even under stress conditions (Levitt 1980; Blum 2005) or regain the recovery of water status in plants after stress is released (Lopes and Reynolds 2011). For doing so, precise alteration in osmotic potential and rigidity in cells of either cell wall or other small cells are required by the crop (Chaves et al. 2003). This strategy is favorable for the plant when stress conditions are either mild or moderate as it helps in maintaining the growth of the plant in such a situation (Ferreira et al. 2017). Besides, it has been bestowed with a number of natural endowments that help in the survival of this crop under abiotic stresses particularly drought. Compensatory ability, allopolyploidy, carbon sequestration, and tolerance to abiotic stress attained by crossing the varieties with *Saccharum spontaneum* are some of these. The compensatory ability is attributed by the compensatory physiologic continuum (tillering potential, development of two types of roots (set and shoot roots), leaves and vast ratooning potential) owed to sugarcane crop. These special properties of sugarcane help the plant in sustaining in abiotic stress conditions (Shrivastava et al. 2015, 2016, 2017). Besides, the requirement of higher favorable temperatures for cane growth, development, and physiological process, with the exception in process of sucrose storage, is another important ability of sugarcane that helps in survival under high temperatures which makes it different from other C_3 crops. Furthermore, as a response of drought stress, several adaptations have been reported for its survival. These are as follows: possessing slender cane stalks (Singh and Ramakrishnan 1977); proper root system of canes; less shoot root ratio; maintenance of moisture content in leaf sheath having greater than 76% (Shrivastava et al. 2003); higher production of osmolytes such as trehalose, free proline, glycine betaine, etc. (Ho et al. 1988; Guimarães et al. 2008; Shrivastava et al. 2016); higher release of drought-induced electrolytes (Viqueira et al. 1985); leaves of the cane possessing higher N and K content while in stalks, higher P and K content; and an elevated amount of hydrolysis of starch in stalks when under water-deficient conditions (Sales et al. 2012).

11.4 Traits for Identifying Drought-Tolerant Cane Varieties

Exposure of sugarcane to drought conditions causes several alterations in physiological and biochemical attributes. Some alterations in attributes cause growth retardation and loss in crop yield, while some attributes help in maintaining the physiological processes as under normal conditions. On this basis, tolerant and susceptible cane varieties for drought conditions are differentiated and identified. Accumulation of proline and low production of glycine betaine in such a situation are common physiological attributes seen in cane as well as in other plants as a tolerant character for drought conditions (Verbruggen and Hermans 2008; Chen and Murata 2011; Giri 2011). Direct association of higher tillering, smaller leaf area, and erect narrow leaves with drought resistance have been reported in variety CP 43/33. Besides, deeper supra stomatic activities, higher amount of sclerenchymatous fibers, thicker external epidermal cell wall and cuticle, copious parenchyma cells, and lower number of stomata have been considered to be the indicators for drought resistance in sugarcane plant (Bajwa and Ahmad 2002). Furthermore, some cane varieties were revealed to avoid drought conditions by adopting the avoidance mechanism, for example, in variety BF-162, rolling of leaves was observed, but in Co 1148 the leaf curling was slight (Venkataramana et al. 1984). Also, studies have revealed the various indicators for tolerance to such conditions in sugarcane. Silva et al. (2008) had shown the association of increased production and yield with an increase in stalk number and stalk height along with an increase in weight under drought conditions. Several studies have reported that there are also indirect means of selection criteria for selecting tolerant cane varieties. These are leaf chlorophyll index, leaf and canopy temperature, stomatal conductance, canopy conductance, rate of photosynthesis, and transpiration (Silva et al. 2007; Endres et al. 2010; da Silva et al. 2012; Basnayake et al. 2015). Thomas and Howarth (2000) and Blum (2005) revealed that higher green leaf area under drought stress is another trait that is considered for the selection of drought-tolerant varieties. Besides, leaf photochemical capacity is also one selecting trait for drought tolerance (Guo and Tan 2015). Furthermore, an important selection criterion is the root characteristics of sugarcane crop which generally a crop adapts for survival under drought condition (Songsri et al. 2008; Wang et al. 2009). In sugarcane crop, the development of deep large roots, as well as higher root length density, may be considered as a selection criterion for drought stress (Tardieu et al. 1992; Smith et al. 2005; Blum 2005; Tardieu 2012). Also, thickness of main or tertiary roots under drought conditions is another important selection factor for a tolerant or resistant variety. The study had shown that there are certain cane hybrids obtained by crossing sugarcane with sorghum and Co 213 possessing deep roots which help in withstanding such conditions (Shrivastava and Srivastava 2006). Singh and Ramakrishnan (1977) had also revealed root-pulling resistance as another trait for drought tolerance whose value ranged from 285.3 to 331.3 kg/plant. Besides, relative turgidity of the cells in cane varieties was decreased rapidly under such a situation as observed in Co 285 and Co 312 due to higher evaporation leaf surface area. Naidu and Bhagyalakshmi (1973 and 1987) had shown that almost 90% of stomata get closed on the fifth day, while in case of drought-susceptible cane varieties, relative turgidity decreases less rapidly (76% in *Badilla* and 0% in *Kaludai Boothan*) after the drought conditions ended, while only 50% of stomata get closed on the seventh day. Sang-Wu et al. (1984) had reported that decrease in the rate of respiration can be a distinguishing feature for tolerant and susceptible cane varieties. Proline content in water-deficient conditions had also been considered as a physiological parameter for identifying drought resistance in sugarcane (Chen et al. 1995). Furthermore, there are certain attributes such as trehalose and glycine betaine in sugarcane crop that helps in tolerating not only the drought stress but also other multiple abiotic stresses (Shrivastava and Srivastava 2012; Shrivastava et al. 2015, 2017). Some of the indigenous varieties developed in India had tolerance characteristics toward drought were Hemja and Katha which were known for adaption under drought stress (Srinivasan 2004).

Development of drought-tolerant varieties has always been an area of attraction for breeders; however, the problem of climate change has made the cane breeders work even more on it. Various sugarcane drought-tolerant varieties have been developed, like Serato, F 134, CP 72/2083, ISD 37, etc. In India, All India Co-ordinated Research Project on Sugarcane (AICRP(s)) works on identification of drought-tolerant varieties, and Central Varietal Release Committee (CVRC) releases some of the identified varieties for commercial purposes, such as Co 8371, Co 87,263, Co 98,014, CoLk 94,184, Co 0218, Co 99,004, etc. (Sinha 2015).

11.5 Response of Sugarcane Crop Under Drought Conditions

Drought is the most important problem of concern for the production of any crop. The emerging problem of climate change has added fuel to fire as this has caused the frequent occurrence of droughts, thereby causing huge losses in agriculture (Boyer 1982; Shrivastava et al. 2016; Arora 2019; Mall et al. 2020). Hall et al. (1961) and Campos et al. (2004) had revealed that crop responses to drought stress are dependent on the species and more frequently on the genotype of the crop. Also, there are other factors such as duration and severity of drought as well as age and phase of crop growth at which drought occurred, on which the response of crop depends (Yang et al. 1993; Chimenti et al. 2006; Pinheiro and Chaves 2011). Even if the time of drought exposure to sugarcane crop is very less, causing no visible symptoms, cells can still lose turgidity and growth of cane may retard to some extent. Cane yield, quality, and sugar recovery are the main parameters of concern under drought stress conditions. Even a small-scale drought may cause a reduction in yield if it occurs at critical stages of growth (Shrivastava et al. 2016). Drought in terms of soil implies that there is less moisture content in it, and this causes yield loss of 0.39 mt/day when there was a loss in moisture content for 75–90 days, but there was a loss in cane yield of 0.256 mt/day after 90–105 days of such conditions (Shrivastava and Srivastava 2006). The experiments conducted on effect of drought on plant and ratoon canes Misra et al. (2016) revealed that post-harvest sucrose losses increased after 240 h of harvest when sugarcane was grown under drought conditions in comparison to the ones grown under normal conditions.

In sugarcane crop, alterations in the morphological and physiological response under drought stress conditions were seen (Mall et al. 2019). Under drought stress condition, the thickness of the sugarcane plant was revealed to be dependent on the magnitude of drought. Cane varieties possessing thinner stalks and mother shoot showed higher green leaves, tillers, and rate of leaf emergence and increased leaf area, in comparison to ones having relatively thicker stalks and taller mother shoots. Higher rate of tillering, leaf number, and ratio of tiller/root weight were also related to drought (Shrivastava and Srivastava 2006). Higher tiller production was seen when drought occurred during the early stages of the growth phase of sugarcane, particularly in several parts of subtropical region of India, which implies that canes having thinner stalks may prosper well than the thicker ones (Lal et al. 1968). Misra et al. (2020a) showed a comparison of the morphological differences in sugarcane under drought and waterlogging stress conditions. They revealed that number of internodes and root weight decrease under drought conditions, but the average internodal length increases. Singh and Reddy (1980) showed that under drought stress there was no significant difference in leaf area and leaf weight; however, there was a decrease in leaf area index. Physiological processes such as leaf temperature, photochemical electron transport, stomatal conductance, rate of transpiration, photosynthesis, respiration, photo-assimilate partitioning were reported to be altered under drought conditions (Negi et al. 1971). The severity of drought also causes effect on the water content of cane leaves (Uprety et al. 1999). From mild to severe drought conditions, Chen et al. (1995) had revealed that water content of the leaves (free water content and relative water content) deteriorated progressively. The situation was observed to be worst in drought-sensitive cane cultivars such as Co 6304 and Yuetang 57–423 as compared to drought-tolerant ones like Guitang 11 and NCo 310 (Ru Kai et al. 1996). Li-Huan et al. (2010) had revealed that water potential had a dominating effect on drought resistance. Besides, a reduction in photosynthesis rate was also seen under such a condition due to the closure of stomata. Reduction in nitrate reductase activities has also been reported under such a situation (Sairam 1994). Venkataramana et al. (1986) had revealed that during the formative phase of cane growth, an increase in stomatal diffusive resistance (K_s) on both sides (abaxial and adaxial) of the sugarcane leaf has been reported when the stress ranges from 60 to 150 days. Furthermore, uptake of nutrients in such a condition has been reported to be reduced due to lower extension of roots in the soil. Also, several metabolic processes were hampered due to reduction in growth hormones such as cytokines, gibberellins, etc. Besides, an increase in accumulation of malondialdehyde (MDA) and permeability of plasma membrane (PMP) was seen during such a situation. Other physiological indexes such as PS II photochemical efficiency (F_v/F_m), PS II potential efficiency (F_v/F_o), variable fluorescence quenching rate ($\Delta F_v/F_o$), and efficiency of potential quantum conversion ($\Delta F_v/F_t$) were decreased in such a situation. A positive relationship between PMP and MDA and the negative relationship between PMP and F_v/F_m as well as between F_v/F_o and $\Delta F_v/F_o$ had been revealed (Gao et al. 2006). Furthermore, the high fiber content in ratoon canes was also seen under partially or unirrigated condition (Zhang et al. 1996).

11.5.1 Management Strategies for Sugarcane Under Drought Conditions

For minimizing or overcoming the drought effect on sugarcane crop, management measures have been recommended. The management strategies used for coping with drought condition in case of sugarcane are listed below.

11.5.1.1 Cultural Practices as Management Strategy for Drought Stress

As it is known that under drought condition soil moisture content (SMC) gets deteriorated and so as to avoid further loss in SMC, lower leaves of cane stalks, which get dried in standing canes in fields may be removed to be used in the inter-row spaces available between the clumps. Earthing up is another way which may be performed in plant and ratoon crop, irrespective of autumn and spring season planting. Also, for checking loss of moisture at surface level, hoeing as cultural practice in the inter-row spacing of sugarcane may be performed.

11.5.1.2 Chemical Application for Mitigating Drought Stress Effect

Exogenous application of nitric oxide enhances the tolerance power in canes for facing the drought (Silveira et al. 2017). Misra et al. (2020b) showed that usage of formulation of benzalkonium chloride and sodium metasilicate was effective in managing the post-harvest sucrose losses in sugarcane under drought stress condition. In drought condition, canes have often been infected with insects and pests, and so as to control it, usage of chlorpyrifos 20EC @ 5 lha^{-1} in 1200 L water as spray in cane rows has been preferred for controlling termites, while for black bug and rodents, application of endosulfan (35EC @670 mL/ha in 1000 L water) and pre-baiting followed by zinc phosphide baiting thereafter bromoiodide cake baiting, respectively, have been known to be effective (Shrivastava and Srivastava 2006).

11.5.1.3 Use of Plant Growth-Promoting Rhizobacteria (PGPR) in Mitigating Drought Stress in Sugarcane

The application of PGPR in sugarcane has also been reported as another effective management strategy for coping up with drought stress (Pereira et al. 2019). PGPR are important to plants due to several beneficial effects. The production of antimicrobials, phytohormones, and siderophores and rise in nutrient levels are the reasons behind their beneficial impact (Someya and Akutsu 2005; Ribeiro and Cardoso 2012; Bal et al. 2013; Paungfoo-Lonhienne et al. 2014; Verma et al. 2019). In drought stress conditions, these rhizobacteria have the capability of increasing the tolerance power of plants by modulating nutrient uptake and phytohormone levels and improving water uptake by secreting metabolites such as exopolysaccharides (Arora et al. 2020). Change in hormonal content helps in rising the capability of water absorption in sugarcane (Aroca and Ruiz-Lozano 2009). Kour et al. (2020) reported phosphate-solubilizing drought-tolerant rhizobacteria with ability to enhance the productivity under drought stress. Pereira et al. (2019) showed that microbial communities present in rhizosphere positively affected sugarcane under drought stress conditions with role of soil bacteria such as *Bacillus*, *Falsibacillus*, *Streptomyces*, *Arthrobacter*, and *Microbacterium*. These bacteria help in coping up with drought stress conditions in sugarcanes. Chandra et al. (2018) reported that two rhizobacterial strains (BSSC11 and BMSE7) caused an improvement in cane growth and an increase in cane tolerance capability under drought stress conditions. Furthermore, application of *Azospirillum* under drought stress conditions (in sugarcane) increases the nitrogen, phosphorus, and potassium content in plants. The study further showed that this interaction varies from genotype to genotype under drought stress conditions. The cultivar M 1176/77 showed a 15% rise in shoot growth and 75% in root dry mass under drought stress when inoculated by strain of *Azospirillum* (Moutia et al. 2010). Chandra et al. (2021) reported that *Bacillus* BS87, a strain from sugarcane rhizosphere, is a preferred option as a bio-fertilizer that can aid to augment the cane growth of the plant and ratoon crop particularly during abiotic stress conditions.

11.6 Future Prospects

Though sugarcane is far superior in tolerating drought conditions through its natural endowments, still there is a need for evaluation of cane germplasm under such a situation. Furthermore, the development of varieties by using *Saccharum* or other related genera having peculiar characteristic (physiological and biochemical) for tolerance toward drought or multiple abiotic stress conditions through breeding is also another aspect to be worked upon. Molecular work on this aspect will help in identifying the genes responsible for helping in tolerating such conditions and also identifying the tolerant cane varieties. Although the first genetically modified (GM) sugarcane has been developed in Indonesia using bet A gene as a tolerant variety to drought conditions, its field performance is still lacking which needs to be worked upon. Also, more transgenic varieties are needed to be developed which could help in managing the problem of low production and productivity in drought conditions. Scanty information is available on application of PGPR and their mechanisms in sugarcane under drought stress conditions which needs to be further investigated. New strains of PGPR tested under drought stress condition need to be inculcated in bio-fertilizers for further investigation under drought-prone ecosystems.

11.7 Conclusion

Sugarcane is a major producer of sugar and ethanol and has also gained much more importance in the present scenario. The natural abilities (low photorespiration, reduced water loss, high temperature tolerance, and several physiological adaptations) possessed by this crop are the characteristic features helping under drought stress conditions. But still, the present scenario of unpredictable weather as a result of climate change is causing cane production more prone to drought stress. Heavy losses have also been reported in sugar production under drought conditions from different parts of the world. However, management of drought stress in sugarcane by various strategies are now known, and these need to be researched upon further so as to tackle the impact of drought and climate change for this very important crop.

References

Anand A (2004) Drought-like situation in 122 districts. The Statesman, New Delhi. Accessed 17 July

Anjum SA, Xie XY, Wang LC, Saleem MF, Man C, Lei W (2011) Morphological, physiological and biochemical responses of plants to drought stress. Afr J Agric Res 6(9):2026–2032

Anjum SA, Ashraf U, Tanveer M, Khan I, Hussain S, Zohaib A, Abbas F, Saleem MF, Wang L (2017) Drought tolerance in three maize cultivars is related to differential osmolyte accumulation, antioxidant defense system, and oxidative damage. Front Plant Sci 8:69

Anonymous (2002) Pakistan News Channel, Hyderabad. Accessed 24 Sept

Apel K, Hirt H (2004) Reactive oxygen species: metabolism, oxidative stress, and signal transduction. Annu Rev Plant Biol 55:373–399

Aroca R, Ruiz-Lozano J (2009) Sustainable agriculture reviews. In: Lichtfouse E (ed) Climate change, intercropping, pest control and beneficial microorganisms, sustainable agriculture reviews, vol 2. Springer, Berlin, pp 229–232

Arora NK (2019) Impact of climate change on agriculture production and its sustainable solutions. Environ Sustain 2(2):95–96

Arora NK, Fatima T, Mishra J, Mishra I, Verma S, Verma R, Verma M, Bhattacharya A, Verma P, Mishra P, Bharti C (2020) Halo-tolerant plant growth promoting rhizobacteria for improving productivity and remediation of saline soils. J Adv Res 26:69–82

Bajji M, Lutts S, Kinet JM (2000) Physiological changes after exposure to and recovery from polyethylene glycol-induced water deficit in callus cultures issued from durum wheat (*Triticum durum* Desf.) cultivars differing in drought resistance. J Plant Physiol 156(1):75–83

Bajwa MA, Ahmad N (2002) Studies on moisture economy and drought tolerance of different cane varieties. Pak Sugar J 17(3):18–21

Bal HB, Da S, Dangar TK, Adhya TK (2013) ACC deaminase and IAA producing growth promoting bacteria from the rhizosphere soil of tropical rice plants. J Basic Microbiol 53(12): 972–984

Basnayake J, Jackson PA, Inman Bamber NG, Lakshmanan P (2015) Sugarcane for water limited environments: variation in stomatal conductance and its genetic correlation with crop productivity. J Exp Bot 66(13):3945–3958

Blum A (2005) Drought resistance, water-use efficiency, and yield potential—are they compatible, dissonant, or mutually exclusive? Crop Pasture Sci 56(11):1159–1168

Boyer JS (1982) Plant productivity and environment. Science 218(4571):443–448

Buckley TN (2005) The control of stomata by water balance. New Phytol 168(2):275–292

Bueno LCS, Mendes ANG, Carvalho SP (2006) Melhoramento Genético de Plantas: Princípios e Procedimentos. UFLA, pp 213–219

Campos H, Cooper A, Habben JE, Edmeades GO, Schussler JR (2004) Improving drought tolerance in maize: a view from industry. Field Crop Res 90(1):19–34

Chandra P, Tripathi P, Chandra A (2018) Isolation and molecular characterization of plant growth-promoting Bacillus spp. and their impact on sugarcane (*Saccharum spp.* hybrids) growth and tolerance towards drought stress. Acta Physiol Plant 40(11):1–15

Chandra A, Chandra P, Tripathi P (2021) Whole genome sequence insight of two plant-growth promoting bacteria (*B. subtilis* BS87 and *B. meagterium* BM89) isolated and characterized from sugarcane rhizosphere depicting better crop yield potentiality. Microbiol Res 247:126733

Chaves MM, Maroco JP, Pereira JS (2003) Understanding plant responses to drought—from genes to the whole plant. Funct Plant Biol 30(3):239–264

Chen TH, Murata N (2011) Glycinebetaine protects plants against abiotic stress: mechanisms and biotechnological applications. Plant Cell Environ 34(1):1–20

Chen RK, Zhang MQ, Lu YB (1995) Studies on the effect of water deficits on physiology in sugarcane (*Saccharum officinarum* L.). J 甘蔗 1

Chimenti CA, Marcantonio M, Hall AJ (2006) Divergent selection for osmotic adjustment results in improved drought tolerance in maize (*Zea mays* L.) in both early growth and flowering phases. Field Crop Res 95(2–3):305–315

Cornic G, Massacci A (1996) Leaf photosynthesis under drought stress. In: Baker NR (ed) Photosynthesis and the environment. Springer, Dordrecht, pp 347–366

da Silva PP, Soares L, da Costa JG, da Silva VL, de Andrade JCF, Gonçalves ER, Neto CER (2012) Path analysis for selection of drought tolerant sugarcane genotypes through physiological components. Ind Crop Prod 37(1):11–19

de Oliveira LAR, Cardoso MN, de Oliveira ACA, Machado CDA, Cardoso BT, MUNIZ ADS, Ledo ADS (2018) Effects of in vitro drought stress on growth, proline accumulation and antioxidant defense in sugarcane. J Agric Sci 10(5):135–149

Endres L, Silva JV, Ferreira VM, Barbosa GVS (2010) Photosynthesis and water relations in Brazilian sugarcane. Open Agric J 4(1):31–37

Ferreira THS, Tsunada MS, Bassi D, Araujo P, Mattiello L, Guidelli GV, Righetto GL, Goncalves VR, Lakshmanan P, Menossi M (2017) Sugarcane water stress tolerance mechanisms and its implications on developing biotechnology solutions. Front Plant Sci 8:1077

Gao SJ, Luo J, Zhang H, Chen RK, Lin Y (2006) Physiological and biochemical indexes of drought resistance of sugarcane (*Saccharum spp.*). Chin J Appl Ecol 17(6):1051–1054

Giri J (2011) Glycine betaine and abiotic stress tolerance in plants. Plant Signal Behav 6(11): 1746–1751

Guimarães ER, Mutton MA, Mutton MJR, Ferro MIT, Ravaneli GC, Silva JAD (2008) Free proline accumulation in sugarcane under water restriction and spittlebug infestation. Sci Agric 65(66): 628–633

Guo Y, Tan J (2015) Recent advances in the application of chlorophyll a fluorescence from photosystem II. Photochem Photobiol 91(1):1–14

Hall AS, Holowenko AR, Laughlin HG (1961) Theory and problems of machine design; Schaums Outline Series. McGraw-Hill, New York

Ho ST, Liu WF, Chen YH, Chen WS (1988) Relationship between free proline accumulation in leaves and yields of sugarcane varieties under water stressed conditions. Rep Taiwan Sugar Res Inst

Kelkar-Khambete A (2014) Drought in Maharashtra: lack of management or vagaries of climate change? http://www.indiawaterportal.org. Accessed 26 Sept 2015

Kooyers NJ (2015) The evolution of drought escape and avoidance in natural herbaceous populations. Plant Sci 234:155–162

Kour D, Rana KL, Yadav AN, Sheikh I, Kumar V, Dhaliwal HS, Saxena AK (2020) Amelioration of drought stress in foxtail millet (*Setaria italica* L.) by P-solubilizing drought-tolerant microbes with multifarious plant growth promoting attributes. Environ Sustain 3:23–34

Lal KN, Mehrotra DN, Tandon JN (1968) Growth behaviour, root extension and juice characters of sugarcane in relation to nutrient deficiency and drought resistance. Indian J Agric Sci 38(5): 790–804

Levitt J (1980) Responses of plants to environmental stresses. Water, radiation, salt, and other stresses, vol 2. Academic Press, New York, NY

Lewis JT (2014) Fire and drought scar Brazilian sugar crop. Wall Street J. https://www.wsj.com/articles/fire-and-drought-scar-brazilian-sugar-crop-1411487285

Li YR, Solomon S (2003) Ethephon: a versatile growth regulator for sugar cane industry. Sugar Tech 5(4):213–223

Li-huan Z, Yongxiu X, Litao Y (2010) Effects of water stress on leaf water and chlorophyll fluorescence parameters of sugarcane seedling. Agric Sci Technol 23

Lisar SYS, Motafakkerazad R, Hossain MM, Rahman MM (2012) Water stress in plants: causes, effects and responses. In: Rahman IMM (ed) Water stress. InTech, London. ISBN: 978-953-307-963-9

Lopes MS, Reynolds MP (2011) Drought adaptive traits and wide adaptation in elite lines derived from re-synthesized hexaploid wheat. Crop Sci 51(4):1617–1626

Lutts S, Kinet JM, Bouharmont J (1996) Effects of salt stress on growth, mineral nutrition and proline accumulation in relation to osmotic adjustment in rice (*Oryza sativa* L.) cultivars differing in salinity resistance. Plant Growth Regul 19(3):207–218

Mall AK, Misra V, Kumar A, Pathak AD (2019) Drought: status, losses and management in sugarcane crop. In: Bharti PK, Ray J (eds) Farm and development: technological perspectives. Discovery Publishing House, New Delhi, pp 165–187

Mall AK, Misra V, Singh BD, Kumar M, Pathak AD (2020) Drought tolerance: breeding efforts in sugarcane. In: Hasaanzuman M (ed) Agronomic crops. Springer, Singapore, pp 157–172

Misra V, Solomon S, Ansari MI (2016) Impact of drought on post-harvest quality of sugarcane crop. Adv Life Sci 20(5):9496–9505

Misra V, Solomon S, Mall AK, Prajapati CP, Hassan A, Allah EFE, Ansari MI (2020a) Morphological assessment of water stressed sugarcane: a comparison of waterlogged and drought affected crop. Saudi J Biol Sci 27(5):1228–1236

Misra V, Solomon S, Hassan A, Allah EFE, Al-Arjani AF, Mall AK, Prajapati CP, Ansari MI (2020b) Minimization of post-harvest sucrose losses in drought affected sugarcanes using chemical formulation. Saudi J Biol Sci 27(1):309–317

Molina MG (2002) Environmental constraints on agricultural growth in 19th century Granada (southern Spain). Ecol Econ 41(2):257–270

Moutia JFY, Saumtally S, Spaepen S, Vanderleyden J (2010) Plant growth promotion by *Azospirillum* sp. in sugarcane is influenced by genotype and drought stress. Plant Soil 337(1):233–242

Naidu KM, Bhagyalakshmi KV (1973) Relative turgidity and stomatal movement of sugarcane varieties in relation to their drought resistance. Indian J Agric Sci 43(11):1016–1018

Naidu KM, Bhagyalakshmi KV (1987) Stomatal movement in relation to drought resistance in sugarcane. Curr Sci 36(20):555–556

National News Agency of Thailand (2014) Biosugar Int Mag. http://www.jornalcana.com.br/biosugar-en/drought-devastates-sugarcane-in-thailand/. Accessed 20 Sept 2014

Negi OP, Naithani SP, Poddar S (1971) Root studies of outstanding sugarcane varieties of Bihar, India. Proc Int Soc Sugar Cane Technol 14:733–738

Paungfoo-Lonhienne C, Lonhienne TGA, Yeoh YK, Webb RI (2014) A new species of Burkholderia isolated from sugarcane roots promotes plant growth. J Microbial Biotechnol 7(2):142–154

Pereira LB, Andrade GS, Meneghin SP, Vicentini R, Ottoboni LMM (2019) Prospecting plant growth promoting bacteria isolated from the rhizosphere of sugarcane under drought stress. Curr Microbiol 76(11):1345–1354

Pimentel C (2004) A Relação da Planta com a Água. Seropédica, Rio de Janeiro, p 171

Pinheiro C, Chaves MM (2011) Photosynthesis and drought: can we make metabolic connections from available data? J Exp Bot 62(3):869–882

Ram B (2017) Status of sugarcane agriculture and sugar industry. In: Proceeding of international symposium on sugarcane research since co 205: 100 years and beyond (Sucrosym-2017), 16–21st September, Coimbatore, pp i–xxvi

Ramesh P (2000) Sugarcane Breeding Institute, Coimbatore, India effect of different levels of drought during the formative phase on growth parameters and its relationship with dry matter accumulation in sugarcane. J Agron Crop Sci 185(2):83–89

Ribeiro CM, Cardoso EJBN (2012) Isolation, selection and characterization of root-associated growth promoting bacteria in Brazil pine (*Araucaria angustifolia*). Microbiol Res 167(2):69–78

Ru Kai C, Muqing Z, Yibo L (1996) Drought-avoidant difference in the 4 cultivars of sugarcane (*S. officinarum* L.). J Sugarcane 3(1):1–4

Sairam RK (1994) Effect of moisture stress on physiological activities of two contrasting wheat genotypes. Indian J Exp Biol 32:594–597

Sales CRG, Ribeiro RV, Machado DFSP, Machado RS, Dovis VL, Lagôa AMMA (2012) Gas exchange and carbohydrate balance in sugarcane plants under root stressful conditions. Bragantia 71(3):319–327

Sang-Wu L, Ming-Muh K, Shiue JJ, Liu WC, Wang TSC (1984) Pre-submergence in relation to soil microbial balance and sugarcane production. Taiwan Sugar 31(1):13–11

Sanji G, Jun L, Rukai C, Muqing Z, Daren P (2002) Photosynthetic physiology indexes of the drought resistance of sugarcane and its comprehensive evaluation. Zuo Wu Xue Bao 28(1): 94–98

Saranga Y, Jiang CX, Wright RJ, Yakir D, Paterson AH (2002) Genetic dissection of cotton physiological responses to arid conditions and their interrelationships with productivity. Plant Cell Environ 27(3):47–58

Shao HB, Chu LY, Jaleel CA, Manivannan P, Panneerselvam R, Shao MA (2009) Understanding water deficit stress-induced changes in the basic metabolism of higher plants-biotechnologically and sustainably improving agriculture and the ecoenvironment in arid regions of the globe. Crit Rev Biotechnol 29(2):131–151

Shrivastava AK, Srivastava MK (2006) Abiotic stresses affecting sugarcane: sustaining productivity. International Book Distributing Co, Lucknow, p 32

Shrivastava AK, Srivastava S (2012) Sugarcane: physiological and molecular approaches for improving abiotic stress tolerance and sustaining crop productivity. In: Improving crop resistance to abiotic stress, pp 885–921

Shrivastava AK, Srivastava DC, Solomon S, Srivastava MK, Singh I (2003) Physiological characters imparting resistance to biotic and abiotic stresses in sugarcane. Sugar Tech 5(3): 105–120

Shrivastava AK, Solomon S, Rai RK, Singh P, Chandra A, Jain R, Shukla SP (2015) Physiological interventions for enhancing sugarcane and sugar productivity. Sugar Tech 17(3):215–226

Shrivastava AK, Srivastava TK, Srivastava AK, Misra V, Shrivastava S, Singh VK, Shukla SP (2016) Climate change induced abiotic stresses affecting sugarcane and their mitigation. ICAR-Indian Institute of Sugarcane Research, Lucknow. (ISBN: 978-93-5265-983-8)

Shrivastava A, Pathak AD, Misra V, Srivastava S, Swapna M, Shukla SP (2017) Sugarcane crop: it's tolerance towards abiotic stresses. In: Minhas P, Rane J, Pasala R (eds) Abiotic stress management for resilient agriculture. Springer, Singapore, pp 375–397

Silva M, Soares RAB, Landell MGA, Campana MP (2007) Agronomic performance of sugarcane families in response to water stress. In: Proceedings of ISSCTC, vol 26, pp 613–623

Silva MDA, Silva JAGD, Enciso J, Sharma V, Jifon J (2008) Yield components as indicators of drought tolerance of sugarcane. Sci Agric 65(6):620–627

Silveira NM, Hancock JT, Frungillo L, Siasou E, Marcos FCC, Salgado I, Machado EC, Ribeiro RV (2017) Evidence towards the involvement of nitric oxide in drought tolerance of sugarcane. Plant Physiol Biochem 115:354–359

Singh S, Ramakrishnan (1977) Early growth attributes of thick-stalked versus thin-stalked varieties in relation to drought resistance in sugarcane. Sugar Cane Breed Newsl 40:1–4

Singh S, Reddy M (1980) Growth, yield and juice quality performance of sugarcane varieties under different soil moisture regimes in relation to drought resistance. Proc Int Soc Sugar Cane Technol 17(1):544–555

Sinha OK (2016) Forty-five years of AICRP on sugarcane. All India Coordinated Research Project on Sugarcane. ICAR-IISR, Lucknow, pp 16–20

Smith D (1978) Cane sugar world. Palmer, New York, p 240

Smith D, Inman-Bamber N, Thorburn (2005) Growth and function of the sugarcane root system. Field Crop Res 92(2–3):169–183

Someya N, Akutsu K (2005) Biocontrol of plant diseases by genetically modified microorganisms: current status and future prospects. In: Siddiqui ZA (ed) PGPR: biocontrol and biofertilization. Springer, Dordrecht, pp 297–312

Songsri P, Jogloy S, Vorasoot N, Akkasaeng C, Patanothai A, Holbrook CC (2008) Root distribution of drought-resistant peanut genotypes in response to drought. J Agron Crop Sci 194(2): 92–103

Srinivasan TV (2004) Improving indigenous sugarcane of India. Sugar Tech 6(3):107–111

Taiz L, Zeiger E (2006) Stress physiology. Plant Physiol 4

Tardieu F (2012) Any trait or trait-related allele can confer drought tolerance: just design the right drought scenario. J Exp Bot 63(1):25–31

Tardieu F, Bruckler L, Lafolie F (1992) Root clumping may affect the root water potential and the resistance to soil-root water transport. Plant Soil 140(2):291–230

Thomas H, Howarth CJ (2000) Five ways to stay green. J Exp Bot 51(1):329–337

Uprety KK, Murti GSR, Bhatt RM (1999) Response of French bean cultivars to water deficits: changes in endogenous hormones, praline and chlorophyll. Biol Plant 40(3):381–388

Venkataramana S, Sundaram SS, Naidu KM (1984) Growth behavior of field grown sugarcane in relation to environmental parameters and soil moisture stress. Agric For Meteorol 31(3–4): 251–260

Venkataramana S, Rao PNG, Naidu K M (1986) The effects of water stress during the formative phase on stomatal resistance and leaf water potential and its relationship with yield in ten sugarcane varieties. Field Crop Res 13:345–353

Verbruggen N, Hermans C (2008) Proline accumulation in plants: a review. Amino Acids 35(4): 753–759

Verma M, Mishra J, Arora NK (2019) Plant growth-promoting rhizobacteria: diversity and applications. In: Sobti RC, Arora NK, Kothari R (eds) Environmental biotechnology: for sustainable future. Springer, Singapore, pp 129–173

Viqueira L, Acosta I, Rodriguez CR (1985) Study on drought resistance in relation to nitrate reductase and proline content in sugarcane. Rev INICA 2(1):64–77

Wahid A (2007) Physiological implications of metabolite biosynthesis for net assimilation and heat-stress tolerance of sugarcane (*Saccharum officinarum*) sprouts. J Plant Res 120(2):219–228

Wang H, Siopongco J, Wade LJ, Yamauchi A (2009) Fractal analysis on root systems of rice plants in response to drought stress. Environ Exp Bot 65(2–3):338–344

Yang CM, Fan MJ, Hsiang WM (1993) Growth and yield responses of maize (*Zea mays* L.) to soil water deficits. II. Effects of water deficit timing and strength. J Agric Res China 42:173–186

Zarabi S (2004) Bad monsoon may upset growth prospects. Hindustan Times (Lucknow), July 22, pp 13
Zargar A, Sadiq R, Naser B, Khan FI (2011) A review of drought indices. Environ Rev 19:333–349
Zhang MQ, Chen RK, Yu SL (1996) Changes of polyamine metabolism in drought-stressed sugarcane leaves and their relation to drought resistance. Acta Phytophysiol Sin 22(3):327–332
Zlatev Z, Lidon FC (2012) An overview on drought induced changes in plant growth, water relations and photosynthesis. Emir J Food Agric 24(1):57–72

12 The Beneficial Effects of Arbuscular Mycorrhizal Fungi and Compost on Plant Tolerance to Drought and Salinity Stresses: A Study on Date Palm and Alfalfa

Mohamed Ait-El-Mokhtar, Raja Ben-Laouane, Abderrahim Boutasknit, Mohamed Anli, Fatima El-Amerany, Salma Toubali, Soufiane Lahbouki, Toshiaki Mitsui, Said Wahbi, Marouane Baslam, and Abdelilah Meddich

Abstract

Phoenix dactylifera L. and *Medicago sativa* L. are two important crops, which play economic, social, and ecological roles in oasis ecosystems. However, the ecosystems where these plants grow are subjected to drought and salinity, which are the main limiting factors for their productivity. The present study was carried out to evaluate the date palm and alfalfa response to drought and salinity stresses and to study the effect of arbuscular mycorrhizal fungi (AMF) and compost in

M. Ait-El-Mokhtar
Laboratory of Agro-Food, Biotechnologies and Valorization of Vegetal Bioresources, Department of Biology, Faculty of Sciences Semlalia, Cadi Ayyad University, Marrakech, Morocco

Department of Biology, Faculty of Sciences and Techniques Mohammedia, Hassan II University, Casablanca, Morocco

R. Ben-Laouane · A. Boutasknit · M. Anli · F. El-Amerany · S. Toubali · S. Lahbouki · S. Wahbi · A. Meddich (✉)
Laboratory of Agro-Food, Biotechnologies and Valorization of Vegetal Bioresources, Department of Biology, Faculty of Sciences Semlalia, Cadi Ayyad University, Marrakech, Morocco
e-mail: a.meddich@uca.ma

T. Mitsui
Laboratory of Biochemistry, Faculty of Agriculture, Niigata University, Niigata, Japan University, Niigata, Japan

Department of Life and Food Sciences, Graduate School of Science and Technology, Niigata University, Niigata, Japan

M. Baslam
Laboratory of Biochemistry, Faculty of Agriculture, Niigata University, Niigata, Japan University, Niigata, Japan

N. K. Arora, B. Bouizgarne (eds.), *Microbial BioTechnology for Sustainable Agriculture Volume 1*, Microorganisms for Sustainability 33,
https://doi.org/10.1007/978-981-16-4843-4_12

enhancing the drought and salt tolerance. Date palm seedlings were grown under well-watered (75% field capacity: FC) and drought stress (25% FC) conditions with and without AMF inoculation on the one hand. On the other hand, date palm and alfalfa seedlings were grown under non-saline and saline conditions (0 and 240 mM NaCl; 0 and 120 mM NaCl respectively) with and without AMF inoculation and compost addition. Growth, mineral, physiological, as well as biochemical traits were evaluated in this study. Plant growth traits especially shoot and root dry weight were negatively affected by drought and salt stress. Nevertheless, mycorrhizal and composted plants showed higher growth parameters under drought and salt stress conditions compared to the untreated plants. AMF and compost mitigated the decline of P, K^+, and Ca^{2+} concentrations induced by drought and salinity. Besides, AMF inoculation improved physiological traits through elevating photosynthetic efficiency, stomatal conductance, and leaf water potential under the two abiotic stresses. Furthermore, drought and salt stresses caused an increment in the antioxidant enzymes activities (superoxide dismutase, catalase and peroxidase, as well as ascorbate peroxidase). However, AMF inoculation induced further increase of these activities. In conclusion, colonization with AMF and compost addition could protect date palm and alfalfa plants against the negative drought and salt stress effects by mitigating their oxidative damage and improving their nutrient acquisition.

Keywords

Arbuscular mycorrhizal fungi · Compost · Drought stress · Salt stress · Growth · Oxidative stress · Tolerance · Date palm · Alfalfa

12.1 Introduction

The production of many crops in the world is confronted with problems caused by drought and salinity than by any other characteristics of climate (Ashraf et al. 2013; Gupta and Huang 2014; Ait-El-Mokhtar et al. 2020a). Drought and salinity are two widespread environmental abiotic stress factors in more than three quarters of the land surface of the globe (Li et al. 2007). These constraints impair almost all plant processes including morphological, physiological, and biochemical aspects, causing cell death, reduction in sizes of fruits and leaves, stomatal closure, photosynthetic perturbations, osmotic and oxidative stress, and membrane permeability (Nawaz et al. 2010; Fahad et al. 2017; Ait-El-Mokhtar et al. 2020b; Ben-Laouane et al. 2020a). In order to mitigate their effects on agriculture, the cultivation of plants that have the ability to survive in dry and saline environments is encouraged.

Date palm and alfalfa are among the drought- and salinity-resistant plants cultivated in dry and quasi-dry land (Yaish and Kumar 2015; Mouradi et al. 2018; Ait-El-Mokhtar et al. 2020c). Palm trees are known as a symbol of life in the desert due to their diverse benefits. Its cultivation offers fruits which contain essential nutrients (e.g., carbohydrates, amino acids, fibers, and minerals) as well as bioactive molecules (e.g., vitamins and antioxidants) (Vayalil 2012; Kamal-Eldin and Ghnimi 2018). Regarding alfalfa, it is considered as an important forage crop for hundreds of species of herbivores and prey animals and for industry due to its nutritional composition (i.e., protein) and to its high productivity (Flajoulot et al. 2005). Even though they have the ability to resist, they still suffer from severe damage and are going to extinct slowly over time.

Nowadays, many techniques are provided to manage abiotic stresses. The appropriate irrigation practices and plant breeding technologies are mostly used to overcome these issues (Hasegawa and Bressan 2000; Munns 2002). However, they are very expensive and consume more energy. Thus, the use of biological sources, like mutualistic microorganisms (e.g., arbuscular mycorrhizal fungi (AMF)) and organic amendments (e.g., compost), might be an alternative solution to increase either soil fertility or plant growth during stressful environments (e.g., drought and saline conditions) (Rady et al. 2016; Santander et al. 2019). AMF are symbiont fungi that belong to the phylum *Glomeromycota* (Schubler et al. 2001). They can confer many advantages to host plants as inducing plant growth, mineral acquisition water uptake, and increasing plant resistance to biotic and abiotic stresses (Porcel et al. 2012; Giri 2017; Meddich et al. 2018). Otherwise, compost is a bio-organic compound that is generated by recycling different organic wastes, such as leaves, grass, fruit, vegetable, manure, and olive mill waste (Azim et al. 2017). The application of compost to the soil is considered as one of the most important and effective strategies that still applies in the present times. Compost promotes plant growth due to its richness in humic acid, hormones, and beneficial microorganisms (Keeling et al. 1995) and its provision of essential micro- and macronutrients (El Amerany et al. 2020).

The interaction of roots with AMF and compost could mitigate and reduce water and salt stress severity by improving photosynthetic activity and chlorophyll content (Hidri et al. 2016), reducing the absorption of sodium (Na^+) and chloride (Cl^-) (Daei et al. 2009), and increasing the level of their antioxidation compounds such as polyphenol oxidase, superoxide dismutase, and catalase (Lakhdar et al. 2008; Meddich et al. 2015; Ye et al. 2019; Ait-El-Mokhtar et al. 2019, 2020a). Many crops could potentially benefit from compost as well as AMF; however, the benefit might be different and depends on either crop species, compost doses, or AMF strains used. Here, we will discuss the beneficial effects of AMF as well as compost, produced locally, on the physiology and biochemistry of date palm and alfalfa plants subjected to water and salt stresses.

12.2 Materials and Methods

12.2.1 Preparation of AMF and Seedlings

The used AMF were *Glomus monosporus*, *Glomus clarum*, and *Glomus deserticola* strains and Aoufous mycorrhizal fungi consortium (obtained from the palm grove of Tafilalet, southeast of Morocco) (Meddich et al. 2015). This consortium consisted of a mixture of native species as follows: *Glomus* sp. (15 spores g^{-1} soil), *Sclerocystis* sp. (9 spores g^{-1} soil), and *Acaulospora* sp. (1 spore g^{-1} of soil).

Date palm (Boufeggous *cv.*) and alfalfa (*Medicago Sativa* L.) seeds were disinfected using sodium hypochlorite (10%) for 10 min and were rinsed multiple times using sterile distilled water. The germination was carried out at 38 °C for 3 weeks for date palm seeds and at 28 °C for 1–2 days for alfalfa seeds.

12.3 Experimental Design and Growth Conditions

12.3.1 Experiment 1

Two-month *Phoenix dactylifera* cv. Boufeggous seedlings were planted in 5 L black plastic buckets filled with 4 kg of sterilized (3 h at 180 °C) sand–peat mixture (2:1 v/v). The methodology for applying water stress was described by Tobar et al. (1994) and Meddich et al. (2000). The pots were then placed in a greenhouse. Five mycorrhizal treatments (non-inoculated control, Aoufous mycorrhizal consortium, *G. deserticola*, *G. clarum*, and *G. monosporus*) and two water regimes (25% and 75% FC) were applied.

12.3.2 Experiment 2

Young seedlings of alfalfa were transplanted into plastic pots containing sterile sand, mixed with sterile compost (autoclaved at 200 °C for 1 h for 3 successive days) at a rate of 5%. The used compost was characterized as follows: 0.270 mg/L P (phosphorus), $0.03 \cdot 10^{-3}$ mg/g NH_4^+ (ammonium), $0.07.10^{-3}$ mg/g NO_3^- (nitrate), 0.44 NH_4^+/NO_3^- ratio, 7.86 pH, 30.65% TOC (total organic carbon), 2.19% N (nitrogen), 14 C/N ratio, and 49% ashes. This experiment was performed in a greenhouse (with a 16/8 h day/night cycle and average temperature of 25.5 °C, relative humidity average of 68.5%, and light of 410 $\mu m^{-2}\ s^{-1}$) for 2 months, at the Faculty of Science Semlalia, Cadi Ayyad University Marrakesh, Morocco. The experiment consisted of two levels of salt 0 and 120 mM NaCl with and without Aoufous consortium application (+AMF and −AMF). The soil salinization was gradually carried out to prevent osmotic shock. The bags were irrigated as needed with distilled water and different salt solutions to keep the target salt levels. Two treatments were applied: uninoculated plants (control), and Aoufous consortium (AMF-treated plants) under 0 and 120 mM NaCl conditions. Many agro-physiological and biochemical traits were evaluated after 2 months of cultivation.

12.3.3 Experiment 3

After 2 months of germination, the palm plants were transplanted into plastic pots containing sterile sand. Seedlings were irrigated as needed with distilled water for

5 months after germination, and then they were subjected to 0 and 240 mM NaCl conditions. The experiment was performed in a greenhouse for 10 months under the same conditions as in the experiment 2. The experiment was designed in four treatments with two levels of salt in the soil (0 and 240 mM of NaCl) with and without Aoufous consortium inoculation (+AMF and −AMF). Ten replicates were performed for each treatment for a total of 40 completely randomized block pots.

12.3.4 Experiment 4

After 2 months of germination, date palm seedlings were transplanted into plastic pots containing a mixture of sterile sand and compost. The organic amendment was prepared from green waste and was the same compost used in the experiment 2. The compost was added in a proportion of 5% (w/w) when transplanting seedlings. Five months after transplantation, two salinity levels were applied (0 and 240 mM NaCl) as describe in experiment 3.

12.4 Measured Parameters

12.4.1 Assessment of Root Colonization, Growth, and Plant Biomass Accumulation

At harvest, the root part was separated from the shoot one and the biometrical traits (plant height, number of leaves, leaf area, and dry weight of shoots and roots) were measured. The roots were cut into 1 cm segments and were stained for mycorrhizal colonization assessment as described by Phillips and Hayman (1970). The percentage of mycorrhizal colonization was measured using the formula described by McGonigle et al. (1990).

12.4.2 Water Content, Relative Water Content, and Leaf Water Potential

The water content (WC) of the aerial part was expressed in grams of water per gram of the dry matter (g/g DW). The relative water content (RWC) was determined on the leaf at the same level according to Barrs and Weatherly (1962). The leaf water potential (LWP) was determined using a pressure chamber as described by Scholander et al. (1965).

12.4.3 Stomatal Conductance, Chlorophyll Fluorescence, and Electrolyte Leakage

Stomatal conductance (gs) was measured at harvest using a porometer system (Leaf Porometer, Decagon Device, Inc., Washington, USA) following the user manual instructions. Chlorophyll fluorescence was determined using a chlorophyll fluorometer (Opti-sciences OSI 30 p) after 30 min of leaf dark adaptation. Three parameters were recorded: initial (F_0), maximal (F_m), and variable (F_v) fluorescence. The efficiency of photosystem II was expressed as the F_v/F_m ratio. Electrolyte leakage (EL) measurements were performed according to Lutts et al. (1996).

12.4.4 Photosynthetic Pigments, Hydrogen Peroxide, Malondialdehyde, and Phenolic Contents

Photosynthetic pigments were extracted as described by Smith and Benitez (1955), and the content of chlorophyll a, chlorophyll b, carotenoids, and total chlorophyll was estimated according to Arnon's (1949) formula. Malondialdehyde (MDA) and hydrogen peroxide concentrations in leaves and roots were evaluated following the methods of Dhindsa et al. (1981) and Velikova et al. (2000), respectively. The phenolic compound extraction was carried following the method of El Hadrami et al. (1997), and their concentrations were evaluated using the Folin–Ciocalteu reagent protocol.

12.4.5 Antioxidant Enzymatic Activities

Assay of enzymes was performed according to Ait-El-Mokhtar et al. (2019), and the supernatants were kept at −20 °C for subsequent biochemical assays. The activity of superoxide dismutase (SOD, EC 1.15.1.1) was determined following Beyer and Fridovich's (1987) method. The catalase (CAT, EC 1.11.1.6), ascorbate peroxidase (APX, EC 1.11.1.11), and peroxidase (POD, EC 1.11.1.7) activities were measured according to Aebi (1984), Amako et al. (1994), and Polle et al.'s (1994) protocols respectively. For measurement of PPO activity (EC 1.14.18.1), it was performed following the method of Hori et al. (1997). The protein concentration in the different extracts was determined as described by Bradford (1976) using bovine serum albumin (BSA) as a standard.

12.4.6 Nutrient Analysis

The mineral analyses were carried out on powdered oven-dried shoots and roots attacked with 98% H_2SO_4 and 30% H_2O_2. P concentrations were measured according to Olsen and Dean (1965). Na^+, K^+, and Ca^{2+} concentrations were assessed by flame photometry (JENWAY, PFP7) following the method of Wolf (1982).

12.4.7 Statistical Analysis

The results data are means based on five replicates ± standard error (S.E.) per treatment. The statistical analysis was carried out using COSTAT software (version 6.3). Data were subjected to analysis of variance (ANOVA). Differences between means were assessed by Duncan's multiple range test (DMRT) at $P < 0.05$.

12.5 Results

12.5.1 Effects of AMF on Improving Tolerance of Date Palm to Drought Stress

12.5.1.1 Effect of Water Stress on Palm Root Colonization with AMF

The frequency of infection (F) of date palm root system by AMF was lightly affected by the deficiency of soil water (Fig. 12.1a). For the two imposed water regimes, F stayed high (>60%) for all the tested fungi. Aoufous consortium, *G. monosporus*, and *G. clarum* were more infectious ($F > 90\%$) than *G. deserticola* whatever the applied water regime. The intensity of palm roots colonization by different AMF decreased with increasing soil water deficiency (Fig. 12.1b). Mycorrhizal intensity was higher for Aoufous consortium, *G. monosporus*, and *G. Clarum* than for *G. deserticola*.

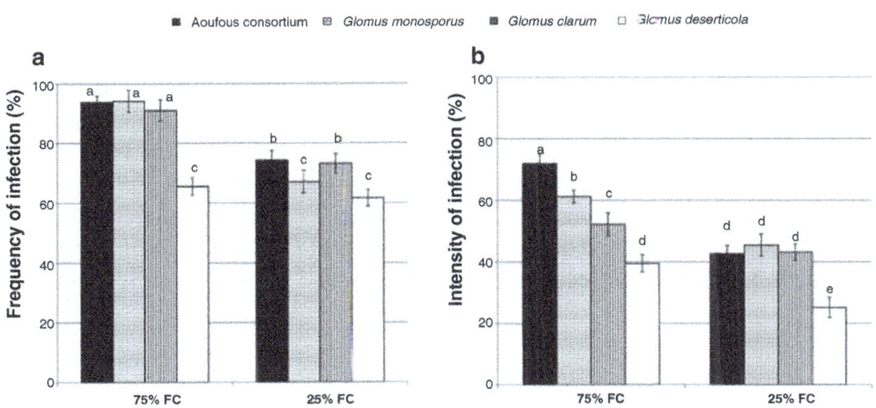

Fig. 12.1 Effect of water deficit on frequency (**a**) and intensity (**b**) of mycorrhizal infection after 9 months from AMF inoculation of date palm roots. Bars of each parameter labeled by different letters are significantly different ($P < 0.05$) as assessed by Duncan's test

12.5.1.2 Effect of AMF and Water Stress on Growth and Mineral Nutrition of Date Palm

The obtained results showed that mycorrhizal palms showed greater number of leaves compared to the non-mycorrhizal plants and regardless of the applied water regime (Table 12.1). The palms inoculated with Aoufous consortium, *G. monosporus*, and *G. clarum* had significantly improved leaf area compared to the control or *G. deserticola* mycorrhizal plants, while leaf numbers and leaf area were significantly reduced under water stress. In addition, shoot dry matter accumulation was significantly enhanced with the application of AMF and regardless of the applied water regime.

The Aoufous consortium, *G. monosporus*, and *G. clarum* application produced more shoot dry matter under severe water regime. Furthermore, mycorrhizal plant shoots recorded higher concentrations of P, K, and Ca compared to the non-mycorrhizal plants regardless of the water treatment (Table 12.2). In the same vein, Na concentrations were significantly higher in inoculated plants under favorable water regime (75% FC).

12.5.1.3 Effect of AMF and Water Stress on the Water Status of Date Palm

For both water regimes, mycorrhizal plants recorded a similar WC to the control (Table 12.3). In contrast, inoculated plants with Aoufous consortium and *G. monosporus* showed an increased relative WC compared to the stressed control. Moreover, AMF-treated palms recorded higher LWP values in comparison to the control under both water treatments. *G. monosporus* and Aoufous consortium inoculated plants had the highest LWP, while the stomatal resistance was significantly enhanced in non-inoculated palms compared to the inoculated ones.

12.5.1.4 Effect of Water Deficit and AMF on Peroxidase and Polyphenol Oxidase Activities and Phenolic Content

Peroxidase (POX) and polyphenol oxidase (PPO) activities were significantly increased in the presence of water stress (Table 12.4). In the presence of water stress, POX activity was higher in *G. deserticola*-treated palms, while PPO activity was significantly greater in Aoufous consortium and *G. monosporus* inoculated plants compared to the non-inoculated plants. Under 25% FC, soluble phenol values were increased in inoculated and control plants except *G. deserticola*-treated palms. The highest content was recorded in *G. clarum* and *G. monosporus* inoculated plants.

Table 12.1 Effect of water deficit on the growth parameters of mycorrhizal and non-mycorrhizal date palm after 9 months from AMF inoculation

Water regime		Number of leaves	Plant height (cm)	Leaf area (cm^2)	Shoot dry weight ($g\ plant^{-1}$)
75% FC	Control	3.89 ± 0.22c	43.22 ± 1.89de	27.54 ± 1.50d	1.21 ± 0.19f
	Aoufous consortium	5.13 ± 0.30a	42.31 ± 2.48c	49.67 ± 1.53a	2.77 ± 0.23c
	Glomus monosporus	4.60 ± 0.32b	41.23 ± 2.39ab	48.71 ± 2.63a	3.48 ± 0.28a
	Glomus clarum	4.35 ± 0.27bc	40.72 ± 2.38ab	34.25 ± 1.63b	2.45 ± 0.15d
	Glomus deserticola	4.06 ± 0.23bc	33.64 ± 1.97 cd	29.75 ± 1.75 cd	3.17 ± 0.26b
25% FC	Control	2.65 ± 0.15e	27.97 ± 1.64e	22.16 ± 0.75e	0.81 ± 0.09 g
	Aoufous consortium	4.43 ± 0.26bc	37.37 ± 2.20bc	28.68 ± 1.68d	1.94 ± 0.08e
	G. Monosporus	4.16 ± 0.14bc	36.48 ± 2.14bcd	32.29 ± 0.98bc	1.82 ± 1.03e
	G. clarum	4.08 ± 0.14bc	34.53 ± 2.03 cd	27.56 ± 0.96d	1.89 ± 0.11e
	G. deserticola	3.09 ± 0.09d	28.86 ± 1.69e	22.30 ± 1.30e	1.18 ± 0.13f

Values of each parameter labeled by different letters indicate significant differences ($P < 0.05$) as assessed by Duncan's test

Table 12.2 Effect of water deficit on the mineral content of mycorrhizal and non-mycorrhizal date palm after 9 months from AMF inoculation

Water regime		P (mg/plant)	K (mg/plant)	Ca (mg/plant)	Na (mg/plant)
75% FC	Control	3.61 ± 0.21f	1.68 ± 0.05f	1.19 ± 0.07 h	0.20 ± 0.01f
	Aoufous consortium	9.90 ± 0.58c	4.67 ± 0.27c	2.58 ± 0.15e	0.53 ± 0.03b
	G. monosporus	18.11 ± 1.06a	7.86 ± 0.46a	1.91 ± 0.11f	0.72 ± 0.04a
	G. clarum	10.99 ± 0.64b	2.57 ± 0.15e	3.77 ± 0.22a	0.46 ± 0.02c
	G. deserticola	11.10 ± 0.65b	5.67 ± 0.33b	3.15 ± 0.18c	0.72 ± 0.04a
25% FC	Control	1.75 ± 0.10 g	1.10 ± 0.04 g	1.11 ± 0.06 h	0.22 ± 0.01f
	Aoufous consortium	11.56 ± 0.68b	3.66 ± 0.21d	1.63 ± 0.09 g	0.35 ± 0.02d
	G. monosporus	4.91 ± 0.29e	3.83 ± 0.22d	3.42 ± 0.20b	0.35 ± 0.02d
	G. clarum	8.77 ± 0.51d	1.96 ± 0.11f	2.90 ± 0.17d	0.45 ± 0.02c
	G. deserticola	5.39 ± 0.31 g	1.10 ± 0.04 g	1.03 ± 0.06 h	0.29 ± 0.01e

Values of each parameter labeled by different letters indicate significant differences ($P < 0.05$) as assessed by Duncan's test

Table 12.3 Effect of water deficit on water status of mycorrhizal and non-mycorrhizal date palm after 9 months from AMF inoculation

Water regime		Water content (WC) (g/g dw)	Relative water content (RWC) (%)	Leaf water potential (LWP) (MPa)	Stomatal resistance (s/cm)
75% FC	Control	2.99 ± 0.23a	78.27 ± 1.42b	−30.37 ± 0.70d	2.35 ± 0.10c
	Aoufous consortium	2.80 ± 0.07a	79.33 ± 2.08b	−27.57 ± 0.40c	2.25 ± 0.20c
	G. monosporus	3.21 ± 0.47a	82.65 ± 2.32a	−23.97 ± 0.89a	2.17 ± 0.12c
	G. clarum	2.84 ± 0.12a	76.51 ± 2.18b	−27.20 ± 0.23c	2.42 ± 0.05c
	G. deserticola	2.83 ± 0.13a	71.96 ± 2.09c	−28.52 ± 0.96c	2.43 ± 0.19c
25% FC	Control	2.88 ± 0.08a	78.27 ± 1.42e	−35.03 ± 1.72f	3.21 ± 0.06a
	Aoufous consortium	2.94 ± 0.04a	79.33 ± 2.08d	−28.61 ± 0.77c	2.15 ± 0.14c
	G. monosporus	2.99 ± 0.21a	82.65 ± 2.32d	−25.46 ± 0.63b	2.13 ± 0.02c
	G. clarum	2.68 ± 0.17a	76.51 ± 2.18e	−30.67 ± 0.15d	2.25 ± 0.10c
	G. deserticola	3.03 ± 0.27a	71.96 ± 2.09e	−33.09 ± 1.08e	2.51 ± 0.15b

Values of each parameter labeled by different letters indicate significant differences ($P < 0.05$) as assessed by Duncan's test

Table 12.4 Effect of water deficit on peroxidase (POD) and polyphenol oxidase (PPO) activities and phenol content of mycorrhizal and non-mycorrhizal date palm after 9 months from AMF inoculation

Water regime		POD (Unit/g fw)	PPO (Unit/g fw)	Phenols content (Catechine equivalent/g fw)
75% FC	Control	281.67 ± 6.34i	366.67 ± 3.71i	660.32 ± 3.36 h
	Aoufous consortium	418.89 ± 6.23 h	606.67 ± 5.08 h	952.38 ± 7.11 g
	G. monosporus	750.00 ± 6.58e	736.67 ± 3.92 g	615.87 ± 6.09i
	G. clarum	723.33 ± 9.68f	850.00 ± 3.25f	1401.97 ± 7.48c
	G. deserticola	576.11 ± 6.92 g	1070.00 ± 6.58e	1177.78 ± 5.53e
25% FC	Control	1427.50 ± 7.74b	1176.67 ± 6.79d	1347.30 ± 7.45d
	Aoufous consortium	1124.17 ± 6.65c	1816.67 ± 6.79a	1022.22 ± 2.23f
	G. monosporus	1428.33 ± 4.10b	1588.33 ± 8.55b	1629.52 ± 4.88b
	G. clarum	1060.00 ± 6.96d	1181.67 ± 6.88d	2101.59 ± 5.67a
	G. deserticola	1780.00 ± 3.31a	1290.00 ± 9.33c	1026.03 ± 7.36f

Values of each parameter labeled by different letters indicate significant differences ($P < 0.05$) as assessed by Duncan's test

12.5.2 Effects of AMF on Improving Tolerance of Date Palm to Salt Stress

12.5.2.1 Effect of Salt Stress and AMF on Growth Parameters of Date Palm

The salt stress significantly reduced the growth performances including shoot height, number of leaves, leaf area, and shoot and root dry weight (Table 12.5). The greatest reductions were recorded for root dry weight and leaf area (41 and 39%, respectively). The application of AMF diminished the deleterious impact of salt on the growth traits. Indeed, inoculated palms under stressed conditions had significant high plant height, number of leaves, and leaf area. Plants' dry weights were greater in mycorrhizal palms than in non-mycorrhizal plants.

No AMF colonization was observed in non-inoculated date palm roots. In the absence of salinity, AMF inoculation induced important date palm root colonization, while the application of NaCl stress significantly reduced the colonization of the root system of inoculated seedlings. Effectively, mycorrhizal frequency and intensity declined by 43 and 63%, respectively, under 240 mM NaCl (Table 12.5).

12.5.2.2 Effect of Salt Stress and AMF on Water and Physiological Traits of Date Palm

The application of salinity significantly reduced LWP and stomatal conductance (33 and 38%, respectively) (Table 12.6). However, AMF application significantly improved water and physiological traits in comparison to non-inoculated palms under 240 mM NaCl. The highest improvements were recorded for stomatal conductance and F_v/F_m (112 and 116%, respectively).

Table 12.5 Effect of salinity on mycorrhizal colonization and growth parameters of mycorrhizal and non-mycorrhizal date palm after 8 months from AMF inoculation

NaCl treatment		Plant height (cm)	Number of leaves	Leaf area (cm^2)	Shoot dry weight (g $plant^{-1}$)	Root dry weight (g $plant^{-1}$)	AMF infection frequency (Fa) %	AMF infection intensity (Ma) %
0 mM	−AMF	30.6 ± 0.39c	3.2 ± 0.42c	18.95 ± 0.68c	1.55 ± 0.06c	1.11 ± 0.10c	–	–
	+AMF	35.3 ± 0.54a	5.2 ± 0.42a	32.40 ± 0.65a	4.99 ± 0.89a	3.36 ± 0.53a	100 ± 0.00a	74 ± 4.16a
240 mM	−AMF	27.32 ± 0.40d	3.1 ± 0.32c	11.52 ± 1.20d	1.09 ± 0.14d	0.66 ± 0.09d	–	–
	+AMF	31.54 ± 0.34b	4.1 ± 0.32b	20.62 ± 0.55b	3.51 ± 0.47b	1.68 ± 0.22b	100 ± 0.00a	27.38 ± 4.44b

Values of each parameter labeled by different letters indicate significant differences ($P < 0.05$) as assessed by Duncan's test

Table 12.6 Effect of salinity on water and physiological parameters of mycorrhizal and non-mycorrhizal date palm after 8 months from AMF inoculation

NaCl treatment		Relative water content (RWC) %	Leaf water potential (LWP) (bar)	Stomatal conductance (gs) (mmol m^{-2} s^{-1})	Photosynthetic efficiency (F_v/F_m)
0 mM	−AMF	73.72 ± 1.35b	−23.70 ± 0.98c	17.63 ± 1.66c	0.47 ± 0.03c
	+AMF	82.81 ± 1.78a	−15.96 ± 1.44a	45.76 ± 4.31a	0.78 ± 0.03a
240 mM	−AMF	50.03 ± 0.76d	−31.47 ± 2.81d	10.87 ± 1.72d	0.34 ± 0.01d
	+AMF	65.92 ± 1.14c	−21.97 ± 1.05b	23.00 ± 1.20b	0.73 ± 0.01b

Values of each parameter labeled by different letters indicate significant differences ($P < 0.05$) as assessed by Duncan's test

12.5.2.3 Effect of Salt Stress and AMF on Nutrient Contents of Date Palm

Plants' phosphorus content was significantly reduced by salt stress (43%), while Na uptake was increased two times in stressed palms in comparison to the control. In addition, K and Ca contents in plants grown in the presence of salinity were lowered by 45 and 31%, respectively compared to the control. In contrast, AMF inoculation improved the P, K, and Ca uptake and reduced Na concentrations in stressed plants (Fig. 12.2).

12.5.2.4 Effect of Salt Stress and AMF on Photosynthetic Pigments of Date Palm

Photosynthetic pigments content was significantly lowered under 240 mM NaCl (Fig. 12.3). Application of NaCl stress reduced Chl a by 54%, Chl b by 43%, carotenoids by 37%, and total chlorophyll content by 50%, respectively, in comparison to the non-stressed date palm. In contrast, mycorrhizal plants showed an improvement of pigments under salt stress.

12.5.2.5 Effect of Salt Stress and AMF on MDA, H_2O_2, and Protein Contents of Date Palm

The results in Table 12.7 indicate that salt stress induced a significant enhancement of MDA and H_2O_2 concentrations in shoots and roots of date palms in comparison to the control. However, salinity significantly reduced protein concentration in all date palms. AMF inoculation showed a decline in MDA and H_2O_2 contents, while it recorded an increase in protein content (34%).

12.5.2.6 Effect of Salt Stress and AMF on Antioxidant Enzyme Activity of Date Palm

Salt stress induced a significant enhancement of antioxidant enzyme (SOD, CAT, POD, and APX) activity both in shoots and roots (Figs. 12.4 and 12.5). This activity was greater in roots compared to the shoots. Furthermore, AMF application further increased the activity of these enzymes in stressed and non-stressed date palms.

12.5.3 Effects of AMF on Improving Tolerance of Alfalfa to Salt Stress

12.5.3.1 Effect of Salt Stress and AMF on Alfalfa Growth Traits

Salt stress application significantly reduced shoot and root dry weights of *M. sativa* (Fig. 12.6). In addition, dry biomass production was improved in shoot and root by AMF treatment compared to the control under salt stress. Under normal conditions, shoot dry weight was increased by AMF compared to non-inoculated plants.

12.5.3.2 Effect of Salt Stress and AMF on Alfalfa Physiological Traits

The stomatal conductance values in *M. sativa* plant were decreased under salt stress with no significant difference between AMF inoculated and non-inoculated plants

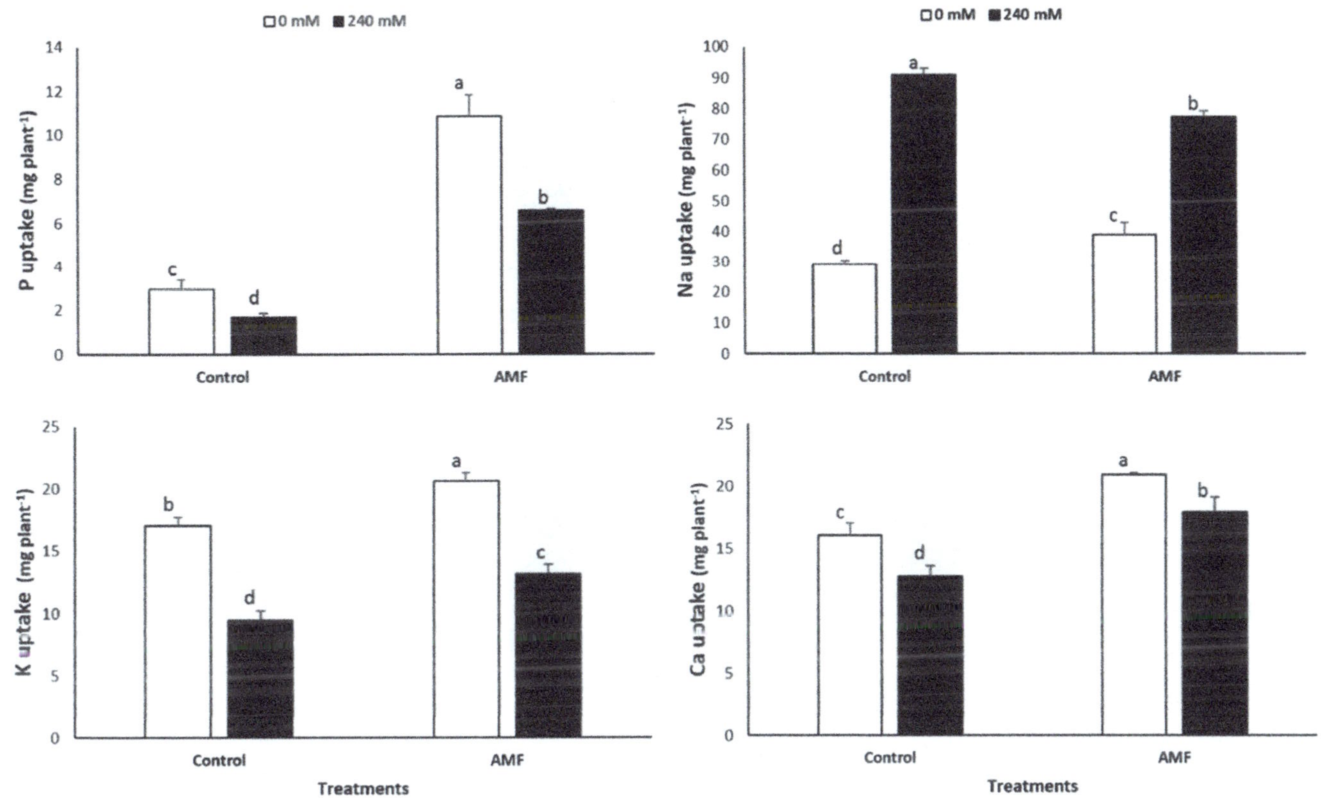

Fig. 12.2 Effect of salinity on mineral uptake of mycorrhizal and non-mycorrhizal date palm after 8 months from AMF inoculation. Bars of each parameter labeled by different letters are significantly different ($P < 0.05$) as assessed by Duncan's test

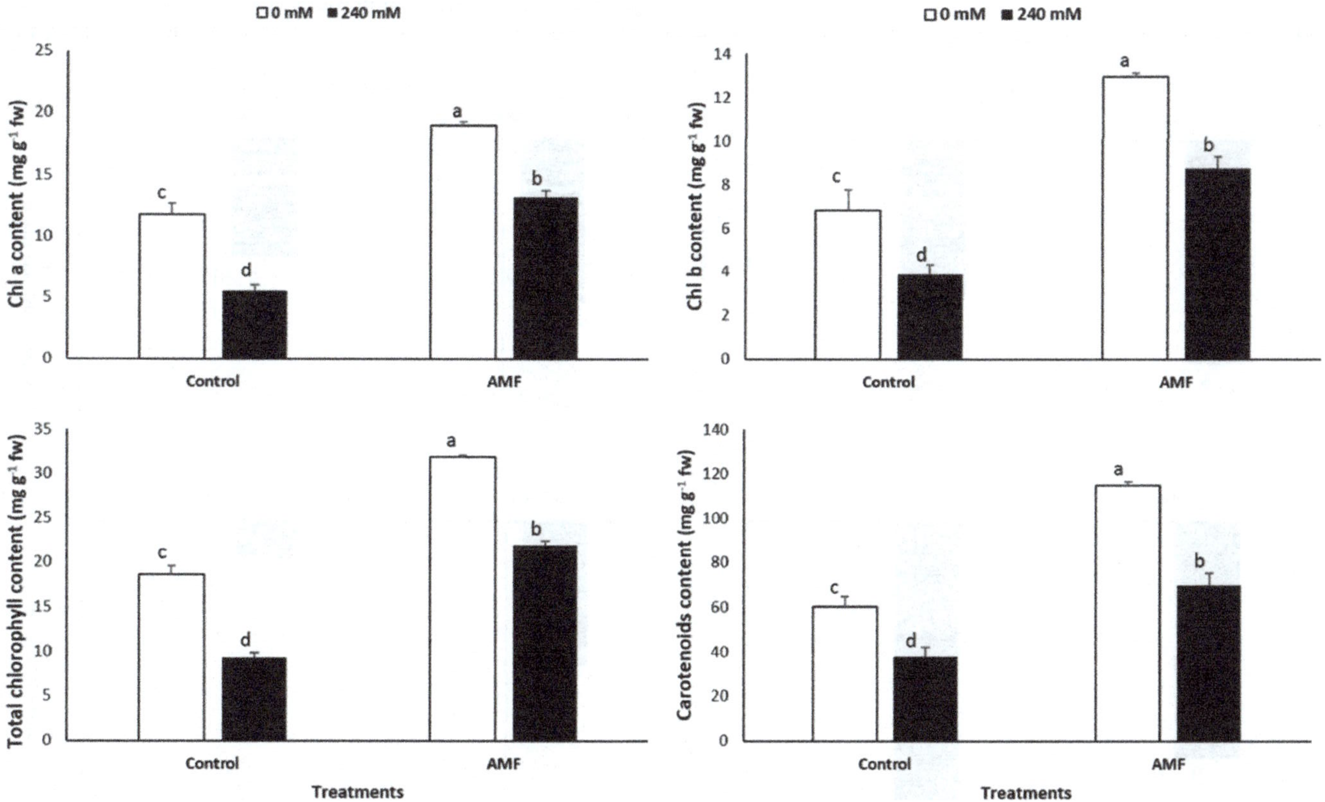

Fig. 12.3 Effect of salinity on photosynthetic pigment contents of mycorrhizal and non-mycorrhizal date palm after 8 months from AMF inoculation. Bars of each parameter labeled by different letters are significantly different ($P < 0.05$) as assessed by Duncan's test

Table 12.7 Effect of salinity on the content of hydrogen peroxide, MDA, and protein contents in the shoot and roots of mycorrhizal and non-mycorrhizal date palm after 8 months from AMF inoculation

NaCl treatment		Shoot			Root		
		H_2O_2 nmol g^{-1} fw	MDA μmol g^{-1} fw	Protein g^{-1} fw	H_2O_2 nmol g^{-1} fw	MDA μmol g^{-1} fw	Protein g^{-1} fw
0 mM	−AMF	17.85 ± 0.47b	23.05 ± 1.30b	21.23 ± 0.75b	36.83 ± 2.01c	11.58 ± 0.64b	11.67 ± 0.47c
	+AMF	11.82 ± 1.41c	16.69 ± 0.79	26.57 ± 1.27a	24.86 ± 1.99d	6.88 ± 0.79c	14.00 ± 0.41a
240 mM	−AMF	24.81 ± 2.40a	27.18 ± 0.30a	16.04 ± 1.02d	51.33 ± 2.21a	14.29 ± 0.30a	8.61 ± 0.57d
	+AMF	19.47 ± 1.07b	21.56 ± 1.37b	20.08 ± 0.31c	41.72 ± 1.40b	8.43 ± 0.30c	12.38 ± 0.25b

Values of each parameter labeled by different letters indicate significant differences ($P < 0.05$) as assessed by Duncan's test

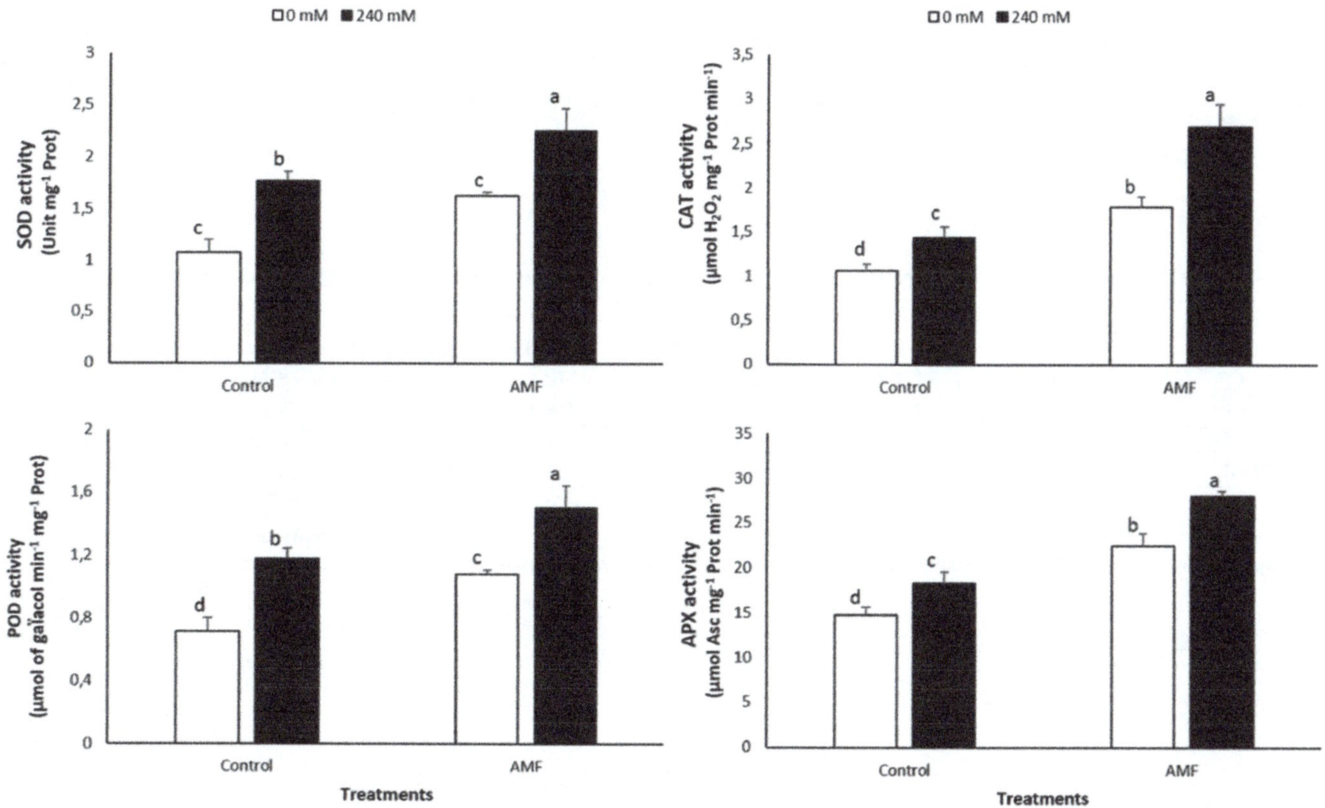

Fig. 12.4 Effect of salinity on the activity of ascorbate peroxidase (APX), catalase (CAT), peroxidase (POD), and superoxide dismutase (SOD) in the shoot of mycorrhizal and non-mycorrhizal date palm after 8 months from AMF inoculation. Bars of each parameter labeled by different letters are significantly different ($P < 0.05$) as assessed by Duncan's test

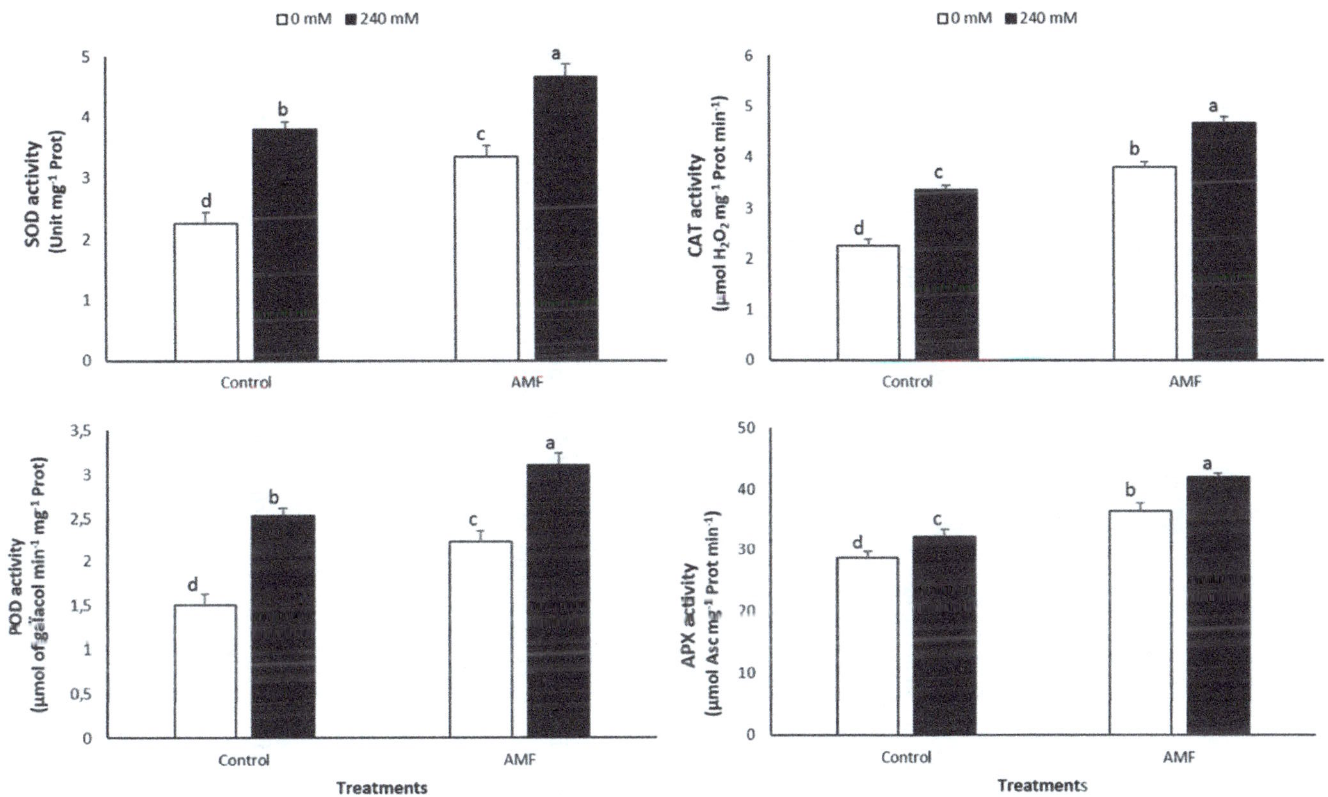

Fig. 12.5 Effect of salinity on the activity of ascorbate peroxidase (APX), catalase (CAT), peroxidase (POD), and superoxide dismutase (SOD) in the root of mycorrhizal and non-mycorrhizal date palm after 8 months from AMF inoculation. Bars of each parameter labeled by different letters are significantly different ($P < 0.05$) as assessed by Duncan's test

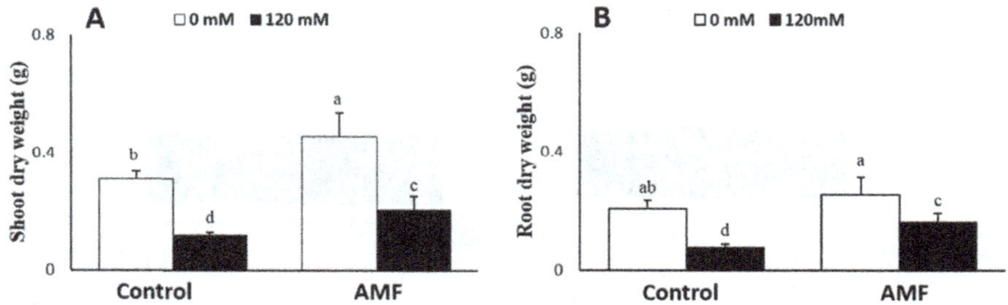

Fig. 12.6 Effect of salinity on shoot (**a**) and root (**b**) dry weight of mycorrhizal and non-mycorrhizal alfalfa after 2 months from AMF inoculation. Bars of each parameter labeled by different letters are significantly different ($P < 0.05$) as assessed by Duncan's test

(Fig. 12.7a). However, the photosynthetic efficiency (F_v/F_m) has shown no significant difference in all treatments under normal and salt stress conditions (Fig. 12.7b). This parameter was enhanced when plants were inoculated with AMF under salt stress and non-stress conditions. In addition, salt stress significantly decreased LWP values, in treated and untreated AMF plants (Fig. 12.7c). However, no significant difference has been observed in mycorrhizal plants under salt stress compared to the control. The highest LWP value was reported in AMF inoculated alfalfa under normal conditions. As shown in Fig. 12.7d, inoculation with AMF was significantly effective to decrease electrolyte leakage. The decrease of this parameter in mycorrhizal plants was significant under saline and non-saline conditions compared to non-inoculated plants.

12.5.3.3 Effect of Salt Stress and AMF on Alfalfa Biochemical Traits

PPO- as well as POD-specific activities were distinctly increased under salt stress (Table 12.8). Nevertheless, these activities were significantly pronounced in the root than in the shoot tissue. Under salt stress conditions and in the root part, the highest PPO activity values were observed in plants inoculated with AMF. However, the obtained results under stress showed that POD activity was higher in mycorrhizal plants compared to non-mycorrhizal plants in shoot and root tissues.

12.5.4 Effects of Compost on Improving Tolerance of Date Palm to Salt Stress

12.5.4.1 Effect of Compost Application and Salinity Stress on Growth Performances of Date Palm

As shown in Table 12.9, the salinity stress significantly decreased biometrical parameters (plant height, leaf numbers, leaf area, and shoot and root dry weight). The most affected parameter by salt stress was root dry weight, which showed a decrease of 45% followed by leaf area with a decrease of 39%. The application of compost significantly enhanced the growth parameter in the presence of 240 mM

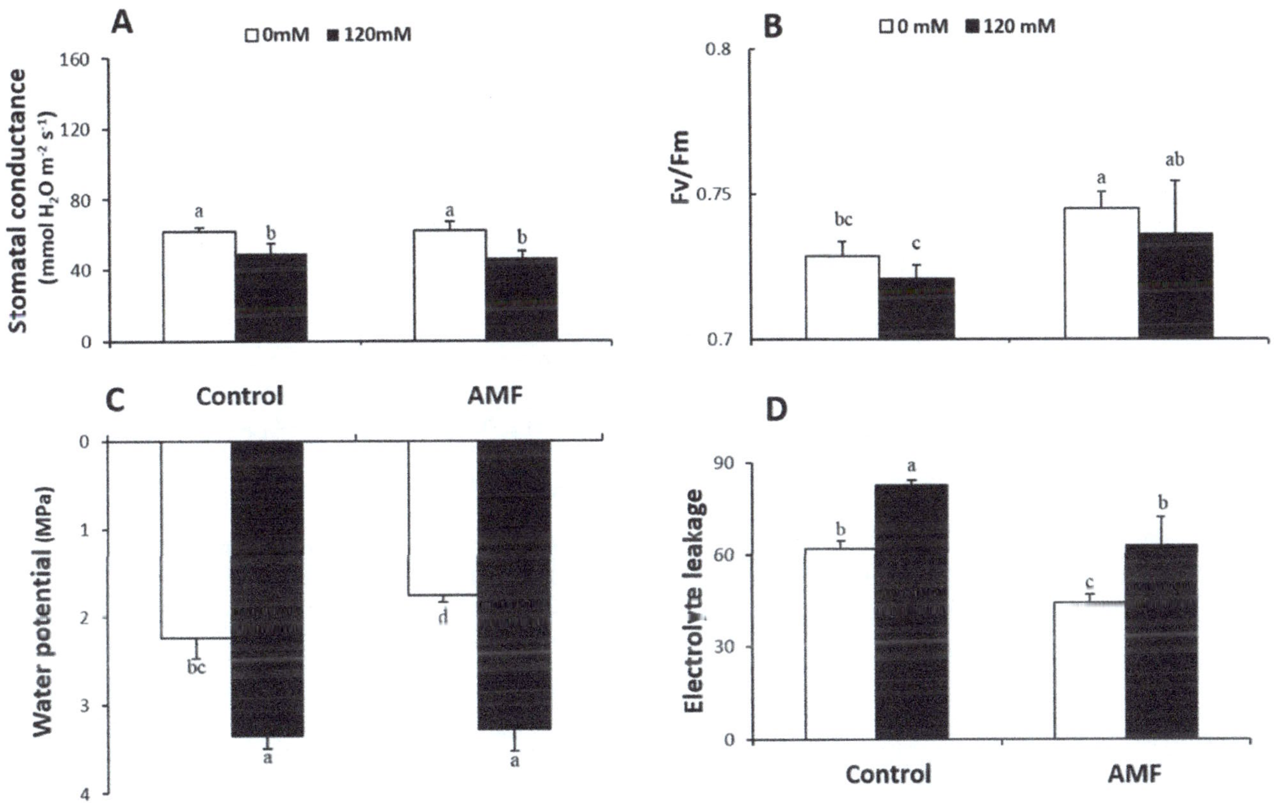

Fig. 12.7 Effect of salinity on stomatal conductance (**a**), photosynthetic efficiency (F_v/F_m) (**b**), water potential (**c**), and electrolyte leakage (**d**) of mycorrhizal and non-mycorrhizal alfalfa after 2 months from AMF inoculation. Bars of each parameter labeled by different letters are significantly different ($P < 0.05$) as assessed by Duncan's test

Table 12.8 Effect of salinity on the polyphenol oxidase (PPO) and peroxidase (POD) activity in shoot and root of mycorrhizal and non-mycorrhizal alfalfa after 2 months from AMF inoculation

NaCl treatment		Shoot		Root	
		PPO µmol of catechol min^{-1} mg^{-1} protein	POD µmol guaiacol mg^{-1} protein min^{-1}	PPO µmol of catechol min^{-1} mg^{-1} protein	POD µmol guaiacol mg^{-1} protein min^{-1}
0 mM	−AMF	0.59 ± 0.03de	0.34 ± 0.10 cd	1.43 ± 0.05d	8.64 ± 0.23e
	+AMF	0.66 ± 0.00 cd	0.24 ± 0.10d	1.86 ± 0.03b	11.74 ± 0.84 cd
120 mM	−AMF	1.02 ± 0.02b	0.46 ± 0.09bc	1.78 ± 0.06bc	14.21 ± 0.72c
	+AMF	0.97 ± 0.03b	0.48 ± 0.06b	2.28 ± 0.11a	21.93 ± 1.91a

Each parameter values labeled by different letters indicate significant differences ($P < 0.05$) as assessed by Duncan's test

Table 12.9 Effect of salinity on growth paramcters of amended and non-amended 10 month-old date palms

NaCl treatment		Plant Height (cm)	Number of leaves	Leaf area (cm^2)	Shoot dry weight ($g\ plant^{-1}$)	Root dry weight ($g\ plant^{-1}$)
0 mM	−Compost	30.43 ± 0.45c	3.4 ± 0.55c	18.81 ± 0.88c	1.55 ± 0.08c	1.10 ± 0.12c
	+Compost	33.7 ± 0.6a	5.0 ± 0.00a	31.72 ± 0.37a	3.60 ± 0.52a	1.98 ± 0.13a
240 mM	−Compost	27.27 ± 0.55d	3.2 ± 0.48c	11.37 ± 1.57d	1.06 ± 0.16d	0.62 ± 0.11d
	+Compost	29.87 ± 0.78b	4.2 ± 0.45b	16.84 ± 0.70b	2.60 ± 0.70b	1.21 ± 0.39b

Values of each parameter labeled by different letters indicate significant differences ($P < 0.05$) as assessed by Duncan's test

NaCl. Date palms grown in the presence of compost under salt stress showed an increase of plant height, number of leaves, leaf area, and dry weight compared to non-amended seedlings under the same conditions.

12.5.4.2 Effect of Compost Application and Salinity Stress on Mineral Nutrition of Date Palm

Figure 12.8 shows that shoot phosphorus content was significantly decreased by salinity (41%), while Na content was enhanced five times in stressed plants compared to the control. In addition, aerial Ca and K concentrations of salt-affected palms decreased by 37 and 31%, respectively, in comparison to the control. Date palm seedlings amended with compost showed an increase in the concentration of K, P, and Ca and a decrease in Na content in the shoot of palms grown in the presence of salt stress.

12.6 Discussion

12.6.1 Effects of AMF on Improving Tolerance of Date Palm to Water Stress

Drought is one of the main constraints that affects plant growth and development, consequently hampering productivity (Fernández-Lizarazo and Moreno-Fonseca 2016). Exposure of plants to drought is identified as the most harmful environmental stress for plants, as it reduces their growth and yield more than any environmental constraint (Izanloo et al. 2008; Osakabe et al. 2014). Besides, date palm is considered a resistant species to drought because of its many adaptive strategies such as the optimization of water use (Meddich et al. 2015), thus affecting the plant morphology, physiology, and biochemical factors (Baslam et al. 2014; Benhiba et al. 2015). The use of AMF can offer a possibility to alleviate drought and other environmental constraints, by forming symbiotic associations with most terrestrial vascular plants. These fungi are famous to improve plant growth and health by enhancing mineral uptake and by improving tolerance to different environmental stresses (Clark and Zeto 2000; Turnau and Haselwandter 2002; Ben-Laouane et al. 2019). Many studies showed that AMF symbiosis could protect the date palm plant against harmful effects of drought by a better capacity to colonize the root system (Baslam et al. 2014; Meddich et al. 2015; Paymaneh et al. 2019). Our study reported that the AMF inoculation has strongly enhanced the host plants' mycorrhization parameters under drought stress. The strong level of root colonization under low soil WC is linked to the best adapted and/or effective AMF-plant interaction under drought (Lee et al. 2012). The root colonization can enhance host plant growth under abiotic stress conditions (Meddich et al. 2018; Ait-El-Mokhtar et al. 2019; Ben-Laouane et al. 2019; Boutasknit et al. 2020). However, the application of AMF benefited date palm height, number of leaves formed, leaf surface, and shoot and root dry matter production compared to the control plants under water stress (Goicoechea et al. 2014; Meddich et al. 2015). The improvement of growth and biomass of mycorrhizal

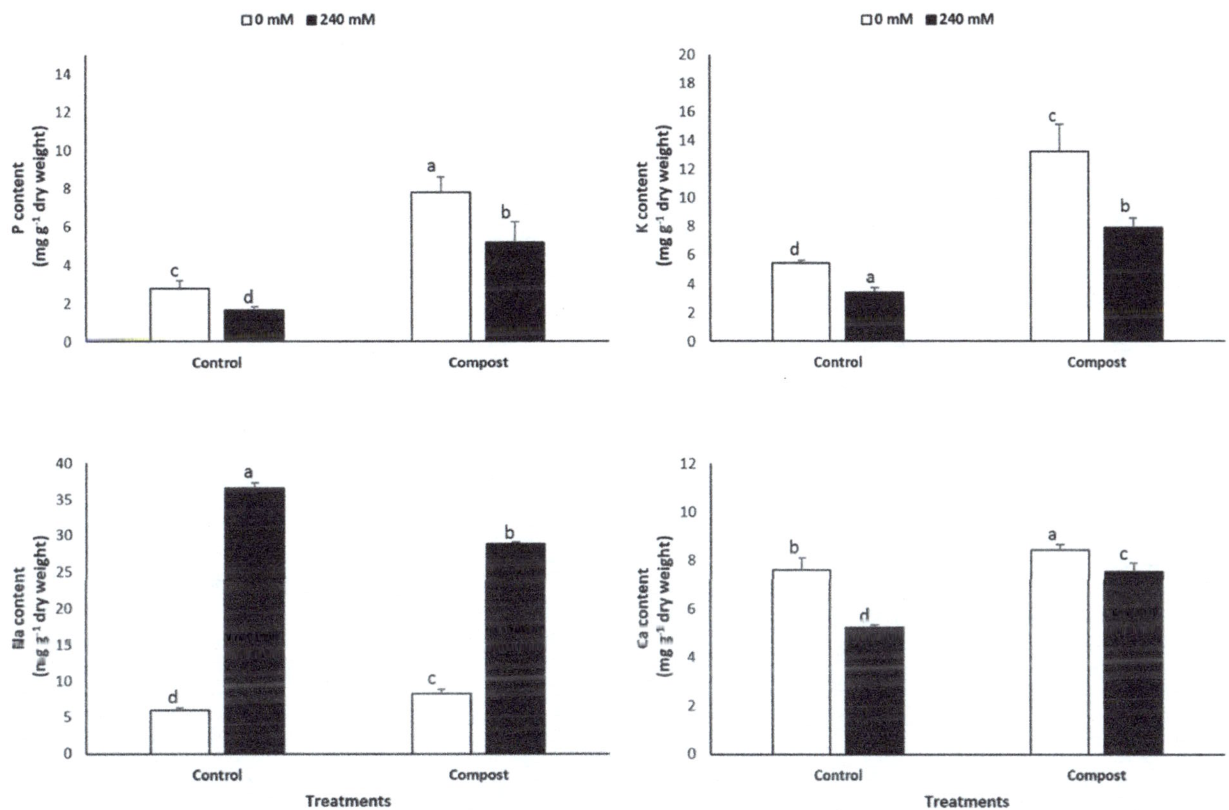

Fig. 12.8 Effect of salinity on aerial mineral uptake of amended and non-amended 10 month-old date palms. Bars of each parameter labeled by different letters are significantly different ($P < 0.05$) as assessed by Duncan's test

plants under water stress indicates a positive role of mycorrhizal symbiosis in host plant growth. In addition, the beneficial effect of AMF on date palm plants growth, under low soil water, may be related to the better uptake of low mobility nutrients from the substrate. In fact, a previous study of our team showed that date palm inoculated with AMF accumulated more mineral nutrition in leaves than non-AMF plants whatever the soil moisture regime is (Baslam and Goicoechea 2012; Baslam et al. 2014; Meddich et al. 2015). The possible mechanism of the stimulation activity of AMF on the performance of date palm is the modulation of the root system architecture and a better absorption of water and poorly mobile macro- and micronutrients (Rouphael et al. 2015; Meddich et al. 2015). It is well known that AMF can enhance the water status of host plants under water stress conditions. Meddich et al. (2015) reported higher leaf WC and LWP in mycorrhizal than in non-AMF plants under drought. This enhanced water status of AMF plants may be due to external hyphal extraction of soil water (Ruiz-Lozano and Azcón 1996), higher stomatal conductance (Zhu et al. 2012; Meddich et al. 2015), indirect effect of increased phosphate and other nutrient uptake (Meddich et al. 2020), greater osmotic adjustment (Boutasknit et al. 2020), as well as greater root hydraulic conductivity (Augé 2001) than non-AMF plants under water stress conditions. Anli et al. (2020) observed that better plant nutrition due to mycorrhizal symbiosis can help date palm to improve water absorption, mitigate the negative impact of water stress and maintain growth. In addition, a better water circulation in the plant may partially explain this tolerance in the presence of AMF. In fact, plant molecular analysis demonstrated the expression of genes coding for synthesis of aquaporins or water channel proteins in the roots of AMF plants under drought stress conditions (Zézé et al. 2007). Besides, AMF hyphae may reward the inhibition of aquaporin activity in AMF plants under water deficit (Quiroga et al. 2019). It is known that prevention of dehydration during drought stress is the result of a close-fitting balance between the ability of root to uptake and distribute water throughout plant tissues (Wahbi et al. 2005; Quiroga et al. 2019). Other studies reported that AMF plants often present greater stomatal conductance compared to non-AMF plants under drought stress (Augé et al. 2014; Boutasknit et al. 2020). Many studies found that plants inoculated with AMF showed an enhancement in the stomatal conductance under drought, such as rosemary (*Rosmarinus officinalis* L.) (Sánchez-Blanco et al. 2004), tangerine (*Citrus tangerine*) (Wu et al. 2008), date palm (*P. dactylifera* L.) (Meddich et al. 2015), and rice (*Oryza sativa* L.) (Ruíz-Sánchez et al. 2011). The increase in the stomatal conductance of mycorrhizal date palms under water stress can be attributed to a better water potential of the leaves. The water potential of plant leaves increased under drought stress according to preexisting literature (Chitarra et al. 2016; Mo et al. 2016; Anli et al. 2020). However, the LWP of mycorrhized date palm was lower than that of the control group under drought stress, which could be due to the fact that AMF helped plants to adapt to drought stress.

Drought stress leads obviously to an overproduction of reactive oxygen species (ROS) which can threaten cells by inducing destructive physiological interference to plants (Cruz De Carvalho 2008). Nevertheless, ROS accumulation may inhibit photosynthesis and respiration and may affect membrane permeability, gene expression, and protein functions (Erice et al. 2007). Plants could protect themselves against oxidative damage caused by stress through enzymatic and non-enzyme detoxification mechanisms of the ROS (Yang et al. 2008; Guo et al. 2010). Caravaca et al. (2005) showed that non-AMF plants under drought stress conditions increased their SOD and POX activities as compared to AMF plants because these plants had a lesser oxidative stress. This demonstrates that during long periods of drought, AMF are crucial to help date palm plants to cope with the oxidative stress (Baslam et al. 2014). As a result, the accumulation of ROS in plants under drought stress was kept at a relatively low level by strengthening the components of the scavenging enzyme system, thereby reducing the degree of cell membrane peroxidation and mitigating the severe damage caused by drought stress. Mycorrhizal plants under water stress have been shown to accumulate soluble sugars that are known to improve their ability to tolerate drought conditions (Boutasknit et al. 2020; Anli et al. 2020). Numerous studies have shown that soluble sugar concentrations in mycorrhizal plants were higher than those in non-mycorrhizal plants under drought stress (Meddich et al. 2015; Huang et al. 2020; Boutasknit et al. 2020).

12.6.2 Effects of AMF on Improving Tolerance of Date Palm and Alfalfa to Salt Stress

As expected, salinity negatively affects alfalfa and date palm growth parameters (shoot and root biomass, plant elongation, and leaf area). It has been reported that root dry weight is generally less damaged by excess salts than shoot dry weight as observed here in alfalfa plants (Munns and Tester 2008). However, in date palm seedlings, root biomass was more critically affected by salt stress than shoot biomass. A previous study explains this by the fact that roots are the first to be exposed to excess soil salinity where they take minerals and water and transport them to the leaves (Garg and Pandey 2016). On the other hand, roots play a key role in salt tolerance, as they are the first area of defense against NaCl and are able to directly reduce or block sodium absorption (Rana et al. 2019) Salt generally affects root length and structure by decreasing cell size and division and affecting differentiation pathways (Bernstein 2013). The decrease in growth due to salinity is reported by previous studies on date palm (Diatta et al. 2014; Sperling et al. 2014) and alfalfa (Elboutahiri et al. 2008; Ashrafi et al. 2014) cultivars. Previous research has indicated that the reduction of plant establishment and growth by salinity is owing to both the osmotic and ionic phases. During the first phase, growth inhibition is mainly due to the difficulty for the plant to absorb soil water. Second phase is due to the toxic effect of the NaCl ions in the plant, which is observed with the higher salt concentration (Porcel et al. 2012; Evelin et al. 2019).

Mycorrhizal status, measured here as mycorrhizal frequency and intensity, was significantly declined under saline conditions. This could be due to delayed germination of spores and reduced growth and extension of AMF hyphae (Juniper and Abbott 2006; Elboutahiri et al. 2010). Salinity might also increase H_2O_2 level in the

AMF roots, which can eventually lead to the deterioration of arbuscules (Fester and Hause 2005).

In this study, AMF inoculation with indigenous Aoufous consortium seemed to improve host tolerance salinity by mitigating the negative effects of salt stress on plant growth and development. Indeed, AMF improved different growth parameters in the salt-stressed alfalfa and date palm plants. The enhanced growth of AMF-treated plants under salt stress may be linked to improved mineral nutrition of the host plant through AMF (Evelin et al. 2019). Previous studies reported that plants treated with AMF in the presence of NaCl stress showed a significant improvement in the uptake of P, K^+, and Ca^{2+}, versus a significant reduction in Na^+ uptake which could promote biomass production (Sallaku et al. 2019). Inorganic phosphate transporters were found on *Glomus versiforme* hyphae that implicated in the direct uptake of P. In addition, a glutamine synthase gene was discovered in *Rhizophagus intraradices*, which increases the probability of N metabolism at the hyphae level that can be transported later to the plant (Salvioli et al. 2012). Felicia et al. (2017) showed that improved nitrogen absorption in salt-stressed AMF plants is related to higher expression of nitrate and ammonium transporters. The improvement of P and N absorption through AMF inoculation could improve the tolerance of alfalfa and date palms to salt stress as demonstrated in our data. This may be due to the role of P in preserving the integrity of vacuolar membrane and facilitating the Na^+ compartmentalization in the vacuoles (Miransari 2011; Bothe 2012). Furthermore, potassium ions are essential to the plant because of their implication in the osmotic equilibrium, have a vital role in the stomata opening/closing, and are a main element in protein biosynthesis (Evelin et al. 2019). On the other hand, calcium plays the role of a second messenger in the nuclear region of root hairs, and during salinity, its level is enhanced to transmit signals (Evelin et al. 2009). Minimizing the storage of Na^+ and Cl^- ions in tissues is a mechanism used by plants to fight against salt stress. Miransari (2011) showed that AMF could alleviate salt stress on plant growth by accumulating Na^+ in their vacuoles as strategies of tolerance to salt stress. The increases in K^+ and Ca^{2+} uptake are other strategies that allowed AMF plants to better tolerate salinity.

AMF application might improve the soil physicochemical and biological traits, which indirectly could positively influence plants' growth and tolerance to abiotic stress. Production of glomalin is the key factor of the contribution of AMF to soil aggregation. This glycoprotein substance improves soil structure by retaining soil particles in aggregates and keeping them stable, which enhances soil aeration and drainage and generates greater microbial activity (Garcia et al. 2019; Ben-Laouane et al. 2020b). Glomalin also plays a critical role in capturing several toxic elements, such as Na^+ ions (Kohler et al. 2010).

Salt stress also disrupted and damaged the photochemical reactions of photosynthesis, particularly PSII, as shown by the decrease in the F_v/F_m ratio in alfalfa and date palm plants. In addition, salinity had an adverse effect on chlorophyll concentration. This negative effect may be related to the disruption of chloroplast by salinity (Parida and Das 2005; Parihar et al. 2015) and specific enzymes responsible for the photosynthetic pigment synthesis (Murkute et al. 2006). A reduction in mineral uptake necessary for chlorophyll biosynthesis may also decrease the chlorophyll concentration. However, the present study showed an increment in chlorophyll content, photosystem II efficiency measured as F_v/F_m ratio, as well as stomata conductance in alfalfa plants and date palm seedlings inoculated with AMF under salt stress. These enhancements may be related to a better photosynthesis capacity as a result of improved CO_2 assimilation and can be linked to mineral uptake enhancement particularly magnesium, an essential constituent of chlorophyll molecule (Sheng et al. 2008). The improvement in the activity of PSII in AMF plants under salt stress was most probably linked to an accumulation of certain metabolites such as proline, sugar, and glycine betaine and enhancement of the antioxidant activity, which may protect the PSII complex and the CO_2-fixing enzymes (Sheng et al. 2008). Similarly, AMF plants and seedling had greater capacity to assimilate CO_2 due to high stomatal conductance. Augé (2000) proposed that enhanced sink strength of AMF roots could be a cause of the observed favorable effect of AMF inoculation on stomatal conductance (gs). In addition, AMF colonization leads to a higher LWP under salinity as a result of changing root morphology (Augé et al. 2008). Sheng et al. (2008) reported that AMF application improved plant WC versus non-AMF plants. This can be explained by the increase of hydraulic conductivity and water absorption ability mediated by AMF application on the plant roots even under salt stress (Kapoor et al. 2008). The enhancement of water uptake could contribute to the maintaining of a larger leaf area and better gs and therefore better CO_2 assimilation. Campanelli et al. (2013) showed the importance of AMF in increasing water availability to the plant, possibly by improving root density, which reduce saline stress in alfalfa. Previous research has shown that AMF association affects plant water regulation by enhancing hydraulic conductivity via the positive regulation of root aquaporin genes, by boosting osmolytes or by activating hormonal signals like abscisic acid (ABA) that mediates stomatal conductance (Xie et al. 2018).

Under stressful conditions, the reduction in photosynthetic performance lets plants absorb more light energy than can be consumed by photosynthetic carbon fixation. This excessive accumulation of energy will lead to generation of ROS in the leaves of plants (Hossain and Dietz 2016). Our data showed that salt stress induced oxidative damage through increased H_2O_2 content, lipid peroxidation, and electrolyte leakage in date palm and alfalfa plants. The same results were reported by Lamaoui et al. (2019) in *Argania spinosa*, Rasool et al. (2013) in *Cicer arietinum* L., Alqarawi et al. (2014) in *Ephedra aphylla*, and Sandhu et al. (2017) in *M. sativa*. Roots are the organs that are directly affected by saline conditions; therefore, they have a higher ROS accumulation than shoots. ROS interact with unsaturated lipid membranes leading to the loss of its integrity (Pedranzani et al. 2016). Our results demonstrated that AMF inoculation protected alfalfa and date palm plants from ROS as indicated by the decrease of electrolyte leakage, MDA and H_2O_2 content in treated plants compared to the control. Stress-induced oxidative damage results from the increase of ROS such as H_2O_2, leading to lipid peroxidation and then membrane dysfunction. ROS have the potential to diffuse through membrane aquaporin for affecting the cells over longer distance (Lamaoui et al. 2019). AMF may have restricted the diffusion of radicals by eliminating them at the sites of production. The reduced ROS accumulation and lower membrane damage in AMF plants may be related to the significant enhancement of antioxidant activities and phosphate metabolism (He et al. 2007).

In the present report, higher activity of antioxidant enzymes such as SOD, CAT, POD, PPO, and APX in AMF plants compared with non-AMF plants under salt stress was combined with the decrease of H_2O_2 accumulation and lipid peroxidation as well as electrolyte leakage reduction. This may indicate a less oxidative damage in the AMF plants. The same results were reported by Abdel Latef and Chaoxing (2011) in *Solanum lycopersicum*, Wu et al. (2010) in *Poncirus trifoliate*, and Hidri et al. (2016) in *Sulla carnosa*. It is known that SOD is the first enzyme that catalyzes the transformation of O_2^- into H_2O_2 and O_2. APX and CAT interact with these active forms of oxygen and keep them at low levels. They act to detoxify H_2O_2 by transforming it in water and oxygen (Mittler 2002). Thus, reduced H_2O_2 and MDA levels in inoculated plants might be due to significant increase in these antioxidant activities (He et al. 2007). Therefore, inoculated date palm and alfalfa plants are more tolerant to salt stress than non-treated ones as the result of the higher antioxidant activity, which may explain their higher biomass growth.

12.6.3 Effects of Compost on Improving Tolerance of Date Palm to Salt Stress

Many researches have reported that organic amendments confer various positive changes in plant growth and physiological and biochemical traits under salinity stress (Tartoura et al. 2014; Beykkhormizi et al. 2018). Compost application significantly alleviated the negative effects of salinity which can be explained by the fact that plants are able to absorb macro- and micro-nutriments provided by the composted waste (Cavagnaro 2014). Moreover, compost amendment can improve soil physicochemical and biological parameters (Weber et al. 2014) and enhance the liberation of nutrients in the soil, making them available to the plant (Akhzari et al. 2015).

The deleterious effects of salinity on plant growth and development may be linked to nutrient imbalances (Evelin et al. 2019). In our report, the P, K^+, and Ca^{2+} uptake in date palm plants decreased under saline condition. In addition, Na^+ uptake was higher in seedlings grown under 240 mM NaCl, inducing the nutrient imbalance and thus disrupting photosynthesis and other metabolic processes (Ramoliya et al. 2004). Compost application has been reported to affect plant physiological functions, particularly mineral uptake, and thus plant tolerance to salinity conditions (Chaichi et al. 2017; Mbarki et al. 2018). Chaichi et al. (2017) reported that tomato plants grown under salt stress treated with compost tea enriched with AMF presented an improvement of ion uptake, such as K^+, Ca^{2+}, and Mg^{2+}. In addition, Mbarki et al. (2018) demonstrated that the organic matter application enhances N and P contents in plants grown in salt stress. The increase of K^+, P, and Ca^{2+} acquisition could be among strategies of salt stress tolerance. The enhancement of date palm tolerance to soil salinity through enhanced P uptake could be related to the role of this element in conserving the cell membrane integrity and in helping the compartmentalization of toxic ions in vacuoles (Bothe 2012). Furthermore, the limitation of Na^+ uptake by the decrease of Na^+ mobility in compost-treated soil could be considered as another strategy to improve plant adaptation in salinity conditions (Mbarki et al. 2018). Moreover, organic amendment has been reported to be directly linked to soil Na removal including Na leaching and Na adsorption ratio (Chaganti and Crohn 2015; Sun et al. 2015).

12.7 Future Prospects

Since the last few decades, the scientific community has highlighted the importance of the use of AMF and compost to overcome various effects of environmental constraints on plant performances in the context of the changing climate. The present study opens new avenues for promising biotechnological techniques to mitigate the deleterious effects of drought and salinity on crop development and productivity. AMF and compost play a pivotal role in plant growth promotion and improve plant tolerance to drought and salt stress, which are closely linked to climate change. The combined effect of AMF and compost should be investigated with different formulations including various tolerant AMF strains and consortia (to drought and salt) and various composts prepared from different organic materials. Furthermore, molecular approaches could present new insights into the underlying mechanisms involved in the alleviation of these abiotic stresses on plant growth and development to promote their productivity in hostile environments.

12.8 Conclusion

Our study showed that inoculation with AMF or compost improved date palm and alfalfa growth under salinity or drought conditions. The native Aoufous consortium was the most effective in increasing the growth, physiological traits, as well as antioxidant enzymes under drought conditions. These indigenous AMF have also been effective in promoting the tolerance of date palm and alfalfa to salt stress. Drought and salt stress significantly reduced growth of date palm and alfalfa plants due to their evident effects on physiological as well as biochemical parameters. These constraints enhanced hydrogen peroxide and lipid peroxidation production inducing membrane integrity loss and simultaneously decreased the essential nutrient uptake. AMF inoculation and application of compost alleviated the harmful impact of drought and salinity on plant growth performances, by enhancing relative water content, LWP, photosynthetic efficiency, pigment content, antioxidant enzymes, and P, K^+, and Ca^{2+} contents. In addition, AMF and compost application mitigated the salt stress-induced changes by reducing the excess uptake of Na^+ under saline conditions. Based on these findings, our results support the view that plants inoculated with AMF or amended with compost can alleviate the deleterious effects under drought and salt stresses and stimulate plant development, mainly growth, physiological traits, accumulation of antioxidant enzymes, and reduction of cell membrane damage. Thus, these indigenous biofertilizers like AMF and compost could be an encouraging biological tool to improve plant performance and development and to alleviate drought and salt stress damages.

References

Abdel Latef AAH, Chaoxing H (2011) Effect of arbuscular mycorrhizal fungi on growth, mineral nutrition, antioxidant enzymes activity and fruit yield of tomato grown under salinity stress. Sci Hortic 127:228–233

Aebi H (1984) Catalase in vitro. Methods Enzymol 105:121–126

Ait-El-Mokhtar M, Ben Laouane R, Anli M, Boutasknit A, Wahbi S, Meddich A (2019) Use of mycorrhizal fungi in improving tolerance of the date palm (*Phoenix dactylifera* L.) seedlings to salt stress. Sci Hortic 253:429–438

Ait-El-Mokhtar M, Baslam M, Ben-Laouane R, Anli M, Boutasknit A, Mitsui T et al (2020a) Alleviation of detrimental effects of salt stress on date palm (*Phoenix dactylifera* L.) by the application of arbuscular mycorrhizal fungi and/or compost. Front Sustain Food Syst 4:131

Ait-El-Mokhtar M, Boutasknit A, Ben-Laouane R, Anli M, El Amerany F, Toubali S et al (2020b) Vulnerability of oasis agriculture to climate change in Morocco. In: Karmaoui A, Barrick K, Reed M, Baig M (eds) Impacts of climate change on agriculture and aquaculture. IGI Global, Hershey, PA, pp 76–106

Ait-El-Mokhtar M, Fakhech A, Anli M, Ben-Laouane R, Boutasknit A, Wahbi S et al (2020c) Infectivity of the palm groves arbuscular mycorrhizal fungi under arid and semi-arid climate and its edaphic determinants towards efficient ecological restoration. Rhizosphere 15:100220

Akhzari D, Attaeian B, Arami A, Mahmoodi F, Aslani F (2015) Effects of vermicompost and arbuscular mycorrhizal fungi on soil properties and growth of *Medicago polymorpha* L. Compost Sci Utili 23:142–153

Alqarawi AA, Abd Allah EF, Hashem A (2014) Alleviation of salt-induced adverse impact via mycorrhizal fungi in *Ephedra aphylla* Forssk. J Plant Interact 9:802–810

Amako K, Chen GX, Asada K (1994) Separate assays specific for ascorbate peroxidase and Guaiacol peroxidase and for the Chloroplastic and cytosolic isozymes of ascorbate peroxidase in plants. Plant Cell Physiol 35:497–504

Anli M, Baslam M, Tahiri A, Raklami A, Symanczik S, Boutasknit A et al (2020) Biofertilizers as strategies to improve photosynthetic apparatus, growth, and drought stress tolerance in the date palm. Front Plant Sci 11:516818

Arnon DI (1949) Copper enzymes in isolated chloroplasts. Polyphenoloxidase in *Beta vulgaris*. Plant Physiol 24:1–15

Ashraf M, Shahbaz M, Ali Q (2013) Drought-induced modulation in growth and mineral nutrients in canola (*Brassica napus* L.). Pak J Bot 45:93–98

Ashrafi E, Zahedi M, Razmjoo J (2014) Co-inoculations of arbuscular mycorrhizal fungi and rhizobia under salinity in alfalfa. Soil Sci Plant Nutr 60:619–629

Augé RM (2000) Stomatal behavior of arbuscular mycorrhizal plants. In: Arbuscular mycorrhizas: physiology and function, pp 201–202

Augé RM (2001) Water relations, drought and vesicular-arbuscular mycorrhizal symbiosis. Mycorrhiza 11:3–42

Augé RM, Toler HD, Sams CE, Nasim G (2008) Hydraulic conductance and water potential gradients in squash leaves showing mycorrhiza-induced increases in stomatal conductance. Mycorrhiza 18:115–121

Augé RM, Toler HD, Saxton AM (2014) Arbuscular mycorrhizal symbiosis alters stomatal conductance of host plants more under drought than under amply watered conditions: a meta-analysis. Mycorrhiza 25:13–24

Azim K, Komenane S, Soudi B (2017) Agro-environmental assessment of composting plants in southwestern of Morocco (Souss-Massa region). Int J Recycl Org Waste Agric 6:107–115

Barrs HD, Weatherly PE (1962) A re-examination of the relative turgidity technique for estimating water deficits in leaves. Aust J Biol Sci 15:413–428

Baslam M, Goicoechea N (2012) Water deficit improved the capacity of arbuscular mycorrhizal fungi (AMF) for inducing the accumulation of antioxidant compounds in lettuce leaves. Mycorrhiza 22(5):347–359

Baslam M, Qaddoury A, Goicoechea N (2014) Role of native and exotic mycorrhizal symbiosis to develop morphological, physiological and biochemical responses coping with water drought of date palm, *Phoenix dactylifera*. Trees Struct Funct 28:161–172

Benhiba L, Fouad MO, Essahibi A, Ghoulam C, Qaddoury A (2015) Arbuscular mycorrhizal symbiosis enhanced growth and antioxidant metabolism in date palm subjected to long-term drought. Trees 29(6):1725–1733

Ben-Laouane R, Meddich A, Bechtaoui N, Oufdou K, Wahbi S (2019) Effects of arbuscular mycorrhizal fungi and rhizobia symbiosis on the tolerance of *Medicago sativa* to salt stress. Gesunde Pflanzen 71(2):135–146

Ben-Laouane R, Baslam M, Ait-El-Mokhtar M, Anli M, Boutasknit A, Ait-Rahou Y et al (2020a) Potential of native arbuscular mycorrhizal fungi, rhizobia, and/or green compost as alfalfa (*Medicago sativa*) enhancers under salinity. Microorganisms 8(11):1695

Ben-Laouane R, Ait-El-Mokhtar M, Anli M, Boutasknit A, Ait Rahou Y, Raklami A et al (2020b) Green compost combined with mycorrhizae and rhizobia: a strategy for improving alfalfa growth and yield under field conditions. Gesunde Pflanzen:1–15

Bernstein N (2013) Effects of salinity on root growth plant roots. In: Eshel A, Beeckman T (eds) Plant roots: the hidden half, vol 36. CRC, Boca Raton, FL, pp 1–18

Beyer WF, Fridovich I (1987) Assaying for superoxide dismutase activity: some large consequences of minor changes in conditions. Anal Biochem 161:559–566

Beykkhormizi A, Sarghein SH, SarafrazArdakani MR, Moshtaghioun SM, Mousavi Kouhi SM (2018) Alleviation of salinity stress by Vermicompost extract: a comparative study on five fennel landraces. Commun Soil Sci Plant Anal 49:2123–2130

Bothe H (2012) Arbuscular mycorrhiza and salt tolerance of plants. Symbiosis 58:7–16

Boutasknit A, Baslam M, Ait-el-mokhtar M, Anli M (2020) Arbuscular mycorrhizal Fungi mediate drought tolerance and recovery in two contrasting carob (*Ceratonia siliqua* L.) ecotypes by regulating stomatal, water relations, and (in) organic adjustments. Plants 9(1):80

Bradford MM (1976) A rapid and sensitive method for the quantitation of microgram quantities of protein utilizing the principle of protein-dye binding. Anal Biochem 72:248–254

Campanelli A, Ruta C, Mastro GD, Morone-Fortunato I (2013) The role of arbuscular mycorrhizal fungi in alleviating salt stress in *Medicago sativa* L. var. icon. Symbiosis 59:65–76

Caravaca F, Alguacil MM, Hernández JA, Roldán A (2005) Involvement of antioxidant enzyme and nitrate reductase activities during water stress and recovery of mycorrhizal *Myrtus communis* and *Phillyrea angustifolia* plants. Plant Sci 169:191–197

Cavagnaro TR (2014) Impacts of compost application on the formation and functioning of arbuscular mycorrhizas. Soil Biol Biochem 78:38–44

Chaganti VN, Crohn DM (2015) Evaluating the relative contribution of physiochemical and biological factors in ameliorating a saline–sodic soil amended with composts and biochar and leached with reclaimed water. Geoderma 259:45–55

Chaichi MR, Keshavarz-Afshar R, Lu B, Rostamza M (2017) Growth and nutrient uptake of tomato in response to application of saline water, biological fertilizer, and surfactant. J Plant Nutr 40: 457–466

Chitarra W, Pagliarani C, Maserti B, Lumini E, Siciliano I, Cascone P et al (2016) Insights on the impact of arbuscular mycorrhizal symbiosis on tomato tolerance to water stress. Plant Physiol 171:1009–1023

Clark RB, Zeto SK (2000) Mineral acquisition by arbuscular mycorrhizal plants. J Plant Nutr 23: 867–902

Cruz De Carvalho MH (2008) Drought stress and reactive oxygen species: production, scavenging and signaling. Plant Signal Behav 3:156–165

Daei G, Ardekani MR, Rejali F, Teimuri S, Miransari M (2009) Alleviation of salinity stress on wheat yield, yield components, and nutrient uptake using arbuscular mycorrhizal fungi under field conditions. J Plant Physiol 166:617–625

Dhindsa RS, Plumbdhindsa P, Thorpe TA (1981) Leaf senescence—correlated with increased levels of membrane-permeability and lipid-peroxidation, and decreased levels of superoxide-dismutase and catalase. J Exp Bot 32:93–101

Diatta ILD, Kane A, Agbangba CE, Sagna M, Diouf D, Aberlenc-Bertossi F et al (2014) Inoculation with arbuscular mycorrhizal fungi improves seedlings growth of two sahelian date palm cultivars (*Phoenix dactylifera* L., cv. Nakhla hamra and cv. Tijib) under salinity stresses. Adv Biosci Biotechnol 05:64–72

El Amerany F, Rhazi M, Wahbi S, Taourirte M, Meddich A (2020) The effect of chitosan, arbuscular mycorrhizal fungi, and compost applied individually or in combination on growth, nutrient uptake, and stem anatomy of tomato. Sci Hortic 21:109015

El Hadrami I, Ramos T, El Bellaj M, El Idrissi-Tourane A, Macheix JJ (1997) A Sinapic derivative as an induced Defence compound of date palm against *fusarium oxysporum* f. sp. *albedinis*, the agent causing Bayoud disease. J Phytopathol 145:329–333

Elboutahiri N, ThamiAlami I, Briz M, Alfaiz C (2008) The effect of salinity and high temperature on biomass production of some alfalfa landraces. Options Méditérr 79:293–297

Elboutahiri N, Thami-Alami I, Udupa SM (2010) Phenotypic and genetic diversity in *Sinorhizobium meliloti* and *S. medicae* from drought and salt affected regions of Morocco. BMC Microbiol 10(1):1–3

Erice G, Aranjuelo I, Irigoyen JJ, Sánchez-Díaz M (2007) Effect of elevated CO2, temperature and limited water supply on antioxidant status during regrowth of nodulated alfalfa. Physiol Plant 130:33–45

Evelin H, Kapoor R, Giri B (2009) Arbuscular mycorrhizal fungi in alleviation of salt stress: a review. Ann Bot 104:1263–1280

Evelin H, Devi TS, Gupta S, Kapoor R (2019) Mitigation of salinity stress in plants by arbuscular mycorrhizal Symbiosis: current understanding and new challenges. Front Plant Sci 10:470

Fahad S, Bajwa AA, Nazir U, Anjum SA, Farooq A, Zohaib A (2017) Crop production under drought and heat stress: plant responses and management options. Front Plant Sci 8:1–16

Felicia V, Russia P, Ingraffia R, Giambalvo D, Frenda AS, Martinelli F (2017) Arbuscular mycorrhizal symbiosis mitigates the negative effects of salinity on durum wheat. PLoS One 12:e0184158

Fernández-Lizarazo JC, Moreno-Fonseca LP (2016) Mecanismos de tolerancia al estréspor déficit hídrico en plantas inoculadas con hongos micorrízicos arbusculares. Una Rev Agron Colomb 34:179–189

Fester T, Hause G (2005) Accumulation of reactive oxygen species in arbuscular mycorrhizal roots. Mycorrhiza 15:373–379

Flajoulot S, Ronfort J, Baudouin P, Barre P, Huguet T, Huyghe et al (2005) Genetic diversity among alfalfa (*Medicago sativa*) cultivars coming from a breeding program, using SSR markers. Theor Appl Genet 111:1420–1429

Garcia L, Dattamudi S, Chanda S, Jayachandran K (2019) Effect of salinity stress and microbial inoculations on glomalin production and plant growth parameters of snap bean (*Phaseolus vulgaris*). Agronomy 9:545

Garg N, Pandey R (2016) High effectiveness of exotic arbuscular mycorrhizal fungi is reflected in improved rhizobial symbiosis and trehalose turnover in Cajanuscajan genotypes grown under salinity stress. Fungal Ecol 21:57–67

Giri B (2017) Mycorrhizal dependency and growth response of *Gliricidia sepium* (Jacq.) Kunth ex Walp. Under saline condition. Plant Sci Today 4:154

Goicoechea N, Baslam M, Erice G, José J (2014) Increased photosynthetic acclimation in alfalfa associated with arbuscular mycorrhizal fungi (AMF) and cultivated in greenhouse under elevated CO2. J Plant Physiol 171:1774–1781

Guo J, Yang Y, Wang G, Yang L, Sun X (2010) Ecophysiological responses of *Abies fabri* seedlings to drought stress and nitrogen supply. Physiol Plant 139:335–347

Gupta B, Huang B (2014) Mechanism of salinity tolerance in plants: physiological, biochemical, and molecular characterization. Int J Genomics 2014:701596

Hasegawa PM, Bressan RA (2000) Plant cellular and molecular responses to high salinity. J Sci Food Agric 91:463–499

He Z, He C, Zhang Z, Zou Z, Wang H (2007) Changes of antioxidative enzymes and cell membrane osmosis in tomato colonized by arbuscular mycorrhizae under NaCl stress. Colloids Surf B Biointerfaces 59:128–133

Hidri R, Barea JM, Ben MOM, Abdelly C, Azcón R (2016) Impact of microbial inoculation on biomass accumulation by *Sulla carnosa* provenances, and in regulating nutrition, physiological and antioxidant activities of this species under non-saline and saline conditions. J Plant Physiol 201:28–41

Hori K, Wada A, Shibuta T (1997) Changes in Phenoloxidase activities of the galls on leaves of *Ulmus davidana* formed by *Tetraneura fuslformis* (Homoptera: Eriosomatidae). Appl Ent Zool 26:365–371

Hossain MS, Dietz KJ (2016) Tuning of redox regulatory mechanisms, reactive oxygen species and redox homeostasis under salinity stress. Front Plant Sci 7:548

Huang D, Ma M, Wang Q, Zhang M, Jing G, Li C et al (2020) Arbuscular mycorrhizal fungi enhanced drought resistance in apple by regulating genes in the MAPK pathway. Plant Physiol Biochem 149:245–255

Izanloo A, Condon AG, Langridge P, Tester M, Schnurbusch T (2008) Different mechanisms of adaptation to cyclic water stress in two south Australian bread wheat cultivars. J Exp Bot 59: 3327–3346

Juniper S, Abbott LK (2006) Soil salinity delays germination and limits growth of hyphae from propagules of arbuscular mycorrhizal fungi. Mycorrhiza 16:371–379

Kamal-Eldin A, Ghnimi S (2018) Classification of date fruit (*Phoenix dactylifera*, L.) based on chemometric analysis with multivariate approach. J Food Meas Charact 12:1020–1027

Kapoor R, Sharma D, Bhatnagar AK (2008) Arbuscular mycorrhizae in micropropagation systems and their potential applications. Sci Hortic 116:227–239

Keeling AA, Griffiths BS, Ritz K, Myers M (1995) Effects of compost stability on plant growth, microbiological parameters and nitrogen availability in media containing mixed garden-waste compost. Bioresour Technol 54:279–284

Kohler J, Caravaca F, Roldán A (2010) An AM fungus and a PGPR intensify the adverse effects of salinity on the stability of rhizosphere soil aggregates of *Lactuca sativa*. Soil Biol Biochem 42: 429–434

Lakhdar A, Hafsi C, Rabhi M, Ouerghi Z, Abdelly C, Montemurro F et al (2008) Application of municipal solid waste compost reduces the negative effects of saline water in *Hordeum maritimum* L. Bioresour Technol 99:7160–7167

Lamaoui M, Chakhchar A, Benlaouane R, El Kharrassi Y, Farissi M, Wahbi S et al (2019) Uprising the antioxidant power of *Argania spinosa* L. callus through abiotic elicitation. C R Biol 342:7–17

Lee BR, Muneer S, Jung WJ, Avice JC, Ourry A, Kim TH (2012) Mycorrhizal colonization alleviates drought-induced oxidative damage and lignification in the leaves of drought-stressed perennial ryegrass (*Lolium perenne*). Physiol Plant 145:440–449

Li LF, Zhang Y, Zhao ZW (2007) Arbuscular mycorrhizal colonization and spore density across different land-use types in a hot and arid ecosystem. Southwest China J Plant Nutr Soil Sci 170: 419–425

Lutts S, Kinet JM, Bouharmont J (1996) NaCl-induced senescence in leaves of rice (*Oryza sativa* L.) cultivars differing in salinity resistance. Ann Bot 78:389–398

Mbarki S, Cerdà A, Zivcak M, Brestic M, Rabhi M, Mezni M et al (2018) Alfalfa crops amended with MSW compost can compensate the effect of salty water irrigation depending on the soil texture. Process Saf Environ Prot 115:8–16

McGonigle TP, Miller MH, Evans DG, Fairchild GL, Swan JA (1990) A new method which gives an objective measure of colonization of roots by vesicular- arbuscular mycorrhizal Fungi. New Phytol 115:495–501

Meddich A, Oihabi A, Abbas Y, Bizid E, Meddich A, Oihabi A et al (2000) Rôle des champignons mycorhiziens à arbuscules de zones arides dans la résistance du trèfle (*Trifolium alexandrinum* L.) au déficit hydrique. Botany 93:369–377

Meddich A, Jaiti F, Bourzik W, El A, Hafidi M (2015) Use of mycorrhizal fungi as a strategy for improving the drought tolerance in date palm (*Phoenix dactylifera*). Sci Hortic 192:468–474

Meddich A, Ait El Mokhtar M, Bourzik W, Mitsui T, Baslam M, Hafidi M (2018) Optimizing growth and tolerance of date palm (*Phoenix dactylifera* L.) to drought, salinity, and vascular fusarium-induced Wilt (*Fusarium oxysporum*) by application of arbuscular mycorrhizal fungi (AMF). In: Root biology, pp 239–250

Meddich A, Oufdou K, Boutasknit A, Raklami A, Tahiri A, Ben-Laouane R et al (2020) Use of organic and biological fertilizers as strategies to improve crop biomass, yields and physico-chemical parameters of soil. In: Meena R (ed) Nutrient dynamics for sustainable crop production. Springer, Singapore, pp 247–288

Miransari M (2011) Interactions between arbuscular mycorrhizal fungi and soil bacteria. Appl Microbiol Biotechnol 89:917–930

Mittler R (2002) Oxidative stress, antioxidants and stress tolerance. Trends Plant Sci 7:405–410

Mo YL, Wang YQ, Yang RP, Zheng JX, Liu CM, Li H et al (2016) Regulation of plant growth, photosynthesis, antioxidation and osmosis by an arbuscular mycorrhizal fungus in watermelon seedlings under well watered and drought conditions. Front Plant Sci 7:644

Mouradi M, Farissi M, Bouizgaren A, Lahrizi Y, Qaddoury A, Ghoulam C (2018) Alfalfa and its Symbiosis responses to osmotic stress. N Perspect Forage Crop 17:149–168

Munns R (2002) Comparative physiology of salt and water stress. Plant Cell Environ 25:239–250

Munns R, Tester M (2008) Mechanisms of salinity tolerance. Annu Rev Plant Biol 59:651–681

Murkute AA, Sharma S, Singh SK (2006) Studies on salt stress tolerance of citrus rootstock genotypes with arbuscular mycorrhizal fungi. Hortic Sci 33:70–76

Nawaz K, Hussain K, Majeed A, Khan F, Afghan S, Ali K (2010) Fatality of salt stress to plants: morphological, physiological and biochemical aspects. Afr J Biotechnol 9:5475–5480

Olsen SR, Dean LA (1965) Phosphorus 1. Methods of soil analysis. Part 2. Chemical and microbiological properties, pp 1035–1049

Osakabe Y, Osakabe K, Shinozaki K, Tran LSP (2014) Response of plants to water stress. Front Plant Sci 5:1–8

Parida AK, Das AB (2005) Salt tolerance and salinity effects on plants: a review. Ecotoxicol Environ Saf 60:324–349

Parihar P, Singh S, Singh R, Singh VP, Prasad SM (2015) Effect of salinity stress on plants and its tolerance strategies: a review. Environ Sci Pollut Res 22:4056–4075

Paymaneh Z, Sarcheshmehpour M, Bukovská P, Jansa J (2019) Could indigenous arbuscular mycorrhizal communities be used to improve tolerance of pistachio to salinity and/or drought? Symbiosis 79(3):269–283

Pedranzani H, Rodríguez-Rivera M, Gutiérrez M, Porcel R, Hause B, Ruiz-Lozano JM (2016) Arbuscular mycorrhizal symbiosis regulates physiology and performance of *Digitaria eriantha* plants subjected to abiotic stresses by modulating antioxidant and jasmonate levels. Mycorrhiza 26:141–152

Phillips JM, Hayman DS (1970) Improved procedures for clearing roots and staining parasitic and vesicular-arbuscular mycorrhizal fungi for rapid assessment of infection. Trans Br Mycol Soc 55:158–IN18

Polle A, Otter T, Seifert F (1994) Apoplastic peroxidases and lignification in needles of Norway spruce (*Picea abies* L.). Plant Physiol 106:53–60

Porcel R, Aroca R, Ruiz-Lozano JM (2012) Salinity stress alleviation using arbuscular mycorrhizal fungi: a review. Agron Sustain Dev 32:181–200

Quiroga G, Erice G, Aroca R, Chaumont F, Ruiz-Lozano JM (2019) Contribution of the arbuscular mycorrhizal symbiosis to the regulation of radial root water transport in maize plants under water deficit. Environ Exp Bot 167:103821

Rady Santander MM, Semida WM, Hemida KA, Abdelhamid MT (2016) The effect of compost on growth and yield of *Phaseolus vulgaris* plants grown under saline soil. Int J Recycl Org Waste Agric 5:311–321

Ramoliya PJ, Patel HM, Pandey AN (2004) Effect of salinization of soil on growth and macro- and micro-nutrient accumulation in seedlings of *Salvadora persica* (Salvadoraceae). For Ecol Manage 202:181–193

Rana MM, Takamatsu T, Baslam M, Kaneko K, Itoh K, Harada N et al (2019) Salt tolerance improvement in rice through efficient SNP and marker-assisted selection coupled with speed-breeding. Int J Mol Sci 20:2585

Rasool S, Ahmad A, Siddiqi TO, Ahmad P (2013) Changes in growth, lipid peroxidation and some key antioxidant enzymes in chickpea genotypes under salt stress. Acta Physiol Plant 35:1039–1050

Rouphael Y, Franken P, Schneider C, Schwarz D, Giovannetti M, Agnolucci M et al (2015) Arbuscular mycorrhizal fungi act as biostimulants in horticultural crops. Sci Hortic 196:91–108

Ruiz-Lozano JM, Azcón R (1996) Viability and infectivity of mycorrhizal spores after long term storage in soils with different water potentials. Appl Soil Ecol 3:183–186

Ruíz-Sánchez M, Armada E, Muñoz Y, García de Salamone IE, Aroca R, Ruíz-Lozano JM et al (2011) *Azospirillum* and arbuscular mycorrhizal colonization enhance rice growth and physiological traits under well-watered and drought conditions. J Plant Physiol 168:1031–1037

Sallaku G, Sandén H, Babaj I, Kaciu S, Balliu A, Rewald B (2019) Specific nutrient absorption rates of transplanted cucumber seedlings are highly related to RGR and influenced by grafting method, AMF inoculation and salinity. Sci Hortic 243:177–188

Salvioli A, Zouari I, Chalot M, Bonfante P (2012) The arbuscular mycorrhizal status has an impact on the transcriptome profile and amino acid composition of tomato fruit. BMC Plant Biol 12(1): 1–2

Sánchez-Blanco MJ, Ferrández T, Morales MA, Morte A, Alarcón JJ (2004) Variations in water status, gas exchange, and growth in *Rosmarinus officinalis* plants infected with *Glomus deserticola* under drought conditions. J Plant Physiol 161:675–682

Sandhu D, Cornacchione MV, Ferreira JFS, Suarez DL (2017) Variable salinity responses of 12 alfalfa genotypes and comparative expression analyses of salt-response genes. Sci Rep 7: 1–18

Santander Schubler C, Sanhueza M, Olave J, Borie F, Valentine A, Cornejo P (2019) Arbuscular mycorrhizal colonization promotes the tolerance to salt stress in lettuce plants through an efficient modification of ionic balance. J Soil Sci Plant Nutr 19(2) 321–331

Scholander PF, Hammel HT, Bradstreet ED, Hemmingsen EA (1965) Sap pressure in vascular plants. Science (80) 148:339–346

Schubler A, Schwarzot D, Walke C (2001) A new fungal phylum, the Glomeromycota: phylogeny and evolution. Mycol Res 105:1413–1421

Sheng M, Tang M, Chen H, Yang B, Zhang F, Huang Y (2008) Influence of arbuscular mycorrhizae on photosynthesis and water status of maize plants under salt stress. Mycorrhiza 18:287–296

Smith JH, Benitez A (1955) Chlorophylls: analysis in plant materials. In: Paech K, Tracey MV (eds) Modern methods of plant analysis/Moderne Methoden der Pflanzenanalyse. Modern methods of plant analysis/Moderne Methoden der Pflanzenanalyse, vol 4. Springer, Berlin, Heidelberg, pp 142–196

Sperling O, Lazarovitch N, Schwartz A, Shapira O (2014) Effects of high salinity irrigation on growth, gas-exchange, and photoprotection in date palms (*Phoenix dactylifera* L., cv. Medjool). Environ Exp Bot 99:100–109

Sun Y, Zhang X, Wu Z, Hu Y, Wu S, Chen B (2016) The molecular diversity of arbuscular mycorrhizal fungi in the arsenic mining impacted sites in Hunan Province of China. J Environ Sci 39:110–118

Tartoura KAH, Youssef SA, Tartoura ESAA (2014) Compost alleviates the negative effects of salinity via up-regulation of antioxidants in *Solanum lycopersicum* L. plants. Plant Growth Regul 74:299–310

Tobar RM, Azcon R, Barea JM (1994) The improvement of plant N acquisition from an ammonium-treated, drought-stressed soil by the fungal symbiont in arbuscular mycorrhizae. Mycorrhiza 4:105–108

Turnau K, Haselwandter K (2002) Arbuscular mycorrhizal fungi, an essential component of soil microflora in ecosystem restoration. Mycorrhizal Technol Agric:137–149

Vayalil PK (2012) Date fruits (*phoenix dactylifera* Linn): an emerging medicinal food. Crit Rev Food Sci Nutr 52:249–271

Velikova V, Yordanov I, Edreva A (2000) Oxidative stress and some antioxidant systems in acid rain-treated bean plants. Plant Sci 151:59–66

Wahbi S, Wakrim R, Aganchich B, Tahi H, Serraj R (2005) Effects of partial rootzone drying (PRD) on adult olive tree (*Olea europaea*) in field conditions under arid climate: I. physiological and agronomic responses. Agric Ecosyst Environ 106:289–301

Weber J, Kocowicz A, Bekier J, Jamroz E, Tyszka R, Debicka M et al (2014) The effect of a sandy soil amendment with municipal solid waste (MSW) compost on nitrogen uptake efficiency by plants. Eur J Agron 54:54–60

Wolf B (1982) Communications in Soil Science and Plant Analysis a comprehensive system of leaf analyses and its use for diagnosing crop nutrient status. Commun Soil Sci Plant Anal 13:1035–1059

Wu QS, Xia RX, Zou YN (2008) Improved soil structure and citrus growth after inoculation with three arbuscular mycorrhizal fungi under drought stress. Eur J Soil Biol 44:122–128

Wu QS, Zou YN, He XH (2010) Contributions of arbuscular mycorrhizal fungi to growth, photosynthesis, root morphology and ionic balance of citrus seedlings under salt stress. Acta Physiol Plant 32:297–304

Xie W, Hao Z, Zhou X, Jiang X, Xu L, Wu S et al (2018) Arbuscular mycorrhiza facilitates the accumulation of glycyrrhizin and liquiritin in *Glycyrrhiza uralensis* under drought stress. Mycorrhiza 28:285–300

Yaish MW, Kumar PP (2015) Salt tolerance research in date palm tree (*Phoenix dactylifera* L.), past, present and future perspectives salt tolerance research in date palm tree (*Phoenix dactylifera* L.), past, present, and future perspectives. Front. Plant Sci 6:348

Yang Y, Chen Y, Li W (2008) Arbuscular mycorrhizal fungi infection in desert riparian forest and its environmental implications: a case study in the lower reach of Tarim River. Prog Nat Sci 18: 983–991

Ye L, Zhao X, Bao E, Cao K, Zou Z (2019) Effects of arbuscular mycorrhizal Fungi on watermelon growth, elemental uptake, antioxidant, and photosystem II activities and stress-response gene expressions under salinity-alkalinity stresses. Front Plant Sci 10:1–12

Zézé A, Brou YC, Meddich A, Marty F (2007) Molecular characterisation of a mycorrhizal inoculant that enhances *Trifolium alexandrium* resistance under water stress conditions. Afr J Biotechnol 6:1524–1528

Zhu XC, Song FB, Liu SQ, Liu TD, Zhou X (2012) Arbuscular mycorrhizae improves photosynthesis and water status of *Zea mays* L. under drought stress. Plant Soil Environ 58(4):186–191

Phosphate-Solubilizing Actinomycetes as Biofertilizers and Biopesticides: Bioformulations for Sustainable Agriculture

13

Brahim Bouizgarne

Abstract

Currently, there is a growing need of biofertilizers and biopesticides for sustainable agriculture, especially those formulated by using microorganisms or their products. Actinobacteria (actinomycetes) could display positive interaction with plants by acting as rhizosphere colonizers, endophytes, or symbionts. Numerous studies focused on their beneficial effects toward plants as growth promoters by mechanisms such as phosphate solubilization, production of phytohormones, decreasing ethylene production, or protection against phytopathogens (by production of siderophores, antibiotics, and lytic enzymes or by inducing systemic resistance). Thus, efficient strategies were developed to explore their potential in order to enhance plant fitness and crop yield. Since the last few decades, interest in actinomycetes able to solubilize insoluble phosphate forms is particularly increasing. However, these works are still not sufficient compared to that on the Gram-positive *Bacillus* and the Gram-negative *Pseudomonas* bacteria. In addition, actinomycetes with antagonistic effects against phytopathogens are largely isolated and their effects proved in greenhouses and/or field experiments. Although phosphate-solubilizing actinomycetes display high diversity, versatility and adaptation to harsh conditions, high metabolite production potential, and suitability to formulations, rare commercial compounds are available in markets. This chapter presents an overview of the diversity and importance of actinomycetes in natural and agricultural soils. It also presents recent knowledge on beneficial traits of phosphate-solubilizing actinomycetes to plants and their application as biofertilizers and/or biopesticides with emphasis on

B. Bouizgarne (✉)
Laboratory of Plant Biotechnology (Biotechnologies Végétales), Faculty of Sciences, University Ibn Zohr, Agadir, Morocco
e-mail: b.bouizgarne@uiz.ac.ma

N. K. Arora, B. Bouizgarne (eds.), *Microbial BioTechnology for Sustainable Agriculture Volume 1*, Microorganisms for Sustainability 33,
https://doi.org/10.1007/978-981-16-4843-4_13

actinobacteria-based formulations and obstacles that hinder their development and commercialization.

Keywords

Phosphate-solubilizing actinomycetes · Diversity · Biofertilizers · Biopesticides · Sustainable agriculture

13.1 Introduction

After nitrogen (N), phosphorus (P) is the most essential nutrient for plant growth and represents one of the major growth-limiting macronutrients. It accounts for between 0.2 and 0.8% of the dry weight of plants, and it is contained within nucleic acids, enzymes, co-enzymes, nucleotides, and phospholipids. P is essential for plant growth and development. Adequate P availability is required in many physiological and biochemical plant activities including cell division, photosynthesis, development of aerial and root systems, formation of flowers and seeds and maturity, N fixation in legumes, and starch storage.

In soil, P exists as insoluble inorganic phosphorus and insoluble organic phosphorus. In general, most soils are poor regarding available P (orthophosphate), and its deficiency severely restricts plant growth and yields. In fact, availability of soluble forms of P in the soils is limited because phosphorus is largely present as complexed, insoluble forms of iron (Fe), aluminum (Al), and calcium (Ca) (respectively calcium phosphate ($Ca_3(PO_4)_2$), in calcareous soils and aluminum phosphate (AlPO) and ferrous phosphate (FePO) in acidic soils. In addition, soluble forms are rapidly converted in complex forms. Fe and Al at high pH and Ca at low pH fix the soluble form into insoluble form rendering it unavailable to the plants (Johnson and Loepper 2006). This conversion is also responsible for the loss of about 75% of chemical P fertilizer, added to soils each year that rapidly becomes unavailable for plants.

Currently, constantly increasing global food demand and need to improve agricultural productivity led to the use of excess chemical containing P fertilizers. The Food and Agriculture Organization (FAO) of the United Nations (2017) reports that the world's population is expected to grow to almost ten billion by 2050, and global food security could be in jeopardy, due to mounting pressures on natural resources and to climate change. Each year, huge amounts of P based fertilizers are applied in conventional agriculture. However, only a low percentage of P contained in these fertilizers (30%) can be used by plants, while most of it is either leached or converted into insoluble phosphates. Thus, conventional agriculture causes soil and groundwater pollution including eutrophication and accumulation of toxic compounds. In addition, direct application of rock phosphate as fertilizer is not effective in most soils as plants could not extract P from these forms. During the last decades, the search for alternative methods such as the isolation and application of phosphate-solubilizing soil microbia gained much attention. Their use for plant growth promotion is a promising ecofriendly strategy for improving crop productivity while preserving environmental quality. Indeed, increasing P use efficiency is a major challenge in intensive agricultural production systems that could be addressed efficiently by these bacteria as it could result in enhancing the growth and yield of economically important crops. Phosphate-solubilizing soil microbia can effectively solubilize insoluble rock phosphate forms or mineralize organic phosphate forms into soluble form, consequently making soluble P bioavailable to plants. They also prevent the released P from being immobilized again (Richardson et al. 2011). Among soil microorganisms, phosphate-solubilizing actinomycetes are of great interest. Nowadays, attempts to develop their application for biological plant growth promotion as alternative strategy is ongoing, and efforts are made toward better understanding of phosphate-solubilizing actinomycetes' functional diversity, their root and rhizosphere colonizing ability, as well as their solubilizing mechanisms and other growth-promoting traits. In addition, actinomycetes are well known for their production of lytic enzymes and secondary metabolites involved in their antagonistic effects against phytopathogens. Among secondary metabolites, antibiotics are often biodegradable, making them a useful tool for the development of biopesticides with minimal side effects toward the environment (Law et al. 2017). In order to make actinomycetes efficient components of sustainable agriculture, development of bioformulations with phosphate-solubilizing actinomycetes is considered an open research field which is still poorly considered. Indeed, one proposed strategy is to combine rock phosphate with slow-release actinomycetes as a sustainable solution for agriculture. In this chapter, we focused on diversity of soil actinobacteria, their ability to solubilize P, and their potential as biofertilizers. The mechanisms of inorganic phosphate solubilization and of organic P mineralization by these actinomycetes are highlighted, and the potential of their formulations and examples of their efficient use as biofertilizers and/or biopesticides in a sustainable agriculture framework are discussed. Finally, their potential as suitable components for diverse formulations is discussed with emphasis on obstacles that hinder development of such products.

13.2 Diversity of Actinobacteria and Their Importance in Soil

Actinomycetes are Gram-positive, filamentous bacteria with a high Shargaff coefficient (GC content greater than 70%). They are aerobic bacteria but some are facultative anaerobes. These bacteria are heterotrophic or chemotrophic but most are chemo-heterotrophs. They are able to use a wide variety of energy sources including recalcitrant complex polymers (Lacombe-Harvey et al. 2018; Castañeda-Cisneros et al. 2020). One of the common features to all actinomycetes is the presence of mycelial structures with frequent formation of aerial mycelium and conidia. These characters bring them closer to fungi. However, these microorganisms are indeed bacteria because they exhibit all the characteristics of prokaryotic microorganisms, in particular the absence of a nuclear membrane surrounding the nucleus. The diameter of their filaments is also similar to that of bacteria; the filaments of species belonging to the genus *Streptomyces* have a diameter of the order of 0.5–2.0 µm (Loria et al. 2003). Actinomycetes were at the origin of the discovery of actinomycin by Wacksman in 1940 from a culture of *Streptomyces antibioticus* (Wacksman and Woodruff 1940) and streptomycin in *Streptomyces griseus* (Barka et al. 2016). To date, these bacteria hold great importance in antibiotic biotechnology despite progress in chemical synthesis. In fact, 75% of the known antibiotics are naturally derived from actinomycetes and more particularly from the genus *Streptomyces*. Taxonomically, actinomycetes belong to the kingdom prokaryotes, Class *Actinomycetes*, and order *Actinomycetales* (Ludwig et al. 2012). Most actinomycetes have vegetative mycelium, also named substrate mycelium and an aerial mycelium. The aerial mycelium forms a cottony or powdery structure. The structure of vegetative mycelium varies depending on the composition of the culture medium, the growth conditions particularly the temperature, and the presence of stimulating or inhibiting substances. These mycelia can show pigmentation variable: white, gray, yellow, orange, etc., and often spores allow them to survive under unfavorable conditions and the resumption of vegetative growth when conditions become less hostile. Actinomycetes are adapted to various ecological environments, so they can be found in natural soils, in fresh or saline waters, and in the air. However, they are more abundant in the soil than other environments. In natural soils, they represent one of the most abundant microorganism groups among the soil microbial community in comparison with other telluric microorganisms. Species belonging to the genus *Streptomyces* can account for almost 95% of the total micropopulation of soil actinomycetes (Barka et al. 2016). Their vertical distribution in soil ranges from the surface to more than 2 m deep (Breton et al. 1989) where their density ranges from about 10^6 to 10^9 cells per gram (Barka et al. 2016). The abundance of actinomycetes at soil level is influenced by several physicochemical parameters of the soil. Indeed, the temperature, the pH, and the humidity of the soils govern their growth. They can be preserved for long periods, thanks to the spores, which germinate in contact with exogenous nutrients (Goodfellow and Williams 1983). Some actinomycetes are pathogenic such as *Streptomyces scabies*, *S. acidiscabies*, *S. brasiliscabiei*, *S. caviscabies*, *S. europaeiscabiei*, *S. ipomea*, *S. stelliscabiei*, *S. turgidiscabies*, and *S. reticuliscabies* (Hudec et al. 2021; Corrêa et al. 2021). Other actinomycetes belonging to the genus *Frankia* are symbiotic of higher plants forming associations called actinorhiza and allowing N fixation (Bouizgarne et al. 2015). However, most known actinomycetes are saprophytes or rhizosphere colonizers. Their saprophyte lifestyle is well studied as they are involved in decomposition of most recalcitrant organic compounds in polymeric form, in particular keratin, chitin, cellulose, and lignins (Gopalakrishnan et al. 2014; Bhatti et al. 2017; Lacombe-Harvey et al. 2018; Gong et al. 2020). By their ability to fix N and to decompose dead organic matter and organic matter recycling, they contribute to agricultural soil fertility.

The diversity among the actinomycetes could be evaluated by studying classical morphological and cultural characteristics as described by the International Streptomyces Project (ISP) (Shirling and Gottlieb 1966) and by more modern molecular methods (gene sequencing) (for review see Bouizgarne and Ait Ben Aoumar 2014).

13.2.1 Phosphate-Solubilizing Actinobacteria as Biofertilizers

In order to apply actinomycetes as efficient plant growth promoters in greenhouses and fields, they should be efficient root colonizers and show good rhizosphere competence (El-Tarabily et al. 2008; Bouizgarne 2013). The dynamics of root colonization ability of actinomycetes was well-studied by Merzaeva and Shirokikh (2006) and Franco et al. (2007). Actinomycetes show various plant growth-promoting (PGP) mechanisms. Mechanisms of action of actinomycetes could be either direct or indirect. Direct mechanisms include N fixation and production of phytohormones and siderophores (Fe chelators), providing plants with soluble phosphate and increasing nutrient uptake, while indirect stimulation of plant growth includes protection against phytopathogens either by antagonistic effects (production of secondary metabolites, competition, parasitism) or induced resistance (Bouizgarne 2013). Figure 13.1 summarizes the mechanisms involved in the beneficial effects of actinobacteria.

Numerous actinomycete genera and species including *Streptomyces*, *Micromonospora*, *Actinopolyspora*, *Actinomadura*, *Kitasatospora*, *Nocardioides*, *Kibdelosporangium*, *Thermobifida*, and *Goirdonia* possess phosphate-solubilizing potential (Hoberg et al. 2005; Barreto et al. 2008; Abdulla 2009; Franco-Correa et al. 2010; Satyaprakash et al. 2017). Phosphate-solubilizing actinomycetes exhibit various mechanisms in order to make P available for plants and contribute to their plant growth promotion abilities. These mechanisms are as follows.

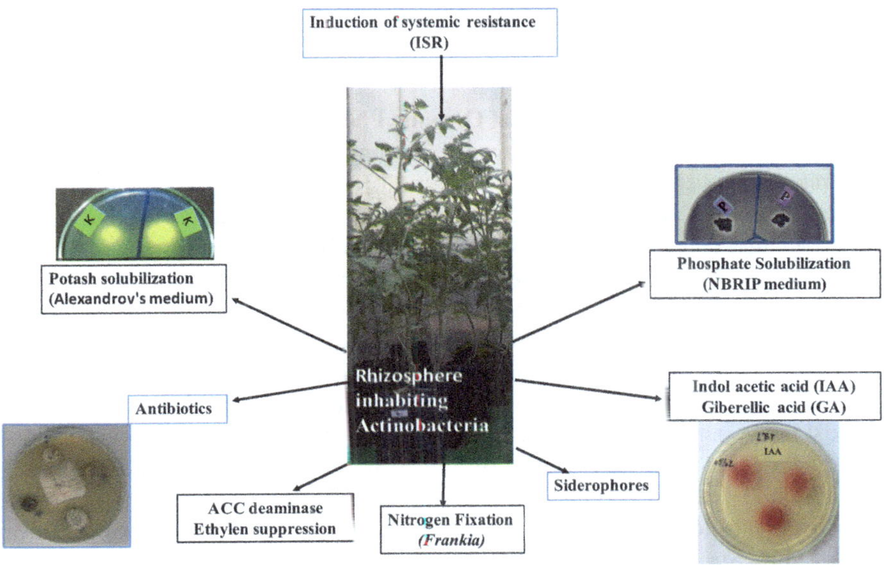

Fig. 13.1 Mechanisms deployed by actinobacteria as plant growth promoting rhizobacteria (PGPR) and biocontrol agents

Solubilization by Lowering Soil pH and Cation Chelation Organic acid production is the principal mechanism for solubilization of soil P. This is allowed by microbial production of organic acids or the release of protons (Hoberg et al. 2005; Abdulla 2009; Kumar et al. 2018). In alkaline soils, solubilization of complex insoluble calcium rock phosphate forms is rendered easy by decreasing soil pH by producing organic acids. Organic acids are the products of glucose which are used as carbon sources (Satyaprakash et al. 2017). The ability of actinomycetes to increase P availability by organic acid release was shown by Hoberg et al. (2005), Abdulla (2009), Farhat et al. (2015), and Vargas Hoyos et al. (2021). Abdulla (2009), investigating actinomycetes originating from St. Katherine, Egypt, found that organic acids were produced by 97% of the isolated actinobacteria. Organic acids such as gluconic and 2-ketogluconic acid were found in culture extracts (Vargas Hoyos et al. 2021). However, gluconic acid is the most commonly reported and most efficient solubilizing agent (Jog et al. 2014; Farhat et al. 2015; Vargas Hoyos et al. 2021). Work by Farhat et al. (2015) demonstrated that solubilizing activity of a *Streptomyces* was proved to be concomitant with a lowering in pH due to the secretion of gluconic acid. Moreover, the bacterium showed presence of the gene *gdh* that encodes the glucose dehydrogenase (GDH) responsible for gluconic acid production from glucose and *pqq* (involved in biosynthesis of the pyrroloquinoline quinone cofactor of GDH). Another work by Jog et al. (2014) aimed to investigate the key glyoxylate cycle enzymes isocitrate dehydrogenase (IDH), isocitrate lyase (ICL), and malate synthase (MS) in the culture supernatant of a strain of *Streptomyces* with high phosphate solubilization potential. Results showed a significant increase in gene expression of the two enzymes (ICL and MS) and their activities that resulted in malate overproduction, responsible for lowered pH during stationary phase.

Solubilization by Mineralization of Organic Phosphate Mineralization occurs in soils where large amount of organic P compounds such as nucleic acids, phospholipids, sugar phosphates, phytic acid, and phosphonates are converted. Mineralization of organic phosphates through the secretion of phosphatases (enzymes that convert organic compounds containing P) is well known. Alkaline phosphatases (Juma and Tabatabai 1988) or acidic phosphatase enzymes (Richardson et al. 2011) were found in various actinomycete genera such as *Actinomadura*, *Micromonospora*, *Streptosporangium*, *Nocardia*, and *Thermobifida* (Franco-Correa et al. 2010; Bhatti et al. 2017). Moreover, Tao et al. (2008) found that solubilization by acidification and mineralization of organic phosphate may coexist in the same actinobacterial strain. Mineralization by both acidic and alkaline phosphatases (enzymes able to hydrolyze phosphoric esters) was found in *Streptomyces* and *Thermobifida* (Franco-Correa et al. 2010), while phytates (enzymes able to convert inositol phosphate) were also found to be another existing mechanism deployed by *Streptomyces* (Reza-Ghorbani-Nasrabadi et al. 2012).

In laboratory, the ability of selected actinobacterial isolates to solubilize phosphate from the insoluble tricalcium phosphate ($Ca_3 (PO_4)_2$), both qualitatively and qualitatively, is usually tested by using solid and liquid media (Nautiyal 1999).

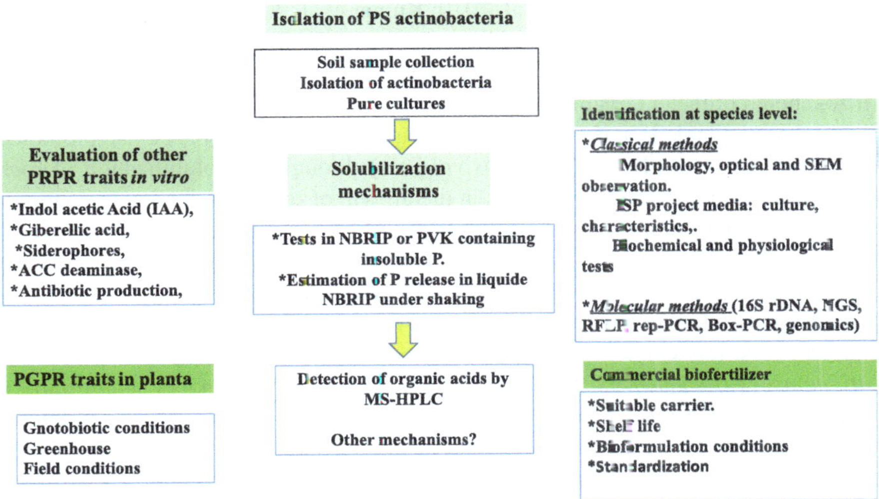

Fig. 13.2 Protocol for isolating phosphate-solubilizing actinobaceria and development of commerial biofertilizers

Plating the bacterial isolates on NBRIP (National Botanical Research Institute's phosphate) containing 10 g/L de Glucose, 5 g/L de $MgCl_2.6H_2O$, 0.25 g/L de $MgSO_4.7H_2O$, 0. 2 g/L de KCl, 0.1 g/L $(NH4)_2SO_4$, 15 g/L d'agar, et 5 g/L de $Ca_3(PO_4)_2$, or Pikovskaya medium (PVK) is used. After incubation, isolates which showed clear halos around their colonies were considered as phosphate solubilizers. Diameter of the halos are measured for each isolate by subtracting colony diameter from the total diameter. The NBRIP medium was successfully used for checking phosphate solubilization by *Actinomyces* and *Streptomyces* (Farhat et al. 2015). Quantitative estimation of phosphate solubilization could be carried out according to Mehta and Nautiyal (2001) on liquid NBRIP medium under shaking and measurements of phosphate concentrations according to Fiske and Subbarow (1925). Figure 13.2 presents protocol scheme for isolation, characterization of phosphate-solubilizing actinobacteria, characterization of their PGP traits in vitro and in presence of plants, as well as development of commercial phosphate-solubilizing-based biofertilizers.

Phosphate-solubilizing microorganisms are group of beneficial microorganisms that could improve plant growth by making soluble P available to them. Among this group, phosphate-solubilizing actinomycetes could enhance plant growth by improving P acquisition efficiency, thereby converting the insoluble forms of P to soluble form. Consequently, they could represent an efficient promising approach for managing P deficiency in agricultural soils and also as biofertilizers. P use efficiency in agricultural lands can be improved through inoculation of soil or seed and seedling treatment by phosphate-solubilizing actinobacteria. Several studies demonstrated that plant growth promotion relies on the ability of the actinomycetes to solubilize phosphate (El-Tarabily et al. 2008; Dicko et al. 2018). Actinomycetes belonging to genera *Streptomyces*, *Streptoverticillium*, and *Micromonospora*

(El-Tarabily et al. 2008; Gupta et al. 2010; Kumar et al. 2018) were studied for their ability to solubilize phosphates. In general, their solubilizing effects were particularly showed in deficient phosphate soils. El-Tarabily et al. (2008) reported that the rhizosphere-competent actinobacterium *Micromonospora endolithica* induced increase in available P in the soil, in greenhouse experiments. This isolate promoted the growth of bean plants (*Phaseolus vulgaris*) through the solubilization of rock phosphate (as evidenced by increases in the growth of roots and shoots) in comparison with another, non-phosphate-solubilizing *Micromonospora* species. In a work by Jog et al. (2014), strains of *Streptomyces* mhcr 0816 and mhce 0811 showing phosphate solubilization activity significantly improved wheat (*Triticum aestivum*) growth concomitant to an increase in mineral nutrients (Fe, manganese [Mn], and P). Similar work was conducted on sorghum by using strains taxonomically close to *Streptomyces tsusimaensis*, *Streptomyces caviscabies*, *Streptomyces setonii*, *Streptomyces africanus*, and an unidentified *Streptomyces* spp. in greenhouse conditions, showing significant enhancement in plant height, leaf area, stem weight, leaf weight, root length, root surface area, root volume, and root dry weight. Experiments conducted in field conditions showed that those isolates resulted in enhancement in rice growth and yield (Gopalakrishnan et al. 2013). More recent work by Boubekri et al. (2021) and Vargas Hoyos et al. (2021), reported that phosphate-solubilizing actinomycete strains enhanced wheat (*T. aestivum*) and soybean growth. Combination of phosphate-solubilizing actinomycetes in a tripartite association with plants and other beneficial microorganisms could result in enhanced plant growth effects by synergy between microorganisms. Co-inoculation of *Streptomyces* MCR9, *Streptomyces* MCR 26, and *Thermobifida* MCR24, with the arbuscular and mycorrhizal fungi *Glomus mosseae*, resulted in stimulated fungal spore germination and fungal development from spores. In pot experiment, this combination led to significantly increased total mycorrhizal root length of *Glomus* inoculated plants, shoot biomass enhancement, and an increased N and P uptake (Franco-Correa et al. 2010). In addition to phosphate solubilization, actinomycetes could also act by other mechanisms including phytohormones such as indole acetic acid (Tokala et al. 2002; Hasegawa et al. 2008) as well as increasing other nutrients' availability such as N. *Streptomyces lydicus* strain WYEC108 showed to increase the nodulation frequency in pea and could produce siderophores allowing improved uptake of iron in pea nodules (Tokala et al. 2002). Table 13.1 presents plant growth promotion exhibited by selected phosphate-solubilizing actinomycetes in experiments under greenhouse or in natural conditions.

13.2.2 Phosphate-Solubilizing Actinomycetes as Biopesticides

The search for antagonists among soil microorganisms is a preferred way to control plant diseases and is considered as a good alternative to chemicals which create imbalances in the microbial community structures and have detrimental effects on the environment. Among natural compounds produced by actinomycetes, antibiotics and enzymes could play a key role in inhibition of other soil microorganisms.

Table 13.1 Plant growth-promoting traits exhibited by phosphate-solubilizing actinobacteria

Actinobacteria	Crop	Growth conditions	Main results	References
Streptomyces rishiriensis 3AS4	Soybean	Greenhouse conditions	Plants inoculated with 3AS4 in the presence of rock phosphate (RP) showed increased plant height by 80% and a shoot-root ratio of 30% higher than non-inoculated control without P addition	Vargas Hoyos et al. (2021)
**Streptomyces albovirids* P18 **Streptomyces griseorubens* BC3 **Streptomyces griseorubens* BC10 **Nocardiopsis alba* BC11 with phosphate and rock potassium solubilization abilities	Wheat (*Triticum aestivum* var. Vitron)	Greenhouse conditions	Significant increase in root length by up to 23.84%, in root volume by up to 71.46%, root dry weight by up to 162.41%, shoot length by up to 23.56%, and shoot dry weight by up to 65.68%	Boubekri et al. (2021)
Streptomyces roseocinereus MS1B15	Barley (*Hordeum vulgare*)	Pots in greenhouse condition	Increased shoot and ear length. Increase in available phosphorus in ears and leaves. Increase in P and N contents in the soil	Chouyia et al. (2020)
**Streptomyces* A20 **Streptomyces* 5.1 **Streptomyces* 7.1	Rice (Fedearroz 733 (and Fedearroz 60)	Gnotobiotic and Greenhouse experiments	Improved growth of rice	Suárez-Moreno et al. (2019)
Strains of *Actinomycetes* sp. H7, O19 and AHB12	Maize	Seed treatments, Pot and greenhouse experiment	*Actinomycetes* sp. H7 significantly increased the fresh and dry biomass of the aerial part Enhancement in seed yield with the combination of the tree strains	Dicko et al. (2018)
**Streptomyces tusimaensis* **Streptomyces caviscabies* **Streptomyces setonii* **Streptomyces africanus* *Streptomyces* sp.	Sorghum	Greenhouse conditions	Increase in plant height, leaf area, stem weight, and leaf weight. Enhancement in root length and dry weight and in the root volume and root surface area	Gopalakrishnan et al. (2013)

(continued)

Table 13.1 (continued)

Actinobacteria	Crop	Growth conditions	Main results	References
**Streptomyces rochei* IDWR19, **Streptomyces thermolilacinus* IDWR81	Wheat (var. Lokwan)	Pot experiments under natural sunlight and temperature	An increase of 12.2 and 24.5% in shoot length of plants, respectively, inoculated with *S. rochei* and *S. thermolilacinus*. Increases in biomass of 1·eight- and 2·threefold were recorded for the two isolates, respectively	Jog et al. (2012)
Streptomyces sp.	Wheat (cv. Chamran)	Greenhouse experiments	In normal and saline conditions: Increased plant growth and development. Increases in germination rate, percentage and uniformity, shoot length, and dry weight. Increased concentrations of N, P, Fe, and Mn in wheat shoots	Sadeghi et al. (2012)
Co-inoculated of arbuscular mycorrhizal fungus *Glomus mosseae* with *Streptomyces* MCR9 *Thermobifida* MCR24 *Streptomyces* MCR 26	Clover plants (*Trifolium repens*)	Pot experiment	Significant plant growth enhancement Increased N and P acquisition by plants. Increase in total mycorrhizal root length of AMF inoculated plants	Franco-Correa et al. (2010)
Microbacterium sp. F10a	Wheat (*Triticum aestivum*)	Pot experiments with uncontaminated and PAH-contaminated soils	Significant increase in wheat growth. In uncontaminated soil, root increased by 36%. In PAH-contaminated soils inoculated by F10A, root dry masses increased by 87% compared with soil inoculated by dead bacteria	Sheng et al. (2009)
Micromonospora endolithica	Bean (*Phaseolus vulgaris*)	Greenhouse	In soil amended with either single super-phosphate or powdered rock phosphate, *M. endolithica* promoted the growth of roots and shoots of bean plants concomitant to an increase in available P in the soil and in N, P, K, S, Mg, and Fe in the roots and shoots	El-Tarabily et al. (2008)

Around 23,000 bioactive secondary metabolites are reported to be produced by microorganisms of which 45% (10,000 compounds) are produced by actinomycetes (Berdy 2005). The genus *Streptomyces* is the largest producer of known secondary metabolites of actinomycetes with approximately 70–80% bioactive natural products (Berdy 2005). Since the discovery of the first bacterial antibiotic by Selman Abraham Waksman (awarded with The Nobel Prize in Physiology or Medicine 1952), a huge progress has been made on developing researches on the discovery of new antibiotics. By their ability to colonize rhizospheric environments, actinomycetes can compete with several telluric microorganisms. Competition could involve production of siderophores, which allow Fe to chelate and thus deprive other microorganisms, particularly phytopatogens. Some actinomycetes are parasites of fungi and produce enzymes such as chitinases and β-1,3 glucanases that allow them to degrade the wall of fungal cells. The ability to produce siderophores and chitinolytic and glucanolytic enzymes, involved in the antagonism against phytopathogens, are well documented (Hong et al. 2002; de Vasconcellos and Cardoso 2009; Khamna et al. 2009; Gopalakrishnan et al. 2014).

Due to their antagonistic abilities against phytopathogens, actinomycetes have been the subject of several studies aimed at the biocontrol of phytopathogenic diseases. For example, the strain *S. lydicus* WYEC108 with mycoparasitic properties toward *Pythium ultimum* and *Rhizoctonia solani* (Yuan and Crawford 1995) acts by the production of chitinases (Mahadevan and Crawford 1997), while works by Hong et al. (2002) demonstrated the involvement 1,3-β-glucanase in antagonism by *Streptomyces* against several fungal pathogens including *Pythium aphanidermatum*, *Colletotrichum higginsianum*, *Acremonium lactucum*, and *Fusarium oxysporum*. Currently, antibiotic-producing actinomycetes constitute a valuable tool in biological control of plant pathogens. The strategy of searching antagonistic actinomycetes had been the subject of many studies mostly in natural soils and rhizospherical soils of various plants. In order to select high effective antagonists, some considerations regarding soil sampling are to be taken. Linderman et al. (1983) and Cook et al. (1987) recommended searching antagonistic microorganisms in healthy plants in a field with diseased plants of the same species. In addition, antagonistic effect of actinomycetes may be linked to the production of antibiotics in rhizospherical soils, which proved to provide more actinomycetes than non-rhizosphere soils (Crawford et al. 1993). For example, population densities of *Streptomyces griseoviridis* were higher in rhizosphere soil of *Brassica rapa* sp. oleifera and *Daucus carota* than in non-rhizosphere soil (Kortemaa et al. 1994, 1997). Thus, it's recommended to search effective actinobacteria in samples of rhizosphere soils (Bouizgarne 2013). At laboratory scale, suitable techniques for screening for effective antagonistic strains are to be considered. In vitro, actinomycete activities could be monitored by using various techniques: direct confrontation, agar disc method, or activities of culture filtrates or culture extracts.

The ability of several actinomycetes to produce antifungal compounds which inhibit in vitro bacterial and fungal growth has been largely studied. This antagonistic effect might allow them to inhibit the microbial growth of several bacterial and fungal phytopathogens in greenhouse and field conditions and thus reduce or eliminate the impact of most plant diseases of soil-borne origin (Hamby and Crawford 2000; El-Tarabily and Sivasithamparam 2006; de Vasconcellos and Cardoso 2009; Jog et al. 2014; Bouizgarne and Ouhdouch 2017). The antagonistic properties of *Streptomyces* against phytopathogens belonging to bacteria and fungi such as *Alternaria*, *Aspergillus*, *Clavibacter*, *Colletotrichum*, *Curvularia*, *Fusarium*, *Penicillium*, *Rhizoctonia*, *Sclerotium*, or *Xanthomonas* are well established (Khamna et al. 2009; de Oliveira et al. 2010; Poovarasan et al. 2013; Jog et al. 2014; Tamreihao et al. 2016). In addition to suppression of bacterial and fungal plant diseases, species of *Streptomyces* also showed antagonistic effects against the plant-parasitic nematode *Meloidogyne incognita* (Nimnoi and Ruanpanun 2020). However, few studies demonstrated that actinobacteria could release their antibiotics in soil. *Streptomyces hygroscopicus* var. geldanus used to control pea root rot caused by *R. solani* produces geldanamycin (an antibiotic with antibacterial, antiprotozoal, antifungal activities), which was detected at soil level (Rothrock and Gottlieb 1984). Nigericin was also detected in soil by Trejo-Estrada et al. (1998). Thus, several actinomycetes or their antifungals have been used in biological control of plant diseases. *Streptomyces griseus* var. autotrophicus or its antifungal metabolite, faeriefungin (Nair et al. 1989), has been used to control *Fusarium* wilt vascular disease of asparagus (*A. officinalis* L) (Smith et al. 1990). Geldanamycin produced by *Streptomyces hygroscopicus* var. geldanus was applied to suppress *R. solani* in soil (Rothrock and Gottlieb 1984). Fungichromine from cultures of *Streptomyces padanus* have been demonstrated effective in reducing damping-off caused by *R. solani* in cabbage (Shih et al. 2003). The effects of the antibiotics blasticidin S, kasugamycin, and oligomycin A have demonstrated to control rice blast caused by *Magnaporthe oryzae* (Law et al. 2017). Importance of antibiotics in biocontrol was also demonstrated, for example, by Beausejour et al. (2001) who showed that a mutant of *Streptomyces* strain defective in the production of geldanamycin lost the ability to control common scab disease. In addition, antagonism by antibiotics could also be involved in plant growth promotion as indirect mechanism. De Vasconcellos and Cardoso (2009) reported that *Streptomyces kasugaensis* able to inhibit the growth of *Fusarium* and *Armillaria* pine rot resulted in plant promotion in growth of *Pinus taeda* under greenhouse experiment. Tokala et al. (2002) demonstrated that *S. lydicus* WYEC 108 induced increased root nodulation and nitrogenase activity and nodular assimilation in pea (*Pisum sativum*). Among the well-known commercialized products, antagonistic *Streptomyces violaceusniger* YCED9, an antifungal producer (Hamby and Crawford 2000), also showed carrot growth promotion under gnotobiotic conditions. Another indirect mechanism by which actinomycetes could enhance growth of plants is protection by induction of systemic resistance toward subsequent challenge by phytopathogens. It was reported that cucumber showed increased disease resistance toward the phytopathogen *Sclerotinia sclerotiorum* by modulated salicylic and jasmonic acid pathways in response to a *Streptomyces* application (Mun et al. 2020).

Although phosphate-solubilizing actinomycetes were isolated in laboratories and their antagonistic effects against phytopathogens were demonstrated in vitro, only few researches demonstrated their protective effects in greenhouse and/or field conditions. It is obvious that strains with phosphate solubilizing abilities (and eventually other PGP) traits) in combination with biocontrol effects could result in more plant fitness and health. It was showed that species of phosphate-solubilizing *Streptomyces* could inhibit phytopathogens of rice (Debananda et al. 2009) or tomato (de Oliveira et al. 2010). Efficient biocontrol effects by phosphate-solubilizing actinomycetes were shown, for instance, against *Pythium aphanidermatum* and *F. oxysporum* f. sp. radicis-lycopersici, respectively, causal agents of damping-off and foot and root rot of tomato (Postma et al. 2013), against *Fusarium* wilt of chickpea (Gopalakrishnan et al. 2011), and against sheath blight disease caused by *Rhizoctonia solani* in rice (Tamreihao et al. 2016). Table 13.2 presents selected examples of phosphate-solubilizing actinomycetes with antagonism toward plant pathogens and, when achieved, their biocontrol abilities of plant diseases.

13.3 Bioformulations of Actinobacteria: Present and Future Perspectives

Sustainable agriculture could rely on soil management, genetic engineering of plants, and application of bioinoculants in order to increase nutrient amounts available to plants or protect them against phytopathogens (Arora 2018). Plant growth-promoting bacteria (PGPB)-based bioinoculants could be applied as bioformulations. Bioformulation is the process of mixing microorganisms (with nutritional supplements) or their products with inert materials to enhance their activity and viability during storage and field application.

Bioformulation of phosphate-solubilizing microbes as inoculants has been the focus of several studies in the last few decades, as it could represent an efficient sustainable, ecofriendly strategy. In fact, the use of less-soluble P sources such as rock phosphate in combination with actinomycetes or microbial consortia (actinobacteria and AMF, or other eubacteria) presents some advantages over phosphate fertilizers: (1) low cost of culture conditions including cheap byproducts and formulations including cost-effective carriers, (2) lower to no environmental detrimental impact, and (3) slow release of P from the rock phosphate to the soil, which could contribute to limit the rate of P adsorption and precipitation. The later aspect has attracted attention, and research is currently directed toward developing smart fertilizers with gradual release of nutrients. Other traits regarding quality and safety standards for registration could be summarized as follows (1) availability of the microorganism in viable and active form and adequate inoculum density, (2) effectiveness as plant growth promoter and/or biocontrol agent against plant pathogens, (3) no toxicity to humans or animals is observed, (4) rapid biodegradation, and (5) no genetic transfer possibility to other related microorganisms.

The carrier plays the role to guarantee long shelf life and to make formulation application to plants easiest possible. Shelf life could be evaluated as the ability to

Table 13.2 Examples of phosphate-solubilizing actinobacteria with antagonism toward plant pathogens and biocontrol abilities

Phosphate-solubilizing actinobacteria	Source of isolation	Pathogen (disease)	Test plant (crop)	Antagonistic or biocontrol effect	Reference
Streptomyces corchorusii UCR3–16	Rice (var. Thesa) rhizosphere soil	*_Bipolaris oryzae_ (brown spot) *_Pyricularia oryzae_ (blast disease) *_Rhizoctonia solani_ (sheath blight) *_Rhizoctonia oryzae-sativae_ (aggregate sheath blight) *_Fusarium oxysporum_ (root rot) *_Curvularia oryzae_ (leaf spot)	Rice	Enhancement of vigor index of inoculated seedlings in rice treated seeds under pathogen-challenged conditions Significantly reduced the sheath blight disease in rice leaves	Tamreihao et al. (2016)
Streptomyces pluricolorscens R18(6)	Tomato (_Lycopersicon esculentum_) (endophytic)	Phytopathogens including *_Xanthomonas campestris_ *_Clavibacter michiganensis_ subsp. _michiganensis_ *_Ralstonia solanacearum_ *_Fusarium. oxysporum_ f. sp. lycopersici	Not tested	Growth inhibition of all tested phytopathogens	de Oliveira et al. (2010)
Streptomyces mhcr0817	Wheat rhizosphere soil _(Triticum aestivum)_	*_Aspergillus niger_ *_Penicillium chrysogenum_ *_Microsporum gypseum_	Not tested	– Inhibition of fungal growth – Production of antifungal metabolite	Jog et al. (2014)
Streptomyces pseudovenezuelae 13.4.2	Marine clay soil	*_Pythium aphanidermatum_ (damping-off) *_Fusarium oxysporum_ f. sp. radicis lycopersici: FORL (crown and root rot)	Tomato	Significant control of _P. aphanidermatum_ and FORL	Postma et al. (2013)
Streptomyces vinaceusdrappus LSCH-10C	Lake sediment	*_Curvularia oryzae_ *_Pyricularia oryzae_ *_Bipolaris oryzae_ *_Fusarium oxysporum_	Not tested	Growth inhibition, chitinase activities	Ningthoujam et al. (2009)

maintain viability at a high population density. In addition, the carrier should support the viability of the actinomycetes during its application. It is obvious that the ability of actinobacteria, particularly those of the genus *Streptomyces*, to form desiccant-resistant spores makes them ideal bioinoculants as they can be formulated in dry formulations. This trait is not found in Gram-negative bacteria. Indeed, dry formulations (granules or powders) are generally preferred because of extended shelf life and easy transportation and storage, which can be suspended in liquid (oil, water, or oil in water emulsion) at the time of application (Sabaratnam and Traquair 2002). Peat, rice, talcum powder, starch granules, Kaolin clay, Zeolite, powdered milk, charcoal, and alginate beads are examples of good carriers (Trejo-Estrada et al. 1998: Sabaratnam and Traquair 2002; Soares et al. 2007; Postma et al. 2013). For instance, biopesticide such as *Streptomyces melanosporofaciens* EF-76, used to control potato's scab disease, was formulated in chitosan-polyphosphate beads (Jobin et al. 2005). *Streptomyces pseudovenezuelae* strain 13.4.2, a biocontrol agent for *F. oxysporum* f. sp. radicis lycopersici (causal agent of tomato foot and root rot disease), was formulated with animal bone charcoal (ABC) (Postma et al. 2013). Among works on bioformulations of phosphate-solubilizing actinobacteria, Tamreihao et al. (2016) used a talcum powder formulation for *Streptomyces corchorusii* UCR3–16 for application as biofertilizer for rice plants.

Actinomycete-based commercial bioformulations with biocontrol activities are Mycostop (a fungicide containing *S. griseoviridis* K61, formulated as wettable powder), Actinovate (a fungicide containing *S. lydicus*, formulated as soluble powder), Nogall (a bactericide containing *Agrobacterium radiobacter* strain K1026), Mykocide (a fungicide containing *Streptomyces colombiensis*), Actofit (an insecticide containing *Streptomyces avermitilis*), Bactophil (a product for controlling seed diseases containing *Streptomyces albus*), Actin (a fungicide containing *Streptomyces atrovirens* as liquid formulation), and Bialaphos (a herbicide containing *S. hygroscopicus* and *Streptomyces viridochromogenes* as powder formulation) (Vurukonda et al. 2018; Gosal et al. 2020). In addition, available are some active substances derived from Streptomyces BLA-S (a fungicide containing blasticidin S), Kasumin and Kasurab (fungicides containing kasugamycin), and Agrimycin (a bactericide containing streptomycin) (Vurukonda et al. 2018). Mycostop, a well-known and largely used biofungicide, based on the mycelium and spores of *S. griseoviridis*, was proved to be effective against *Fusarium* wilt of carnation, damping-off of cruciferous plants, and root rot of cucumber (Tahvoven 1988). In addition, two products with both biocontrol and PGP activities are available: Action Iron® and Micro 108®, both containing *S. lydicus* WYEC108, but to our knowledge, no product with phosphate-solubilizing actinomycetes is available despite of their abovementioned advantages. In the light of such a fact, we can say that while the knowledge of the mechanisms by which actinomycetes are acting as phosphate solubilizers and/or biopesticides, and although numerous research papers are published on development of bioinoculants containing formulations, few actinomycete-based products are currently available in markets, and rare (if not available at all) are those with proved P-solubilizing activities. For instance, Mycostop is the only *Streptomyces*-based plant protection product registered in the EU, Canada, and the USA (Vurukonda et al. 2018), others in Asia and Ukraine. However, most of them are registered as products for plant protection. Among constraints that hinder development and commercialization of bioformulations, we can cite:

- Difficulties regarding isolation procedures, identification of the microbial inoculant, and in developing cost-effective, optimal growing conditions for inoculum production as well as avoiding possible contaminations.
- Suitable carrier for bioformulation to ensure long shelf life and suitable formulation form (powder, liquid, or granule) and method of application (soil inoculation as mixture or by irrigation, seed inoculation, and application to leaves or roots of seedlings) (Law et al. 2017).
- Problems regarding ascertaining efficiencies in field conditions, as in case of many selected microorganisms, usually in vitro does not correlate with their in vivo behavior and consequently expected goals are not reached.
- Problems regarding discrepancies in efficiency caused by ecological parameters interference as bioformulations might be affected by physical and chemical factors such as organic matter, pH, nutrient level, and soil moisture which might result in failure in field conditions.
- Difficulties to meet farmers' needs, particularly in terms of ease of application method and concerns about comparative costs and efficiency between chemical and biological products. In this context, biological products should be attractive.
- Concerns regarding regulatory standards: quality control and regulation of biofertilizers and biopesticides (Arora et al. 2016).

13.4 Conclusion

Numerous studies addressed the search and application of plant growth-promoting bacteria or rhizobacteria (respectively, PGPB and PGPR). Actinomycetes could constitute a group of great importance due to mechanisms that benefit plant growth. Thus, sustainable agriculture, based on environmentally friendly techniques by using actinomycetes as growth promoters and/or disease biocontrol agents, could reduce the use of chemicals. The objective is to reduce negative impact on environment while developing smart, slow P-releasing biofertilizer formulations. Though numerous researches concerned actinomycetes as a particular soil-adapted taxon, few researches addressed the application of phosphate-solubilizing actinomycetes as PGPB for enhancing plant growth and/or as biopesticides at field scale. The advantages of using actinomycetes are numerous: (1) they are adapted to most soils with ability to thrive in different environmental conditions, (2) their high rhizosphere competence and adaptation to root conditions of most crops, (3) their ability to produce resisting spores, (4) their remarkable production of diverse metabolites with antibiotic activity, especially by *Streptomycetes* strains, (5) their biocontrol abilities toward plant pathogens, and (6) their suitability to formulations as they can be formulated as dry or liquid formulations with relatively long shelf life. When applied in fields, several of them might display other PGP mechanisms (phytohormone production, siderophore productions, etc.) that can contribute largely to their effects in soils. Thus, adequate understanding of the mechanisms involved in enhancing plant growth effects and their interaction with biotic and abiotic factors should be taken into consideration.

In addition to lytic enzymes and antibiosis, eliminating deleterious phytopathogens could also be due to systemic resistance induction that confers protection to plants. Despite their remarkable traits, actinomycetes are under-explored for plant growth promotion, as compared to other bacteria such as *Pseudomonas*, *Bacillus*, or fungi (such as Phosphomax, Symbion-P, or Biophos). It is noticed that only few commercial efficient products based on actinomycetes are available for use in agriculture and no phosphate-solubilizing bioformulations of actinomycetes are available. Abovementioned obstacles are to be overcome, and researchers are encouraged to make actinomycetes available to farmers, by producing multifunctional formulations with phosphate-solubilizing actinomycetes as biofertilizers and as biopesticides. Researches in this topic are worth to be encouraged, and deep investigations should be performed for species not yet studied. Species able to produce various PGP mechanisms including phosphate solubilization, lytic enzymes, show antibiosis and induction of systemic resistance in plants are considered great candidates for successful formulations and sustainable agriculture.

References

Abdulla H (2009) Bioweathering and biotransformation of granitic rock minerals by actinomycetes. Microb Ecol 58:753–761

Arora NK (2018) Agricultural sustainability and food security. Environ Sustain 1(3):1–3

Arora NK, Verma M, Prakash J, Mishra J (2016) Regulation of biopesticides: global concerns and policies. In: Arora NK, Mehnaz S, Balestrini R (eds) Bioformulations: for sustainable agriculture. Springer, New Delhi, pp 283–299

Barka EA, Vatsa P, Sanchez L, Gaveau-Vaillant N, Jacquard C, Meier-Kolthoff JP, Klenk HP, Clément C, Ouhdouch Y, van Wezel GP (2016) Taxonomy, physiology, and natural products of Actinobacteria. Microbiol Mol Biol Rev 80:1–43

Barreto TR, da Silva ACM, Soares ACF, de Souza JT (2008) Population densities and genetic diversity of actinomycetes associated to the rhizosphere of *Theobroma cacao*. Braz J Microbiol 39:464–470

Beausejour J, Agbessi S, Beaulieu C (2001) Geldanamycin producing strains as biocontrol agents against common scab of potato. Can J Microbiol 23:194

Berdy J (2005) Bioactive microbial metabolites. J Antibiot 58:1–26

Bhatti AA, Haq S, Bhat RA (2017) Actinomycetes benefaction role in soil and plant health. Microb Pathog 111:458–467

Boubekri K, Soumare A, Mardad I, Lyamlouli K, Hafidi M, Ouhdouch Y, Kouisni L (2021) The screening of potassium- and phosphate-solubilizing Actinobacteria and the assessment of their ability to promote wheat growth parameters. Microorganisms 9:470

Bouizgarne B (2013) Bacteria for plant growth promotion and disease management. In: Maheshwari D (ed) Bacteria in agrobiology: disease management. Springer, Berlin, Heidelberg, pp 15–47

Bouizgarne B, Ait Ben Aoumar A (2014) Diversity of plant associated Actinobacteria. In: Maheshwari DK (ed) Bacterial diversity in sustainable agriculture, vol 1. Springer, Cham, pp 41–99

Bouizgarne B, Ouhdouch Y (2017) Management of tomato foot and root rot (TFRR) by biocontrol agents with emphasis on factors affecting its effectiveness. In: Kumar V, Kumar M, Sharma S, Prasad R (eds) Probiotics and plant health. Springer, Singapore, pp 1–19

Bouizgarne B, Oufdou K, Ouhdouch Y (2015) Actinorhizal and Rhizobial–legume symbioses for alleviation of abiotic stresses. In: Arora NK (ed) Plant microbes symbiosis: applied facets. Springer, New Delhi, pp 273–295

Breton A, Theilleux J, Sanglier JJ, Viobis G (1989) Organismes producteurs: biologie, taxonomie et écologie. In: Larpent JP, Sanglier JJ (eds) Biotechnologies des antibiotiques. Masson, Paris, pp 33–70

Castañeda-Cisneros YE, Mercado-Flores Y, Anducho-Reyes MA, Álvarez-Cervantes J, Ponce-Lira B, Evangelista-Martínez Z, Téllez-Jurado A (2020) Isolation and selection of *Streptomyces* species from semi-arid agricultural soils and their potential as producers of xylanases and cellulases. Curr Microbiol 77:3460–3472

Chouyia FE, Romano I, Fechtali T, Fagnano M, Fiorentino N, Visconti D, Idbella M, Ventorino V, Pepe O (2020) P-solubilizing *Streptomyces roseocinereus* MS1B15 with multiple plant growth-promoting traits enhance barley development and regulate rhizosphere microbial population. Front Plant Sci 11:1137

Cook RJ, Sitton JW, Haglund WA (1987) Influence of soil treatments on growth and yield of wheat and implications for control of Pythium root rot. Phytopathology 77:1192–1198

Corrêa DBA, do Amaral DT, da Silva MJ, Destéfano SAL (2021) *Streptomyces brasiliscabiei*, a new species causing potato scab in South Brazil. Antonie Van Leeuwenhoek. https://doi.org/10.1007/s10482-021-01566-y

Crawford DL, Lynch JM, Whipps JM, Ousley MA (1993) Isolation and characterization of Actinomycetes antagonists of a fungal root pathogen. Appl Environ Microbiol 59:3899–3905

de Oliveira MIF, da Silva MG, Van Der Sand ST (2010) Anti-phytopathogen potential of endophytic actinobacteria isolated from tomato plants (*Lycopersicon esculentum*) in southern Brazil, and characterization of *Streptomyces* sp. R18(6), a potential biocontrol agent. Res Microbiol 161:565–572

de Vasconcellos RLF, Cardoso EJBN (2009) Rhizospheric streptomycetes as potential biocontrol agents of *fusarium* and *Armillaria* pine rot and as PGPR for *Pinus taeda*. BioControl 54:807–816

Debananda SN, Suchitra S, Salam N (2009) Screening of actinomycete isolates from niche habitats in Manipur for antibiotic activity. Am J Biochem Biotech 5(4):221–225

Dicko AH, Babana AH, Kassogué A, Fané R, Nantoumé D, Ouattara D, Dao S (2018) A Malian native plant growth promoting Actinomycetes based biofertilizer improves maize growth and yield. Symbiosis 75(3):267–275

El-Tarabily KA, Sivasithamparam K (2006) Non-streptomycete Actinomycetes as biocontrol agents of soil-borne fungal plant pathogens and as plant growth promoters. Soil Biol Biochem 38:1505–1520

El-Tarabily KA, Nassar AH, Sivasithamparam K (2008) Promotion of growth of bean (*Phaseolus vulgaris* L.) in a calcareous soil by a phosphate solubilizing rhizosphere-competent isolate of *Micromonospora endolithica*. Appl Soil Ecol 39:161–171

Farhat MB, Boukhris I, Chouayekh H (2015) Mineral phosphate solubilization by *Streptomyces* sp. CTM396 involves the excretion of gluconic acid and is stimulated by humic acids. FEMS Microbiol Lett 362(5):fnv008

Fiske CH, Subbarow Y (1925) The colorimetric determination of phosphorus. J Biol Chem 66:375–400

Food and Agriculture Organization of the United Nations, FAO (2017) The future of food and agriculture—trends and challenges. FAO, Rome. ISBN 978-92-5-109551-5

Franco C, Michelsen P, Percy N, Conn V, Listiana E, Moll S, Loria R, Coombs J (2007) Actinobacterial endophytes for improved crop performance. Aust Plant Pathol 36:524–531

Franco-Correa M, Quintana A, Duque C, Suarez C, Rodríguez MX, Barea JM (2010) Evaluation of actinomycete strains for key traits related with plant growth promotion and mycorrhiza helping activities. Appl Soil Ecol 45:209–217

Gong X, Xiang W, Cao XYY, Hao Y, Li L, Wang Q, Zou H, Qian C (2020) Microbispora cellulosiformans sp. nov., a novel actinomycete with cellulase activity isolated from soil in the cold region. Antonie Van Leeuwenhoek 113:2053–2062

Goodfellow M, Williams ST (1983) Ecology of actinomycetes. Annu Rev Microbiol 37:189–216

Gopalakrishnan S, Pande S, Sharma M, Humayun P, Kiran BK, Sandeep D, Vidya MS, Deepthi K Rupela O (2011) Evaluation of *actinomycete* isolates obtained from herbal vermicompost for biological control of *fusarium* wilt of chickpea. Crop Prot 30:1070–1078

Gopalakrishnan S, Srinivas V, Vidya SM, Rathore A (2013) Plant growth-promoting activities of *Streptomyces* spp. in sorghum and rice. SpringerPlus 2:574

Gopalakrishnan S, Vadlamudi S, Bandikinda P, Sathya A, Vijayabharathi R, Rupela O, Himabindu K, Katta K, Varshney RK (2014) Evaluation of *Streptomyces* strains isolated from herbal vermicompost for their plant growth-promotion traits in rice. Microbiol Res 169:40–48

Gosal SK, Kaur J, Kaur J (2020) Microbial biotechnology: a key to sustainable agriculture. In: Kumar M, Kumar V, Prasad R (eds) Phyto-microbiome in stress regulation. Springer, Singapore, pp 219–243

Gupta N, Sahoo D, Basak U (2010) Evaluation of *in vitro* solubilization potential of phosphate solubilising *Streptomyces* isolated from phyllosphere of *Heritiera fomes* (mangrove). Afr J Microbiol Res 4:136–142

Hamby MK, Crawford DL (2000) The enhancement of plant growth by selected Streptomyces species. In: American Society for Microbiology, 100th General Meeting, Los Angeles, CA. Abstract no: 567

Hasegawa S, Meguro A, Shimizu M, Nishimura T, Toyoda K, Shiraishi T, Kunoh H (2008) Two bioassay methods to evaluate root-accelerating activity of *Streptomyces* sp. MBR52 metabolites. Actinomycetologica 22:42–45

Hoberg E, Marschner P, Lieberei R (2005) Organic acid exudation and pH changes by *Gordonia* sp. and *Pseudomonas fluorescens* grown with P adsorbed to goethite. Microbiol Res 160 177–187

Hong T-Y, Cheng C-W, Huang J-W, Meng M (2002) Isolation and biochemical characterization of an endo-1,3-β-glucanase from *Streptomyces sioyaensis* containing a C-terminal family 6 carbohydrate-binding module that binds to 1,3-β-glucan. Microbiology 148:1151–1159

Hudec C, Novinscak A, Filion M (2021) Diversity and virulence of *Streptomyces* spp. causing potato common scab in Prince Edward Island, Canada. Phytopathology 111:617–626

Jobin G, Couture G, Goyer C, Brzezinski R, Beaulieu C (2005) Streptomycete spores entrapped in chitosan beads as a novel biocontrol tool against common scab of potato. Appl Microbiol Biotechnol 68:104–110

Jog R, Nareshkumar G, Rajkumar S (2012) Plant growth promoting potential and soil enzyme production of the most abundant *Streptomyces* spp. from wheat rhizosphere. J Appl Microbiol 113:1154–1164

Jog R, Pandya M, Nareshkumar G, Rajkumar S (2014) Mechanism of phosphate solubilization and antifungal activity of *Streptomyces* spp. isolated from wheat roots and rhizosphere and their application in improving plant growth. Microbiology 160:778–788

Johnson SE, Loepper RH (2006) Role of organic acids in phosphate mobilization from iron oxide. Soil Sci Soc Am J 70:222–234

Juma NG, Tabatabai MA (1988) Phosphatase activity in corn and soybean roots: conditions for assay and effects of metals. Plant and Soil 107:39–47

Khamna S, Yokota A, Lumyong S (2009) Actinomycetes isolated from medicinal plant rhizosphere soils: diversity and screening of antifungal compounds, indole-3-acetic acid and siderophore production. World J Microbiol Biotechnol 25:649–655

Kortemaa H, Rita H, Haahtela K, Smolander A (1994) Root colonization ability of antagonistic *Streptomyces griseoviridis*. Plant and Soil 163:77–83

Kortemaa H, Pennanen A, Smolander A, Haahtela K (1997) Distribution of antagonistic *Streptomyces griseoviridis* in rhizosphere and non-rhizosphere sand. J Phytopathol 145:137–143

Kumar A, Kumar A, Patel H (2018) Role of microbes in phosphorus availability and acquisition by plants. Int J Curr Microbiol App Sci 7:1344–1347

Lacombe-Harvey M-È, Brzezinski R, Beaulieu C (2018) Chitinolytic functions in actinobacteria: ecology, enzymes, and evolution. Appl Microbiol Biotechnol 102:7219–7230

Law JW-F, Ser H-L, Khan TM, Chuah L-H, Pusparajah P, Chan K-G, Goh B-H, Lee L-H (2017) The potential of *Streptomyces* as biocontrol agents against the rice blast fungus, *Magnaporthe oryzae* (*Pyricularia oryzae*). Front Microbiol 8:3

Linderman RG, Moore LW, Baker KF, Cooksey DA (1983) Strategies for detecting and characterizing systems for biological control of soilborne plant pathogens. Plant Dis 67:1058–1064

Loria R, Coombs J, Yoshida M, Kers J, Bukhalid R (2003) A paucity of bacterial root diseases: *Streptomyces* succeeds where others fail. Physiol Mol Plant Pathol 62:65–72

Ludwig W, Euzéby J, Whitman WB (2012) Taxonomic outline of the phylum Actinobacteria. In: Goodfellow M, Kämpfer P, Busse H-J, Trujillo ME, Suzuki K, Ludwig W, Whitman WB (eds) Bergey's manual of systematic bacteriology, vol 5, 2nd edn. Springer, New York, pp 29–31

Mahadevan B, Crawford DL (1997) Properties of the chitinase of the biocontrol agent *Streptomyces lydicus* WYEC108. Enzyme Microb Technol 20:489–493

Mehta S, Nautiyal CS (2001) An efficient method for qualitative screening of phosphate-solubilizing bacteria. Curr Microbiol 43:51–56

Merzaeva OV, Shirokikh IG (2006) Colonization of plant rhizosphere by actinomycetes of different genera. Microbiology 75:226–230

Mun B-G, Lee W-H, Kang S-M, Lee S-U, Lee S-M, Lee DY, Shahid M, Yun B-W, Lee I-N (2020) *Streptomyces* sp. LH 4 promotes plant growth and resistance against *Sclerotinia sclerotiorum* in cucumber via modulation of enzymatic and defense pathways. Plant and Soil 448:87–103

Nair MG, Putnam AR, Mishra SK, Mulks MH, Taft WH, Keller JE, Miller JR, Zhu PP, Meinhart JD, Lynn DG (1989) Faeriefungin: a new broad-spectrum antibiotic from *Streptomyces griseus* var. *autotrophicus*. J Nat Prod 52:797–809

Nautiyal CS (1999) An efficient microbiological growth médium for screening phosphate solubilizing microorganisms. FEMS Microbiol Lett 170:265–270

Nimnoi P, Ruanpanun P (2020) Suppression of root-knot nematode and plant growth promotion of chili (*Capsicum flutescens* L.) using co-inoculation of *Streptomyces* spp. Biol Control 145: 104244

Ningthoujam DS, Sanasam S, Tamreihao K, Nimaichand S (2009) Antagonistic activities of local actinomycete isolates against rice fungal pathogens. Afr J Microbiol Res 3:737–742

Poovarasan S, Mohandas S, Paneerselvam P, Saritha B, Ajay KM (2013) Mycorrhizae colonizing actinomycetes promote plant growth and control bacterial blight disease of pomegranate (*Punica granatum* L. cv Bhagwa). Crop Prot 53:175–181

Postma J, Clematis F, Nijhuis EH, Someus E (2013) Efficacy of four phosphate-mobilizing bacteria applied with an animal bone charcoal formulation in controlling *Pythium aphanidermatum* and *fusarium oxysporum* f. sp. *radices lycopersici* in tomato. Biol Control 67:284–291

Reza-Ghorbani-Nasrabadi GR, Alikhani HA, Hamedi J (2012) Identification and determination of extracellular phytate-degrading activity in actinomycetes. World J Microbiol Biotechnol 28: 2601–2608

Richardson AE, Lynch JP, Ryan PR, Delhaize E, Smith FA, Smith SE, Harvey PR, Ryan MH, Veneklaas EJ, Lambers H, Oberson A, Culvenor RA, Simpson RJ (2011) Plant and microbial strategies to improve the phosphorus efficiency of agriculture. Plant and Soil 349:121–156

Rothrock CS, Gottlieb D (1984) Role of antibiosis in antagonism of *Streptomyces hygroscopicus* var *geldanus* to *Rhizoctonia solani* in soil. Can J Microbiol 30:1440–1447

Sabaratnam S, Traquair JA (2002) Mechanism of antagonism by *Streptomyces griseocarneus* (strain Di944) against fungal pathogens of greenhouse-grown tomato transplants. Can J Plant Pathol 37:197–211

Sadeghi A, Karimi E, Dahaji PA, Javid MG, Dalvand Y, Askari H (2012) Plant growth promoting activity of an auxin and siderophore producing isolate of *Streptomyces* under saline soil conditions. World J Microbiol Biotechnol 28:1503–1509

Satyaprakash M, Nikitha T, Reddi EUB, Sadhana B, Vani SS (2017) A review on phosphorous and phosphate solubilising bacteria and their role in plant nutrition. Int J Curr Microbiol App Sci 6: 2133–2144

Sheng XF, He LY, Zhou L, Shen YY (2009) Characterization of *Microbacterium* sp. F10a and its role in polycyclic aromatic hydrocarbon removal in low-temperature soil. Can J Microbiol 55: 529–535

Shih H-D, Liu Y-C, Hsu F-L, Mulabagal V, Dodda R, Huang J-W (2003) Fungichromin: a substance from *Streptomyces padanus* with inhibitory effects on *Rhizoctonia solani*. J Agric Food Chem 51:95–99

Shirling E, Gottlieb D (1966) Methods for characterization of *Streptomyces* species. Int J Syst Bacteriol 16:313–340

Smith J, Putnam A, Nair M (1990) In vitro control of *fusarium* diseases of *Asparagus officinalis* L. with a *Streptomyces* or its polyene antibiotic, Faeriefungin. J Agric Food Chem 38:1729–1733

Soares ACF, Da Silva C, Da Silva M, Perez JO (2007) Production of streptomycete inoculum in sterilized rice. Sci Agric 64:641–644

Suárez-Moreno ZR, Vinchira-Villarraga DM, Vergara-Morales DI, Castellanos L, Ramos FA, Guarnaccia C, Degrassi G, Venturi V, Moreno-Sarmiento N (2019) Plant-growth promotion and biocontrol properties of three *Streptomyces* spp. isolates to control bacterial rice pathogens. Front Microbiol 10:290

Tahvoven R (1988) Microbial control of plant diseases with *Streptomyces* spp. EPPO Bull 18:55–59

Tamreihao K, Ningthoujam DS, Nimaichand S, Singh ES, Reena P, Singh SH, Nongthomba U (2016) Biocontrol and plant growth promoting activities of a *Streptomyces corchorusii* strain UCR3-16 and preparation of powder formulation for application as biofertilizer agents for rice plant. Microbiol Res 192:260–270

Tao G, Tian S, Cai M, Xie G (2008) Phosphate solubilizing and mineralizing abilities of bacteria isolated from soils. Pedosphere 18:515–523

The Nobel Prize in Physiology or Medicine (1952). NobelPrize.org. Nobel Media AB 2021. https://www.nobelprize.org/prizes/medicine/1952/summary/

Tokala RK, Strap JL, Jung MC, Crawford DL, Salove MH, Deobald LA, Bailey FJ, Morra MJ (2002) Novel plant-microbe rhizosphere interaction involving *Streptomyces lydicus* WYEC108 and the pea plant (*Pisum sativum*). Appl Environ Microbiol 68:2161–2171

Trejo-Estrada S, Paszczynski A, Crawford D (1998) Antibiotics and enzymes produced by the biocontrol agent *Streptomyces violaceusniger* YCED-9. J Ind Microbiol Biotechnol 21:81–90

Vargas Hoyos HA, Chiaramonte JB, Barbosa-Casteliani AG, Fernandez Morais J, Perez-Jaramillo JE, Nobre Santos S, Nascimento Queiroz SC, Soares Melo I (2021) An *Actinobacterium* strain from soil of cerrado promotes phosphorus solubilization and plant growth in soybean plants. Front Bioeng Biotechnol 22(9):579906

Vurukonda SSKP, Giovanardi D, Stefani E (2018) Plant growth promoting and biocontrol activity of *Streptomyces* spp. as endophytes. Int J Mol Sci 19:952

Wacksman SA, Woodruff HB (1940) The soil as a source of microorganisms antagonistics to disease producing bacteria. In: Bergey's manual of systematic bacteriology, vol 4. Williams & Wilkins, pp 2333–2648

Yuan WM, Crawford D (1995) Characterization of *Streptomyces lydicus* WYEC108 as a potential biocontrol agent against fungal root and seed rots. Appl Environ Microbiol 61:3119–3128

Active Role of Yeast in Environmental Sustainability

14

Ketaki Ramani, Manan Desai, Disha Changela, and Kruti Dangar

Abstract

The present global economy with an increasing number of government policies and technological innovations fosters an increasing urge toward environmentally sustainable solutions. For centuries, yeasts have been used in fermentation processes. The unique physiology and versatile growth conditions make yeasts a promising tool for bioproduction of various products ranging from wines to jet fuels and, more recently, substituting certain plastic materials. Metabolic engineering tools and genome sequencing have seen tremendous growth in recent years, enhancing the production rates and decreasing the overall cost. Yeasts possess the potential to contribute to sustainability and a green economy. This chapter covers recent developments related to yeasts in white biotechnology.

Keywords

Saccharomyces cerevisiae · Sustainable fermentation · Biofuels · Biomaterials · White biotechnology · Food security · Biofertilizers · Bioagents

K. Ramani
Department of Microbiology, School of Science, RK University, Rajkot, Gujarat, India

M. Desai
Department of Biochemistry, The Institute of Science, Mumbai, Maharashtra, India

D. Changela · K. Dangar (✉)
Department of Biochemistry, Saurashtra University, Rajkot, Gujarat, India

N. K. Arora, B. Bouizgarne (eds.), *Microbial BioTechnology for Sustainable Agriculture Volume 1*, Microorganisms for Sustainability 33,
https://doi.org/10.1007/978-981-16-4843-4_14

14.1 Introduction

The environment is essential for all the living beings from humans to microorganisms. Throughout ages, both human activities and natural causes have had a significant negative impact on the ecosystem (Fig. 14.1). It is, therefore, essential to adopt more sustainable practices to prevent further degradation and with the hope to repair the existing damage. The concept of sustainability was realized early during the eighteenth to nineteenth century by the German forestry industry to address the detrimental effects of the over-exploitation of forests (Vornholz 1994). It has been a highly discussed topic because of its global impact, but now in the post-COVID-19 pandemic period, the approach towards sustainable environmental development requires serious consideration.

However, there is no single sustainable development approach since it deals with multifaceted development problems. Broadly, industrial sustainability requires economic viability, environmental compatibility, and social responsibility. The aim of developing sustainable industries is to promote economic growth without environmental degradation. When used responsibly, biotechnology can help equilibrate biotic and abiotic relationships by providing solutions for various anthropogenic activities, propelling research toward socioeconomic and environmental benefits through renewable energy sources and other industrial applications. Developing countries such as India and Brazil are now eyed as potential markets mainly for agricultural (green) and industrial (white) biotechnologies (Barcelos et al. 2018).

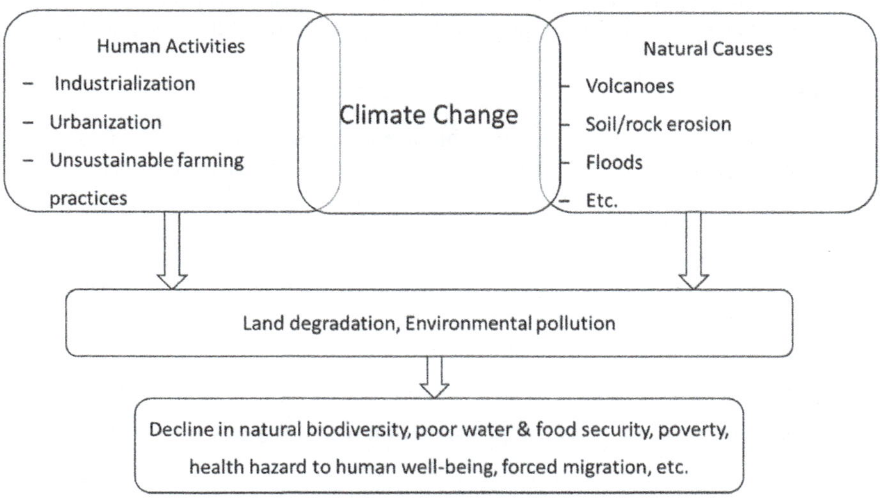

Fig. 14.1 Impact of growth and development on the environment. Rapid and uncontrolled growth of human activities and natural phenomena form a wicked interconnected cascade of events that have depleted natural resources and land quality for cultivation and polluted nearly all ecological niches. This skewed ecological balance has affected global climate and urgently requires more sustainable solutions

Table 14.1 Different color codes assigned to biotechnology and their applications in various industries (Kafarski 2012; Matyushenko et al. 2016; Barcelos et al. 2018; Graciano et al. 2019; Steiner 2020)

Color code	Industry
White	Industrial biotechnology
Green	Agricultural biotechnology, geomicrobiology, biofuels, biofertilizers/pesticides, and bioremediation
Blue	Coastal, marine, and freshwater biotechnology
Red	Biomedicine, biopharmaceutics, molecular diagnostics, vaccines and therapeutics
Yellow	Food and nutritional biotechnology
Brown	Desert and arid zone biotechnology
Gold	Bioinformatics and nanotechnology
Violet	Ethical, moral, and legal issues related to patents, inventions, and intellectual property rights
Grey	Waste management biotechnology
Dark	Abuse of biotechnology through bioterrorism and biowarfare

The "green" and "white" colors are among the other color codes (Table 14.1) assigned to identify the different areas of biotechnology.

Industrial or white biotechnology is based on modifying biological organisms, processes, products, and systems for producing goods and services. An industrial alliance where the by-product of one becomes the starting material for another would allow sustainable industrial ecology. Although bioproducts are less toxic and considered renewable, biodegradable, and recyclable, optimizing the efficiency and specificity of bioprocesses and the specific characteristics of bioproducts would not only increase its profitability and capacity but also reduce its environmental footprint. Among the various microorganisms used, one has provided humans with bread, beer, and wine since antiquity—the yeasts. These versatile microbes have applications in various biotechnological enterprises. This chapter briefly describes the role of yeasts in sustainable biotechnology (Ivanov and Hung 2010; Gavrilescu 2010; Ezeonu et al. 2012).

14.2 Yeasts and Yeast Engineering

A quick Internet search will describe yeasts as a single-cell eukaryotic microorganisms constituting about 1% of all described fungal species belonging to the kingdom fungi. These versatile chemoorganotrophs are broadly categorized into two phyla, ascomycetes and basidiomycetes (non-saccharomyces yeasts). Taxonomically diverse yeast communities are of ecological importance, and their dynamics have been mainly studied in managed soils like vineyards or agricultural fields (Kurtzman and Fell 2006; Yurkov 2018; Li et al. 2020). A popular member of the ascomycetes phylum is the Baker's yeast or *Saccharomyces cerevisiae*. Through gene duplication and metabolic rewiring, it has gained an evolutionary advantage

Table 14.2 Use of yeast and yeast-based products in industries that are currently being focused on for sustainable development

Industry	Application	References
Biopharmaceuticals	Protein production	Love et al. (2018)
Vaccine development	Hepatitis virus and human papillomavirus vaccines	Kumar and Kumar (2019)
Food industry	Valuable metabolites production and food waste valorization	Karim et al. (2020)
	Terpenoid and polyphenol production as natural food preservatives	Lyu et al. (2019)
Agriculture and animal husbandry	Sustainable protein feed for aquaculture	Øverland and Skrede (2017)
	Animal feed additives	Shurson (2018)

known as the Crabtree effect, allowing a competitive dominance by outgrowing other microorganisms (Hagman et al. 2013; Dashko et al. 2014; Sabater-Muñoz et al. 2020).

Since the mapping of the yeast genome in 1996, *S. cerevisiae* has become a well-established cell factory in industrial biotechnology. Various strains have been used, produced in vitro, or subjected to adaptive evolutionary methods to develop a library based on the product and other parameters such as pH, temperature, and salt concentrations to aid in various industrial processes; some recent applications are summarized in Table 14.2. Despite its successful domestication and contribution to the developing bioeconomy, the current approach is considered unsustainable. A potential solution is a systems-based approach where metabolic engineering and the available gene manipulation tools can help increase the productivity and usability of a given strain (Nandy and Srivastava 2018). In the post-genomic era, with an increase in genetic and biochemical data, the focus of metabolic engineering has now shifted to manipulating the entire cell and not just a single pathway—systems metabolic engineering. This approach is an advantage over conventional genetic engineering techniques by expanding the metabolic engineering toolbox enabling control of gene expression and modulating regulatory networks throughout the cell, facilitating overproduction of secondary metabolites (Yadav et al. 2012; Borodina and Nielsen 2014).

14.2.1 Fermentation

There appears a classical association between fermented foods and beverages and the human diet that has been part of the local gastronomic tradition. Fermentation is a very diverse process involving a series of enzymatic biotransformation of major and minor food components determined by the substrate, microorganisms, and the product. The most commonly used microorganisms include yeasts and several bacterial species such as *Acetobacter*, lactic acid bacteria, *Propionibacterium*, and *Bacillus* spp. (Kårlund et al. 2020). Engineered *S. cerevisiae* strains produced by rewiring different metabolic pathways now provide innovative approaches for developing sustainable and novel yeast-based processes. This allows the optimization of the fermentation process in *S. cerevisiae* using episomal plasmids to overexpress selected genes, reconstruct heterologous pathways and develop robust genome-scale metabolic models (Hou et al. 2012). For instance, an important pathway target for strain engineering is the Ehrlich pathway allowing the production of short branched-chain fatty acids that serve as intermediates for the production of industrially important chemicals (Yu et al. 2016).

A successful and advantageous implication of metabolic engineering is observed in vitiviniculture. For a more sustainable approach, the wine industry has resorted to microorganisms that influence various stages from grape harvest to bottling and waste treatment, including fermentation. Microbial resources exploitable for sustainability improvements include a wide array of genera and species comprising yeasts (both non-*Saccharomyces* and *Saccharomyces*), fungi, and bacteria. Early inoculation of non-Saccharomyces yeasts such as *Debaryomyces vanrijiae* or *Torulaspora delbrueckii* has been shown to improve the aroma profile or contribute to the fruity notes by increasing the volatile thiol content, respectively. In addition, they provide must (young wine; freshly crushed grape juice) bioprotection by producing antimicrobial and antifungal compounds. They are often mixed with the commercial starters of *S. cerevisiae* early during vinification (Curiel et al. 2017; Nardi 2020). Another example is of a biocontrol agent, *Metschnikowia pulcherrima*. It produces an insoluble red compound with antifungal activity, pulcherrimin, and is effective against several pathogenic yeasts such as *Candida albicans*. However, *M. pulcherrima* does not show antifungal activity against *S. cerevisiae* and is also compatible with alcoholic fermentation, making it a suitable starter prior to *Saccharomyces* inoculation. The positive influence of non-*Saccharomyces* yeasts allows controlled mixed fermentation by considering a mixed starter of selected non-*Saccharomyces* strains and *S. cerevisiae* to enhance the wine quality. They also help prevent the accumulation of unwanted compounds such as ochratoxin A or ethyl carbamate, affecting food safety. Apart from improving the quality of the wine, a proper combination of yeasts and/or bacterial strains such as lactic acid bacteria along with relevant modification in processing and technology can help in energy savings as well as in reduced use of additives like sulfites (Padilla et al. 2016; Benito et al. 2019; Nardi 2020). Like wine, beer brewing also requires yeast fermentation. The term beer refers to a broad group of fermented beverages with a different taste and aroma. In general, beer is a product of fermented cereals or hopped drink obtained from liquified starch. The most commonly used yeast for beer production, in particular lager-type beer, is *Saccharomyces pastorianus*. As with wine, genome editing and evolutionary engineering have allowed the designing of hybrid yeasts to enhance flavors and aroma, exploring new opportunities for its industrial application (Denby et al. 2018; Iattici et al. 2020).

Beer brewing is closely connected with the microbial fermentation of cereals. The most abundant by-product produced during beer manufacturing is the brewers' spent grain (BSG), which is mainly barley grain husks. This protein and fiber-rich bulk constitute about 85% of the total by-products obtained. In large-scale beer-producing countries like the United States, Brazil, or Germany, enormous amounts of BSG are produced and require proper disposal. One of the main applications of BSG has been as an animal feed or disposed of in landfills. However, it may serve an alternative purpose. Given its beneficial attributes and bioactive contents, BSG is being incorporated into food products such as ready-to-eat snacks or cookies, promoting different and sustainable applications and potential reusability of this agro-industrial by-product. Furthermore, it can also be reused as a growth medium for yeasts by replacing current yeast culture media sustainably and economically (Lynch et al. 2016; Cooray et al. 2017; Iattici et al. 2020).

Fermentation is not restricted only to brewing and wine production. Fermented food products, in particular traditional and indigenous products, have a significant sociocultural impact and are an archaic and inexpensive way to prepare food products with longer shelf life, better nutritive properties, and a good source of probiotic microorganisms such as different genera of lactic acid bacteria (Rezac et al. 2018). Various microorganisms used, including yeasts, impart characteristic flavor to the fermented food. Several milk products such as cheese and yogurts as well as different flavored milk exploit a variety of yeast strains. The correct selection of microorganisms and raw materials can have a significant impact on the final product. For instance, yeast strains of *Kluyveromyces lactis* and *Debaryomyces hansenii* are responsible for the characteristic aroma of Camembert cheese. Similarly, the addition of yeast strains *Saccharomyces boulardii* and *S. cerevisiae* in acidophilus-yeast fermented milk produced by a bacterial strain *Lactobacillus acidophilus* increases the antioxidant properties of the milk as well as enhances the viability of the bacteria (Nandy and Srivastava 2018). More recently, yeast fermentation of coffee beans has been shown to improve the quality of the coffee. Fermentation of green coffee beans with *Wickerhamomyces anomalu* (formerly *Pichia anomala*) increased the antioxidant property and reduced the total tannin content of fermented coffee (Haile and Kang 2020).

Food sustainability aims to provide food security, optimal use of natural and human resources, and respect to biodiversity and ecosystems (Tamang et al. 2020). Fermented food products have high nutritive content and cater to food security issues in countries like Nigeria, thereby providing subsistence to the regional economy and boosting rural livelihood (Adesulu and Awojobi 2014). Another example is that of cereals, which are an important part of the human diet. During the processing stages, by-products such as bran and BSG are generated in large quantities. Despite being rich in nutrients, these by-products are used as feed, fuel, substrates for biorefinery, or waste and are only partially recycled. Fermentation along with tailored enzymatic treatments of these by-products provides a feasible approach for innovative enhancements in the bioprocessing technology, thereby encouraging sustainable green biotechnologies to revalorize the high quantity of by-products (Verni et al. 2019; Tlais et al. 2020).

14.2.2 Food Security

Food security and insecurity are serious global issues, especially in developing and underdeveloped nations. Security refers to the availability access, and utilization of food, whereas insecurity is associated with disruption of food due to the climate, inconsistent cropping patterns, or unemployment (Challa et al. 2019). The concept of food security originated in the mid-1970s to address the global food crisis and combat global hunger. The classical and widely accepted definition of food security "when all people, at all times, have physical and economic access to sufficient, safe, nutritious food that meets their dietary needs and food preferences for an active and healthy life" is predominantly governed by two key factors, price and affordability. The high cost of nutritious food consumes nearly the entire earnings of low-income people, making it unaffordable for many, especially those from underdeveloped countries like Nigeria (Herforth et al. 2020). Population growth has led to an increased demand for healthy human food and animal feed, particularly in developing countries, steering the research toward more sustainable nonconventional sources. Figure 14.2 compares the impact on food insecurity through the prevalence of undernourishment and the number of undernourished people in developing and underdeveloped countries like India and Nigeria, respectively. According to the Food and Agriculture Organization (FAO) of the United Nations, developing nations like India have taken a positive step toward food security and are likely to reduce the prevalence of undernourishment in the future, whereas countries like Nigeria require more support in dealing with food insecurity (FAO 2020). Microorganisms provide an elegant solution to address this issue, and their properties such as fermentation (previous section) have been exploited through biotechnology to effectively improve the nutritive quality of food and provide a potential solution for global food shortage. Apart from being associated with fermentation, yeast *S. cerevisiae* has proven to be a promising and inexpensive source for producing proteins from microbial biomass, often referred to as single-cell proteins (SCPs) (Razzaq et al. 2020). There are also commercially available products from *Saccharomyces*, *Fusarium*, and *Torulopsis* such as Marmite® or Quorn™. The latter contains a mycoprotein from a filamentous fungus *Fusarium venenatum* to manufacture food-grade SCP and is used as a meat substitute (Upcraft et al. 2020).

Traditionally, as discussed in the previous section, fermentation has been used to make various food products (like cheese) and beverages (wine or beer) and preserve food improving their storage life. With the increasing need for healthy and nutritive food products, this age-old technique has been revisited to improve the nutritive value of the food product and ensuring food security. This has led to the popularization of indigenous fermented products that have had traditional and cultural significance. For example, yeasts like *Pichia kudriavzevii* and *S. cerevisiae* are predominantly present in nearly all indigenous fermented food and beverages of sub-Saharan African countries like Ethiopia, producing compounds like folate and providing probiotic properties (Johansen et al. 2019; Koricha et al. 2020). Similarly, when yeast is used as a starter in microbial fermentation of cocoa beans, it produces functional volatile compounds and increases the nutritive value quality, and safety

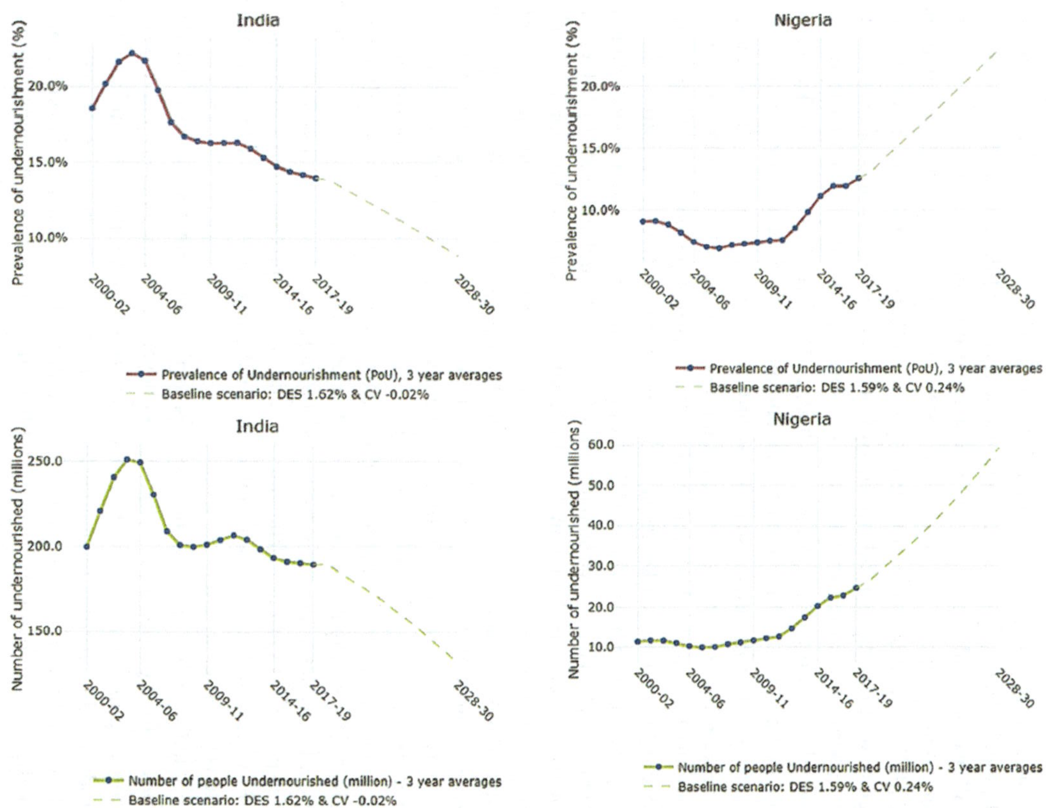

Fig. 14.2 Prevalence of undernourishment and number of undernourished in India and Nigeria. Developing countries like India are seriously considering the issue of food security and are working toward reducing the prevalence of undernourishment (PoU) and the number of undernourished people (NoU). Compared

to the years 2003–2005, PoU has decreased from 22.2% to 14% in 2017–2019. Also, the NoU has decreased from 251.1 million in 2003–2005 to 189.2 million in 2017–2019. In contrast, in underdeveloped countries like Nigeria, the prevalence of undernourishment has seen a gradual rise since 2011–2013 from 12.0% to 13.8% (2018–2020) and is expected to reach 23% by 2028–2030. Similarly, Nigeria has also seen a gradual increase in the number of undernourished since 2005–2007 from ten million to 24.6 million (2017–2019) and is expected to reach 59.2 million by 2028–2030. This projection is represented and predicted by the Prevalence of Undernourishment (PoU) projection tool based on the FAO data (FAO 2020) of the product; this has attracted the interest of the chocolate industry (Mota-Gutierrez et al. 2019).

The probiotic properties of fermented products have added another layer of eubiotics to food security. By definition, eubiotics refer to the use of feed additives such as probiotics, prebiotics, essential oils, and organic acids to ensure the hygienic and healthy living of the livestock. The increasing global demands for food have resulted in the industrialization of animal productions and administration of antimicrobial drugs to livestock. Such uncontrolled and prolonged use of antibiotics has contributed to antibiotic resistance in animals, thus increasing the risk of zoonotic diseases. Pathogenic bacteria such as *Salmonella* or *Campylobacter* use animal-derived food products (milk, meat, or eggs) as reservoirs and cause foodborne infections in humans (Santovito et al. 2018). Therefore, a healthy relationship between humans and animals is required to prevent another pandemic (The Lancet 2020). Prebiotic feed additives can modulate the poultry gastrointestinal tract by limiting the establishment of campylobacter, thereby increasing bird growth rates and preventing human foodborne infections (Ricke et al. 2020). Yeasts and yeast-derived products have recently been explored and found to control the spread of zoonoses. Yeast cell wall-based products provide a sustainable commercial alternative to antimicrobial growth products and improve host immunity. In addition, *Saccharomyces*-derived prebiotic refined functional carbohydrates have been found to reduce human foodborne pathogens in poultry (Santovito et al. 2018; Froebel et al. 2019). Yeasts and yeasts-derived products are versatile and serve as a potential economic tool to combat malnutrition in developing countries. Future studies and research are required to explore and develop innovative applications of yeasts related to food security.

14.2.3 Biofertilizers and Biocontrol Agents

Accumulation of industrial effluents, pesticides, and other agro-chemicals degrade the quality and nutritive value of the land, resulting in poor crop yield. Food security and crop productivity depend on soil characteristics, which in turn are modulated by the dynamic chemistry of soil microbes that have co-evolved with plants (Arora 2018). Agroecological practice of biofertilization appears as a promising method of revitalizing the soil through microorganisms (Gouda et al. 2018; Arora et al. 2020). Soil arbuscular mycorrhizal fungi (AMF) are microbial symbionts of plants, which are now being used to improve yields of important crops. AMF form an intricate network with plant roots and aid in increasing the overall biomass productivity, crop performance, and soil quality. As a result, AMF is being considered as a natural alternative to chemical fertilizers to prevent further soil degradation, promote sustainable agriculture, and boost global food security. A practical example of this can be seen in the countries of the Great Mekong Region that are slowly adopting biofertilization as opposed to conventional practices leading to increased production of biofertilizers in China. However, this technology is in its infancy and requires extensive research (Rodriguez and Sanders 2015; Igiehon and Babalola 2017;

Atieno et al. 2020; Gujre et al. 2021). Among other microbes, yeasts have shown a potential to be bioagents capable of managing sustainable agriculture. Yeasts have shown activity against several diseases affecting fruits, vegetables, and plants. Various yeast species play diverse roles in sustainable agriculture. For example, *Cryptococcus sp.* serve as biocontrol agents inhibiting growth of pathogens and reducing growth of lesions and necrosis in plants. Alvarez et al. (2019) reported yeast strain *Candida sake* 41E as biocontrol agent against *Penicillium expansum*, a common post-harvest pathogen of apple. Brewers' yeast waste was found to enhance the growth and nutrient content of sugarcane and tomato plants (Lonhienne et al. 2014). Another example is of *S. cerevisiae* strain PSY-4 that is known to promote plant growth, especially for corn plants. Similarly, yeast strains *D. hansenii* Dh-67, *Lachancea thermotolerans* Lt-69, and *S. cerevisiae* Sc-5 have shown to promote seedling growth and development by increasing its dry weight and chlorophyll content. Yeasts also help control and prevent post-harvest diseases. *Candida* sp. prevent growth of pathogenic stains like *Penicillium digitatum* and *Botrytis cinerea* in fruits like grapes, apples, peaches, and citrus fruits. Recently, psychrophilic Antarctic yeasts such as *Metschnikowia fructicola*, *Pichia guilliermondii*, and *Saccharomycopsis schoenii* among others were found to prevent growth of phytopathogens at low temperatures and being considered as bio-control agents used to prevent post-harvest diseases among fruits when stored in cold (Ferreira et al. 2019; Fernandez-San Millan et al. 2020; Mukherjee et al. 2020). Despite these observations and reports, the use of yeasts and yeast waste as biofertilizers requires more innovative studies to further probe into this aspect and help promote sustainable agriculture.

14.2.4 Biofuels

The global economy is dependent on carbon-based fuels. With depleting petroleum reservoirs and associated geopolitical issues, a need for more sustainable fuel sources has increased, leading to a paradigm shift directed towards biofuels. Presently, the most widely used biofuel is ethanol which is produced from sugar cane, corn, or wheat. Other than ethanol, advanced alcohols such as propanol, 1-butanol, and 3-methyl 1-butanol are also gaining popularity among alcohol obtained from renewable lignocellulosic biomass processed from food and agricultural wastes (Hegde and Trabold 2018). High concentrations of ethanol can be produced by exploiting the make-accumulate-consume strategy during alcoholic fermentation (Hagman et al. 2013). This remarkable evolutionary trait has proven beneficial in the framework of a sustainable industry.

Bioethanol is the most widely used biofuel with a distinguished history dating back to the early 1900s. Popularly, it has been blended with gasoline (gasohol) to reduce the emission of exhaust gases. Several sources such as lignocellulosic biomass, or agricultural and forest residues, can be used to produce bioethanol. Apart from these, even newspaper has been considered for bioethanol production to minimize municipal solid wastes. Yeast-based bioethanol production depends on the conversion rate of carbohydrates to ethanol and the industrial conditions maintained during production. Yeast strains such as *Pichia stipitis* NRRL-Y-7124, *S. cerevisiae* RL-11, and *Kluyveromyces fragilis* Kf1, in particular *S. cerevisiae*, have shown to be efficient bioethanol producers from spent coffee grounds (Mussatto et al. 2012). Although *S. cerevisiae* shows tolerance to ethanol, the sustainability of fermentation reaction can be significantly affected by high alcoholic stress (at levels >15%). To overcome this, directed evolution and metabolic engineering have led to the development of ethanol-tolerant *S. cerevisiae* mutant strains. However, since the commercial yeast, *S. cerevisiae*, cannot ferment pentose sugars, ethanol-tolerant and thermotolerant *K. marxianus*, capable of fermenting both hexose and pentose sugars, can be used. Furthermore, modifications such as cellular immobilization are advantageous by improving cell density, high substrate conversion, and resistance toward inhibitors for sustainable commercialization of bioethanol production. Depending upon the feedstock, pretreatment, hydrolysis, and fermentation parameters used for *S. cerevisiae* and *K. marxianus*, the bioethanol production can vary from 0.21 to 1.2 g/L/h with an ethanol concentration of 20.6 to 69.2 g/L (Stanley et al. 2010; Ruriani et al. 2012; Byadgi and Kalburgi 2016; Azhar et al. 2017). Interestingly, ethanol productivity can be increased to 6.55 g/L/h using immobilized *S. cerevisiae* on MCM-41 zeolite with alginate embedding to ferment sugar molasses (Zheng et al. 2012). However, under favorable conditions, higher concentrations of bioethanol (21.45 g/L) can also be obtained from stems of the shrub *Prosopis juliflora* using *K. marxianus* (Sivarathnakumar et al. 2019).

Nevertheless, *S. cerevisiae* remains an attractive option for the biofuel industry. It is now metabolically engineered to meet the growing demand for advanced biofuels: butanols (Kuroda and Ueda 2016), terpenes such as farnesene (Tippmann et al. 2016), bisabolene (Peralta-Yahya et al. 2011), amorphadiene as a jet fuel substitute (Buijs et al. 2013), and fatty acid ethyl esters (FAEEs), as an alternative for biodiesel (Zhou et al. 2016). Furthermore, metabolic engineering has also allowed improved tolerance towards short-chain alcohols and industrial production of long-chain alkanes in *S. cerevisiae* (Swidah et al. 2015; Buijs et al. 2015). Besides *S. cerevisiae*, several other species are exploited by the biofuel industry. *Yarrowia lipolytica* is a nonconventional oleaginous yeast found in dairy products and oily foods. It is a non-pathogenic obligate aerobic yeast capable of consuming hydrophobic substrates (different types of oils, fats, and hydrocarbons) and other renewable carbon sources such as lignocellulosic sugars. Oleaginous yeasts, in general, have unique physiology allowing them to produce lipids and oleochemicals. Metabolically engineered *Y. lipolytica* is used for manufacturing sesquiterpenes like farnesene, which, in turn, is used in biodiesel production. Other oleaginous species such as *Rhodosporidium toruloides*, *Lipomyces starkey*, *Rhodotorula glutinis*, *Trichosporon fermentans*, and *Cryptococcus curvatus* are emerging as promising hosts for the biofuel industry (You et al. 2017; Darvishi et al. 2017; Adrio 2017; Liu et al. 2019). Recently, Martinez-Silveira et al. (2019) reported optimization and production of biodiesel by oleaginous strain of *Rhodotorula graminis* S1/2R.

Another promising source of biofuel is biohydrogen. Hydrogen gas is considered one of the cleanest fuels produced from food wastes. Several microorganisms produce biohydrogen via dark fermentation of organic feedstocks and biomass, making it an attractive alternative for biofuel (Cabrol et al. 2017; Hegde and Trabold 2018). Alternately, yeast-based biohydrogen can be produced using bioengineered *S. cerevisiae* or by appropriate pretreatment of *S. cerevisiae* or edible fungus *Gymnopus contrarius* J2, to increase the production of biohydrogen (Waks and Silver 2009; Sheng et al. 2018; Antonopoulou 2020). Despite its potential, there are no standardized protocols for the industrial production of biohydrogen. Currently, biohydrogen is being integrated with fungal biofuels (biodiesel and bioethanol) for improved energy recovery (Ghosh and Roy 2019; Srivastava et al. 2020). Even though biohydrogen and bioethanol can serve as an alternative to conventional fuels, more research is required, in particular, for yeast-based production to explore its usability for sustainable, eco-friendly applications.

14.2.5 Biomaterials

Biomaterials are characterized as natural or synthetic materials comprising entirely or partly from biological sources capable of interacting with living systems. These are used extensively in medicine, tissue engineering, and (bio)material science among others. Yeast is the latest addition to the already available biomaterials. The filamentous nature of *S. cerevisiae* has recently allowed innovative applications related to material fabrication. It has found applications in the textile and packing industries. The type of fungus and fabrication factors affect the nature of mycelium-based composite materials. For instance, heat-pressed mycelium materials have density and elastic modulus similar to natural materials such as cork or wood. These lignocellulosic-based composites are renewable, hydrophobic, nonflammable, and biodegradable, therefore less environmentally toxic compared to synthetic materials. Mycelium composites from *Trametes versicolor*, when integrated with flax, hemp, or straw, exhibit good insulation properties than conventional. As a result, fungal composites have found innovative applications in the textile and automotive industries (Cerimi et al. 2019; Appels et al. 2019; Elsacker et al. 2019). Fungal mycelium composites have also found a very unexpected application. To reduce exposure to formaldehyde emission from certain wood materials like medium-density fiberboard or particle boards, mycelium-based biopolymer composites have gained attention. The mycelia digest and bind to the surface of lignocellulosic materials forming an entangled network providing additional mechanical strength to the wood panels along with fire resistance and acoustical absorption properties. Moreover, mycelium modification reduces water absorption and thickness swelling of the hybrid composites, making it a promising system for the packing and furniture industries (Sun et al. 2019).

Yeasts have also carved a niche for themselves in the electronics industry, where yeast-derived cathode material has shown to be a promising material for energy conversion and storage in lithium-oxygen batteries (Zhu et al. 2018). Another exemplary application developed by VTT Technical Research Centre of Finland and Aalto University, Finland, is the unique Koravaa fungal headphones made of various bioengineered materials: a 3D-printed headband made from yeast-produced lactic acid, ear padding made of a fungal hydrophobin and plant cellulose, the imitation leather-like material from fungus *Phanerochaete chrysosporium*, speaker mesh from biosynthetic spider silk produced by microbes and the outer foam of mycelium-cellulose composite (www.korvaa.com).

Yeast-based biomaterials have also found uses in tissue engineering. Although the application of yeasts in this field has been poorly explored, several yeast species, especially those used in the food industry and with the "Generally Regarded As Safe" (GRAS) status, can have potential application in tissue engineering (Kyle et al. 2009). The bioengineered yeast-derived nanovacuoles used to enhance drug penetration in tumor xenografts without eliciting an immune response (Gujrati et al. 2016) or chromosomal integration for collagen production in *Pichia pastoris* and *Hansenula polymorpha* highlight some of the applications of yeasts related to tissue engineering (de Bruin et al. 2000; Werten et al. 2001).

14.3 Conclusion and Future Prospects

Yeasts have immeasurable applications in biotechnology with a global focus on developing new sustainable products. The omics era has revolutionized the commercial applications of this microorganism. The systems-level understanding of yeasts has further expanded the horizons for future explorations and innovation in yeast biotechnology. The newly developed CRISPR/Cas9-based gene editing has been used to modify several *Saccharomyces* and numerous non-Saccharomyces yeast genomes. It was used to create genetically modified (GM) yeast strains of enological relevance to enhance the flavor and manufacture safer wines (van Wyk et al. 2020; Vilela 2021). However, this technology is still in its infancy and requires standardization and ethical approval for using GM strains for commercial applications. *S. cerevisiae* is the best-studied and one of the most widely used eukaryotes in several industrial processes. Despite its efficient adaptation, as the omics era progresses toward untangling the yeast genome, there is still a great potential of either optimizing existing strains or exploiting the immense natural reservoir of environmental isolates. Apart from the industrially dominating *S. cerevisiae*, several species such as *Kluyveromyces* sp. and others are being explored and emerging as promising candidates ushering sustainable white biotechnology. With more insight and understanding of yeast dynamics and interactions with other microorganisms, newer, innovative, and sustainable applications can be expected in the future.

References

Adesulu AT, Awojobi KO (2014) Enhancing sustainable development through indigenous fermented food products in Nigeria. Afr J Microbiol Res 8(12):1338–1343

Adrio JL (2017) Oleaginous yeasts: promising platforms for the production of oleochemicals and biofuels. Biotechnol Bioeng 114(9):1915–1920

Alvarez A, Gelezoglo R, Garmendia G, González ML, Magnoli AP, Arrate E, Cavaglieri LR, Vero (2019) Role of Antarctic yeast in biocontrol of *Penicillium expansum* and patulin reduction of apples. Environ Sustain 2:277–283

Antonopoulou G (2020) Designing efficient processes for sustainable bioethanol and bio-hydrogen production from grass Lawn waste. Molecules 25(12):2889

Appels FVW, Camere S, Montalti M, Karana E, Jansen KMB, Dijksterhuis J et al (2019) Fabrication factors influencing mechanical, moisture- and water-related properties of mycelium-based composites. Mater Des 161:64–71

Arora NK (2018) Agricultural sustainability and food security. Environ Sustain 1(3):217–219

Arora NK, Fatima T, Mishra I, Verma S (2020) Microbe-based inoculants: role in next green revolution. In: Shukla V, Kumar N (eds) Environmental concerns and sustainable development. Springer, Singapore, pp 191–246

Atieno M, Herrmann L, Nguyen HT, Phan HT, Nguyen NK, Srean P et al (2020) Assessment of biofertilizer use for sustainable agriculture in the great Mekong region. J Environ Manage 275: 111300

Azhar H, Abdulla R, Jambo S, Marbawi H, Jualang Azlan G, Mohd Faik AA et al (2017) Yeasts in sustainable bioethanol production: a review. Biochem Biophys Rep 10:52–61

Barcelos M, Lupki FB, Campolina GA, Nelson DL, Molina G (2018) The colors of biotechnology: general overview and developments of white, green and blue areas. FEMS Microbiol Lett 365(21)

Benito S, Ruiz J, Belda I, Kiene F, Beisert B, Navascués E et al (2019) Application of non-saccharomyces yeasts in wine production. In: Sibirny A (ed) Non-conventional yeasts: from basic research to application. Springer, Berlin, pp 75–89

Borodina I, Nielsen J (2014) Advances in metabolic engineering of yeast *Saccharomyces cerevisiae* for production of chemicals. Biotechnol J 9(5):609–620

Buijs NA, Siewers V, Nielsen J (2013) Advanced biofuel production by the yeast *Saccharomyces cerevisiae*. Curr Opin Chem Biol 17(3):480–488

Buijs NA, Zhou YJ, Siewers V, Nielsen J (2015) Long-chain alkane production by the yeast *Saccharomyces cerevisiae*. Biotechnol Bioeng 112(6):1275–1279

Byadgi SA, Kalburgi PB (2016) Production of bioethanol from waste newspaper. Proc Environ Sci 35:555–562

Cabrol L, Marone A, Tapia-Venegas E, Steyer JP, Ruiz-Filippi G et al (2017) Microbial ecology of fermentative hydrogen producing bioprocesses: useful insights for driving the ecosystem function. FEMS Microbiol Rev 41(2):158–181

Cerimi K, Akkaya KC, Pohl C, Schmidt B, Neubauer P (2019) Fungi as source for new bio-based materials: a patent review. Fungal Biol Biotechnol 6(1):17

Challa S, Dutta T, Neelapu NRR (2019) Fungal white biotechnology applications for food security: opportunities and challenges. In: Yadav A, Singh S, Mishra S, Gupta A (eds) Recent advancement in white biotechnology through fungi. Perspective for value-added products and environments. Fungal biology, vol 2. Springer, Cham, pp 119–148

Cooray ST, Lee JJL, Chen WN (2017) Evaluation of brewers' spent grain as a novel media for yeast growth. AMB Express 7(1):117

Curiel JA, Morales P, Gonzalez R, Tronchoni J (2017) Different non-saccharomyces yeast species stimulate nutrient consumption in *S. cerevisiae* mixed cultures. Front Microbiol 8:2121

Darvishi F, Fathi Z, Ariana M, Moradi H (2017) *Yarrowia lipolytica* as a workhorse for biofuel production. Biochem Eng J 127:87–96

Dashko S, Zhou N, Compagno C, Piškur J (2014) Why, when, and how did yeast evolve alcoholic fermentation. FEMS Yeast Res 14(6):826–832

de Bruin EC, de Wolf FA, Laane NCM (2000) Expression and secretion of human α1 (I) procollagen fragment by *Hansenula polymorpha* as compared to *Pichia pastoris*. Enzyme Microb Technol 26(9):640–644

Denby CM, Li RA, Vu VT, Costello Z, Lin W, Chan LJG et al (2018) Industrial brewing yeast engineered for the production of primary flavor determinants in hopped beer. Nat Commun 9(1): 965

Elsacker E, Vandelook S, Brancart J, Peeters E, De Laet L (2019) Mechanical, physical and chemical characterisation of mycelium-based composites with different types of lignocellulosic substrates. PLoS One 14(7):e0213954

Ezeonu CS, Tagbo R, Anike EN, Oje OA, Onwurah INE (2012) Biotechnological tools for environmental sustainability: prospects and challenges for environments in Nigeria—a standard review. Biotechnol Res Int 2012:450802

FAO (2020) Prevalence of undernourishment (PoU) projection tool. Food and Agriculture Organization of the United Nations, Rome

Fernandez-San Millan A, Farran I, Larraya L, Ancin M, Arregui LM, Veramendi J (2020) Plant growth-promoting traits of yeasts isolated from Spanish vineyards: benefits for seedling development. Microbiol Res 237:126480

Ferreira EMS, Resende DA, Vero S, Pimenta RS (2019) The use of psychrophilic Antarctic yeast in the biological control of post-harvest diseases of fruits stored at low temperatures. In: Rosa L (ed) Fungi of Antarctica. Springer, Cham, pp 243–263

Froebel LK, Jalukar S, Lavergne TA, Lee JT, Duong T (2019) Administration of dietary prebiotics improves growth performance and reduces pathogen colonization in broiler chickens. Poult Sci 98(12):6668–6676

Gavrilescu M (2010) Environmental biotechnology: achievements, opportunities and challenges. Dynam Biochem Process Biotechnol Mol Biol 4:1–36

Ghosh S, Roy S (2019) Novel integration of biohydrogen production with fungal biodiesel production process. Bioresour Technol 288:121603

Gouda S, Kerry RG, Das G, Paramithiotis S, Shin HS, Patra JK (2018) Revitalization of plant growth promoting rhizobacteria for sustainable development in agriculture. Microbiol Res 206: 131–140

Graciano R, Ribeiro J, Macêdo A, Lavareda P, Oliveira P, Netto J et al (2019) Recent patents applications in red biotechnology: a mini-review. Recent Pat Biotechnol 13(3):170–186

Gujrati V, Lee M, Ko YJ, Lee S, Kim D, Kim H et al (2016) Bioengineered yeast-derived vacuoles with enhanced tissue-penetrating ability for targeted cancer therapy. Proc Natl Acad Sci U S A 113(3):710–715

Gujre N, Soni A, Rangan L, Tsang DCW, Mitra S (2021) Sustainable improvement of soil health utilizing biochar and arbuscular mycorrhizal fungi: a review. Environ Pollut 268:115549

Hagman A, Säll T, Compagno C, Piskur J (2013) Yeast "make-accumulate-consume" life strategy evolved as a multi-step process that predates the whole genome duplication. PLoS One 8(7): e68734–e68734

Haile M, Kang WH (2020) Antioxidant properties of fermented green coffee beans with *Wickerhamomyces anomalus* (strain KNU18Y3). Fermentation 6:18

Hegde S, Trabold TA (2018) Chapter 5. Sustainable waste-to-energy technologies: fermentation. In: Trabold TA, Babbitt CW (eds) Sustainable food waste-to-energy systems. Academic Press, San Diego, CA, pp 69–88

Herforth A, Bai Y, Venkat A, Mahrt K, Ebel A, Masters WA (2020) Cost and affordability of healthy diets across and within countries. Background paper for the state of food security and nutrition in the world. In: FAO agricultural development economics technical study no. 9. FAO, Rome

Hou J, Tyo KEJ, Liu Z, Petranovic D, Nielsen J (2012) Metabolic engineering of recombinant protein secretion by *Saccharomyces cerevisiae*. FEMS Yeast Res 12(5):491–510

Iattici F, Catallo M, Solieri L (2020) Designing new yeasts for craft brewing: when natural biodiversity meets biotechnology. Beverages 6:1

Igiehon NO, Babalola OO (2017) Biofertilizers and sustainable agriculture: exploring arbuscular mycorrhizal fungi. Appl Microbiol Biotechnol 101(12):4871–4881

Ivanov V, Hung YT (2010) Applications of environmental biotechnology. In: Wang L, Ivanov V, Tay JH (eds) Environmental biotechnology. Handbook of environmental engineering, vol 10. Humana Press, Totowa, NJ, pp 1–17

Johansen PG, Owusu-Kwarteng J, Parkouda C, Padonou SW, Jespersen L (2019) Occurrence and importance of yeasts in indigenous fermented food and beverages produced in sub-Saharan Africa. Front Microbiol 10:1789

Kafarski P (2012) Rainbow code of biotechnology. Chemik 66:814–816

Karim A, Gerliani N, Aïder M (2020) *Kluyveromyces marxianus*: an emerging yeast cell factory for applications in food and biotechnology. Int J Food Microbiol 333:108818

Kårlund A, Gómez-Gallego C, Korhonen J, Palo-Oja OM, El-Nezami H, Kolehmainen M (2020) Harnessing microbes for sustainable development: food fermentation as a tool for improving the nutritional quality of alternative protein sources. Nutrients 12(4):1020

Koricha AD, Han DY, Bacha K, Bai FY (2020) Diversity and distribution of yeasts in indigenous fermented foods and beverages of Ethiopia. J Sci Food Agric 100(9):3630–3638

Kumar R, Kumar P (2019) Yeast-based vaccines: new perspective in vaccine development and application. FEMS Yeast Res 19(2):foz007

Kuroda K, Ueda M (2016) Cellular and molecular engineering of yeast *Saccharomyces cerevisiae* for advanced biobutanol production. FEMS Microbiol Lett 363(3):fnv247

Kurtzman CP, Fell JW (2006) Yeast systematics and phylogeny—implications of molecular identification methods for studies in ecology. In: Péter G, Rosa C (eds) Biodiversity and ecophysiology of yeasts. The yeast handbook. Springer, Berlin, Heidelberg, pp 11–30

Kyle S, Aggeli A, Ingham E, McPherson MJ (2009) Production of self-assembling biomaterials for tissue engineering. Trends Biotechnol 27(7):423–433

Li AH, Yuan FX, Groenewald M, Bensch K, Yurkov AM, Li K et al (2020) Diversity and phylogeny of basidiomycetous yeasts from plant leaves and soil: proposal of two new orders, three new families, eight new genera and one hundred and seven new species. Stud Mycol 96: 17–140

Liu Y, Jiang X, Cui Z, Wang Z, Qi Q, Hou J (2019) Engineering the oleaginous yeast *Yarrowia lipolytica* for production of α-farnesene. Biotechnol Biofuels 12(1):296

Lonhienne T, Mason M, Ragan M, Philip H, Schmidt S, Paungfoo-Lonhienne C (2014) Yeast as a biofertilizer alters plant growth and morphology. Crop Sci 54:785

Love KR, Dalvie NC, Love JC (2018) The yeast stands alone: the future of protein biologic production. Curr Opin Biotechnol 53:50–58

Lynch KM, Steffen EJ, Arendt EK (2016) Brewers' spent grain: a review with an emphasis on food and health. J Inst Brewing 122(4):553–568

Lyu X, Lee J, Chen WN (2019) Potential natural food preservatives and their sustainable production in yeast: terpenoids and polyphenols. J Agric Food Chem 67(16):4397–4417

Martinez-Silveira A, Pereyra V, Garmendia G, Refo C, Vero S (2019) Optimization of culture conditions of *Rhodotorula graminis* S1/2R to obtain saponifiable lipids for the production of second-generation biodiesel. Environ Sustain 2:419–428

Matyushenko I, Sviatukha I, Grigorova-Berenda L (2016) Modern approaches to classification of biotechnology as a part of NBIC—technologies for bioeconomy. BJEMT 14:1–14

Mota-Gutierrez J, Barbosa-Pereira L, Ferrocino I, Cocolin L (2019) Traceability of functional volatile compounds generated on inoculated cocoa fermentation and its potential health benefits. Nutrients 11(4):884

Mukherjee A, Verma JP, Gaurav AK, Chouhan GK, Patel JS, Hesham AEL (2020) Yeast a potential bio-agent: future for plant growth and postharvest disease management for sustainable agriculture. Appl Microbiol Biotechnol 104(4):1497–1510

Mussatto SI, Machado EMS, Carneiro LM, Teixeira JA (2012) Sugars metabolism and ethanol production by different yeast strains from coffee industry wastes hydrolysates. Appl Energy 92: 763–768

Nandy SK, Srivastava RK (2018) A review on sustainable yeast biotechnological processes and applications. Microbiol Res 207:83–90

Nardi T (2020) Microbial resources as a tool for enhancing sustainability in winemaking. Microorganisms 8:4

Øverland M, Skrede A (2017) Yeast derived from lignocellulosic biomass as a sustainable feed resource for use in aquaculture. J Sci Food Agric 97(3):733–742

Padilla B, Gil JV, Manzanares P (2016) Past and future of non-saccharomyces yeasts: from spoilage microorganisms to biotechnological tools for improving wine aroma complexity. Front Microbiol 7:411

Peralta-Yahya PP, Ouellet M, Chan R, Mukhopadhyay A, Keasling JD, Lee TS (2011) Identification and microbial production of a terpene-based advanced biofuel. Nat Commun 2(1):483

Razzaq ZU, Khan MKI, Maan AA, Rahman S (2020) Characterization of single cell protein from *Saccharomyces cerevisiae* for nutritional, functional and antioxidant properties. J Food Meas Charact 14(5):2520–2528

Rezac S, Kok CR, Heermann M, Hutkins R (2018) Fermented foods as a dietary source of live organisms. Front Microbiol 9:1785

Ricke SC, Lee SI, Kim SA, Park SH, Shi Z (2020) Prebiotics and the poultry gastrointestinal tract microbiome. Poult Sci 99(2):670–677

Rodriguez A, Sanders IR (2015) The role of community and population ecology in applying mycorrhizal fungi for improved food security. ISME J 9(5):1053–1061

Ruriani EKA, Sunarti TC, Meryandini A (2012) Yeast isolation for bioethanol production. HAYATI J Biosci 19(3):145–149

Sabater-Muñoz B, Mattenberger F, Fares MA, Toft C (2020) Transcriptional rewiring, adaptation, and the role of gene duplication in the metabolism of ethanol of *Saccharomyces cerevisiae*. MSystems 5(4):e00416–e00420

Santovito E, Greco D, Logrieco AF, Avantaggiato G (2018) Eubiotics for food security at farm level: yeast Cell Wall products and their antimicrobial potential against pathogenic bacteria. Foodborne Pathog Dis 15(9):531–537

Sheng T, Zhao L, Gao L, Liu W, Wu G, Wu J et al (2018) Enhanced biohydrogen production from nutrient-free anaerobic fermentation medium with edible fungal pretreated rice straw. RSC Adv 8(41):22924–22930

Shurson GC (2018) Yeast and yeast derivatives in feed additives and ingredients: sources, characteristics, animal responses, and quantification methods. Anim Feed Sci Technol 235: 60–76

Sivarathnakumar S, Jayamuthunagai J, Baskar G, Praveenkumar R, Selvakumari IAE, Bharathiraja B (2019) Bioethanol production from woody stem *Prosopis juliflora* using thermo tolerant yeast *Kluyveromyces marxianus* and its kinetics studies. Bioresour Technol 293:122060

Srivastava RK, Shetti NP, Reddy KR, Aminabhavi TM (2020) Biofuels, biodiesel and biohydrogen production using bioprocesses: a review. Environ Chem Lett 18(4):1049–1072

Stanley D, Bandara A, Fraser S, Chambers PJ, Stanley GA (2010) The ethanol stress response and ethanol tolerance of *Saccharomyces cerevisiae*. J Appl Microbiol 109(1):13–24

Steiner U (2020) Biotechnology. In: Fachenglisch Für BioTAs Und BTAs, pp 1–80

Sun W, Tajvidi M, Hunt CG, McIntyre G, Gardner DJ (2019) Fully bio-based hybrid composites made of wood, fungal mycelium and cellulose nanofibrils. Sci Rep 9(1):3766

Swidah R, Wang H, Reid PJ, Ahmed HZ, Pisanelli AM, Persaud KC et al (2015) Butanol production in *S. cerevisiae* via a synthetic ABE pathway is enhanced by specific metabolic engineering and butanol resistance. Biotechnol Biofuels 8(1):97

Tamang JP, Cotter PD, Endo A, Han NS, Kort R, Liu SQ et al (2020) Fermented foods in a global age: east meets west. Compr Rev Food Sci Food Safety 19(1):184–217

The Lancet (2020) Zoonoses: beyond the human–animal–environment interface. Lancet 396(10243):1

Tippmann S, Scalcinati G, Siewers V, Nielsen J (2016) Production of farnesene and santalene by *Saccharomyces cerevisiae* using fed-batch cultivations with RQ-controlled feed. Biotechnol Bioeng 113(1):72–81

Tlais AZA, Fiorino GM, Polo A, Filannino P, Di Cagno R (2020) High-value compounds in fruit, vegetable and cereal byproducts: an overview of potential sustainable reuse and exploitation. Molecules 25(13):2987

Upcraft T, Johnson R, Finnigan T, Hallett J, Guo M (2020) Protein from renewable resources: mycoprotein production from agricultural residues. In: Pierucci S, Manenti F, Bozzano G, Manca D (eds) 30 European symposium on computer aided process engineering, vol 48. Elsevier, Amsterdam, pp 985–990

van Wyk N, Kroukamp H, Espinosa MI, von Wallbrunn C, Wendland J, Pretorius IS (2020) Blending wine yeast phenotypes with the aid of CRISPR DNA editing technologies. Int J Food Microbiol 324:108615

Verni M, Rizzello CG, Coda R (2019) Fermentation biotechnology applied to cereal industry by-products: nutritional and functional insights. Front Nutr 6:42

Vilela A (2021) An overview of CRISPR-based technologies in wine yeasts to improve wine flavor and safety. Fermentation 7:1

Vornholz G (1994) The sustainable development approach. Intereconomics 29(4):194–198

Waks Z, Silver PA (2009) Engineering a synthetic dual-organism system for hydrogen production. Appl Environ Microbiol 75(7):1867–1875

Werten MWT, Wisselink WH, Jansen-van den Bosch TJ, de Bruin EC, de Wolf FA (2001) Secreted production of a custom-designed, highly hydrophilic gelatin in *Pichia pastoris*. Protein Eng Des Sel 14(6):447–454

Yadav VG, De Mey M, Lim CG, Ajikumar PK, Stephanopoulos G (2012) The future of metabolic engineering and synthetic biology: towards a systematic practice. Metab Eng 14(3):233–241

You S, Yin Q, Zhang J, Zhang C, Qi W, Gao L et al (2017) Utilization of biodiesel by-product as substrate for high-production of β-farnesene via relatively balanced mevalonate pathway in *Escherichia coli*. Bioresour Technol 243:228–236

Yu AQ, Pratomo Juwono NK, Foo JL, Leong SSJ, Chang MW (2016) Metabolic engineering of *Saccharomyces cerevisiae* for the overproduction of short branched-chain fatty acids. Metab Eng 34:36–43

Yurkov AM (2018) Yeasts of the soil—obscure but precious. Yeast 35(5):369–378

Zheng C, Sun X, Li L, Guan N (2012) Scaling up of ethanol production from sugar molasses using yeast immobilized with alginate-based MCM-41 mesoporous zeolite composite carrier. Bioresour Technol 115:208–214

Zhou YJ, Buijs NA, Zhu Z, Qin J, Siewers V, Nielsen J (2016) Production of fatty acid-derived oleochemicals and biofuels by synthetic yeast cell factories. Nat Commun 7(1):11709

Zhu X, Yan Y, Wan W, Wang Y, Wu Y, He X et al (2018) Yeast-derived active carbon as sustainable high-performance electrodes for lithium–oxygen batteries. Mater Lett 215:71–74

Suppressive Effect of Root Knot Nematode *Meloidogyne* spp. During Composting of Tomato Residues

15

Azim Khalid, Soudi Brahim, Périssol Claude, Imane Thami-Alami, and Roussos Sevastianos

Abstract

The suppression of pathogens in general, as well as phytopathogenic agents in particular, by the composting process is quite oftenly attributed to the sanitation effect of the thermophile phase. This ensures hygienization and sanitation of compost against plant pathogens and phytoparasites to prevent their dissemination in horticultural production. The root knot nematode is one of the most economically damaging genera of plant-parasitic nematodes on horticultural crops. Composting horticultural wastes should ensure compost hygienization from root knot nematodes. Thus, the objectives of this work are to produce hygienic compost based on tomato residues and to study the suppression kinetic of *Meloidogyne* spp. during composting. Three levels of initial C/N ratio were tested (20, 30, and 40), combined with three levels of initial moisture (30, 50, and 80%). Several parameters were monitored such as temperature, oxygen, as well as the quality parameters of the composts produced. The results obtained showed

A. Khalid (✉)
Research Unit of Integrated Crop Production, Regional Center of Agronomy Research, INRA-CRRA, B.P.124, Inzegane, Agadir, Morocco

S. Brahim
Department of Environment and Natural Resources, IAV Hassan II-Rabat. B.P. 6202, Madinat Al Irfane, Morocco

P. Claude · R. Sevastianos
Institut Méditerranéen de Biodiversité et d'Ecologie marine et continentale (IMBE) Aix-Marseille Université, CNRS, IRD, Avignon Université, Campus de l'Etoile, Centre Saint-Jérôme, Marseille Cedex, France

I. Thami-Alami
Laboratory of Microbiology, Regional Center of Agronomy Research, INRA-CRRA Avenue Mohamed Belarbi Alaoui B.P 6356-Rabat Institute, Inzegane, Agadir, Morocco

N. K. Arora, B. Bouizgarne (eds.), *Microbial BioTechnology for Sustainable Agriculture Volume 1*, Microorganisms for Sustainability 33,
https://doi.org/10.1007/978-981-16-4843-4_15

that despite the thermophile phase at the start of the composting marked by a temperature reaching 60 °C, 10% of the root knot nematodes' larvae were not eliminated. This shows that temperature alone as a sanitizing factor does not allow the compost hygienization. However, at the end of the process, root knot larvae were completely suppressed. This is probably owed to the contribution of the mesophilic flora which colonizes the substrate after the thermophile phase.

Keywords

Compost hygienization · *Meloidogyne* spp. · C/N ratio · Mesophile phase · Crop damage

15.1 Introduction

Cultivated plants are often exposed to many critical conditions which favor pathologies inception following different forms of metabolic disturbance. These abiotic stressors can be climatic (high or low temperatures) or anthropogenic in origin through the application of substances at high doses of toxicity such as pesticides or heavy metals. According to Singla and Krattinger (2016), biotic stress involves a living organism qualified as a pathogenic agent (fungus, bacteria, mycoplasma, virus, and viroid) or other agents such as mites, insects, mollusks, nematodes, and herbivores (Bonnemain and Chollet 2003). Several parasites attack crops under Moroccan conditions, such as phytoparasitic nematodes (*Meloidogyne* spp.) that represent one of the most important and damaging parasites with regard to cultivated plants and more particularly vegetable crops (Karajeh 2008; Asif et al. 2020).

Compost is a hygienic product, stable in composition, and rich in humic substances, resulting from a biochemical conversion by aerobic biodegradation of organic matter (Azim et al. 2017). The suppressive ability of composts to plant diseases has been attributed to the chemical composition of composts, availability of nutrients, and microbial composition due both to the supply of microorganisms and to the stimulation of those in the soil (Pane et al. 2019). All composts, if correctly processed and applied, can suppress telluric phytopathogens in amended soils and ensure healthy soil to the crop. However, not all composts have the same ability to effectively inhibit plant pathogens. Yet, soils in general are not suppressive, and disease management is based mainly on chemical control (Tian et al. 2002). Consequently, interest in the use of biological control means was increasingly considered in recent times with growing concern for sustainable agriculture that respects the environment (Barratt et al. 2017).

Nowadays, the market for compost and organic soil conditioners seems to be promising for the agriculture, home gardening, landscaping, horticulture, and construction sectors. In fact, the overall compost sector is expected to reach an estimated $8.9 billion by 2025, with a CAGR (Compound Annual Growth Rate) of 5% to 7% from 2020 to 2025 (Fig. 15.1). The main precursors of this market are the increasing

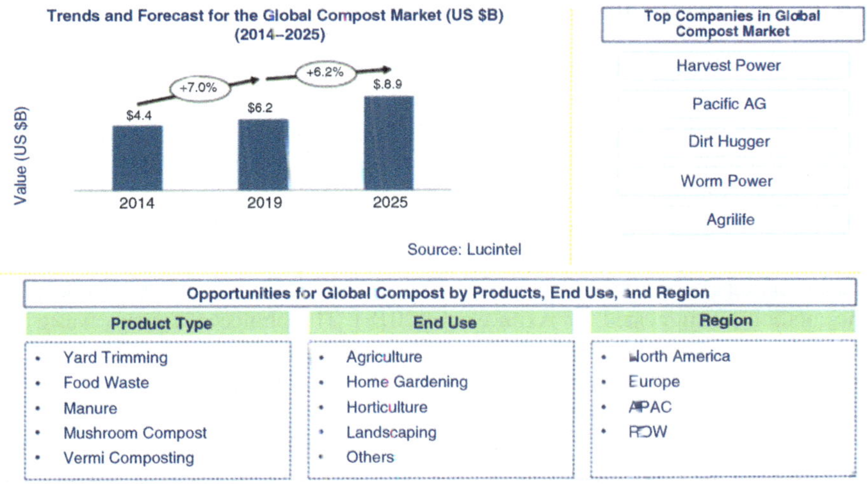

Fig. 15.1 Growth of global compost market and the main opportunities for compost market segment (Lucintel 2020)

search for organic products and the increased awareness of the dangers associated with chemical fertilizers and pesticides (Lucintel 2020).

Recently, studies have focused on the use of composted organic matter for the biological protection of plants against nematodes (Hallmann and Meressa 2018). Indeed, the use of compost has been able to significantly reduce the nematode population on plants to keep it below an acceptable threshold (Atandi et al. 2017). This has oriented research toward the promotion of other forms of composts. On the other hand, the suppressive effects on phytopathogenic diseases and phytoparasitic nematodes are often observed during the composting process, opening up entirely new perspectives. Noble and Roberts (2004) have claimed that a sufficient exposure time of thermophile temperature is enough to eradicate nematodes during composting. Menke and Grossmann (1971) reported that for the sanitation of *Meloidogyne incognita* inoculated in the organic residues of pepper and tomato, it is necessary to go through a thermophile phase with an average temperature of 57 °C for at least 19 h in pile composting. In contrast, for sanitation of *M. incognita* inoculated to green waste in the form of egg masses, a maximum temperature of 74 °C during 4 days was required in dynamic free heap composting (Noble and Roberts 2004). However, it is only progressively that researchers begin to understand under what conditions root knot nematodes can resist during composting. The objective of this work is to produce a hygienic compost based on tomato residues and to study the suppression kinetic of root knot nematodes during composting with different initial C/N ratios and moisture.

15.2 Materials and Methods

15.2.1 Studied Factors

Composting was carried out in Melk Zher experimental center (INRA-Agadir) 45 km south of Agadir city (coordinate: N30.0440758,W-9.548735). This study focused on composting green waste from tomatoes, sheep manure, and sawdust. During the biodegradation process, microorganisms use carbon as an energy source and nitrogen for the production of proteins necessary for their development. For this reason, initial C/N ratios of the mixtures (20, 30 and 40) were calculated according to the method as suggested by Azim et al. (2018). The choice of tomato waste was motivated by the fact that horticultural producers consider it to be a source of inoculum for root knot nematodes (*Meloidogyne* spp.). As a consequence, the composting of this high risky waste for vegetable crops requires rigorous monitoring throughout the process, while acting on two parameters: temperature and competition between microorganisms.

Composting was carried out in 220 L biocomposters ("Biolane" brand, Finland) with a patented natural ventilation system, equipped with a thermal probe and an evacuation hatch. Moistening is done in a way to start with three moisture levels (30, 50, and 80% v/v) for each level of C/N ratio. The factors studied were the initial C/N ratio with three levels (20, 30, and 40) and the initial moisture with three levels (30, 50, and 80% (v/v)). All the treatments were carried out in biocomposters in number of nine having received the combination of the two factors studied (three levels of C/N * three levels of RH% = 9 combinations).

15.2.2 Formulation of the Mixtures

The method presented below allowed to carry out relatively balanced mixtures with respect to their C/N (Table 15.1). It is based for each category of waste, on their total carbon and total nitrogen contents. The raw materials were brewed, in particular those of tomato crop, to obtain a particle size of less than 3 cm. The constituents were mixed manually (Picture 15.1a) and transferred to the biocomposters (Picture 15.1b). The amount of water applied was adjusted with a probe measuring the moisture of the mixture in order to guarantee a starting threshold in accordance with the levels of the factor studied.

15.2.3 Observations and Measured Parameters

15.2.3.1 Physical Parameters

The temperature reflects the level of microbial activity. A multiparameter probe (CompostManager®V3.11.f- [SLC], UK) was carefully placed 80 cm deep in the biocomposter in order to measure the actual temperature of the compost pile. This probe also measures relative moisture (v/v %), O_2 and CO_2 level instantly with a

Table 15.1 Composition of mixtures with initial values of total carbon and total nitrogen

Treatments	C/N20			C/N30			C/N40		
Organic wastes	Sheep manure	Saw dust	Tomato waste	Sheep manure	Saw dust	Tomato waste	Sheep manure	Saw dust	Tomato waste
Ci (%)	35.41	59.4	42.2	35.41	59.4	42.2	35.41	59.4	42.2
Ni (%)	3.19	0.09	3.15	3.19	0.09	3.15	3.19	0.09	3.15
(C/N) i	11.1	660	13.39	11.1	660	13.39	11.1	660	13.39
% (p/p)	20	25	55	18	47	35	20	60	20
(C/N) mean	19.3 ≈ 20			29.7 ≈ 30			40.60 ≈ 40		

Ci total carbon, *Ni* total nitrogen, *C/Ni* total carbon to total nitrogen ratio, *C/NM* total carbon to total nitrogen ratio of the mixture

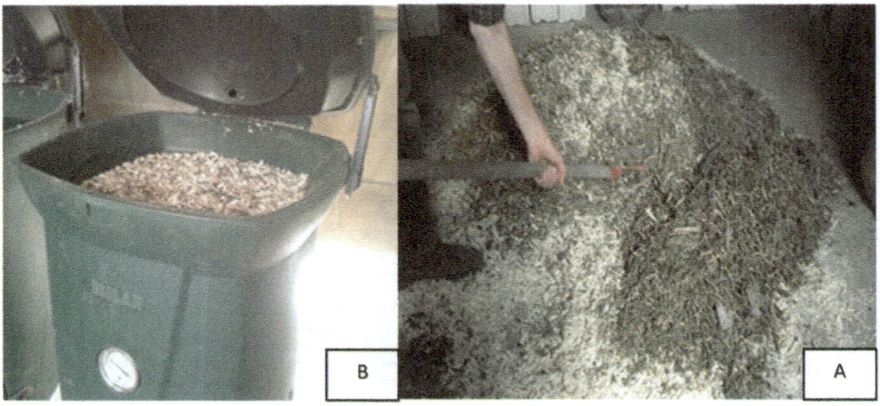

Picture 15.1 Manual mixing of the constituents (**a**). Biocomposter filled with mixed ram materials (**b**)

correction of the O_2 content as a function of temperature. Relative moisture content was measured before watering the biocomposters to avoid values biasing. The measurements were taken weekly for the first 2 months and then every 10 days until the end of the composting process.

15.2.3.2 Physicochemical Parameters

All analyzes were performed in three repetitions for each sample taken from the nine combinations throughout the duration of the experiment. The sampling consisted in taking three composite samples of the biocomposter from the center of the pile. A reversal is carried out every 2 weeks to homogenize the heap.

The pH was measured using an InoLab pH meter (at room temperature) in a 1/10 (w/v) aqueous extract (compost/distilled water) after stirring for 30 min. The electrical conductivity (EC) was measured on the other hand (at room temperature) using a Bioblock Scientific EC meter in an aqueous extract 1/5 (w/v) (compost/distilled water) after an hour of agitation.

The organic matter (MO) was obtained by difference between the weighed mass of the dry sample (M1) and the mass of the same sample calcined at 480 °C (M2) at a constant weight for more than 6 h. The incinerator used was that of NaberTherm. The total organic carbon (TOC) of the samples was estimated from MO using the following formula:

$$\text{TOC} = \text{OM}/1.724$$

All analyses were performed in triplicates for each sample taken from the nine biocomposters during the experiment. The sampling consists in taking three composite samples of the biocomposter from the center of the pile. Biomass turning was carried out every 2 weeks to homogenize the heap. Analysis of the total nitrogen content was performed using the standard Kjeldahl procedure (Gerhardt: Kjeldatherm-Vapodest 20, Germany) as described by Jackson (1967).

15.2.3.3 Microbial Parameters

Determination of the Bacterial Load of the Mesophilic and Thermophile Flora
To follow the evolution of the total mesophilic and thermophile microflora, microbial analyses were carried out according to the method described by Saidi et al. (2009). The work was carried out under aseptic conditions under the hood. The material and the culture medium were sterilized under 120 °C for 20 min:

- *Sample extraction and dilutions*: A series of dilutions were carried out using 10 g of compost and 90 mL of sterile distilled water as the mother solution was stirred for 2 h. Using a sterile pipette, 1 mL of each dilution was spread on the petri dishes containing TSA (Tryptic Soy Agar) as medium. The petri dishes were incubated at 22 °C for mesophile flora and 45 °C for thermophile flora for 5 days. After incubation, a colony count was performed only on dishes with a number of colonies between 30 and 300.
- *Results expression*: The determination of the bacterial load is made by counting the colonies, and the results are expressed in CFU (colony-forming units)/1 g of compost: according to the mathematical formula below:

$$N = \sum C/(n_1 + 0.1\, n_2) * d * V$$

where

- N: Number of CFUs per gram of compost.
- Σ: colonies: Sum of colonies of interpretable boxes.
- V: Volume of spread solution (1 mL).
- $n1$: Number of dishes considered at the first dilution selected,
- $n2$: Number of dishes considered at the second dilution used,
- $d1$: Factor of the first dilution selected.

15.2.3.4 Nematode Extraction and Counting

The extraction of root knot nematodes (*Meloidogyne* spp.) was carried out in the laboratory from the roots of infested tomato plants. Since tomato waste is one of the essential components of this trial, it is important to monitor the suppressiveness effect of root knot nematodes during composting:

- *Selection of infested roots*: Tomato roots were separated from the soil by shaking them lightly or by passing them under a water shower or after soaking. The roots were then placed on a paper towel to absorb excess of water. They were then cut to extract the inoculum.

Nematode extraction: Baermann (1917) modified method was used for this step by collecting the infested tomato roots with nematodes, rinsing, and cutting them finely with a scissor. The cut roots were ground with a plastic hammer avoiding excessive pressure. The ground paste was sieved successively in sieves of different diameters ranging from 1 mm to 40 μm. Nematodes were then recovered in a beaker by rinsing the sieve with a jet of water. A bleach drop was added to the recovered nematodes to burst the egg masses keeping them in suspension by shaking for 30 min. Nematode counting was performed in a small grid plate under 40× magnification of a microscope and reported to gram of roots.

The inoculation of the biocomposters with the nematodes was carried out with a suspension of 120,000 J2/5 L/biocomposter (5 L of fresh water containing larvae J2 of *Meloidogyne* spp. at a concentration of 24 J2/mL). The inoculation density was set to 545 J2/L of biocompost with useful volume of 220 L. The extraction of the larvae was complemented by the presence of egg masses in the aqueous suspension. Each egg mass contains about 200 to 300 J2 larvae.

15.2.4 Cress Test (Compost Quality)

This test assesses the maturity of a compost in terms of its anti-germination effect on watercress seeds. The test consists of watercress seeds germinating on a mixture of compost with black peat (40% to 60% by volume) and counting the number of seedlings emerged at 2 and 7 days. This test is carried out in a phytotron under controlled conditions (25 °C and 80% of humidity). For each of the nine treatments, three repetitions were carried out in addition to control (black peat) at 100%.

15.2.5 Statistical Analysis

All statistical analyses were performed using IBM SPSS Statistics 25. Multiple comparisons of means were confirmed by the Tukey test at $p < 0.05$.

15.3 Results and Discussion

The evolution of the composting process was determined by monitoring some physicochemical and microbial parameters for 2 months. All of these parameters provide information on the transformations that have occurred during composting. Likewise, they make it possible to better assess the degree of degradability of the initial substrates and to determine the maturity of the final products.

15.3.1 Physical Parameters

15.3.1.1 Temperature

The temperature evolution during composting is a first index which reflects the degree of degradability of the compost. In fact, temperature is a result of the microbial activity during fermentation, resulting from the lysis of carbon-carbon bonds, which is an exothermic reaction (Chang et al. 2019). However, the compost temperature curve studied at a classic pace of composting was characterized by two important phases (fermentation phase and maturation phase). The fermentation phase is characterized by a thermophile cycle with a rise in temperature up to 70 °C (Fig. 15.2). This increase is the result of intense microbial activity which results from the degradation of the simple molecules present in the substrate (Azim et al. 2018).

The maturation phase is characterized by a decrease in temperature to ambient one by involving the fungi in the degradation of complex molecules (cellulose, hemicellulose, lignin, etc.). The most significant temperatures resulted from the treatment C/N20 ratio. This can be explained by the presence of nitrogen which promotes the development of microbial activity more intensely compared to other treatments C/N30 and C/N40. Likewise, C/N20 entered the cooling phase early compared to the other two treatments. The carbon content in the C/N30 and C/N40 treatments is high, which explains the continuity of the microbial activity of degradation even after 42 days of mixture.

15.3.1.2 Oxygen Consumption

Figure 15.3 shows the evolution of the oxygen consumption of the different combinations. The values are expressed by difference between the ambient oxygen content (21%) and the oxygen content in the pile's center. The initial oxygen

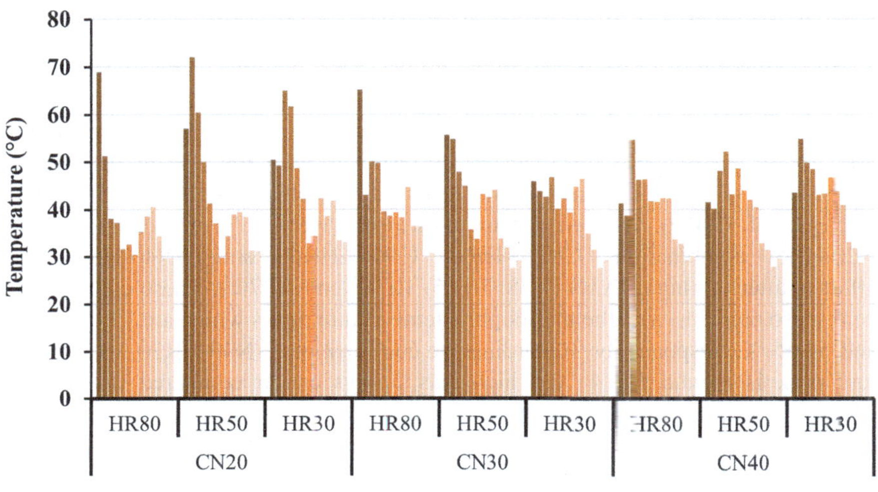

Fig. 15.2 Temperature evolution of the different treatments during the composting period

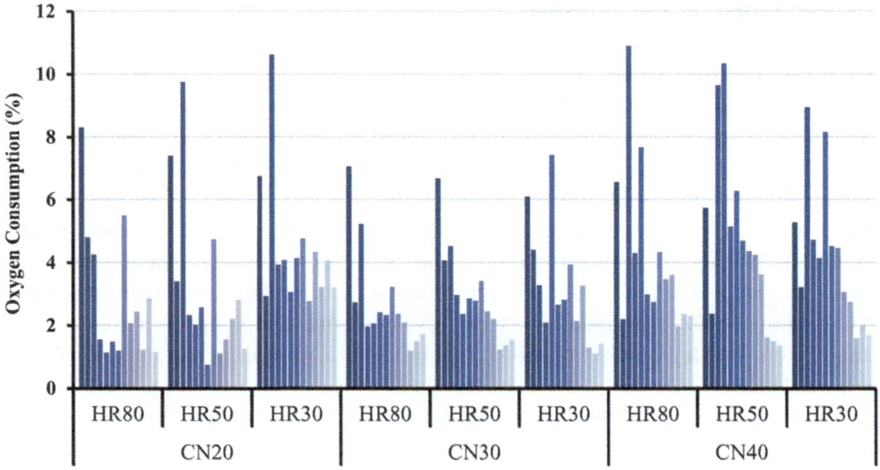

Fig. 15.3 Evolution of the oxygen consumption of the different treatments during the composting period

Table 15.2 Pearson correlation value (R^2) between oxygen consumption and initial moisture values

	C/N20	C/N30	C/N40
Pearson correlation value (R^2)	0.9993	0.9459	0.9976

consumption values of the C/N20 treatments, C/N30, and C/N40 are positively correlated (Table 15.2) with the initial moisture contents (30, 50, and 80%). In fact, the higher the initial moisture of the pile is, the more intense microbial activity becomes while maintaining adequate ventilation and porosity. This result confirms those found by Mustin (1987) and Khater (2015).

The evolution of oxygen consumption does not follow the same trend between treatments. However, it appears clear that the most intense oxygen consumption was recorded by the C/N20 treatment with 7.5% against 6.62 and 5.87%, respectively, for C/N30 and C/N40. This demand corresponds with the rise in thermophile temperature (Fig. 15.2) which is the result of microbial activity as reported by Chang et al. (2019). Oxygen consumption peaks were only recorded for the C/N20-HR50 combinations, C/N20-HR30, C/N40-HR80, and C/N40-HR30 only 10 days after mixing (DAM) (9.78, 10.65, 10.92, and 8.97%). These consumption peaks also coincide with the thermophile temperature peaks which are 72.03, 65.07, and 54.65 °C, respectively. Regarding the combination C/N40-HR30, its temperature peak is around 54.84 °C which was 2 days earlier compared to the other abovementioned combinations. It should also be noted that the amplitude of O_2 consumption is higher for C/N20 and C/N40 treatments than for that of C/N30, despite the persistence of mesophilic temperatures during the composting period. The O_2 consumption of the C/N40 treatment is very variable because the percentage of sawdust is high, which implies a high porosity and subsequently drainage of the moist water. These conditions are unfavorable to microbial growth and development, which is hampered by low moisture that does not exceed 30% (v / v).

15.3.2 Physicochemical Parameters

15.3.2.1 pH

Figure 15.4 shows the evolution of the pH of the different combinations during the composting period. At the beginning of the composting process, there is an initial acidic pH (6.4). The initial acidic pH of the mixtures is due to the nature of these raw materials, in particular those of tomatoes which have a pH of 5.94. After 20 days of mixing, the pH has increased to be alkaline around 7.5. This increase in pH can be explained by the development of the thermophile flora which hydrolyzes protein and organic nitrogen; this generates a rise in ammonium nitrogen and subsequently a rise in pH. Between 20 and 40 DAM, the pH is slightly stabilized for C/N20 and C/N40. The steady trend is maybe due to the buffering capacity of the humus during the maturation phase (Chaher et al. 2020). Concerning C/N30, there was a slight acidification probably due to the production of organic acids and the dissolution of CO_2 in the medium and/or the degradation products of simple molecules (carbohydrates and lipids Fig. 15.4). The pH is stabilized to a slightly alkaline value, which promotes the development of actinomycetes and alkaline bacteria (Mustin 1987). Statistical analysis revealed no significant differences between the different initial C/N ratios or between the values of the initial RH.

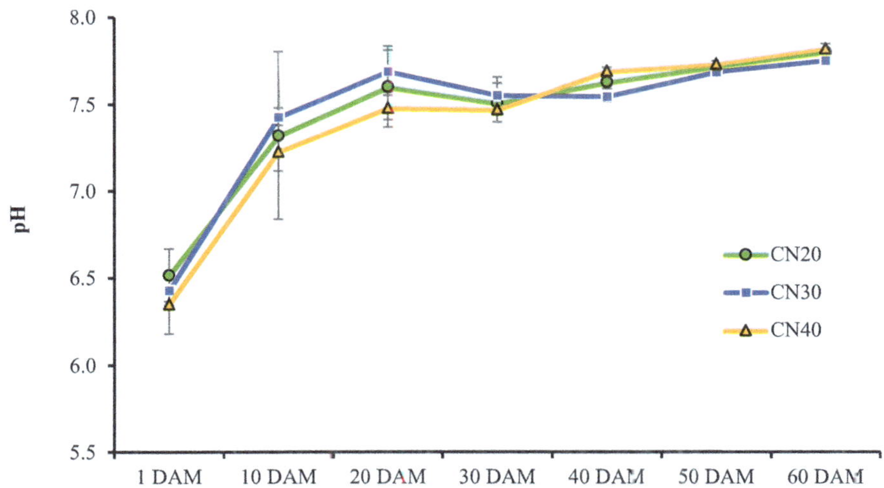

Fig. 15.4 Evolution of the pH of the different treatments during the composting period

15.3.2.2 Electrical Conductivity

Figure 15.5 shows the evolution of the EC of the different combinations during the composting period. The EC for the ratios C/N20, C/N30, and C/N40 had an initial value between 10 and 14 dS cm^{-1}. After 10 days of composting, a decrease in the salinity rate is observed, reaching values between 2 and 6 dS cm^{-1} which is maybe due to the humification of organic materials which was transformed salt and micromolecular organic acids to macromolecular humus (Sanchez-Monedero et al. 2018). The successive moistening of the biocomposters was done with low-salinity water (0.9 dS cm^{-1}) without turning. After 2 months of the process, the EC increases very slightly for C/N20 and C/N30 composts. This can be explained by the temperature and humidity conditions which are favorable for the degradation of organic matter and its mineralization and thus an increase in EC (Chaher et al. 2020). There is a significant difference between the EC values of the C/N20, C/N30, and C/N40 treatments. The salinity is proportional to the high-salinity tomato waste (15.75 dS cm^{-1}). It should be noted that by decreasing the C/N ratio, the proportion of tomato waste increases, and subsequently the EC increases.

15.3.2.3 C/N Ratio

Figure 15.6 illustrates the evolution of the C/N ratio of the different combinations during composting period. After the first 10 days of composting, the nitrogen content decreased from values between 1.31 and 1.28% to 1.25 and 1.19%. This decrease is mainly linked to nitrogen losses in the form of volatile ammonia during thermophile phase (Jiang et al. 2019). The initial C/N ratios decreased from 20.68, 28.98, and 39.60 to 19.16, 23.39, and 27.08, respectively, for the three treatments C/N20, C/N30, and C/N40 at the end of the process. The decrease in C/N ratio is generally linked to the loss of organic carbon by biodegradation of organic matter and the

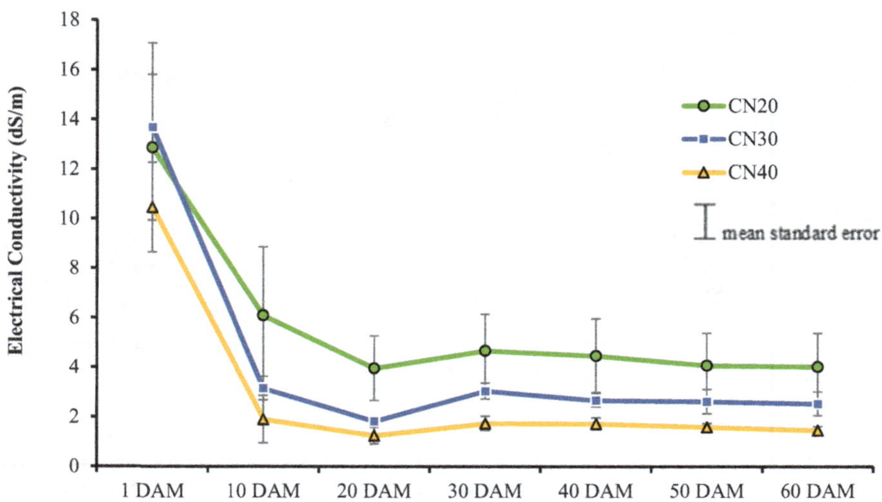

Fig. 15.5 Evolution of the EC of the different treatments during the composting period

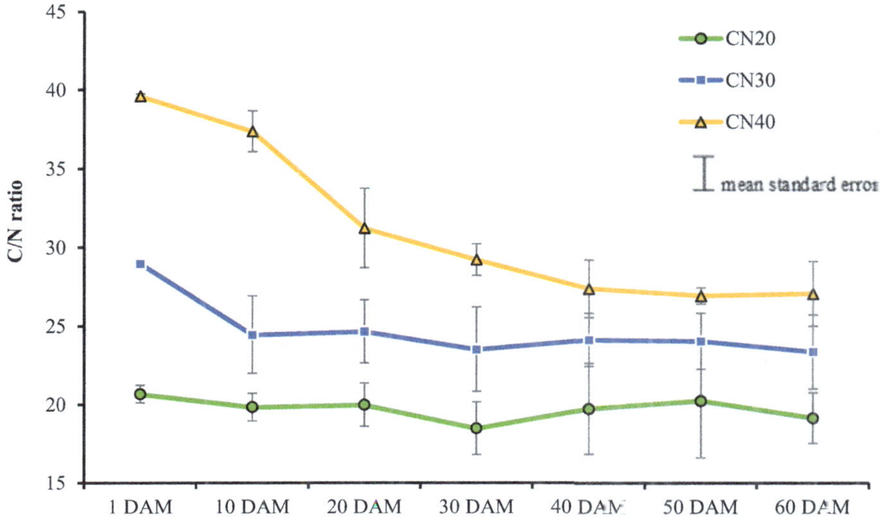

Fig. 15.6 Evolution of the C/N ratio of the different combinations during the composting period

release of CO_2 (Meng et al. 2018). The final values attained by C/N20 and C/N30 treatments are around 20 after 2 months of composting, affirming the good maturity of the final compost. According to Barje (2010), a compost is considered mature if it has a C/N ratio between 10 and 20. The decrease in C/N in the C/N40 treatment is more marked than those of C/N20 and C/N30 treatments. This may be due to the temperature which continued to be in the mesophilic ranges for the C/N40 treatment and therefore, a continuation of the degradation of carbon by the mesophilic fungi (Jiang et al. 2019).

15.3.3 Biological Parameters

15.3.3.1 Thermophilic and Mesophilic Microflora

The diversity of the existing microflora is linked to the different levels of biodegradability of the organic constituents. Different communities of microorganisms will thus succeed one another according to their specialty of decomposition and their enzymatic arsenal. This succession of decomposition explains the succession of the different phases of composting (Liu et al. 2018). However, the total thermophilic microflora allows better control of the thermophile phase which constitutes the critical phase during the composting process.

Figures 15.7 and 15.8 respectively show the evolution of the mesophilic and thermophilic flora as a function of all the treatments. At the start of the process, the total mesophilic and thermophilic flora are equal. After 30 DAM, a significant increase in the total mesophilic flora (Fig. 15.8) was observed and reached a value

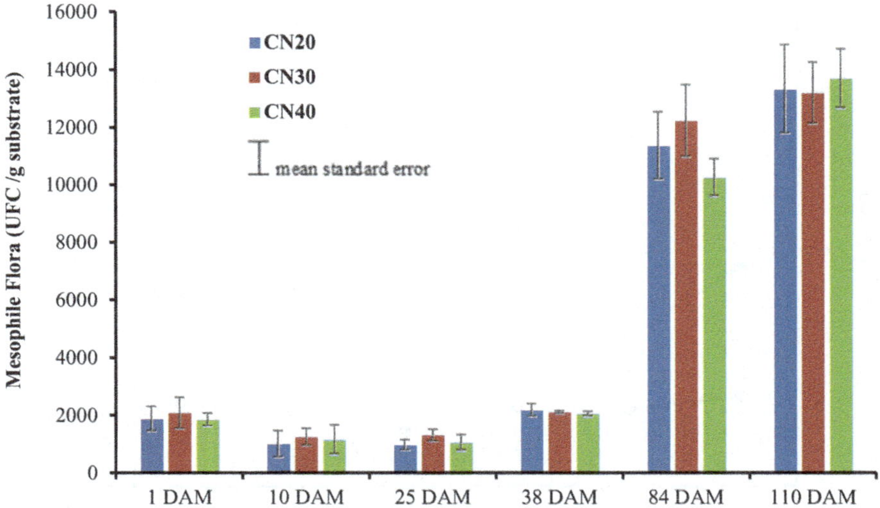

Fig. 15.7 Evolution of the mesophilic flora according to the three treatments during composting

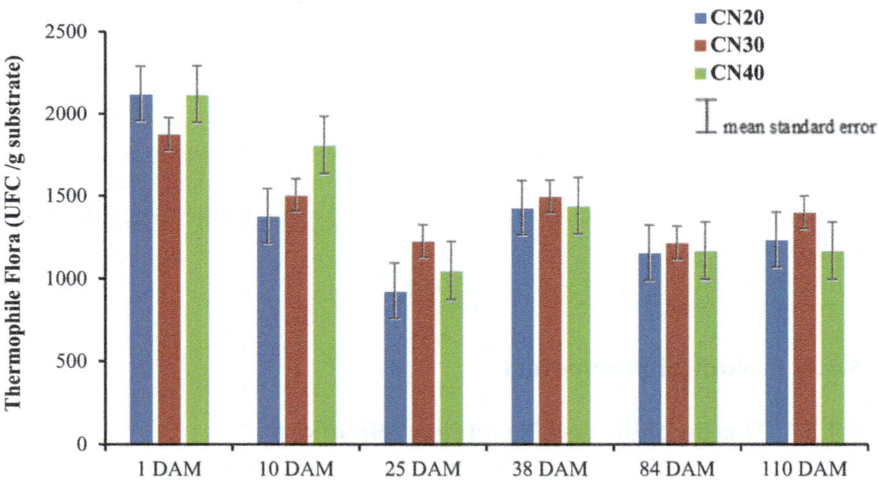

Fig. 15.8 Evolution of the thermophilic flora according to the three treatments during composting

of 1.2 10^4 CFU/g compost. The proliferation of this flora indicates the predominance of mesophilic bacteria responsible for the degradation of labile organic matter (Saidi et al. 2009). This predominance also indicates an early maturity of the compost with the development of antagonistic microorganisms (Fuchs 2002), which gives the compost the ability to self-sanitize and thereby protect plants against telluric diseases transmitted through the compost (Pérez-Piqueres et al. 2006). The richness of compost in beneficial microorganisms improves the biodiversity of overexploited soils. Indeed, the amendment of soils with compost induces an increase in the populations of soil microorganisms by a factor of 1000 (Lazarovits 2001), which clearly shows the benefits of compost compared to peat which, in several studies, has been found to be much less rich in microorganisms (than composts) (Larbi 2006). Statistical analysis revealed no significant difference between all treatments.

The thermophilic flora (Fig. 15.8) in the compost studied experienced high values, showing that the compost is progressing well. The thermophile phase which coincides with the rise of the thermophilic microflora was indeed short (25 DAM). Near the end of the process, the thermophilic flora stabilized at the same time as the mesophilic flora began to increase in 84 DAM. Mustin (1987) indicated that microorganisms in the compost, at any time, create the conditions for their own destruction, which are optimal for the following microbial populations engaged in composting. Composting also transforms organic matter into a substrate that becomes hostile for the growth and survival of most pathogens (Deloraine et al. 2002). According to Soudi (2005), these are either antagonisms of the populations of saprophytes developed in the compost, or the toxicity of the substances generated by the metabolism of this microflora or even unfavorable conditions found at certain times of the composting (pH, moisture, C / N ratio or others).

15.3.3.2 Evolution of Root Knot Nematode Populations

Figure 15.9 shows the evolution of the population of root knot nematodes (*Meloidogyne* spp.) according to the different treatments throughout the process. The initial values at inoculation were equal in terms of J2 larva (545 J2/L of biocomposter giving 54.5 J2/100 mL substrate). With the presence of egg masses and the favorable conditions to their hatching (temperature around 35 °C and RH of 60%), part of the egg masses hatched to directly give J2 larvae. This explains the difference in initial density at sampling between the different combinations. The rise in temperature during composting normally suppresses the majority of pathogenic germs. This "hygienization" of the compost is generally carried out for composting taking place in good conditions (turning, aerobic, sufficient duration, humidity). Noble and Roberts (2004) have shown that almost all germs and parasites are destroyed during composting, as long as the temperatures are between 40 and 70 °C. With regard to root knot nematodes *M. incognita*, the lethal temperature is 57 °C (Menke and Grossmann 1971; Duval 1992).

On the third sampling, the population densities dropped to record an average value of 114.44 J2/100 mL (substrate) with a standard deviation of 34.68 J2/100 mL (substrate). This decrease corresponded the thermophilic phase of the different combinations and reached its end 14 DAM. During the mesophilic phase, the densities of the nematodes increased again toward the fourth sample for all combinations. This increase may be due to the hatching of residual egg masses after the first inoculation. The temperature conditions (mean of 40.05 ± 5.02 °C) and moisture content (mean of 27.06 ± 13.01%) were in fact slightly favorable. This clearly shows that the egg masses can withstand thermophilic temperatures which have reached 68.91, 72.03, and 65.22 °C, respectively, for the combinations C/N20 M80, C/N20 M50, and C/N30 M80. These thermophile temperatures are in

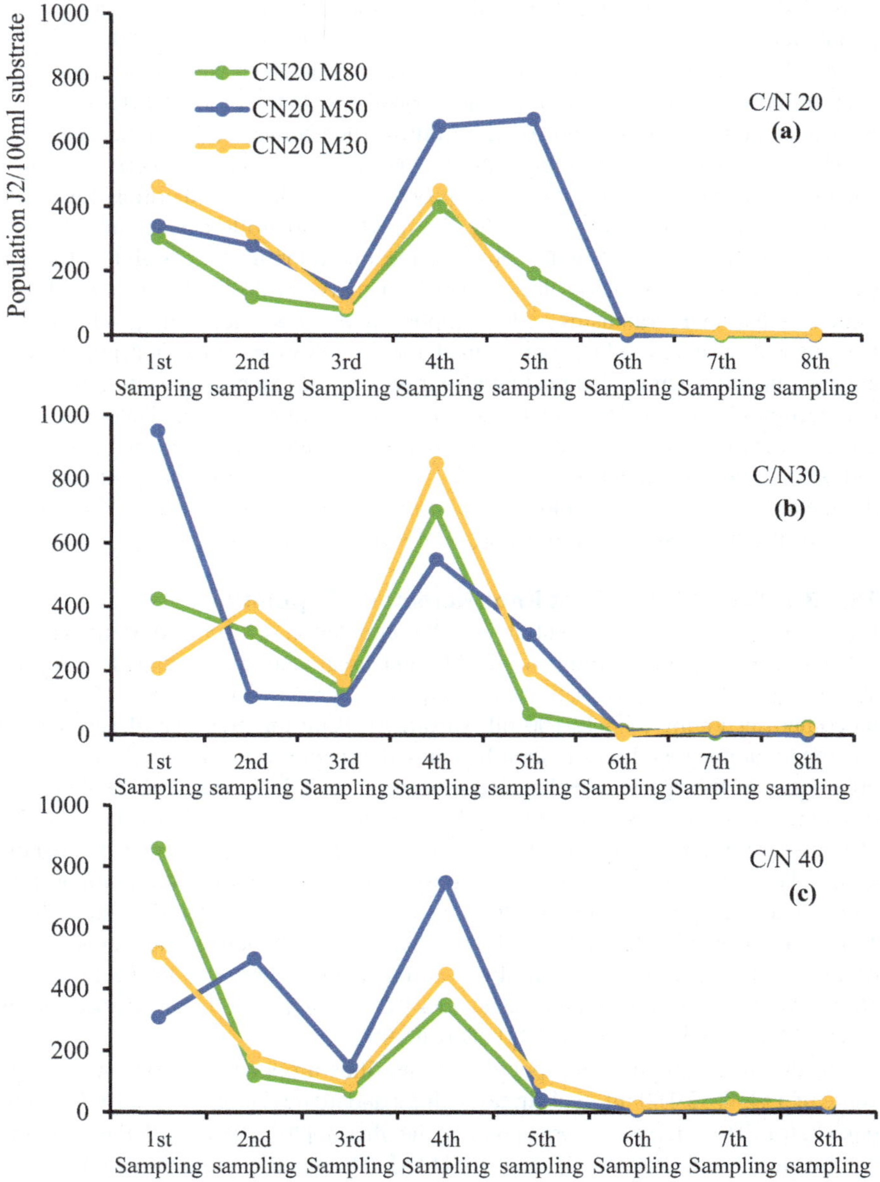

Fig. 15.9 Evolution of root knot nematodes (*Meloidogyne* spp.) according to the different treatments throughout the process. (**a**) stands for the treatment with the initial C/N ratio of 20, (**b**) for treatment with initial C/N ratio 30, and (**c**) for treatment with initial C/N ratio 40

fact maximum temperatures reached in the substrate at the sampling locations. This result, therefore, overturned what was proved by Duval (1992).

On the fifth and the sixth samplings, all nematode populations decreased drastically toward values close to zero except for the combination C/N20 M50, which was kept stable between the fourth and the fifth sampling. The thermal profile remained mesophilic, in particular for the C/N40 treatment with these three variants (average 42.78 ± 0.95 °C) and the C/N30 treatment (average 41.89 ± 3.28 ° C). On the other hand, the thermal profile of C/N20 treatment cooled to record an average of 34.71 ± 0.52 °C, which partly explains the less drastic drop in the densities of the J2 larvae.

For the last two samplings (seventh and eighth samplings), the nematode densities remained close to zero until the end of the process. Despite the mesophilic temperature between 20 and 35 °C, root-knot nematodes were almost eliminated, which can be explained by the oriented antagonism of the mesophilic flora against root-knot nematodes. Indeed, in some cases, the time/temperature combination is not sufficient to eradicate the pathogens, and yet they are eliminated. Thus, for the same organism, the temperature necessary for its destruction is lower in a compost heap than in the laboratory (Lopez-Real and Foster 1985). During composting, microbial antagonisms, along with the toxic substances formed or even unfavorable conditions that occur at certain times of composting (pH or others), are all associated to be suppressive against pathogens (Duval 1992; Soudi 2005; Pane et al. 2019). Fungi, bacteria, nematodes, and viruses that cause crop damage are very vulnerable to stress from composting. Even organisms that form sclerotia and are very persistent in the soil are destroyed (Soudi 2005). The hygienization process of J2 larvae takes place by thermal remediation of fermentable waste during the thermophile phase. In contrast, the survival of egg masses of *Meloidogyne* spp. is influenced in a lesser extent by thermophile temperature (Hallmann and Meressa 2018). At the end of the process, only the combinations C/N20 M80, C/N20 M50, and C/N30 M50 are free from root knot nematode larvae in their respective samples. This confirms that these combinations have been effective for the eradication of *Meloidogyne* spp. inoculated as larvae and egg masses.

15.3.4 Compost Quality Tests

Figure 15.10 shows the results of the cress test on the produced composts according to the C/N treatments. Statistical analysis revealed no significant difference for the initial C/N factor compared to a control consisting of 100% black peat. The cress test has shown that the composts produced are not phytotoxic since the average germination rate exceeds 50%. However, it should be noted that composts with a germination index of less than 60% should not be amended before 90 days have elapsed since the crop was set up (Soudi 2005). This is the case for composts C/N20 (GI = 58.89% ± 0.51) and C/N40 (IG = 56.67% ± 1.2), while that for C/N30 can be amended to the soil even just before transplanting or sowing the crop.

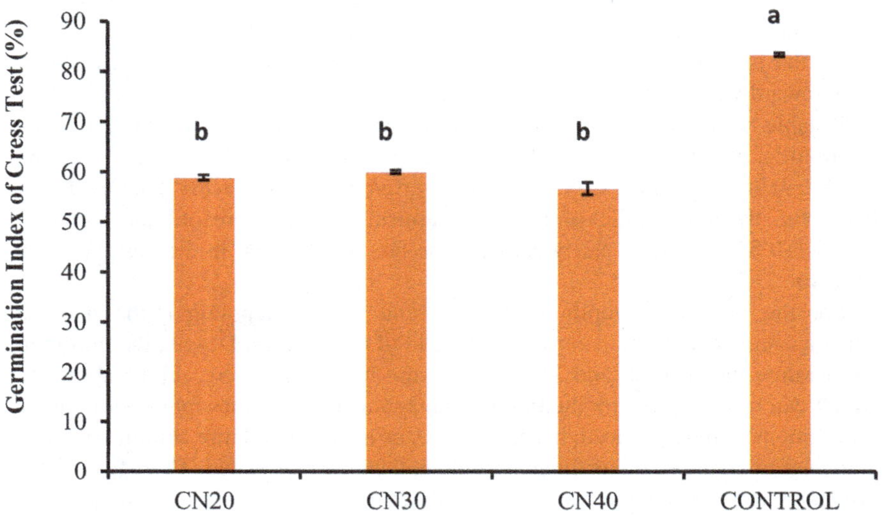

Fig. 15.10 Cress test results on compost produced according to C/N20, C/N30, and C/N40 treatments

15.4 Future Prospects

While increasing compost production and demand, attention must be paid to the compost quality and plant safety. As far as compost is amended on the agricultural soils, stakeholders should avoid making it a potential dissemination agent of plant diseases and parasites. Therefore, plant and crop safety should be at higher standards of quality check for the plant diseases, parasite control and suppression during composting. There is a wide research area to determine the optimal conditions to suppress phytopathogens and parasites during composting at multiscale level.

According to the Sustainable Development Goals (SDG) of the United Nations, carbon sequestration is one of the main actions that could be taken in to consideration especially under the SGD N°13 (Climate Action). Regrettably, carbon sequestration is less considered than carbon emission attenuation. Besides and since compost market is expending, policy makers and stakeholders could take advantage of carbon trade and finance to boost compost production and tackle organic waste issue. This objective could be achieved through subsidizing circular economy (compost plant and machinery), regulation improvement, government institutions procurement of compost, community composting, funding R&D for startups creation, networking of compost stakeholders, and education and extension for farmers and future generations.

15.5 Conclusion

The results of this study indicate a good biodegradability of the raw materials, in particular the treatments C/N20 and C/N30 which have shown good results in terms of maturity and optimization. On the other hand, the C/N40 treatment will take an additional period of time to arrive at a final C/N ratio of less than 15. With regard to the kinetics of suppression of root knot nematodes, a slight resistance of the nematodes was observed after the thermophile phase, despite the thermophilic temperature up to 73 °C, where 90% of the J2 larvae were eradicated. This shows that the thermophilic temperature alone during the composting process does not completely eliminate J2 larvae of *Meloidogyne* spp. At the end of the process, the nematodes are probably destroyed by the antagonism of the mesophilic flora. Increasing initial C/N ratio increases the duration of composting and subsequently its production cost. C/N20 and C/N30 treatments appear to be the most suitable for direct soil amendment. At the same time, the sanitation of the compost vis-à-vis the nematodes cannot be carried out exclusively by thermophilic temperature (70 °C); the results have clearly shown that the combined action of thermophilic temperature (reduction of the population), with the action of the mesophilic flora (antagonistic action), was finally able to sanitize the compost of this hazardous phytoparasites. It is recommended to carry out other validation tests in order to detect which groups of microbes have a nematicidal and antagonistic action against root knot nematodes during the mesophile phase.

References

Asif M, Khan F, Ansari T, Ahmad F, Tariq M, Khan A, Siddiqui MA, Khair S, Ahmed W (2020) Elicitation of resistance through the exploration of acibenzolar-*S*-methyl (ASM) against *Meloidogyne incognita* in tomato under salt stress condition. Environ Sust 3:313–318

Atandi J, Haukeland S, Kariuki G, Coyne D, Karanja E, Musyoka M et al (2017) Organic farming provides improved management of plant parasitic nematodes in maize and bean cropping systems. Agric Ecosyst Environ 247:265–272

Azim K, Soudi B, Boukhari S, Perissol C, Roussos S, Thami Alami I (2017) Composting parameters and compost quality: a literature review. Org Agric 8(2):141–158

Azim K, Faissal Y, Soudi B, Perissol C, Roussos S, Thami Alami I (2018) Elucidation of functional chemical groups responsible of compost phytotoxicity using solid-state 13C NMR spectroscopy under different initial C/N ratios. Environ Sci Pollut Res 25(4):3408–3422

Baermann G (1917) Ein einfache Methode zur Auffindung von Anklyostomum (Nematoden) Larven in Erdproben. Geneeskd Tijdschr Ned Indie 57:131–137

Barje F (2010) Biotransformation du mélange des déchets d'huileries d'olive - ordures ménagères: Approche physico-chimique, suivi biochimique, bilan humique et qualité agronomique. Thèse de 3ème cycle, Université Cadi Ayyad, Faculté des Sciences Semlalia, Marrakech

Barratt B, Moran V, Bigler F, van Lenteren J (2017) The status of biological control and recommendations for improving uptake for the future. BioControl 63(1):155–167

Bonnemain JL, Chollet JF (2003) L'arsenal phytosanitaire face aux ennemis des plantes. Considérations générales. C R Biol 326(1):1–7

Chaher N, Chakchouk M, Engler N, Nassour A, Nelles M, Hamdi M (2020) Optimization of food waste and biochar in-vessel co-composting. Sustainability 12(4):1356

Chang R, Guo Q, Chen Q, Bernal M, Wang Q, Li Y (2019) Effect of initial material bulk density and easily-degraded organic matter content on temperature changes during composting of cucumber stalk. J Environ Sci 80:306–315

Deloraine A, Hedreville L, Arthus C (2002) Etude bibliographique sur l'évaluation des risques liés aux bio-aérosols générés par le compostage des déchets. Contrat ADEMe/Careps n° 0075038, Rapport n° 317

Duval J (1992) L'élimination des phytopathogènes par le compostage. Ecological agriculture projects, McGill University (Macdonald campus). AGRO-BIO-310-03. http://eap.mcgill.ca/agrobio/ab310-03.htm

Fuchs JG (2002) Practical use of quality compost for plant health and vitality improvement. In: Insam H, Riddech N, Klammer S (eds) Microbiology of composting. Springer, Berlin, Heidelberg, pp 435–444

Hallmann J, Meressa BH (2018) Nematode parasites of vegetables. In: Sikora RA, Coyne D, Hallmann J, Timper P (eds) Plant parasitic nematodes in subtropical and tropical agriculture. CAB International and USDA, p 898

Jackson ML (1967) Soil chemical analysis *in*: Karlen DL, Varvel GE, bullock DG, cruse RM (1994) crop rotations for the 21st century. Adv Agron 53:1–45

Jiang J, Pan Y, Yang X, Liu J, Miao H, Ren Y et al (2019) Beneficial influences of pelelith and dicyandiamide on gaseous emissions and the fungal community during sewage sludge composting. Environ Sci Pollut Res 26(9):8928–8938

Karajeh M (2008) Interaction of root-knot nematode (*Meloidogyne javanica*) and tomato as affected by hydrogen peroxide. J Plant Prot Res 48(2):181–187

Khater ESG (2015) Some physical and chemical properties of compost. Int J Waste Resour 5:1

Larbi M (2006) Influence de la qualité des composts et de leurs extraits sur la protection des plantes contre les maladies fongiques. [Influence of the quality of composts and compost extracts on their capacity to protect plants against fungal diseases.] Thesis. Institut de recherche de l'agriculture biologique FiBL. CH-Frick; l' Université de Neuchâtel

Lazarovits G (2001) Management of soil-borne plant pathogens with organic soil amendments: a disease control strategy salvaged from the past. Can J Plant Pathol 23(1):1–7

Liu L, Wang S, Guo X, Zhao T, Zhang B (2018) Succession and diversity of microorganisms and their association with physicochemical properties during green waste thermophile composting. J Waste Manag 73:101–112

Lopez-Real J, Foster M (1985) Plant pathogen survival during composting of agricultural organic wastes. In: Gasser JKR (ed) Composting of agricultural wastes. Elsevier Applied Science, London, pp 291–299

Lucintel (2020) Compost Market Report: Trends, Forecast and Competitive Analysis. Value Market Report (ID: 4846238). Lucintel, p 211. https://www.lucintel.com/compost-market.aspx

Meng L, Zhang S, Gong H, Zhang X, Wu C, Li W (2018) Improving sewage sludge composting by addition of spent mushroom substrate and sucrose. Bioresour Technol 253:197–203

Menke G, Grossmann F (1971) Einfluss der Schnellkompostierung von Mull auf Ereger von Pflanzenkrankheiten. Z Pflanzenkr 71:75–84

Mustin M (1987) Le compost: La gestion de la matière organique. Francois Dubusc, Paris, 954p

Noble R, Roberts SJ (2004) Eradication of plant pathogens and nematodes during composting: a review. Plant Pathol 53(5):548–568

Pane C, Spaccini R, Piccolo A, Celano G, Zaccardelli M (2019) Disease suppressiveness of agricultural greenwaste composts as related to chemical and bio-based properties shaped by different on-farm composting methods. Biol Control 137:104026

Pérez-Piqueres A, Edel-Hermann V, Alabouvette C, Steinberg C (2006) Response of soil microbial communities to compost amendments. Soil Biol Biochem 38(3):460–470

Saidi N, Kouki S, M'hiri F, Jedidi N, Mahrouk M, Hassen A et al (2009) Microbiological parameters and maturity degree during composting of *Posidonia oceanica* residues mixed with vegetable wastes in semi-arid pedo-climatic condition. J Environ Sci 21(10):1452–1458

Sanchez-Monedero M, Cayuela M, Roig A, Jindo K, Mondini C, Bolan N (2018) Role of biochar as an additive in organic waste composting. Bioresour Technol 247:1155–1164
Singla J, Krattinger S (2016) Biotic stress resistance genes in wheat. In: Reference module in food science. https://doi.org/10.1016/B978-0-08-100596-5.00229-8
Soudi B (2005) The composting of cattle manure and crop residues from greenhouse production system. In: Newsletter of Transfer of Technology in Agriculture, PNTTA No. 129. http://www.agrimaroc.net/129.pdf
Tian S, Fan Q, Xu Y, Liu H (2002) Biocontrol efficacy of antagonist yeasts to gray mold and blue mold on apples and pears in controlled atmospheres. Plant Dis 86(8):848–853

Analyzing the Capabilities of *Actinobacteria* for Remediation Through Metagenomic Analysis of Contaminated Habitats

16

Kruti Dangar, Ketaki Ramani, and Disha Changela

Abstract

Industrialization has increased the number of pollutants in air, water, and land to toxic levels. Effectively removing these pollutants from the environment is a challenge faced by many countries around the world. Heavy metals such as lead (Pb), cadmium (Cd), and zinc (Zn) and organic pollutants such as polycyclic aromatic hydrocarbons (PAHs) are taken up by organisms at various levels of ecological chains causing deteriorating effects on all the organisms. The bioremediation process has long been known to sustainably detoxify environmental pollutants with the help of microorganisms. Polluted areas often harbor microorganisms that are not cultivable but essential for the bioremediation process. Metagenomic studies help to identify such organisms and present the array of microorganisms apt for bioremediation of the contaminated site. Advances in sequencing techniques help identify microorganisms faster. It also helps with the identification of specific genes, proteins, and enzymes that can hasten the remediation process. *Actinobacteria* are soil dwellers and have a unique set of enzyme systems with the ability to take up these pollutants effectively. This chapter presents the various metagenomic approaches and the *Actinobacteria* reported possessing the bioremediation ability.

Keywords

Bioremediation · Metagenomics · *Actinobacteria* · Heavy metals · Organic pollutants

K. Dangar (✉) · D. Changela
Department of Biochemistry, Saurashtra University, Rajkot, Gujarat, India

K. Ramani
Department of Microbiology, School of Science, RK University, Rajkot, Gujarat, India

N. K. Arora, B. Bouizgarne (eds.), *Microbial BioTechnology for Sustainable Agriculture Volume 1*, Microorganisms for Sustainability 33,
https://doi.org/10.1007/978-981-16-4843-4_16

16.1 Introduction

Bioremediation is a process that deals with microorganisms to degrade and detoxify environmental contaminants (Lovley 2003; Chakraborty et al. 2012; Arora 2018). Mainly three approaches are used to classify the process of bioremediation. The first approach is natural attenuation that is based on the native microbial communities which degrade or detoxify contamination naturally. Biostimulation (second approach) deals with native organisms along with some addition of nutrients to increase the rate of biodegradation. The third approach is bioaugmentation which requires the addition of non-native organisms into the contaminated site to enhance the process of bioremediation (Salanitro et al. 2000; Payne et al. 2011). The population of bacteria in the soil, ocean, and human body are approximately 2.6×10^{29}, 3.6×10^{29}, and $10^{12–14}$ cells, respectively (Torsvik and Ovreas 2002; Sogin et al. 2006; Tremaroli and Bäckhed 2012; De Mandal et al. 2014). *Actinobacteria* are Gram-positive bacteria with high GC content in DNA (Ventura et al. 2007; Hamedi et al. 2013). The genera belonging to *Actinobacteria* show variation in taxonomic affiliation, physiology, morphology, and biocatalytic potential. (Atlas 1997; Singh et al. 2013). *Actinobacteria*, *Acidimicrobiia*, *Coriobacteriia*, *Nitriliruptoria*, *Rubrobacteria*, and *Thermoleophilia* are the six major classes of the phylum *Actinobacteria* (Ventura et al. 2007). *Actinobacteria* are an important source for antimicrobial (Wang et al. 2021), antitumor (Kamala et al. 2020), anti-inflammatory (Nair and Abraham 2020), and antioxidant compounds (Law et al. 2020). Today, two-thirds of the world's antibiotics are derived from *Actinobacteria* (Mast et al. 2020). Huge numbers of researchers have documented the role of *Actinobacteria* as soil amendment agents (Yadav et al. 2021) as formulations of biofertilizers and biopesticides (Shivlata and Satyanarayana 2017). *Actinobacteria* are also reported to improve soil health by nitrogen fixation (Sathya et al. 2017) and phosphate solubilization (Yadav et al. 2018; Lunggani and Suprihadi 2019), along with increasing uptake of iron and reduce biotic and abiotic stresses (Prathyusha and Bramhachari 2018). The diverse pollutants such as pesticides, heavy metals, dyes, hydrocarbons/polyaromatic hydrocarbons, and plastic are degraded or decomposed / detoxified by *Actinobacteria*. The list of few species belonging to phylum *Actinobacteria* along with their ability to degrade pollutants is enlisted in Table 16.1. It clearly shows a multipotential facet of *Actinobacteria* in various habitats.

16.2 Metagenomics

The major uphill battle is to identify and know the ecological role of 90–99% of organisms that cannot be cultivated under standard laboratory conditions (Dangar et al. 2017). The combination of polymerase chain reaction (PCR) and next-generation sequencing (NGS) technologies has helped a lot in measuring microbial diversity along with the discovery of genes, enzymes, pathways, and diverse metabolic products (Felczykowska et al. 2015). Various molecular methods such as

Table 16.1 Species of actinobacteria reported to degrade or detoxify/remove pollutants

Actinobacteria	Pollutant	Reference
Actinomadura sp. *AF-555*	Poly (3-hydroxybutyrate-co-3-hydroxyvalerate	Shah et al. (2010)
Aminobacter sp. *MSH1*	Benzonitrile herbicides	Frkova et al. (2014)
Amycolatopsis and *Streptomyces species*	Heavy metals such as Zn, Pb	El Baz et al. (2015)
Amycolatopsis oliviviridis sp. nov.	Polylactic acid bioplastic	Penkhrue et al. (2018)
Amycolatopsis orientalis	Azo dye (amido black)	Chengalroyen and Dabbs (2013)
Arthrobacter sp. *AK-YN-10*	s-triazine	Sagarkar et al. (2016)
Arthrobacter sp. *U3*	Mercury	Giovanella et al. (2015)
Arthrobacter sulfureus BZ73	Petroleum hydrocarbons	Margesin et al. (2013)
Cellulomonas variformis GPM2609	Hydrocarbons	Brito et al. (2006)
Dietzia sp. *PD1*	Azo dyes (Congo red and indigo carmine)	Das et al. (2016)
Dietzia sp. *strain GS-1*	Disodium terephthalate	Sugimori et al. (2000)
Frankia alni ACN149	Atrazine	Rehan et al. (2014)
Gordonia alkanivorans HKI0136^T	Tar	Kummer et al. 1999
Gordonia sp. *JAAS1*	Chlorpyrifos	Abraham et al. (2013)
Janibacter limosus P3-3-X1	Phenols and its derivatives	Su et al. (2019)
Janibacter sp. *KAS23*	Pentachlorophenol	Khessairi et al. (2014)
Kibdelosporangium aridum JCM 7912	Poly (-lactide)	Jarerat et al. (2003)
Kocuria indica DP-K7	Methyl red	Kumaran et al. (2020)
Kocuria sp. *CL2*	Pentachlorophenol	Karn et al. (2011)
Microbacterium mangrovi MUSC 115T	Starch	Ser et al. (2018)
Micrococcus glutamicus NCIM 2168	Sulfonated diazo dye C.I. reactive Green 19A	Saratale et al. (2009)
Mycobacterium chlorophenolicum PCP-1	Pentachlorophenol	Wittmann et al. (1998)
Nocardiopsis sp.	Azo dye reactive Orange-16	Chittal et al. (2019)
Nocardiopsis sp. *mrinalini9*	Polythene, plastic, and diesel	Singh and Sedhuraman (2015)
Pimelobacter simplex BZ91	n-alkanes, phenol	Margesin et al. (2013)
Rhodococcus equi TB-60	Toluene-2,4-dicarbamic acid dibutyl ester (TDCB)	Akutsu-Shigeno et al. (2006)
Rhodococcus erythropolis	Oil	Zvyagintseva et al. (2001)

(continued)

Table 16.1 (continued)

Actinobacteria	Pollutant	Reference
Rhodococcus qingshengii JB301	Triphenylmethane dyes: Crystal violet and methyl violet	Li et al. (2014)
Rhodococcus sp. *BCH2*	Atrazine	Kolekar et al. (2014)
Saccharothrix aerocolonigenes TE5	Azo dyes: Reactive Red 1, Reactive Orange 107, Reactive black 5	Rizwana Parvez and Palempalle (2015)
Streptomyces aureus HP-S-01	Deltamethrin	Chen et al. (2011)
Streptomyces chrestomyceticus S20	Malachite green	Vignesh et al. (2020)
Streptomyces diastaticus (PA2)	Cypermethrin	Janarthanan et al. (2018)
Streptomyces sp. *AC1–6*	Diazinon	Briceño et al. (2015)
Streptomyces sp. *Hlh1*	Petroleum	Baoune et al. (2019)
Streptomyces sp. *S501*	Petroleum	Chen et al. (2020)
Streptomyces sp. *strain AH-B*	Quinclorac	Lang et al. (2018)
Thermobifida alba AHK119	Aliphatic-aromatic co-polyester film	Hu et al. (2010)
Zhihengliuella sp. *ISTPL4*	Phenanthrene	Mishra et al. (2019)

genetic fingerprinting, metagenomics, metaproteomics, and metatranscriptomics have resolved the limitation of non-cultivable approaches (Kumar et al. 2020; Abiraami et al. 2020; Mukherjee and Reddy 2020).

The term metagenomics was given by Jo Handelsman in 1998. Metagenomics is based on the study of genetic material directly collected from environmental samples followed by sequencing and analysis of data to comprehend the occurrence and adaptation of microbes (Handelsman et al. 1998). The extraction of total genomic DNA, amplification, sequencing, and analysis of sequencing data are four milestones of sequence-based metagenomics (Kamble et al. 2020; Bøifot et al. 2020; Grieb et al. 2020; Dhanjal et al. 2020). The quality and quantity of final results rely on various key decisions such as sample collection, storage, protocols for DNA extraction, and amplification of DNA. Sequencing platform and annotation tools via bias gene are reported by various researchers (Garrido-Cardenas and Manzano-Agugliaro 2017; Berini et al. 2017; Garlapati et al. 2019). Figure 16.1 shows the overview of the process of sequence-based metagenomics.

Sequencing technology is growing rapidly and has transcended from classical technology to NGS. The various sequencing technologies such as Applied Biosystem's SOLiD, Ion Torrent, Pacific Biosciences (PacBio), single-molecule real-time (SMRT) sequencing, 454/Roche, and Illumina/Solexa are reported in a huge number of research papers (Thomas et al. 2012; Fakruddin and Chowdhury 2012; Cao et al. 2017; Nguyen et al. 2020). The assembly of sequences obtained by sequencing is carried out by de novo and reference-based. The DNA sequences are

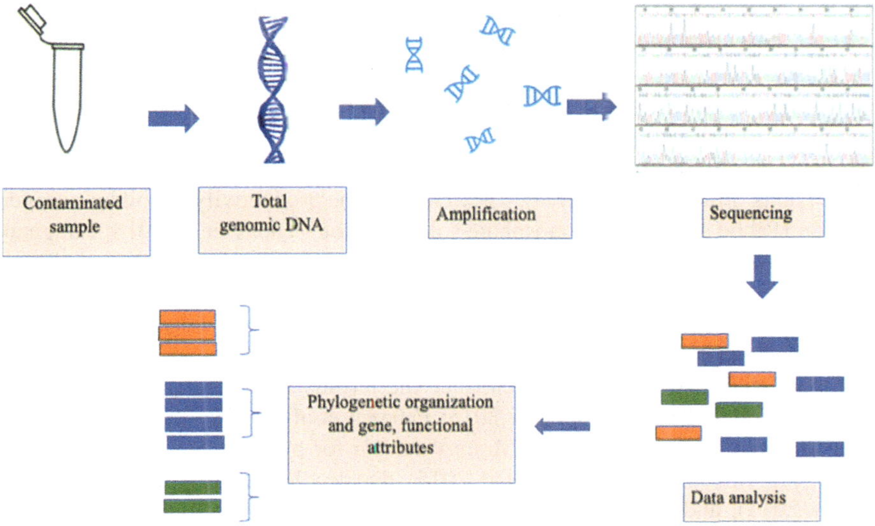

Fig. 16.1 Overview of process of sequence-based metagenomics analysis

sorted into a group which is referred to as binning. These groups can further be used to portray an organism which is called as operational taxonomic unit (OTU). Binning of sequences is converted into groups (Thomas et al. 2012; Edet et al. 2017).

The iMicrobe (http://imicrobe.us) (Youens-Clark et al. 2019), Viral Informatics Resource for Metagenome Exploration (VIROME) (http://virome.dbi.udel.edu/) (Wommack et al. 2012), Metagenomics Rapid Annotation using Subsystem Technology (MR-RAST) (https://www.mg-rast.org/) (Meyer et al. 2008), European Bioinformatics Institute (EBI) (http://www.ebi.ac.uk/metagenomics) (Mitchell et al. 2020), and Kyoto encyclopedia of genes and genomes (KEGG) (https://www.genome.jp/kegg/) (Ogata et al. 1999; Kanehisa and Goto 2000) databases are used in metagenomics for identification and functional gene diversity of residing microbes (Kim et al. 2013). The functional aspects of metagenomics opened the key for the identification of desirable genes, enzymes, and proteins. The total genomic DNA is isolated directly from the environment and is inserted into the appropriate host. This is followed by observation of the expression of desirable genes, proteins, and enzymes that have been reported (Rondon et al. 2000: Guazzaroni et al. 2015; Singer et al. 2020). Hence, using the tool of metagenomics, a diverse array of *Actinobacteria* have been discovered which is a prospective way for bioremediation.

16.3 Role of *Actinobacteria* in Soil and Importance of Metagenomic Analysis

Actinobacteria include a diverse group of significant species, the majority of which possess the properties to promote plant growth and decontaminate pollutants (Shivlata and Satyanarayana 2017). *Actinobacteria* are also reported as soil alteration agents (Fracchia et al. 2006). They enhance the productivity of soil by nitrogen fixation (Ribbe et al. 1997), phosphate solubilization (Yadav et al. 2018; Lunggani and Suprihadi 2019), and positive support to plant mycorrhizal growth, and their association (Sathya et al. 2017) boosts the iron accessibility by siderophore production (Valencia-Cantero et al. 2007. *Actinobacteria* reduce the growth of soilborne pathogens and plant pathogens by the production of antagonistic enzymes, decrease soil biotic and abiotic stress, and enhance the fertility of soil (Errakhi et al. 2007; Srivastava et al. 2014; Heng et al. 2015; Faheem et al. 2015). Several researchers have reported the potential of soil *Actinobacteria* for biodegradation (Frkova et al. 2014), detoxification (Giovanella et al. 2015), dye decolorization (Chengalroyen and Dabbs 2013), and bioplastic degradation (Penkhrue et al. 2018). It indicates that *Actinobacteria* act as a prospective solution to clean up the polluted sites and protect the soil ecosystem from pollutants. One gram of soil contains approximately 10^8 bacterial cells (Raynaud and Nunan 2014); however, owing to the limitations of culture-dependent approaches, only 1–10% of bacteria can be cultivated and maintained under the standard laboratory conditions (Handelsman 2004). The remaining 90–99% non-cultivable bacteria with hidden potential can be identified by metagenomic and other omic approaches (Riesenfeld et al. 2004; Alves et al. 2018). It is important to identify the *Actinobacteria* communities involved in various significant functions in soil as degradation, detoxification, and/or removal rate of pollutants. Metagenomic studies assist to identify the diversity and functional attributes of non-cultivable bacteria. Metagenomic analysis also increases our knowledge about the total microbial ecology, soil ecology at genes, and genome level. Furthermore, such metagenomic approaches contribute toward a better understanding of the *Actinobacteria* from various contaminated and/or non-contaminated habitats with extreme complexity.

16.4 Metagenomic Studies of *Actinobacteria* Involved in Bioremediation of Heavy Metals

In the last three decades, due to massive industrialization and urbanization, there has been an increase in pollution as well as the dissemination of heavy metals (Iyer et al. 2005). Several studies demonstrated the existence of bacteria in polluted habitats with metal resistance genes (MRGs), degradation genes, and antibiotic resistance genes (Liu et al. 2020: Guo et al. 2020; Ma et al. 2020). The outcome of the identification of bacteria by culture-dependent methods and culture-independent methods has shown huge variance (Dangar et al. 2017). Recent developments in molecular biology and NGS approaches have greatly enhanced our understanding of the microbial communities involved in various bioremediation processes for environmental cleanup (Techtmann and Hazen 2016; Arora et al. 2018; Bharagava et al. 2019). Literature has revealed the presence of key microbes in some specific contaminated sites and has been reported for bioremediation. For example, *Brevundimonas* spp., *Rhodanobacter*, and *Rhodocyclaceae* were shown to have the ability to reduce nitrate and uranium respectively, in contaminated sites (Kavitha et al. 2009; Green et al. 2012). The residing bacterial communities at various heavy metal [cadmium (Cd), lead (Pb), zinc (Zn), chromium (Cr), and arsenic (As)]-contaminated soils were captured by pyrosequencing of 16S rRNA gene fragments (Hur et al. 2011; Sheik et al. 2012; Chodak et al. 2013). Water was collected from New Mexico, USA, and the DNA was extracted by using Geneclean® Turbo-Kit and the sequencing platform IlluminaIIx used for the metagenomic study. The metagenomic-based taxonomic profile showed the presence of *Firmicutes*, *Proteobacteria*, and *Actinomycetes* in an aquifer contaminated with tetrachloroethene (Reiss et al. 2016). The water samples were collected from the hot springs of Reshi and Yumthang, Sikkim, India. The metagenomic analysis of these two sites has shown the presence of *Actinobacteria* with a relative abundance of 22.68% and 98.1%, respectively. Further, the metadata was analyzed by BacMetScan V.1.0 which demonstrated the presence of metal resistance genes. The gene for Zn tolerance was found in the Yumthang hot spring, and resistance of copper (Cu), Cd, Hg, and As was detected in Reshi hot spring (Najar et al. 2020).

Three soil samples (S1, S2, and S3) contaminated with Pb and Zn were collected from Bama mine, Iran. S1 S2 and S3 contained low, moderate, and high concentrations of heavy metals correspondingly. Each soil sample was subjected to extraction of total genomic DNA using PowerSoil DNA isolation kit. 16Sr DNA from soil was amplified using 515F and 806R primers. PCR products were sequenced by MiSeq Illumina platform, and sequencing data were analyzed by the Quantitative Insights into Microbial Ecology software package (QIIME). As a result of the analysis of metagenomic data, total 2252 operational taxonomic units were observed in three samples. The second most abundant percentage of phyla was *Actinobacteria* in S1 (19.92%,), S2 (19.66%), and S3 (17.28%). In heavy metal site S3, the most abundant genus was *Solirubrobacter* (16.73%) which belonged to the phylum *Actinobacteria* (Hemmat-Jou et al. 2018).

The metagenomic approach was used to understand the structural and functional diversity present in the Ganges Brahmaputra Delta aquifer system. An arsenic-contaminated groundwater of the abovementioned site was subjected to metagenomic analysis. The metagenomic DNA was isolated by PowerWater® DNA Isolation Kit for whole-genome shotgun (WGS) sequencing. The metagene data was analyzed through the MG-RAST (Metagenomic Rapid Annotations Using Subsystems Technology) server. The results demonstrated that the most abundant bacteria were *Proteobacteria* (62.6%) and *Actinobacteria* (3.7%) (Das et al. 2017).

Soil samples contaminated with Pb, Zn, and Cr were explored for identification of bacterial communities by metagenomics. The metagenomic DNA was extracted by a modified method as given by Zhou et al. (1996) and Deja-Sikora (2012). To identify the microbial community of metagenomic DNA, the amplification of the V3 and V4 region of the 16S rRNA gene was performed by using primers 357F and 786R. PCR products were subjected to pyrosequencing on GS FLX and GS Junior instruments. Deja-Sikora (2012) reported that the most abundant phylum was *Proteobacteria*, while *Actinobacteria* were also observed. Along with the most abundant class of *Actinobacteria*, other classes such as *Alphaproteobacteria*, *Acidobacteria*, *Gammaproteobacteria*, *Sphingobacteria*, and *Betaproteobacteria* were also observed in polluted soil. The study presented novel findings of a more abundant uncultured genus of *Solirubrobacter* (*Actinobacteria*, *Solirubrobacteraceae*) in the Olkusz soils (Gołębiewski et al. 2014).

Soil contaminated with vanadium was collected from vanadium smelting plant, Panzhihua City, China. Metagenomic DNA was collected by FastDNA Spin Kit and amplified by PCR primers 338F and 806R for amplification of 16S rRNA genes. The 16S rRNA gene amplified product was sequenced by using Illumina MiSeq platform. Metadata was analyzed and converted into OUTs (operational taxonomic units), and for identification of bacterial community, UPARSE version 7.0 (http://drive5.com/uparse/) and Ribosomal Database Project (RDP) classifier (http://rdp.cme.msu.edu/) were used. From metadata of each sample, 24 phyla were observed, among them the most dominant phyla was *Actinobacteria* with a relative abundance of 16.27–50.57%. To support this finding, Cao et al. (2017) have described *Actinobacteria* along with other phyla such as *Proteobacteria*, *Acidobacteria*, and *Firmicutes* commonly residing in the soils contaminated with vanadium. The presence of *Actinobacteria* was higher and *Firmicutes* was lower with increasing the depth of the smelting plant (Wang et al. 2021).

The soil samples were collected from Korea (36°44′05″N, 127°12′12″E) and Belgium (51°12′41″N; 5°14′32″E) which were contaminated with various heavy metals. PowerSoil DNA Isolation Kit was used for the isolation of total genomic DNA. DNA was amplified with 519F-816R primers, and amplified products were sequenced by using MiSeq™ platform. Metadata was further analyzed by QIIME. In samples *Proteobacteria*, *Actinobacteria*, *Verrucomicrobia*, *Bacteroidetes*, *Chloroflexi*, *Planctomycetes*, *Firmicutes*, and *Acidobacteria* phyla were existing with more than 1% of relative abundance. Faoro et al. (2010) and Janssen (2006) observed that the most dominant phyla, *Proteobacteria* and *Actinobacteria*, were observed in soil collected from most parts of the world. *Proteobacteria* and *Actinobacteria* act as pivotal microorganisms in soil by performing various activities such as biological nitrogen fixation, oxidation of iron and methane, production of plant growth hormones, and solubilization of various metals (Miyashita 2015; Itävaara et al. 2016; Yadav et al. 2018). *Nocardia* and *Arthrobacter* belong to the phylum *Actinobacteria* and showed the highest abundance in Korean soil (Ofek et al. 2012; Igalavithana et al. 2019). Both genera increase the plant growth, whereas *Nocardia* enhances the process of degradation of phenolic compounds (Malarczyk et al. 1994; Yadav et al. 2018; Igalavithana et al. 2019).

16.5 Metagenomic Studies of *Actinobacteria* Involved in Bioremediation of Organic Pollutants

The degradation of polycyclic aromatic hydrocarbons (PAHs) and volatile aromatics collectively indicated as benzene, toluene, ethylbenzene, and xylene (BTEX) were reported in soil samples by using various biological techniques (Andreoni and Gianfreda 2007). The cultivable strains of *Rhodococcus* and *Arthrobacter* belonging to *Actinobacteria* were found in BTEX polluted soils (Stapleton and Sayler 2000; Hendrickx et al. 2006). The phenanthrene degradation genes were observed in *Nocardioides* sp. KP7 (Saito et al. 1999; Kanaly et al. 2000; Daane et al. 2001). In diesel-contaminated Canadian high Arctic soils, metagenomics was coupled to reverse transcriptase real-time PCR (RT-qPCR) analysis. The results showed the presence of hydrocarbon-degrading genes, which mainly belonged to *Gammaproteobacteria*, *Alphaproteobacteria*, and *Actinobacteria* (Yergeau et al. 2012). The high abundance of *Actinobacteria* was observed along with cytochrome P450 alkane hydroxylase genes associated with increased degradation of hydrocarbon in diesel-contaminated sites (Schneiker et al. 2006; Yergeau et al. 2012). The addition of heavy metals and hydrocarbons in an environment creates disturbance of residing microbial communities. Over the time, some microbes develop an ability to degrade, resist, and/or utilize heavy metals and hydrocarbons (Girvan et al. 2005; Van Dorst et al. 2014; Pessoa-Filho et al. 2015). Most studies were conducted on tapping the potential of such microbial communities to enhance the productivity of bioremediation (Bell et al. 2013). By performing the metagenomic analysis of natural and diesel-contaminated soils, the change in the shift of the dominant phylum from *Proteobacteria* to *Actinobacteria* was observed (Jung et al. 2016). The genus such as *Rhodococcus*, *Nocardioides*, and *Gordonia* belonging to *Actinobacteria* were found to possess alkane 1-monooxygenase enzyme (Van Dorst et al. 2014; Jung et al. 2016). The E26, E29, and E44 sites of petroleum seepage near marine sediments were collected from the Eastern Gulf of Mexico. Metagenomic analysis depicted the dominance of *Actinobacteria* with high relative abundance in a sample collected from E44 (Dong et al. 2019). Pyrosequencing proved that the main shifts of bacterial communities were caused by the addition of a mobilizing agent. It strongly represented the decrease in proliferation of *Actinobacteria* (*Mycobacteriaceae*) and *Bacteroidetes* (*Chitinophagaceae*) leading to the slowdown of PAH degradation. The study also revealed the occurrence of *Actinobacteria* in PAH-polluted soil by metagenomics (Lladó et al. 2015). The petroleum muck sample was collected from Kandla, India. DNA isolation kit and Ion Torrent platform were used for isolation of metagenomic DNA and sequencing, respectively. A total of 2,228,423 sequences were annotated using this link http://metagenomics.nmpdr.org. The most dominant phylum observed was *Proteobacteria* with 99.09% of relative abundance, while *Actinobacteria* was 0.70%. It showed that *Actinobacteria* can reside in polluted sites (Joshi et al. 2014).

16.6 Conclusion and Future Prospects

The ecological role of *Actinobacteria* in the polluted ecosystems is still largely unknown. Hence, it would be interesting to explore vast genetic resources particularly represented by non-cultivable *Actinobacteria* using the recently developed approaches of metagenomics. The development of new algorithms, standardization of protocols, and integration of functional and sequence analysis of metagenomic, transcriptomics, and proteomics studies have provided a clearer understanding of *Actinobacteria*-based bioremediation pathways. The development of cloning and expression systems of metagenomic DNA can be used for the production of enzymes in large quantities and would certainly add to the possibilities of altering the enzyme for desired traits so as to enhance the removal of pollutants from polluted habitats. Continued efforts to characterize soil *Actinobacteria* diversity and functional attributes will add to our understanding of the potential utility of these bacteria as a source of useful products for biotechnology.

The world is facing a challenge to tackle the enormous amount of increasing pollutants in the environment. Organic pollutants and heavy metals are difficult to degrade and are contaminating soil as well as water. Biological approaches are a sustainable way to address the pollution problem. Microorganisms have been isolated from contaminated sites and used for the bioremediation process. However, there lie a plethora of microorganisms that cannot be cultivated under lab conditions and possess the ability to degrade the pollutants present in the contaminated sites. Metagenomic approaches are robust and comprehensive allowing identification of the microorganism and metabolic pathways from samples, thereby their potential for bioremediation can be harnessed. Actinobacteria have been identified from soil and water samples at several contaminated sites using metagenomic techniques. Advances in metagenomics and the ability of *Actinobacteria* toward bioremediation of several pollutants show promise towards achieving a clean environment.

References

Abiraami TV, Singh S, Nain L (2020) Soil metaproteomics as a tool for monitoring functional microbial communities: promises and challenges. Rev Environ Sci Biotechnol 19:73–102

Abraham J, Shanker A, Silambarasan S (2013) Role of Gordonia sp JAAS 1 in biodegradation of chlorpyrifos and its hydrolysing metabolite 3, 5, 6-trichloro-2-pyridinol. Lett Appl Microbiol 57(6):510–516

Akutsu-Shigeno Y, Adachi Y, Yamada C, Toyoshima K, Nomura N, Uchiyama H et al (2006) Isolation of a bacterium that degrades urethane compounds and characterization of its urethane hydrolase. Appl Microbiol Biotechnol 70(4):422

Alves LD, Westmann CA, Lovate GL, de Siqueira GM, Borelli TC, Guazzaroni ME (2018) Metagenomic approaches for understanding new concepts in microbial science. Int J Genomics 2018:2312987

Andreoni V, Gianfreda L (2007) Bioremediation and monitoring of aromatic-polluted habitats. Appl Microbiol Biotechnol 76(2):287–308

Arora NK (2018) Bioremediation: a green approach for restoration of polluted ecosystems. Environ Sustain 1:305–307

Arora NK, Fatima T, Mishra I, Verma M, Mishra J, Mishra V (2018) Environmental sustainability: challenges and viable solutions. Environ Sustain 1:309–340

Atlas R (1997) Principles of microbiology. WCB McGrill-Hill, New York

Baoune H, Aparicio JD, Pucci G, El Hadj-Khelil AO, Polti MA (2019) Bioremediation of petroleum-contaminated soils using *Streptomyces sp.* Hlh1. J Soil Sediment 19(5):2222–2230

Bell TH, Yergeau E, Juck DF, Whyte LG, Greer CW (2013) Alteration of microbial community structure affects diesel biodegradation in an Arctic soil. FEMS Microbiol Ecol 85:51–61

Berini F, Casciello C, Marcone GL, Marinelli F (2017) Metagenomics: novel enzymes from non-culturable microbes. FEMS Microbiol Lett 364(21):fnx211

Bharagava RN, Purchase D, Saxena G, Mulla SI (2019) Applications of metagenomics in microbial bioremediation of pollutants: from genomics to environmental cleanup. In: Das S, Dash HR (eds) Microbial diversity in the genomic era. Academic Press, Elsevier, London, pp 459–477

Bøifot KO, Gohli J, Moen LV, Dybwad M (2020) Performance evaluation of a new custom, multi-component DNA isolation method optimized for use in shotgun metagenomic sequencing-based aerosol microbiome research. Environ Microbiome 15(1):1–23

Briceño G, Fuentes MS, Rubilar O, Jorquera M, Tortella G, Palma G et al (2015) Removal of the insecticide diazinon from liquid media by free and immobilized *Streptomyces* sp. isolated from agricultural soil. J Basic Microbiol 55(3):293–302

Brito EM, Guyoneaud R, Goñi-Urriza M, Ranchou-Peyruse A, Verbaere A, Crapez MA et al (2006) Characterization of hydrocarbonoclastic bacterial communities from mangrove sediments in Guanabara Bay, Brazil. Res Microbiol 157(8):752–762

Cao Y, Fanning S, Proos S, Jordan K, Srikumar S (2017) A review on the applications of next generation sequencing technologies as applied to food-related microbiome studies. Front Microbiol 8:1829

Chakraborty R, Wu CH, Hazen TC (2012) Systems biology approach to bioremediation. Curr Opin Biotechnol 23:483–490

Chen S, Lai K, Li Y, Hu M, Zhang Y, Zeng Y (2011) Biodegradation of deltamethrin and its hydrolysis product 3-phenoxybenzaldehyde by a newly isolated *Streptomyces aureus* strain HP-S-01. Appl Microbiol Biotechnol 90(4):1471–1483

Chen C, Chen S, Zhang W, Yuan F, Yu J, Liu Q (2020) *Streptomyces sp.* S501, a marine petroleum-degrading actinobacterium isolated from sediment of Yalujiang Estuary, northern China, and its genome annotation. Curr Microbiol 77(11):3643–3650

Chengalroyen MD, Dabbs ER (2013) Identification of a gene responsible for amido black decolorization isolated from *Amycolatopsis orientalis*. World J Microbiol Biotechnol 29(4):625–633

Chittal V, Gracias M, Anu A, Saha P, Rao KB (2019) Biodecolorization and biodegradation of azo dye reactive orange-16 by marine Nocardiopsis sp. Iran J Biotechnol 17(3):18–26

Chodak M, Gołębiewski M, Morawska-Płoskonka J, Kuduk K, Niklińska M (2013) Diversity of microorganisms from forest soils differently polluted with heavy metals. Appl Soil Ecol 64:7–14

Daane LL, Harjonio I, Zylstra GJ, Hagblom MM (2001) Isolation and characterization of polycyclic aromatic hydrocarbon-degrading bacteria associated with the rhizosphere of salt marsh plants. Appl Environ Microbiol 67:2683–2691

Dangar KG, Raiyani NM, Pandya RD, Singh SP (2017) Uncultivated lineages and host–microbe interaction in saline environment. In: Singh R, Kothari R, Koringa P, Singh S (eds) Understanding host-microbiome interactions- an omics approach. Springer, Singapore, pp 13–28

Das P, Banerjee P, Zaman A, Bhattacharya P (2016) Biodegradation of two azo dyes using *Dietzia* sp. PD1: process optimization using response surface methodology and artificial neural network. Desalin Water Treat 57(16):7293–7301

Das S, Bora SS, Yadav RN, Barooah M (2017) A metagenomic approach to decipher the indigenous microbial communities of arsenic contaminated groundwater of Assam. Genomics Data 12:89–96

De Mandal S, Sanga Z, Nachimuthu SK (2014) Metagenomic analysis of bacterial community composition among the cave sediments of indo-Burman biodiversity hotspot region. Policies Procedures 2:631

Deja-Sikora E (2012) Search for bacterial cadmium, zinc, lead, copper and chromium resistance genes in metagenomes of soils polluted with heavy metals. PhD thesis, Nicolaus Copernicus University, Torun, Poland

Dhanjal DS, Chopra C, Chopra RS (2020) Metagenomic DNA sequencing: technological advances and applications. In: Chopra RS, Chopra C, Sharma NR (eds) Metagenomics: techniques, applications, challenges and opportunities. Springer, Singapore, pp 37–53

Dong X, Greening C, Rattray JE, Chakraborty A, Chuvochina M, Mayumi D et al (2019) Metabolic potential of uncultured bacteria and archaea associated with petroleum seepage in deep-sea sediments. Nat Commun 10(1):1–2

Edet UO, Antai SP, Brooks AA, Asitok AD, Enya O, Japhet FH (2017) An overview of cultural, molecular and metagenomic techniques in description of microbial diversity. J Adv Microbiol 19:1–9

El Baz S, Baz M, Barakate M, Hassani L, El Gharmali A, Imziln B (2015) Resistance to and accumulation of heavy metals by actinobacteria isolated from abandoned mining areas. Sci World J 2015:761834

Errakhi R, Bouteau F, Lebrihi A, Barakate M (2007) Evidence of biological control capacities of *Streptomyces* spp. against *Sclerotium rolfsii* responsible for damping-off disease in sugar beet (*Beta vulgaris* L.). World J Microbiol Biotechnol 23:1503–1509

Faheem M, Raza W, Zhong W, Nan Z, Shen Q, Xu Y (2015) Evaluation of the biocontrol potential of *Streptomyces goshikiensis* YCXU against *fusarium oxysporum* f. sp. *niveum*. Biol Control 81: 101–110

Fakruddin MD, Chowdhury A (2012) Pyrosequencing-an alternative to traditional sanger sequencing. Am J Biochem Biotechnol 8(1):14–20

Faoro H, Alves AC, Souza EM, Rigo LU, Cruz LM, Al-Janabi SM et al (2010) Influence of soil characteristics on the diversity of bacteria in the southern Brazilian Atlantic Forest. Appl Environ Microbiol 76:4744–4749

Felczykowska A, Krajewska AK, Zielinska S, Los JM (2015) Sampling, metadata and DNA extraction—important steps in metagenomics studies. Acta Biochim Pol 62(1):151–160

Fracchia L, Dohrmann AB, Martinotti MG, Tebbe CC (2006) Bacterial diversity in a finished compost and vermicompost: differences revealed by cultivation-independent analyses of PCR-amplified 16S rRNA genes. Appl Microbiol Biotechnol 71:942–952

Frkova Z, Badawi N, Johansen A, Schultz-Jensen N, Bester K, Sørensen SR, Karlson UG (2014) Degradation of three benzonitrile herbicides by *Aminobacter* MSH1 versus soil microbial communities: pathways and kinetics. Pest Manag Sci 70(8):1291–1298

Garlapati D, Charankumar B, Ramu K, Madeswaran P, Murthy MR (2019) A review on the applications and recent advances in environmental DNA (eDNA) metagenomics. Rev Environ Sci Biotechnol 18(3):389–411

Garrido-Cardenas JA, Manzano-Agugliaro F (2017) The metagenomics worldwide research. Curr Genet 63(5):819–829

Giovanella P, Costa AP, Schäffer N, Peralba MC, Camargo FA, Bento FM (2015) Detoxification of mercury by bacteria using crude glycerol from biodiesel as a carbon source. Water Air Soil Pollut 226(7):224

Girvan MS, Campbell CD, Killham K, Prosser JI, Glover LA (2005) Bacterial diversity promotes community stability and functional resilience after perturbation. Environ Microbiol 7:301–313

Gołębiewski M, Deja-Sikora E, Cichosz M, Tretyn A, Wróbel B (2014) 16S rDNA pyrosequencing analysis of bacterial communities in heavy metals polluted soils. Microb Ecol 67(3):635–647

Green SJ, Prakash O, Jasrotia P, Overholt WA, Cardenas E, Hubbard D et al (2012) Denitrifying bacteria from the genus *Rhodanobacter* dominate bacterial communities in the highly contaminated subsurface of a nuclear legacy waste site. Appl Environ Microbiol 78:1039–1047

Grieb A, Bowers RM, Oggerin M, Goudeau D, Lee J, Malmstrom RR et al (2020) A pipeline for targeted metagenomics of environmental bacteria. Microbiome 8(1):1–7

Guazzaroni ME, Silva-Rocha R, Ward RJ (2015) Synthetic biology approaches to improve biocatalyst identification in metagenomic library screening. J Microbial Biotechnol 8(1):52–64

Guo XP, Sun XL, Chen YR, Hou L, Liu M, Yang Y (2020) Antibiotic resistance genes in biofilms on plastic wastes in an estuarine environment. Sci Total Environ 745:140916

Hamedi J, Mohammadipanah F, Ventosa A (2013) Systematic and biotechnological aspects of halophilic and halotolerant actinomycetes. Extremophiles 17(1):1–3

Handelsman J (2004) Metagenomics: application of genomics to uncultured microorganisms. Microbiol Mol Biol Rev 68(4):669–685

Handelsman J, Rondon MR, Brady SF, Clardy J, Goodman RM (1998) Molecular biological access to the chemistry of unknown soil microbes: a new frontier for natural products. Chem Biol 5(10):R245–R249

Hemmat-Jou MH, Safari-Sinegani AA, Mirzaie-Asl A, Tahmourespour A (2018) Analysis of microbial communities in heavy metals-contaminated soils using the metagenomic approach. Ecotoxicology 27(9):1281–1291

Hendrickx B, Junca H, Vosahlova J, Lindner A, Ruegg I, Faber F et al (2006) Alternative primer sets for PCR detection of genotypes involved in bacterial aerobic BTEX degradation: distribution of the genes in BTEX degrading isolates and in subsurface soils of a BTEX contaminated industrial site. J Microbiol Methods 64:230–265

Heng JLS, Shah UKM, Rahman NAA, Shaari K, Hamzah H (2015) *Streptomyces ambofaciens* S2—a potential biological control agent for *Colletotrichum gloeosporioides* the causal agent for anthracnose in red chili fruits. J Plant Pathol Microbiol S1:006

Hu X, Thumarat U, Zhang X, Tang M, Kawai F (2010) Diversity of polyester-degrading bacteria in compost and molecular analysis of a thermoactive esterase from *Thermobifida alba* AHK119. Appl Microbiol Biotechnol 87(2):771–779

Hur M, Kim Y, Song HR, Kim JM, Choi YI, Yi H (2011) Effect of genetically modified poplars on soil microbial communities during the phytoremediation of waste mine tailings. Appl Environ Microbiol 77:7611–7619

Igalavithana AD, Kim KH, Jung JM, Heo HS, Kwon EE et al (2019) Effect of biochars pyrolyzed in N_2 and CO_2, and feedstock on microbial community in metal (loid) s contaminated soils. Environ Int 126:791–801

Itävaara M, Salavirta H, Marjamaa K, Ruskeeniemi T (2016) Geomicrobiology and metagenomics of terrestrial deep subsurface microbiomes. Adv Appl Microbiol 94:1–77

Iyer A, Mody K, Jha B (2005) Biosorption of heavy metals by a marine bacterium. Mar Pollut Bull 50(3):340–343

Janarthanan R, Gopikrishnan V, Kavitha K, Murugan A, Balagurunathan R (2018) Biodegradation of cypermethrin metabolites using terrestrial actinobacterium, *Streptomyces diastaticus* (PA2) and its GC-MS analysis. Int J ChemTech Res 11(05):509–520

Janssen PH (2006) Identifying the dominant soil bacterial taxa in libraries of 16S rRNA and 16S rRNA genes. Appl Environ Microbiol 72:1719–1728

Jarerat A, Tokiwa Y, Tanaka H (2003) Poly (L-lactide) degradation by *Kibdelosporangium aridum*. Biotechnol Lett 25(23):2035–2038

Joshi MN, Dhebar SV, Bhargava P, Pandit AS, Patel RP et al (2014) Metagenomic approach for understanding microbial population from petroleum muck. Genome Announc 2(3):e00533

Jung J, Philippot L, Park W (2016) Metagenomic and functional analyses of the consequences of reduction of bacterial diversity on soil functions and bioremediation in diesel-contaminated microcosms. Sci Rep 6:23012

Kamala K, Sivaperumal P, Kamath SM, Thilagaraj WR, Rajaram R (2020) Marine Actinobacteria as a source for emerging biopharmaceuticals. In: Kim SK (ed) Encyclopedia of marine biotechnology, vol 4. Wiley, Hoboken, NJ, pp 2095–2105

Kamble A, Sawant S, Singh H (2020) 16S ribosomal RNA gene-based metagenomics: a review. Biomed Res J 7(1):5

Kanaly RA, Bartha R, Watanabe K, Harayama S (2000) Rapid mineralization of benzo [a] pyrene by a microbial consortium growing on diesel fuel. Appl Environ Microbiol 66:4205–4211

Kanehisa M, Goto S (2000) KEGG: Kyoto encyclopedia of genes and genomes. Nucleic Acids Res 28(1):27–30

Karn SK, Chakrabarti SK, Reddy MS (2011) Degradation of pentachlorophenol by *Kocuria* sp. CL2 isolated from secondary sludge of pulp and paper mill. Biodegradation 22:63–69

Kavitha S, Selvakumar R, Sathishkumar M, Swaminathan K, Lakshmanaperumalsamy P, Singh A et al (2009) Nitrate removal using *Brevundimonas diminuta* ATCC 8486 from groundwater. Water Sci Technol 60:517–524

Khessairi A, Fhoula I, Jaouani A, Turki Y, Cherif A, Boudabous A et al (2014) Pentachlorophenol degradation by *Janibacter* sp., a new actinobacterium isolated from saline sediment of arid land. Biomed Res Int 2014:296472

Kim M, Lee KH, Yoon SW, Kim BS, Chun J, Yi H (2013) Analytical tools and databases for metagenomics in the next-generation sequencing era. Genomics Inform 11(3):102–113

Kolekar PD, Phugare SS, Jadhav JP (2014) Biodegradation of atrazine by *Rhodococcus* sp. BCH2 to N-isopropylammelide with subsequent assessment of toxicity of biodegraded metabolites. Environ Sci Pollut Res 21(3):2334–2345

Kumar AM, Ravindran B, Sarsaiya S, Chen H, Wainaina S, Singh E et al (2020) Metagenomics for taxonomy profiling: tools and approaches. Bioengineered 11(1):356–374

Kumaran S, Ngo AC, Schultes FP, Tischler D (2020) Draft genome sequence of *Kocuria indica* DP-K7, a methyl red degrading actinobacterium. 3 Biotech 10:175

Kummer C, Schumann P, Stackebrandt E (1999) *Gordonia alkanivorans* sp. nov., isolated from tar-contaminated soil. Int J Syst Evol Microbiol 49(4):1513–1522

Lang Z, Qi D, Dong J, Ren L, Zhu Q, Huang W et al (2018) Isolation and characterization of a quinclorac-degrading Actinobacteria *Streptomyces* sp. strain AH-B and its implication on microecology in contaminated soil. Chemosphere 199:210–217

Law JW, Letchumanan V, Tan LT, Ser HL, Goh BH, Lee LH (2020) The rising of "modern Actinobacteria" era. Prog Microbe Mol Biol 3(1):a0000064

Li G, Peng L, Ding Z, Liu Y, Gu Z, Zhang L et al (2014) Decolorization and biodegradation of triphenylmethane dyes by a novel *Rhodococcus qingshengii* JB301 isolated from sawdust. Ann Microbiol 64(4):1575–1586

Liu JL, Yao J, Zhu X, Zhou DL, Duran R, Mihucz VG et al (2020) Metagenomic exploration of multi-resistance genes linked to microbial attributes in active nonferrous metal (loid) tailings. Environ Pollut 273:115667

Lladó S, Covino S, Solanas A, Petruccioli M, Annibale A, Viñas M (2015) Pyrosequencing reveals the effect of mobilizing agents and lignocellulosic substrate amendment on microbial community composition in a real industrial PAH-polluted soil. J Hazard Mater 283:35–43

Lovley DR (2003) Cleaning up with genomics: applying molecular biology to bioremediation. Nat Rev Microbiol 1:35–44

Lunggani AT, Suprihadi A (2019) Characterization phosphate-solubilizing marine actinobacteria associated with *Sargassum* sp from Menjangan kecil island, Indonesia. In: J Phys: conference series, volume 1217, the 8th international seminar on new paradigm and innovation on natural science and its application, Central Java, Indonesia, vol 1. IOP, p 012161

Ma C, Lo PK, Xu J, Li M, Jiang Z, Li G et al (2020) Molecular mechanisms underlying lignocellulose degradation and antibiotic resistance genes removal revealed via metagenomics analysis during different agricultural wastes composting. Bioresour Technol 314:123731

Malarczyk E, Rogalski J, Leonowicz A (1994) Transformation of ferulic acid by soil bacteria *Nocardia* provides various valuable phenolic compounds. Acta Biotechnol 14:235–241

Margesin R, Moertelmaier C, Mair J (2013) Low-temperature biodegradation of petroleum hydrocarbons (n-alkanes, phenol, anthracene, pyrene) by four actinobacterial strains. Int Biodeter Biodegr 84:185–191

Mast Y, Stegmann E, Lu Y (2020) Regulation of antibiotic production in Actinomycetes. Front Microbiol 11:1566

Meyer F, Paarmann D, D'Souza M, Olson R, Glass EM, Kubal M et al (2008) The metagenomics RAST server–a public resource for the automatic phylogenetic and functional analysis of metagenomes. BMC Bioinform 9(1):1–8

Mishra A, Rathour R, Singh R, Kumari T, Thakur IS (2019) Degradation and detoxification of phenanthrene by actinobacterium *Zhihengliuella sp.* ISTPL4. Environ Sci Pollut Res 27:27256–27267

Mitchell AL, Almeida A, Beracochea M, Boland M, Burgin J, Cochrane G et al (2020) MGnify: the microbiome analysis resource in 2020. Nucleic Acids Res 48(D1):D570–D578

Miyashita NT (2015) Contrasting soil bacterial community structure between the phyla Acidobacteria and Proteobacteria in tropical southeast Asian and temperate Japanese forests. Genes Genet Syst 90:61–77

Mukherjee A, Reddy MS (2020) Metatranscriptomics: an approach for retrieving novel eukaryotic genes from polluted and related environments. 3 Biotech 10(2):71

Nair S, Abraham J (2020) Natural products from Actinobacteria for drug discovery. In: Patra J, Shukla A, Das G (eds) Advances in pharmaceutical biotechnology. Springer, Singapore, pp 333–363

Najar IN, Sherpa MT, Das S, Das S, Thakur N (2020) Diversity analysis and metagenomic insights into the antibiotic resistance and metal resistances among Himalayan hot spring Bacteriobiome-insinuating inherent environmental baseline levels of antibiotic and metal tolerance. J Global Antimicrob Resist 21:342–352

Nguyen LN, Commault AS, Kahlke T, Ralph PJ, Semblante GU, Johir MA et al (2020) Genome sequencing as a new window into the microbial community of membrane bioreactors—a critical review. Sci Total Environ 704:135279

Ofek M, Hadar Y, Minz D (2012) Ecology of root colonizing *Massilia* (Oxalobacteraceae). PLoS One 7(7):e40117

Ogata H, Goto S, Sato K, Fujibuchi W, Bono H, Kanehisa M (1999) KEGG: Kyoto encyclopedia of genes and genomes. Nucleic Acids Res 27(1):29–34

Payne RB, May HD, Sowers KR (2011) Enhanced reductive dechlorination of polychlorinated biphenyl impacted sediment by bioaugmentation with a dehalorespiring bacterium. Environ Sci Technol 45:8772–8779

Penkhrue W, Sujarit K, Kudo T, Ohkuma M, Masaki K, Aizawa T et al (2018) *Amycolatopsis oliviviridis* sp. nov. a novel polylactic acid-bioplastic-degrading actinomycete isolated from paddy soil. Int J Syst Evol 68(5):1448–1454

Pessoa-Filho M, Barreto CC, dos Reis Junior FB, Fragoso RR, Costa FS, de Carvalho MI et al (2015) Microbiological functioning, diversity, and structure of bacterial communities in ultramafic soils from a tropical savanna. Antonie Van Leeuwenhoek 107(4):935–949

Prathyusha AM, Bramhachari PV (2018) Novel perspectives of biotic and abiotic stress tolerance mechanism in Actinobacteria. In: Singh BP, Gupta VK, Passari AK (eds) New and future developments in microbial biotechnology and bioengineering. Elsevier, Amsterdam, pp 235–244

Raynaud X, Nunan N (2014) Spatial ecology of bacteria at the microscale in soil. PLoS One 9(1): e87217

Rehan M, Kluge M, Fränzle S, Kellner H, Ullrich R, Hofrichter M (2014) Degradation of atrazine by *Frankia alni* ACN14a: gene regulation, dealkylation, and dechlorination. Appl Microbiol Biotechnol 98(13):6125–6135

Reiss RA, Guerra P, Makhnin O (2016) Metagenome phylogenetic profiling of microbial community evolution in a tetrachloroethene-contaminated aquifer responding to enhanced reductive dechlorination protocols. Stand Genomic Sci 11(1):88

Ribbe M, Gadkari D, Meyer O (1997) N2 fixation by *Streptomyces thermoautotrophicus* involves a molybdenum-dinitrogenase and a manganese-superoxide oxidoreductase that couple N_2 reduction to the oxidation of superoxide produced from O_2 by a molybdenum-CO dehydrogenase. J Biol Chem 272:26627–26633

Riesenfeld CS, Schloss PD, Handelsman J (2004) Metagenomics: genomic analysis of microbial communities. Annu Rev Genet 38:525–552

Rizwana Parvez S, Palempalle UM (2015) Decolourisation and detoxification of reactive azo dyes by *Saccharothrix Aerocolonigenes* TE5. J Appl Environ Microbiol 3(2):58–62

Rondon MR, August PR, Bettermann AD, Brady SF, Grossman TH, Liles M et al (2000) Cloning the soil metagenome: a strategy for accessing the genetic and functional diversity of uncultured microorganisms. Appl Environ Microbiol 66(6):2541–2547

Sagarkar S, Bhardwaj P, Storck V, Devers-Lamrani M, Martin-Laurent F, Kapley A (2016) S-triazine degrading bacterial isolate *Arthrobacter sp.* AK-YN10, a candidate for bioaugmentation of atrazine contaminated soil. Appl Microbiol Biotechnol 100(2):903–913

Saito A, Iwabuchi T, Harayama S (1999) Characterization of genes for enzymes involved in the phenanthrene degradation in *Nocardioides* spp. KP7. Chemosphere 38:1331–1337

Salanitro JP, Johnson PC, Spinnler GE, Maner PM, Wisniewski HL, Bruce C (2000) Field-scale demonstration of enhanced MTBE bioremediation through aquifer bioaugmentation and oxygenation. Environ Sci Technol 34:4152–4162

Saratale RG, Saratale GD, Chang JS, Govindwar SP (2009) Ecofriendly degradation of sulfonated diazo dye CI reactive Green 19A using *Micrococcus glutamicus* NCIM-2168. Bioresour Technol 100(17):3897–3905

Sathya A, Vijayabharathi R, Gopalakrishnan S (2017) Plant growth-promoting actinobacteria: a new strategy for enhancing sustainable production and protection of grain legumes. 3 Biotech 7(2):102

Schneiker S, dos Santos VAPM, Bartels D, Bekel T, Brecht M et al (2006) Genome sequence of the ubiquitous hydrocarbon-degrading marine bacterium *Alcanivorax borkumensis*. Nat Biotechnol 24:997–1004

Ser HL, Tan WS, Cheng HJ, Yin WF, Chan KG, Ab Mutalib NS et al (2018) Draft genome of starch-degrading actinobacterium, *Microbacterium mangrovi* MUSC 115T isolated from inter-tidal sediments. Prog Drug Discov Biomed Sci 1(1):a0000005

Shah AA, Hasan F, Hameed A (2010) Degradation of poly (3-hydroxybutyrate-co-3-hydroxyvalerate) by a newly isolated *Actinomadura* sp. AF-555, from soil. Int Biodeter Biodegr 64(4):281–285

Sheik CS, Mitchell TW, Rizvi FZ, Rehman Y, Faisal M, Hasnain S et al (2012) Exposure of soil microbial communities to chromium and arsenic alters their diversity and structure. PLoS One 7(6):e40059

Shivlata L, Satyanarayana T (2017) Actinobacteria in agricultural and environmental sustainability. In: Singh J, Seneviratne G (eds) Agro-environmental sustainability. Springer, Cham, pp 173–218

Singer GA, Shekarriz S, McCarthy A, Fahner N, Hajibabaei M (2020) The utility of a metagenomics approach for marine biomonitoring. bioRxiv. https://doi.org/10.1101/2020.03.16.993667

Singh MJ, Sedhuraman P (2015) Biosurfactant, polythene, plastic, and diesel biodegradation activity of endophytic *Nocardiopsis sp. mrinalini9* isolated from *Hibiscus rosasinensis* leaves. Bioresour Bioprocess 2:2

Singh SP, Thumar JT, Gohel SD, Kikani B, Shukla R, Sharma A, Dangar K (2013) Actinomycetes from marine habitats and their enzymatic potential. In: Trincone A (ed) Marine enzymes for biocatalysis. Woodhead Publishing, Elsevier, London, pp 191–214

Sogin ML, Morrison HG, Huber JA, Welch DM, Huse SM, Neal PR, Arrieta JM, Herndl GJ (2006) Microbial diversity in the deep sea and the underexplored "rare biosphere". Proc Natl Acad Sci 103:12115–12120

Srivastava S, Patel JS, Singh HB, Sinha A, Sarma BK (2014) *Streptomyces rochei* SM3 induces stress tolerance in chickpea against *Sclerotinia sclerotiorum* and NaCl. J Phytopathol 163:583–592

Stapleton RD, Sayler G (2000) Changes in subsurface catabolic gene frequencies during natural attenuation of petroleum hydrocarbons. Environ Sci Technol 34:1991–1999

Su S, Liao L, Yu Y, Zhang J, Chen B (2019) Genomic data mining of an Antarctic deep-sea actinobacterium, *Janibacter limosus* P3-3-X1. Mar Genomics 43:100684

Sugimori D, Dake T, Nakamura S (2000) Microbial degradation of disodium terephthalate by alkaliphilic *Dietzia* sp. strain GS-1. Biosci Biotechnol Biochem 64(12):2709–2711

Techtmann SM, Hazen TC (2016) Metagenomic applications in environmental monitoring and bioremediation. J Ind Microbiol Biotechnol 43(10):1345–1354

Thomas T, Gilberts J, Meyer F (2012) Metagenomics—a guide from sampling to data analysis. Microb Inform Exp 2(3):1–12

Torsvik V, Ovreas L (2002) Microbial diversity and function in soil: from genes that ecosystems. Curr Opin Microbiol 5:240–245

Tremaroli V, Bäckhed F (2012) Functional interactions between the gut microbiota and host metabolism. Nature 489(7415):242–249

Valencia-Cantero E, Hernandez-Calderon E, Velazquez-Becerra C, Lopez-Meza JE, Alfaro-Cuevas R, Lopez-Bucio J (2007) Role of dissimilatory fermentative iron-reducing bacteria in Fe uptake by common bean (*Phaseolus vulgaris* L.) plants grown in alkaline soil. Plant Soil 291:263–273

Van Dorst J, Siciliano SD, Winsley T, Snape I, Ferrari BC (2014) Bacterial targets as potential indicators of diesel fuel toxicity in sub Antarctic soils. Appl Environ Microbiol 80:4021–4033

Ventura M, Canchaya C, Tauch A, Chandra G, Fitzgerald GF, Chater KF, van Sinderen D (2007) Genomics of Actinobacteria: tracing the evolutionary history of an ancient phylum. Microbiol Mol Biol Rev 71(3):495–548

Vignesh A, Manigundan K, Santhoshkumar J, Shanmugasundaram T, Gopikrishnan V, Radhakrishnan M et al (2020) Microbial degradation, spectral analysis and toxicological assessment of malachite green by *Streptomyces chrestomyceticus* S20. Bioprocess Biosyst Eng 43:1457–1468

Wang T, Li F, Lu Q, Wu G, Jiang Z, Liu S et al (2021) Studies on diversity, novelty, antimicrobial activity, and new antibiotics of cultivable endophytic actinobacteria isolated from psammophytes collected in Taklamakan Desert. J Pharm Anal 11(2):241–250

Wittmann C, Zeng AP, Deckwer WD (1998) Physiological characterization and cultivation strategies of the pentachlorophenol-degrading bacteria *Sphingomonas chlorophenolica* RA2 and *Mycobacterium chlorophenolicum* PCP-1. J Ind Microbiol Biotechnol 21(6):315–321

Wommack KE, Bhavsar J, Polson SW, Chen J, Dumas M, Srinivasiah S et al (2012) VIROME: a standard operating procedure for analysis of viral metagenome sequences. Stand Genomic Sci 6(3):421–433

Yadav AN, Verma P, Kumar S, Kumar V, Kumar M, Sugitha TC et al (2018) Actinobacteria from rhizosphere: molecular diversity, distributions, and potential biotechnological applications. In: Singh BP, Gupta VK, Passari AK (eds) New and future developments in microbial biotechnology and bioengineering. Elsevier, Amsterdam, pp 13–41

Yadav KK, Shrivastava N, Solanki AC, Upadhyay S, Trivedi M (2021) Actinobacteria interventions in plant and environment fitness. In: Solanki MK, Kashyap PL, Ansari RL, Kumari B (eds) Microbiomes and plant health. Academic Press, Elsevier, New York, pp 397–427

Yergeau E, Sanschagrin S, Beaumier D, Greer CW (2012) Metagenomic analysis of the bioremediation of diesel-contaminated Canadian high Arctic soils. PLoS One 7(1):e30058

Youens-Clark K, Bomhoff M, Ponsero AJ, Wood-Charlson EM, Lynch J, Choi I et al (2019) iMicrobe: tools and data-driven discovery platform for the microbiome sciences. Giga Sci 8(7): giz083

Zhou J, Bruns MA, Tiedje JM (1996) DNA recovery from soils of diverse composition. Appl Environ Microbiol 62(2):316–322

Zvyagintseva IS, Poglazova MN, Gotoeva MT, Belyaev SS (2001) Effect on the medium salinity on oil degradation by nocardioform bacteria. Microbiology 70(6):652–656

Biodegradation and Remediation of Pesticides in Contaminated Agroecosystems: Special Reference to Glyphosate and Paraquat

17

Mohamed Maldani, Laila Nassiri, and Jamal Ibijbijen

Abstract

The usage of pesticides in agricultural practices contributes to an improvement in food production through monitoring insects, weeds, and crop diseases, with the aim of ensuring food sustainability to meet the needs of increasing population. However, the widespread application of these substances has numerous harmful consequences on the soil and on both environmental and human health. Moreover, in developed nations, the nonrational usage of chemicals, as well as banned forms, presents a major danger and increases contaminated agricultural lands at alarming rates, as well as polluting surface and groundwater. Glyphosate [*N*-(phosphonomethyl) glycine] is a broad-spectrum systemic herbicide that blocks the enzyme required by plants to produce amino acids and proteins. At present, in all fields (environment, agriculture, toxicology), it has become important to talk about paraquat (1,1′-dimethyl-4,4′-bipyridium) when it comes to glyphosate. Similarly, in recent years, these two herbicides have been the focus of many toxicities and biodegradation studies. Bioremediation by microbial biotechnology is also one of the most recommended solutions to alleviate the impact of these contaminants and is known to be an environmentally safe soil and water remediation technology. The principle of bioremediation is the modification and removal of pesticides in the form of nontoxic compounds used as nutrients for plants. Several approaches are commonly used, such as biostimulation and bioaugmentation. The discovery of potent microbial strains and the screening of degradation genes are currently a challenge for scientific researchers. In this context, this chapter highlights and summarizes contaminants, their

M. Maldani · L. Nassiri · J. Ibijbijen (✉)
Environment, Valorization of Microbial and Plant Resources Unit, Faculty of Sciences, Moulay Ismail University of Meknes, Meknes, Morocco
e-mail: j.ibijbijen@fs.umi.ac.ma

N. K. Arora, B. Bouizgarne (eds.), *Microbial BioTechnology for Sustainable Agriculture Volume 1*, Microorganisms for Sustainability 33,
https://doi.org/10.1007/978-981-16-4343-4_17

environmental implications, and the biotechnological use of bacteria that may be used for bioremediation in order to remediate polluted areas.

Keywords

Pesticides · Contamination · Glyphosate · Paraquat · Bioremediation · Biotechnology

17.1 Introduction

In the frenzy of sustainable production with better yields, agriculture has been modernized by means of mechanization to meet the needs of both humans and animals. In the face of a scarcity of arable land, this modernity has implemented new methods that are constantly improving. One of the main aspects of plant development is the management and eradication of weeds. Since the industrial revolution and in the face of the rampant demography of humanity, most modern crop protection activities have called for the use of phytosanitary products of various natures operating on the entire growth cycle of unwanted plants, widely known as weeds. If it is agreed that phytosanitary products of all kinds contribute to the improvement of agricultural speculation, it must be understood that their imprudent use constitutes a direct and indirect threat to humans, animals, and the functioning of the agrosystems concerned.

Soil is probably the most exposed environmental factor and definitely the most susceptible to pesticide contamination. Its weakness can be clarified in the sense that it is rapidly polluted, while its remediation may take several decades after the removal of toxins prior to its rehabilitation. Humans are either directly or indirectly impacted by the harm and effects of the use of pesticides. According to the World Health Organization (WHO 2017), the rise in the number of diseases associated with chemical intoxications, such as colorectal cancer, leukemia, allergies, Parkinson's, is primarily attributed to the use of photoproducts and chemicals (e.g. pesticide) that cause health problems and death in many parts of the world. Similar findings have been verified at the national level by the Center for Antipoison and Pharmacovigilance of Morocco (CAPM 2010). To ensure the long-term viability of plant defense products, many countries have banned or restricted the use of pesticides in agriculture. The European Union has just voted to ban the use of glyphosate for 5 years. This chapter highlights and summarizes contaminants, their environmental implications, and the biotechnological use of bacteria that may be used for bioremediation in order to remediate polluted areas. The purpose of this chapter is as follows: what is a pesticide? and what are its derivatives, its modes of action, and its relevance to agriculture? several other questions that have made us interested in this research area, to which this chapter offers answers.

17.2 Background on Pesticides

17.2.1 Definitions

There are many definitions of pesticides according to scientific or commercial meanings in the literature. Any of these meanings are given according to the objective sought.

17.2.1.1 Definitions of "Pesticide" According to Researchers

Pesticide is the act of killing a pest. It is a chemical compound or mixture used to control pests including insects, fungi, molds, and weed species. These substances are also widely referred to as plant protection products (Allsop et al. 2015). The pesticide is a term composed of "pest" insect harmful to the plant and "cide" from the Latin "caedere" which means, to attack, slaughter, and kill (Rainaud 2013). Generally, the term pesticide is a common name for all molecules or formulations of products that kill harmful species, whether used in the agricultural field or in other applications.

17.2.1.2 Definition According to the CAPM

According to Centre Antipoison du Maroc (CAPM), pesticides refer to all chemical or biological products intended to kill living elements considered harmful (microbe, animal, or plant) or intended to prevent their growth, including unwanted species of plants or animals responsible for damage during or injurious to the cultivation, processing, storage or marketing of food, agricultural products, wood, vectors of human or animal diseases and harmful organisms of materials, premises, and habitats. Pesticides shall also designate an active substance, a commercial specialty, or formulation consisting of one or more active substances as well as a variety of adjuvants, solvents, inert ingredients, residual substances, and metabolites that occur during degradation of the product (CAPM 2010).

17.2.1.3 Definition According to the FAO and the IARC

A pesticide is a substance or combination of chemical or biological compounds, which is intended to repel, kill, or combat harmful organisms or to be used as a plant growth regulator. According to the International Agency for Research on Cancer (IARC), pesticides are substances used to prevent, control, or remove organisms considered to be harmful. It also talks about phytosanitary products or plant protection products (WHO 2017).

17.2.1.4 Definition According to the Profanes

A plant protection product, in the case of profanes, is a product used to treat or prevent plant disease organisms, is frequently used to increase the yield of treated crops or to suppress the growth of certain plants, is sometimes used to ensure better survival of seeds and fruits, and may be dependent on one or more active substances or microorganisms. These active ingredients may be mineral or organic, may be of natural origin, or may be derived from synthetic chemistry (Crop Life Maroc 2018).

17.2.1.5 Definition According to the US Environmental Protection Agency (USEPA)

A pesticide is a substance or combination of substances intended for the following purposes: prevent, kill, repel, or reduce any harmful organism. Besides, pesticide products contain both active and inert ingredients: an "active ingredient" prevents, kills, repels, or attenuates a parasite or is a plant regulator, defoliating, dehydrating, or stabilizing nitrogen. In regards to, "inert ingredients", which are any substance other than an active ingredient, intentionally included in a pesticide product which is important for the performance and usability of products (USEPAO 2004).

17.2.2 Pesticides Roles

Pesticides for agricultural use are among the deciding factors for pest control in the production of crop. Certainly, for a long time, the main means of struggle were physical: collecting of larvae, eggs, adult insects, destruction of diseased plants with fire, and manual and mechanical weeding (Calvet 2005). However, in their search for better yields, humans had no choice but to use pesticides intensively. Bearing in mind that without the use of pesticides the production will be poor, the use of pesticides in agriculture has become inevitable in order to ensure good safety and better crop production. More concisely, these items are used, on the one hand, to combat crop pests and, on the other hand, to ensure the supply of nutrients required for the production and growth of plants (Zhang 2018).

The combination of protection provided to crops by pesticides and the availability of nutrients guarantees a significant yield. The lack of arable land due to desertification and salinization leaves little choice but the use of pesticides and GMOs to feed humanity and combat famine. This position given to pesticides is beneficial; farmers are able to maximize agricultural production. Pesticides have a variety of beneficial effects. These include protecting crops, preserving food and resources, and preventing vector-borne diseases and other species that may pose a threat to public health and the economy (National Research Council 1993). According to Boulet (2005), pesticides are primarily used in agriculture for about 90% of uses but may also be used in other professional sectors or are intended for domestic or therapeutic usage. All of this demonstrates the significant role that pesticides perform in fields other than agriculture.

17.2.3 Historical Overview and Evolution of Pesticides

The pesticide understanding provides a brief overview of pesticide history. Historically, men have often learned how to adapt to nature through simple or complex evolutionary practices. Around 4000 years ago, farmers in the Tigris and Euphrates used ancient methods to improve the productivity of their soil through basic methods of weed eradication and the elimination of rats. In his 2005 novel, Calvet explains that in ancient times, man's genius led him to primitive activities that were basically physical: the elimination of larvae, eggs, and adult insects, elimination of diseased plants with burning, and mechanical and manual weeding. Gradually, the genius of man has often grown. According to Calvet (2005), citing Homer and Pliny assumes in his book that the usage of chemicals is comparatively ancient. This seniority is demonstrated by the usage of sulfur (after Homer) and arsenic (according to Pliny). It is estimated that arsenic has been used as an insecticide since the end of the seventeenth century, as well as nicotine, the toxic properties of which were discovered by Jean de la Quintinie (1626–1688) stated by Calvet (2005).

Armed with accumulated expertise in the elimination of crop pests, humans have become creative and developed guidelines for the management of insecticides, fungicides, and herbicides, which became quite popular in the nineteenth century. Over the ages, man has made an attempt to monitor the rivals of civilizations. One of the symbols of neutralizing forces and devastating crops is the usage of pesticides in a variety of ways. These chemicals have been used to a small degree since antiquity, the key component of which is insecticides. Historically, around 4500 years ago, Sumerians were considered to be the first to use insecticides to destroy mosquitoes and mites. Sulfur is the primary composition of these insecticides. In the fifteenth century, the Chinese started to use mercury and arsenic to monitor insects in the garden. While the seventeenth century was characterized by the usage of tobacco as a contact insecticide according to the IARC, it emerged in the 1991 edition.

Plant and animal products have been more widely used and readily available as active ingredients in pesticides. As for the eighteenth and early nineteenth centuries, only a few chemical substances were usable, and their usage was not quite common. Pyrethrum, extracted from the dried flowers of *Chrysanthaemum cineraria folium* "pyrethrum daisies," has been used as an insecticide for over 2000 years. From 1870 onward, the number of compounds available for use increased gradually and diversified (George and Shukla 2011). In the twentieth century, coinciding with the Second World War, the advancement of technology provided considerable momentum to the many methods used to produce biological weapons from such pesticide ingredients. In the end, this initiative contributed to the development of synthetic pesticides. Expansion of synthetic pesticides intensified in the 1940s by finding the insecticidal properties of DDT (dichlorodiphenyl trichloroethane), BHC (β-benzene hexachloride), aldrin, dieldrin, endrin, chlordane, parathion, captan, and 2,4-D. Among these, DDT has been the most common due to its wide-spectrum action and low mammalian toxicity and has also reduced insect-borne diseases such as malaria, yellow fever, and typhus, as stated in the November 2000 Pesticide History Blog (Kinkela 2011). In the same bulletin in pesticide history, almost generalized usage of DDT has been documented, culminating in resistance trends found in house flies in 1946. In addition to product tolerance, according to some sources, DDT has caused harm to nontarget plants and animals.

Another research shows the harm incurred by the abuse of chemicals. But new-generation pesticides seemed very healthy, particularly compared to the types of arsenic that killed people in the 1920s and 1930s (Martini and Phillips 2009). In addition, issues with the spontaneous usage of these pesticides have been illustrated by Rachel Carson in her book Silent Spring (Carson 2002). In recognition of the adverse consequences and hazards of these goods, an educational program has increased awareness of the health hazards that may be linked with the unregulated usage of pesticides and paved the way for cleaner and more environmentally sustainable products. Work started in the 1970s and 1980s on the manufacture of safe pesticides has persisted. Research has culminated in the creation of well-developed herbicide groups, the best-selling and most effective in the world of glyphosate (Duke 2018), low-use sulfonylurea and imidazolinone herbicides, as well as dinitroanilines and aryloxyphenoxypropionate and cyclohexanediones.

As with insecticides, the 3rd generation of pyrethroids was synthesized, with the advent of avermectins, benzoylureas, and *Bacillus thuringiensis* as a spray remedy. The fungicide families triazole, morpholine, imidazole, pyrimidine, and dicarboxamide have appeared during this time. In this desperate sprint to sell the agrochemical product lines, these substances had a single mode of operation, rendering them more selective, and the emergence of resistance issues on the other side of the coin contributed to consideration of management techniques to reduce negative impact (George and Shukla 2011). According to George and Shukla's (2011) study on pesticides and cancer, research efforts in the modern age of the 1990s centered on finding new members of established families with higher selectivity and enhanced environmental and toxicological profiles. In addition, modern agrochemical families have been launched in the market; such as triazolopyrimidine, triketone, and isoxazole herbicides; strobilurin and azolon fungicides; and chloronicotinyl, spinosyn, fiprole, and diacylhydrazine insecticides. Many of these modern agrochemicals are used in grams rather than in kilograms per hectare. Advances in genetics have made it possible to set up a pest management scheme requiring the use of genetically engineered crops with exposure to a large variety of herbicides or pests. These involve herbicide-tolerant crops such as soya beans, rice, canola, cotton, rice hybrids, European corn borer, and wheat wart (Benbrook 2012; Abubakar et al. 2020). In addition to the genetic factor, the usage of advanced pest control strategies is rapidly common, discouraging the production of pest species and reducing the use of agrochemicals. These developments have altered the essence of insect management and have the ability to minimize and/or modify the essence of the agrochemicals used. New concept techniques and methods of implementation minimize the likelihood of pesticide poisoning and cancer risk (George and Shukla 2011).

17.3 Classification of Pesticides

About 100 chemical groups of pesticides and nearly 10,000 industrial formulas, comprising of the active components and adjuvants, are described in numerous types (liquids, solids, granules, powders) in the literature. Pesticides presently accessible in the market are distinguished by a range of chemical structures, functional classes, and behaviors that their description is complicated (WHO 2009). In general, pesticide consists of two types of constituents: the active ingredients which give it its herbicide, insecticidal, or fungicidal activity and the formulants which complete the formulation. These two components shall decide the family to which each pesticide individual belongs. The formulations are normally composed of one or more active substances and one or more adjuvants (Abubakar et al. 2020). It improves the performance, protection, and ease of use of the product (Leong et al. 2020). The wording often refers to the physical shape in which the plant safety commodity is placed on the market. Commonly, active substances can be categorized either by the existence of the organisms to be combated (1st classification system) or by the molecular nature of the primary active agent that makes them up (2nd classification system) (Fig. 17.1).

17.3.1 Biological Classification

17.3.1.1 Herbicides

They are the most commonly used pesticides in the world. Herbicides are meant to restrict the adventitious plant species vying with plants to be covered by slowing their development by suppressing their photosynthesis phenomenon. They can be limited or absolute (Kaur 2019). Industrial herbicides have several common pathways for operation. Several enzymes involved in the biosynthesis of amino

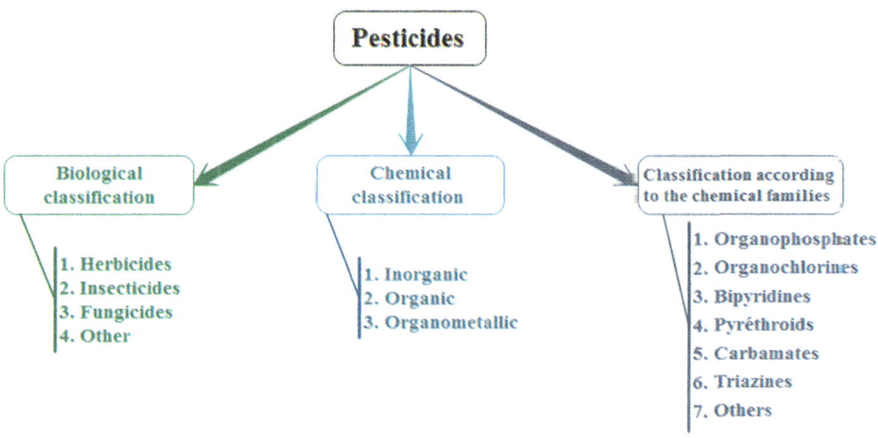

Fig. 17.1 Classification of pesticides (Calvet 2005)

acids are active sites for herbicides (e.g., glyphosate). Many different herbicide groups inhibit photosynthesis by binding to the quinone binding protein (e.g., paraquat), D-1, to avoid the movement of photosynthetic electrons (Duke 1990; Dayan et al. 2019).

17.3.1.2 Insecticides

Insecticides are chemicals engineered to destroy or reduce dangerous insects. Insecticides are used to defend plants from insects. They interfere by killing them or stopping their life cycle. There are various types: neurotoxins, insecticides that work on respiration, and insecticides that interact with the establishment of the cuticle. Organochlorines, organophosphates, and carbamates are the most widely found families. Insect repellents such as diethyltoluamide (DEET) (Nauen et al. 2019) are also listed in this group.

17.3.1.3 Fungicides

Fungicides are toxins that destroy or suppress the fungi and molds responsible for some diseases. In these species, they may behave differently. A distinction is rendered between respiratory inhibitors, cell division inhibitors, sterol biosynthesis inhibitors, amino acid, protein biosynthesis fungicides, and finally carbohydrate and polyol metabolism fungicides (Zubrod et al. 2019).

17.3.1.4 Other Pesticides

These include algaecides (against algae), rodenticides (against mice, rats, moles, and other rodents), and acaricides (against mites).

17.3.2 Chemical Classification

17.3.2.1 Inorganic Pesticides

Present inorganic pesticides have reasonably low toxicity and no effects on the ecosystem. Pesticides in this category, such as borates, silicates, and sulfuric, are minerals derived from the soil. There are no more inorganic insecticides. The only herbicide being used today as a complete weed killer is sodium chlorate. Most of the inorganic pesticides are sulfur and copper-based fungicides. These are used for vines, fruit trees, and many vegetable crops (Calvet 2005).

17.3.2.2 Organometallic Pesticides

This category comprises fungicides; the molecule is made up of a metal complex such as zinc and manganese and the organic anion dithiocarbamate. Examples of such pesticides are mancozeb (with zinc) and maneb (with manganese) (Calvet 2005; Amghar 2019).

17.3.2.3 Organic Pesticides

These are pesticides that include organic compounds and are usually categorized into three major groups: organochlorine, organophosphorus, and carbamate.

17.3.3 Classification According to the Chemical Families

Pesticides can also be categorized according to the chemical family to which the active substances belong. The major pesticides commonly used belong to a few broad chemical families: organophosphates, organochlorines, carbamates, pyrethroids, and triazines, are the key classes of pesticides that have been reported to be carcinogenic in different models (George and Shukla 2011; Kaur 2019).

17.3.3.1 Organochlorines

Organochlorine pesticides are organic compounds, and the oldest synthetic pesticides. These are molecules comprising at least one carbon-chlorine bond formed by chlorination of different unsaturated hydrocarbons (Mawussi 2008; Rani et al. 2017). Organochlorine molecules are lipophilic and very persistent, distinguished by low water solubility but strong organic solvent solubility, immune to biological and chemical degradation (Singh and Singh 2017; Pierre et al. 2018). In comparison, the impact of ultraviolet radiation is usually more harmful and derivatives are even stable than the original compounds (Yang et al. 2004). Organochlorine pesticides include DDT, hexachlorobenzene, aldrin, dieldrin, and lindane. The usage of pesticides with organochlorine is officially outlawed in many temperate nations, but many are still used in marine ecosystems. Furthermore, they appear to be used in certain tropical countries (Gao et al. 2008; Brodskiy et al. 2016; Qu et al. 2017).

17.3.3.2 Organophosphates

Organophosphates are synthetic substances that degrade quite rapidly in the atmosphere but have neurotoxic effects on vertebrates (Testud and Grillet 2007; Poirier et al. 2017). Organophosphates comprise both phosphorus-containing insecticides (chlorpyrifos) and herbicides (glyphosate). They are both produced from phosphoric acid and are usually the most harmful of all pesticides (Eto 2018). These compounds are a class of pesticides that function on the enzyme acetylcholinesterase. They function by irreversibly blocking acetylcholinesterase, which is important for nerve transmission in insects, humans, and most animals (Poirier et al. 2017). The capacity to block acetylcholinesterase (and thus toxicity) can differ greatly from one compound to another. Organophosphorus compounds are destroyed by sunlight, climate, and soil, while limited quantities can remain and end up in food and water (Sharifi et al. 2015; Khedr et al. 2019). However, while organophosphates degrade faster, they are more toxic and pose a danger to consumers of these compounds (Eto 2018).

17.3.3.3 Bipyridines

Bipyridines are a family of organic, heterocyclic, and crude formula $(C_5H_4N)_2$ chemical compounds, formed by two coupled pyridines, which are used as precursors in herbicides and insecticides. Paraquat and diquat are the major herbicides in the chemical family (Khan 1974).

17.3.3.4 Pyrethroids

Pyrethroids are neurotoxic insecticides used, on the one hand, for crop safety in cultivation and, on the other, for the management of harmful insects (Kadala 2011). There are natural pyrethroids present in chrysanthemum flowers (Li et al. 2020a). Its presence dates back to the 1970s, after the ban on pesticides created by organochlorine, which accumulated in the atmosphere and the human body. Synthetic pyrethroids are more robust, particularly with regard to light (Qiang et al. 2002). They are therefore more lipophilic and thus better cross the cuticle of insects and are inserted into biological membranes (Ensley 2018; Pedersen et al. 2020). The lethal poisonings connected with these agents remain extraordinary. Their accelerated hepatic metabolism greatly decreases human toxicity. They were also an alternative to older molecules (organochlorines, organophosphates, carbamates, etc.) (Ramchandra et al. 2019).

17.3.3.5 Carbamates

Carbamates used as insecticides were found in the 1950s. Carbamates are used as insecticides and fungicides on a very significant variety of insects, aphids, and mites, as well as on nematodes and mammals owing to their capacity to inactivate the enzyme acetylcholinesterase (Testud and Grillet 2007).

17.3.3.6 Triazines

Triazines include a broad variety of applications. Many of them are employed as selective herbicides. The selective existence of triazines is attributable to the fact that certain plants can metabolize these compounds while others may not (Hall et al. 2018). They monitor several broadleaf weeds and grasses (D'Archivio et al. 2018). Mainly used for maize production, these pesticides are often used for citrus fruits, nuts, sugar cane, sorghum, and cotton (Villamizar and Brown 2016; Le Bars et al. 2020). Triazines are among the oldest herbicides, and studies into these substances began in the 1950s, with triazine herbicides triggering cancer in humans is also under consideration.

17.4 Pesticides and the Environment

17.4.1 Transport Mechanisms of Pesticide in the Environment

A large number of pesticides penetrate plants in different forms (stomata, cuticle, root, etc.). The remaining quantity is distributed into the atmosphere and does not only reside in the position where it has been used. Even a pesticide is detected in places far from where it is used. They actually move from one location to another (Wang et al. 2019). The pesticide transport may be carried out in a variety of forms. In this segment, we shall restrict ourselves to transport by runoff, infiltration, groundwater drainage, air drift, and volatilization (Boulet 2005; Sharma et al. 2020) (Fig. 17.2).

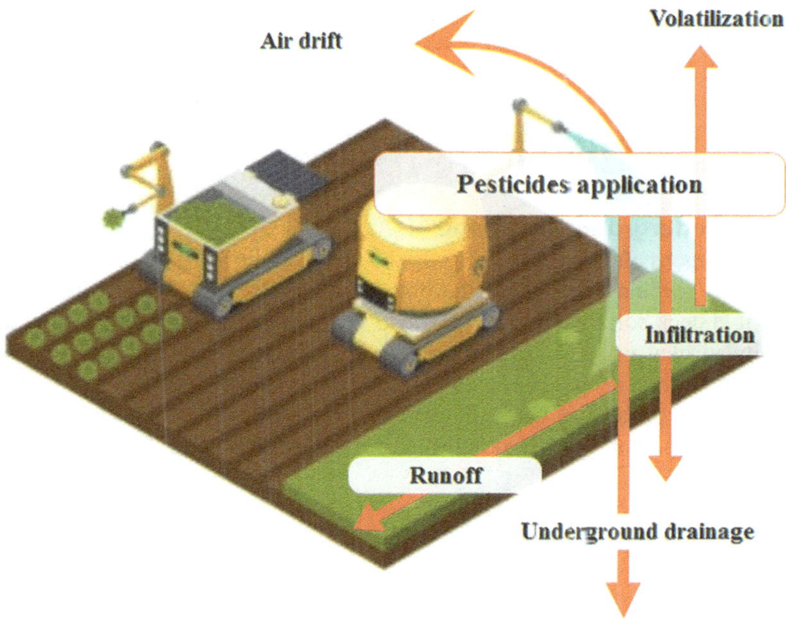

Fig. 17.2 Pesticides transport in the environment (Calvet 2005)

17.4.1.1 Runoff

Rainwater runoff happens anytime the rain is heavy enough or lengthy enough for the surface layer of the soil to be fully saturated. Due to the permeability strongly related to the soil texture, extra rain cannot reach the soil. In these conditions, the water then rises to the surface in a limited concentration of time to form gullies or by utilizing land drainage routes (ditches and underwater drains) to enter the watercourses. This water takes everything in its course, including pesticides. Pesticide exposure by water runoff is caused by the slope of the land, the quality of the plant cover and its value, the form of land, the planting methods, the severity of the rainfall, the physicochemical characteristics of each pesticide, and the period between the application of the pesticide and the rain following the application. Pesticides are present in bodies of water, lakes, and reservoirs through this method of transport. These storage sites will gradually penetrate the soil (Calvet 2005; Chen et al. 2019) (Fig. 17.2).

17.4.1.2 Infiltration

Water soaking in the soil may contribute to dissolved pesticides. This penetration in areas of stagnation of water or in soils with a sandy texture is distinguished by rapid permeability, taking into account the high porosity rate (Barriuso and Calvet 1992). In the upper layers, where soil microorganism activity is essential, pesticides can undergo some degradation. Chemical and physical processes can often convert or dilute the existence of pesticides. These various mechanisms usually make pesticides less harmful (Ren et al. 2018). The amount of water penetration depends on the porosity of the soil. Porous soil is composed of hard products such as sand and gravel. Water infiltrates easily in this form of soil and may penetrate the phreatic zone. Consequently, the latter is susceptible to pollution. Conversely, fine-textured soil with a batting crust, such as clay, is less permeable to pollution, so water is more steadily infiltrating (Calvet 2005) (Fig. 17.2).

17.4.1.3 Underground Drainage

Drainage happens after the water has crossed the upper horizons of the surface and is expressed by seeping to create trickles of water. Generally, it happens when clay soil is present at deep soil thicknesses (Barriuso and Calvet 1992). In agricultural areas, especially in clay or fine-textured soils, excess water is drained by groundwater drainage to watercourses. This hypodermic flow allows the transport of pesticide contaminants to the sites of deposition or to the waterways. Pesticides with rainwater penetrate into the soil reach groundwater drainage system and discharged into the streams (Singh et al. 2019) (Fig. 17.2).

17.4.1.4 Air Drift

Air drift allows the wind to scatter droplets of pesticide vapors beyond the goal area. It can be a significant cause of pollution for the surrounding areas of the treated areas. The key variables that affect drift are the scale of the sprayed pesticide, wind direction, and spray height. Big droplets easily settle in the treated region, but fine droplets may remain longer in the air and be carried by the wind over long distances (Fang et al. 2017) (Fig. 17.2).

17.4.1.5 Volatilization

During and after spraying, various pesticides begin to volatilize. These vapors grow above the treated region and are then carried by air currents over long distances. The phase of volatilization is accentuated by the sun. These pesticides ultimately are transported by runoff, infiltration, and underground drainage. The pesticide residues are detected somewhere else before their disappearance or reduction (Houbraken et al. 2016) (Fig. 17.2).

17.4.2 Behavior of Pesticides in the Soil

The paradigm of sustainable agriculture with the establishment of gigantic latifundium-type farms has little choice but to return to the use, often huge, of pesticides in order to increase yields. These pesticides interact with terrestrial environments by their residues, having produced the intended results. These residues become unnecessary environmental contaminants (Khan 1980; Fang et al. 2017). Residues may or may not have an impact on the environment, and their existence in the soil requires a number of physical processes and physicochemical, biochemical, and chemical reactions. These various soil associations somehow decide the behavior of the residues and their destiny (Khan 1980). Transport in and out of the soil system is governed by physical and physicochemical processes, though, chemical and biochemical reactions influence pesticide transformations, resulting in molecular alterations and degradation toward complete mineralization (Ren et al. 2018).

17.4.2.1 Retention

Some pesticide molecules can be susceptible to temporary or long-term sequestration in the soil. This condition is linked to the accumulation or immobilization of pesticide residues (Calvet 2005). This retention phenomenon, according to Guimont (2005), is a mechanism that immobilizes the pesticide molecules or their transformation or degradation components, regardless of their dissolved or gaseous condition. By means of this concept, it is clear to understand that the pesticide pathway is first rendered by going through the unsaturated zone of the soil until it comes into touch with the solid process, which is, of course, its end. This movement is not at all accidental; it is attributed to the sequence of many physicochemical and to biological phenomena.

The final result of the preservation mechanism is the release of molecules after changes to the configuration of the solid phase or the nature of the liquid phase. According to Guimont (2005), while preservation is due to adsorption, the release is due to desorption. According to the same source, where preservation is due to stabilization, release occurs by molecular diffusion alone or through hydrolysis reactions or shifts in the arrangement of solid constituents. It concludes that, in both situations, the release takes the molecules into the liquid process where they become mobile and can be transferred and consumed by living organisms.

17.4.2.2 Adsorption

In addition to the retention, the pesticide activity in the soil may follow or is affected by the adsorption mechanism (Khan 1980). In their studies, Kah et al. (2007); Wang et al. 2020; Wang et al. 2020) estimated that adsorption can be defined as a surface phenomenon that leads to pesticide aggregation, ionized or not, at the interface of two immiscible phases (a solid phase produced by clay particles and organic matter and a liquid phase (the soil solution) in which pesticide molecules are contained in solution). This condition is commonly found in the soil's solid/liquid interface. Adsorption can occur in this interface, unrolling as often on the outer surface of the soil particle facing the liquid as on the inner surface of the pores in contact with the soil solution. Adsorption does not arise in the soil in the same manner, since it is closely associated with the existence of the substrate. In cultivated soils, pesticide adsorption at the liquid/solid soil interface is the most significant as opposed to uncultivated soils (Khan 1980). In addition, Barriuso and Calvet (1992) demonstrate that the process of adsorption depends on the quality of the soil. They find that soils abundant in organic matter consume all nonionic pesticides in vast amounts. In this regard, two pesticide families are characterized by their loads of markedly different

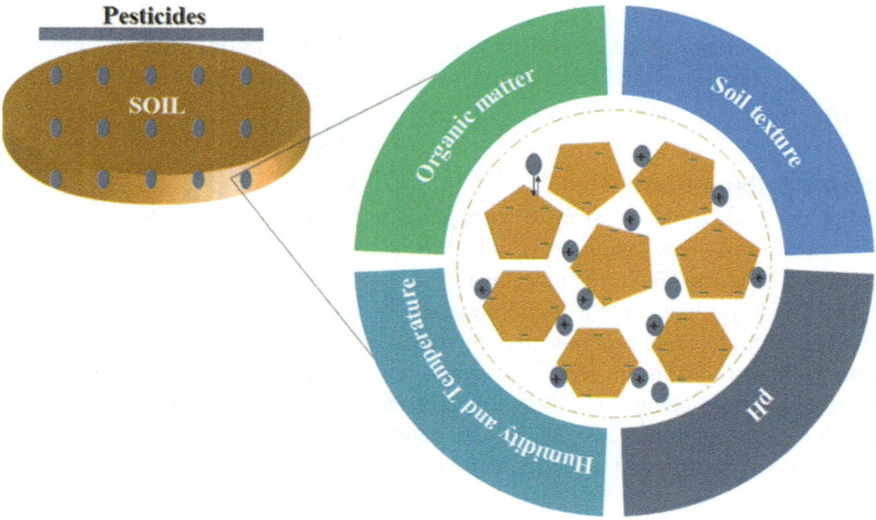

Fig. 17.3 Parameters that affect pesticide adsorption in soil

adsorption characteristics: (1) cationic pesticides are highly absorbed by soils containing a lot of clay minerals (especially montmorillonite and vermiculite) and organic matter; (2) anionic pesticides are highly consumed by soils abundant in metal oxides and hydroxides (Khan 1980; Barriuso and Calvet 1992; Calvet 2005).

In addition to the existence of the soil structure, there are other influences that may greatly affect the absorption of pesticides in the soil. These involve temperature, soil and water pH, particle size distribution, surface area of solids, soil/water ratio, and adsorbent physical properties (Fig. 17.3). This entire set of variables and parameters combined allows it possible to assess the adsorption potential of the soil.

Adsorption-desorption Process

Adsorption-desorption mechanisms in the soil play a crucial role in all physical processes influencing pesticide residue actions in the crop-plant environment, such as volatilization in the air, soil water movement resulting from soil movement and pesticides leaching to the surface or groundwater, and absorption from the soil by plants or fauna. In addition, biological activity and decay are often impaired by adsorption. Adsorption is commonly characterized as the adhesion or attraction of one or more ion or molecular layers to the surface. In soil, the amount of all forms of ion or molecule fixation on or in the solid process is termed adsorption (Barriuso and Calvet 1992; Yavari et al. 2020).

Pesticides may be adsorbed by soil organic matter and inorganic soil fractions, and the interaction between adsorption at all sites relies on the soil properties as well as the chemical composition of the pesticide. Adsorption of inorganic and organic matter is not attributed to a single pesticide-soil interface process. There is a spectrum of potential adsorption interactions, beginning with fixed site adsorption and finishing with a distinction of three-dimensional stages, such as aqueous solution and soil organic matter. Links differ between those that are reversible and those that are derived from so-called attached residues (Calvet 2005; Sharifi et al. 2015; Sarkar et al. 2020; Wang et al. 2020; Wang et al. 2020). Pesticide adsorption may be quantitatively defined by a linear adsorption coefficient (KD) or by adsorption constants extracted from the Langmuir or Freundlich adsorption equation. The linear KD is the concentration quotient in the solid phase of the soil divided by that of the liquid phase of the soil. It is relevant to the classification of adsorption by dilute aqueous solutions where the distinction between the solid and liquid phases of the soil is more or less independent of the concentration (Calvet 2005). Numerous experiments have shown that adsorption is strongly associated with organic soil carbon (García-Delgado et al. 2020), at least in the case of nonionic pesticides.

17.4.2.3 Mobilization

Pesticide mobility in soil is related, on the one hand, to adsorption and, on the other, to the mass flow of dissolved fractions. Adsorption slows down the mass wave, which consists of diffusion, convection, and dispersion, and is also a function of removing pesticides from the solution through biological and chemical reactions (Calvet 2005; Elbana et al. 2019; Holten et al. 2019). Thereby the diffusion is a physical mechanism by which molecules, atoms, and ions–because of their thermal mobility–go from higher concentrations to lower concentrations. This means of transport is independent of the flow of water. Whereas, convection is the passive movement of solutes in moving water. It is a function of the volume flow velocity of water. On the other hand, the dispersion is the diffusion or combination of solutes in flowing pore water arising from the varying flow speeds of each volume of water. Like convection, it is a connection to the mass movement of soil water, which relies on the current. Based on these processes influencing the mobility of pesticides in the soil, the related soil parameters (apart from adsorption coefficients and microflora biodegradation) are texture, porosity, density, layer humidity, flow velocity, and aggregate stability (Calvet 2005).

17.4.3 Transfer of Pesticides from Soil to Water

Surface runoff, tile absorption, and incorporation through groundwater are the primary mechanisms of pesticides travelling through water (Climent et al. 2019). These systems are also possible causes of pesticide pollution of rivers, reservoirs, and groundwater (Wang et al. 2019). Contamination of groundwater and drinking water with pesticides has been of major interest in recent years. Forecasting leaching trends for surface pesticides are also becoming extremely relevant in determining potential concentrations in groundwater and drinking water in the future.

Given the extraordinary value of adsorption properties and degradation rates for soil pesticide leaching, soil texture may be definitive in specific cases. Organic soil matter influences the mobility and leaching of pesticides not only by adsorption but also favorably by producing greater aggregate stability (Kah and Brown 2007; Zhu et al. 2019). Furthermore, organic matter improves pesticide mobility and leaching by forming water-soluble additives between pesticides that are poorly soluble in water and humic compounds that are easily soluble in water. The temperature also causes the leaching of pesticides in the soil. Interactions between different chemicals is linked to soil mobility and leaching can take place under the mutual control of their solubility in water or competition for soil adsorption sites (Vryzas 2018; Larson 2019; Sarkar et al. 2020).

The interaction of all the parameters examined above and others not explored may contribute to the occurrence of pesticide residues in groundwater. The amounts of residues of pesticides found in groundwater from agricultural usage include those for which theoretical and experimental models have not been predicted to enter groundwater. In previous years, chlorinated pesticides such as aldrin and DDT have become the pesticides most widely contaminating drinking water (Grung et al. 2015; Meftaul et al. 2020). Their existence was not anticipated due to their poor solubility in water; but, due to their biological and chemical durability, they were able to enter groundwater. Currently, pesticides found in groundwater after agricultural usage belong to a number of chemical groups, including carbamates, triazines, acetanilides, uracils, dinitrophenols, and others (Nantia et al. 2017; Zhao et al. 2017). Traces of readily biodegradable pesticides can be identified in groundwater at an analytically measurable degree if their mobility is large.

17.4.4 Detection and Quantification of Pesticides in Environmental Matrices

Analysis, when pesticides are identified or quantified, is important both for food security and protection and for the ecosystem. Many testing units and laboratories are interested in pesticide detection in various matrices: soil, fruit, water, and blood. Abundant literature on this topic demonstrates that researchers more or less adopt the approach by changing that information according to needs (matrices, pesticide form, etc.).

Pesticides are found in different environmental components in the form of complicated mixtures, and, in general, at low concentrations, their study involves a preparation to simplify and reconcentrate samples that must conform with requirements such as viability, matrix, and respect for the ecosystem, etc. Pesticide extraction is a critical step until the study of the above; extraction is carried out according to the matrix: (1) for strong matrices (such as soils and sediments) one of the initial methods to be used for the extraction of solid matrices has been the extraction by Soxhlet. However, it is a lengthy technique involving massive quantities of solvents. Today, the most widely employed methods are microwave-assisted extraction and extraction accelerated by a solvent at high temperature and pressures (Cruz 2016); (2) for liquid matrices liquid-liquid extraction is one of the first methods used to separate organic compounds from liquid matrices. This is an

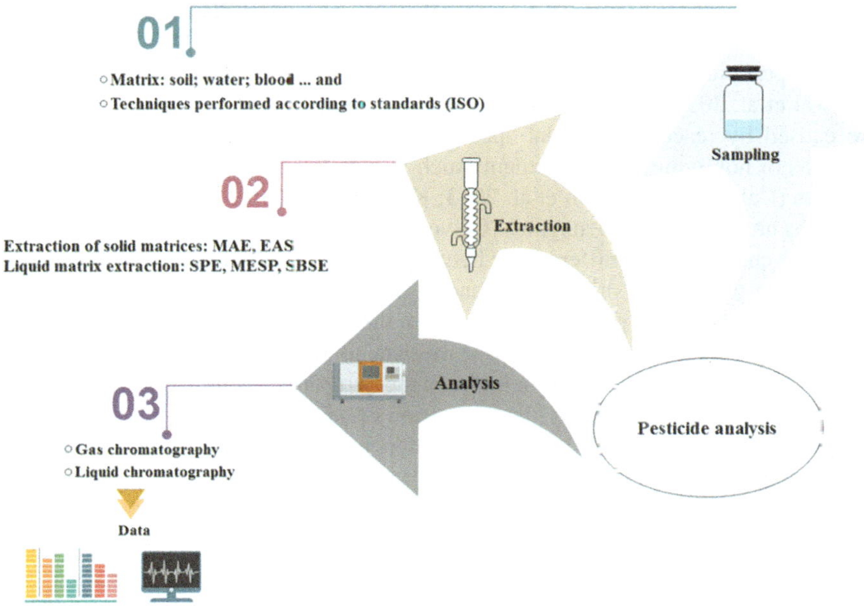

Fig. 17.4 Pesticide analysis steps (Van der Hoff and van Zoonen 1999; Reemtsma et al. 2013; Moreno-González et al. 2017)

efficient method because Soxhlet extraction has many disadvantages (use of massive quantities of solvents, lengthy handling, multiple measures, etc.) (Tankiewicz et al. 2011). Improvements have been made to this methodology, in particular, the production of liquid-liquid micro-extraction, which requires just a few microliters of solvents and in other methods based on other concepts, such as solid-phase extraction (SPE), micro-extraction on solid phase (MESP), and stir-bar sorptive extraction (SBSE). At present, with the latest technologies, it is possible to evaluate water samples without previous extraction while retaining sensitivity consistent with the study of a non-highly diluted natural medium (Reemtsma et al. 2013) (Fig. 17.4).

Depending on the type of the pesticides being examined, samples that have been extracted using one of the techniques mentioned above may be analyzed in order to classify the pesticides present. Two analytical separation techniques are typically used for their detection and quantification: gas chromatography (GC) (Van der Hoff and van Zoonen 1999) and liquid chromatography (LC) (Moreno-González et al. 2017) (Fig. 17.4). These methods are combined with a basic or universal detector. Thus, while having accuracy, the mass spectrometer has the benefit of being a virtually universal detection instrument.

17.4.5 Transformation and Degradation of Pesticides

In soil, pesticides experience separate reactions to biotic and abiotic transition (Jaiswal et al. 2017). Abiotic reactions involve all those that are not enzymatic but are caused by reactive chemical species or molecular function in the soil or by catalysis of nonliving soil constituents such as metal oxides and organic or mineral surfaces (Calvet 2005; Cycoń et al. 2013; Khedr et al. 2019). In comparison, biotic reactions are catalyzed by enzymes (Calvet 2005; Kumar et al. 2018). To date, it is also quite challenging to differentiate between abiotic and biotic triggers of chemical reactions in the soil. In different instances, the same reaction substance originates partly from the enzyme phase and partly from the abiotic phase (Martin-Laurent and Therond 2017). The transition from one mechanism to another is neither sudden nor clearly established. The distinction between the two pathways utilizing sterilized and unsterilized soil does not imply the proportion of abiotic and biotic reactions found in natural soil, since sterilization, in addition to preventing enzymatic behavior, induces improvements in soil structure as in free radicals in soil or singlet oxygen in the aqueous process (Mamouni et al. 2018). Therefore, the abiotic portion of the conversion products derived from sterilized soil should be treated as a minimum limit of possible abiotic reactions.

Pesticide mineralization involves the full breakdown into tiny inorganic molecules such as CO_2, CO, H_2O, NH_3, H_2S, and Cl. This is the only way to remove xenobiotic substances from the atmosphere (Andreu and Picó 2004; Calvet 2005). A final product arising from the co-metabolic operation of some organisms can be further degraded by other classes of organisms or may be exposed to an abiotic attack. So, soil mineralization is a dynamic mechanism involving multiple biotic and abiotic causes. The overall degradation quantification of international soil compounds is very relevant for their ecotoxicological assessment. The physicochemical properties of the pesticide decide the behavior and biological activity of the pesticide in the soil. The main properties of pesticides include molecular size, ionization, lipophilicity, polarization, and volatility, but usually one or two of them have a dominant effect (Calvet 2005; Guimont 2005).

The fate of pesticides in the soil ecosystem typically depends on the abiotic transformation associated with physical, chemical, and photochemical processes, as well as the biological transformation associated with the behavior of living organisms, in particular microorganisms (Omakor et al. 2001; Jaiswal et al. 2017). The pesticide behavior in soils is defined by the following processes: (1) volatilization, (2) plant absorption, (3) leaching and runoff, (4) soil sorption and binding, (5) chemical degradation, and (6) soil microorganism degradation (Calvet 2005). Volatilization is the main cause of pesticide loss in target areas, especially when pesticides are added to the soil or plant surface. The rate of failure always exceeds that of chemical degradation, runoff, or leaching (Houbraken et al. 2016). The absorption of pesticides by the roots of plants is another process that decreases the volume of pesticides in the soil. This process is potentially the primary cause of bioaccumulation in the food chain and the primary path of human and animal exposure (Navarro et al. 2007).

Owing to the different processes that pesticides undergo in the soil, their longevity in this setting is generally limited, although in certain situations they may leave residual residues in the soil for weeks, months, or even years (Testud and Grillet 2007; Du et al. 2017; Oudejans et al. 2020). The half-life of a chemical depends on the variety of variables, including climate, soil texture, pH, temperature, humidity, minerals, organic material, form, and quality of the substance and its single or multiple applications (Calvet 2005). Further tests, including different physical and chemical properties and soil parameters, are therefore needed to evaluate the pesticide activity in this environment.

17.5 Microbial Degradation of Pesticides: Biotic Degradation

In the preceding parts and paragraphs, the issue and function of pesticides and their physicochemical characteristics have been discussed. There is also a question of the assimilation or deterioration of the pesticide in the soil following usage and in what shape it is degraded. It is evident that pesticides are not assimilated by plants and contaminants are present in the soil owing to the spraying and recovery process. Their effect on vulnerable microfauna and groundwater pollution is increasingly understood. In order to minimize or alleviate their harmful effects on the ecosystem in general and, in particular, on soils and waterways, consideration or enforcement of the degradation protocol for these pesticides is important and crucial. The bibliography dealing with this topic illustrates the biologic degradation of pesticides.

17.5.1 Mode of Pesticides Degradation in the Soil

In order to keep the soil fertile, it is important to concentrate on the pesticide fate in the soil. If soil scientists are persuaded that restoring soil fertility by pedogenesis is a long-term method, some researchers still believe that microorganism degradation remains a fascinating process. The microbiological degradation of pesticides in soils has been extensively documented. Degradation originates from the action of microorganisms, meaning that the pesticide is used as a nutritional substrate and is degraded by enzymes (Khan et al. 1976). To this, it should be added that, in general, biodegradation is more essential than abiotic degradation. According to Calvet (2005), the mode of degradation is exceptional in the so-called recalcitrant pesticides. The dynamics of of microorganisms is found in the upper horizons of the soil where the conditions are optimum due to the existence of organic matter, oxygen and humidity, and the iso-humic complex (Liu et al. 2018).

17.5.2 Pesticides Biodegradation

Particular focus was put on the dominant mechanism of biodegradation in the pesticide degradation in the surface layers of the soil. The observations made by Calvet (2005) make it possible to understand how microorganisms work to remove pesticides from the soil. He concludes that biodegradation is due to various chemical transformations that change the composition and structure of the molecules brought to the soil. This research adds that these modifications may be restricted to the plain and simple removal of a functional group leading to different intermediates or transformation products or to the complete degradation of mineral molecules (mineralization). Throughout this method of biodegradation and mineralization, a whole series of intermediate molecules between the initial molecule and the final molecules can thus be generated and made available in the soil (McAuliffe et al. 1990; Moneke et al. 2010; Obuotor et al. 2016; Poirier et al. 2017).

It is important to realize that the intermediate molecules are transitory. According to Barriuso et al. (1996), it may be concluded that they have features that vary from primary molecules since, more frequently, they do not have the same retention and toxicity as the original molecule, but are not inherently less polluting and less harmful, which requires recognizing them and understanding their environmental properties and destiny. It should be remembered that only the process of pesticide mineralization contributes to their complete removal from the natural environment (Andreu and Picó 2004). It is also certain that mineralization is the safest approach to neutralize pesticides by extracting them from the environment and preserving soil fertility.

17.5.3 Microorganisms' Role in Biotic Degradation

Microflora activity is often used to generate substances that benefit the natural environment. This activity is created by biotic degradation in soils with the collective or individual involvement of fungi, algae, protozoa, and bacteria. However, pesticide-degrading microorganisms are typically primarily composed of bacteria and fungi, since they are mostly present and identified during degradation processes involving microorganisms (Jacob et al. 1988; Benslama 2014; Kryuchkova et al. 2014; Obuotor et al. 2016). The phenomena favoring the involvement of bacteria and fungi as formidable agents in the biotic degradation of pesticide molecules is the existence of ideal conditions for their growth. They should adapt to the temperature and humidity levels in order to take action. The biodegradation dependent on pH, carbon, and mineral content is determined by the presence of clay in the soil. Calvet (2005) also considers that the size of microbial biomass is equally significant. It is an average of 200 mg C/kg in sandy soils low in organic matter and can be around 900 mg C/kg in rich soils. This microflora is not made up of a single species but of a variety of species. In order to differentiate its activity from other microorganisms operating in the biotic degradation of pesticides, its role is defined as degrading. The activity of microflora in the process of degradation depends closely, at the same time, on its enzymatic activity, which defines the nature of the chemical reactions, and on its environment, which affects its growth and survival (Aislabie and Lloyd-Jones 1995). There are three components: physical (temperature, water content, and soil structure), chemical (organic compounds (phenolic acids, amino acids), inorganic (amino nitrate, orthophosphate) dissolved in soil solution, oxygen, pH), and biological (other organisms, plant roots, soil animals) (Aislabie and Lloyd-Jones 1995; Calvet 2005; Kah et al. 2007).

17.5.4 Nature and Reactions Involved in the Biodegradation Action

The process of biodegradation will begin when the environmental conditions are optimum. However, it is important to note that the basic feature of degradation due to microorganisms is the catalysis of chemical reactions by enzymes. In order to achieve this, pesticides must be dissolved during the soil liquid phase, which is necessary to activate the process. According to (Calvet 2005), there are mainly two situations: (1) chemical reactions catalyzed by intracellular enzymes; in this situation, pesticides must be absorbed in order to be processed; (2) chemical reactions are triggered by extracellular enzymes. In this situation, pesticide absorption is not required. Both conditions tend to ensure the degradation of the pesticide compounds, based on whether or not they have to be absorbed in advance. A minimum of environmental conditions are needed for microorganisms to begin their pesticide molecule degrading activities. Now we need to consider the processes by which these microorganisms are able to degrade pesticides. Calvet (2005) and Gupta et al. (2016) consider that microorganisms can be active in pesticide degradation under different mechanisms of action.

17.5.4.1 Direct Metabolism

Microorganisms require nutrients that are critical to their development (N, P, K, S, and trace elements), water, and energy for their growth and maintenance of activity. From a biochemical point of view, in the absence of absorption of light energy, particularly from the sun, the production of energy results from the exchange of electrons between the electron donors (organic and inorganic oxidizing compounds) and the electron receptors (molecular oxygen, nitrates, sulfates, ferric compounds, and organic compounds) (Calvet 2005; Wang et al. 2020; Wang et al. 2020). This direct metabolism is the consequence of the use of pesticides as an energy source. Numerous pesticides may be substantial or complementary sources of energy for microorganisms by various catabolic and enzyme-catalyzed chemical reactions. By utilizing this energy from pesticide degradation, the bacteria that make up the larger part of the microorganisms are able to conduct all their chemical reactions by targeting the original molecule and inorganic molecules by mineralization. On the other hand, other microorganisms may only conduct part of the transition process. This circumstance involves the aggregation of joint efforts through the presence of many organisms to achieve mineralization, each utilizing successive chemical transformations as a source of energy (Pino and Peñuela 2011; Teerakun et al. 2020).

The beneficial function of the consortium in degradation is not just due to the existence of all the enzymes required for degradation but may also benefit from the sharing of nutrients and the removal of substances that prevent degrading microorganisms (Calvet 2005). Microorganisms are even able to destroy several pesticides belonging to various chemical families by a direct metabolism pathway (Abraham et al. 2014; Ramya and Vasudevan 2020). Depending on the existence of the functional groups, the degradation of pesticides by these microorganisms is more or less simple. The processes contribute to the production of inorganic molecules (carbon dioxide, ammonia, water, sulfate anions, and phosphates).

According to Fournier et al. (1996), kinetics presents a delay step that precedes a fast mineralization phase. The subsequent application of the pesticide would eliminate the lag step. It is also possible to transfer the mineralization potential from treated soil to untreated soil. The soil where mineralization happens includes unique bacteria capable of generating it. However, the speed of mineralization is improved by successive applications. As described above, mineralization is complete degradation. This mineralization is a pathway for direct metabolism. It happens in a series of chemical reactions, the first of which is hydrolysis, or oxidation, reduction, or addition, accompanied by other forms of transformation (Calvet 2005).

17.5.4.2 Co-metabolism

This degradation mechanism was first described by Leadbetter and Forster in 1959. According to Dalton and Stirling (1982) co-metabolism is a process in which microorganisms ensure their conservation and proliferation at the expense of the organic substrate by degrading pesticides without becoming a source of energy and nutrients for them, the initial reactions of which are catalyzed by non-specific. In general, as opposed to direct metabolism, co-metabolism does not lead to complete pesticide degradation, particularly where a single strain concerns the initial pesticide molecule. On the other hand, it is likely that many strains can function together in sequences to achieve degradation and to produce metabolites, or even to use others as energy substrates and mineralize them (Teerakun et al. 2020). The metabolites produced have properties different from those of the original pesticide, in particular those linked to their transport and toxicity. They are also more polar and thus water-soluble and therefore more poisonous.

17.5.4.3 Conjugation and Condensation

The particularity of the conjugation process is that it refers to the unification of two molecules. Methylation and acetylation are two conjugation reactions in soil microflora (Schrack et al. 2009). This type of mechanism leads to the union of about two to five molecules and polycondensates of larger molecular size when a greater number of molecules are brought together. The reactions of polycondensates induced by microorganisms play an important role in the incorporation of pesticides into humic substances in the soil and contribute to the formation of bound residues (Calvet 2005).

17.5.5 Herbicides Studied

17.5.5.1 Glyphosate

Definition and Nomenclature

A variety of definitions are given for glyphosate, but the one that seems fairly concise and complete to us is the definition given by Yu et al. (2007), Zhan spil; et al. (2018) and by Steinrücken and Amrhein (1980) that "Glyphosate (N-phosphonomethyl) glycine) is a broad-spectrum systemic herbicide that blocks an enzyme that the plant requires to produce amino acids and proteins." Through its consistency and performance, glyphosate is the most commonly used herbicide in the world, by individuals as well as by professionals. Its effectiveness at low doses makes it commonly used. According to Maldani et al. (2017, 2021), it is the best-selling herbicide in Morocco and around the world. Its crucial role is much more valued in the production and development of genetically modified plants in order to tolerate herbicides (Duke and Powles 2009). The products of glyphosate have been sold worldwide. These products have been sold under a variety of titles, including Abundit Extra, Credit, Xtreme, Glifonox, Glyphogan, Ground-Up, Rodeo, Roundup, Touch Down, Tragli, Wipe Out, and Yerbimat (IARC 2018).

Historical Overview on Glyphosate

By observing the chronology of all subjects addressed in the scientific literature concerning glyphosate, the first to be developed was Henri Martin in Schaffhouse (Switzerland) in 1950, without understanding the scope of the possibilities provided by this pesticide. It was perceived to be a result of secondary importance. Two decades later, the work of John E. Franz, a chemist in the Monsanto Company (1970), applied herbicide properties and showed that glyphosate, a systemic herbicide, destroys plants by blocking the enzyme required to produce proteins (Duke and Powles 2008). Subsequently, glyphosate was patented under the trading mark Roundup. Despite this significant finding, the most notable findings of its use were that glyphosate was sold in Malaysia for rubber cultivation and then in the United Kingdom for wheat cultivation beginning in 1974. Also, in the same year, it was first authorized by the American authorities in the United States, for weeding in nonagricultural zones. After the discovery of its fascinating herbicide properties, glyphosate has achieved worldwide success. In spite of its widespread application, Monsanto embarked on a wide-ranging campaign to develop genetically modified organisms (GMOs) for greater resistance to Roundup in the 1990s. From 1996 onward, the company marketed Roundup Ready crops based on corn and soy GMOs in 2011. Massive areas have been sown by these two speculations, and more than 650,000 tons of glyphosate have been applied worldwide.

Glyphosate has attracted a number of companies following the expiry of its American patent in 2000 and provided the world-famous popularity and the craze of individuals and professionals. It is sold under a number of brand names by more than 40 firms (Nandula 2010; Duke 2018). The market leader is Monsanto, which supplies all of the herbicides produced worldwide. Other companies such as Syngenta, Bayer, BASF, and DOW also offer their own glyphosate products for the majority of the market share. The evolution of glyphosate did not happen like a peaceful river. Some have begun work on environmental and public health effects. In 2012, the research by Professor Gilles-Eric Séralini on the long-term toxicity of Roundup created a stir. Then, in March 2015, the WHO-dependent IARC classified glyphosate as likely to be carcinogenic to humans; consequently, a range of countries have taken precautionary action on their use. France, for its part, has agreed to prohibit its usage within 3 years (Duke 2018; Lautre 2019). These fair timeframes consist of enabling farmers and researchers to identify substitutes that are less toxic than glyphosate. Glyphosate is currently accepted in the European Union until 15 December 2022 (EFSA 2021).

Characteristics

Durand (2007) classifies glyphosate (Fig. 17.5) in the poor organic acid family as very soluble in water: 10.5 g/L at 20 °C at a pH of 2. This acid is approximately ten times less soluble than organic solvents. In addition to this active ingredient, other glyphosate herbicides include different chemical additives, such as surfactants, which promote the absorption of glyphosate by the plant in order to achieve its impact. In order to help describe glyphosate at the physical and chemical levels, Table 17.1 points forth the key characteristics.

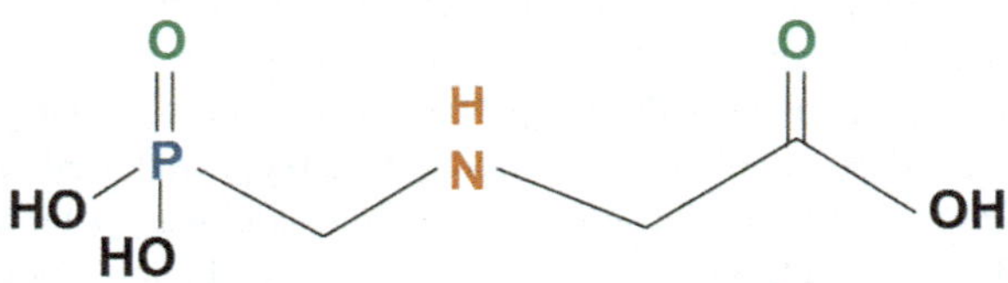

Fig. 17.5 Structure of the glyphosate molecule

Table 17.1 Physicochemical characteristics of glyphosate

Chemical name	N-(phosphonomethyl) glycine
Chemical family	Amino acid
Chemical formula	$C_3H_8NO_5P$
Density	1.71 (g/mL)
Molecular weight	169.1 g/mol
Melting point	200 °C
Solubility in water	10.5 g/L at 20 °C et pH de 2 ; 11.6 g/L at 25 °C

Mode of Action

In the process of weed neutralization, herbicides follow two modes of action, preemergence herbicide which is a substance that works before weeds are formed and the postemergence herbicide. In the case of postemergence herbicides, a distinction must be made between those which establish a contact action and those which manifest a systemic reaction. As with other herbicide classes, glyphosate is consumed by the leaves and branches and is then transferred to the roots and the aerial portion by the plant's vascular system. According to Steinrücken and Amrhein (1980), it works by preventing the biosynthesis of aromatic amino acids (tyrosine, phenylalanine, and tryptophan) that are necessary for the growth and survival of plants. The deficiency of aromatic amino acids contributes to the termination of protein production and the development of some phenolic compounds. Durand (2007) considers that the blocked enzyme is 5-enolpyruvylshikimate-3-phosphate synthase (EPSPS), which is localized in chloroplasts and catalyzes the synthesis of 5-enolpyruvylshikimate-3-phosphate (EPSP), phosphoenolpyruvate (PEP), and shikimate-3-phosphate; glyphosate molecules compete with phosphoenolpyruvate (PEP) and not with shikimate-3-phosphate. It does not bind directly to the active site of the enzyme but to a potential allosteric location, creating a structural difference that would keep PEP from binding. Inhibition of EPSP triggers the interruption of three aromatic amino acids (phenylalanine, tyrosine, and tryptophan) which are important for plant synthesis (Giesy et al. 2000) (Fig. 17.6).

Glyphosate Adjuvants

Adjuvants boost the physicochemical properties and the biological performance of the active ingredients; they can be solvents, dispersants, emulators, or surfactants. These compounds do not have herbicide action (Rainaud 2013). In some nations, such as France, all formulations containing glyphosate produce surfactants such as adjuvants (Rainaud 2013). This raises the contact area between the sprayed liquid and the leaves for improved action by encouraging the penetration of the product through the cuticle, pore, or plant stomata. Even if their function is limited to a subordinate intervention to promote contact, they are therefore necessary for optimal herbicide activity. It is projected that just 10% of the volume of glyphosate sprayed will penetrate the plant (Leaper and Holloway 2000). It should be remembered that there is a multitude of surfactant-type adjuvants available. Various forms of surfactants are ethoxylated fatty amines, phosphate esters, alkyl polyglucosides, and propoxylated quaternary ammonium salts. Each of the surfactants has its own peculiarities and physical, chemical, biological, and toxicological effects (Kaczmarek et al. 2019).

In glyphosate formulations, primarily surfactantsare in the first two families. The Roundup variety primarily uses fatty ethoxylated tallow amines; the most widely used is polyoxyethylene amine (POEA), a detergent that facilitates the distribution of spray droplets on the leaves and is found to be approximately 15%, and studies have shown that it is 10,000 times more harmful to human cells than glyphosate (Mesnage

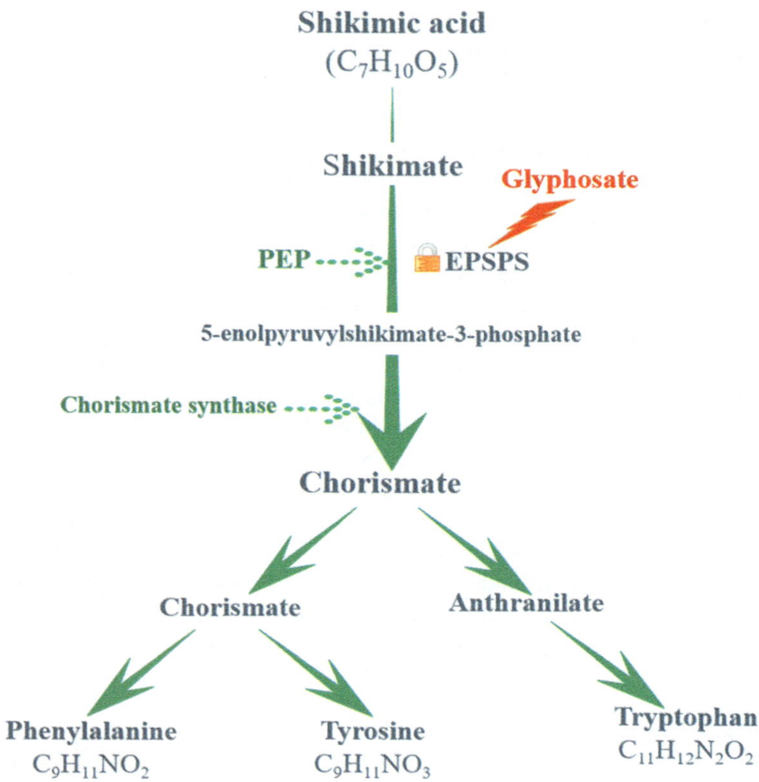

Fig. 17.6 Mechanism of action of glyphosate (Dill 2005)

et al. 2013). Thus, the separate products in the Roundup variety composed of 14.5% to 75% glyphosate salts, whereas the remainder of the formulation comprises of a dozen large adjuvants, the structure of which is mostly held confidential (Deschomets 2001; Rainaud 2013).

Glyphosate in Soil

The behavior and transformation of glyphosate in soil depends on the different interactions with the herbicide and soil under specific environmental conditions such as precipitation, temperature, organic matter, etc. and are affected by numerous soil factors and processes (Gimsing and Borggaard 2001; Bento et al. 2016; Khan 1980; Hermansen et al. 2020; Meftaul et al. 2020). Application rates and formulations can also have an effect on pesticide activity (Weber et al. 1965; Weber 2018). In their experiments, Bento et al. (2016) assessed the dissipation kinetics of glyphosate and its metabolite AMPA in the soil as a feature of temperature and soil moisture and found that, under biotic conditions, glyphosate dissipation and AMPA formation/dissipation are mainly influenced by temperature and humidity and glyphosate was 30 times more persistent in cold and dry weather than in hot and humid weather.

Adsorption

Glyphosate is active and quickly absorbed in soils, therefore not quite mobile (Zhao et al. 2009), and in certain situations, primarily sandy or limestone soils, adsorption is more moderate (Sprankle et al. 1975; Piccolo et al. 1994). Adsorption of glyphosate relies on many factors; pH highly affects glyphosate adsorption (Ololade et al. 2019); typically, adsorption rises as the pH reduces. Glyphosate is negatively charged in highly alkaline pHs, as are the surfaces of clay minerals, oxides, and organic matter, reducing adsorption. Glyphosate is attached to soil constituents by its phosphonic group, and this results in standard interactions with the occurrence of competition with phosphate at adsorption sites (Sprankle et al. 1975; Pessagno et al. 2008; Pereira et al. 2019). Gimsing and Borggaard (2001) have shown that on goethite, phosphate can also remove glyphosate from its sorption sites. According to Laitinen et al. (2008), the impact of the concentration of phosphorus in two soil types (clay and sandy loam) on glyphosate adsorption decreased with a rise in phosphorus concentrations (Fig. 17.7).

Several studies have reported that soil physicochemical properties such as organic matter quality, acidity, and texture largely affect the irreversible or reversible reaction of glyphosate in soil (Gómez Ortiz et al. 2017; Silva et al. 2019; Meftaul et al. 2020). As glyphosate is a small molecule of carboxyl, amino, and phosphonate groups, it is strongly absorbed by minerals in soils (Gimsing et al. 2007). As a consequence, glyphosate is heavily adsorbed to clay and organic matter and largely

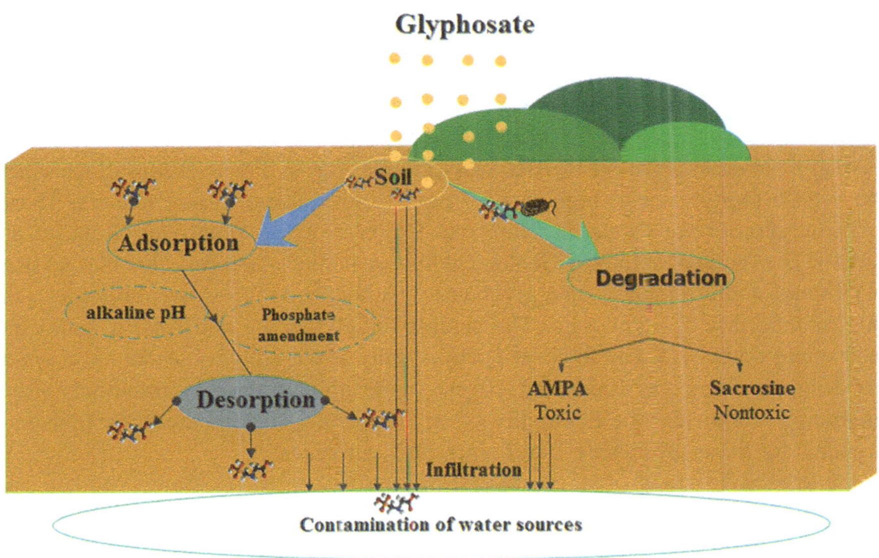

Fig. 17.7 Glyphosate behavior in soil

immobile in the soil, rendering it quite inaccessible to microbial degradation and accumulation in soil over time. As a consequence, glyphosate and AMPA can survive for more than a year in clay soils and wash rapidly in sandy soils (Okada et al. 2019). On the other hand, glyphosate can bind to humic substances through its phosphonic acid group and react with polyvalent cations adsorbed to organic matter and clay (Piccolo et al. 1994; Gómez Ortiz et al. 2017).

Glyphosate Mobility in Soil

Glyphosate mobility in various compartments is potentially restricted due to its strong soil sorption capacity (Piccolo et al. 1994; Padgette et al. 1995; Zhao et al. 2009; Padilla and Selim 2020). Dose, occurrence, preferential flow, sorption and degradation, and soil type affect the amount of glyphosate usable for leaching (Vereecken 2005; Norgaard et al. 2014; Duke 2020). According to Borggaard and Gimsing (2008), glyphosate leaching is typically caused by precipitation and soil composition, whereas sorption and degradation impair the mobility of glyphosate in soils.

Glyphosate can be transported in dissolved or particulate form and can enter water bodies (Meftaul et al. 2020). Owing to its low mobility, soil volatilization and low vapor pressure (9.8×10^{-8} mmHg at 25 °C) of glyphosate in the atmosphere is marginal (Giesy et al. 2000). In addition, runoff is another movement from agricultural fields and urban areas such as waterproof and linked paved surfaces (Grandcoin et al. 2017). As a consequence, the usage of glyphosate in certain impermeable and linked surfaces in some European countries is banned (Meftaul et al. 2020).

Presence of Glyphosate and Its Metabolites in Water Sources

Glyphosate traces have been identified in many surface and groundwater systems (Peruzzo et al. 2008; Toss et al. 2017). Persistence is generally 170 days, with a half-life of between 45 and 60 days (Peruzzo et al. 2008). Some experiments have also demonstrated a half-life of many years (Carlisle and Trevors 1988). Nowadays, glyphosate and its main metabolite AMPA have also been found in surface water and also groundwater (Okada et al. 2020). Glyphosate and AMPA levels in surface water have been found to be as high as 10^{-6} μg/L in several European countries. These concentrations were above the European Union drinking water cap (Botta et al. 2009). The findings of the US Geological Survey in nine states revealed that AMPA was found more regularly than glyphosate; its concentration was equivalent to or higher than that of glyphosate. The analysis of glyphosate and AMPA pathways across the water cycle and their association with various soils is mandatory in order to determine their environmental effect. While glyphosate is a commonly used herbicide worldwide, the possibility of environmental contamination by transport systems for these substances is still not well documented.

Impact of Glyphosate on Soil Microflora

Studies on the ecotoxicity of glyphosate and its metabolites to soil fauna are also inconclusive (Dominguez et al. 2016). Indeed, experiments have shown that glyphosate does not greatly impact soil microflora when used at normal doses. Glyphosate toxicity to soil microorganisms is usually deemed negligible as glyphosate undergoes accelerated microbial degradation and immobilization in soils (Giesy et al. 2000). On the other hand, various experiments have reported chronic effects on soil microflora as a whole: worms, microbes, fungi, etc. (Druille et al. 2013; Dominguez et al. 2016; Iori et al. 2020; Romina et al. 2020). Glyphosate functions by inhibiting the function of 5-enolpyruvylshikimate-3-phosphate synthase (EPSPS), an enzyme occurring in plants, microbes, and fungi (Padgette et al. 1995), some of which play a key role in the soil nutrient cycle (Zhan et al. 2018).

Impact of Glyphosate on Human Health

According to the WHO classification, glyphosate is graded as "U" (presents an unlikely or acute danger of daily use) (WHO 1994). Several authors have described the clinical existence of glyphosate and its accumulation in blood, urine, and breast milk (Zouaoui et al. 2013; Bus 2015; Niemann et al. 2015; Tsao et al. 2016; Woźniak et al. 2020). Glyphosate toxicity is defined by multiple symptoms such as gastrointestinal signs, impaired consciousness, hypotension, respiratory impairment, metabolic acidosis, and renal failure (Chang et al. 1999; Eddleston 2020). The death rate from glyphosate toxicity is estimated to be 3.2% in the research, including 601 patients with acute glyphosate poisoning. In the same report, the authors suggested that mortality was closely correlated with older age, increased intake, and increased intake of plasma glyphosate (>734 mg/mL) (Rainaud 2013).

The glyphosate composition includes surfactants that are expected to improve its toxicity. According to Zhang et al. (2019), glyphosate exposure may raise the risk of non-Hodgkin lymphoma by 40% and thus trigger cancer. Several reports have demonstrated that glyphosate influences human health; it can induce endocrine, nervous system, and leukemia disorders (De Roos et al. 2005; Mahoney 2018; Gastiazoro et al. 2020; Martinez et al. 2020). Enzyme disruptions have also been hypothesized and shown in human cell cultures (disruption of cytochrome P450, DNA structure in human mammary epithelial and placental cells) (Benachour et al. 2007) (Fig. 17.8).

Degradation of Glyphosate

After application, most pesticides reach the soil due to direct usage and/or leaf leaching (Torstensson 1985; Xiang et al. 2020). Pesticide-containing plants can return to the soil during the growing cycle (leaf senescence) or after harvest as crop residues. Glyphosate degradation in the soil is primarily biological with laboratory half-lives (DT50) varying from less than 1 day to 40 days (Mamy and Barriuso 2005). The predominant metabolite of glyphosate AMPA is more persistent: the DT50 ranged from 25 to 75 days (Feng and Thompson 1990; Mamy and Barriuso 2005; Singh et al. 2020). Glyphosate degradation relies on soil composition and properties, climatic factors, and management practices for phosphate fertilizers (Gimsing et al. 2007; Helander et al. 2012; Munira et al. 2016). The key product arising from the metabolism of glyphosate is AMPA, which has a strong

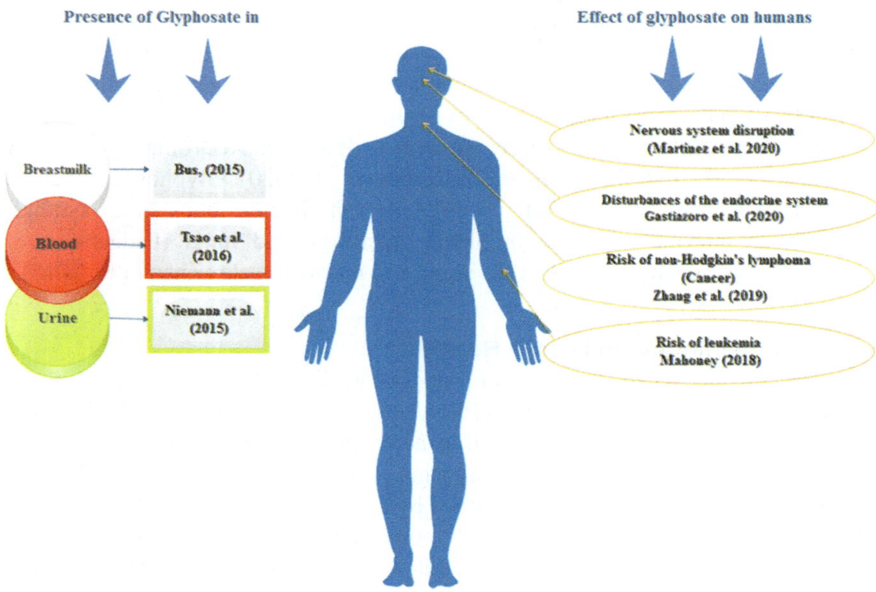

Fig. 17.8 Glyphosate effects on human health

environmental persistence and high soil mobility (Simonsen et al. 2008) and is therefore poisonous to nontarget species (Mañas et al. 2009; Singh et al. 2020).

Glyphosate-Degrading Microorganisms

Microorganisms perform a vital function in the transformation and mineralization of organophosphates. Thus, there is a broad variety of bacteria in the literature that degrade glyphosate. Microorganisms play a crucial role in the transformation and mineralization of organophosphates. Thus, in the literature there is a wide range of bacteria which degrade glyphosate; Table 17.2 summarizes in chronological order the bacteria which degrade glyphosate.

Mechanism of Glyphosate Biodegradation

In soil as well as in water, glyphosate degrades primarily through biotic processes under the action of microorganisms (Sharifi et al. 2015); glyphosate biodegradation processes have been extensively studied. In reality, glyphosate is effectively biodegraded by soil microorganisms. During the biodegradation of glyphosate, two metabolic pathways were identified, the first leading to the intermediate formation of glycine and the second leading to the formation of AMPA (Durand 2007). The rupture of the C-P lyase bond contributes to the development of sarcosine (*N*-methylglycine) and inorganic phosphorus. Sarcosine, in turn, is transformed into glycine and formaldehyde. Subsequently, glycine is used by bacteria to synthesize protein, nucleic acid, and amino acids such as serine, cysteine, methionine, and histidine (Fig. 17.9). Jacob et al. (1985) and (1987) described for the first time the

Table 17.2 Microorganism-degrading glyphosate

Microorganisms	Source	Gram status	Glyphosate detected metabolite	Glyphosate use	References
Serratia marcescens	Baghdad soil, Iraq	–	No data	Utilization of glyphosate as sole carbon source	Radif and Albaayit (2019)
Providencia rettgeri GDB 1	Glyphosate-contaminated soil, Taiwan	–	AMPA	Use of glyphosate with a concentration of 10,000 mg / L	Xu et al. (2019)
Ochrobactrum intermedium Sq20	Soil contaminated with glyphosate	–	Sarcosine and glycine	Utilization of glyphosate as sole carbon source	Firdous et al. (2018)
Achromobacter sp. MPK 7A	Glyphosate-contaminated soil	–	Sarcosine	Utilization of glyphosate as sole phosphorus source	Ermakova et al. (2017)
Comamonas odontotermitis P2	Glyphosate-contaminated soil in Australia	–	No data	Utilization of glyphosate as sole carbon and phosphorus source	Firdous et al. (2017a)
Ochrobactru`m intermedium Sq20	Glyphosate-contaminated indigenous soil	–	Sarcosine and glycine	Utilization of glyphosate as sole carbon source	Firdous et al. (2017b)
Enterobacter zibanae sp.	Sandy soil from Algeria	–	No data	Utilization of glyphosate as sole phosphorus source	Benslama and Boulahrouf (2016)
Pseudomonas putida	Sandy soil from Algeria	–	No data	Utilization of glyphosate as sole phosphorus source	Benslama and Boulahrouf (2016)
Salinicoccus spp.	Qom Hoz-e Soltan Lake, Iran	+	AMPA	Utilization of glyphosate as sole phosphorus source	Sharifi et al. (2015)
Bacillus subtilis Bs-15	Rhizospheric soil of a paper plant	+	No data	Utilization of glyphosate as a sole carbon and phosphorus source	Yu et al. (2015)
Bacillus subtilis	Rice field soil contaminated by glyphosate	+	No data	Utilization of glyphosate as sole phosphorus source	Nwokoro et al. (2015)
Pseudomonas sp.	Rice field soil contaminated by glyphosate	–	No data	Utilization of glyphosate as sole phosphorus source	Nwokoro et al. (2015)

(continued)

Table 17.2 (continued)

Microorganisms	Source	Gram status	Glyphosate detected metabolite	Glyphosate use	References
Enterobacter cloacae K7	Rhizoplane of various plants in Russia	–	Sarcosine and glycine	Utilization of glyphosate as sole phosphorus source	Kryuchkova et al. (2014)
Ochrobactrum sp. GDOS	Soil	–	AMPA	Utilization of glyphosate as sole phosphate source	Hadi et al. (2013)
Ochrobactrum anthropi GPK3	Glyphosate-contaminated soil	–	AMPA, glyoxylate, sarcosine, glycine, and formaldehyde	Utilization of glyphosate as sole phosphorus source	Sviridov (2012)
Bacillus cereus CB4	Glyphosate-polluted soil in the herbicide plant, China	+	AMPA, glyoxylate, sarcosine, glycine, and formaldehyde	Utilization of glyphosate as sole phosphorus source	Fan et al. (2012)
Achromobacter sp. MPS 12A	Methylphosphonic acid-contaminated soil	–	Sarcosine, glycine, and formaldehyde	Utilization of glyphosate as sole phosphorus source	Sviridov et al. (2012)
Acetobacter sp.	Rice field soil, Nigeria	–	No data	Utilization of glyphosate as a sole carbon and phosphorus source	Moneke et al. (2010)
Pseudomonas fluorescens	Rice field soil, Nigeria	–	No data	Utilization of glyphosate as a sole carbon and phosphorus source	Moneke et al. (2010)
Geobacillus caldoxylosilyticus T20	Central heating system water	+	AMPA and glyoxylate	Utilization of glyphosate as sole phosphorus source	Obojska et al. (2002)
Streptomyces sp. StC	Raw sludge from a municipal sewage treatment plant	+	Sarcosine and glycine	Utilization of glyphosate as sole phosphorus and nitrogen	Obojska et al. (1999)
Pseudomonas pseudomallei 22	Soil	–	No data	Utilization of glyphosate as sole phosphorus source	Peñaloza-Vazquez et al. (1995)
Pseudomonas sp. 4ASW	Glyphosate-contaminated soil	–	Sarcosine	Utilization of glyphosate as sole phosphorus source	Dick and Quinn (1995)

Pseudomonas sp. GLC11	Mutant of *Pseudomonas* sp. PAO1 on selective medium	–	No data	Utilization of glyphosate as sole phosphorus source	Selvapandiyan and Bhatnagar (1994)
Rhizobium meliloti 1021	Mutant of *Rhizobiaceae* meliloti induced by transposon Tn5 mutagenesis which is resistant to streptomycin	–	Sarcosine and glycine	Utilization of glyphosate as sole phosphorus source	Liu et al. (1991)
Achromobacter sp. LW9	Activated sludge from glyphosate process waste stream	–	AMPA	Utilization of glyphosate as a sole carbon source in presence of phosphate	McAuliffe et al. (1990)
Agrobacterium radiobacter SW9	Activated sludge from a waste stream	–	AMPA	Utilization of glyphosate as a sole carbon source in presence of phosphate	McAuliffe et al. (1990)
Alcaligenes sp. GL	Non-axenic cultures of the cyanobacterium *Anacystis nidulans*	–	Sarcosine and glycine	Utilization of glyphosate as sole phosphorus source	Lerbs et al. (1990)
Arthrobacter atrocyaneus ATCC 13752	German collection of microorganisms and cell cultures	+	AMPA and CO_2	Utilization of glyphosate as sole phosphorus source	Pipke et al. (1987a)
Arthrobacter sp. GLP-1/Nit-1	Mutant of *Arthrobacter* sp. GLP-1	+	Phosphate	Utilization of glyphosate as sole phosphorus source as well as sole nitrogen source	Pipke et al. (1987b)
Pseudomonas sp. LBr	Activated sludge from glyphosate process waste stream	–	AMPA and glycine	Utilization of glyphosate as sole phosphorus source	Jacob et al. (1988)
Agrobacterium radiobacter	Sludge from water treatment plant in America	–	No data	Utilization of glyphosate as sole phosphorus source	Wackett et al. (1987)
Arthrobacter sp. GLP-1	Accidental contaminant of *Klebsiella pneumoniae*	+	Phosphate and glycine	Utilization of glyphosate as sole phosphorus source	Pipke et al. (1987b)

(continued)

Table 17.2 (continued)

Microorganisms	Source	Gram status	Glyphosate detected metabolite	Glyphosate use	References
Flavobacterium sp. GD1	Monsanto-activated sludges	–	AMPA and phosphate	Utilization of glyphosate as sole phosphorus source	Balthazor and Hallas (1986)
Pseudomonas sp. SG-1	Aerobic digester liquid	–	AMPA	Utilization of glyphosate as sole phosphorus source	Talbot et al. (1984)
Pseudomonas sp. PG2982	*Pseudomonas aeruginosa* ATCC 9027	–	Sarcosine, phosphate, glycine, and formaldehyde	Utilization of glyphosate as sole phosphorus source	Moore et al. (1983)

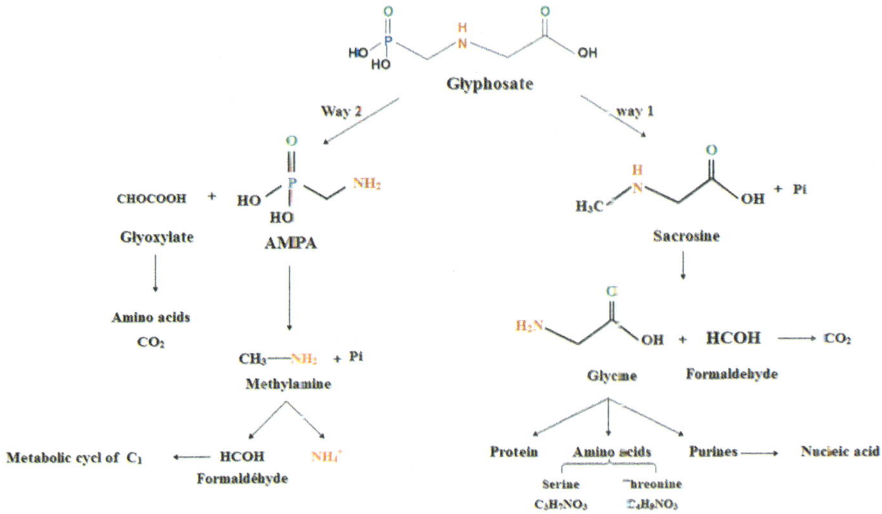

Fig. 17.9 Biodegradation mechanism of glyphosate (Zhan et al. 2013)

metabolic pathway for the degradation of glyphosate by *Pseudomonas* sp. PG2982, isolated from the soil, and Pipke et al. (1987b), worked on the *Arthrobacter* sp. strain, GLP-1 report. The study of *Pseudomonas* sp. PG2982 was the first to show this metabolic pathway for glyphosate by a bacterium. The second degradation mechanism for glyphosate contributes to the development of AMPA; this pathway includes the transformation of glyphosate into AMPA through the loss of a C2 unit (glyoxylate). This compound is then dephosphorylated by C-P lyase and transformed into methylamine, followed by formaldehyde (Singh et al. 2020).

Enzymes Involved in the Degradation of Glyphosate

The common feature of all the reactions involved in microorganism degradation is that they are all catalyzed by enzymes, most of which are intracellular. This means the reactions take place inside microbial bodies and are accompanied by pesticide absorption. In glyphosate biodegradation, glyphosate oxidoreductase (GOX) is the main enzyme for the degradation of glyphosate to AMPA via the cleavage of the C-N bond. Thus, phosphonoacetaldehyde hydrolase, phosphonoacetate hydrolase, and phosphoenolpyruvate hydrolase are the four catabolic C-P-binding enzymes of C-P lyase distinguished by degrading glyphosate bacteria (Villarreal-Chiu 2012). However, only C-P lyase can detach the bond of C-P from glyphosate, whereas the other three enzymes are extremely restricted to their own substrate (Bujacz et al. 1995; Zhan et al. 2018). The C-P bond of glyphosate is stable for hydrolysis and chemolysis and photolysis. The C-P lyase complex with a strong specificity for glyphosate is, therefore, necessary for the degradation of glyphosate via the cleavage of the inactivated C-P bond and the formation of sarcosine. Extremely selective C-P lyase for glyphosate has not been characterized at a genetic and biochemical level (Sviridov et al. 2014).

Extraction and Analysis of Glyphosate

In the background of environmental problems, pesticides to be analyzed, such as glyphosate, can be found in a number of matrices. In general, pesticides are more or less directly related to their constituents, regardless of the model. They are surrounded by a variety of organic and inorganic molecules from the media themselves or by adjuvants employed throughout the application of the products.

Pesticide extraction in different environmental matrices is an important and crucial phase in the pesticide analysis (Karasali et al. 2016), the aim of which is twofold: the first is to remove pesticides from their matrices, and the second is to isolate them from other substances in order to obtain solutions for their chemical analysis. The extraction shall be carried out using solvents selected according to the properties of pesticides and matrices (Narenderan et al. 2019).

The lack of sufficient chemical groups in glyphosate molecules (e.g., chromophores or fluorophores) hinders the detection of glyphosate molecules through traditional detectors. Moreover, due to its ionic character, low volatility, and low mass, it is difficult to decide the low concentrations needed for residue analysis (Ibáñez et al. 2005). Determination of soil residues at a low level of concentration (for example, less than 0.1 mg/kg) is often difficult due to the complexity of this matrix sample (Ibáñez et al. 2005). Owing to the extremely polar existence of glyphosate, the extraction of organic solvents from environmental matrices is almost difficult.

Most of the methods developed for glyphosate and AMPA analysis need derivatization procedures to enable gas chromatography or high-performance liquid chromatography (HPLC) analysis (Wang et al. 2016). GC/MS procedures include the derivatization of various reagents (Ibáñez et al. 2005). An analytical approach has been established and validated for the rapid and rigorous determination of low concentrations of glyphosate and its primary metabolite AMPA in soil by LC-MS/MS. The preparation of the sample involved a 9-fluorenyl-methylchloroformate (FMOC-Cl) derivatization stage until it was determined to improve the chromatographic conduct of glyphosate and AMPA. The process meets the criteria for excellent sensitivity and unmistakable confirmation of residues identified using two MS/MS transitions for each compound (Karasali et al. 2019).

17.5.5.2 Paraquat

Definition and Nomenclature

Paraquat is a weed killer that affects nearly all plant organisms (Tu and Bollen 1968). It prevents the respiration and photosynthesis cycles. This substance is rapidly absorbed by the leaves and transferred by the sap. However, this translocation happens only in the direction of plant organs where the movements of the sap are necessary. The flagship product of Syngenta, Gramoxone, is the most popular trade name for paraquat, although several producers still market herbicides under many different terms.

Historical Overview of Paraquat

Paraquat was first synthesized by Weidel and Russo in 1882 and used as an herbicide in 1955 at the Jealott's Hill research center (Sousa et al. 2020). Paraquat is a herbicidal pesticide developed for commercial purposes as early as 1961 by Imperial Chemical Companies (ICI), which later became Syngenta and was first used in agriculture in 1962 (US EPA 1987). After glyphosate, paraquat is one of the most commonly used herbicides in the world (Maldani et al. 2017). In 2003, the European Union approved paraquat by registering it in Annex I to Directorate 91/414/EEC by Direction 2003/112/EC. Despite its toxicity to humans and the environment marketing authorization was issued. Paraquat is officially outlawed in Europe. The European community has revoked the directive authorizing paraquat, arguing that the authorization protocol did not take adequate account of the correlation between the herbicide and Parkinson's disease.

Characteristics

Paraquat (1,1′-dimethyl-4,4′-bipyridinium) is an active herbicidal agent (photosynthetic inhibitor) belonging to the quarterly ammonium family (Bipyridiles) (Fig. 17.10, Table 17.3). Paraquat is decomposed at about 300 °C. It is soluble in an acidic or neutral solution and unstable in a simple medium since it is rapidly hydrolyzed (Table 17.3).

Mode of Action

Paraquat is a fast-acting, nonselective contact herbicide absorbed by the vegetation. It destroys plant tissue by disrupting photosynthesis and breaking up cell membranes, enabling water to escape, contributing to rapid leaf drying (Fig. 17.11). It may even be transported within the plant, which raises the risk of

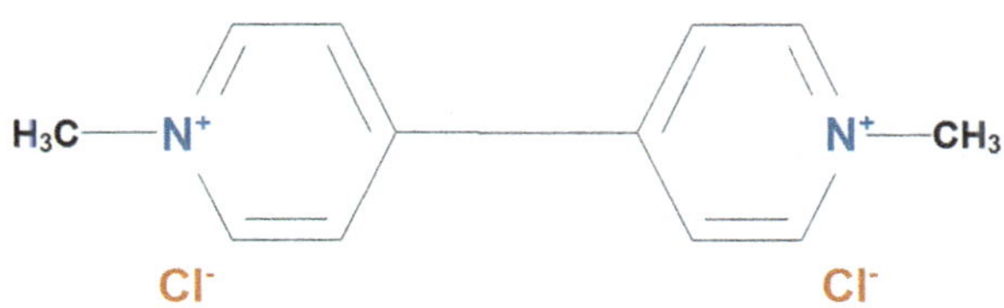

Fig. 17.10 Structure of the paraquat molecule

Table 17.3 Physicochemical characteristics of paraquat

Chemical name	1,1′-Dimethyl-4,4′-bipyridinium
Chemical family	Bipyridiles/bipyridines
Chemical formula	$C_{12}H_{14}C_{12}N_2$
Density	1.25 g/cm^3
Molecular weight	257.15 g/mol
Melting point	300 °C
Solubility	Water, alcohol

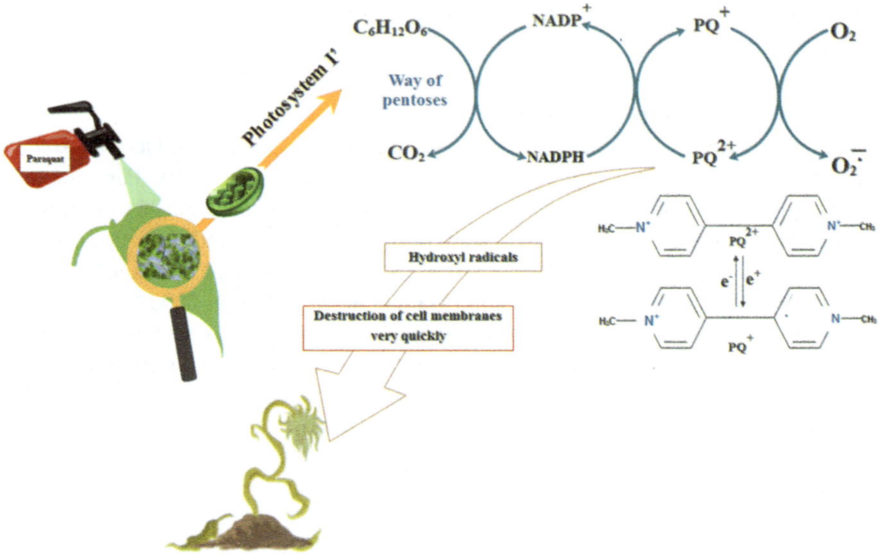

Fig. 17.11 Mechanism of action of paraquat (Bonneh-Barkay et al. 2005)

residues (Pomeroy 2012). Studies have shown that paraquat molecules transmit electrons from iron-sulfur centers to chloroplast photosystem I. As chlorophyll captures energy from sunlight, it is transmitted as an electron flow through the "Photosystem I." Reduced paraquat molecules combine with oxygen to create superoxides (O_2) that generate hydroxyl radicals that invade the plant's cell membranes (Bonneh-Barkay et al. 2005). Generally, paraquat diverts this flow, creating extremely reactive free radicals that easily kill cell membranes, reverse their material, and manifest themselves as yellowing and drying (Cohen and Doherty 1987). The leaves turn brown, often less than 30 min after application, and in the case of extreme sunlight, complete drying is observed after a few days.

Paraquat in Soil

Paraquat and Soil Texture

In soil pesticide behavior, a specific parameter, the texture of the soil, must be tested. Pesticides and sorption do not occur in the same way depending on soil composition. Paraquat is no exception to this state of affairs. In their studies on the impact of activated clay particles on the adsorption of paraquat from aqueous solution, Tsai et al. (2003) observed that the size of the clay particles played a key role in the adsorption potential and the rate of adsorption. It seems obvious that the mobility of the paraquat remains closely associated with the texture of the soil. Thus, in soil with a prevailing silt grain, paraquat is not or is not a tiny mobile. This condition is often found in soils with a silky texture and clay loam silt. On the other hand, it is mildly mobile in sandy soil (Damanakis et al. 1970; Amondham et al. 2006; Sarkar et al. 2020).

Sorption and Desorption of Paraquat

Despite mobility and behavior, the sorption and desorption of paraquat in soils depend on the quantity and relative quality of clay minerals and organic matter (Knight and Denny 1970; Gondar et al. 2012). It is useful to realize that the behavior of the adsorption exhibits variability based on the quality of the clay that makes up the clay-humic complex. Khan (1974) and Gondar et al. (2012) argued that load-dependent controls absorb soil pH. Findings of Weber et al. (1965), indicate that, in the case of kaolinite, adsorption happens on the edges or faces of clay particles rather than in intermediate locations. The absorption of paraquat can rely on other factors in the soil. Thus, Weber et al. (1965) and Amondham et al. (2006) showed that oxidized minerals, such as iron oxides, are the other factors likely to impact the adsorption/desorption of paraquat with comparatively little effect. Weber and Weed (1968) showed that the temperature and duration of exposure did not greatly impact kaolinite or montmorillonite sorption of paraquat.

The correlation between the absorption of paraquat and the texture and some elements of the soil is not often known. Indeed, in their numerous experiments, Pateiro-Moure et al. (2009) have shown that the existence of organic matter does not inherently improve the adsorption of paraquat in some soils rich in clay. They concluded that the adsorption of paraquat by agricultural soils improved when organic matter was eliminated by soil treatment with H_2O_2. It follows from this finding that there is a simple explanation based on the fact that, in the presence of an organic matter-clay mineral complex, organic matter obstructs some of the possible sites offered by clay minerals for the binding of paraquat.

Paraquat Transport in Soil

It is important to look at the conditions of movement on the ground. Studies on soil adsorption/desorption have shown that there are major repercussions for the fate of paraquat in soil and surface waters. Research of the 1980s outlined the numerous fundamental processes for the fate and transport of paraquat in soils. Studies explained the reliance of these mechanisms on soil organic matter and clay content. This association between transport pathways for paraquat and unique soil parameters, such as soil organic matter, has been verified by Khan (1974), Gondar et al. (2012), Zbair et al. (2019), and Sarkar et al. (2020). The importance of transporting paraquat through the soil profile has been the subject of several experiments and has been questioned by several authors. Khan et al. (1975) observed that herbicide movement would occur as it is adsorbed to colloidal clays or as organic matter filled with paraquat. To further understand this phenomenon, studies performed by Vinten et al. (1983) on a soil column found that the suspension of Li-montmorillonite transported more than 50% of 14C-paraquat added to sandy loam soils at a depth of 12 cm. Leonard et al. (1979) experiments in Georgian Piedmont concentrated on the mechanical and physical properties of paraquat. The author's studies have shown that saltwater or wind erosion of paraquat-laden sediments is the only possible method for transferring paraquat to surface waters. Through a synthesis of studies and experience explaining the transport of paraquat in the soil (Vinten et al. 1983; Amondham et al. 2006; Zbair et al. 2019; Sarkar et al. 2020), it can be shown that the nature of the soil is fundamental, which explains the optimal conditions for the transport of the herbicide in the substrate by the complexity of the colloid bonds and the interaction between them.

Presence of Paraquat and Its Metabolites in Water Sources

The great flexibility of paraquat has contributed to overexploitation in a variety of regions. While paraquat is typically inactivated by photodegradation and nonreversible adsorption of clays or organic materials, several studies have documented the existence of this herbicide in water supplies (Ardiwinata et al. 2019; Li et al. 2020b). Water contamination can occur, for example, by vertical transport of dissolved paraquat colloids collected through the soil. Moreover, its excellent hydrolysis resistance and its high water solubility greatly lead to an elevated likelihood of water leakage (Badroo and Nandurkar 2019; Goel and Singh 2019). The involvement of paraquat in drinking water supplies may pose a serious threat to human health due to its acute toxicity, which has been the key reason for its ban in many countries. Inexpensive and highly efficient materials capable of extracting this herbicide from the water have become increasingly important. It is one of the key ways of exposing humans to paraquat.

Impact of Paraquat on Soil Microflora

The presence of bacteria in the soil leads to the fertility of the soil and is an important connection in pedogenesis. The existence of pesticides such as paraquat may be seriously influenced by the bacterial community and the dynamism of these bacteria as a major factor in soil productivity (Maldani et al. 2018). Paraquat can have detrimental consequences for the organisms responsible for the nutrient cycle. In order to elucidate this effect of herbicides on the soil trophic chain, Gadkari (1988) examined the effects of paraquat on the mixed culture of nitrifying bacteria in the aqueous laboratory method. This researcher observed that paraquat-induced total inhibition of ammonium and nitrate oxidation. To obtain this effect, two doses of 5 μg/mL and 10 μg/mL were used. These two concentrations had the benefit of completely inhibiting ammonium oxidation for more than 40 days. In addition to inhibition of ammonium, it has been observed that nitrate oxidation has also been completely inhibited at the same concentrations for 28 days. This impact persisted at lower doses of 1 μg/mL, where ammonium and nitrate oxidation were found to be prevented for more than 18 days. This experience demonstrates that the application of paraquat, even at concentrations often deemed poor, has a substantial effect on the proper functioning of the trophic chain of the soil. Its usage must be justified so as not to affect the soil's vitality and fertility. Other experiments also studied the impact of paraquat on soil microorganisms. According to Martin et al. (2001), who examined the effect of paraquat on symbiotic nitrogen-fixing bacteria (*Rhizobium* sp.). Only 17.14% of the strains evaluated were immune to paraquat up to 400 ppm. Other experiments have shown that paraquat does not impact soil microflora (Dennis et al. 2018).

Impact of Paraquat on Human Health

Paraquat poisoning as a result of oral ingestion was first described in 1966. It tends to be highly harmful to humans. In this regard, various instances of disease and/or death have been recorded (Hayes et al. 1991). The dosage of lethal ingestion of paraquat in humans is 35 mg/kg (Fonger et al. 2014). A limit of 3.5 mg/h can be ingested through the skin or respiratory tract without injury (Wagner 1983). Ingestion of paraquat frequently induces burning sensations in the mouth and throat, accompanied by inflammation of the gastrointestinal tract, culminating in stomach discomfort, lack of appetite, nausea, vomiting, and diarrhea. Paraquat is drained by the digestive tract within 2 h of digestion. Next, paraquat accumulates in body organs, such as the liver, kidneys, lungs, and central nervous system (Gil et al. 2014). Some indicators of toxicity include thirst, shortness of breath, and fast heart rate. Such toxic symptoms include respiratory dysfunction, lung sores, and liver injury. Any symptoms may not develop before a few days after exposure (Bourée and Ensaf 2016). The most harmful results of paraquat are typically found in workers exposed to this herbicide. The effects examined are moderate liver injury, renal tubular necrosis, and pulmonary fibrosis. Above everything, pulmonary fibrosis is liable for mortality, and unintentional intake of paraquat may even contribute to serious systemic toxicity (Garnier et al. 2003). The oxidative and free radical processing properties of paraquat will contribute to oxidative stress in several organs, which would facilitate the onset of diseases such as cancer or pulmonary inflammation (Tyagi and Singh 2020). According to the findings of Ahmadian et al. (2020) on lymphocytes, paraquat may be involved in peripheral lymphocyte apoptosis via a cascade of cellular events that are closely related to oxidative stress. Several reports have demonstrated that there is a clear link between paraquat and Parkinson's disease (Tanner et al. 2011; Tamano et al. 2019) (Fig. 17.12).

Paraquat Degradation

Paraquat can be degraded by exposure to natural or chemical influences or also biological factors. As exposed to ultraviolet light, sunlight, and soil microorganisms, paraquat degrades to fewer harmful products than the parent compound (Huang et al. 2019). However, since paraquat quickly absorbs soil particles, it prevents sunlight decomposition. The US Environmental Protection Agency (US EPA 1987) report showed that paraquat sprayed on sandy soil surfaces would not decay until the soil had been irradiated with natural sunlight for 2 years.

Microorganisms Degrading Paraquat

In addition to the degradation by light, it is important to realize that this process may also be achieved in a biological way. Based on several studies, a broad variety of bacteria present in soil solution can slowly degrade the bioavailable paraquat (Table 17.4). Other laboratory and field studies and incubation experiments have verified and concluded that microbial degradation induces generally slow

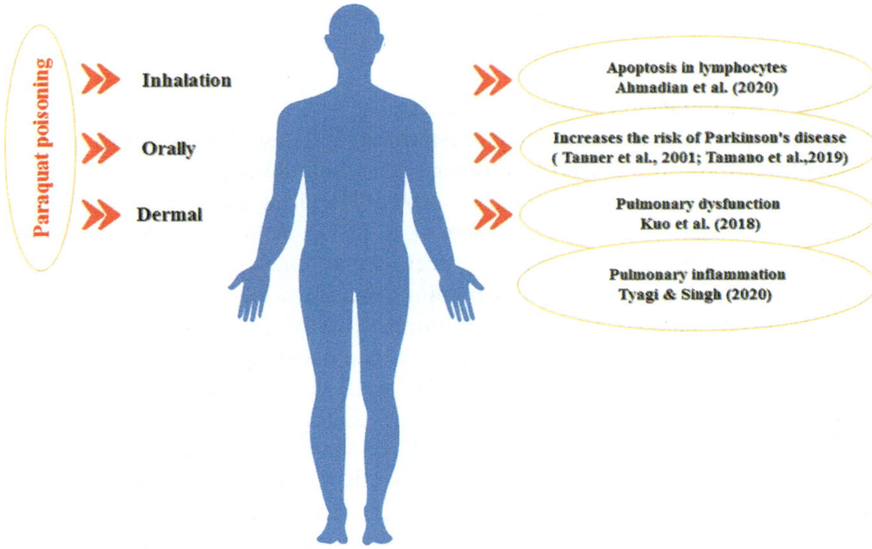

Fig. 17.12 Effects of paraquat on human health

degradation of paraquat in its bioavailable shape, as shown by the researchers (Funderburk and Bozarth 1967) in their numerous studies on the mechanisms of degradation of paraquat. The findings of the experiment by Teerakun et al. (2020) aimed at testing the optimization conditions of paraquat biodegradation by the microbial consortium showed that the consortium had bacteria having similarity with *Sphingomicrobium marinum* (97%), *Ferrovibrio xuzhouensis* (93%), *Azospirillum lipoferum* (93%), *Altererythrobacter xinjiangensis* (94%), and *Xanthobacter autotrophicus* (92%).

Biodegradation Mechanism of Paraquat

The mechanism for biodegradation of paraquat has been identified as seen in Fig. 17.13 (Huang et al. 2019); the first step in transforming paraquat is demethylation to monoquat by microbial action. A new oxidative cleavage of the monoquat cycle forms the 4-carboxy-1-methylpyridinium ion in the next step. Pyridinium carbons are emitted as CO_2 through 14C marking, and 4-carboxy-1-methylpyridinium ion is readily converted to methylamine and CO_2 through microbial activity in soils. Methylamine can be used as a source of nitrogen and carbon for microbial growth. During paraquat biodegradation, ring fragmentation compounds identified as three carboxylic acids were oxalate, formate, and succinate, while methylamine and carbon dioxide were identified as the ultimate metabolites. However, paraquat degradation enzymology is seldom reported, and no other intermediate has been described. Microorganisms that degrade paraquat can use the product of pathways upstream of cellular energy through glycolytic and tricarboxylic acid pathways. Further studies on the biodegradation of paraquat are required to detect

Table 17.4 Paraquat-degrading microorganisms

Microorganisms	Source	Gram	Paraquat use/results	References
Providencia rettgeri	Soil, Nigeria	–	Use of paraquat as nutrients	Obuotor et al. (2016)
Pseudomonas sp.	Soil, Nigeria	–	Use of paraquat as nutrients	Obuotor et al. (2016)
Clostridium prazmowski	Soil contaminated with paraquat	+	80.3% degradation of paraquat after 3 days	Han et al. (2014)
Sporohalobacter orenetal	Soil contaminated with paraquat	–	86.2% degradation of paraquat after 3 days	Han et al. (2014)
Enterobacter cloacae PQ02	Soil, city Sanya, China	–	95% degradation of paraquat (50 mg/L) in 7 days in the presence of an additional electron donor such as sucrose	Wu et al. (2013)
Pseudomonas putida	Industrial Biotechnology Innovation Center, Colombia	–	47.3% of paraquat (69.76 mg/L) was degraded in 3 days in the presence of 25% of nutrients, about 95% of degradation of paraquat (69.76 mg/L) in 3 days in the presence of 15 g/L of activated charcoal	Kopytko et al. (2002)
Aerobacter aerogenes	Soils	–	Use of paraquat as the sole source of carbon and nitrogen	Tu and Bollen (1968)
Agrobacterium tumefaciens	Soils	–	Use of paraquat as the sole source of carbon and nitrogen	Tu and Bollen (1968)
Pseudomonas fluorescens	Soils	–	Use of paraquat as the sole source of carbon and nitrogen	Tu and Bollen (1968)
Bacillus cereus	Soils	–	Use of paraquat as the sole source of carbon and nitrogen	Tu and Bollen (1968)

the typical metabolites and enzymes responsible for the transformation of the various intermediates. Degradation by biological means or not causes one to raise concerns regarding the lifetime of paraquat in the soil. The most widely recorded half-life values vary from roughly 7 years (Hance et al. 1980).

Extraction and Analysis of Paraquat

Several studies have centered on validation and creation of methods for measuring paraquat in various matrices (Shivhare and Gupta 1991; Maya et al. 2011; Pizzutti et al. 2016; Rashidipour et al. 2019). Taking into account their physicochemical

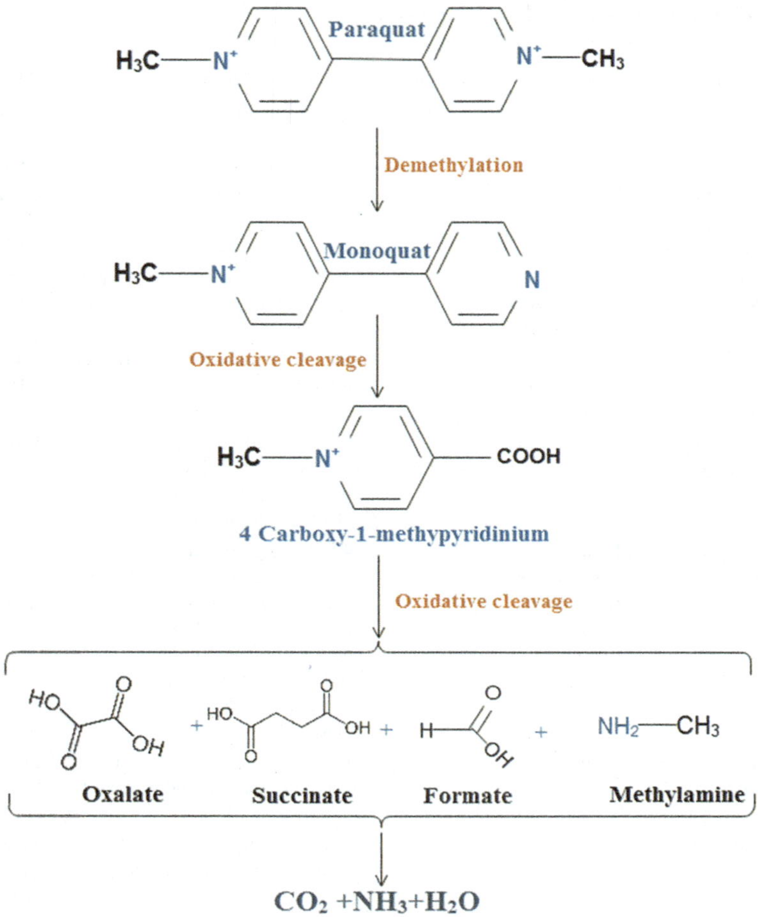

Fig. 17.13 Biodegradation metabolism of paraquat (Huang et al. 2019)

properties, which preclude their determination by multi-residue methods, several methods and analytical techniques for the determination of paraquat have been documented in recent years (Pizzutti et al. 2016). Spectrophotometric, spectrofluorimetric, potentiometric (Vytvras and Šimĭčkovă-Štajerova 1989; Maya et al. 2011; Yao et al. 2013), and methods using GC with solid-phase extraction and solid-phase micro-extraction (Almeida and Yonamine 2007), have been used. Most of the experiments applied concerned environmental matrices, such as water (Maya et al. 2011; Shawish et al. 2012; Yao et al. 2013) and biological matrices, such as blood and urine (Almeida and Yonamine 2007; Shawish et al. 2012). LC coupled with mass spectrometry (LC-MS/MS) and ionization is now the preferred method owing to parameters such as specificity and selectivity (Pizzutti et al. 2016).

17.6 Conclusion

Soil is one of the most susceptible ecosystems to contamination by chemical inputs, such as pesticides. It is easily polluted, although it can take many years to restore again. Glyphosate and paraquat are the most powerful and commonly used herbicides in the world. Unfortunately, they are among the phytosanitary products that induce contamination of natural resources, such as soil and water. Generally, the scientific work on these two herbicides is restricted to laboratory studies owing to analytical difficulties in the field. Bioremediation or biological decontamination of areas contaminated by a pesticide is primarily focused on the isolation of strains with bioremediation ability. Biodegradation optimization is area of focus along with the studies on the impact of abiotic factors on microbes. Evaluating the ecological threats of a herbicide requires monitoring the evolution of the active molecule in tandem with the microbial biomass of contaminated soil

References

Abraham J, Silambarasan S, Logeswari P (2014) Simultaneous degradation of organophosphorus and organochlorine pesticides by bacterial consortium. J Taiwan Inst Chem Eng 45:2590–2596

Abubakar Y, Tijjani H, Egbuna C, Adetunji CO, Kala S, Kryeziu TL et al (2020) Pesticides, history, and classification. In: Egbuna C, Sawicka B (eds) Natural remedies for pest, disease and weed control. Elsevier, Amsterdam, pp 29–42

Ahmadian E, Eftekhari A, Kavetskyy T, Khosroushahi AY, Turksoy VA, Khalilov R (2020) Effects of quercetin loaded nanostructured lipid carriers on the paraquat-induced toxicity in human lymphocytes. Pestic Biochem Physiol 167:1–8

Aislabie J, Lloyd-Jones G (1995) A review of bacterial-degradation of pesticides. Soil Res 33:925–942

Allsop M, Huxdorff C, Johnston P, Santillo D, Thompson K (2015) Pesticides and our health—a growing concern. Greenpeace Research Laboratories (RLG), School of Biosciences, Innovation Centre Phase 2, University of Exeter, Exeter, p 56

Almeida RM, Yonamine M (2007) Gas chromatographic—mass spectrometric method for the determination of the herbicides paraquat and diquat in plasma and urine samples. J Chromatogr B 853:260–264

Amghar D (2019) Contribution à l'étude de l'influence d'un herbicide, le Glyphosate et d'un fongicide, le Mancozèbe, sur la germination, la croissance et la physiologie de deux céréales: *Hordeumvulgare L. et Avenasativa L.* PhD Thesis, Mouloud Mammeri University, Tizi Ouzou, Algeria

Amondham W, Parkpian P, Polprasert C, Delaune RD, Jugsujinda A (2006) Paraquat adsorption, degradation, and remobilization in tropical soils of Thailand. J Environ Sci Health Pt B 41:485–507

Andreu V, Picó Y (2004) Determination of pesticides and their degradation products in soil: critical review and comparison of methods. Trends Anal Chem 23:772–789

Ardiwinata AN, Harsanti ES, Kurnia A, Sulaeman E, Fauriah R, Paputri DMW (2019) Contamination of paraquat residues in soil and water from several provinces in Indonesia. In: Proceeding of AIP Conference 2120, p 40024

Badroo IA, Nandurkar HP (2019) Paraquat dichloride induced carbohydrate metabolism alterations in fresh water Teleost fish *Channa punctatus* (Bloch, 1793) after acute Exposure. In: Proceeding of National Conference Special Issue, p 42

Balthazor TM, Hallas LE (1986) Glyphosate-degrading microorganisms from industrial activated sludge. Appl Environ Microbiol 51(2):432–434

Barriuso E, Calvet R (1992) Soil type and herbicides adsorption. Int J Environ Anal Chem 46:117–128

Barriuso E, Calvet R, Schiavon M, Soulas G (1996) Les pesticides et les polluants organiques des sols. Etude Gest des sols 3:279–296

Benachour N, Sipahutar H, Moslemi S, Gasnier C, Travert C, Seralini GE (2007) Time-and dose-dependent effects of roundup on human embryonic and placental cells. Arch Environ Contam Toxicol 53:126–133

Benbrook CM (2012) Impacts of genetically engineered crops on pesticide use in the US-the first sixteen years. Environ Sci Eur 24:24

Benslama O (2014) Isolement et caractérisation des bactéries capables de dégrader l'herbicide Glyphosate et optimisation des conditions de culture pour une dégradation plus efficace. PhD Thesis, Constantine 1 University, Algeria

Benslama O, Boulahrouf A (2016) High-quality draft genome sequence of Enterobacter sp. Bisph2, a glyphosatedegrading bacterium isolated from a sandy soil of Biskra, Algeria. Genomics Data 8:61–66

Bento CPM, Yang X, Gort G, Xue S, van Dam R, Zomer P et al (2016) Persistence of glyphosate and aminomethylphosphonic acid in loess soil under different combinations of temperature, soil moisture and light/darkness. Sci Total Environ 572:301–311

Bonneh-Barkay D, Reaney SH, Langston WJ, Di Monte DA (2005) Redox cycling of the herbicide paraquat in microglial cultures. Mol Brain Res 134:52–56

Borggaard OK, Gimsing AL (2008) Fate of glyphosate in soil and the possibility of leaching to ground and surface waters: a review. Pest Manag Sci Former Pest Sci 64:441–456

Botta F, Lavison G, Couturier G, Alliot F, Moreau-Guigon E, Fauchon N et al (2009) Transfer of glyphosate and its degradate AMPA to surface waters through urban sewerage systems. Chemosphere 77:133–139

Boulet A (2005) Lutte contre la pollution des eaux par les pesticides utilisés en zones non agricoles: analyse et synthèse des actions engagées et recommandations. PhD Thesis, Inst Natl Agron Paris, France

Bourée P, Ensaf A (2016) Guyane: intoxication fatale au paraquat. Rev Francoph des Lab 478:67–70

Brodskiy ES, Shelepchikov AA, Feshin DB, Agapkina GI, Artukhova MV (2016) Content and distribution pattern of dichlorodiphenyltrichloroethane (DDT) in soils of Moscow. Moscow Univ Soil Sci Bull 71:27–34

Bujacz B, Wieczorek P, Krzysko-Lupicka T, Golab Z, Lejczak B, Kavfarski P (1995) Organophosphonate utilization by the wild-type strain of *Penicillium notatum*. Appl Environ Microbiol 61:2905–2910

Bus JS (2015) Analysis of Moms Across America report suggesting bioaccumulation of glyphosate in US mother's breast milk: implausibility based on inconsistency with available body of glyphosate animal toxicokinetic, human biomonitoring, and physico-chemical data. Regul Toxicol Pharmacol 73:758–764

Calvet R (2005) Les pesticides dans le sol: conséquences agronomiques et environnementales. France Agricole Editions, 637 p

CAPM (2010) Centre Anti Poison du Maroc. Toxicologie Maroc. vol 4. 1er trimestre. 15 p

Carlisle SM, Trevors JT (1988) Glyphosate in the environment. Water Air Soil Pollut 39:409–420

Carson R (2002) Silent spring. Houghton Mifflin Harcourt en English—40th anniversary ed. 1st Mariner Books ed, 287 p

Chang CY, Peng YC, Hung DZ, Hui WH, Yang DY, Lin TJ (1999) Clinical impact of upper gastrointestinal tract injuries in glyphosate-surfactant oral intoxication. Hum Exp Toxicol 18: 475–478

Chen C, Guo W, Ngo HH (2019) Pesticides in stormwater runoff—a mini review. Front Environ Sci Eng 13:72

Climent MJ, Coscollà C, López A, Barra R, Urrutia R (2019) Legacy and current-use pesticides (CUPs) in the atmosphere of a rural area in central Chile, using passive air samplers. Sci Total Environ 662:646–654
Cohen GM, Doherty MA (1987) Free radical mediated cell toxicity by redox cycling chemicals. Br J Cancer Suppl 8:46
Crop Life Maroc (2018) Article sur le marché des produits phytosanitaires au Maroc. http://www.croplife.ma/Pdf/Article_sur_le_marche_phytosanitaire_au_Maroc.mai_2011.pdf. Accessed 7 Oct 2018
Cruz JM (2016) Etude de la contamination par les pesticides des milieux eau , air et sols : developpement de nouveaux outils et application à l 'estuaire de la Gironde. PhD Thesis, University Of Bordeaux, Bordeaux, France
Cycoń M, Zmijowska A, Wójcik M, Piotrowska-Seget Z (2013) Biodegradation and bioremediation potential of diazinon-degrading *Serratia marcescens* to remove other organophosphorus pesticides from soils. J Environ Manage 117:7–16
D'Archivio AA, Maggi MA, Odoardi A, Santucci S, Passacantando M (2018) Adsorption of triazine herbicides from aqueous solution by functionalized multiwall carbon nanotubes grown on silicon substrate. Nanotechnology 29:65701
Dalton H, Stirling DI (1982) Co-metabolism. Philos Trans R Soc Lond B Biol Sci 297:481–496
Damanakis M, Drennan DSH, Fryer JD, Holly K (1970) The adsorption and mobility of paraquat on different soils and soil constituents. Weed Res 10:264–277
Dayan FE, Barker A, Bough R, Ortiz M, Takano H, Duke SO (2019) Herbicide mechanisms of action and resistance. Plants 8:341
De Roos AJ, Blair A, Rusiecki JA, Hoppin JA, Svec M, Dosemeci M et al (2005) Cancer incidence among glyphosate-exposed pesticide applicators in the Agricultural Health Study. Environ Health Perspect 113:49–54
Dennis PG, Kukulies T, Forstner C, Orton TG, Pattison AB (2018) The effects of glyphosate, glufosinate, paraquat and paraquat-diquat on soil microbial activity and bacterial, archaeal and nematode diversity. Sci Rep 8:2119
Deschomets G (2001) Les adjuvants du glyphosate. Phytoma, la défense des végétaux 541:14–16
Dick RE, Quinn JP (1995) Control of glyphosate uptake and metabolism in Pseudomonas sp. 4ASW. FEMS Microbiol Lett 134(2–3):177–182
Dill GM (2005) Glyphosate-resistant crops: history, status and future. Pest Manag Sci 61(3): 219–224
Dominguez A, Brown GG, Sautter KD, De Oliveira CMR, De Vasconcelos EC, Niva CC et al (2016) Toxicity of AMPA to the earthworm *Eisenia andrei* Bouché, 1972 in tropical artificial soil. Sci Rep 6:19731
Druille M, Cabello MN, Omacini M, Golluscio RA (2013) Glyphosate reduces spore viability and root colonization of arbuscular mycorrhizal fungi. Appl Soil Ecol 64:99–103
Du J, Gridneva Z, Gay MC, Trengove RD, Hartmann PE, Geddes DT (2017) Pesticides in human milk of Western Australian women and their influence on infant growth outcomes: a cross-sectional study. Chemosphere 167:247–254
Duke SO (1990) Overview of herbicide mechanisms of action. Environ Health Perspect 87:263–271
Duke SO (2018) The history and current status of glyphosate. Pest Manag Sci 74:1027–1034
Duke SO (2020) Glyphosate: environmental fate and impact. Weed Sci 68:201–207
Duke SO, Powles SB (2008) Glyphosate: a once-in-a-century herbicide. Pest Manag Sci Former Pest Sci 64:319–325
Duke SO, Powles SB (2009) Glyphosate-resistant crops and weeds: now and in the future. AgBioForum 12:346–357
Durand C (2007) Les Biotechnologies au feu de l'éthique. L'Harmattan, Paris
Eddleston M (2020) Poisoning by pesticides. Medicine 48:214–217
EFSA, European Food Safety Authority (2021) Glyphosate: application for renewal. Available via DIALOG. https://www.efsa.europa.eu/en/topics/topic/glyphosate. Accessed 01 May 2021

Elbana T, Gaber HM, Kishk FM (2019) Soil chemical pollution and sustainable agriculture. In: El-Ramady H, Alshaal T, Bakr N, Elbana T, Mohamed E, Belal AA (eds) The soils of Egypt. World soils book series. Springer, Cham, pp 187–200

Ensley SM (2018) Pyrethrins and pyrethroids. In: Gupta RC (ed) Veterinary toxicology. Academic Press, Elsevier, London, pp 515–520

Ermakova IT, Shushkova TV, Sviridov AV, Zelenkova NF, Vinokurova NG, Baskunov BP, Leontievsky AA (2017) Organophosphonates utilization by soil strains of Ochrobactrum anthropi and Achromobacter sp. Arch Microbiol 199(5):665–675. https://doi.org/10.1007/s00203-017-1343-8

Eto M (2018) Organophosphorus pesticides. CRC, Boca Raton, FL, 399 p

Fan J, Yang G, Zhao H, Shi G, Geng Y, Hou T, Tao K (2012) Isolation, identification and characterization of a glyphosate-degrading bacterium, Bacillus cereus CB4, from soil. J Gen Appl Microbiol 58(4):263–271

Fang Y, Nie Z, Die Q, Tian Y, Liu F, He J et al (2017) Organochlorine pesticides in soil, air, and vegetation at and around a contaminated site in southwestern China: concentration, transmission, and risk evaluation. Chemosphere 178:340–349

Feng JC, Thompson DG (1990) Fate of glyphosate in a Canadian forest watershed. 2. Persistence in foliage and soils. J Agric Food Chem 38:1118–1125

Firdous S, Iqbal S, Anwar S (2017a) Optimization and modeling of glyphosate biodegradation by a novel Comamonas odontotermitis P2 through response surface methodology. Pedosphere. https://doi.org/10.1016/S1002-0160(17)60381-3

Firdous S, Iqbal S, Anwar S, Jabeen H (2017b) Identification and analysis of 5-enolpyruvylshikimate-3-phosphate synthase (EPSPS) gene from glyphosate resistant Ochrobactrum intermedium Sq20. Pest Manag Sci 74:1184–1196. https://doi.org/10.1002/ps.4624

Firdous S, Iqbal S, Anwar S, Jabeen H (2018) Identification and analysis of 5 enolpyruvylshikimate 3 phosphate synthase (EPSPS) gene from glyphosate resistant Ochrobactrum intermedium Sq20. Pest Manag Sci 74(5):1184–1196

Fonger GC, Hakkinen P, Jordan S, Publicker S (2014) The National Library of Medicine's (NLM) Hazardous Substances Data Bank (HSDB): background, recent enhancements and future plans. Toxicology 325:209–216

Fournier JC, Soulas G, Parekh NR (1996) Pesticide degradation in soils. In: Tarradellas J, Bitton G, Rossel D (eds) Soil ecotoxicology. CRC, p 85

Funderburk HH, Bozarth GA (1967) Review of the metabolism and decompostion of diquat and paraquat. J Agric Food Chem 15:563–567

Gadkari D (1988) Effects of atrazine and paraquat on nitrifying bacteria. Arch Environ Contam Toxicol 17:443–447

Gao J, Liu L, Liu X, Lu J, Zhou H, Huang S et al (2008) Occurrence and distribution of organochlorine pesticides—lindane, p, p′-DDT, and heptachlor epoxide—in surface water of China. Environ Int 34:1097–1103

García-Delgado C, Marín-Benito JM, Sánchez-Martín MJ, Rodríguez-Cruz MS (2020) Organic carbon nature determines the capacity of organic amendments to adsorb pesticides in soil. J Hazard Mater 390:1–9

Garnier R, Bazire A, Chataigner D (2003) Effets sur la santé de l'utilisation professionnelle du paraquat. Arch des Mal Prof médecine du Trav 64:310–324

Gastiazoro MP, Durando M, Milesi MM, Lorenz V, Vollmer G, Varayoud J et al (2020) Glyphosate induces epithelial mesenchymal transition-related changes in human endometrial Ishikawa cells via estrogen receptor pathway. Mol Cell Endocrinol 510:110841

George J, Shukla Y (2011) Pesticides and cancer: insights into toxicoproteomic-based findings. J Proteomics 74:2713–2722

Giesy JP, Dobson S, Solomon KR (2000) Ecotoxicological risk assessment for Roundup®herbicide. In: Ware GW (ed) Reviews of environmental contamination and toxicology. Reviews of environmental contamination and toxicology, vol 167. Springer, New York, pp 35–120

Gil H, Hong JR, Jang SH, Hong SY (2014) Diagnostic and therapeutic approach for acute paraquat intoxication. J Korean Med Sci 29:1441–1449

Gimsing AL, Borggaard OK (2001) Effect of KCl and $CaCl_2$ as background electrolytes on the competitive adsorption of glyphosate and phosphate on goethite. Clays Clay Miner 49:270–275

Gimsing AL, Szilas C, Borggaard OK (2007) Sorption of glyphosate and phosphate by variable-charge tropical soils from Tanzania. Geoderma 138:127–132

Goel A, Singh O (2019) Herbicides poisoning: Paraquat and Diquat. In: Singh O, Juneja D (eds) Principles and practice of critical care toxicology. Jaypee Brothers Medical, New Delhi, p 312

Gómez Ortiz AM, Okada E, Bedmar F, Costa JL (2017) Sorption and desorption of glyphosate in mollisols and ultisols soils of Argentina. Environ Toxicol Chem 36:2587–2592

Gondar D, López R, Antelo J, Fiol S, Arce F (2012) Adsorption of paraquat on soil organic matter: effect of exchangeable cations and dissolved organic carbon. J Hazard Mater 235:218–223

Grandcoin A, Piel S, Baures E (2017) AminoMethylPhosphonic acid (AMPA) in natural waters: its sources, behavior and environmental fate. Water Res 117:187–197

Grung M, Lin Y, Zhang H, Steen AO, Huang J, Zhang G et al (2015) Pesticide levels and environmental risk in aquatic environments in China—a review. Environ Int 81:87–97

Guimont S (2005) Devenir des pesticides dans les sols en fonction de l'état d'humidité et du mode de circulation de l'eau dans le sol. PhD thesis, National Polytechnic Institute of Lorraine, Lorraine, France

Gupta M, Mathur S, Sharma TK, Rana M, Gairola A, Navani NK, Pathania R (2016) A study on metabolic prowess of Pseudomonas sp. RPT 52 to degrade imidacloprid, endosulfan and coragen. J Hazard Mater 301:250–258

Hadi F, Mousavi A, Noghabi KA, Tabar HG, Salmanian AH (2013) New bacterial strain of the genus Ochrobactrum with glyphosatedegrading activity. J Environ Sci Health B 48(3):208–213

Hall JC, Donnelly-Vanderloo MJ, Hume DJ (2018) Triazine-resistant crops: the agronomic impact and physiological consequences of chloroplast mutation. In: Duke SO (ed) Herbicide-resistant crops. CRC, Boca Raton, pp 107–126

Han X, Yuan R, Wang GQ, Zhang CJ (2014) Isolation of paraquat degrading bacteria and identification of degradation characteristics. Anhui Agricu Sci Bull 20:38–39. https://doi.org/10.16377/j.cnki.issn1007-7731.2014.08.004

Hance RJ, Byast TH, Smith PD (1980) Apparent decomposition of paraquat in soil. Soil Biol Biochem 12:447–448

Hayes WJ, Laws ER, Clarkson TW (1991) Handbook of pesticide toxicology; Classes of Pesticides, vol 2. Academic Press, San Diego, p 1576

Helander M, Saloniemi I, Saikkonen K (2012) Glyphosate in northern ecosystems. Trends Plant Sci 17:569–574

Hermansen C, Norgaard T, de Jonge LW, Moldrup P, Müller K, Knadel M (2020) Predicting glyphosate sorption across New Zealand pastoral soils using basic soil properties or Vis—NIR spectroscopy. Geoderma 360:114009

Holten R, Larsbo M, Jarvis N, Stenrød M, Almvik M, Eklo OM (2019) Leaching of five pesticides of contrasting mobility through frozen and unfrozen soil. Vadose Zo J 18:1–10

Houbraken M, van den Berg F, Butler Ellis CM, Dekeyser D, Nuyttens D, De Schampheleire M et al (2016) Volatilisation of pesticides under field conditions: inverse modelling and pesticide fate models. Pest Manag Sci 72:1309–1321

Huang Y, Zhan H, Bhatt P, Chen S (2019) Paraquat degradation from contaminated environments: current achievements and perspectives. Front Microbiol 10:1–9

IARC (2018) Monographs: GLYPHOSATE. file:///C:/Users/pc/Desktop/biblio/Ref/IARCMONOGRAPH.pdf. Accessed 28 Jul 2019

Ibáñez M, Pozo ÓJ, Sancho JV, López FJ, Hernández F (2005) Residue determination of glyphosate, glufosinate and aminomethylphosphonic acid in water and soil samples by liquid chromatography coupled to electrospray tandem mass spectrometry. J Chromatogr A 1081:145–155

Iori S, Dalla Rovere G, Ezzat L, Smits M, Ferraresso SS, Babbucci M et al (2020) The effects of glyphosate and AMPA on the mediterranean mussel *Mytilus galloprovincialis* and its microbiota. Environ Res 182:108984

Jacob GS, Schaefer J, Stejskal EO, McKay RA (1985) Solid-state NMR determination of glyphosate metabolism in a *Pseudomonas* sp. J Biol Chem 260:5899–5905

Jacob GS, Garbow JR, Schaefer J, Kishore GM (1987) Solid–state NMR studies of regulation of N-(Phosphonomethyl)glycine and glycine metabolism in *Pseudomonas* sp. strain PG2982. J Biol Chem 262:1552–1557

Jacob GS, Garbow JR, Hallas LE, Kimack NM, Kishore GM, Schaefer J (1988) Metabolism of glyphosate in *Pseudomonas* sp. strain LBr. Appl Environ Microbiol 54:2953–2958

Jaiswal DK, Verma JP, Yadav J (2017) Microbe induced degradation of pesticides in agricultural soils. In: Singh S (ed) Microbe-induced degradation of pesticides. Environmental science and engineering. Springer, Cham, pp 167–189

Kaczmarek DK, Rzemieniecki T, Marcinkowska K, Pernak J (2019) Synthesis, properties and adjuvant activity of docusate-based ionic liquids in pesticide formulations. J Ind Eng Chem 78: 440–447

Kadala PA (2011) Action des pyréthrinoïdes sur le canal sodique activé par le potentiel des neurones du système olfactif de l'abeille domestique *Apis mellifera*. PhD thesis, University Of Avignon, Avignon, France

Kah M, Brown CD (2007) Changes in pesticide adsorption with time at high soil to solution ratios. Chemosphere 68:1335–1343

Kah M, Beulke S, Brown CD (2007) Factors influencing degradation of pesticides in soil. J Agric Food Chem 55:4487–4492

Karasali H, Marousopoulou A, Machera K (2016) Pesticide residue concentration in soil following conventional and low-input crop management in a Mediterranean Agro-Ecosystem, in Central Greece. Sci Total Environ 541:130–142

Karasali H, Pavlidis G, Marousopoulou A (2019) Investigation of the presence of glyphosate and its major metabolite AMPA in Greek soils. Environ Sci Pollut Res 26:36308–36321

Kaur G (2019) Herbicides and its role in induction of oxidative stress—a review. Int J Environ Agric Biotechnol 4:995–1004

Khan SU (1974) Determination of diquat and paraquat residues in soil by gas chromatography. J Agric Food Chem 22:863–867

Khan SU (1980) Pesticides in the soil environment. Elsevier, New York, p 248

Khan SU, Marriage PB, Saidak WJ (1975) Residues of paraquat in an orchard soil. Can J Soil Sci 55(1):73–75

Khan M, Gassman M, Haque R (1976) Biodegradation of pesticides. Chemtech 6:62–69

Khedr T, Hammad AA, Elmarsafy AM, Halawa E, Soliman M (2019) Degradation of some organophosphorus pesticides in aqueous solution by gamma irradiation. J Hazard Mater 373: 23–28

Kinkela D (2011) DDT and the American century: global health, environmental politics, and the pesticide that changed the world. The University of North Carolina Press, Chapel Hill, p 272

Knight BAG, Denny PJ (1970) The interaction of paraquat with soil: adsorption by an expanding lattice clay mineral. Weed Res 10:40–48

Kopytko M, Chalela G, Zauscher F (2002) Biodegradation of two commercial herbicides (Gramoxone and Matancha) by the bacteria Pseudomonas putida. Electron J Biotechnol 5(2): 0–1

Kryuchkova YV, Burygin GL, Gogoleva NE, Gogolev YV, Chernyshova MP, Makarov OE et al (2014) Isolation and characterization of a glyphosate-degrading rhizosphere strain, *Enterobacter cloacae* K7. Microbiol Res 169:99–105

Kumar S, Kaushik G, Dar MA, Nimesh S, Lopez-Chuken UJ, Villarreal-Chiu JF (2018) Microbial degradation of organophosphate pesticides: a review. Pedosphere 28:190–208

Laitinen P, Siimes K, Rämö S, Jauhiainen L, Eronen L, Oinonen S et al (2008) Effects of soil phosphorus status on environmental risk assessment of glyphosate and glufosinate-ammonium. J Environ Qual 37:830–838

Larson SJ (2019) Pesticides in surface waters: distribution, trends, and governing factors, vol 3. CRC, Boca Raton, FL, 400 p

Lautre Y (2019) Demandons l interdiction des pesticides à base de glyphosate: La France doit respecter son engagement de sortir du glyphosate en 3 ans, au plus tard fin 2020. http // yonnelautre.fr/spip.php?article4073 Accessed 23 Sept 2020

Le Bars M, Sidibe F, Mandart E, Fabre J, Le Grusse P, Diakite CH (2020) Évaluation des risques liés à l'utilisation de pesticides en culture cotonnière au Mali. Cah Agric 29:4

Leadbetter ER, Foster JW (1959) Incorporation of molecular oxygen in bacterial cells utilizing hydrocarbons for growth. Nature 184:1428–1429

Leaper C, Holloway PJ (2000) Adjuvants and glyphosate activity. Pest Manag Sci Former Pest Sci 56:313–319

Leonard RA, Langdale GW, Fleming WG (1979) Herbicide runoff from upland piedmont watersheds—data and implications for modeling pesticide transport. J Environ Qual 8:223–229

Leong WH, Teh SY, Hossain MM, Nadarajaw T, Zabidi-Hussin Z, Chin SY et al (2020) Application, monitoring and adverse effects in pesticide use: the importance of reinforcement of Good Agricultural Practices (GAPs). J Environ Manage 260:109987

Lerbs W, Stock M, Parthier B (1990) Physiological aspects of glyphosate degradation in Alcaligenes sp. strain GL. Arch Microbiol 153(2):146–150

Li W, Lybrand DB, Xu H, Zhou F, Last RL, Pichersky E (2020a) A trichome-specific, plastid-localized *Tanacetum cinerariifolium* Nudix protein hydrolyzes the natural pyrethrin pesticide biosynthetic intermediate trans-Chrysanthemyl Diphosphate. Front Plant Sci 11:482

Li Y, Liu L, Kuang H, Xu C (2020b) Preparing monoclonal antibodies and developing immunochromatographic strips for paraquat determination in water. Food Chem 311:125897

Liu Y, Lonappan L, Brar SK, Yang S (2018) Impact of biochar amendment in agricultural soils on the sorption, desorption, and degradation of pesticides: a review. Sci Total Environ 645:60–70

Liu CM, McLean P, Sookdeo C, Cannon F (1991) Degradation of the herbicide glyphosate by members of the family rhizobiaceae. Appl Environ Microbiol 57(6):1799–1804

Mahoney K (2018) The Impact of Glyphosate Exposure on Leukemia and Lymphoma Cases in the State of Iowa, Thesis, Carthage College, Kenosha, Wisconsin, United States

Maldani M, Dekaki EM, Nassiri L, Ibijbijen J (2017) State of art on the use of pesticides in Meknes Region, Morocco. Am J Agric Sci 4:138–148

Maldani M, Ben Messaoud B, Nassiri L, Ibijbijen J (2018) Influence of paraquat on four rhizobacteria strains: *Pantoea agglomerans, Rhizobium nepotum, Rhizobium radiobacter* and *Rhizobium tibeticum*. Open Environ Sci 10:48–55

Maldani M, Aliyat FZ, Cappello S, Morabito M, Giarratana F, Nassiri L, Ibijbijen A (2021) Effect of glyphosate and paraquat on seed germination, amino acids, photosynthetic pigments and plant morphology of *Vicia faba, Phaseolus vulgaris* and *Sorghum bicolor*. Environ Sust. https:// doi.org/10.1007/s42398-021-00170-0

Mamouni A, Mansour M, Schmitt P (2018) Abiotic degradation pathways of selected pesticides in the presence of oxygen species in aqueous solutions. In: Mansour M (ed) Fate and prediction of environmental chemicals in soils, plants, and aquatic systems. CRC, Boca Raton, FL, pp 253–274

Mamy L, Barriuso E (2005) Glyphosate adsorption in soils compared to herbicides replaced with the introduction of glyphosate resistant crops. Chemosphere 61:844–855

Mañas F, Peralta L, Raviolo J, Ovando HG, Weyers A, Ugnia L et al (2009) Genotoxicity of AMPA, the environmental metabolite of glyphosate, assessed by the Comet assay and cytogenetic tests. Ecotoxicol Environ Saf 72:834–837

Martin MJ, Host GE, Lenz KE, Isebrands JG (2001) Simulating the growth response of aspen to elevated ozone: a mechanistic approach to scaling a leaf-level model of ozone effects on photosynthesis to a complex canopy architecture. Environ Pollut 115 425–436

Martinez MA, Rodríguez JL, Lopez-Torres B, Martínez M, Martínez-Larrañaga MR, Maximiliano JE et al (2020) Use of human neuroblastoma SH-SY5Y cells to evaluate glyphosate-induced effects on oxidative stress, neuronal development and cell death signaling pathways. Environ Int 135:105414

Martini SA, Phillips M (2009) Nutrition and food commodities in the 20th century. J Agric Food Chem 57:8130–8135

Martin-Laurent F, Therond O (2017) Comment les sols atténuent-ils naturellement les pesticides? Biofutur 36:54

Mawussi G (2008) Bilan environnemental de l'utilisation de pesticides organochlorés dans les cultures de coton, café et cacao au Togo et recherche d'alternatives par l'évaluation du pouvoir insecticide d'extraits de plantes locales contre le scolyte du café (*Hypothenemus hampei Ferrari*). PhD Thesis, University Of Toulouse, Toulouse, France

Maya F, Estela JM, Cerdà V (2011) Improved spectrophotometric determination of paraquat in drinking waters exploiting a Multisyringe liquid core waveguide system. Talanta 85:588–595

McAuliffe KS, Hallas LE, Kulpa CF (1990) Glyphosate degradation by*Agrobacterium radiobacter* isolated from activated sludge. J Ind Microbiol 6:219–221

Meftaul IM, Venkateswarlu K, Dharmarajan R, Annamalai P, Megharaj M (2020) Pesticides in the urban environment: a potential threat that knocks at the door. Sci Total Environ 711:134612

Mesnage R, Bernay B, Séralini GE (2013) Ethoxylated adjuvants of glyphosate-based herbicides are active principles of human cell toxicity. Toxicology 313:122–128

Moneke AN, Okpala GN, Anyanwu CU (2010) Biodegradation of glyphosate herbicide in vitro using bacterial isolates from four rice fields. Afr J Biotechnol 26:4067–4074

Moore JK, Braymer HD, Larson AD (1983) Isolation of a Pseudomonas sp. which utilizes the phosphonate herbicide glyphosate. Appl Environ Microbiol 46(2):316–320

Moreno-González D, Pérez-Ortega P, Gilbert-López B, Molina-Díaz A, García-Reyes JF, Fernández-Alba AR (2017) Evaluation of nanoflow liquid chromatography high resolution mass spectrometry for pesticide residue analysis in food. J Chromatogr A 1512:78–87

Munira S, Farenhorst A, Flaten D, Grant C (2016) Phosphate fertilizer impacts on glyphosate sorption by soil. Chemosphere 153:471–477

Nandula VK (2010) Glyphosate resistance in crops and weeds: history, development, and management. Wiley, Hoboken, NJ, 321 pp

Nantia EA, Moreno-González D, Manfo FP, Sonchieu J, Moundipa PF, García-Campaña AM et al (2017) Characterization of carbamate pesticides in natural water from Cameroon. Anal Lett 50: 1397–1409

Narenderan ST, Meyyanathan SN, Karri VVSR (2019) Experimental design in pesticide extraction methods: a review. Food Chem 289:384–395

National Research Council (1993) Managing global genetic resources: agricultural crop issues and policies. National Academies Press, Washington, DC, p 480

Nauen R, Slater R, Sparks TC, Elbert A, Mccaffery A (2019) IRAC: insecticide resistance and mode-of-action classification of insecticides. Mod Crop Prot Compd 3:995–1012

Navarro S, Vela N, Navarro G (2007) An overview on the environmental behaviour of pesticide residues in soils. Spanish J Agric Res:357–375

Niemann L, Sieke C, Pfeil R, Solecki R (2015) A critical review of glyphosate findings in human urine samples and comparison with the exposure of operators and consumers. J für Verbraucherschutz und Leb 10:3–12

Norgaard T, Moldrup P, Ferré TP, Olsen P, Rosenbom AE, de Jonge LW (2014) Leaching of glyphosate and aminomethylphosphonic acid from an agricultural field over a twelve-year period. Vadose Zo J 13:1–18

Nwokoro O, Dibua MEU (2015) Degradation of soil cyanide by single and mixed cultures of Pseudomonas stutzeri and Bacillus subtilis. Arch Ind Hyg Toxicol 65(1):113–119. https://doi.org/10.2478/10004-1254-65-2014-2449

Obojska A, Lejczak B, Kubrak M (1999) Degradation of phosphonates by Streptomycete isolates. Appl Microbiol Biotechnol 51(6):872–876

Obojska A, Ternan NG, Lejczak B, Kafarski P, McMullan G (2002) Organophosphonate utilization by the thermophile Geobacillus caldoxylosilyticus T20. Appl Environ Microbiol 68(4): 2081–2084

Obuotor TM, Yahaya PB, Akinloye OA, Sakariyau AO (2016) Biodegradation of paraquat dichloride-contaminated soil with fermented corn steep. Uniosun J Sci 1:67–73

Okada E, Costa JL, Bedmar F (2019) Glyphosate dissipation in different soils under no-till and conventional tillage. Pedosphere 29:773–783

Okada E, Allinson M, Barral MP, Clarke B, Allinson G (2020) Glyphosate and aminomethylphosphonic acid (AMPA) are commonly found in urban streams and wetlands of Melbourne, Australia. Water Res 168:1–8

Ololade OO, Aiyesanmi AF, Okoronkwo AE, Ololade IA, Adanigbo P (2019) Influence of electrolyte composition and pH on glyphosate sorption by cow-dung amended soil. J Environ Sci Health Pt B 54:758–769

Omakor JE, Onyido I, Buncel E (2001) Mechanisms of abiotic degradation and soil--water interactions of pesticides and other hydrophobic organic compounds. Part 3. Nucleophilic displacement at the phosphorus centre of the pesticide fenitrothion [O, O-dimethyl O-(3-methyl-4-nitrophenyl) phosph]. J Chem Soc Perkin Trans 2:324–330

Oudejans L, Mysz A, Snyder EG, Wyrzykowska-Ceradini B, Nardin J, Tabor D et al (2020) Remediating indoor pesticide contamination from improper pest control treatments: persistence and decontamination studies. J Hazard Mater 397:1–11

Padgette SR, Kolacz KH, Delannay X, Re DB, LaVallee BJ, Tinius CN et al (1995) Development, identification, and characterization of a glyphosate-tolerant soybean line. Crop Sci 35 1451–1461

Padilla JT, Selim HM (2020) Environmental behavior of glyphosate in soils. In: Donald LS (ed) Advances in Agronomy, vol 159. Elsevier, Amsterdam, pp 1–34

Pateiro-Moure M, Pérez-Novo C, Arias-Estévez M, Rial-Otero R, Simal-Gándara J (2009) Effect of organic matter and iron oxides on quaternary herbicide sorption—desorption in vineyard-devoted soils. J Colloid Interface Sci 333:431–438

Pedersen KE, Pedersen NN, Meyling NV, Fredensborg BL, Cedergreen N (2020) Differences in life stage sensitivity of the beetle *Tenebrio molitor* towards a pyrethroid insecticide explained by stage-specific variations in uptake, elimination and activity of detoxifying enzymes. Pestic Biochem Physiol 162:113–121

Peñaloza-Vazquez A, Mena GL, Herrera-Estrella L, Bailey AM (1995) Cloning and sequencing of the genes involved in glyphosate utilization by Pseudomonas pseudomallei. Appl Environ Microbiol 61(2):538–543

Pereira RC, Anizelli PR, Di Mauro E, Valezi DF, da Costa ACS, Zaia CTB et al (2019) The effect of pH and ionic strength on the adsorption of glyphosate onto ferrihydrite. Geochem Trans 20:1–14

Peruzzo PJ, Porta AA, Ronco AE (2008) Levels of glyphosate in surface waters, sediments and soils associated with direct sowing soybean cultivation in north pampasic region of Argentina. Environ Pollut 156:61–66

Pessagno RC, Sánchez RMT, dos Santos AM (2008) Glyphosate behavior at soil and mineral—water interfaces. Environ Pollut 153:53–59

Piccolo A, Celano G, Arienzo M, Mirabella A (1994) Adsorption and desorption of glyphosate in some European soils. J Environ Sci Health Pt B 29:1105–1115

Pierre M, Jean-Baptiste AA, Sébastien DD (2018) Organochlorine pesticides contamination in human milk in Abidan (Côte d'Ivoire). J Toxicol Environ Health Sci 10:10–14

Pino N, Peñuela G (2011) Simultaneous degradation of the pesticides methyl parathion and chlorpyrifos by an isolated bacterial consortium from a contaminated site. Int Biodeter Biodegr 65:827–831

Pipke R, Amrhein N, Jacob GS, Schaefer J, Kishore GM (1987a) Metabolism of glyphosate in an Arthrobacter sp. GLP-1. FEBS J 165(2):267–273

Pipke R, Schulz A, Amrhein N (1987b) Uptake of glyphosate by an *Arthrobacter* sp. Appl Environ Microbiol 53(5):974–978

Pizzutti IR, Vela GME, de Kok A, Scholten JM, Dias JV, Cardoso CD et al (2016) Determination of paraquat and diquat: LC-MS method optimization and validation. Food Chem 209:248–255

Poirier L, Jacquet P, Elias M, Daudé D, Chabrière E (2017) La décontamination des organophosphorés: vers de nouvelles alternatives. In: Annales pharmaceutiques francaises, vol 75(3). Elsevier, Masson, 226, p 209

Pomeroy A (2012) Biochemical mechanisms of paraquat toxicity. Academic Press, London, 240 pp

Qiang H, Zhong X, Jing R, Jin G (2002) Progress of pyrethroids insecticides. J Qingdao Inst Chem Technol 1

Qu C, Albanese S, Lima A, Li J, Doherty AL, Qi S et al (2017) Residues of hexachlorobenzene and chlorinated cyclodiene pesticides in the soils of the Campanian Plain, southern Italy. Environ Pollut 231:1497–1506

Radif HM, Albaayit SF (2019) Isolation, identification and role of glyphosate-degrading bacteria from soils of Baghdad. Eurasia Proc Sci Technol Eng Math (EPSTEM) 6:135–137

Rainaud PL (2013) Evaluation des risques à long terme des herbicides à base de glyphosate sur la santé humaine. PhD Thesis, University of Picardy Jules Verne, Amiens, France

Ramchandra AM, Chacko B, Victor PJ (2019) Pyrethroid poisoning. Indian J Crit Care Med 23(4): 267–271

Ramya K, Vasudevan N (2020) Biodegradation of synthetic pyrethroid pesticides under saline conditions by a novel halotolerant *Enterobacter ludwigii*. Desalin Water Treat 173:255–266

Rani M, Shanker U, Jassal V (2017) Recent strategies for removal and degradation of persistent & toxic organochlorine pesticides using nanoparticles: a review. J Environ Manage 190:208–222

Rashidipour M, Heydari R, Maleki A, Mohammadi E, Davari B (2019) Salt-assisted liquid–liquid extraction coupled with reversed-phase dispersive liquid–liquid microextraction for sensitive HPLC determination of paraquat in environmental and food samples. J Food Meas Charact 13: 269–276

Reemtsma T, Alder L, Banasiak U (2013) Multimethod for the determination of 150 pesticide metabolites in surface water and groundwater using direct injection liquid chromatography–mass spectrometry. J Chromatogr A 1271:95–104

Ren X, Zeng G, Tang L, Wang J, Wan J, Liu Y et al (2018) Sorption, transport and biodegradation—an insight into bioavailability of persistent organic pollutants in soil. Sci Total Environ 610:1154–1163

Romina SM, Laura LV, Lozano VL, Vera MS, Ferraro M, Izaguirre I et al (2020) New findings on the effect of glyphosate on autotrophic and heterotrophic picoplankton structure: a microcosm approach. Aquat Toxicol 222:1–10

Sarkar B, Mukhopadhyay R, Mandal A, Mandal S, Vithanage M, Biswas JK (2020) Sorption and desorption of agro-pesticides in soils. In: Prasad MNV (ed) Agrochemicals detection, treatment and remediation. Butterworth-Heinemann, Elsevier, pp 189–205

Schrack D, Coquil X, Ortar A, Benoît M (2009) Persistence of pesticides in water from farm plots recently converted to organic farming. Carrefours l'Innovation Agron 4:259–268

Selvapandiyan A, Bhatnagar RK (1994) Isolation of a glyphosatemetabolising Pseudomonas: detection, partial purification and localisation of carbon-phosphorus lyase. Appl Microbiol Biotechnol 40(6):876–882

Sharifi Y, Pourbabaei AA, Javadi A, Abdolmohammad MH, Saffari M, Morovvati A (2015) Biodegradation of glyphosate herbicide by *Salinicoccus* spp isolated from Qom Hoze-soltan lake, Iran. Environ Health Eng Manag J 2:31–36

Sharma PK, Mayank M, Ojha CSP, Shukla SK (2020) A review on groundwater contaminant transport and remediation. ISH J Hydraul Eng 26:112–121

Shawish HMA, Ghalwa NA, Hamada M, Basheer AH (2012) Modified carbon paste electrode for potentiometric determination of diquat dibromide pesticide in water and urine samples. Mater Sci Eng C 32(2):140–145

Shivhare P, Gupta VK (1991) Spectrophotometric method for the determination of paraquat in water, grain and plant materials. Analyst 116:391–393

Silva V, Mol HGJ, Zomer P, Tienstra M, Ritsema CJ, Geissen V (2019) Pesticide residues in European agricultural soils—a hidden reality unfolded. Sci Total Environ 653:1532–1545

Simonsen L, Fomsgaard IS, Svensmark B, Spliid NH (2008) Fate and availability of glyphosate and AMPA in agricultural soil. J Environ Sci Health Pt B 43:365–375

Singh T, Singh DK (2017) Phytoremediation of organochlorine pesticides: concept, method, and recent developments. Int J Phytoremediation 19:834–843

Singh S, Singh SK, Chaubey J, Arora H, Bhandari K, Vyas A (2019) Inter-state mobility of pesticide leachate in underground water. AGUFM 2019:H41J

Singh S, Kumar V, Datta S, Wani AB, Dhanjal DS, Romero R et al (2020) Glyphosate uptake, translocation, resistance emergence in crops, analytical monitoring, toxicity and degradation: a review. Environ Chem Lett:1–40

Sousa S, Maia ML, Correira-Sá L, Fernandes VC, Delerue-Matos C, Calhau C et al (2020) Chemistry and toxicology behind insecticides and herbicides. In: Rakhimol KR, Thomas S, Volova T, Jayachandran K (eds) Controlled release of pesticides for sustainable agriculture. Springer, Cham, pp 59–109

Sprankle P, Meggitt WF, Penner D (1975) Adsorption, mobility, and microbial degradation of glyphosate in the soil. Weed Sci 23:229–234

Steinrücken HC, Amrhein N (1980) The herbicide glyphosate is a potent inhibitor of 5-enolpyruvylshikimic acid-3-phosphate synthase. Biochem Biophys Res Commun 94:1207–1212

Sviridov A (2012) Enzyme systems of organophosphonate catabolism of soil bacteria Achromobacter sp. and Ochrobactrum anthropi GPK3. PhD thesis (in Russian). Pushchinoa 152:120–132

Sviridov AV, Shushkova TV, Zelenkova NF, Vinokurova NG, Morgunov IG, Ermakova IT, Leontievsky AA (2012) Distribution of glyphosate and methylphosphonate catabolism systems in soil bacteria Ochrobactrum anthropi and Achromobacter sp. Appl Microbiol Biotechnol 93(2):787–796

Sviridov AV, Shushkova TV, Ermakova IT, Ivanova EV, Leontievsky AA (2014) Glyphosate: safety risks, biodegradation, and bioremediation. In: Cao G, Orrù R (eds) Current environmental issues and challenges. Springer, Dordrecht, pp 183–195

Tamano H, Nishio R, Morioka H, Furuhata R, Komata Y, Takeda A (2019) Paraquat as an environmental risk factor in Parkinson's disease accelerates age-related degeneration via rapid influx of extracellular Zn^{2+} into Nigral dopaminergic neurons. Mol Neurobiol 56:7789–7799

Tankiewicz M, Fenik J, Biziuk M (2011) Solventless and solvent-minimized sample preparation techniques for determining currently used pesticides in water samples: a review. Talanta 86:8–22

Talbot HW, Johnson LM, Munnecke DM (1984) Glyphosate utilization by Pseudomonas sp. and Alcaligenes sp. isolated from environmental sources. Curr Microbiol 10(5):255–259

Tanner CM, Kamel F, Ross GW, Hoppin JA, Goldman SM, Korell M et al (2011) Rotenone, paraquat, and Parkinson's disease. Environ Health Perspect 119:866–872

Teerakun M, Reungsang A, Chaowarat M, Saraphirom P (2020) Optimization of Paraquat degradation with microbial consortium from contaminated soil using statistic method. Int J 18:73–79

Testud F, Grillet JP (2007) Insecticides organophosphorés, carbamates, pyréthrinoïdes de synthèse et divers. Encycl Med Chir 2:1–24

Torstensson L (1985) Behaviour of glyphosate in soils and its degradation. In: Herbic glyphosate, pp 137–150

Toss V, Leito I, Yurchenko S, Freiberg R, Kruve A (2017) Determination of glyphosate in surface water with high organic matter content. Environ Sci Pollut Res 24:7880–7888

Tsai WT, Lai CW, Hsien KJ (2003) Effect of particle size of activated clay on the adsorption of paraquat from aqueous solution. J Colloid Interface Sci 263:29–34

Tsao YC, Lai YC, Liu HC, Liu RH, Lin DL (2016) Simultaneous determination and quantitation of paraquat, diquat, glufosinate and glyphosate in postmortem blood and urine by LC--MS-MS. J Anal Toxicol 40:427–436

Tu CM, Bollen WB (1968) Interaction between paraquat and microbes in soils. Weed Res 8:38–45

Tyagi N, Singh R (2020) Paraquat-induced oxidative stress and lung inflammation. In: Chakraborti S, Parinandi N, Ghosh R, Ganguly N, Chakraborti T (eds) Oxidative stress in lung diseases. Springer, Singapore, pp 245–270

US EPA (1987) Unfinished business: a comparative assessment of environmental problems, EPA report EPA-230/2-87-025. US Environmental Protection Agency, Washington, DC. https://nepis.epa.gov/Exe/ZyPDF.cgi?Dockey=2000CRX3.PDF

USEPAO (2004) Overview of the ecological risk assessment process in the Office of Pesticide Programs, US Environmental Protection Agency: Endangered and Threatened Species Effects Determinations. DIANE Publishing, Washington, DC. 92 p

Van der Hoff GR, van Zoonen P (1999) Trace analysis of pesticides by gas chromatography. J Chromatogr A 843:301–322

Vereecken H (2005) Mobility and leaching of glyphosate: a review. Pest Manag Sci Former Pest Sci 61:1139–1151

Villamizar ML, Brown CD (2016) Modelling triazines in the valley of the River Cauca, Colombia, using the annualized agricultural non-point source pollution model. Agric Water Manag 177: 24–36

Villarreal-Chiu JF (2012) Organic phosphonate metabolism by marine bacteria. PhD Thesis, Queen's University Belfast, Belfast, Royaume-Uni

Vinten AJA, Yaron B, Nye PH (1983) Vertical transport of pesticides into soil when adsorbed on suspended particles. J Agric Food Chem 31:662–664

Vryzas Z (2018) Pesticide fate in soil-sediment-water environment in relation to contamination preventing actions. Curr Opin Environ Sci Health 4:5–9

Vytvras K, Šimĭčkovă-Štajerova B (1989) Potentiometric ion-pair formation titrations of bisquaternary cations: determination of diquat and paraquat. Anal Chim Acta 226:177–182

Wackett LP, Shames SL, Venditti CP, Walsh CT (1987) Bacterial carbonphosphorus lyase: products, rates, and regulation of phosphonic and phosphinic acid metabolism. J Bacteriol 169(2):710–717

Wagner SL (1983) Clinical toxicology of agricultural chemicals. Noyes Data Corporation, Park Ridge, NJ, p 306

Wang S, Liu B, Yuan D, Ma J (2016) A simple method for the determination of glyphosate and aminomethylphosphonic acid in seawater matrix with high performance liquid chromatography and fluorescence detection. Talanta 161:700–706

Wang R, Yuan Y, Yen H, Grieneisen M, Arnold J, Wang D et al (2019) A review of pesticide fate and transport simulation at watershed level using SWAT: current status and research concerns. Sci Total Environ 669:512–526

Wang T, Yu C, Chu Q, Wang F, Lan T, Wang J (2020) Adsorption behavior and mechanism of five pesticides on microplastics from agricultural polyethylene films. Chemosphere 244:1–12

Wang XQ, Liu J, Zhang N, Yang H (2020) Adsorption, mobility, biotic and abiotic metabolism and degradation of pesticide exianliumi in three types of farmland. Chemosphere 254:1–12

Weber JB (2018) Properties and behavior of pesticides in soil. In: Richard CH, Daniel JS (eds) Mechanisms of pesticide movement into ground water. CRC, Boca Raton, pp 15–42

Weber JB, Weed SB (1968) Adsorption and desorption of diquat, paraquat, and prometone by montmorillonitic and kaolinitic clay minerals. Soil Sci Soc Am J 32(4):485–487

Weber JB, Perry PW, Upchurch RP (1965) The influence of temperature and time on the adsorption of paraquat, diquat, 2, 4-D and prometone by clays, charcoal, and an anion-exchange resin 1. Soil Sci Soc Am J 29:678–688

Weidel H, Russo M (1882) Studies on the pyridine. Monthly notebooks for chemistry and related parts of other sciences 3(1):850-885

WHO (1994) Glyphosate. International Program on Chemical Safety. 67 p

WHO (2009) The WHO recommended classification of pesticides by hazard and guidelines to classification 2009. World Health Organization, International Programme on Chemical Safety, Geneva

WHO (2017) Manuel sur l'élaboration et l'utilisation des spécifications FAO/OMS pour les pesticides, Rome, 318 p

Woźniak E, Reszka E, Jabłońska E, Mokra K, Balcerczyk A, Huras B (2020) The selected epigenetic effects of aminomethylphosphonic acid, a primary metabolite of glyphosate on human peripheral blood mononuclear cells (in vitro). Toxicol In Vitro 66:1–8

Wu CY, Liu JK, Chen SS, Deng X, Li QF (2013) Isolation and characterization of paraquat-degrading extracellular humus-reducing bacteria from vegetable field. Adv Mater Res 807–809: 1026–1030. https://doi.org/10.4028/www.scientific.net/amr.807-809.1026

Xiang Y, Lu X, Yue J, Zhang Y, Sun X, Zhang G et al (2020) Stimuli-responsive hydrogel as carrier for controlling the release and leaching behavior of hydrophilic pesticide. Sci Total Environ 722:137811

Xu B, Sun QJ, Lan JCW, Chen WM, Hsueh CC, Chen BY (2019) Exploring the glyphosate-degrading characteristics of a newly isolated, highly adapted indigenous bacterial strain, Providencia rettgeri GDB 1. J Biosci Bioeng 128(1):80–87

Yang Y, Guo Y, Hu C, Wang Y, Wang E (2004) Preparation of surface modifications of mesoporous titania with monosubstituted Keggin units and their catalytic performance for organochlorine pesticide and dyes under UV irradiation. Appl Catal A Gen 273:201–210

Yao F, Liu H, Wang G, Du L, Yin X, Fu Y (2013) Determination of paraquat in water samples using a sensitive fluorescent probe titration method. J Environ Sci 25:1245–1251

Yavari S, Sapari NB, Malakahmad A, Razali MAB, Gervais TS, Yavari S (2020) Adsorption–desorption behavior of polar imidazolinone herbicides in tropical paddy fields soils. Bull Environ Contam Toxicol 104:121–127

Yu Q, Cairns A, Powles S (2007) Glyphosate, paraquat and ACCase multiple herbicide resistance evolved in a *Lolium rigidum* biotype. Planta 225:499–513

Yu XM, Yu T, Yin GH, Dong QL, An M, Wang HR, Ai CX (2015) Glyphosate biodegradation and potential soil bioremediation by Bacillus subtilis strain Bs-15. Genet Mol Res 14(4) 14717–14730

Zbair M, Anfar Z, Ahsaine HA, Khallok H (2019) Kinetics, equilibrium, statistical surface modeling and cost analysis of paraquat removal from aqueous solution using carbonated jujube seed. RSC Adv 9:1084–1094

Zhan H, Feng Y, Fan X, Chen S (2018) Recent advances in glyphosate biodegradation. Appl Microbiol Biotechnol 102:5033–5043

Zhang W (2018) Global pesticide use: profile, trend, cost/benefit and more. Proc Int Acad Ecol Environ Sci 8:1–27

Zhang L, Rana I, Shaffer RM, Taioli E, Sheppard L (2019) Exposure to glyphosate-based herbicides and risk for non-Hodgkin lymphoma: a meta-analysis and supporting evidence. Mutat Res Mutat Res 781:186–206

Zhao B, Zhang J, Gong J, Zhang H, Zhang C (2009) Glyphosate mobility in soils by phosphate application: laboratory column experiments. Geoderma 149:290–297

Zhao F, Wang S, She Y, Zhang C, Zheng L, Jin M et al (2017) Subcritical water extraction combined with molecular imprinting technology for sample preparation in the detection of triazine herbicides. J Chromatogr A 1515:17–22

Zhu X, Liu W, Chen J, Bruijnzeel LA, Mao Z, Yang X et al (2019) Reductions in water, soil and nutrient losses and pesticide pollution in agroforestry practices: a review of evidence and processes. Plant and Soil:1–42

Zouaoui K, Dulaurent S, Gaulier JM, Moesch C, Lachâtre G (2013) Determination of glyphosate and AMPA in blood and urine from humans: About 13 cases of acute intoxication. Forensic Sci Int 226:20–25

Zubrod JP, Bundschuh M, Arts G, Brühl CA, Imfeld G, Knäbel A et al (2019) Fungicides: an overlooked pesticide class? Environ Sci Technol 53:3347–3365

Microbial Degradation of Polychlorinated Biphenyls (PCBs): Usage of Bacteria and Fungi

18

Ningthoujam Samarendra Singh

Abstract

Microbial communities are almost omnipresent in various ecosystems all over the world and can be utilized for degrading pollutants from the environment matrices. Among the pollutants, polychlorinated biphenyls (PCBs) are the most hazardous chemicals that cause profound deleterious effects on human health and biota communities. The present chapter attempts to summarize the biodegradation of polychlorinated PCBs by bacteria and fungi in laboratory and field conditions. Microbial degradation is often regarded to be the best alternative to traditional physical and chemical approaches due to sustainability and cost-effective benefits. This chapter concludes with forward directions as challenges and prospects in PCBs microbial degradation.

Keywords

Biodegradation · Bioremediation · Bacteria · Environmental pollution · Fungi · Polychlorinated biphenyl

18.1 Introduction

Polychlorinated biphenyls (PCBs) are anthropogenic and aromatic chlorinated-organic chemicals that exhibit high persistent, bioaccumulative nature and toxic effect on human health and ecosystems (Erickson 2001; Vasilyeva and Strijakova 2007; Erickson and Kaley 2011; Grimm et al. 2015; Chirsaz et al. 2020). The

N. S. Singh (✉)
Division of Agricultural Chemicals, ICAR-Indian Agricultural Research Institute, New Delhi, Delhi, India

N. K. Arora, B. Bouizgarne (eds.), *Microbial BioTechnology for Sustainable Agriculture Volume 1*, Microorganisms for Sustainability 33,
https://doi.org/10.1007/978-981-16-4343-4_18

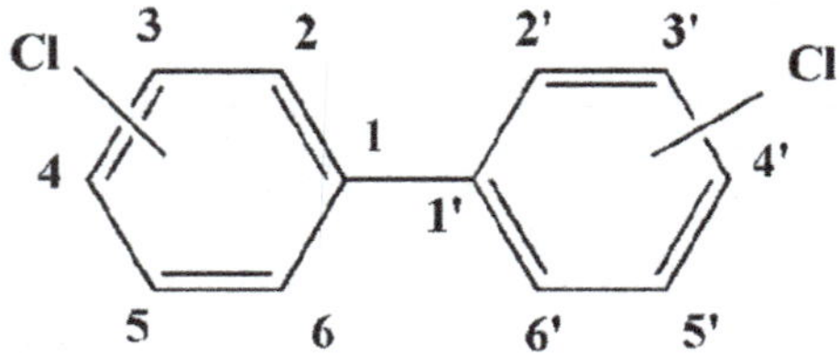

Fig. 18.1 Chemical structure of PCBs

biphenyl backbone structure of PCBs has 1–10 chlorine atoms, forming at least 209 congeners configuration (Ballschmiter and Zell 1980; Wiegel and Wu 2000) (Fig. 18.1). Generally, PCBs are synthesized by catalytic chlorination of biphenyls nucleus, resulting in a complex mixture of different congeners (Borja et al. 2005). Some of the PCBs properties are largely governed by the chlorination level onto the biphenyl nucleus (Elangovan et al. 2019) (Table 18.1). For example, the lower-chlorinated PCBs are volatile, oily liquids, colorless, and hydrophilic, while other higher-chlorinated PCBs act differently (Borja et al. 2005). Owing to high boiling point, high flashpoint, nonflammable, and electrical insulating properties, PCBs are often used for many commercial and industrial purposes, such as pesticides, capacitors, dyes, lubricants, rubber products, transformers, pigments, heat transfer fluids, plasticizers, coolants, carbonless copy paper, and hydraulic fluids (Erickson 2001; Erickson and Kaley 2011; Pentyala et al. 2011; Grimm et al. 2015; IARC 2016; Sharma et al. 2018). They are mainly sold in markets under trade names like Aroclor in the USA, Clophen in Germany, and Kanechlor in Japan (IARC 2016; Sadañoski et al. 2020).

Despite the PCBs restriction imposed by the Stockholm Convention on uses and production, it was estimated in 1929 that about 1.3–1.5 million metric tons of PCBs produced worldwide (of which more than 70% are tri-, tetra-, and penta-chlorinated biphenyls) have contributed contamination in various environmental matrices like soil, sediment, air, and water (Breivik et al. 2002; Vasilyeva and Strijakova 2007; Pieper and Seeger 2008; Grimm et al. 2015; IARC 2016) (Fig. 18.2). For instance, the presence of considerable PCBs congener ranging from 16.15 to 477.85 μg/ kg (dry weight) in river sediments of Pearl River Delta, China (Lai et al. 2015; Wang et al. 2019b). In fact, a majority of PCBs contamination sources are associated with accidental discharges, improper disposal, poorly maintained transformers, leachate, hazardous waste sites, uncontrolled landfills, atmospheric deposition, and incinerators (Breivik et al. 2002).

Lammel and Stemmler (2012) studied on global atmosphere-ocean circulation model from 1950 to 2010 periods and reported that PCB congeners are widely distributed in various environments. Moreover, it was also established that PCBs in air, water, and soil could persist over a long time (Sharma et al. 2018). PCBs deposition in the soil is considered to be the ultimate source and sink due to adsorption on soil organic carbon (IARC 2016). This adsorption is influenced by sorbate properties, sorbent properties, and sorption mechanism (Robinson and Lenn 1994). PCBs are found to be detected in the Ny-Alesund soil, Arctic, and even Antarctica, transported by atmospheric, riverine, ice, and biota system (Zhang et al.

Table 18.1 Physical and chemical properties of polychlorinated biphenyls (PCBs)

PCB congener	Molecular weight	Melting point (°C)	Log K_{ow}	Water solubility (μg/L)	Boiling point (°C)	Vapor pressure (Pa at 25 °C)	No. of isomers	Structure
PCB 10 (2, 4-dichlorobiphenyl)	223.09 g/mol)	24.4–149	–	–	312	0.24	12	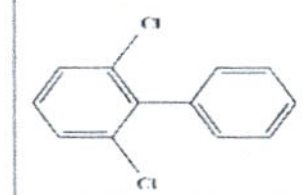
PCB 28 (2, 4, 4′-trichlorobiphenyl)	257.54 g/mol	28–87	–	–	337	0.054	24	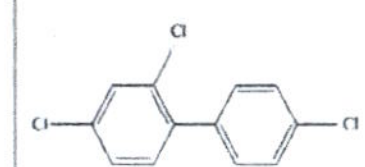
PCB 52 (2, 2′, 5, 5′-tetrachlorobipheny)	291.98 g/mol	47–180	–	–	360	0.012	42	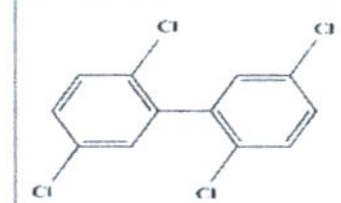
PCB 138 (2, 2′, 3, 4, 4′,5′-hexachlorobipheny)	360.87 g/mol	77–200	6.5–7.4	15.9	400	5.8×10^{-4}	46	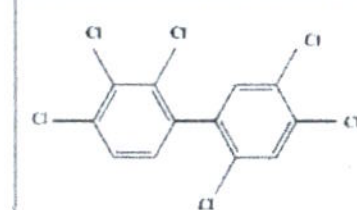
PCB 153 (2, 2′, 4, 4′, 5, 5′-hexachlorobipheny)	360.87 g/mol	77–200	6.7	0.9	400	5.8×10^{-4}	46	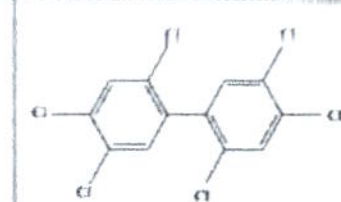
PCB 180 (2, 2′, 3, 4, 4′, 5, 5′-heptachlorobiphenyl)	395.32 g/mol	83–149	6.7–7.2	0.2	417	1.3×10^{-4}	24	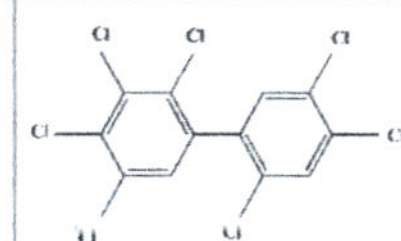

Source: IARC (2016)

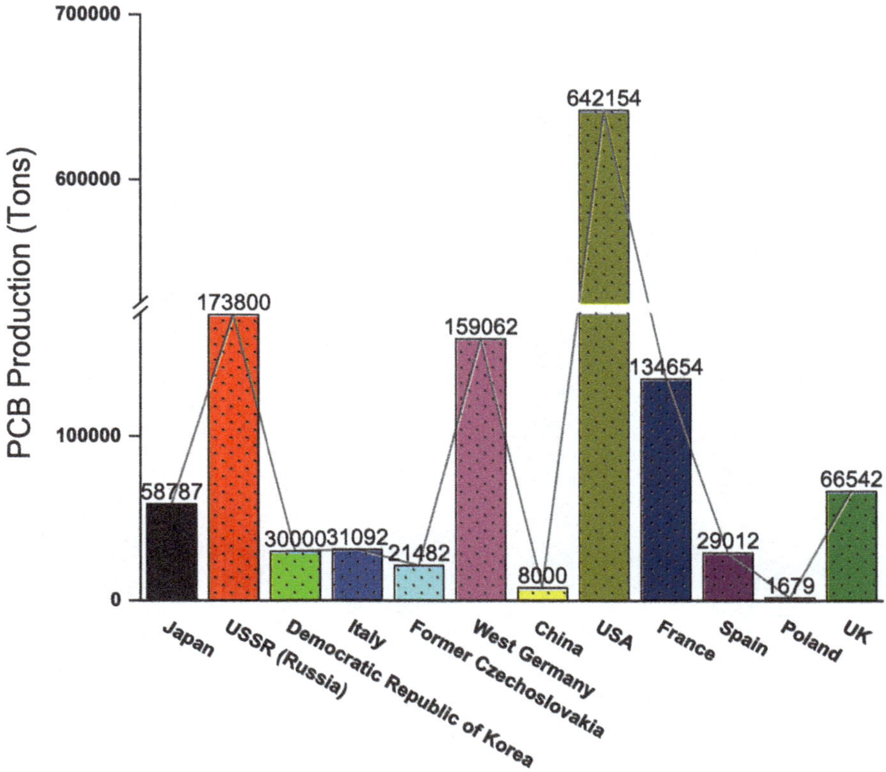

Fig. 18.2 Estimated global production of PCB congeners between 1930 and 2012. Source: Breivik et al. (2002) and IARC (2016)

2014; Grimm et al. 2015; Hung et al. 2016; IARC 2016; Sharma et al. 2018; Lo Giudice et al. 2019). Moreover, the fate and persistence of individual PCBs in the environment differ greatly according to the type of congeners (Breivik et al. 2002). Bioaccumulation of PCBs in the bodies of phytoplankton, zooplankton, plants, fish, and other organisms occurs through the uptake of contaminated water (Jan Campfens 1997; Gomes et al. 2013; Ododo and Wabalo 2019). Similarly, the bioaccumulation of PCBs in the leaves, roots, and seeds of the mangal plant results from contaminated sediments (Alegria et al. 2016; Kelly et al. 2007). Additionally, PCBs are known to cause adverse effects on humans health, wildlife, and ecological system (Maervoet et al. 2004; Arora 2018). The exposure of PCBs could lead to multiple health problems in humans, for example, developmental problem, cancer, adverse immune system, tumor initiation and promotion, skin diseases, abnormal behavioral effects, reproductive failure, ill-reproductive system, nervous system problems, genotoxicity, and endocrine system disruption (Pieper and Seeger 2008; Grossman 2013; Elnar et al. 2015; Chakraborty and Das 2016; Faroon and Ruiz 2016; Balejčíková 2018).

18.2 Analytical Methodology for the Determination of PCBs

In microbial degradation study, gas chromatography (GC) is the most commonly used instrument for quantifying the PCBs congeners and their metabolites. Study by Su et al. (2019) employed GC 2010 Plus equipped with an ECD and Sil capillary column (60 m × 0.25 mm × 0.25 μm) film thickness to measure PCB 77 concentrations. The carrier gas was used as helium at a flow rate of 4.3 mL/min. The oven temperature was set as 80 °C for 3 min, followed by 80 °C to 160 °C at a rate of 8 °C/min, 160 °C to 270 °C at a rate of 4 °C/min, and 270 °C for 25 min. The detector and injector temperatures were programmed at 250 °C and 270 °C, respectively. Another study examined the 33 commercial paint pigments from local paint stores using GC-MS/MS and found 50 PCB congeners, especially azo and phthalocyanine pigments (Hu and Hornbuckle 2010; Grossman 2013). Mouhamadou et al.'s (2013) study in fungal biodegradation has also demonstrated PCBs analysis by using GC equipped with a DB-XLB capillary column (0.25 mm × 60 m × 0.25 mm) and ECD detector, Agilent Technologies, USA. Hydrogen was used as the carrier gas for the analysis. The parameter of the GC instrument was maintained as follows: injector temperature at 250 °C, detector temperature at 310 °C, and the column temperature at 50 °C. Moreover, Beaudette et al. (1998) investigated the PCBs degradation with Perkin-Elmer 8500 GC coupled with ECD detector and capillary column (15 m × 0.53 mm × 0.5 mm). The instrument employed ultrapure nitrogen as the carrier gas, where the injection volume is 1 mL with injection port temperature 250 °C. The initial oven temperature was at 120 °C for 25 min, raised to 210 °C at 10 °C/min, and held at 210 °C for 9 min. The detector temperature was maintained at 300 °C.

It was reported by Sangely et al. (2009) that PCBs were extracted from 500 mL aqueous assays using n-hexane through liquid-liquid extraction. It was monitored using high-resolution GC-MS (HRGC-LRMS), Thermo Finnigan Trace 2000 series fitted with Restek RTX-5MS columns (30 m × 0.25 mm × 0.25 μm) on electron impact mode. Helium was used as carrier gas at the rate of 1.2 mL/min and an injection volume of 1 μL sample in split-less mode at 250 °C. It is also regarded to offer the most sensitive and reproducible ionization technique. The GC-MS was kept at 250 °C. The oven temperature was programmed initially at 50 °C for 1 min, ramped up 20 °C/min to 310 °C, and maintained for 6 min. The parameters of GC running full-scan electron impact as follows: 5 min for the solvent delay, 70 eV of electron impact energy, source temperature at 200 °C, 150 μA for emission current, and 350 V at detector voltage. In another research work, Sun et al. (2016) reported that the quantitative analysis of the PCBs and MeO-PCBs was carried out using GC/MS instrument, Agilent Technologies, USA, fitted with a DB-5 MS capillary column (30 m × 0.25 mm × 0.25 μm) that operated by selected ion monitoring (SIM) mode with an electron impact source, using helium as the carrier gas at 1.0 mL/min flow rate. The initial oven temperature was kept at 80 °C, ramped to 140 °C at 10 °C/min, and raised to 300 °C at 2.5 °C/min. The selected ion monitoring (SIM) mode was used for quantitative determination. Similarly, Zhang et al. (2016) also developed a field analysis approach for the bioremediation study of PCBs like

Aroclor 1016, 1221, 1232, 1242, 1248, 1254, and 1260 from the contaminated soil and sediment. In this approach, the soil sample (0.5 g) was extracted by $KMnO_4$ (2 mL, 0.2 M) and H_2SO_4 (0.5 mL, 6 M) under the headspace solid-phase microextraction with polydimethylsiloxane (100 mm) fiber samples and analyzed using a portable scale GC-MS instrument equipped with ECD detector. Moreover, Yin et al. (2011) reported that PCB analysis using GC-MS of Agilent 7890-A coupled with Agilent 5973 mass-selective detector, HP-5 capillary column (30 m × 0.25 mm ID × 0.25 μm), and Agilent autosampler 7890-A injector.

It has been shown that PCBs and their metabolites from the composite soil sample were extractions using hexane/acetone (1:1, v/v) and 1% acetic acid in n-hexane/acetone (1:1, v/v) in the Dionex 200 ASE system, Palaiseau, France. It was analyzed using GC-MS, Varian, CA, fitted ion trap detector and DB-5MS column, Agilent, Czech Republic (30 m × 0.25 mm × 0.25 mm). The split/split-less injector was maintained at 240 °C. The oven temperature was programmed initially at 60 °C and held for 1 min in the split-less mode, opened splitter with ratio 1:50, heated up to 120 °C at a rate of 25 °C/min, and then finally ramped up to 240 °C at a rate of 2.5 °C/min and held for 28 min. Besides, the instrument was set with solvent delay time at 5 min and transfer line temperature at 240 °C under electron impact at 70 eV (Stella et al. 2015). Additionally, Sangely et al. (2009) demonstrated that PCBs analysis was done using HRGC-LRMS fitted with Restek RTX-5MS column (5% diphenyl; 95% dimethylpolysiloxane; 30 m × 0.25 mm × 0.25 μm) on EI mode. The sample (1 μL) was injected in split-less mode at 250 °C. It was operated using helium as carrier gas at a 1.2 mL/min flow rate. The oven temperature initially set up at 50 °C for 1 minute, ramped up to 310 °C at 20 °C/min, and held at 310 °C for 6 min. Another study conducted by Chakraborty and Das (2016) determined PCBs and their degradation products using Thermo Scientific Trace GC Ultra with model Polaris Q MS equipped with TR-1MS capillary column (30 m × 0.25 ID 0.25 μm). The ion source was maintained at 230 °C, and the transfer line temperature was continued at 280 °C. The oven temperature program was set up at 50 °C for 2 min, ramped up to 80 °C at 10 °C/min for 5 min, increased to 280 °C at 15 °C/min for 5 min, and then held at 300 °C at 10 °C/min for 5 min. The injection volume of the instrument was found to be 1 μL under a slit ratio of 20:1. Helium was used as carrier gas at 1 mL/min with the solvent system as ethyl acetate/water (50–95%, v/v) under a linear gradient for 40 min. The MS operated with IE at 70 eV with a mass range from 40 to 600 a.m.u at three micro-scans. Besides, Chen et al. (2017) reported that PCBs congeners from commercial milk were extracted using the QuEChERS method and then cleaned up by gel permeation chromatography. Then, it was then quantified by GC fitted with Bruker Scion TQ triple quadruple MS using the BR-5MS column (15 m × 0.25 mm × 0.25 μm). The carrier gas was used as helium at a constant flow rate of 1.8 mL/min. The transfer line was maintained at 280 °C and source temperatures were maintained at 250 °C, respectively. The initial oven temperature was started at 90 °C and held for 1 min, raised to 150 °C at a rate of 50 °C/min and held for 1 min, ramped up to 310 °C at a rate of 8 °C/min, and held for 3 min. The MS/MS was employed in EI positive mode at 70 eV. The injection volume was 2 μL under pulsed split-less mode at 250 °C. The individual PCB analytes were quantified from the sum of two MRM transitions and characterized using their abundance ratios in the instrument consoles.

Concurrently, Horváthová et al. (2018) have evaluated the quantitative analysis of seven PCB congeners, namely, PCB 8, PCB 28, PCB 52, PCB 101, PCB 118, PCB 153, and PCB 138 on HP5890A GC-ECD fitted with silica capillary column (30 m × 0.25 mm) with nonpolar stationary phase HP-5MS (0.25 mm). The parameters of the column temperature were maintained as follows: initial temperature raised from 70 °C to 150 °C at 25 °C/min for 2 min, ramped up from 150 °C to 200 °C at 3 °C/min, and increased from 200 °C to 280 °C at 8 °C/min and held for 3 min. Helium was used as carrier gas with a constant flow of 1.5 mL/min. The injection temperature of the instrument was 250 °C using split injection mode. Recently, Olatunji (2019) has shown to determine PCBs congeners in fresh root and leafy vegetables using GC-MS Analysis, Agilent Technologies, quipped with DB-5 capillary column. The sample (1.0 μL) was injected in the split-less mode with helium as carrier gas at a flow rate of 3 mL/min. The injector port was maintained at 250 °C, the initial temperature at 70 °C for 2 min, ramped up to 180 °C at rate 25 °C/ min and held for 3 min, 15 °C/min to 250 °C and held for 2 min, and then 8 °C/ min to 290 °C and held for 5 min. Furthermore, Elangovan et al. (2019) carried out PCBs analysis after sample extraction by using gas/liquid chromatography (GC/LC) tandem with mass spectrometry (MS).

18.3 Microbial Degradation of PCBs

PCBs are comprised of complex mixtures of diverse chemical congeners (Pieper 2005). They can be removed using different approaches such as physical, chemical, and biological methods (Robinson and Lenn 1994). Firstly, the physical method includes destroying the PCBs congeners through incineration at high temperature under thermal oxidation, for example, removing PCBs deposition from soil (Robinson and Lenn 1994). Secondly, the chemical method involves employing dechlorination of PCBs using alkali metal, polyethylene, and glycolate reagents (Robinson and Lenn 1994). However, these physical methods and chemical methods are known to be uneconomical as well as non-environmental friendly in nature (Gangar et al. 2019). Lastly, the biological method relies on PCB's removal by microbial communities and their diverse enzymes through aerobic, anaerobic, bioaugmentation, and bio-stimulation degradation (Robinson and Lenn 1994; Kour et al. 2021) (Fig. 18.3). It usually involves mediation by aerobic and anaerobic microorganisms to degrade to less toxic chemical compounds, such as carbon dioxide, water, and methane through mineralization, transformation, and metabolism process (Robinson and Lenn 1994; Vasilyeva and Strijakova 2007; Pieper and Seeger 2008; Gomes et al. 2013). More specifically, PCBs biodegradation is mostly attributed by diverse enzymes, namely, 2-hydroxyl-6-oxo-6-phenylhexa-2,4-dienoic acid hydrolase, dihydrodiol dehydrogenase, 2,3-dihydroxybiphenyl dioxygenase, and biphenyl dioxygenase (Elangovan et al. 2019). Although bio-stimulation degradation is done by providing natives microbes or consortiums

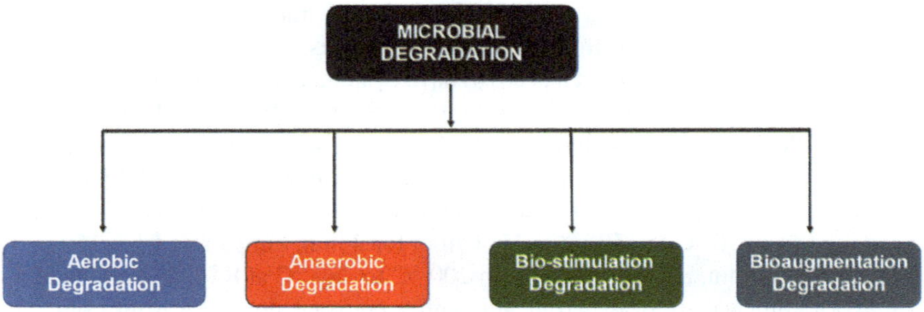

Fig. 18.3 Classification of different approaches to microbial degradation of PCB congeners

with added nutrients, electron donors, or acceptors but bioaugmentation degradation is done by adding selected laboratory-grown microbes to soil or media under experimentation (Hou et al. 2020). Indeed, the biological method is regarded as a cost-effective and eco-friendly approach to remove PCB congeners from the heavily polluted environment (Harms et al. 2011).

There are two major routes in PCBs biodegradation: mineralization and co-metabolism process. In mineralization process, PCBs are utilized by microbes as carbon and energy source and converted into constituent elements (Borja et al. 2005). While in co-metabolism process, the PCBs and another suitable substrate are utilized by microbes as carbon and energy source without complete conversion into the constituent element (Borja et al. 2005). The effectiveness of PCBs biodegradation is directly influenced by carbon sources, microorganisms, solubility, bioavailability, electron acceptors, pH, conductivity, PCBs structure, temperature, light intensity, oxygen content, and concentration (Robinson and Lenn 1994; Borja et al. 2005). For instance, microbial dechlorination of PCB congeners is slower in ortho-substituted PCBs than in meta- and para-substituted PCBs (Wiegel and Wu 2000). As such, the versatility and complex metabolic interactions are an important part of microbial communities (Pieper 2005).

18.3.1 Bacterial Degradation

Earlier PCBs biodegradation work by Ahmed and Focht (1973) reported that *Achromobacter* sp. and *Pseudomonas aeruginosa* could metabolize mono- and di-chlorinated biphenyls degradation as the sole carbon source. Similarly, it was also observed that the *Achromobacter* sp. is responsible for PCBs degradation to chlorinated benzoic acid (Furukawa et al. 1978). Bedard et al. (1987) also found *Alcaligenes eutrophus* H850 and *Pseudomonas putida* LB400 are highly capable of co-metabolizing a broad range of PCB congeners. Yet, Aroclor 1254 contaminated soil from the transformer spill is dechlorinated reductively by more than 50% using mixed microbes and sediment under anaerobic conditions at 12 °C (Tiedje et al. 1993). Another study by Arai et al. (1998) found that *Rhodococcus erythropolis* TA421 can degrade biphenyl using 2,3-dihydroxybiphenyl dioxygenase, biphenyl

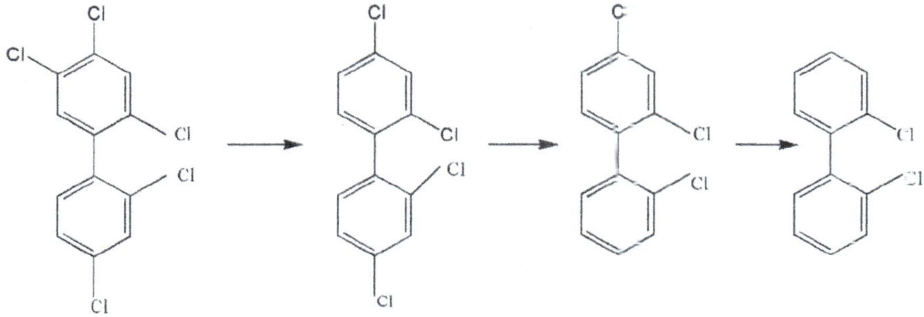

Fig. 18.4 Anaerobic pathway of PCBs biodegradation (Borja et al. 2005)

dioxygenase, and hydrolase. Indeed, *Bacillus* sp. strain JF8 isolated from the compost tends to utilize the biphenyl compounds, such as tetra- and pentachlorobiphenyl for carbon sources (Shimura et al. 1999). As noted, the bioremediation of Aroclor 1254 contaminated soil from the transformer spill results in nearly 50% degradation when mixed with soil, dechlorinating organisms, and river sediment (Robinson and Lenn 1994). *Rhodococcus* sp. strain RHA1 is found to transform Kanechlors 200, 300, 400, and 500, including chlorobenzoic acid within 3 days (Seto et al. 1995). Moreover, Billingsley et al. 1997 studied the PCBs transformation as the carbon source by *Pseudomonas* strain LB400. Many studies have demonstrated that the higher-chlorinated PCBs degrade anaerobically into lower-chlorinated PCBs in sewage sludge, surface soil, aerobic bioreactors, and surface sediments (Ahmed and Focht 1973; Abramowicz 1995; Borja et al. 2005; Pieper and Seeger 2008; Elangovan et al. 2019) (Fig. 18.4). For example, anaerobic dehalogenation of PCBs by *Dehalococcoides* sp. (Borja et al. 2005). At the same time, Wiegel and Wu (2000) revealed that PCBs biodegradation undergoes reductive dichlorination activities in contaminated soil and aquatic sediment by microbial cleavage of meta- and para-position of congeners under anaerobic conditions. Kaya et al. (2018) investigated the anaerobic dechlorination of PCB 118 and Aroclor 1254 by indigenous organohalide-respiring bacteria in the sediment microcosms, namely, Grasse River, Fox River, and Baltimore Harbor. The authors stated that the dechlorination of penta-, hexa-, and hepta-chlorinated PCBs were about 88%, 69%, and 51% in Grasse River; 88%, 87%, and 83% in Fox River; and 66% and 70% in Baltimore Harbor, respectively.

Under anaerobic and aerobic conditions, acclimatized microbes such as *Rhodococcus* sp., *Pseudomonas* sp., *Pseudoxanthomonas* sp., *Agromyces* sp., and *Brevibacillus* sp. isolated from the PCBs containing transformer oil using polyurethane foams as bio-carrier could dechlorinate PCBs regardless of the number of chlorine atoms attached to biphenyl ring (Nabavi et al. 2013). Notably, Davison and Veal (1997) evaluated *Alcaligenes faecalis* type II BPSI-2 and *Sphingomonas paucimobilis* BPSI-3 that mineralize PCBs to CO_2 and water ranging from 24.4%

to 38·1%. Furukawa (2000) also outlined that *Alcaligenes* sp. strain Y42 and *Acinetobacter* sp. strain P6 could co-metabolize PCBs as the sole source of carbon for growth and energy. The author noted that the rate and extent of PCBs dichlorination depends on various factors, namely, electron donors, pH, temperature, carbon sources, and dechlorinating bacteria. *Burkholderia xenovorans* LB400 and *Rhodococcus* sp. RHA1 can mineralize 2-chlorobiphenyl in anaerobic and aerobic biotreatment (Rodrigues et al. 2006). Rayford and Payne et al. (2011) underlined the use of in situ bioaugmentation of dehalorespiring bacterium such as *Dehalobium chlorocoercia* that dechlorinate Aroclor 1260 by about 56% in contaminated sediment of Baltimore Harbor after 120 days. Lu et al. (2018) suggested that *Dehalococcoides mccartyi* CG1 could contribute to enantioselective dichlorination of chiral PCB 174, PCB 149, and PCB 132. *P. aeruginosa* MD2 exhibited biodegradation of PCBs in the wastewater of Notwane sewage treatment plant, Botswana (Mathews 2014). With the use of metagenomics, three *D. mccartyi* strains CG1, CG4, and CG5 that dechlorinate Aroclor 1260 on tetrachloroethene addition can be identified from the studied site (Wang et al. 2014). *Pseudomonas furukawaii* sp. nov isolated from PCB-contaminated soil in Japan showed strong linkage to *Pseudomonas* genus with 97.8% sequence similarities with *Pseudomonas balearica* strain SP1402T (Kimura et al. (2018).

Hitherto, it has been established by Fujihara et al. (2015) that *P. aeruginosa* KF702 isolated from PCBs-contaminated soil could utilize PCBs as the sole source of carbon, due to biphenyl catabolism genes that facilitate the degradation of aromatic chemical compounds. Likewise, *P. aeruginosa* JP-11 isolated from coastal sediments of Odisha, India, degraded more than 98% of PCB congeners, as sole carbon sources, using HOPDA benzoate-catechol metabolism pathway in 72 h period (Chakraborty and Das 2016). Recently, Nuzzo et al. (2017) have isolated two native organohalide-respiring bacteria belonging to genera *Dehalococcoide* from the PCB-contaminated marine sediment of the Venice Lagoon. Other researchers observed that in situ PCBs dechlorination by *D. mccartyi* CG1 in river sediment microcosm could be stimulated through halo priming of tetrachloroethene (Xu et al. 2018). As the chlorination level increases, the efficiency of PCBs dichlorination decreases rapidly (Furukawa 2000). Horváthová et al. (2018) showed the effectiveness of PCBs biodegradation in Delor 103-contaminated river sediments using individual bacterial strains and their consortia, namely, *Achromobacter xylosoxidans*, *Stenotrophomonas maltophilia*, *Ochrobactrum anthropic*, and *Rhodococcus ruber*. The observation further revealed that the consortia of *R. ruber* could result in greater PCBs degradation than individual isolate, establishing a synergistic relationship with other exogenous native microbes that contribute nearly 50% of PCBs biodegradation. Moreover, the resuscitated *Castellaniella* sp. SPC4 could aerobically degrade PCB 77 in 20 mg/l of media at a 0.066/h rate using a bph-encoded biphenyl pathway (Su et al. 2019).

Some of the bacteria genera are highly efficient in degrading PCBs by aerobic oxidation, namely, *Acinetobacter*, *Alcaligenes*, *Achromobacter*, *Janibacter*, *Burkholderia*, *Comamonas*, *Paenibacillus*, *Pseudomonas*, *Sphingomonas*, *Ralstonia*, *Enterobacter*, *Micrococcus*, *Corynebacterium*, *Bacillus*, *Arthrobacter*,

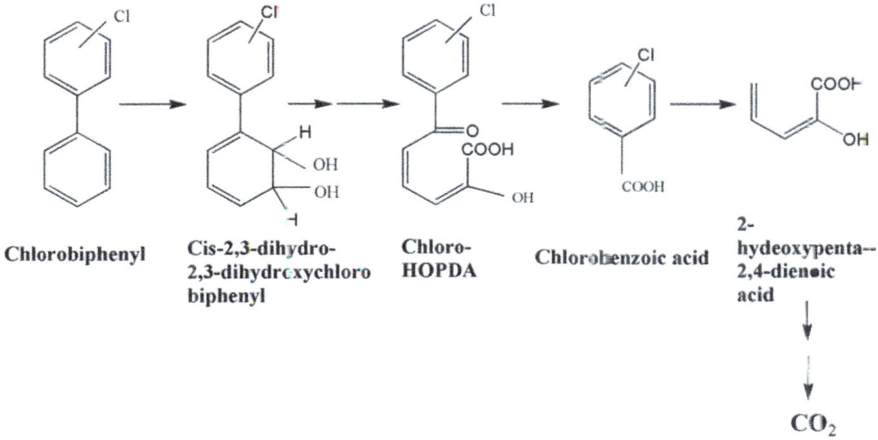

Fig. 18.5 Aerobic pathway of PCBs biodegradation by biphenyl-oxidizing bacteria (Furukawa 2000; Sharma et al. 2018)

and *Rhodococcus* (Furukawa 2000; Borja et al. 2005; Leigh et al. 2006; Elangovan et al. 2019) (Fig. 18.5). Moreover, Ponce et al. (2011) reported that tocopherol and berry extract might enhance the aerobic degradation of PCBs in the soil by *B. xenovorans* strain LB400, suggesting the protective effect of antioxidant compounds. Also, Nwinyi (2011) studied bacteria such as *Pseudomonas*, *Micrococcus*, *Arthrobacter*, and *Acinetobacter* isolated from landfill soil of Edo State, Nigeria, that utilize PCBs containing askarel oil as the sole carbon source through aerobic degradation. Commandeur et al. (1996) found that *Alcaligenes* sp. JBI used aerobic degradation to metabolize the lower-chlorinated PCBs (<5 chlorine atoms) using 3-dihydroxybiphenyl 1,2-dioxygenase and catechol 2,3-dioxygenase. Subsequent studies reported that the lower-chlorinated PCBs might be degraded rapidly by aerobic degradation (Ahmed and Focht 1973; Tiedje et al. 1993; Furukawa 2000; Borja et al. 2005; Bedard 2008; Pieper and Seeger 2008). It is also assumed that the aerobic degradation of PCBs can take place in surface soil and sediments by microbes that utilize PCBs and convert into chlorobenzoic acids through the biphenyl catabolism pathway (Furukawa 2000; Borja et al. 2005; Vasilyeva and Strijakova 2007; Pieper and Seeger 2008). In another study, De et al. (2006) showed that the ability of *Pseudomonas* CH07 to aerobically biodegrade the Clophen A-50 by more than 40% within 40 h period.

According to Jia et al. (2019), at pH 7 and 45 °C, *Arthrobacter* sp. could hydrolyze PCBs by meta-cleavage; and *Alcaligenes* sp. could co-metabolize PCBs by oxidative degradation. Wang et al. (2019a) also studied the removal of PCBs from the contaminated site by *D. mccartyi* strains CG3, *D. mccartyi* strains SG1, *Methanosarcina*, and *Desulfovibrio* using reductive dehalogenase CG3-RD11 and SG1-RD28, where interspecies synergistic interaction in the microbial communities facilitates organohalide respiration. Based on Horváthová et al. (2019) experiment, mixed cultures of *Rhodococcus* sp. and *S. maltophilia* can carry out PCBs biodegradation in minimal mineral water media, resulting in 59% PCBs degradation in contaminated sediment and 65% PCBs degradation in water media. More recently, Samadi et al. (2019) isolated and identified *Lysinibacillus macrolides* and *Bacillus firmus* from PCBs-contaminated soil with the electrical-transformer liquid that removes 2-chlorobiphenyl and 2,4-dichlorobiphenyl by more than 80% within 96 h. The biodegradation of PCB congeners by diverse bacterial species is shown in Table 18.2.

18.3.2 Fungal Degradation

The fungal biodegradation of PCB congeners is considered to be an effective way for remediation of contaminated sites. In 1993, a study by Šašek et al. (1993) found that the white-rot fungi such as *Hnigricans nigricans* strain can achieve more than 50% PCBs degradation in polluted soil. The authors also noted higher PCBs removal in the fungal consortia consisting of *Pleurotus ostreatus*, *Pseudomonas* B2, *H. nigricans*, *Candida boidinii*, *Lentinus edodes*, and *Candida lipolytica* than by individual isolate within the 6 weeks, indicative of the synergistic effect of the microorganisms. In particular, other researchers studied *Phanerochaete chrysosporium* in low-N (ligninolytic) and high-N (non-ligninolytic) media that undergo Aroclors degradation as high as 60.9% of Aroclors-1242, 30.5% of Aroclors-1254, and 17.6% of Aroclors-1260 and observed no difference in degradation in ortho-, meta-, and para-chlorines attached to the biphenyl nucleus (Yadav et al. 1995). It has been found that white-rot fungi is able to degrade PCBs, namely, 2,3–4,4′-dichlorobiphenyl, dichlorobiphenyl, 2,4′,5-trichlorobiphenyl, 2,2′,4,4′-tetrachlorobiphenyl, 2,2′,5,5′-tetrachlorobiphenyl, and 2,2′,4,4′,5,5′-hexachlorobiphenyl. Of these white-rot fungal strains, *Bjerkandera adusta* (UAMH 8258 and UAMH 7308), *P. ostreatus* (UAMH 7964), and *Trametes versicolor* UAMH 8272 show the best results in mineralizing PCBs congeners using lignin peroxidase and manganese peroxidase enzyme. More specifically, it was found that there is a limited abiotic loss of about 3.4 ± 11% of 2,2′,4,4’-TeCB and 17 ± 6.7% of 2,2′,4,4′,5,5’-HCB (Beaudette et al. 1998). Another worker has reported that *Doratomyces nanus*, *Doratomyces purpureofuscus*, *Doratomyces verrucisporus*, *Myceliophthora thermophila*, *Phoma eupyrena*, and *Thermoascus crustaceus* isolated from highly contaminated industrial sites could degrade PCBs between 29 and 85% after 7 d, in which *D. verrucisporus* show maximum PCBs removal (Mouhamadou et al. 2013).

Indeed, autochthonous ascomycetes such as *D. nanus*, *D. purpureofuscus*, *D. verrucisporus*, *M. thermophila*, *P. eupyrena*, and *T. crustaceus* exhibit remarkable capacity to degrade the PCB congeners (18.7–33%) without preference for lower or higher-chlorinated biphenyl using mesocosms in the historically contaminated soil and sediment within 6 months (Sage et al. (2014). Stella et al. (2015) isolated indigenous fungi belonging to orders *Ascomycota*, *Paxillus*, *Phoma*, *Mortierella*, and *Basidiomycota* from aged PCBs-contaminated soil and found to possess a very high potential for microbial degradation of PCBs through PLFA

Table 18.2 Biodegradation of PCB congeners involving bacterial species

PCBs (commercial name or congener types)	Bacteria species	Degradation	References
Aroclor 1242	*Acinetobacter* P6	Over 49 days of treatment, *Acinetobacter* P6 in enriched soil with PCBs reduces 65–75% of PCBs and transforms 48–49% to $^{14}CO_2$ (including soil carbonates), while control reduces only 10% of the PCBs and convert less than 2% to $^{14}CO_2$	Focht and Brunner (1985)
Aroclor 1221	Gram-negative psychrotolerant strains Sag-50A and Sag-50G	After 6 weeks of incubation, the slurries of PCBs-contaminated Arctic soils degrade Aroclor 1221 by about 71–76% at 30 °C	Mohn et al. (1997)
Aroclor 1242	*Burkholderia xenovorans* LB400 and *Rhodococcus* sp. RHA1	The bacterial strains in Aroclor 1242-contaminated river sediment could anaerobically dechlorinate 32% of the chlorine from Aroclor 1242 after 1 year PCBs removal rates are 54–57% for high and low inoculation densities during the aerobic phase	Rodrigues et al. (2006)
Clophen A-50 (coplanar PCBs)	Marine mercury-resistant bacterium: *Pseudomonas* sp. CH07	*Pseudomonas* sp. CH07 can degrade by more than 40% within 40 h	De et al. (2006)
Aroclor 1242 and Aroclor 1260	Aerobic heterotrophic bacteria	The presence of Enzyveba results in marked removal of low-chlorinated biphenyls from the soil by 75% and 135% under slurry and solid-phase conditions, respectively	Di Toro et al. (2006)
PCB 10, 28, 52, 138, 153, and 180	*Staphylococcus xylosus*	In mineral salts medium, PCB 138 and 153 are degraded by 71% and 66%, respectively within 24 h ($p < 0.05$)	Leaes et al. (2006)
Aroclor 1260	*Dehalococcoides* bacteria	The *Dehalococcoides* bacteria dechlorinate 64 PCB congeners carrying four to nine chlorines in Aroclor 1260	Bedard et al. (2007)

(continued)

Table 18.2 (continued)

PCBs (commercial name or congener types)	Bacteria species	Degradation	References
		in the sediment, Housatonic River	
Aroclor 1242	*Pseudomonas aeruginosa* TMU56	*P. aeruginosa* TMU56 following 4 days incubation is capable of degrading 73.3% of Aroclor 1242 (200 ppm) without biphenyl as an inducer and grow on 2,4-, 2,5-, 2,20-, and 4,40-dichlorobiphenyl	Hatamian-Zarmi et al. (2009)
Aroclor 1260	*Dehalobium chlorocoercia*	In mesocosms, bioaugmentation with dehalorespiring bacterium reduces about 56% of penta- and higher-chlorinated PCBs than that of control after 120 days	Rayford and Payne et al. (2011)
2-chlorobiphenyl, 4-CB and 2,4′-chlorobiphenyl	*Burkholderia xenovorans* strain LB400	For 85% removal of 2,4′-CB, bioaugmentation (strain LB400) plus biostimulation (tocopherol or berry extract) takes less time (13 days) than bioaugmentation alone (20 days)	Ponce et al. (2011)
Askarel oil (PCBs blend)	*Pseudomonas* sp., *Arthrobacter* sp., *Micrococcus* sp., and *Acinetobacter* sp.	Following incubation for 7 days, bacterial strains are able to utilized aerobically the askarel oil at 15 μL and 20 μL concentrations of askarel oil in minimal media	Nwinyi (2011)
2,3,4,5,6-pentachlorobiphenyl	*Chloroflexi* group	In Anacostia River microcosms, after 135 days of incubation time, about 10–15% removal of the total chlorines per biphenyl is observed in bioaugmentation treatments with or without electron donors like 1,2,3,4-tetrachlorobenzene and pentachloronitrobenzene	Park et al. (2011)

(continued)

Table 18.2 (continued)

PCBs (commercial name or congener types)	Bacteria species	Degradation	References
PCBs in transformer oil	*Rhodococcus* sp., *Pseudomonas* sp., *Pseudoxanthomonas* sp., *Agromyces* sp., and *Brevibacillus* sp.	Under aerobic or anaerobic conditions, 99.2–99.7% of low-chlorinated congeners were reduced at concentration level ranging between 20 μg/L and 700 μg/L	Nabavi et al. (2013)
Aroclor 1260	*Pseudomonas aeruginosa*	After 96 h of treatment, *P. aeruginosa* could degrade completely the Aroclor1260 in wastewater	Mathews (2014)
PCB 28, PCB 52, PCB 101, PCB 153, and PCB 180	*Pseudomonas* sp. MO2A and *Debaryomyces marasmus* CW36	The co-culture under aerobic conditions removes 92.6% of the PCB congeners with 61.4% dechlorination efficiencies within 14 days	Chen et al. (2015)
Aroclor 1254	*Dehalococcoides*, *Dehalococcoidia*	De-chlorinate PCBs from 150 ± 7 to 380 ± 44 μmol of chlorine kg^{-1} $week^{-1}$ in lagoon sediment	Nuzzo et al. (2017)
PCB180, Aroclor 1260	*Dehalococcoides mccartyi* CG1	The priming of tetrachloroethene (0.05–0.2 mM) in PCBs dichlorination could improve the efficiency of *D. mccartyi* CG1 by about 91% than that of control in pure culture and river sediment microcosm. *Dehalococcoides* is enriched more than 20 times in a river sediment microcosm with the amendment of tetrachloroethene	Xu et al. (2018)
Aroclor 1254, PCB 118	Indigenous organohalide-respiring bacteria	The anaerobic dechlorination rates of PCB 118 in sediment microcosms of Grasse River, Baltimore Harbor, and Fox River are 0.0153, 0.0144, and 0.0048 Cl^{-}/biphenyl/day, respectively. Aroclor 1254 dichlorination rates of	Kaya et al. (2018)

(continued)

Table 18.2 (continued)

PCBs (commercial name or congener types)	Bacteria species	Degradation	References
		Grasse River, Baltimore Harbor, and Fox River are 15.3×10^{-3} Cl^-/day, 14.4×10^{-3} Cl^-/day, and 4.8×10^{-3} Cl^-/day, respectively.	
Delor 103	*Ochrobactrum anthropi*, *Achromobacter xylosoxidans*, *Rhodococcus* sp., *Starkeya novella*, *Brevibacterium* sp., and *Stenotrophomonas maltophilia*	The bacterial mixed cultures of two-membered consisting of *Rhodococcus* sp. and *S. maltophilia* exhibit 70% of PCB congeners (highest biodegradation)	Horváthová et al. (2019)
Aroclor 1248	*Dehalococcoides*, Dehalobacter, Dehalogenimonas, and *Geobacter*	PCB 118 concentration in microcosm decreases by about 70% of the total amount of PCB congener over 430 days. The microcosm dechlorinates penta- and tetra-chlorinated PCB congeners without exogenous electron donor	Ewald et al. (2019)

analysis. Moreover, Tigini et al. (2009) isolated and identified native mitosporic fungal species from heavily Aroclor 1260 PCBs-contaminated sites; such as *Penicillium chrysogenum*, *Scedosporium apiospermum*, *Penicillium digitatum*, and *Fusarium solani* that biodegrade 4,4′-dichlorobiphenyl and 2,2′,5,5-′-tetrachlorobiphenyl regardless of ligninolytic enzymes. Pointing (2001) has inferred that white-rot fungi bioremediate the PCBs using ligninolytic enzymes, glycosylated heme peroxidases, lignin peroxidase, and Mn-dependent peroxidase.

Furthermore, Kamei et al. (2006) found that *Phlebia brevispora* TMIC33929 could biodegrade and transform nearly 50% of coplanar PCBs and 30% of 3,3′,4,4′,5-pentaCB to monomethoxylated PCBs. The study also shows that *P. brevispora* degrades especially non-chlorinated meta-position, and adsorb PCBs onto fungal biomass (Fig. 18.6). Sangely et al. (2009) revealed that *P. chrysosporium* exhibit substantial ability to degrade PCB 28, PCB 52, PCB 101, PCB 118, PCB 138, PCB 153, and PCB 180 ranging from 39% to 49% in water and soil and adsorbed about 42–54% of PCB congeners onto the mycelium. Fungal strains such as *Trametes* sp., *Polyporus* sp., *Nigroporus* sp. F33 and *Nigroporus* sp. U11 isolated from the lower northern region of Thailand tend to degrade more than 90% of 2,4-PCB in culture media after 15 days (Siripong et al. (2009). Similarly, four fungal strains, namely, LC1, LC2, LC3, and LC4, isolated

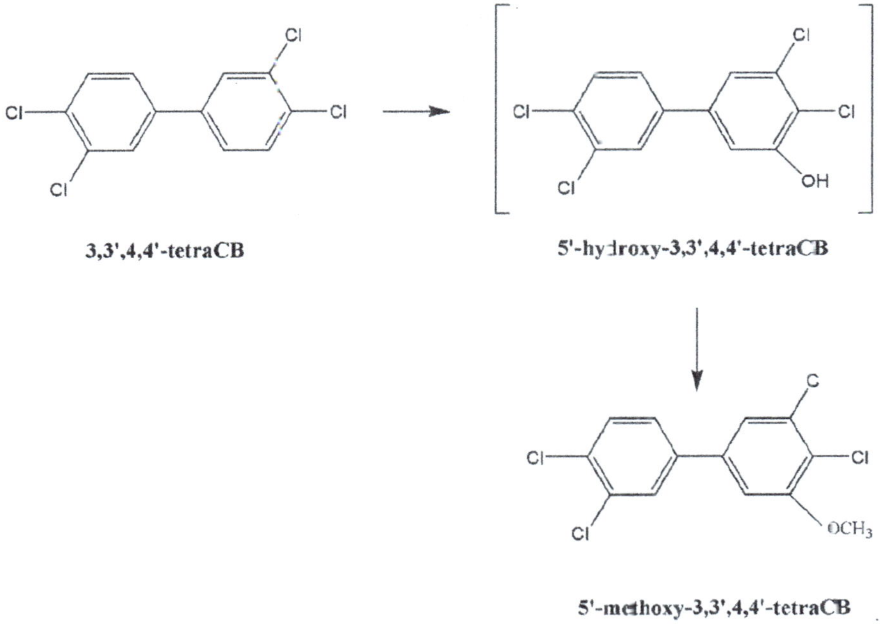

Fig. 18.6 Pathway of PCBs biodegradation by *P. brevispora* TMIC33929 (Kamei et al. 2006)

from contaminated sediment of Licun River could degrade PCBs congener in the presence of sodium dodecyl sulfate: 35% by LC1, 25% by LC2, 38% by LC3, and 15% by LC4 within 35 days (Yin et al. (2011). The researchers also concluded that about 7% of PCBs suffer losses due to volatilization and fungal biosorption. As further noted by Federici et al. (2012) work, the bio-augmentation of *Lentinus tigrinus* with maize stalk in contaminated soil removes Aroclor 1242 and Delor 103 by about 33.6 ± 0.3%.

Gayosso-Canales et al. (2012) study also found that the addition of PCB congener could induce 20-fold more production of laccase by *P. ostreatus*, resulting in enhanced removal of Delor 103/106 from the transformer oil in 30 days. It was reported that *Aspergillus niger* could biodegrade 65% of PCB 1, 38% of PCB 3, and 52% of PCB 38, but not PCB 126 during the study periods. Subsequent work on three fungi isolates revealed that the *A. bisporus* could remove about 31.32–83.91% PCBs; *P. ostreatus* could deplete approximately 37.88–78.29% PCBs; and *L. edodes* could biodegrade about 17.38–75.30% PCB, within 12 weeks, respectively (Gąsecka et al. 2015). Stella et al. (2017) also showed that bioaugmentation of *P. ostreatus* in dumpsite soil could degrade 50% of PCB's aromatic moiety into hydroxylated and methoxylated PCB derivatives such as chlorobenzoates, chlorobenzaldehydes, chlorobenzyl alcohols, and chlorocresol over 12 weeks. During the degradation study, *Eurotiomycetes* and *Sordariomycetes* tend to degrade and adsorb PCBs in low SOC soil more than in high SOC soil (Huang et al. 2018). More recently, Perigon et al. (2019) pointed out that *Thermothelomyces thermophila*,

Schizophyllum commune, *Thermothelomyces heterothallica*, *Thermoascus crustaceus*, and *F. solani* might tolerate and degrade the PCB congener irrespective of their origins and laccase enzymes, for example, 84.79–88.11% of PCBs degradation by *F. solani*, 79.86–86.61% of PCB degradation by *T. heterothallica*, 95.22–95.91% of PCB degradation by *T. crustaceus*, 93.44–94.64% of PCB degradation by *T. thermophila*, and 80.91–86.20% of PCB degradation by *S. commune* under medium liquid experiment. It was established by Sredlova et al. (2020) that PCBs biodegradation in contaminated water achieves removal percentages of about 82, 80, 65, and 50% of di-, tri-, tetra- and penta-chlorinated PCB congeners using spent oyster mushroom substrate in the trickle-bed bioreactor, respectively. As shown in Table 18.3, fungal species biodegrade different PCB congeners in various environment.

18.4 Conclusion and Future Prospects

Many studies have reported the isolation and identification of PCBs-degrading microbes, including their degradation pathways. However, the biodegradation of PCB congeners has remained a great challenge worldwide because of the physical and chemical properties of PCBs, such as low availability, toxicity, hydrophobicity, high sorption, and persistence. It often takes several months, or years, or decades in biodegrading the PCB congeners in various studied environment via complex networks due to physiological, biochemical, ecological, and genetic factors, such as, pH, temperature, tolerance, substrate specificity, kinetics, aerobic or anaerobic conditions, substrates (sediment or soil), other contaminants, etc. Moreover, there is a limitation in upscaling the laboratory findings to fields because in some cases, the selected cultivable bacteria and fungi isolates from laboratory results may not be practical in treating the highly contaminated sites in field conditions. Metagenomic techniques that characterize the composition of microbial communities and their habitat relationships are also susceptible to the common limitations of molecular techniques such as loss of genetic information and provide little information on quantitative physiological characteristics. Although microbial degradation of PCBs is found to suffer from certain drawbacks, it remains the most promising approach due to cost-effectiveness, safety, eco-friendliness, and sustainability. However, a multidisciplinary approach involving chemical, physical, and biological sciences is required to achieve the desired results. Omic approaches like metagenomics have found a new way in identifying un-culturable microbes capable of degrading PCB congeners along with their metabolic pathway. Future research needs to study PCB-degrading microbes for improving their efficiency, survivability, and performance of microbial dynamics using genetic engineering and molecular techniques. Numerous microbes belonging to bacteria and fungi can biodegrade PCBs in aerobic and anaerobic conditions through bioaugmentation and bio-stimulation processes, and there is a need to exploit them efficiently in field conditions.

Table 18.3 Biodegradation of PCB congener involving fungal species

PCBs (commercial name or congener types)	Fungi	Major findings	References
4-chlorobiphenyl and 4,4′-dichlorobiphenyl	*Rhizopus japonicus*	The fungus transforms 4,4-dichlorobiphenyl to hydroxy metabolites	Wallnöfer et al. (1973)
Aroclors 1242, 1254, and 1260	*Phanerochaete chrysosporium*	*P. chrysosporium* degrades 60.9% of Aroclors 1242, 30.5% of 1254, and 17.6% of 1260. It is nonspecific for the position of chlorine substitutions on the biphenyl ring	Yadav et al. (1995)
3,3′,4,4′-tetrachlorobiphenyl, 2,3,3′,4,4′-pentachlorobiphenyl, 2,3′,4,4′,5-pentachlorobiphenyl, 3,3′,4,4′,5-pentachlorobiphenyl, and 2,3′,4,4′,5,5′-hexachlorobiphenyl	*Phlebia brevispora* TMIC33929 and *Phanerochaete sordida* YK624	*P. acerina* biodegrade only 3,3′,4,4′-tetraCB. *P. brevispora* biodegrade 50% of 3,3′,4,4′-tetraCB, 2,3,3′,4,4′-pentaCB, 2,3′,4,4′,5-pentaCB, and 2,3′,4,4′,5,5′-hexaCB	Kamei et al. (2006)
4-Chlorobiphenyl (4-CB) and 4,4′-dichlorobiphenyl (4,4′-DCB)	White-rot fungi: *Phanerochaete chrysosporium* and *Phanerochaete* sp. MZ142	*Phanerochaete* sp. MZ142 could remove 50% of 4,4′-DCB in culture media. *Phanerochaete* sp. MZ142 degrade 4,4′-DCB more than that of *P. chrysosporium*. *P. chrysosporium* transform 4,4′-DCB to 3-hydroxy-4,4′-DCB and 4-hydroxy-3,4′-DCB	Kamei et al. (2006)
2,4-PCB	*Trametes* sp., *Polyporus* sp., *Nigroporus* sp., fungal isolate F33, H11, and *Phanerochaete chrysosporium* IFO 311249	Fungal isolate F33 and *Polyporus* sp. degrade 96.44 ± 2.1% of 2,4-PCB after 15 days of incubation	Siripong et al. (2009)
Aroclor 1260	Mitosporic fungi: *Scedosporium*, *Penicillium*, and *Aspergillus* genera	*Penicillium chrysogenum*, *Scedosporium apiospermum*, *Penicillium digitatum*, and *Fusarium solani* show remarkable PCBs degradation percentages of 17–86% within 30 days of incubation. No ligninolytic enzyme is involved in the degradation	Tigini et al. (2009)

(continued)

Table 18.3 (continued)

PCBs (commercial name or congener types)	Fungi	Major findings	References
Aroclor 1242	*Pleurotus ostreatus*	Using laccase catalyzes, it can achieve 63.8 ± 4.6% PCBs degradation efficiency within 30 days	Gayosso-Canales et al. (2012)
Delor 103	Ligninolytic fungi: *Irpex lacteus* 617/93, *Bjerkandera adusta* 606/93, *Phanerochaete chrysosporium* ME 446, *Phanerochaete magnoliae* CCBAS 134/I, *Pleurotus ostreatus* 3004 CCBAS 278, *Trametes versicolor* 167/93, *Pycnoporus cinnabarinus* CCBAS 595, and *Dichomitus squalens* CCBAS 750	*P. ostreatus* is able to remove PCBs mixture by about 98.4% in complex and by about 99.6% in mineral media	Čvančarová et al. (2012)
PCB 28, 52, 101, 118, 138, 153, and 180	Autochthonous fungal strains: *Doratomyces nanus*, *Doratomyces purpureofuscus*, *Doratomyces verrucisporus*, *Myceliophthora thermophila*, *Phoma eupyrena*, and *Thermoascus crustaceus*	More than 70% degradation irrespective of the number of chlorine substituents on the biphenyl ring. The fungi also show high tolerance toward PCBs	Mouhamadou et al. (2013)
PCB 44, 52, 77, 101, 105, 118, 138, 149, 153, 156, 167, 169, 170, 180, and 194	*Doratomyces nanus*, *Doratomyces purpureofuscus*, *Doratomyces verrucisporus*, *Myceliophthora thermophila*, *Phoma eupyrena*, and *Thermoascus crustaceus*	The fungi reduce 33.3% of the PCBs over 6 months	Sage et al. (2014)
PCB 28, 52, 101, 138, 153 or 180	*Pleurotus ostreatus*, *Lentinula edodes*, and *Agaricus bisporus*	At the end of 12 weeks of incubation, the degree of degradation of a single PCBs by *A. bisporus* ranged between 31.32 ± 1.52 and 83.91 ± 1.07%; by *P. ostreatus* ranged between 37.88 ± 2.54 and 78.29 ± 1.41%; by *L. edodes* ranged 17.38 ± 1.06 and 75.30 ± 1.46%	Gąsecka et al. (2015)

PCB 1, PCB 3, PCB 38, and PCB 126	*Aspergillus niger*	*A. niger* degrades PCB 1, PCB 3, and PCB 38 by 65, 38, 52, and 2% after 12 days, respectively	Kim et al. (2016)
PCB-contaminated soil and Delor 103	White-rot *fungi*: *Pleurotus ostreatus* and *Irpex lacteus*	*P. ostreatus* could remove PCBs from dumpsite soil by 18.5, 41.3, and 50.5% from the bulk, surface, and rhizosphere after 84 days, respectively. Detect mono-hydroxylated PCBs derivatives in biostimulated microcosms	Stella et al. (2017)
Transformer oil: Aroclor 1242, 1254, and 1260	*P. sajor-caju* LBM 105	After 35 days of incubation, *P. sajor-caju* LBM 105 could remove PCBs mixture by 97.7% in complex and 91.7% in mineral media, respectively	Sadañoski et al. (2019)
PCB 28, 52, 101, 118, 138, 153, and 180	*Thermothelomyces thermophila*, *Thermothelomyces heterothallica*, *Thermoascus crustaceus*, and *Fusarium solani*	Remove PCBs by more than 84.79–88.11% in *Fusarium solani*, 79.86–86.61% in *Thermothelomyces heterothallica*, 95.22–95.91% in *Thermoascus crustaceus*, and 93.44–94.64% in *Thermothelomyces thermophila*. *Schizophyllum commune* show resistance to PCBs and their metabolites	Perigon et al. (2019)
Delor 103 (PCBs mixture)	*Pleurotus ostreatus*	*P. ostreatus* removes 82, 80, 65, and 30–50% of di-, tri-, tetra-, and pentachlorinated PCB congeners, respectively	Sredlova et al. (2020)
PCB 28, 52, 101, 118, 138, 153, and 180	*Penicillium chrysogenum*, *P. citreosulfuratum*, *P. canescens*, and *Aspergillus jensenii*	These fungi could biodegrade greater than 70%. Only *P. canescens* able to reduce the toxicity of PCBs and their metabolite	Germain et al. (2021)
Transformer oil (Aroclors 1242, 1254, and 1260)	*Pleurotus pulmonarius* LBM 105	After 21 days, the fungus achieves 65.50 ± 8.09% PCBs degradation using oxidative metabolism by glyceraldehyde-3-phosphate dehydrogenase	Sebastian et al. (2021)

Acknowledgments Mr. Ningthoujam Samarendra Singh acknowledges PG School, ICAR-IARI for extending financial support in the form of a fellowship.

Conflict of Interest: The author declares that there is no conflict of interest regarding the publication of this chapter.

References

Abramowicz DA (1995) Aerobic and anaerobic PCB biodegradation in the environment. Environ Health Perspect 103:97–99

Ahmed M, Focht DD (1973) Degradation of polychlorinated biphenyls by two species of *Achromobacter*. Can J Microbiol 19:47–52

Alegria H, Martinez-Colon M, Birgul A, Brooks G, Hanson L, Kurt-Karakus P (2016) Historical sediment record and levels of PCBs in sediments and mangroves of Jobos Bay, Puerto Rico. Sci Total Environ 573:1003–1009

Arai H, Kosono S, Taguchi K, Maeda M, Song E, Fuji F, Chung SY, Kudo T (1998) Two sets of biphenyl and PCB degradation genes on a linear plasmid in *Rhodococcus erythropolis* TA421. J Ferment Bioeng 86:595–599

Arora NK (2018) Bioremediation: a green approach for restoration of polluted ecosystems. Environ Sust 1:305–307

Balejčíková L (2018) Contamination of water and soil by polychlorinated biphenyl. Acta Hydrol slovaca 2:272–277

Ballschmiter K, Zell M (1980) Analysis of polychlorinated biphenyls (PCB) by glass capillary gas chromatography—composition of technical aroclor- and Clophen-PCB mixtures. Fresenius' Zeitschrift für Anal Chemie 302:20–31

Beaudette LA, Davies S, Fedorak PM, Ward OP, Pickard MA (1998) Comparison of gas chromatography and mineralization experiments for measuring loss of selected polychlorinated biphenyl congeners in cultures of white rot fungi. Appl Environ Microbiol 64:2020–2025

Bedard DL (2008) A case study for microbial biodegradation: anaerobic bacterial reductive dechlorination of polychlorinated biphenyls—from sediment to defined medium. Annu Rev Microbiol 62:253–270

Bedard DL, Haberl ML, May RJ, Brennan MJ (1987) Evidence for novel mechanisms of polychlorinated biphenyl metabolism in *Alcaligenes eutrophus* H850. Appl Environ Microbiol 53:1103–1112

Bedard DL, Ritalahti KM, Löffler FE (2007) The Dehalococcoides population in sediment-free mixed cultures metabolically dechlorinates the commercial polychlorinated biphenyl mixture Aroclor 1260. Appl Environ Microbiol 73:2513–2521

Billingsley KA, Backus SM, Ward OP (1997) Studies on the transformation of selected polychlorinated biphenyl congeners by *Pseudomonas* strain LB 400. Can J Microbiol 43: 782–788

Borja J, Taleon DM, Auresenia J, Gallardo S (2005) Polychlorinated biphenyls and their biodegradation. Process Biochem 40:1999–2013

Breivik K, Sweetman A, Pacyna JM, Jones KC (2002) Towards a global historical emission inventory for selected PCB congeners—a mass balance approach: 1. Global production and consumption. Sci Total Environ 290:181–198

Chakraborty J, Das S (2016) Characterization of the metabolic pathway and catabolic gene expression in biphenyl degrading marine bacterium *Pseudomonas aeruginosa* JP-11. Chemosphere 144:1706–1714. https://doi.org/10.1016/j.chemosphere.2015.10.059

Chen F, Hao S, Qu J, Ma J, Zhang S (2015) Enhanced biodegradation of polychlorinated biphenyls by defined bacteria-yeast consortium. Ann Microbiol 65:1847–1854

Chen X, Lin Y, Dang K, Puschner B (2017) Quantification of polychlorinated biphenyls and polybrominated diphenyl ethers in commercial cows' milk from California by gas chromatography±triple quadruple mass spectrometry. PLoS One 12:1–18

Chitsaz M, Fennell DE, Rodenburg LA (2020) Sources of polychlorinated biphenyls to upper Hudson River sediment post-dredging. Chemosphere 259:127438

Commandeur LC, May RJ, Mokross H, Bedard DL, Reineke W, Govers HA, Parsons JR (1996) Aerobic degradation of polychlorinated biphenyls by *Alcaligenes* sp. JB1: metabolites and enzymes. Biodegradation 7:435–443

Čvančarová M, Křesinová Z, Filipová A, Covino S, Cajthaml T (2012) Biodegradation of PCBs by ligninolytic fungi and characterization of the degradation products. Chemosphere 88:1317–1323

Davison AD, Veal DA (1997) Synergistic mineralization of biphenyl by *Alcaligenes faecalis type* II BPSI-2 and *Sphingomonas paucimobilis* BPSI-3. Lett Appl Microbiol 25:58–62

De J, Ramaiah N, Sarkar A (2006) Aerobic degradation of highly chlorinated polychlorobiphenyls by a marine bacterium, *Pseudomonas* CH07. World J Microbiol Biotechnol 22:1321–1327

Di Toro S, Zanaroli G, Fava F (2006) Intensification of the aerobic bioremediation of an actual site soil historically contaminated by polychlorinated biphenyls (PCBs) through bioaugmentation with a non acclimated, complex source of microorganisms. Microb Cell Factories 5:1–10

Elangovan S, Pandian SBS, Geetha SJ, Joshi SJ (2019) Polychlorinated biphenyls (PCBs): environmental fate, challenges and bioremediation, pp 165–188

Elnar AA, Desor F, Marin F, Soulimani R, Nemos C (2015) Lactational exposure to low levels of the six indicator non-dioxin-like polychlorinated biphenyls induces DNA damage and repression of neuronal activity, in juvenile male mice. Toxicology 328:57–65

Erickson MD (2001) Introduction: PCB properties, uses, occurrence and regulatory history. In: Robertson LW, Hansen LG (eds) PCBs: recent advances in environmental toxicology and health effects. The University Press of Kentucky, Lexington, KY, pp 1–9

Erickson MD, Kaley RG (2011) Applications of polychlorinated biphenyls. Environ Sci Pollut Res 18:135–151

Ewald JM, Humes SV, Martinez A, Schnoor JL, Mattes TE (2019) Growth of *Dehalococcoides* spp. and increased abundance of reductive dehalogenase genes in anaerobic PCB-contaminated sediment microcosms. Environ Sci Pollut Res 27(9):8846–8858

Faroon O, Ruiz P (2016) Polychlorinated biphenyls: new evidence from the last decade. Toxicol Ind Health 32:1825–1847

Federici E, Giubilei M, Santi G, Zanaroli G, Negroni A, Fava F, Petruccioli M, D'Annibale A (2012) Bioaugmentation of a historically contaminated soil by polychlorinated biphenyls with *Lentinus tigrinus*. Microb Cell Factories 11:35

Focht DD, Brunner W (1985) Kinetics of biphenyl and polychlorinated biphenyl metabolism in soil. Appl Environ Microbiol 50:1058–1067

Fujihara H, Yamazoe A, Hosoyama A, Suenaga H, Kimura N, Hirose J, Watanabe T, Futagami T, Goto M, Furukawa K (2015) Draft genome sequence of *Pseudomonas aeruginosa* KF702 (NBRC 110665), a polychlorinated biphenyl-degrading bacterium isolated from biphenyl-contaminated soil. Genome Announc 3(3)

Furukawa K (2000) Biochemical and genetic bases of microbial degradation of polychlorinated biphenyls (PCBs). J Gen Appl Microbiol 46:283–296

Furukawa K, Matsumura F, Ton Omura K (1978) *Alcaligenes* and *Acinetobacter* strains capable of degrading polychlorinated biphenyls. Agric Biol Chem 42:543–548

Gangar T, Bhardwaj KK, Gupta R (2019) Microbes and processes in bioremediation of soil. In: Arora NK, Kumar A, Swati S (eds) Microbes and enzymes in soil health and bioremediation. Microorganisms for sustainability, vol 16. Springer, Singapore, pp 11–37

Gąsecka M, Drzewiecka K, Siwulski M, Sobieralski K (2015) Evaluation of polychlorinated biphenyl degradation through refuse from *Pleurotus ostreatus, Lentinula edodes* and *Agaricus bisporus* production. Folia Hortic 27:135–144

Gayosso-Canales M, Rodríguez-Vázquez R, Esparza-García FJ, Bermúdez-Cruz RM (2012) PCBs stimulate laccase production and activity in *Pleurotus ostreatus* thus promoting their removal. Folia Microbiol (Praha) 57:149–158

Germain J, Raveton M, Binet MN, Mouhamadou B (2021) Screening and metabolic potential of fungal strains isolated from contaminated soil and sediment in the polychlorinated biphenyl degradation. Ecotoxicol Environ Saf 208:111703

Gomes HI, Dias-ferreira C, Ribeiro AB (2013) Overview of in situ and ex situ remediation technologies for PCB-contaminated soils and sediments and obstacles for full-scale application. Sci Total Environ 445–446:237–260

Grimm FA, Hu D, Kania-Korwel I, Lehmler HJ, Ludewig G, Hornbuckle KC, Duffel MW, Bergman Å, Robertson LW (2015) Metabolism and metabolites of polychlorinated biphenyls. Crit Rev Toxicol 45:245–272

Grossman E (2013) Nonlegacy PCBs: pigment manufacturing by-products get a second look. Environ Health Perspect 121:a86–a93

Harms H, Schlosser D, Wick LY (2011) Untapped potential: exploiting fungi in bioremediation of hazardous chemicals. Nat Rev Microbiol 9:177–192

Hatamian-Zarmi A, Shojaosadati SA, Vasheghani-Farahani E, Hosseinkhani S, Emamzadeh A (2009) Extensive biodegradation of highly chlorinated biphenyl and Aroclor 1242 by *Pseudomonas aeruginosa* TMU56 isolated from contaminated soils. Int Biodeterior Biodegrad 63:788–794

Horváthová H, Lászlová K, Dercová K (2018) Bioremediation of PCB-contaminated shallow river sediments: the efficacy of biodegradation using individual bacterial strains and their consortia. Chemosphere 193:270–277

Horváthová H, Lászlová K, Dercová K (2019) Remediation potential of bacterial mixed cultures for polychlorinated biphenyls (PCBs) biodegradation. Acta Chim Slovaca 12:1–7

Hou D, O'Connor D, Igalavithana AD, Alessi DS, Luo J, Tsang DC, Sparks DL, Yamauchi Y, Rinklebe J, Ok YS (2020) Metal contamination and bioremediation of agricultural soils for food safety and sustainability. Nat Rev Earth Environ 1(7):366–381

Hu D, Hornbuckle KC (2010) Inadvertent polychlorinated biphenyls in commercial paint pigments. Environ Sci Technol 44:2822–2827

Huang S, Shan M, Chen J, Penttinen P, Qin H (2018) Contrasting dynamics of polychlorinated biphenyl dissipation and fungal community composition in low and high organic carbon soils with biochar amendment. Environ Sci Pollut Res 25:33432–33442

Hung H, Katsoyiannis AA, Brorström-Lundén E, Olafsdottir K, Aas W, Breivik K, Bohlin-Nizzetto P, Sigurdsson A, Hakola H, Bossi R, Skov H (2016) Temporal trends of persistent organic pollutants (POPs) in Arctic air: 20 years of monitoring under the Arctic monitoring and assessment Programme (AMAP). Environ Pollut 217:52–61

IARC (2016) Polychlorinated biphenyls and Polybrominated biphenyls. IARC Monogr Eval Carcinog risks to humans 107:9–500

Jan Campfens DM (1997) Fugacity-based model of PCB bioaccumulation in complex aquatic food webs. Environ Sci Technol 31:577–583

Jia Y, Wang J, Ren C, Nahurira R, Khokhar I, Wang J, Fan S, Yan Y (2019) Identification and characterization of a meta-cleavage product hydrolase involved in biphenyl degradation from *Arthrobacter* sp. YC-RL1. Appl Microbiol Biotechnol 103:6825–6836

Kamei I, Sonoki S, Haraguchi K (2006) Fungal bioconversion of toxic polychlorinated biphenyls by white-rot fungus, *Phlebia brevispora*. Appl Microbiol Biotechnol 73:932–940

Kamei I, Kogura R, Kondo R (2006) Metabolism of 4,4′-dichlorobiphenyl by white-rot fungi *Phanerochaete chrysosporium* and *Phanerochaete* sp. MZ142. Appl Microbiol Biotechnol 72:566–575

Kaya D, Imamoglu I, Sanin FD, Sowers KR (2018) A comparative evaluation of anaerobic dechlorination of PCB-118 and Aroclor 1254 in sediment microcosms from three PCB-impacted environments. J Hazard Mater 341:328–335

Kelly BC, Ikonomou MG, Blair JD, Morin AE, Gobas FA (2007) Food web-specific biomagnification of persistent organic pollutants. Science 317:236–239

Kim CS, Lim DH, Keum YS (2016) Biodegradation pathways of polychlorinated biphenyls by soil fungus *aspergillus Niger*. Korean J Pestic Sci 20:7–13

Kimura N, Watanabe T, Suenaga H, Fujihara H, Futagami T, Goto M, Hanada S, Hirose J (2018) *Pseudomonas furukawaii* sp. Nov. a polychlorinated biphenyl-degrading bacterium isolated from biphenyl-contaminated soil in Japan. Int J Syst Evol Microbiol 68:1429–1435

Kour D, Kaur T, Devi R, Yadav A, Singh M, Joshi D, Singh J, Sayal DC, Kumar A, Rajput VD, Yadav AN, Sigh K, Singh J, Sayyed RZ, Arora NK, Saxena AK (2021) Beneficial microbiomes for bioremediation of diverse contaminated environments for environmental sustainability: present status and future challenges. Environ Sci Pollut Res 28:24917–24939

Lai Z, Li X, Li H, Zhao L, Zeng Y, Wang C, Gao Y, Liu Q (2015) Residual distribution and risk assessment of polychlorinated biphenyls in surface sediments of the Pearl River Delta, South China. Bull Environ Contam Toxicol 95:37–44

Lammel G, Stemmler I (2012) Fractionation and current time trends of PCB congeners: evolvement of distributions 1950-2010 studied using a global atmosphere-ocean general circulation model. Atmos Chem Phys 12:7199–7213

Leaes FL, Daniel AP, Mello GB, Battisti V, Bogusz S Jr, Emanuell T, Fries LL, Costabeber I (2006) Degradation of polychlorinated biphenyls (PCBs) by *Staphylococcus xylosus* in liquid media and meat mixture. Food Chem Toxicol 44:847–854

Leigh MB, Prouzová P, Macková M, Macek T, Nagle DP, Fletcher JS (2006) Polychlorinated biphenyl (PCB)-degrading bacteria associated with trees in a PCB-contaminated site. Appl Environ Microbiol 72:2331–2342

Lo Giudice A, Caruso G, Rizzo C, Papale M, Azzaro M (2019) Bacterial communities versus anthropogenic disturbances in the Antarctic coastal marine environment. Environ Sust 2:297–310

Lu Q, Qiu L, Xu G (2018) Enantioselective dechlorination of polychlorinated biphenyls in *Dehalococcoides mccartyi* CG1. Appl Environ Microbiol 21:1–12

Maervoet J, Covaci A, Schepens P, Sandau CD, Letcher RJ (2004) A reassessment of the nomenclature of polychlorinated biphenyl (PCB) metabolites. Environ Health Perspect 112: 291–294

Mathews S (2014) Biodegradation of polychlorinated biphenyls (PCBs), Aroclor 1260, in waste-water by isolate MD2 (*Pseudomonas aeruginosa*) from wastewater from Notwane sewage treatment Plant in Gaborone, Botswana. J Bioremed Biodegr 05:3–8

Mohn WW, Westerberg K, Cullen WR, Reimer KJ (1997) Aerobic biodegradation of biphenyl and polychlorinated biphenyls by Arctic soil microorganisms. Appl Environ Microbiol 63:3378–3384

Mouhamadou B, Faure M, Sage L, Marçais J, Souard F, Geremia RA (2013) Potential of autochthonous fungal strains isolated from contaminated soils for degradation of polychlorinated biphenyls potential of autochthonous fungal strains isolated from contaminated soils for degradation of polychlorinated biphenyls. Fungal Biol 117:268–274

Nabavi BF, Nikaeen M, Amin MM (2013) Isolation and identification of aerobic polychlorinated biphenyls degrading bacteria. Int J Environ Health Eng 2:2–5

Nuzzo A, Negroni A, Zanaroli G, Fava F (2017) Identification of two organohalide - respiring Dehalococcoidia associated to different dechlorination activities in PCB—impacted marine sediments. Microb Cell Factories 16:127

Nwinyi OC (2011) Enrichment and identification of Askarel oil (PCB blend) degrading bacteria enriched from landfill sites in Edo state, Nigeria. Agric Biol J North Am 2:89–100

Ododo MM, Wabalo BK (2019) Polychlorinated biphenyls (PCBs) and their impacts on human health: a review, vol 7, pp 73–77

Olatunji OS (2019) Evaluation of selected polychlorinated biphenyls (PCBs) congeners and dichlorodiphenyltrichloroethane (DDT) in fresh root and leafy vegetables using GC-MS. Sci Rep 9:1–10

Park JW, Krumins V, Kjellerup BV, Fennell DE, Rodenburg LA, Sowers KR, Kerkhof LJ, Häggblom MM (2011) The effect of co-substrate activation on indigenous and bioaugmented PCB dechlorinating bacterial communities in sediment microcosms. Appl Microbiol Biotechnol 89:2005–2017

Payne RB, May HD, Sowers KR (2011) Enhanced reductive dechlorination of polychlorinated biphenyl impacted sediment by bioaugmentation with a dehalorespiring bacterium. Env Sci Technol 45:8772–8779

Pentyala SN, Rebecchi M, Mishra S, Rahman A, Stefan R, Rebecchi J, Kodavanti PR (2011) Polychlorinated biphenyls: in situ bioremediation from the environment. J Environ Pollut Hum Health:249–262

Perigon S, Massier M, Germain J, Binet MN, Legay N, Mouhamadou B (2019) Metabolic adaptation of fungal strains in response to contamination by polychlorinated biphenyls. Environ Sci Pollut Res 26:14943–14950

Pieper DH (2005) Aerobic degradation of polychlorinated biphenyls. Appl Microbiol Biotechnol 67:170–191

Pieper DH, Seeger M (2008) Bacterial metabolism of polychlorinated biphenyls. J Mol Microbiol Biotechnol 15:121–138

Pointing SB (2001) Feasibility of bioremediation by white-rot fungi. Appl Microbiol Biotechnol 57: 20–33

Ponce BL, Latorre VK, González M, Seeger M (2011) Antioxidant compounds improved PCB-degradation by *Burkholderia xenovorans* strain LB400. Enzym Microb Technol 49:509–516

Robinson GK, Lenn MJ (1994) The bioremediation of polychlorinated biphenyls (PCBS): problems and perspectives. Biotechnol Genet Eng Rev 12:139–188

Rodrigues JL, Kachel CA, Aiello MR, Quensen JF, Maltseva OV, Tsoi TV, Tiedje JM (2006) Degradation of Aroclor 1242 Dechlorination products in sediments by *Burkholderia xenovorans* LB400 (ohb) and *Rhodococcus* sp. Appl Environ Microbiol 72:2476–2482

Sadañoski MA, Benítez SF, Fonseca MI, Velázquez JE, Zapata PD, Levin LN, Villalba LL (2019) Mycoremediation of high concentrations of polychlorinated biphenyls with *Pleurotus sajor-caju* LBM 105 as an effective and cheap treatment. J Environ Chem Eng 7:103453

Sadañoski MA, Tatarin AS, Barchuk ML, Gonzalez M, Pegoraro CN, Fonseca MI, Levin LN, Villalba LL (2020) Evaluation of bioremediation strategies for treating recalcitrant halo-organic pollutants in soil environments. Ecotoxicol Environ Saf 202:110929

Sage L, Périgon S, Faure M, Gaignaire C, Abdelghafour M, Mehu J, Geremia RA, Mouhamadou B (2014) Autochthonous ascomycetes in depollution of polychlorinated biphenyls contaminated soil and sediment. Chemosphere 110:62–69

Samadi A, Sharifi H, Ghobadi Nejad Z, Yaghmaei S (2019) Biodegradation of polychlorinated biphenyls by *Lysinibacillus macrolides* and *Bacillus firmus* isolated from contaminated soil. Int J Eng Trans B Appl 32:628–633

Sangely M, Sablayrolles C, Vialle C, Strehaiano P, Thannberger L, Montrejaud-Vignoles M (2009) Polychlorinated biphenyls fractioning in aqueous bioremediation assay with *Phanerochaete chrysosporium*. Int J Environ Anal Chem 89:849–856

Šašek V, Volfová O, Erbanová P, Vyas BR, Matucha M (1993) Degradation of PCBs by white rot fungi, methylotrophic and hydrocarbon utilizing yeasts and bacteria. Biotechnol Lett 15:521–526

Sebastian CA, Elizabet AA, Marcela LC, Dario ZP, Isabel FM (2021) Proteomic insight on the polychlorinated biphenyl degrading mechanism of *Pleurotus pulmonarius* LBM 105. Chemosphere 265:129093

Seto M, Kimbara K, Shimura M, Hatta T, Fukuda M, Yano K (1995) A novel transformation of polychlorinated biphenyls by *Rhodococcus* sp. strain RHA1. Appl Environ Microbiol 61:3353–3358

Sharma JK, Gautam RK, Nanekar SV, Weber R, Singh BK, Singh SK, Juwarkar AA (2018) Advances and perspective in bioremediation of polychlorinated biphenyl-contaminated soils. Environ Sci Pollut Res 25:16355–16375

Shimura M, Mukerjee-Dhar G, Kimbara K, Nagato H, Kiyohara H, Hatta T (1999) Isolation and characterization of a thermophilic *Bacillus* sp. JF8 capable of degrading polychlorinated biphenyls and naphthalene. FEMS Microbiol Lett 178:87–93

Siripong P, Oraphin B, Duanporn P (2009) The ability of five fungal isolates from nature to degrade of Polyaromatic hydrocarbons (PAHs) and polychlorinated biphenyls (PCBs) in culture media. J Bas App Sci 3:1076–1082

Sredlova K, Skrob Z, Filipova A, Masin P, Jana Holecova J, Cajtham T (2020) Biodegradation of PCBs in contaminated water using spent oyster mushroom substrate and a trickle-bed bioreactor. Water Res 170:115274

Stella T, Covino S, Burianová E, Filipová A, Křesinová Z, Voříšková J, Větrovský T, Baldrian P, Cajthaml T (2015) Chemical and microbiological characterization of an aged PCB-contaminated soil. Sci Total Environ 533:177–186

Stella T, Covino S, Čvančarová M, Filipová A, Petruccioli M, D'Annibale A, Cajthaml T (2017) Bioremediation of long-term PCB-contaminated soil by white-rot fungi. J Hazard Mater 324: 701–710

Su X, Li S, Cai J, Xiao Y, Tao L, Hashmi MZ, Lin H, Chen J, Mei R, Sun F (2019) Aerobic degradation of 3,3′,4,4′-tetrachlorobiphenyl by a resuscitated strain *Castellaniella* sp. SPC4: kinetics model and pathway for biodegradation. Sci Total Environ 688:917–925

Sun J, Zhu L, Pan L, Wei Z, Song Y, Zhang Y, Qu L, Zhan Y (2016) Detection of methoxylated and hydroxylated polychlorinated biphenyls in sewage sludge in China with evidence for their microbial transformation. Sci Rep 6:1–9

Tiedje JM, Quensen JF, Chee-Sanford J, Schimel JP, Boyd SA (1993) Microbial reductive dechlorination of PCBs. Biodegradation 4:231–240

Tigini V, Prigione V, Di Toro S, Fava F, Varese GC (2009) Isolation and characterisation of polychlorinated biphenyl (PCB) degrading fungi from a historically contaminated soil. Microb Cell Factories 8:5

Vasilyeva GK, Strijakova ER (2007) Bioremediation of soils and sediments contaminated by polychlorinated biphenyls. Microbiology 76(6):639–653

Wallnöfer PR, Engelhardt G, Safe S, Hutzinger O (1973) Microbial hydroxylation of 4-chlorobiphenyl and 4,4′-dichlorobiphenyl. Chemosphere 2:69–72

Wang S, Chng KR, Wilm A, Zhao S, Yang KL, Nagarajan N, He J (2014) Genomic characterization of three unique Dehalococcoides that respire on persistent polychlorinated biphenyls. Proc Natl Acad Sci U S A 111:12103–12108

Wang S, Chen C, Zhao S, He J (2019a) Microbial synergistic interactions for reductive dechlorination of polychlorinated biphenyls. Sci Total Environ 666:368–376

Wang W, Bai J, Zhang G, Jia J, Wang X, Liu X, Cui B (2019b) Occurrence, sources and ecotoxicological risks of polychlorinated biphenyls (PCBs) in sediment cores from urban, rural and reclamation-affected rivers of the Pearl River Delta, China. Chemosphere 218:359–367

Wiegel J, Wu Q (2000) Microbial reductive dehalogenation of polychlorinated biphenyls. FEMS Microbiol Ecol 32:1–15

Xu G, Lu Q, Yu L, Wang S (2018) Tetrachloroethene primes reductive dechlorination of polychlorinated biphenyls in a river sediment microcosm tetrachloroethene primes reductive dechlorination of polychlorinated biphenyls in a river sediment microcosm. Water Res 152:87–95

Yadav JS, Quensen JF, Tiedje JM, Reddy CA (1995) Degradation of polychlorinated biphenyl mixtures (Aroclors 1242, 1254, and 1260) by the white rot fungus *Phanerochaete chrysosporium* as evidenced by congener-specific analysis. Appl Environ Microbiol 61:2560–2565

Yin Y, Guo J, Zheng L, Tian L, Wang X (2011) Capability of polychlorinated biophenyl (PCBs) degrading fungi segregated from sediments. World J Microbiol Biotechnol 27:2567–2574

Zhang P, Ge L, Gao H, Yao T, Fang X, Zhou C, Na G (2014) Distribution and transfer pattern of polychlorinated biphenyls (PCBs) among the selected environmental media of Ny-Ålesund, the Arctic: as a case study. Mar Pollut Bull 89:267–275

Zhang M, Kruse NA, Bowman JR, Jackson GP (2016) Field analysis of polychlorinated biphenyls (PCBs) in soil using solid-phase microextraction (SPME) and a portable gas chromatography-mass spectrometry system. Appl Spectrosc 70:785–793

'he manufacturer's authorised representative in the EU is Springer
'ture Customer Service Centre GmbH, Europaplatz 3, 69115 Heidelberg,
many. If you have any concerns regarding our products, please
'ct ProductSafety@springernature.com

and bound by CPI Group (UK) Ltd, Croydon, CR0 4YY
30/09/2026
02225329-0018